CHEMISTRY OF BIOCONJUGATES

CHEMISTRY OF BIOCONJUGATES

Synthesis, Characterization, and Biomedical Applications

Edited by

RAVIN NARAIN
Department of Chemical and Materials Engineering
University of Alberta
Edmonton, Alberta, Canada

Library of Congress Cataloging-in-Publication Data:

Chemistry of bioconjugates : synthesis, characterization, and biomedical applications / edited by Dr. Ravin Narain, Department of Chemical and Materials Engineering, University of Alberta, Edmonton, Alberta, Canada.
    pages cm
  Includes index.
  ISBN 978-1-118-35914-3 (cloth)
 1. Bioconjugates.   I. Narain, Ravin, editor of compilation.
  QP517.B49C46 2014
  612.1′111–dc23

                                                                                          2013026540

Printed in the United States of America

ISBN: 9781118159248

10 9 8 7 6 5 4 3 2 1

CONTENTS

PREFACE

Combining characteristics of different components into one to generate new molecular systems with unique properties by simply linking one or more (macro)molecules is defined as bioconjugation. The ability to create such biohybrids either covalently or non-covalently has allowed major breakthrough in many industrial and biomedical areas such as bioseparation, targeting, detection, biosensing, biological assays, etc. This book provides a comprehensive account on the chemistries involved in the formation of bioconjugates, followed by an extensive review of all the different types of bioconjugates generated so far from polymers, dendrimers, nanoparticles, carbon nanotubes, hydrogels and so on for different bio-related applications. A section is also devoted to the physicochemical and biochemical properties of bioconjugates. Finally, the book also provides a comprehensive account on the significance of bioconjugation which is lacking in many of the current available resources.

The book begins by providing an overview of the chemistry involved in bioconjugation. Different types of bioconjugation strategies available for the modification of biomolecules (proteins, peptides, carbohydrates, polymers, DNA) are presented. Classical bioconjugation approaches are described first, followed by some recent bioconjugation techniques. This section also provides detailed synthetic protocols for some of the most important strategies for bioconjugation.

Polymer bioconjugates are then discussed separately in three sections, namely polyethylene glycol (PEG), synthetic polymer bioconjugates, and natural polymer bioconjugates. PEG has been extensively used in the development of macromolecular therapeutics and most of the current clinically available therapeutics are PEGylated bioconjugates. PEGylation has been used for proteins, anticancer drugs, and other bioactive molecules such as peptides, antibodies,

oligonucleotides, aptamers, red blood cells, and more recently, viruses. Conjugation of synthetic polymers to biomolecules is an appealing strategy to produce new biomacromolecules with distinctive properties. Typical conjugation strategies are either "*grafting from*" or "*grafting to*" approaches. In "*grafting from*" approach, monomer-functionalized biomolecules are polymerized to produce synthetic polymer bioconjugates. On the other hand, in "*grafting to*" approach, biomolecules are immobilized by reactive coupling reactions. Random and site-specific modifications of natural macromolecules have also been extensively studied and, therefore, an elaborated section has been devoted to this area.

The next section is focused on organic nanoparticle bioconjugates. Different chemical strategies used to couple biomolecules with liposomes, micelles, carbon nanotubes, fullerene, and graphene are discussed. Bioconjugation of biomolecules to those organic nanoparticles has become increasingly important in drug formulation and therapeutic delivery. Choosing the right chemistry between the biomolecule and organic nanoparticle has been the focus of great attention in recent years in view of improving the sustained delivery of these bioconjugates to the targeted site effectively. Carbon nanotubes, fullerene, and graphene have unique properties and their coupling with biomolecules have generated unique materials of high potency in biomedical applications.

Inorganic nanomaterials such as gold, iron oxide, quantum dots, and silica have become key players in the biomedical field. Their unique chemical and physical properties have contributed significantly in further development of these nanomaterials. Their surface properties dictate their colloidal stability and biocompatibility. Therefore, in recent years, several strategies have been developed to conjugate bioactive molecules, targeting ligands and other biologically relevant molecules to broaden the applications of these nanomaterials. This section discusses the different chemistries used in bioconjugation of biomolecules on the surface of these most widely used inorganic nanomaterials.

With the rapid development of the chemistries in bioconjugation, it is now possible to prepare cell-based bioconjugates efficiently. Modifying cell surfaces with bioactive molecules or synthetic polymers has been a versatile way to add advanced features and unique properties to inert cells. Creating a nanoscale layer on a cell surface, for example, significantly improves or even completely changes its biological properties as well as introduces new unique properties, such as chemical functionality, surface roughness, surface tension, morphology, surface charge, surface reflectivity, surface conductivity, and optical properties. Recently, surface modification of living cells has been the subject of study for a variety of biological applications such as imaging, transfection, and control of cell surface interactions. Additionally, microgels and hydrogels have emerged as important materials due to their unique features such as encapsulation, swelling, degradation, and controlled dimensions. Such features are further enhanced by conjugating them chemically or physically with other bioactive molecules. This section reviews different approaches in making those biologically relevant bioconjugates. Subsequently, various conjugation strategies for the preparation of carbohydrate-based vaccines and different types of chemistry used for covalent linkage of the individual vaccine components are discussed. Then, both direct and indirect conjugation techniques, as well as different types of linker molecules used to generate the spacing deemed required between carbohydrate and immunogen are presented.

Finally, once the bioconjugates are synthesized, their structures and function need to be properly characterized to fully understand their properties. Therefore, proper tools are required to fully understand the properties of these complex hybrid biomolecules. The techniques used in the full characterization of these bioconjugates are discussed in detail. This section also focuses on the physicochemical and biochemical properties of bioconjugates. The physical properties of conjugates, including their response to temperature, external field (magnetic field, electric field, ultrasound), and light, are discussed. The chemical properties of conjugates, such as their response to a change in the pH and ionic strength, are also summarized. Additionally, the properties of conjugates in response to glutathione (GSH), hydrogen peroxide (H_2O_2), and glucose are also outlined. These bioconjugates have been implied for a variety of biological applications, including drug and gene delivery applications, biological assays, imaging, and biosensors. The success of these bioconjugates in research laboratories, compared to their precursor biomolecules, has further encouraged their use for industrial applications. Some of these bioconjugates are now used in clinical trials.

Ravin Narain

CONTRIBUTORS

Marya Ahmed, Department of Chemical and Materials Engineering, Alberta Glycomics Centre, University of Alberta, Edmonton, AB, Canada

Keshwaree Babooram, Department of Chemical and Materials Engineering, Alberta Glycomics Centre, University of Alberta, Edmonton, AB, Canada

Mitsuhiro Ebara, Biomaterials Unit, International Center for Materials Nanoarchitectonics (WPI-MANA), National Institute for Materials Science (NIMS), 1-1 Namiki, Tsukuba, Ibaraki, Japan

Kheireddine El-Boubbou, Department of Chemistry, Michigan State University, East Lansing, Michigan, USA; College of Science and Health Professions, King Saud bin AbdulAziz University for Health Sciences, National Guard Health Affairs, Riyadh, Saudi Arabia

Mohammad H. El-Dakdouki, Department of Chemistry, Michigan State University, East Lansing, Michigan, USA; Department of Chemistry, Beirut Arab University, Beirut, Lebanon

Ali Faghihnejad, Department of Chemical and Materials Engineering, Alberta Glycomics Centre, University of Alberta, Edmonton, AB, Canada

Fang Gao, Centre for Blood Research and the Department of Pathology and Laboratory Medicine, Department of Chemistry, University of British Columbia, Vancouver, BC, Canada

Elizabeth R. Gillies, Department of Chemistry and Department of Chemical and Biochemical Engineering, The University of Western Ontario, London, Canada

Anirban Sen Gupta, Department of Biomedical Engineering, Case Western Reserve University, Cleveland, OH, USA

Rachel Hevey, Alberta Glycomics Center and Department of Chemistry, University of Calgary, Calgary, AB, Canada

Lizhi Hong, Shanghai Key Laboratory of Functional Materials Chemistry, East China University of Science and Technology, Shanghai, PR China

Jun Huang, Department of Chemical and Materials Engineering, Alberta Glycomics Centre, University of Alberta, Edmonton, AB, Canada

Xuefei Huang, Department of Chemistry, Michigan State University, East Lansing, Michigan, USA

Muhammad Imran ul-Haq, Centre for Blood Research and the Department of Pathology and Laboratory Medicine, Department of Chemistry, University of British Columbia, Vancouver, BC, Canada

Xiaoze Jiang, State Key Laboratory for Modification of Chemical Fibers and Polymer Materials, College of Material Science and Engineering, Donghua University, Shanghai, PR China

Herbert Kavunja, Department of Chemistry, Michigan State University, East Lansing, Michigan, USA

Jayachandran N. Kizhakkedathu, Centre for Blood Research and the Department of Pathology and Laboratory Medicine, Department of Chemistry, University of British Columbia, Vancouver, BC, Canada

Yohei Kotsuchibashi, Department of Chemical and Materials Engineering, Alberta Glycomics Centre, University of Alberta, Edmonton, AB, Canada

Xue Li, Department of Chemistry, University of Alberta, Edmonton, AB, Canada

Chang-Chun Ling, Alberta Glycomics Center and Department of Chemistry, University of Calgary, Calgary, AB, Canada

Lichao Liu, Shanghai Key Laboratory of Functional Materials Chemistry, East China University of Science and Technology, Shanghai, PR China

Ravin Narain, Department of Chemical and Materials Engineering, Alberta Glycomics Centre, University of Alberta, Edmonton, AB, Canada

Ali Nazemi, Department of Chemistry and Department of Chemical and Biochemical Engineering, The University of Western Ontario, London, Canada

Xiuwei Pan, Shanghai Key Laboratory of Functional Materials Chemistry, East China University of Science and Technology, Shanghai, PR China

Michael J. Serpe, Department of Chemistry, University of Alberta, Edmonton, AB, Canada

Rajesh A. Shenoi, Centre for Blood Research and the Department of Pathology and Laboratory Medicine, Department of Chemistry, University of British Columbia, Vancouver, BC, Canada

Rajesh Sunasee, Department of Chemistry, State University of New York, Plattsburgh, NY, USA

Maria Vamvakaki, Institute of Electronic Structure and Laser, Foundation for Research and Technology, Hellas, Heraklion, Crete, Greece; Department of Materials Science and Technology, University of Crete, Heraklion, Crete, Greece.

Horst A. von Recum, Department of Biomedical Engineering, Case Western Reserve University, Cleveland, OH, USA

Jingguang Xia, Department of Chemistry, Michigan State University, East Lansing, Michigan, USA

Qian Yang, Department of Civil and Environmental Engineering, University of Maryland, College Park, MD, USA

Hongbo Zeng, Department of Chemical and Materials Engineering, Alberta Glycomics Centre, University of Alberta, Edmonton, AB, Canada

Weian Zhang, Shanghai Key Laboratory of Functional Materials Chemistry, East China University of Science and Technology, Shanghai, PR China

Zhenghe Zhang, Shanghai Key Laboratory of Functional Materials Chemistry, East China University of Science and Technology, Shanghai, PR China

Meifang Zhu, State Key Laboratory for Modification of Chemical Fibers and Polymer Materials, College of Material Science and Engineering, Donghua University, Shanghai, PR China

SECTION I

GENERAL METHODS OF BIOCONJUGATION

1

COVALENT AND NONCOVALENT BIOCONJUGATION STRATEGIES

RAJESH SUNASEE[1] AND RAVIN NARAIN[2]

[1]Department of Chemistry, State University of New York, Plattsburgh, NY, USA
[2]Department of Chemical and Materials Engineering, Alberta Glycomics Centre, University of Alberta, Edmonton, AB, Canada

1.1 INTRODUCTION

Bioconjugation—the process of covalently or noncovalently linking a biomolecule to other biomolecules or small molecules to create new molecules—is a growing field of research that encompasses a wide range of science between chemistry and molecular biology. The tremendous achievement of modern synthetic organic chemistry has led to a variety of bioconjugation techniques [1] available for application in research laboratories, medical clinics, and industrial facilities. While bioconjugation involves the fusion of two biomolecules, for example protein–protein, polymer–protein, carbohydrate–protein conjugates, it also involves the attachment of synthetic labels (isotope labels, fluorescent dyes, affinity tags, biotin) to biological entities such as carbohydrates, proteins, peptides, synthetic polymers, enzymes, glycans, antibodies, nucleic acids, and oligonucleotides (ONTs). The product of a bioconjugation reaction is usually termed as a *"bioconjugate"* and synthetic macromolecules produced by bioconjugation approaches are commonly referred to as *biohybrids*, *polymer bioconjugates*, or *molecular chimeras*. Modification of biomolecules is an important technique for modulating the function of biomolecules and understanding their roles in complex biological systems [1a]. However, selective biomolecule modification remains challenging and the ease of generating the desired bioconjugate rapidly under physiological conditions is vital for many applications, such as disease diagnosis, biochemical assays, ligand discovery, and molecular sensing. As applications of bioconjugates continue to grow, an expanded toolkit of chemical methods will be required to add new functionality to specific locations with high yield and chemoselectivity.

The aim of this chapter is to provide a comprehensive review of the different types of bioconjugation methods (covalent and noncovalent approaches) available for the modification of biomolecules (proteins, peptides, carbohydrates, polymers, DNA, etc.). Traditional bioconjugation methods will first be elaborated upon, followed by some modern bioconjugation techniques, particularly the emerging role of bioorthogonal chemistry, where the translation of knowledge of chemical reactions to reactions in living systems can be achieved. While the synthetic aspects of the bioconjugates will be the main focus, a brief description of their applications will also be presented.

1.2 COVALENT BIOCONJUGATION STRATEGIES

The covalent bond is the most common form of linkage between atoms in organic chemistry and biochemistry. The reaction of one functional group with another leads to the formation of a covalent bond via the sharing of electrons between atoms (Figure 1.1).

Covalent bioconjugation strategies are generally categorized as random (modification at multiple sites) or site-specific (modification at a single site) bioconjugation. Traditional covalent bioconjugation strategies preclude control over the regiochemistry of reactions, thereby leading to heterogeneous reaction products and eventually, loss of the

Chemistry of Bioconjugates: Synthesis, Characterization, and Biomedical Applications, First Edition. Edited by Ravin Narain.

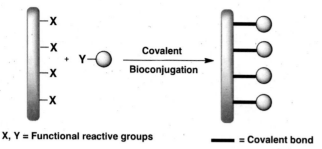

X, Y = Functional reactive groups   ——— = Covalent bond

FIGURE 1.1   Schematic representation of covalent bioconjugation strategy.

biological function of the target biomolecule [1(d)]. However, new methods of bioconjugation that are highly site specific and cause minimal change to the active form of the biomolecule have been developed. For instance, *bioorthogonal* reactions have recently emerged as essential tools for chemical biologists [1(e)]. The following sections survey the covalent modifications of several reactive functional groups (carboxylic acids, aldehydes, ketones, amines, thiols, and alcohols), which are generally present or can be introduced onto macromolecules (proteins, peptides, carbohydrates, nucleic acids, ONTs, etc.).

1.2.1   Carboxyl Modifications

Carboxyl groups are commonly found on the C-terminal ends of proteins and on glutamate (Glu) and aspartate (Asp) amino acid side chains. Carboxylic acids are strong organic acids and the fastest reaction with a nucleophile is removal of the acidic hydrogen to form the carboxylate anion. The resulting anion is resistant to addition reaction with a second nucleophile, and thus makes conjugation through carboxylate group via nucleophilic addition a difficult process. Usually, harsher conditions, acid catalysis, or special reagents are required to promote carboxylic acid-mediated reactions. However, some carboxylate-reactive chemical reactions have been achieved with diazoalkanes and diazoacetyl derivatives (diazoacetate esters and diazoacetamides) and common activating agents such as carbonyldiimidazole (CDI) and carbodiimides to derivatize carboxylic acids. These reactions generate stable covalent linkages namely esters and amides.

1.2.1.1 *Diazoalkanes and Derivatives*   Diazoalkanes, in particular, diazomethane [2] is a powerful reagent for esterification of carboxylic acids. They react instantaneously with

carboxylic acids without the addition of catalysts and may be useful for direct carboxylic acid modification of proteins and synthetic polymers. The reaction mechanism involves nucleophilic attack of the resulting carboxylate anion onto the diazonium ion, followed by an alkylation step to furnish a covalent ester linkage. The driving force of the reaction is the formation of nitrogen, which is a superb leaving group (Scheme 1.1).

Diazomethane, though easily made, is quite toxic, highly explosive, and requires special glassware for reactions. A less explosive and commercially available reagent, trimethylsilyldiazomethane [3], is commonly employed; however, toxicity is still a major concern. In the past, fluorescent diazomethane derivatives have gained much attention for the derivatization of biologically important molecules, especially the nonchromophoric fatty acids [4], bile acids, and prostaglandins. 9-anthryldiazomethane (ADAM) [5, 6] and 1-pyrenyldiazomethane (PDAM) [7, 8] are diazomethane derivatives of the fluorescent dyes anthracene and pyrene, respectively, that have commonly been used as fluorescent labeling reagents for liquid chromatographic determination of carboxylic acids. ADAM and PDAM react readily with carboxylic acids at room temperature in both protic and aprotic solvents. ADAM was found to be unstable and decomposed easily upon storage, while PDAM has a much better chemical stability (a 0.1% (w/v) of PDAM in ethyl acetate solution is stable for 1 week at $\leq -20°C$) [9]. Furthermore, the detection limit for PDAM conjugates (about 20–30 fmol) is reported to be five times better than reported for detection of ADAM conjugates. Fatty acids derivatized with these reagents have been used to measure phospholipase A_2 activity [10].

Protocol for reaction of PDAM with fatty acids [9]:

1. Add 100 µL of 1 mg/mL solution of PDAM in ethyl acetate (ethyl acetate stock solution) to 100 µL of 0.01–10 µg/mL fatty acid solution in methanol.
2. React for 90 minutes at room temperature.
3. Inject 5 µL of reaction mixture into an HPLC column.

1.2.1.2 *Activating Agents*   The direct conversion of a carboxylic acid to an amide with amines is a very difficult process as an acid–base reaction to form a carboxylate ammonium salt occurs first before any nucleophilic substitution reaction happens. As such, amide formation from carboxylic acid is much easier if the acid is first activated (Scheme 1.2)

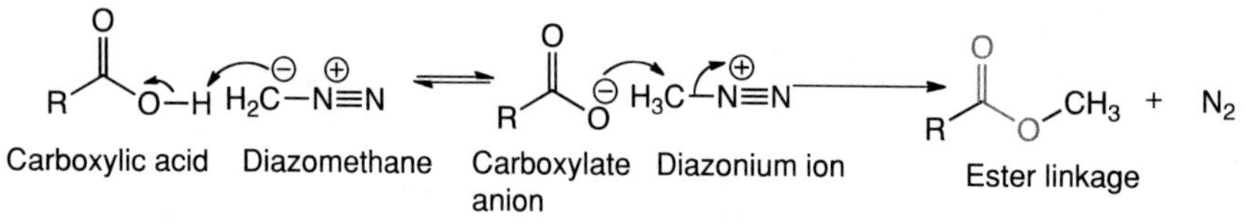

Carboxylic acid   Diazomethane   Carboxylate   Diazonium ion   Ester linkage
                                   anion

SCHEME 1.1   Mechanism of diazomethane esterification reaction.

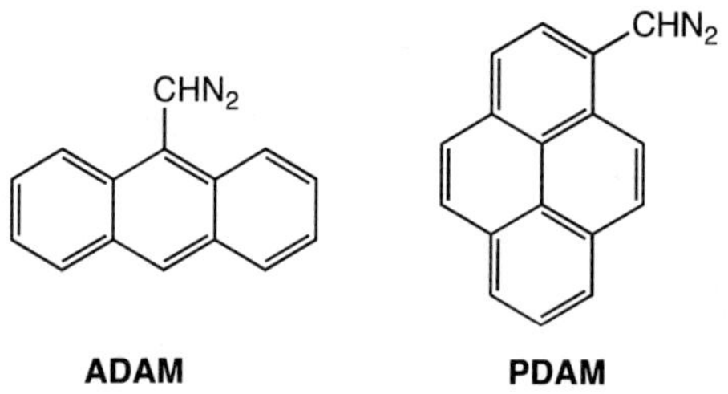

ADAM   **PDAM**

FIGURE 1.2   Fluorescent diazomethane derivatives as labeling reagents.

prior to nucleophilic attack by the amine. This strategy converts the poor carboxy −OH leaving group into a better one. Ester linkages can also be formed using this strategy in the presence of alcohols.

The explosion in the field of peptide chemistry has led to the development of many activating agents that greatly enhance amide formation, but only the most commonly used ones, such as CDI and carbodiimides, will be discussed here (Table 1.1). *N, N′*-Carbonyldiimidazole (CDI) [11] is a white crystalline solid that is useful for activating carboxylic acids to form amide, ester, and thioester linkages. During the reaction, a reactive intermediate, *N*-acylimidazole is formed with liberation of carbon dioxide and imidazole as innocuous side products. The *N*-acylimidazole can then react with amines or alcohols to form stable covalent amide or ester linkages, respectively. CDI is not commonly used in routine peptide synthesis, but nevertheless is quite useful for coupling peptide fragments to form large peptides and small proteins [12]. One unique application of CDI is the synthesis of urea dipeptides [13]. Dicyclohexylcarbodiimide (DCC) and diisopropylcarbodiimide (DIC) are commonly used in organic synthesis for the preparation amides, esters, and acid anhydrides from carboxylic acids. These reagents can also transform primary amides to nitriles, which is a somewhat troublesome side reaction of asparagine and glutamine residues in peptide synthesis. The choice of these carbodiimides depends largely on their solubility properties. DCC was one of the first carbodiimides developed [14] and is widely used in peptide synthesis. It is highly soluble in

dichloromethane, acetonitrile, dimethylformamide (DMF), and tetrahydrofuran, but is insoluble in water. The by-product of a DCC-mediated reaction is dicyclohexylurea, which is nearly insoluble in most organic solvents and precipitates from the reaction mixture as the reaction progresses. Thus, DCC is very useful in solution-phase reactions, but is not appropriate for reactions on resin. Another drawback of DCC-mediated coupling is that trace amounts of dicyclohexylurea remains and are often tedious to remove. DIC was developed as an alternative of DCC since being a liquid, it is easier to handle and also forms a soluble urea by-product, which can easily be removed by simple extraction [15].

1-Ethyl-3-[3-dimethylaminopropyl]carbodiimide (EDC or EDAC) is a versatile modern coupling reagent. It is commonly known as a zero-length cross-linking agent used to conjugate carboxyl groups and amines to form stable covalent amide linkages. Amide bonds typically have a half-life of circa 600 years in neutral solution at room temperature [16], and this extraordinary stability renders amide linkages to be very attractive for bioconjugation. This carbodiimide reagent and its urea by-product are both water soluble; hence, the by-product and any excess reagent are removed by aqueous extraction. EDC reacts with a carboxyl to form an amine-reactive *O*-acylisourea intermediate, which is highly unstable and short-lived in aqueous solution. Thus, hydrolysis is a major competing reaction. It was found that the addition of *N*-hydroxysulfosuccinimide (Sulfo-NHS) stabilizes the amine-reactive intermediate by converting it to a semistable amine-reactive Sulfo-NHS ester (Scheme 1.3), thereby increasing the efficiency of EDC-mediated coupling reactions [17].

Protocol for conjugation of proteins with EDC and Sulfo-NHS [18]:

1. Add EDC (∼2 mM) and Sulfo-NHS (∼5 mM) to protein #1 solution.
2. React for ∼15 minutes at room temperature.
3. Add 2-mercaptoethanol (final concentration of 20 mM) to quench the EDC.
4. Optional step: Separate the protein from excess reducing agent and inactivated cross-linker using a Zeba

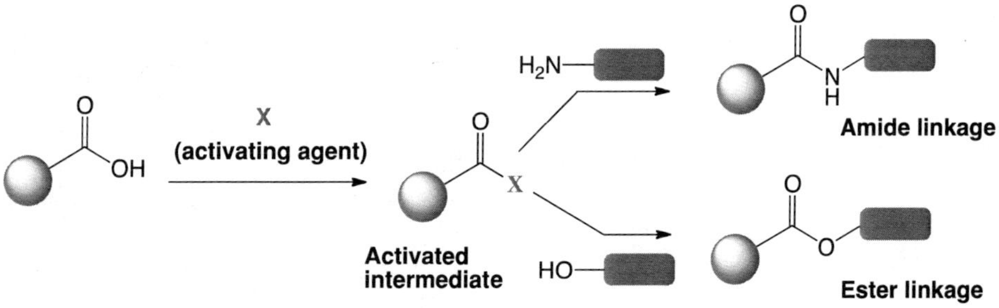

SCHEME 1.2   General strategies for the conjugation of carboxylic acid with amines or alcohols via an activating agent.

TABLE 1.1   Common Activating Agents for Carboxyl-reactive Groups

Activating Agents	Active Intermediates	By-products
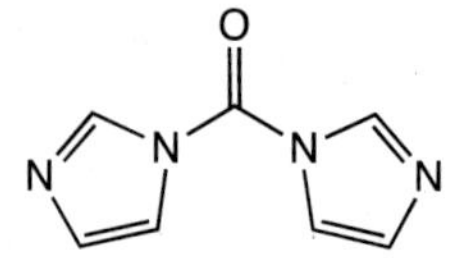 *N, N'*-Carbonyldiimidazole (CDI)	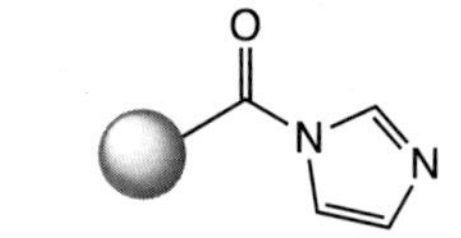 N-Acylimidazole	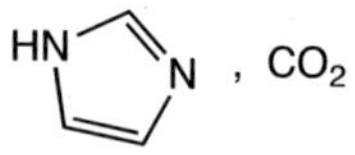 , CO_2 Imidazole
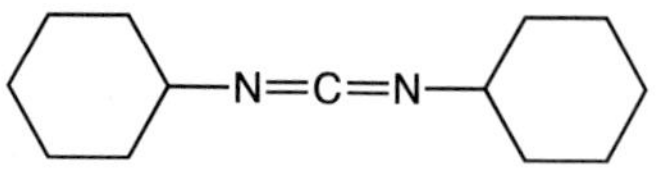 Dicyclohexylcarbodiimide (DCC)	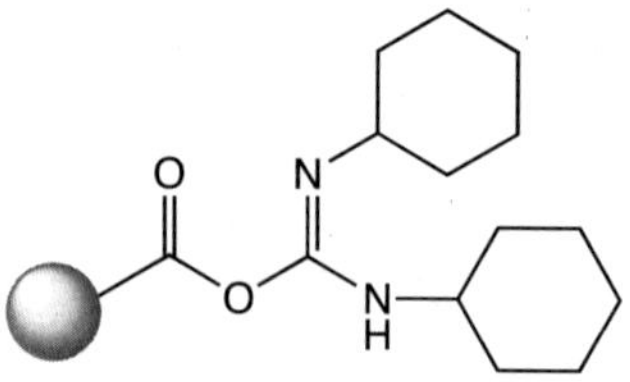 O-Acylisourea	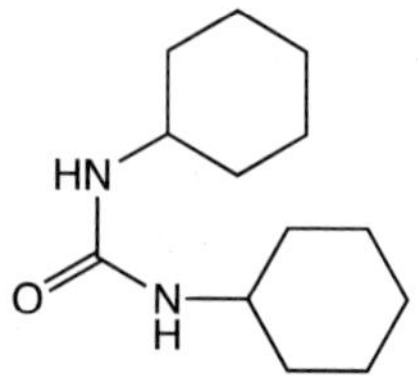 Dicyclohexylurea
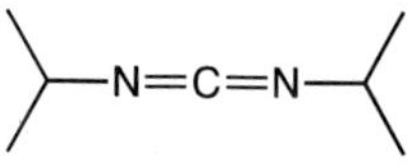 Diisopropylcarbodiimide (DIC)	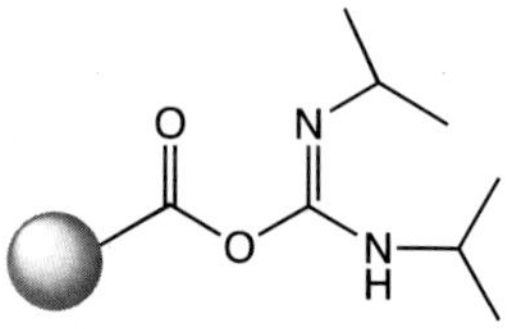 O-Acylisourea ester	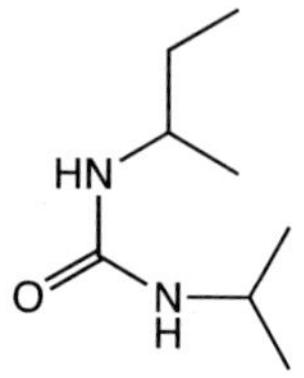 Diisopropylurea
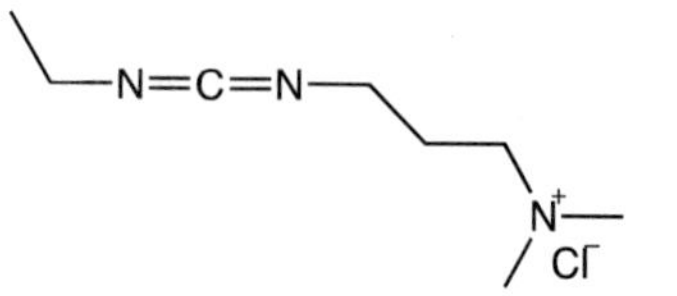 1-ethyl-3-(3-dimethylaminopropyl) carbodiimide hydrochloride (EDC or EDAC)	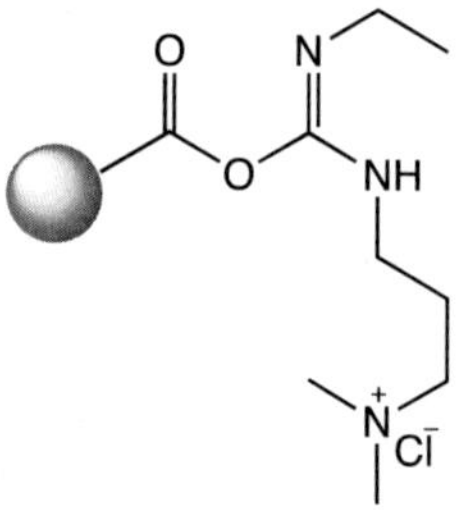 O-Acylisourea ester	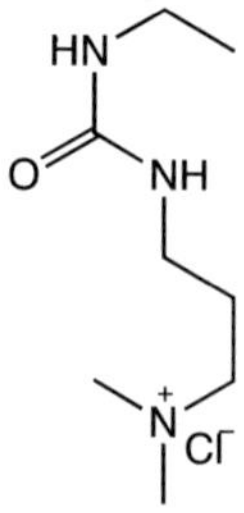 Urea derivative

Desalting Spin Column. Equilibrate the column with activation buffer.

5. Add protein #2 to the reaction mixture or the pooled fractions containing the activated protein at an amount equal to the number of moles of protein #1.

6. React for 2 hours at room temperature.

7. Add hydroxylamine to a final concentration of 10 mM to quench the reaction. (Other means of quenching involve adding 20–50 mM Tris, lysine, glycine, or ethanolamine; however, these primary amine-containing compounds will result in modified carboxyls on protein #1).

8. Remove excess quenching reagent by gel filtration using the same type of column as in Step 4.

A major drawback of carbodiimide activation of amino acid derivatives is that it usually leads to partial racemization of the amino acid. In peptide synthesis, an equivalent of an additive such as triazoles (e.g., 1-hydroxy-7-aza-benzotriazole [19]) is added to minimize this racemization problem. Recently, during the development of prodrugs for the antitumoral agent thiocoraline, a new coupling reagent known as *N,N,N',N'*-tetramethylchloroformamidinium hexafluorophosphate (TCFH) [20] was developed for the coupling of the carboxylic group of an amino acid with the quinolic alcohol to generate an ester linkage (Scheme 1.4). In this case, standard coupling reagents and procedures failed to afford the desired target derivatives. A number of conjugates including PEGylated derivatives with higher solubility were synthesized using the TCFH method.

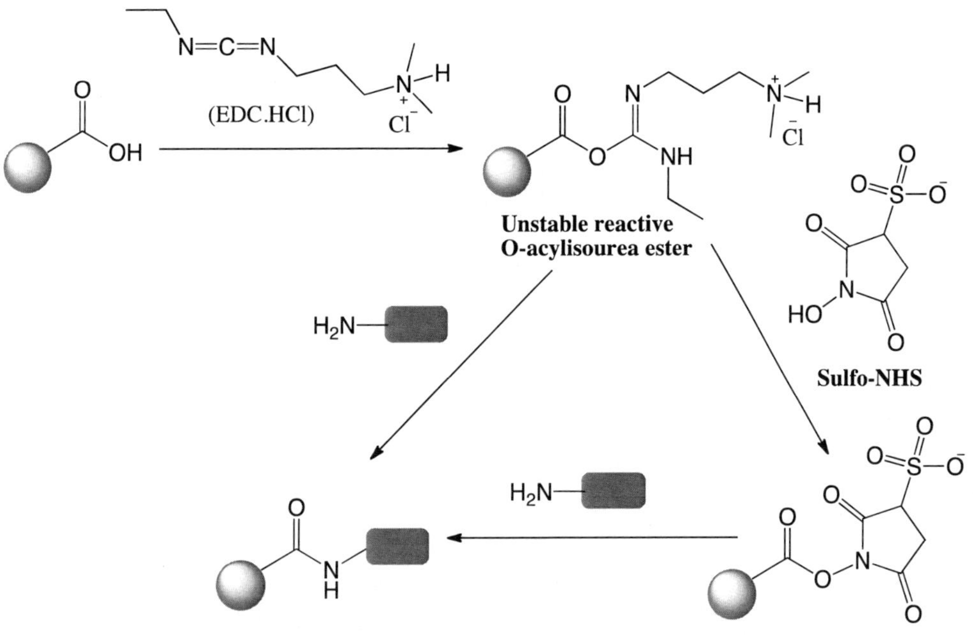

SCHEME 1.3   EDC-mediated protein–carboxylic acid conjugation.

1.2.2   Carbonyl Functional Groups

Aldehydes and ketones are organic compounds that incorporate a carbonyl group (C=O) and are good electrophiles. As such, they undergo nucleophilic addition reactions with various nucleophiles such as amines, N-alkoxyamines (or aminooxy groups), hydrazines, or hydrazide to generate products linked by imine-, oxime-, and hydrazone-reactive groups respectively (Scheme 1.5). The facile synthesis of these carbon–nitrogen double bonds in aqueous solutions at neutral pH makes them attractive for bioconjugation and thus, they have found widespread applications in chemical biology, mainly for the synthesis of nucleic acid conjugates [21]. Aldehydes and ketones are also known as chemical reporters that can tag proteins [22], glycans [23], and other secondary metabolites.

1.2.2.1   Conjugation via Reductive Amination   Reductive amination [24] is a process that transforms a carbonyl

group (typically aldehydes and ketones) into an amine via an intermediate imine (Schiff base). Under acidic conditions, the carbonyl group first reacts with primary amines to form a hemiaminal species, which subsequently loses a water molecule to generate a reversible unstable imine. The imine is then trapped irreversibly with a reducing agent in a one-pot reaction to afford a stable amine product (Scheme 1.6). The overall two-step sequence is called reductive amination. Borohydrides are common reducing agents with sodium cyanoborohydride (NaBH$_3$CN) being the most widely used due to its high selectivity to imines and relative unreactivity with oxo groups [25]. Sodium triacetoxyborohydride, NaBH(OAc)$_3$, was introduced as an alternative mild and nontoxic reducing agent of NaBH$_3$CN [26].

The reaction of a carbonyl group with an amine proceeds with high chemoselectivity and is also compatible with many functional groups present in biomolecules. Carbohydrate–protein conjugates play vital roles in both

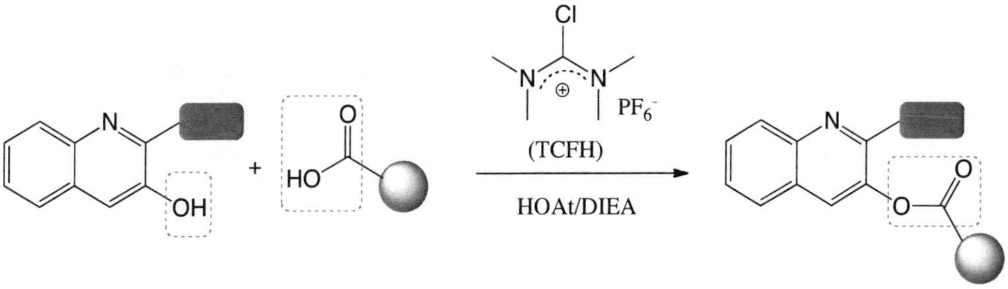

SCHEME 1.4   TCFH-mediated conjugations for ester linkage.

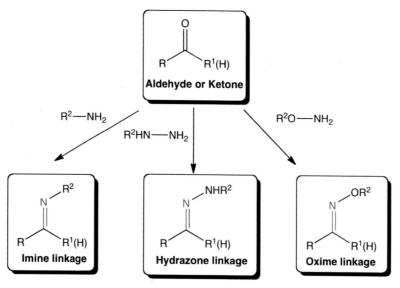

SCHEME 1.5 Bioconjugation via carbon–nitrogen double bonds.

basic and applied research [27] and thus, significant efforts to develop simple and efficient chemical methods for the covalent attachment of carbohydrate molecules to proteins have been investigated [28]. Reductive amination remains one of the key methods for the direct conjugation of carbohydrates to the amino group of proteins especially from unprotected free mono- and oligosaccharides [29–31] (Scheme 1.7).

Typical reductive amination protocol for conjugation of sugars with proteins [30]:

1. Dissolve sugar (60 mM), protein (200 μM), and NaBH$_3$CN (300 mM) in aqueous sodium borate buffer (200 mM, pH 9.0).
2. React with stirring at 37–50°C in a thermostated incubator for 10–24 hours.
3. Dialysis against water followed by lyophilization.

Recently, an improved procedure for direct coupling of carbohydrates to proteins via reductive amination was reported [31]. It was found that the addition of a salt (sodium sulfate, 500 mM) in the reaction mixture highly improved the conjugation efficiency. The improved conditions are compatible with microgram quantities of sugar and afford carbohydrate–protein active conjugates for use in assays and incorporation in microarrays.

1.2.2.2 Conjugation via Hydrazone and Oxime Formation Reaction of carbonyl compounds with primary amines results in formation of unstable reversible Schiff bases, with the equilibrium in water favoring the carbonyl. However, reaction of carbonyl compounds with hydrazides or aminooxy groups leads to formation of Schiff bases (hydrazones and oximes), which are favored in aqueous solution (α-effect nitrogens [32]) as well as being quite stable under physiological conditions [33]. Coupling via hydrazone

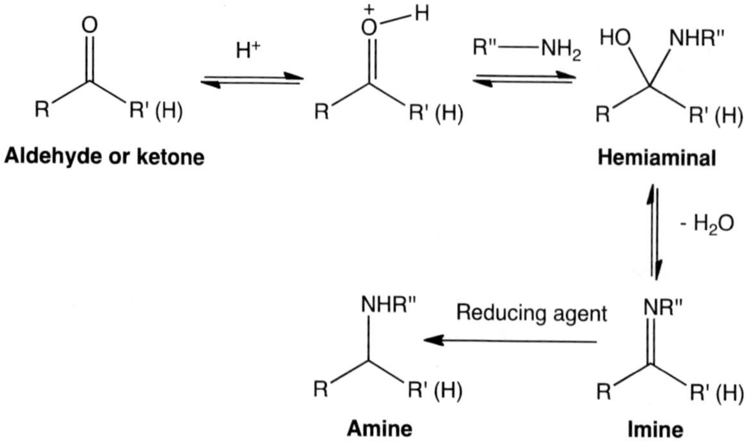

SCHEME 1.6 Reductive amination process.

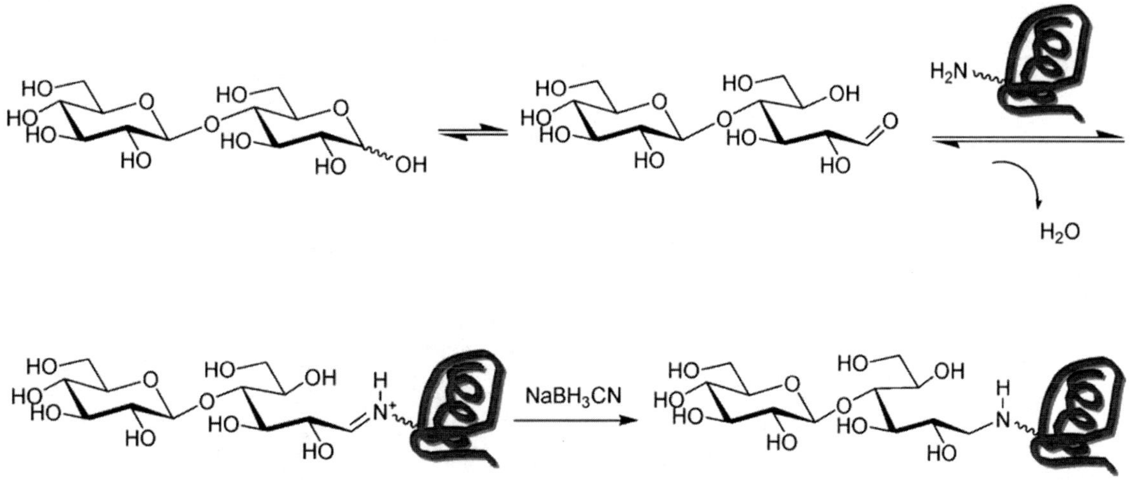

SCHEME 1.7 Bioconjugation of carbohydrates with proteins via reductive amination. Reprinted with permission from Reference 31, Copyright 2008, American Chemical Society.

formation is perhaps the oldest method of bioconjugation. Normally hydrazones are pretty stable; however, in some cases, for example when basic treatment is involved, reduction of the resulting hydrazone linkage with sodium cyanohydride is preferable in order to ensure a stronger covalent linkage. It was observed that the stability of hydrazone depends on the nature of the substituent on the nitrogen and decreases in the following order: aromatic hydrazine > aliphatic hydrazine > hydrazine > hydrazides [34]. Zatsepin et al. used the hydrazone bioconjugation method to synthesize ONT–peptide conjugates (Scheme 1.8) [35]. In this case, the hydrazone formed was prone to hydrolysis in both neutral and basic pH range and hence, reduction was required.

Acylhydrazone linkages were used for the immobilization of peptides to generate peptide microarrays that allowed the sensitive detection of antibodies in blood samples [36]. Aldehydes and ketones react chemoselectively with aminooxy groups to form oxime adducts in mild aqueous solutions. The oxime bond is more stable than hydrazone bond and the reaction proceeds with modest rate in acidic conditions, but are less reactive at pH 7. Recent reports by Dirksen et al. [37] have shown that the rates of imine ligations can be greatly enhanced in the presence of aniline, which behaves as a nucleophilic catalyst. Aniline will first react with the aldehyde to generate an imine. The aniline imine is readily protonated and making it more reactive toward aminooxy reagent. Eventually, loss of aniline leads to formation of the stable oxime product (Scheme 1.9) [38].

Oxime strategies have been used for the glycosylation of peptides, peptoids, proteins, carbohydrates, microarrays, and small molecules of pharmaceutical interest [39]. The aniline strategy has also been applied to accelerate hydrazone conjugation reactions [40]. The method was used to immobilize antibodies onto surfaces for immune-based biosensing platforms [41]. While both hydrazone and oxime are useful conjugates, they do suffer certain limitations, which sometimes restrict their use in biological applications. They are labile to spontaneous hydrolysis of C=N bonds, and alkyl- and acylhydrazones possess short half-lives (about an hour) under physiological conditions [1(d)].

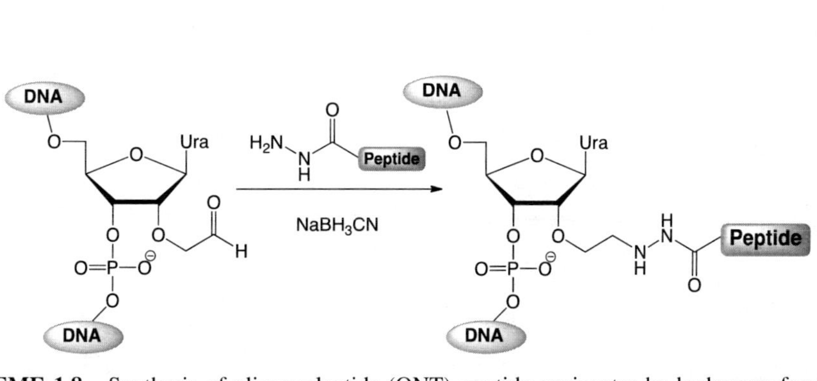

SCHEME 1.8 Synthesis of oligonucleotide (ONT)–peptide conjugates by hydrazone formation followed by reduction. Adapted with permission from Reference 35, Copyright 2002, American Chemical Society.

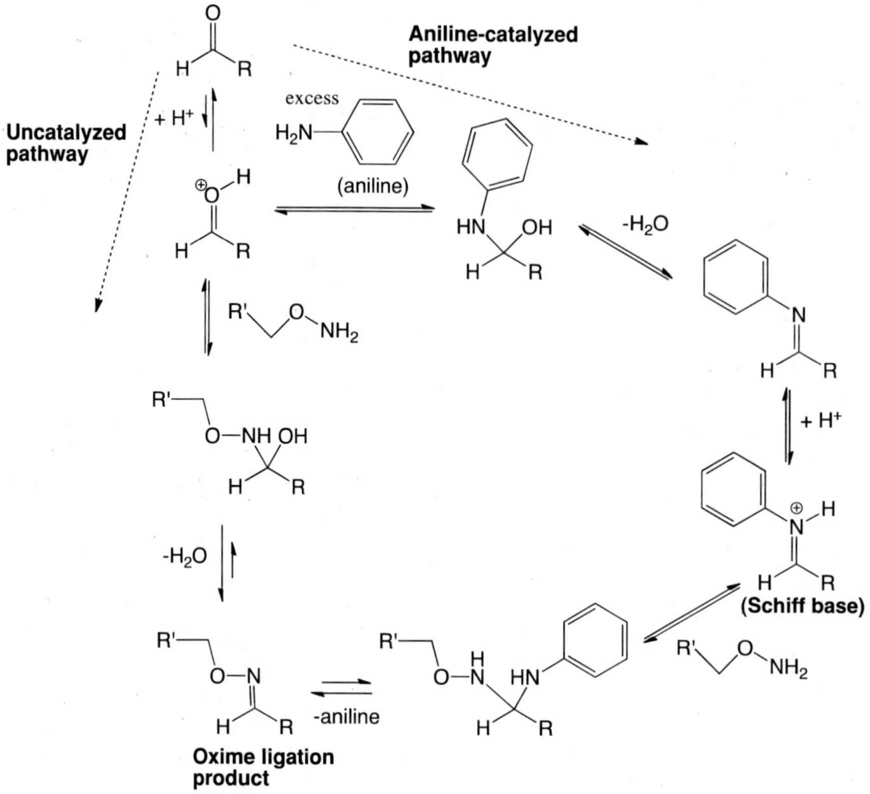

SCHEME 1.9   Aniline-promoted nucleophilic catalysis of oxime conjugation. Adapted with permission from Reference 38, Copyright 2009, John Wiley & Sons, Inc.

1.2.2.3 Conjugation via Mannich and Morita–Baylis–Hillman Reactions   The Mannich reaction is a multicomponent condensation of a nonenolizable aldehyde, like formaldehyde (or ketone), a primary or secondary amine (or ammonia) and a C–H activated compound (aliphatic or aromatic carbonyl compounds or electron-rich aromatic compounds such as phenols) to furnish aminoalkylated products [42]. This powerful reaction discovered in the early 1900s was only recently applied for the synthesis of bioconjugates since it results in stable covalent bonds [43]. A one-step three-component Mannich-type reaction was used for the conjugation of aniline-containing peptides to native tyrosine residues on proteins [44] (Scheme 1.10).

This new bioconjugaton method could be useful in cellular uptake and trafficking studies, protein purification, and materials applications. Recently, Xie et al. used the Mannich reaction to conjugate biomolecules to nanoparticles [45]. Iron oxide nanoparticles functionalized with active hydrogen groups were reacted with amine group-containing cyclic Arginine-Glycine-Aspartic acid (RGD) peptides to develop ultrasmall biocompatible nanoparticles for use as *in vivo* tumor-targeted imaging agents.

Multifunctional bioconjugation by the Morita–Baylis–Hillman (MBH) reaction in aqueous medium was recently developed for the effective modification of oligosaccharides,

peptides, and proteins with fluorescent probes/biotin tags [46]. The MBH reaction involves a carbon–carbon bond formation between the α-position of conjugated carbonyl compounds and carbon electrophiles such as aldehydes or activated ketones. The reaction is usually catalyzed by tertiary amine such as 1,4-diazabicyclo[2.2.2]octane (DABCO) [47]. The resulting β-hydroxy-α-methylene-carbonyl moiety could be further modified via conjugate addition by thiol-based biophysical probes and cysteine-containing peptides and proteins (Scheme 1.11). This mild MBH-based bioconjugation strategy will open up a new direction for the rapid assembly of multifunctional bioconjugates with high structural diversity [46].

1.2.3   Amine Modifications

Amines are organic compounds that possess a basic nitrogen atom with a lone pair. The dominant reactivity of amines is their nucleophilicity and thus most of their reactions involve nucleophilic-to-electrophilic attacks. Virtually all proteins possess lysine moieties, which have a free amine at the N-terminus. Primary amines are especially nucleophilic and this makes them ideal target for conjugation with other reactive groups. Amine bioconjugation strategies have been mainly used to modify proteins, peptides, oligosaccharides,

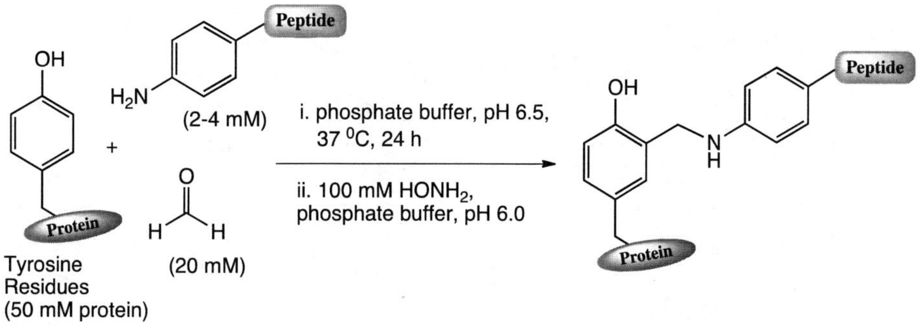

SCHEME 1.10 Three-component Mannich reaction for conjugation of protein and peptide. Adapted with permission from Reference 44, Copyright 2008, American Chemical Society.

ligands, and other biomolecules. The two common reactions of amines are *alkylation* (addition of an alkyl group with loss of a H atom to generate amine linkage) and *acylation* (replacement of a H atom of amino group by an acyl group to form a stable amide linkage), which are typically used for derivatization of the amine-containing side chains of amino acids such as lysine, arginine, and histidine. These side-chain amines behave as good nucleophiles when they are in their unprotonated forms and a moderately basic pH of 8–10 ensures their reactivity. Amine-bearing ligands have been used to conjugate with alkyl halide-bearing surfaces to form stable amine linkages [48]; however, these alkylation reactions proceeded slowly.

Amines also react with other functional groups bearing electrophilic carbon atoms such as isocyanate ($-N=C=O$) and isothiocyanate ($-N=C=S$) to form isourea and isothiourea linkages, respectively. Reactions of isocyanate with amine proceed with good efficiency but are prone to hydrolytic cleavage. Isocyanates deteriorate rapidly upon storage. Alternatively, isothiocyanates react well with amines under alkaline conditions (0.1 M sodium carbonate buffer, pH 9) and are quite stable in water and most solvents. Hydrolysis-resistant diisothiocyanates have been employed for many years for labeling ligands with reporter molecules; however, the thiourea linkage is unstable at lower pH and hence, may be unsuitable for investigations of cell–surface interactions [49]. Antibody conjugates synthesized from fluorescent isothiocyanate have found to degrade over time [50]. Despite all these problems, commercially available fluorescein isothiocyanate (FITC) and tetramethylrhodamine isothiocyanate (TRITC) are still widely used reactive fluorescent dyes for preparing fluorescent bioconjugates (Figure 1.3).

Recently, Roman and Dong [51] prepared fluorescently labeled cellulose nanocrystals by reacting the primary amino group with the isothiocyanate group of FITC to form a thiourea linkage (Scheme 1.12). These fluorescent bioconjugate will be used to study the interaction of cellulose nanocrystals with cells and the biodistribution of cellulose nanocrystals *in vivo*.

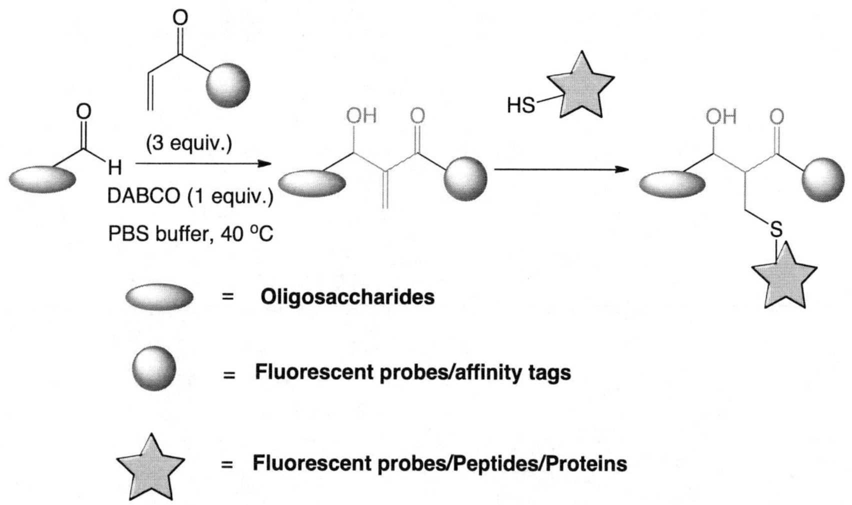

SCHEME 1.11 MBH-based multifunctional bioconjugation strategy. Adapted from Reference 46 with permission from the Royal Society of Chemistry.

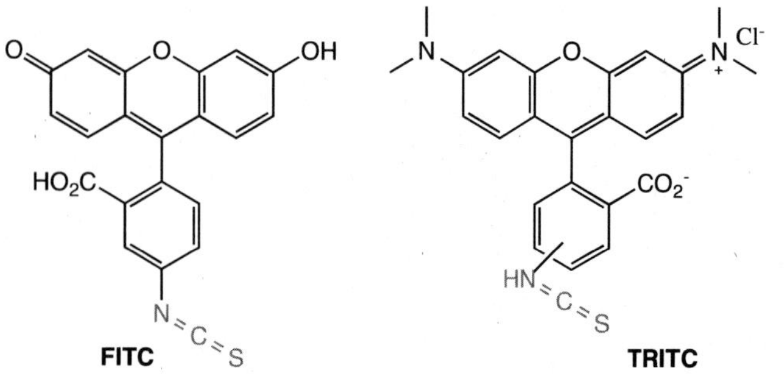

FIGURE 1.3 Structures of fluorescent FITC and TRITC.

Experimental protocol of FITC conjugation to aminated cellulose nanocrystals following the method of Swoboda and Hasselbach [52]:

1. Add FITC (0.32 mmol/g cellulose) to aminated nanocrystals in 50 mM sodium borate buffer solution (50 mL/g cellulose) containing ethylene glycol tetraacetic acid (5 mM), sodium chloride (0.15 M), and sucrose (0.3 M).
2. Stir reaction mixture overnight in the dark.
3. Dialyze for 5 days.
4. Sonicate suspension for 10 minutes, 200 W with ice-bath cooling.
5. Centrifuge for 10 minutes, 4550 G, 25°C and filter through a syringe filter (0.45 m) to remove any aggregates.

Activated esters are also reliable reagents for amine modification since they can form stable amide bond. Typical examples of active esters are succinimidyl (NHS), tetrafluorophenyl (TFP), 4-sulfotetrafluorophenyl (STP), and 4-sulfodichlorophenyl (SDP) esters (Figure 1.4).

Succinimidyl esters display good reactivity with aliphatic amines but low reactivity with aromatic amines, alcohols, and

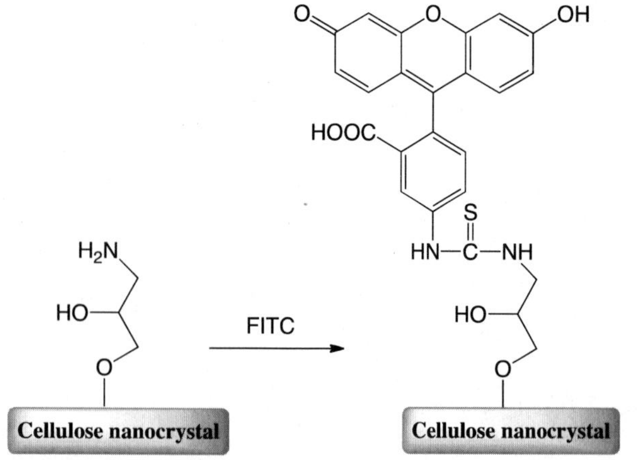

SCHEME 1.12 Synthesis of fluorescently labeled cellulose nanocrystal bioconjugate.

phenols. The major competing reaction of succinimidyl ester conjugation is hydrolysis, which can be minimized if the conjugation is performed below pH 9. Some succinimidyl esters are insoluble in aqueous solution, and as such limit their use for certain specific applications. To overcome this limitation, sulfonated esters such as STP were developed since they are more polar and have better water solubility than simple succinimidyl esters, and can avoid the need of organic solvents in conjugation reactions. STP esters can be prepared from the corresponding phenol (4-sulfo-2,3,5,6-tetrafluorophenol) and are easily purified by chromatography [53]. TFP esters also react smoothly with amines and are more resistant to nonspecific hydrolysis than succinimidyl esters. SDP esters are very hydrolytically stable and have better controlled and consistency in reactions as compared to NHS and TFP esters. Amines react with aldehydes and ketones to generate labile imine linkage, which can be stabilized by reduction with NaBH$_3$CN (reductive amination) to a more stable amine linkage (See Section 1.2.2.1 for more details). Formaldehyde and glutaraldehyde are carbonyl reagents that conjugate with amine via Mannich reactions and/or reductive amination. EDC/NHS-mediated amide bond formation by reaction of an amine with an activated carboxylic compound under physiologic to slightly alkaline conditions (pH 7.2–9) is a very popular and practical conjugation method [54], and this has already been discussed in Section 1.2.1.2 (refer to Scheme 1.3). This method has been commonly used to prepare protein conjugates as well as for the labeling and immobilization of antibodies.

There are several other synthetic chemical groups that will form covalent linkages with amines and this has been well exploited, for example, amine conjugation reactions with epoxides, cyclic anhydrides, imidoesters, carbonates, sulfonyl chlorides, and acyl azides (Table 1.2). Ring-opening reactions of amines with cyclic anhydrides and epoxides are somewhat less reactive and also susceptible to hydrolysis. Imidoesters readily react with amines on proteins with little side reaction such as cross-reactivity with other nucleophilic-reactive groups located on the proteins [55]. The conjugation proceeds well at pH 8–9 and the resulting amidine product does not alter the overall charge of the protein, thereby retaining the native conformation and activity of the protein. Sulfonyl chlorides are very reactive with aliphatic amines at high pH, and protein modification with this reagent is best carried at low temperature. However, they are quite unstable in water, but once conjugated, the sulfonamide linkage is extremely stable [56]. Sulfonyl chlorides are unstable in dimethyl sulfoxide (DMSO) solvent [57], and are not recommended for use in conjugation reactions.

1.2.4 Thiol Modifications

Thiols are organosulfur compounds containing −C–SH or R–SH linkage where R = alkane, alkene, or other

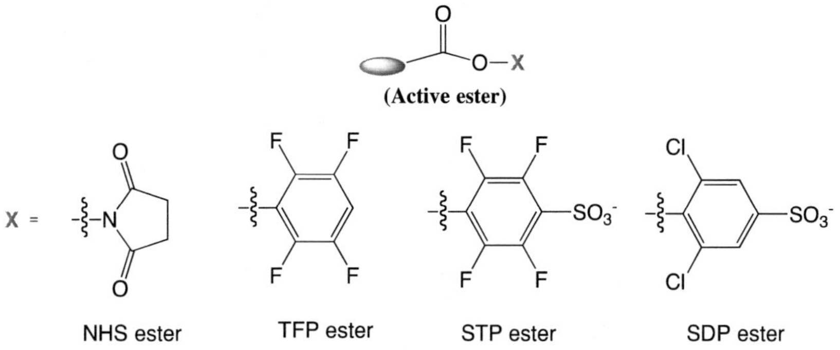

FIGURE 1.4   Chemical structures of active esters for amine modification.

TABLE 1.2   Conventional Amine Covalent Bioconjugation Strategies

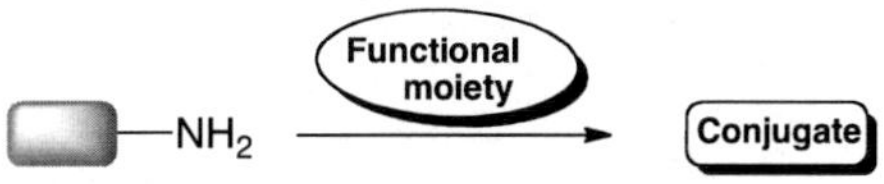

Functional Moiety	Conjugate	References
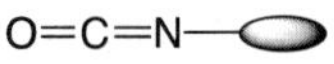 O=C=N—⬭ **Isocyanate**	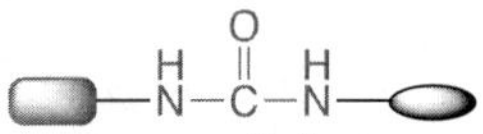 ⬭—N(H)—C(=O)—N(H)—⬭ **Isourea linkage**	58, 59
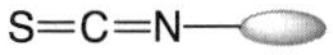 S=C=N—⬭ **Isothiocyanate**	 ⬭—N(H)—C(=S)—N(H)—⬭ **Isothiourea linkage**	49–51, 60
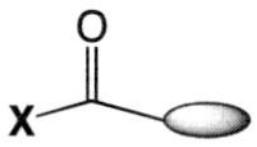 X—C(=O)—⬭ **Active esters** (X = NHS, TFP, STP, SDP)	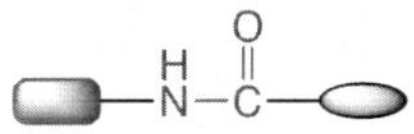 ⬭—N(H)—C(=O)—⬭ **Amide linkage**	53, 54, 61–63
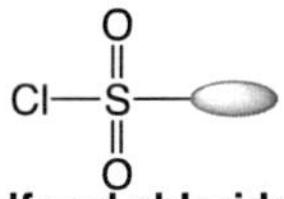 Cl—S(=O)(=O)—⬭ **Sulfonyl chloride**	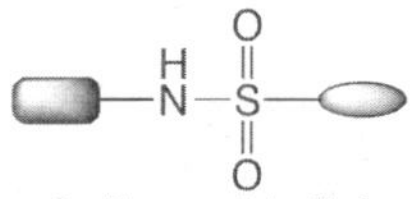 ⬭—N(H)—S(=O)(=O)—⬭ **Sulfonamide linkage**	56, 57
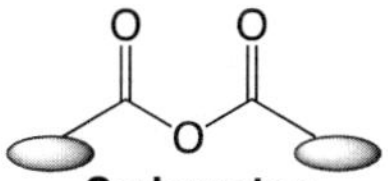 **Carbonates**	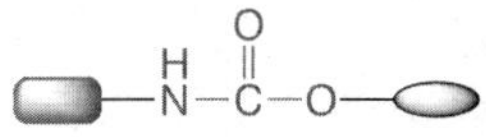 ⬭—N(H)—C(=O)—O—⬭ **Carbamate linkage**	64
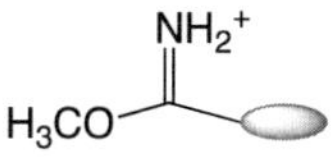 NH_2^+ H₃CO—C—⬭ **Imido esters**	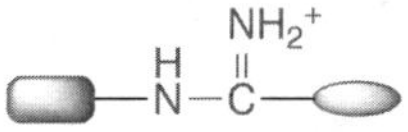 NH_2^+ ⬭—N(H)—C—⬭ **Amidine linkage**	55
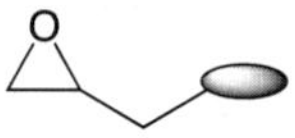 **Epoxide**	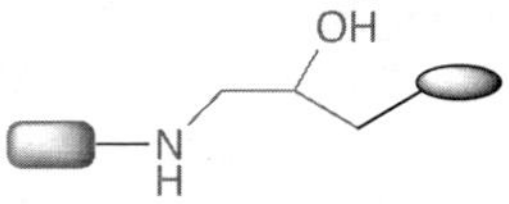 ⬭—N(H)—CH₂—CH(OH)—CH₂—⬭ **Secondary amine linkage**	65, 66
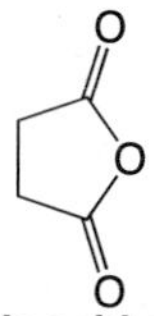 **Cyclic acid anhydride**	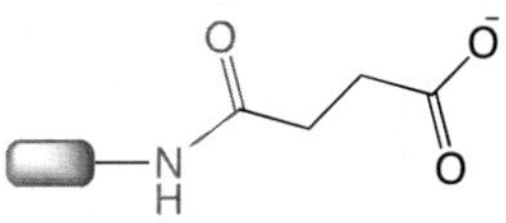 ⬭—N(H)—C(=O)—CH₂—CH₂—C(=O)—O⁻ **Amide linkage**	67
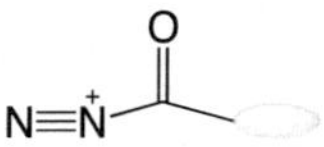 N≡N⁺=N—C(=O)—⬭ **Acyl azides**	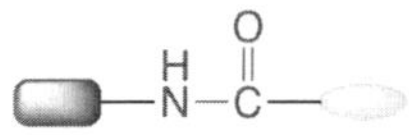 ⬭—N(H)—C(=O)— **Amide linkage**	68, 69

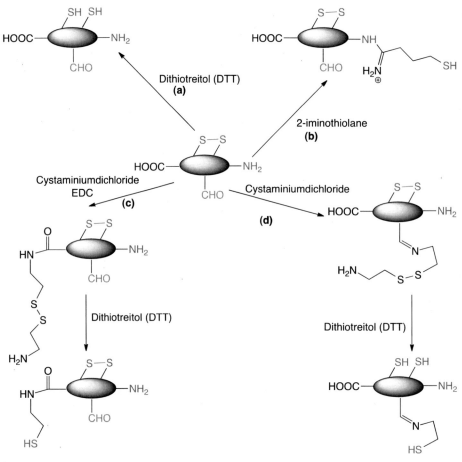

SCHEME 1.13   Incorporation of thiol groups by chemical methods.

carbon-containing groups of atoms. They are also typically known as *sulfhydryls* or *mercaptans*. The thiol group is widely distributed in biological materials and represents an important functional center in biological systems [70]. Thiols play crucial role in maintaining the appropriate oxidation–reduction state of proteins, cells, and organisms. Thiol groups are mainly present in cysteine residues of proteins or they can be generated by chemical methods, for example, reduction of native disulfide bonds with dithiothreitol (DTT) (Scheme 1.13a), coupling of primary amino group with 2-iminothiolane (Traut's reagent) (Scheme 1.13b), cystaminiumdichloride reactions with carboxylic acids (Scheme 1.13c) and aldehydes (Scheme 1.13d) followed by reduction with DTT [71].

However, reformation of disulfide bonds via air oxidation is a common problem during the removal of the reducing agents by dialysis or gel filtration. An alternative powerful reducing agent, namely, tris-(2-carboxyethyl)phosphine (TCEP) [76] was found to prevent disulfide formation and also it does not need to be removed prior to thiol bioconjugation reactions. TCEP is odorless, stable at higher pH and temperature [72] than DTT, and is impermeable to cell membranes and protein hydrophobic core. Depending upon

reaction conditions, TCEP is known to react with thiol-reactive chemical groups, such as iodoacetamides and maleimides [73].

Succinimidyl thiolating reagents have also been developed for incorporating thiol groups into lipids, proteins, and nucleic acids by reaction with amine functional group (Scheme 1.14) [74, 75].

Thiols generally behave as nucleophiles due to the presence of lone pairs of electrons on the sulfur atom; however, the corresponding thiolate anion ($R–S^-$) is a more powerful nucleophile in aqueous solutions. Thus, most bioconjugation reactions involve the thiolate anion. Typical thiol-reactive chemical groups include iodoacetamides (or α-halocarbonyl), maleimides, arylating agents (fluorobenzene), and aziridine derivatives, which react via an *S*-alkylation to form stable thioether linkages (Scheme 1.15)

Iodoacetamides react readily with thiols (usually pH > 7.5) located on proteins, peptides, and thiolated polynucleotides to generate stable thioether bonds [77, 78]. They are more reactive than the corresponding bromoacetamides. However, special care has to be taken when carrying the bioconjugation reaction, as iodoacetamides are very unstable in the presence of light or reducing agents. Reaction must be

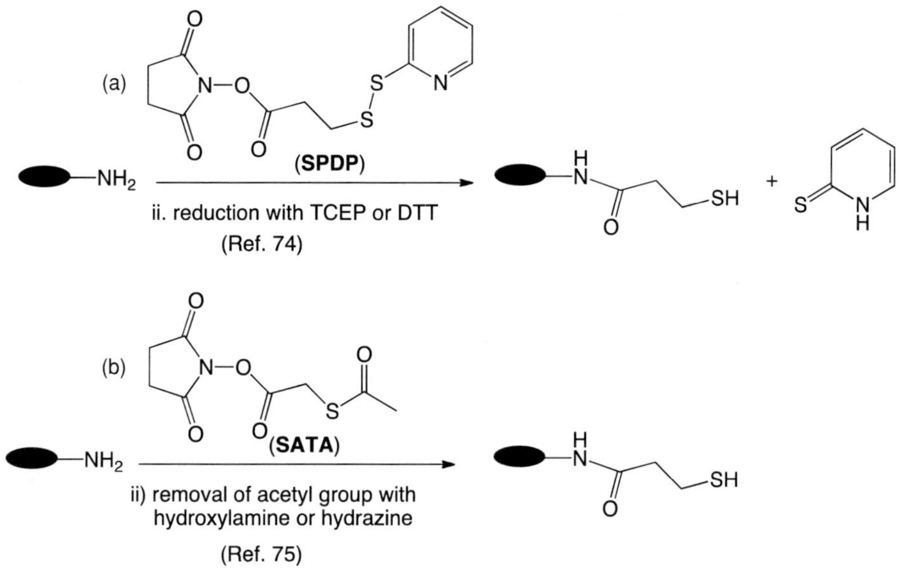

SCHEME 1.14 Thiolation of amine derivatives with succinimidyl 3-(2-pyridyldithio)-propionate (SPDP) and succinimidyl acetylthioacetate (SATA).

carried out in the dark to avoid liberation of free iodine that can react with histidine, tyrosine, and tryptophan residues. Iodoacetamides have been classically used for determining the presence of free cysteine residues in proteins [79] and recently for immobilization and labeling of proteins [80]. Chloroacetamides turn out to have greater specificity than iodoacetamides for cysteine residues [81].

Acryloyl derivatives ($R–CH=CH_2$) possess reactive double bonds that can undergo addition reactions with nucleophilic thiols. A common example of this class of compounds is the maleimide group. Maleimides react irreversibly with thiols in pH range of 6.5–7.5 to afford thioether linkages, which have frequently been used to synthesize neoglycoconjugates [82, 83]. The conjugated double bond of maleimides

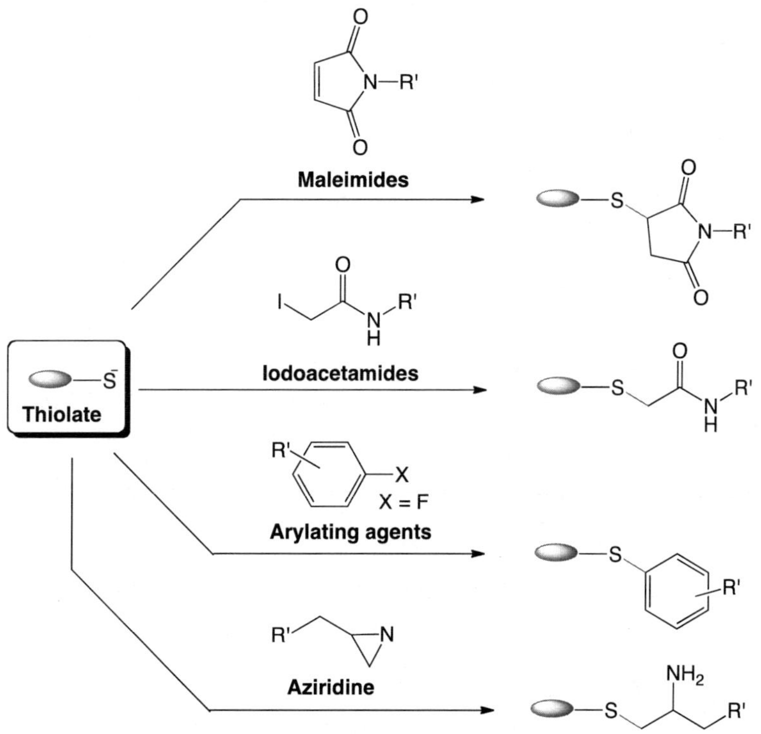

SCHEME 1.15 Formation of thioester bioconjugates via *S*-alkylation-type reactions.

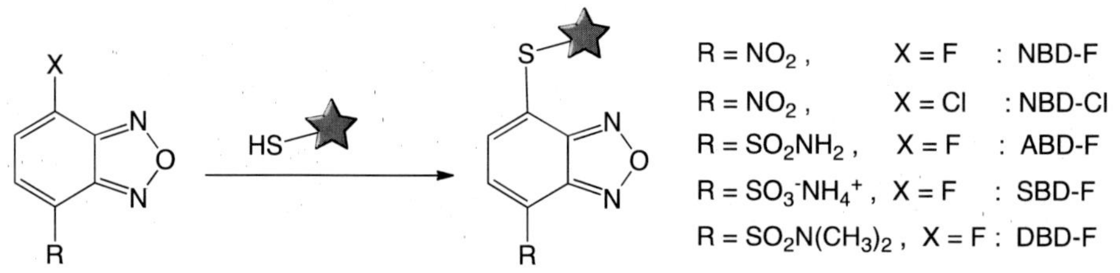

R = NO$_2$,	X = F	: NBD-F
R = NO$_2$,	X = Cl	: NBD-Cl
R = SO$_2$NH$_2$,	X = F	: ABD-F
R = SO$_3^-$NH$_4^+$,	X = F	: SBD-F
R = SO$_2$N(CH$_3$)$_2$,	X = F :	DBD-F

SCHEME 1.16   Conjugation of fluorinated benzoxadiazole derivatives with thiols.

undergoes an alkylation reaction (Michael addition) with thiolates to generate stable succinimidyl thioether bonds. The maleimide/thiol reaction is known to proceed rapidly in neutral aqueous solutions at room temperature, making it ideal for biological applications. Excess maleimides are removed from reaction mixture at the end of the reaction by quenching with free thiols, such as β-mercaptoethanol. Maleimides are more thiol-selective than iodoacetamides and do not react with histidine, tyrosine, or methionine residues. Competitive reactions of maleimides with amines usually require a higher pH than the reaction of maleimides with thiols. Spontaneous hydrolysis of maleimide moiety [84] in basic aqueous media (pH > 8) gives rise to an *N*-acyl derivative that can no longer react with thiols [85] and thus competes significantly with thiol modification. Hydrolysis also produces isomeric succinamic acid thioethers resulting in undesirable heterogeneity, which eventually can alter the activity of bioconjugates. Recent studies have shown that molybdate and chromate can deliberately catalyze imido hydrolysis near neutral pH, thereby providing a strategy to decrease the heterogeneity of maleimide-derived bioconjugates [86]. While both iodoacetamide- and maleimide-thiol bioconjugation strategies have been commonly employed, it has been reported that iodoacetamide conjugates are more toxic while maleimide adducts are less stable during the intracellular reactivity and toxicity studies of haloacetyl and maleimido thiol-reactive probes in HEK 293 cells [87].

Aziridine [88], the nitrogenous analog of epoxides, represents a valuable synthetic block due to its electrophilic nature and thus, strong reactivity toward nucleophiles. Thiolates react readily with aziridines under slightly alkaline conditions via a nucleophilic ring-opening reaction to afford stable thioether linkage and a free amine group. Reactions have been carried out in the presence of a catalytic amount of boron trifluoride etherate [89] or stoichiometric amounts of thiobenzoic acid [90]. Recently, addition of catalytic amount amine base (DBU) has been found to promote thiol/aziridine ligation reactions [91]. This base-promoted aziridine ring-opening strategy was used to conjugate various complex

thiol moieties such as carbohydrates, lipids, and biochemical tags with aziridine-2-carboxylic acid-containing peptides to prepare complex thioglycoconjugates.

Arylating agents react with thiols in a reaction similar to nucleophilic substitution reactions of simple aromatic halides. Reactions proceed rapidly at or below room temperature in the pH range of 6.5–8.0 to yield stable thioether bonds. Some commonly used arylating agents are benzoxadiazole families, such as nitrobenz-2-Oxa-1,3-diazole derivatives (NBD-Cl and the more reactive NBD-F [92], NBD iodoacetate ester (IANBD ester), NBD iodoacetamide (IANBD amide) [93]), 7-fluorobenz-2-oxa-1,3-diazole-4-sulfonamide (ABD-F) [94], Ammonium 7-fluoro-2,1,3-benzoxadiazole-4-sulfonate (SBD-F) and 4-(*N,N*-dimethylaminosulfonyl)-7-fluoro-2,1,3-benzoxadiazole (DBD-F) (Scheme 1.16) [70]. NBD is a functional analog of dinitrophenyl hapten. Thiol conjugates of ABD-F are much more stable in aqueous solution than the NBD thiol conjugates [95]. DBD-F has similar properties to ABD-F and SBD-F; however, the order of reactivities with thiols is as follows: DBD-F > ABD-F > SBD-F [70].

Thiolates also react with disulfide derivatives via a thiol–disulfide interchange reaction. During the process, thiolate attacks one of the sulfur atoms of the disulfide followed by cleavage of the S–S bond toward the formation of a new mixed disulfide derivative (Scheme 1.17). Depending on the amount of thiol used, other mixed disulfide derivatives could be formed.

Among the disulfide derivatives, pyridyl disulfide [96] is the most common one since it reacts with thiols over a broad pH range to form a single mixed disulfide product (Scheme 1.18). Pyridine-2-thione is released as a by-product that helps to monitor the progress of the reaction spectrophotometrically (A$_{max}$ = 343 nm). It can also be removed from the bioconjugates by dialysis or desalting. The disulfide bioconjugation method has been used to covalently conjugate lipids with ONTs [97].

While the classic thiol bioconjugation techniques have been widely used, new methods [98] for selective

SCHEME 1.17   Bioconjugation via disulfide interchange reactions.

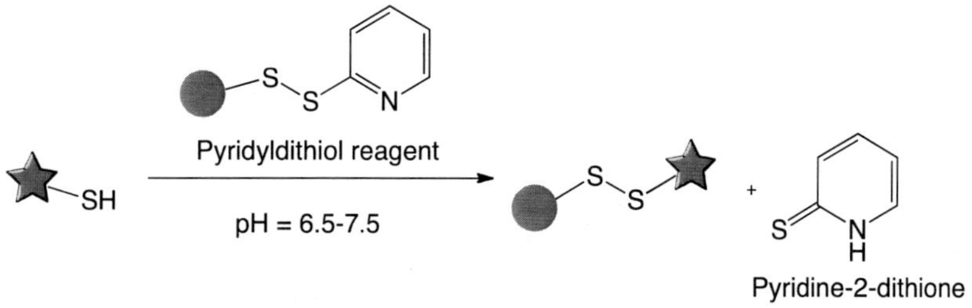

SCHEME 1.18   Chemical conjugation of thiol with pyridyldithiol reagent.

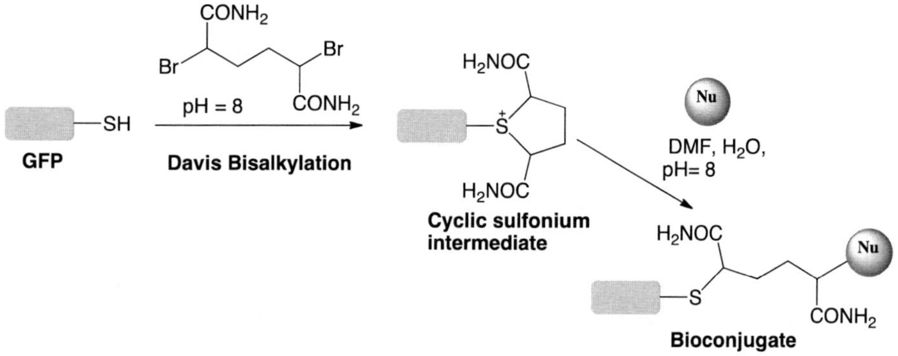

SCHEME 1.19   Bioconjugation via Davis bisalkylation and nucleophilic ring-opening reactions.

modification of thiol continue to be investigated. For instance, Caddick and coworkers [98(a)] recently reported a new thiol bioconjugation strategy of green fluorescent protein (GFP) via an unexpectedly stable cyclic sulfonium intermediate. The reaction involves the treatment of GFP with bisalkylating agents under Davis' conditions [99] followed by treatment with nucleophiles (Scheme 1.19). This new bioconjugation method will open a new entry to functionalized proteins consisting of useful chemical motifs.

Davis and coworkers disclosed a two-step protocol for cysteine modification via an oxidative elimination of cysteine into a dehydroalanine followed by a Michael addition with thiol reagents to afford a stable thioether linkage (Scheme 1.20) [98b]. The oxidative elimination step is induced by *O*-mesitylenesulfonylhydroxylamine (MSH) under alkaline conditions (pH 10–11). The reaction is compatible with methionine residues. However, the

Michael addition step is not stereospecific and thus, the stereochemistry of the cysteine is not preserved. This new bioconjugation method allows the preparation of protein–carbohydrate and protein–peptide conjugates.

1.2.5   Hydroxyl Modifications

The hydroxyl group (−OH) is prevalent in many biologically important molecules, namely carbohydrates, lipids, nucleotides, glycans, peptides/proteins (Thr, Ser, and the phenolic hydroxyl of Tyr), and natural products. In the field of synthetic chemistry, there is an exhaustive list of hydroxyl transformations, and the latter have mainly been applied in the synthesis of natural products. These transformations are generally carried out in an organic solvent and the absence of water. However, hydroxyl bioconjugation reactions turn out to be more challenging as compared to other functional

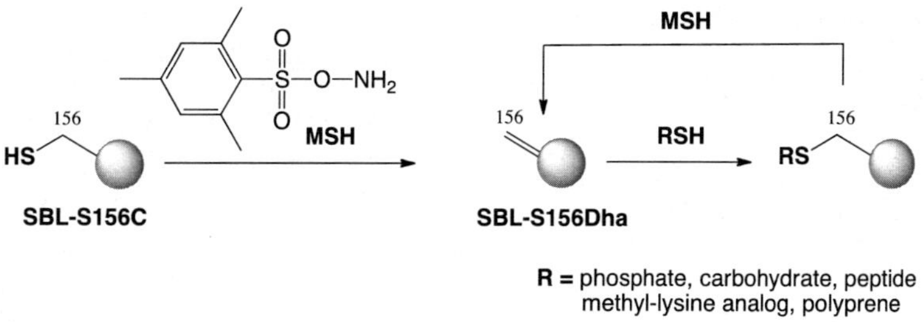

R = phosphate, carbohydrate, peptide
methyl-lysine analog, polyprene

SCHEME 1.20   Davis' two-step method for cysteine modification. Adapted with permission from Reference 98(b), Copyright 2008, American Chemical Society.

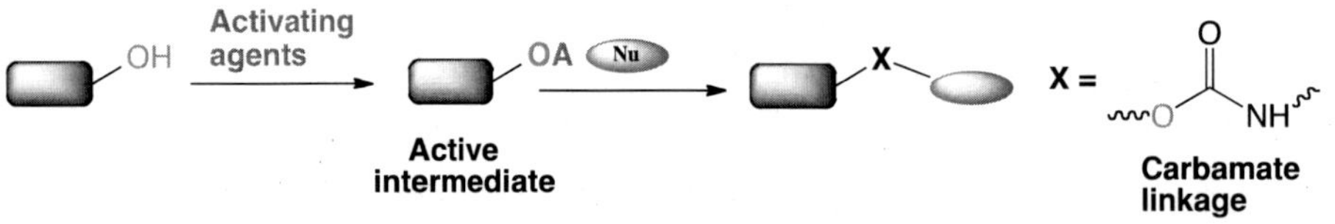

SCHEME 1.21   Bioconjugation via hydroxyl activation and nucleophilic displacement.

groups such as amines, thiols, or carboxylic acids owing to the hydroxyl's relatively low nucleophilicity. Moreover, hydroxyl bioconjugation reactions are often thwarted by the presence of water. Despite these problems, the covalent bioconjugation of two molecular entities via an alcohol has been developed and is generally placed in the following categories:

1.2.5.1   Hydroxyl Activation

The hydroxyl group is transformed to a good leaving group in the presence of an activating agent (A) followed by nucleophilic attack, usually by an amine, to yield a carbamate linkage (Scheme 1.21).

Typical activating agents are *N,N′*-Carbonyldiimidazole (CDI), *N,N′*-Disuccinimidyl carbonate (DSC) and *N*-Hydroxysuccinimidyl chloroformate (Figure 1.5). The active intermediate generated by the reaction of CDI with hydroxyl group can react readily with amines to afford a stable carbamate linkage and the concomitant release of imidazole. This bioconjugation strategy has been applied for immobilization of amine-containing affinity tags [100]. Succinimidyl carbonate was used to activate poly(ethylene) glycol for subsequent coupling with proteins [101]. The major disadvantage with these activating agents is that they are not selective for the hydroxyl group, since they preferentially react with amines and carboxylates.

1.2.5.2   Hydroxyl Oxidation

Hydroxyl groups located on adjacent carbons (commonly known as *cis*-diols) can be easily oxidized by sodium periodate to the corresponding dialdehydes (Scheme 1.22). Periodate oxidation method has been used for conjugation of amine-containing dyes with sugars and polysaccharides and other molecules possessing *cis*-diols. The reactive aldehyde formed can then be used to conjugate with other biomolecules via reductive amination, oxime, or hydrazone [102] methods.

Sodium periodate oxidation has also been used to oxidize adjacent hydroxyl groups on an aromatic ring (for

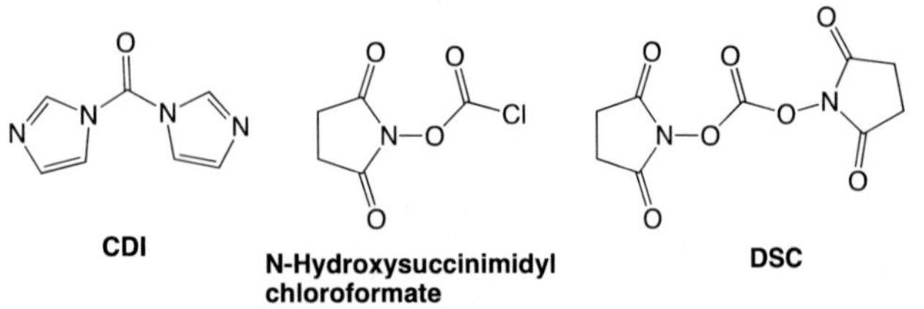

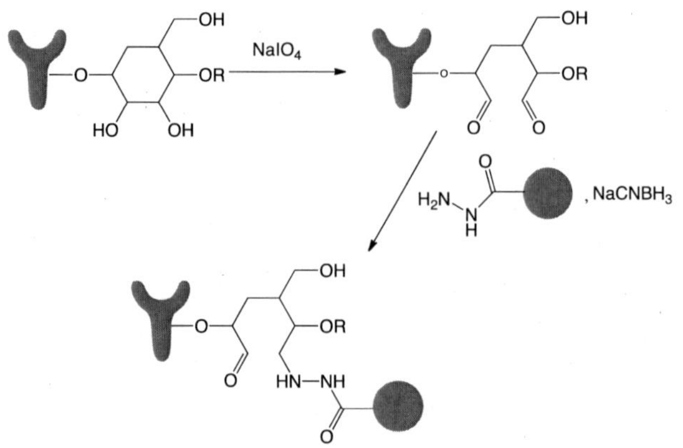

SCHEME 1.22   Sodium periodate oxidation of diols followed by conjugation with hydrazide to generate a covalent hydrazone linkage.

example, *L*-3, 4-dihydroxyphenylalanine derivatives) to generate a reactive *ortho*-quinone intermediate, which can be easily intercepted by a nearby nucleophile through an intramolecular Michael addition (Scheme 1.23) [103].

This novel covalent bioconjugation method was recently extended for the conjugation of protein with a polysaccharide (Scheme 1.24) [104]. However, in this case, after oxidation, the conjugation occurs via an intermolecular Michael addition of the amine from chitosan.

Enzymes such as galactose oxidase have also been exploited for the oxidation of hydroxyl groups on polysaccharide chains [1a].

1.2.5.3   Phenolic Hydroxyl Modifications

The phenolic hydroxyl group of tyrosine is only modestly prevalent and is often buried within the protein structure [105]. As such, it represents an attractive target for bioconjugation. Two main sites of tyrosine residues can be targeted, namely the phenolic hydroxyl group or the electron-rich aromatic ring position *ortho* to the phenolic hydroxyl (Scheme 1.25). In 2004, the Francis group [106] disclosed the first report of a chemoselective method for tyrosine targeting, and further studies established that electron-deficient diazonium salts could be added to tyrosine residues by electrophilic aromatic substitution (Scheme 1.25(A)). Other tyrosine modifications include:

A three-component Mannich-type coupling involving tyrosine, an aldehyde, and an electron-rich aniline to generate an *O*-substituted tyrosine moiety (Scheme 1.25(B)). Strategy has been applied to conjugate proteins with synthetic peptides [44].

FIGURE 1.5   Chemical structures of hydroxyl activating agents.

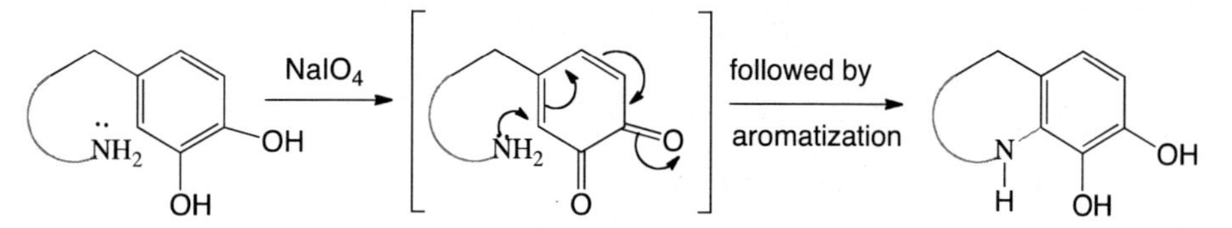

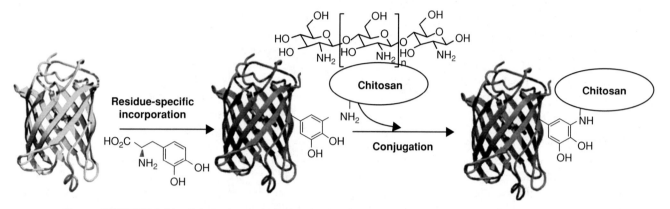

SCHEME 1.23 Ortho-quinone formation by periodate oxidation and intramolecular nucleophilic attack of pendant amine.

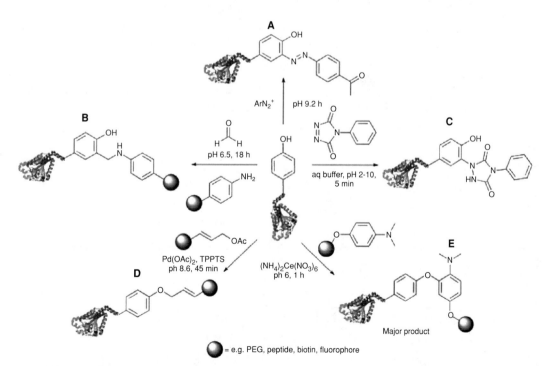

SCHEME 1.24 Schematic representation for the bioconjugation of protein with a polysaccharide. Reprinted with permission from Reference 104, Copyright 2011, American Chemical Society.

An ene-type reaction between tyrosine and a cyclic diazocarboxamide to afford a highly stable 1,2,4-triazoldine-3,5-dione derivative (Scheme 1.25(C)) [107]. This mild, aqueous reaction works well over a broad pH range. An integrin-binding cyclic RGD peptide was conjugated to the therapeutic antibody Herceptin using this method.

A water-compatible selective palladium-catalyzed allylic oxidation of tyrosine residues (Scheme 1.25(D)). Method

SCHEME 1.25 Covalent strategies for tyrosine modifications. Reprinted from Reference 105 with permission from Royal Society of Chemistry.

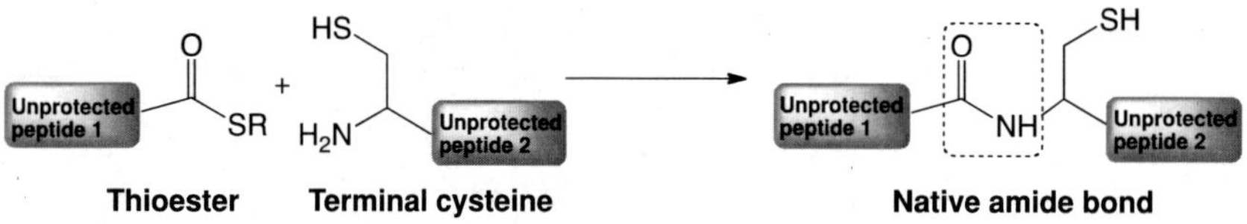

SCHEME 1.26   Schematic representation of NCL.

employs electrophilic π-allyl intermediates derived from allylic acetate and carbamate precursors to modify proteins at room temperature. This transition metal-catalyzed technique provides access to the preparation of synthetic lipoproteins [108].

A cerium-based transition metal-catalyzed strategy recently developed by Francis and coworkers [109]. The electron-rich aromatic ring of tyrosine undergoes an oxidative coupling with electron-rich anilines in the presence of cerium(IV) ammonium nitrate (CAN) as a one-electron oxidant (Scheme 1.24(E)). Attributes of this new bioconjugation strategy are excellent chemoselectivity, mild conditions, low concentration of oxidant and coupling partners, short reaction times, and high yields. Proteins were selectively modified with poly(ethylene)glycol (PEG) and smaller peptides.

1.2.6   Native Chemical Ligation and Expressed Protein Ligation

Native chemical ligation (NCL) is a powerful technique for the linking of two or more unprotected peptides to form large peptide–peptide conjugates. The NCL process involves reaction between a peptide with an N-terminal cysteine and a second peptide having a C-terminal thioester to afford an amide linkage (Scheme 1.26). The key step for the formation of the amide bond is via an S–N acyl shift. Historically, in 1953, Wieland et al. disclosed the chemical foundation of this chemical transformation [110], particularly the S–N acyl transfer step during the synthesis of the dipeptide, valine–cysteine. However, it was not until 1994 that Kent and coworkers [111] at the Scripps Research Institute reported the ligation of thioesters with N-terminal cysteine residues to yield a "native" amide or peptide bond. As such, the reaction was termed NCL.

Mechanistically, the reaction proceeds via a reversible, chemoselective, and regioselective transthioesterification, which connects the peptides through an intermediate thioester. The thioester intermediate undergoes a spontaneous irreversible intramolecular S–N acyl shift rearrangement to form the amide bond at the ligation site and regenerate the cysteine side-chain thiol. While the exact nature of the NCL mechanism is still unclear [112, 113], a generally accepted proposed mechanism is depicted in Scheme 1.27.

Usually, the NCL reaction is enhanced by the presence of thiol catalysts, since the peptide-α-thiolalkyl esters are relatively unreactive under NCL reaction conditions in water at pH = 7. In this case, a thiol–thiolester exchange reaction occurs prior to the transthioesterification step. Common thiol catalysts used for NCL are a benzyl mercaptan (1%)/thiophenol (3%) mixture for chemically synthesized peptide thioesters [114], or 2-mercaptoethanesulfonate sodium salt (MESNa) for recombinant peptide thioesters [115]. However, ligations still usually require long reaction times accompanied with side reactions. A recent study has shown that aryl thiols could be effective catalysts, in particular, (4-carboxymethyl)thiophenol (MPAA) is a highly effective one for NCL [112]. MPAA is water soluble and does not have an offensive odor as one would expect for typical thiol compounds. Chemical ligations are complete within an hour, with high yields.

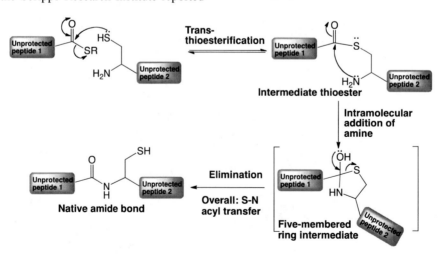

SCHEME 1.27   Proposed mechanism of NCL.

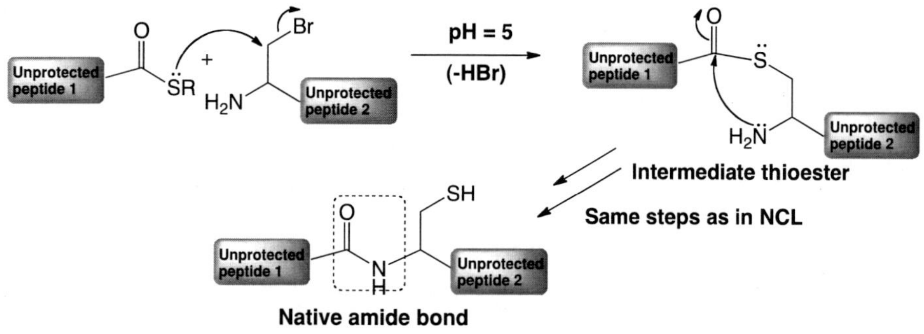

SCHEME 1.28   Tam's modified version of NCL.

A modified version of NCL was reported by Tam and coworkers [116], which employed an N-terminal α-bromoAla whereby reaction with a C-terminal thioester gave the covalent thioester linkage (Scheme 1.28).

NCL reaction is highly chemoselective for the ligation of two diverse functionalized molecules under physiological conditions and does not require any protecting groups. These properties make NCL an attractive and powerful method for modification, synthesis, and semisynthesis of peptides and smaller proteins (chain length < 200 amino acids (aa)). Proteins synthesized by NCL are much larger than the conventional solid-phase peptide synthesis (SPPS) (chain length < 60 aa) [117]. However, larger proteins cannot be easily synthesized by one NCL step. Multi-step NCL of different peptide segments are required [118]. NCL process has been applied to prepare dendrimer–peptide/proteins (GFP) [119] (Scheme 1.29 [120]) and protein–liposome conjugates [121].

Despite the widespread applications of NCL, the chemical synthesis of thioesters has always been the bottleneck in NCL. Most of the thioesters' preparation relied previously on the Boc strategy [111, 116 (b)] using solid-phase peptide synthesis due to the base lability of the thioester. Recent studies have demonstrated that 9-fluorenylmethoxycarbonyl (Fmoc) strategy [122–124] could be used instead of the Boc method to improve the yield of thioester. Several Fmoc-deprotection methods [125–127] were developed in order to liberate the peptide thioester from the resin in high yields.

NCL applications were further enhanced by its marriage with recombinant protein technology toward new powerful approaches to protein semisynthesis. The resulting combined technology is known as *Expressed Protein Ligation* (EPL). EPL [115, 128–131] also known as intein-mediated ligation of expressed proteins [132] allows recombinant and synthetic unprotected polypeptides to be chemoselectively and regioselectively combined together via a native peptide bond under mild aqueous conditions. Since its discovery, EPL's applications have grown significantly in order to address complex biological questions [133]. The overall EPL technique for protein semisynthesis involves the following general steps (Scheme 1.30):

(a) Recombinant protein to be ligated is first expressed as an N-terminal intein, which is fused with a chitin-based domain (CBD) on the C-terminal side of the intein. The CBD aids in the affinity purification process.

(b) NCL is initiated *in situ* by incubating carboxy-terminal Cys peptides with the protein thioesters in the presence of thiol (e.g., thiophenol) in buffer.

(c) Chitin-bound intein is removed by filtration resulting in purified semisynthetic protein.

EPL exploits inteins as a means to form a C-terminal thioester [1e]. N-terminal Cys polypeptides can be obtained

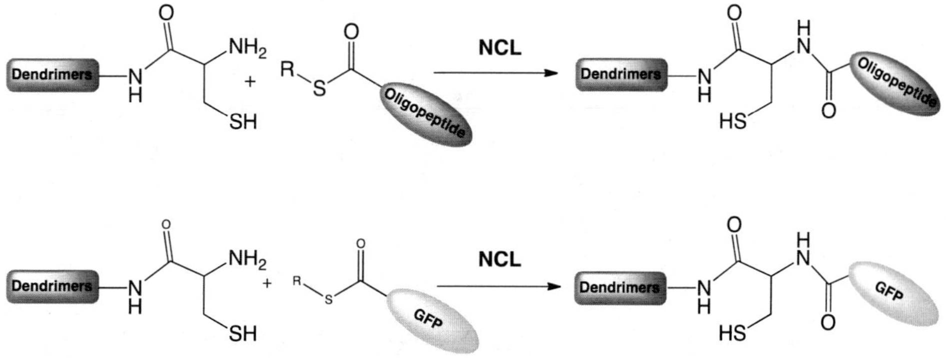

SCHEME 1.29   Peptide–dendrimer and fluorescent protein–dendrimer conjugates via NCL.

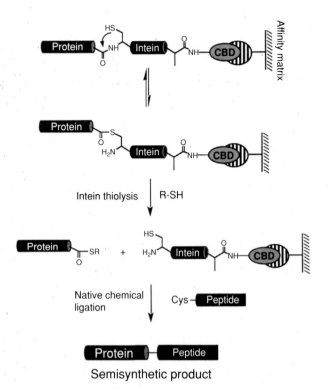

SCHEME 1.30 Schematic representation of EPL process. Reprinted with permission from Reference 115.

recombinantly, using engineered inteins, or by chemical synthesis (SPPS). While the chemical section can be as small as possible, the expressed part is not limited in size. Thus, EPL allows the synthesis of chemically modified proteins [134] of chain length greater than 500 aa, which overcomes the size limitation of NCL. However, in spite of numerous applications of both NCL and EPL, the requirement of a cysteine residue (mimic) at the ligation site is still a major obstacle. In this respect, recent studies have developed to circumvent this limitation. For example, NCL with cysteine mimetics (N-α-(ethanethiol) or an N-α-(oxyethanethiol)) followed by treatment with Zn/H⁺ afforded Gly at the ligation site [135], while NCL combined with desulfurization (Ni/H₂) led to an Ala residue [136] (Scheme 1.31).

A closely related technology to EPL is protein transsplicing (PTS), which is also based on the use of inteins. In PTS, artificially or naturally split inteins are employed to form a new peptide bond between their flanking exteins [137, 138]. PTS has enabled the extension of NCL in living systems for cyclic peptide synthesis [139, 140], *in vivo* semisynthesis of proteins [141], and the study of protein–protein interactions [142].

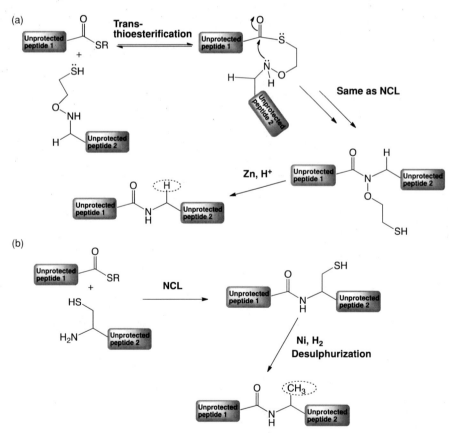

SCHEME 1.31   Alternate ligation methods: (i) Gly and (ii) Ala at ligation sites.

1.2.7 Cross-linking Reagents for Bioconjugation

One of the most useful and ready tools for bioconjugation is perhaps the cross-linking reagents. Cross-linking generally refers to the process of chemically combining two or more molecules via a covalent linkage. These cross-linking reagents (or cross-linkers) possess reactive end-groups that response to specific functional groups such as amines, thiols, carboxyls, and carbonyls. Proteins have many of these functional groups, and thus proteins [143] and peptides are the most studied biomolecules using cross-linking methods; however, cross-linkers are also often used to modify drugs, nucleic acids, and solid surfaces. Nowadays, there is a long, growing list of cross-linkers that are mostly commercially available from many suppliers (Thermo Scientific Pierce, G-Biosciences, Cyanagen, Sigma-Aldrich, etc.). While this makes it easier from a synthetic point of view, it becomes more overwhelming when choosing the correct cross-linker for a particular application. The choice of cross-linkers usually depends on their chemical reactivity and properties with respect to the application. Cross-linkers are normally selected based on the following important features [144]:

(a) *Chemical specificity*: type of reactive groups; whether the reagent possesses the same or different reactive end-groups. Usually, a cross-linker will have a minimum of two reactive groups at either end.

(b) *Spacer arm or connectors*: the length [145] and nature of the spacer arm; the conformational flexibility, hydrophilicity, or hydrophobicity.

(c) *Cell-membrane permeability*: whether the reagents are permeable or impermeable to cells/membranes.

(d) *Chemical reactivity*: whether the reagent will react spontaneously upon addition or it can be activated at a specific time; whether the reagent is photo reactive.

(e) *Cleavability*: whether the cross-linker could be cleaved or reversed when required.

(f) *Important moieties*: whether the reagent contains moieties that can be radiolabeled or tagged with another label.

Note: Readers are directed to the Pierce website (www.piercenet.com) which contains a user-friendly cross-linker selection guide by which the above-listed features may be chosen and a list of those cross-linkers with those selected features will be quickly generated.

As mentioned earlier, cross-linkers have at least two reactive end-groups and those with two reactive groups are usually termed as bifunctional cross-linkers. Bifunctional cross-linkers are further classified as either *homobifunctional* or *heterobifunctional* reagents depending on whether they have the same or different reactive groups (Figure 1.6).

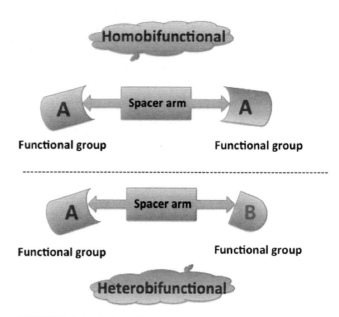

FIGURE 1.6   Schematic representation of homobifunctional (same end-functional groups) and heterobifunctional (different end-functional groups) cross-linkers.

1.2.7.1 Homobifunctional Cross-linkers   Homobifunctional cross-linkers possess similar reactive groups at opposite ends of the cross-linker's spacer arm and are usually symmetrical in design. They have the advantage of being reacted in a one-step chemical cross-linking reaction. However, self-conjugation, polymerization, and intracellular cross-linking are common issues with homobifunctional cross-linking agents. Polymerization can be minimized in a two-step protocol by first allowing the cross-linker to react with one biomolecule, followed by removal of excess cross-linker and by-products. The second biomolecule is then added and allowed to react with the previously activated biomolecule. A common problem in the two-step protocol is that the activated biomolecule intermediate can hydrolyze and degrade rapidly prior to the cross-linking step. Despite these issues, homobifunctional cross-linkers continue to be widely used as they do afford effective bioconjugates. Given the vast number of homobifunctional cross-linkers available, the following tables arrange them with respect to their reactivity toward amino groups, thiols, and other important functional groups. A summary of their attributes and some useful literature references are also provided.

Amines, lysine ε-amines, and N-terminal α-amines are the most abundant groups on proteins, and thus have been the most common target for cross-linking. The two most common amine-reactive groups that have been targeted for cross-linking are the imidoesters and NHS esters (see also Section 1.2.3). Imidoester homobifunctional cross-linkers are among the oldest cross-linkers [146], developed to react with primary amines and afford amidine linkages (Scheme 1.32,

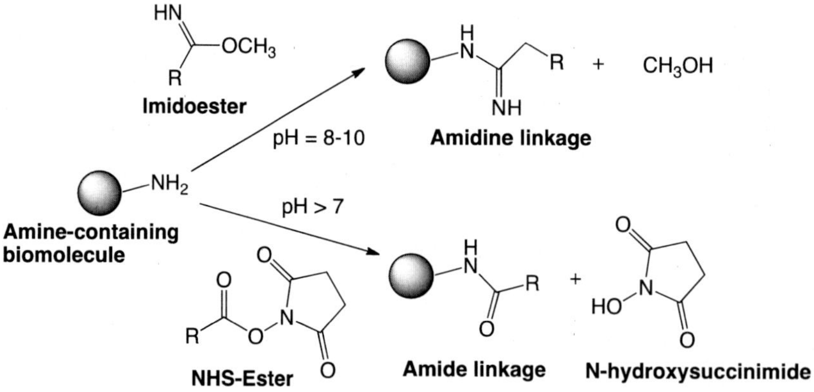

SCHEME 1.32   Conjugation of amines with imidoesters and NHS esters.

Table 1.3). The cross-linking occurs rapidly at pH 10, although amidine formation is favored at pH 8–10. They are highly water soluble but possess short half-lives [147, 148]. The resulting amidine bioconjugate is protonated, and therefore carries a positive charge at physiological pH [149, 150]. Imidoesters can penetrate cell membrane and cross-link proteins within the membrane and thus, imidoester homobifunctional cross-linkers have been used for the study of protein structure, molecular associations in membranes and immobilization of proteins onto solid-phase supports. The amidine linkages formed by imidoester cross-linkers are reversible at high pH and, hence, the more stable and efficient NHS–ester cross-linkers were developed (Table 1.4)[151, 152]. Due to the latter excellent reactivity at physiological

TABLE 1.3   Imidoester Homobifunctional Cross-linkers Reactive toward Amino Groups

Chemical Structures, Names, and Abbreviations	Characteristics/Applications	References
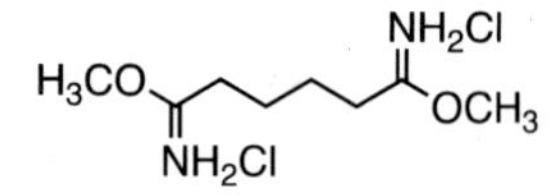 Dimethyladipimidate hydrochloride (DMA)	– Six-atom spacer arm – Noncleavable – Water soluble – Reacts rapidly at pH 8–10 – Reversible at high pH values – Retains charge character – For study of quaternary structure of proteins	155–157
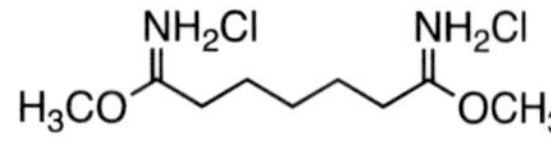 Dimethylpimelimidate hydrochloride (DMP)	– Seven-atom spacer arm – Noncleavable – Water soluble – Reacts at pH 8–10 – Reversible at high pH values – Retains charge character	158–161
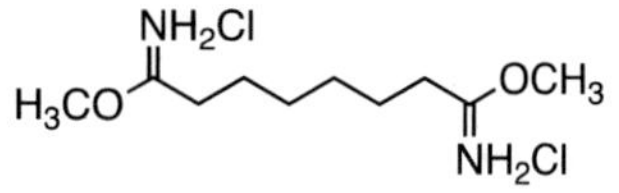 Dimethylsuberimidate hydrochloride (DMS)	– Eight-atom spacer arm – Noncleavable – Water soluble – Reacts at pH 8–10 – Reversible at high pH values – Retains charge character	157, 162
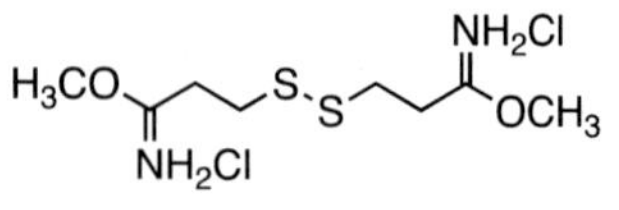 Dimethyl 3,3'-dithiobispropionimidate hydrochloride (DTBP)	– Cleavable, membrane permeable – Water soluble – Reacts at pH 8–10 – Cross-linking can be reversed with reducing agents (DTT) – Net positive charge at physiological pH – Stabilize protein–protein interactions	163–165

TABLE 1.4   NHS-Esters Homobifunctional Cross-linkers Reactive toward Amino Groups

Chemical Structures, Names, and Abbreviations	Characteristics/Applications	References
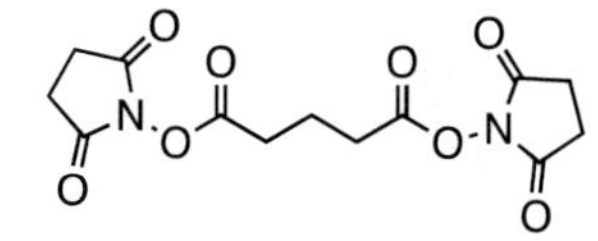 Disuccinimidyl glutarate (DSG)	– Noncleavable – Water insoluble – Reacts at pH 7–9 – For biomarker analysis – Capture of protein interactions on protein array surfaces	166, 167
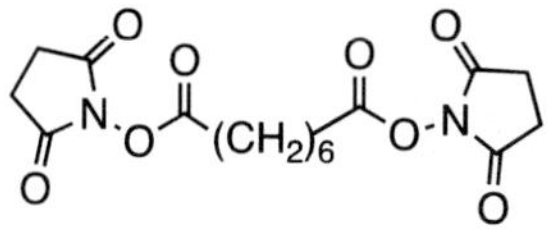 Disuccinimidyl suberate (DSS) 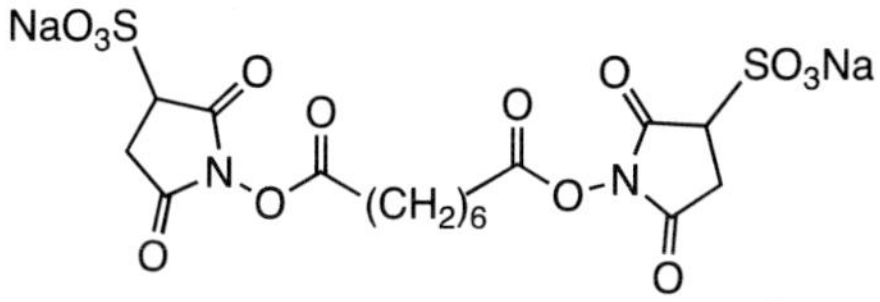 Bis(sulfosuccinimidyl) suberate (BS³)	– Eight-carbon spacer arm – Membrane permeable – Noncleavable – Reacts at pH 7–9 (PBS buffer) – Need to dissolve in organic solvent (DMF or DMSO) – Intracellular conjugations – Protein interaction studies – Membrane impermeable – Water soluble – Noncleavable – Proteomics applications	168–172
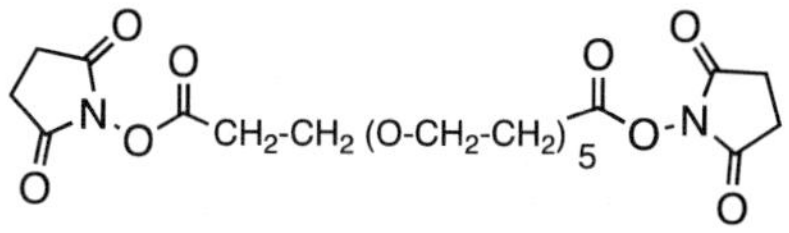 Bis(succinimidyl) penta (ethyleneglycol) BS-(PEG)₅ 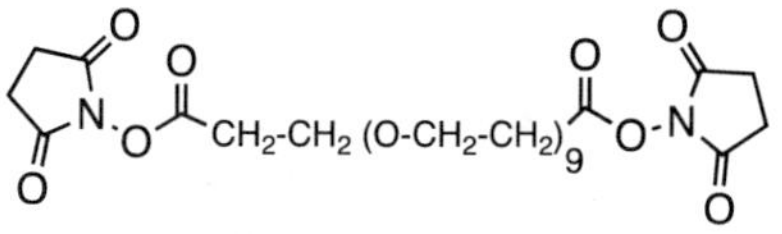 Bis(succinimidyl) nona (ethyleneglycol) BS-(PEG)₉	– Flexible PEG spacer arm – Irreversibly cross-linking – Reacts at pH 7–9 – PEG spacer arm provides increased water solubility, stability – Reduced immunogenicity – Ideal for small molecule or peptide conjugations – Excellent for cross-linking cell surface proteins to stabilize protein interactions	www.piercenet.com
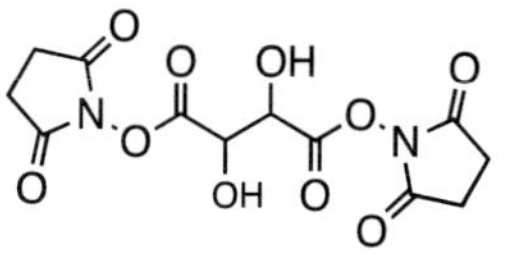 Disuccinimidyl tartrate (DST)	– Four-carbon spacer arm – Cleavable by periodate oxidation – Partially soluble in water – Reacts at pH 7–9 – Sulfo-DST analog also available and more water soluble – Study of protein–lipid complexes	173–175
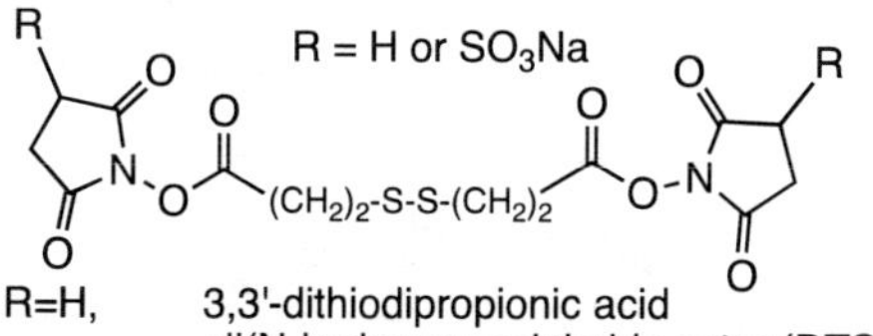 R = H or SO₃Na R=H, 3,3'-dithiodipropionic acid di(N-hydroxysuccinimide ester) (DTSP) R=SO₃Na 3,3'-dithio-bis-(3-sulfo-N-hydroxysuccinimidyl propionate) disodium salt (DTSSP)	– DTSP or DSP: Lomant's reagent—Eight-atom spacer arm – Membrane permeable – Reacts at pH 7–9 (PBS buffer) – Disulfide bond can be cleaved by reducing agents at pH 8.5 – Spacer arm can be cleaved with 5% beta-mercaptoethanol – DTSSP: same as above except – Membrane impermeable – Water soluble – Cell-surface cross-linking possible	176–181

(continued)

TABLE 1.4   (*Continued*)

Chemical Structures, Names, and Abbreviations	Characteristics/Applications	References
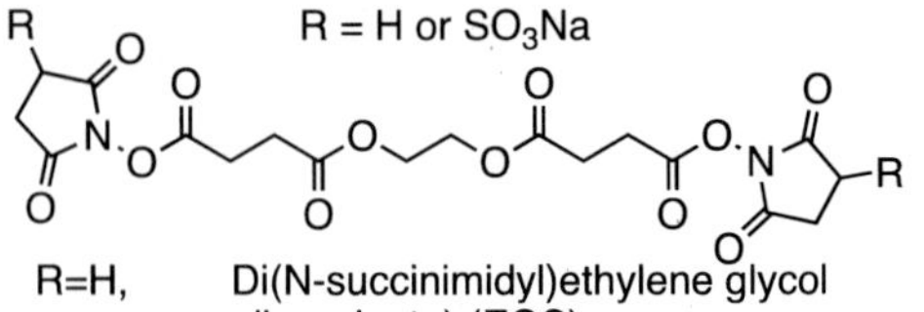 R = H or SO₃Na R=H, Di(N-succinimidyl)ethylene glycol disuccinate) (EGS) R=SO₃Na Ehyleneglycol bis-(3-sulfo-N-hydroxysuccinimylsuccinate) disodium salt (Sulfo-EGS)	– 12-atom spacer arm – EGS is water insoluble – Cleavable by hydroxylamine at pH 8.5 for 3–6 h at 37°C – Sulfo-EGS is water soluble – Reaction conditions same as EGS – Both used for study protein interactions and large protein complexes	182–184
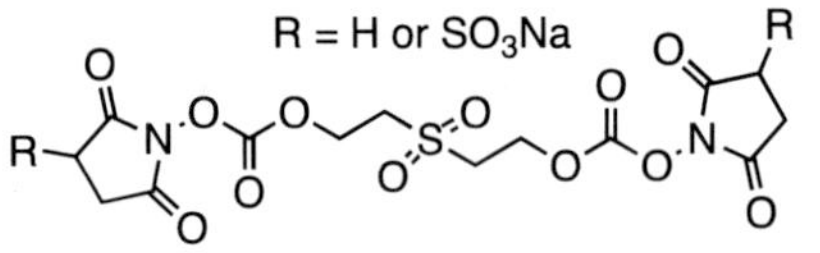 R = H or SO₃Na R=H, Bis[2-(succinimidylcarbonyloxy) ethyl] sulphone (BSOCOES) R=SO₃Na Bis[2-(sulfosuccinimidylcarbonyloxy) ethyl] sulphone (sulfo-BSOCOES)	– BSOCOES is water insoluble – Cleavable by base at pH 8.5 for 2 h at 37°C – Sulfo-BSOCOES is water soluble – Both used for studying cellular and subcellular distribution of the type II vasopressin receptor	185–187

pH, they have slowly replacing imidoester homobifunctional cross-linkers. NHS-ester cross-linkers react with amines in phosphate/carbonate or borate buffers (50–200 mM) to form amide bonds with the release of *N*-hydroxysuccinimide as a by-product (Scheme 1.32). NHSester cross-linkers can be grouped into two types depending on their water-solubility properties; however, they all have almost same reactivity toward amines. Water-insoluble NHS esters usually need to be first dissolved in an organic solvent (DMSO or DMF) prior to addition to the aqueous reaction mixture. Sulfo-NHS esters were introduced as water-soluble and membrane impermeable cross-linkers and can be used when organic solvents are not tolerable to the reaction conditions [153]. Sulfo-NHS cross-linkers possess better half-lives of hydrolysis than the NHS-ester cross-linkers [154]. They have been mainly employed for cell–surface conjugation since they cannot permeate the membrane.

Note: The general precaution for amine conjugation is to avoid buffers containing amines such as Tris or glycine.

Coupling through thiol (sulfhydryl) groups is advantageous since it can be site directed, allow for sequential coupling and yield cleavable products. Usually, a protein in a complex mixture can be specifically labeled if it is the only one with a free thiol group on its surface. The common thiol-reactive groups that have been exploited for the design of cross-linkers are maleimides, iodoacetyl, and pyridyl disulfides (Figure 1.7) (see Section 1.2.4 for more details).

Maleimides have been routinely used for the design of thiol-reactive homobifunctional cross-linkers, which typically react at near neutral pH (6.5–7.5) to afford stable irreversible thioether linkages. Maleimides are unreactive toward tyrosine, histidine, or methionine residues. Thiols must be

excluded from reaction buffers used with maleimides in order to prevent competition for coupling sites. Excess maleimides can be quenched at the end of a reaction by adding free thiols while ethylenediaminetetraacetic acid (EDTA) is often included in the coupling buffer to minimize oxidation of thiols. Many thiol-reactive homobifunctional cross-linkers with different spacer lengths are reported in the literature [188–198]; however, most of these are not easily available and have to be synthesized. The following table 1.5 lists the ones that are commonly employed and commercially available from many suppliers.

Note: The general precaution for thiol conjugation is to remove reducing agents from the conjugation reaction and add a metal chelating agent (EDTA) as an antioxidant.

Other homobifunctional cross-linkers reactive toward thiols that have been reported include pyridyl disulfide (namely 1,4-di-[3′-(2′-pyridyldithio)propionamido]butane (DPDPB)), bisiodoacetamide, and bisepoxide derivatives

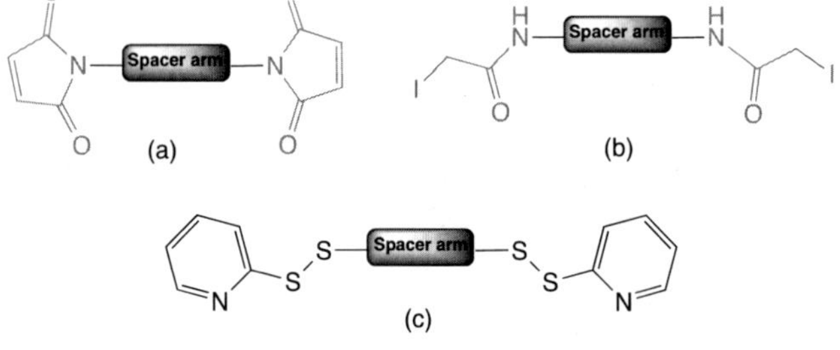

FIGURE 1.7   General representation of thiol-reactive homobifunctional cross-linkers: (a) maleimide, (b) iodoacetyl, (c) pyridyldisulfide.

TABLE 1.5   Maleimides Homobifunctional Cross-linkers Reactive toward Thiol Groups

Chemical Structure, Name, and Abbreviation	Characteristics	References
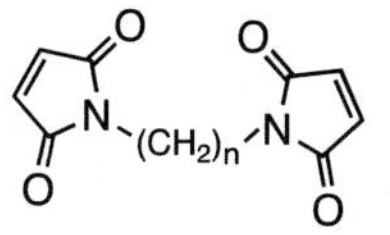 n = 2, 1,2-bis-maleimidoethane (BMOE) n = 4, 1,4-bis-maleimidobutane (BMB) n = 6, 1,6-bis-maleimidohexane (BMH)	– Noncleavable – Water insoluble – Reacts at pH 6.5–7.5—irreversible	199–202
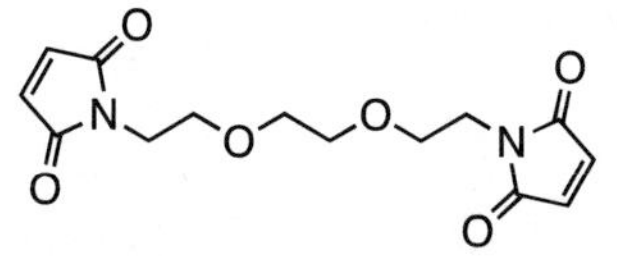 1,8-bis-maleimidotriethylene glycol (BM(PEO)$_3$)	– Polyether spacer – Water soluble – Reacts at pH 6.5–7.5 – Irreversible – Increased stability – Reduced aggregation – Less immunogenicity	203
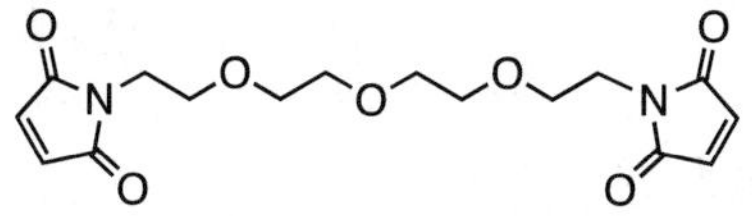 1,11-bis-maleimido-tetraethyleneglycol (BM(PEO)$_3$)		
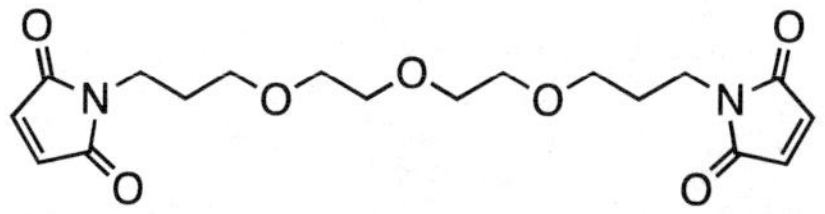 1,13-bis-maleimido-4,7,10-trioxadecane (BM(PEO-PPO)$_2$)		
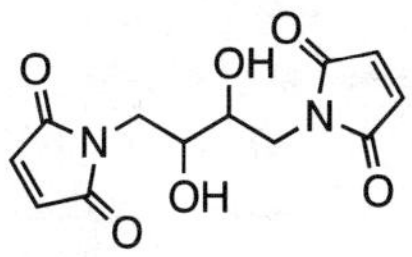 1,4-bis-maleimidyl-2,3-dihydroxybutane (BMDB)	– Mid-length – Reacts at pH 6.5–7.5 – Reversible cross-link – Cleavage by 15mM sodium periodate	204
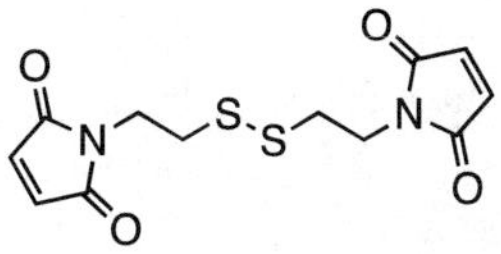 Dithiobismaleimidoethane (DTME)	– Intermediate length – Water insoluble – Reacts at pH 6.5–7.5 – Reversible cross-link – Cleavable by DTT or 2-mercaptoethanol	199, 72(a)

(Figure 1.8). DPDPB reacts with thiols to form disulfide linkages [205, 206]. DPDPD is water insoluble and has to be dissolved in an organic solvent (25 mM DPDPB in DMSO) prior addition to reaction mixture. The disulfide linkage is cleavable with reducing agents. Haloacetamide cross-linkers react with thiols at physiological pH resulting in stable thioether linkages [196]. Nucleophilic ring-opening reactions of thiols with bisepoxides at pH 7.5–8.5 afford thioether bonds and hydroxyl groups [207]. Bisepoxides can also react with nucleophilic amines.

Homobifunctional cross-linking reagents reactive toward carbonyl groups (aldehydes or ketones) are mainly derivatives of hydrazides. Hydrazide-activated cross-linkers will react with aldehydes or ketones at pH 5–7 to generate stable

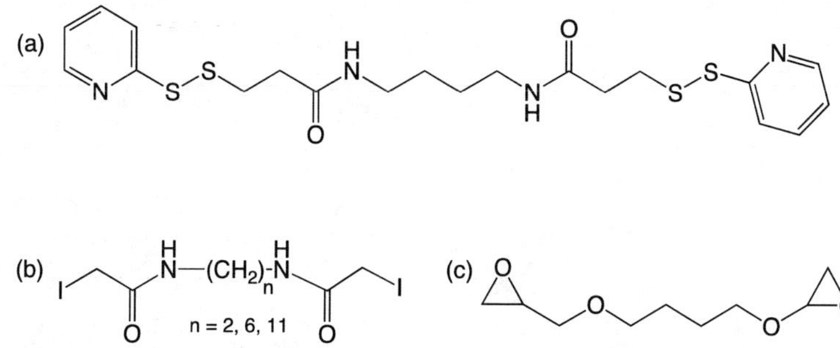

FIGURE 1.8   Homobifunctional cross-linkers reactive toward thiols: (a) DPDPB, (b) bisiodoacetylamide derivatives, and (c) 1,4-butanediol diglycidyl ether (bisepoxides).

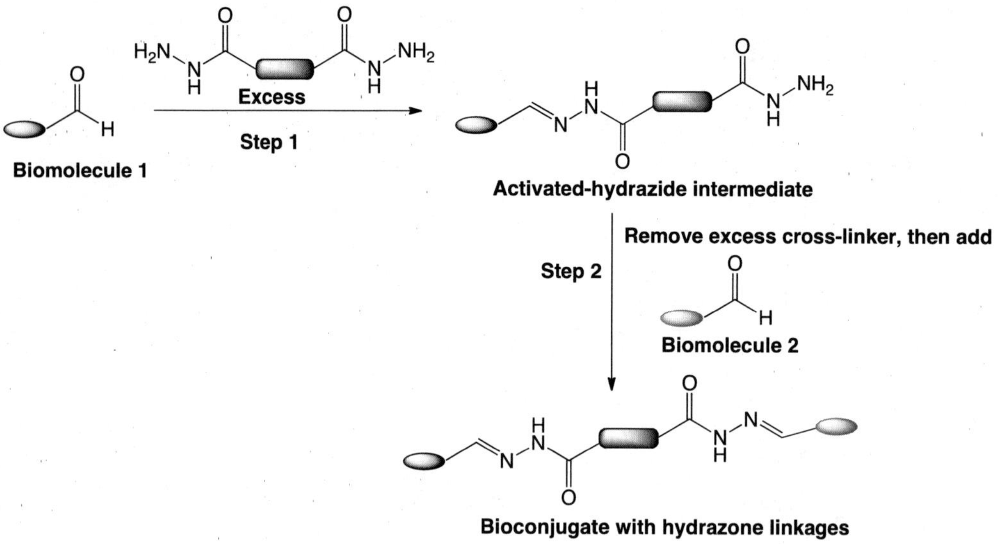

SCHEME 1.33   Conjugation of hydrazide homobifunctional cross-linker with aldehyde-containing biomolecules via a two-step method.

hydrazone linkages. While aldehydes do not readily exist in proteins or macromolecules, they can be readily introduced by mild oxidation of vicinal diols in carbohydrates using sodium meta-periodate. The oxidation has to be carried out in the dark at 0–4°C to avoid side reactions.

These bishydrazide cross-linkers could be used in a single step for the bioconjugation process; however, it is more efficient to perform the cross-linking via a two-step protocol. The two-step protocol involves, first, the addition of a huge excess of hydrazide cross-linker to an aldehyde-containing biomolecule 1 resulting in a hydrazide-activated biomolecule. Excess unreacted hydrazide cross-linker is then removed by desalting followed by addition of biomolecule 2 (Scheme 1.33).

The two most commonly used and commercially available (from Aldrich) hydrazide homobifunctional cross-linkers are carbohydrazide and adipic acid dihydrazide (ADH) (Figure 1.9). Carbohydrazide [208, 209] is a small five-atom spacer homobifunctional cross-linker with excellent water solubility properties. ADH, on the other hand, is a 10-atom spacer homobifunctional cross-linker, which is also water soluble. ADH was introduced in 1980 in glycoconjugate synthesis for the preparation of *Haemophilus influenzae* type b polysaccharide–protein conjugates and since then has been a popular reagent for constructing lattice-like constructs of polysaccharides with proteins [210–213].

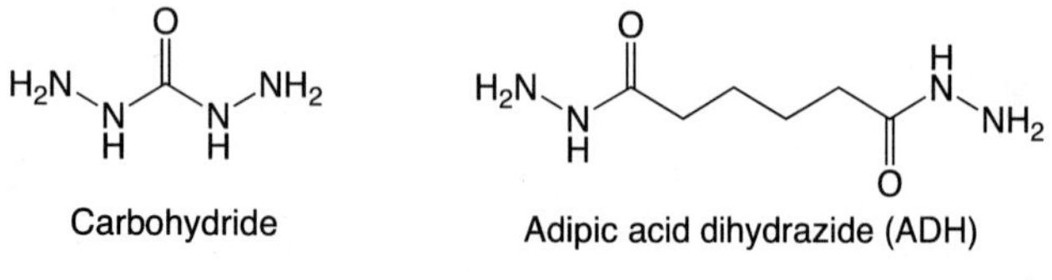

Similarly, the reverse of hydrazide-activation strategy described above, involves the use of dialdehyde derivatives as homobifunctional cross-linking reagents for conjugation of biomolecules. Aldehydes will react with hydrazides and amines at pH 5–7 to form hydrazone and imine linkages respectively. Imine linkage can be subsequently reduced by $NaBH_3CN$ to the more stable amine bond. The reaction with hydrazides is typically faster than with amines, rendering them vital for site-specific cross-linking. Among the dialdehyde homobifunctional cross-linkers (glyoxal, malonaldehyde, succinaldehyde), glutaraldehyde (linear five-carbon spacer) is the most popular and has been extensively used by immunologists in the 1970s for the bioconjugation of an enzyme with an antibody [214–216] (Scheme 1.34). Glutaraldehyde's popularity is mainly due to its commercial availability, low cost, high reactivity, and thermally and chemically stable cross-links [217]. In this case also, the two-step experimental protocol is more efficient than the single-step bioconjugation method [216]. In spite of the widespread use of glutaraldehyde as a cross-linker, it is still unclear about the nature of the mechanism of cross-linking with proteins, which has been a subject of huge debate in the past years. This is mainly because glutaraldehyde can exist in different forms even for specific and controlled reaction conditions [218].

1.2.7.2 Heterobifunctional Cross-linkers
Heterobifunctional cross-linkers have two different reactive groups that allow for sequential (two-stage) conjugations. The major advantage of heterobifunctional cross-linkers is that it helps to minimize undesirable polymerization or intramolecular cross-linking by allowing the least stable group to react first, at such condition where the other group is nonreactive. By

FIGURE 1.9   Chemical structures of carbohydride and ADH.

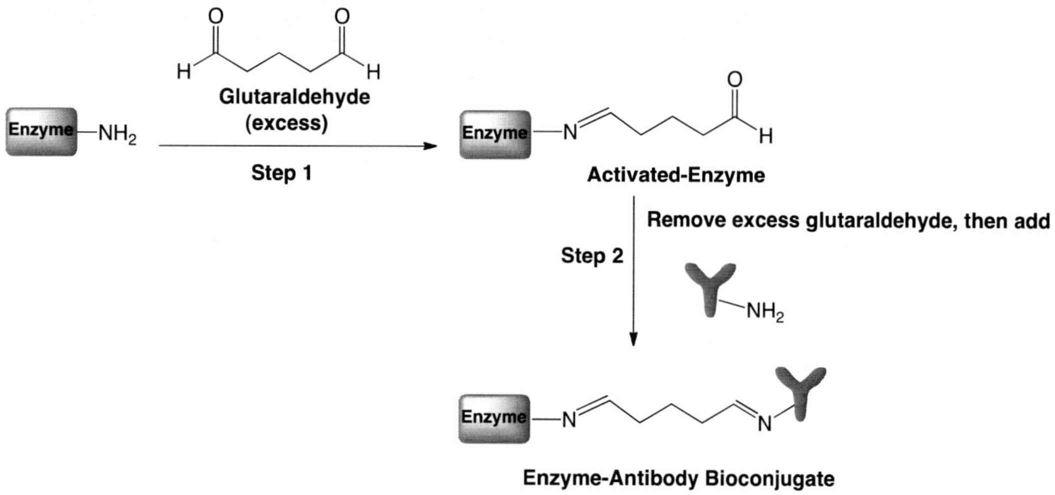

SCHEME 1.34 Conjugation of dialdehyde-homobifunctional cross-linker with amine-containing biomolecules via a two-step method.

far, the most commonly used heterobifunctional cross-linkers are those consisting of an amine-reactive succinimidyl ester (e.g., NHS ester) at one end and a thiol-reactive group on the other end connected by a spacer (Figure 1.10).

The thiol-reactive groups are typically maleimides, pyridyl disulfides, and α-haloacetyls. The NHS-ester reactivity is less stable in aqueous solution and is usually reacted first in sequential cross-linking procedures with amines to afford amide linkages. Table 1.6 lists the various commercially available (Thermo Scientific Pierce) noncleavable heterobifunctional cross-linkers (NHS ester-*spacer*-Maleimide) reactive toward amines and thiols having an aliphatic, aromatic, or a cyclohexane ring as spacer arms. Figure 1.11 depicts heterobifunctional cross-linkers reactive toward amine and thiols possessing different lengths of PEG-based spacers. The general structure is NHS ester-PEG$_n$-Maleimide, which differs in the number of discrete ethylene glycol units ($n = 2, 4, 6, 8, 12,$ or 24). The PEG (or polyethylene oxide (PEO)) spacer arms help in maintaining the solubility of the bioconjugate.

Other advantages include increased stability, highly flexible, nontoxic and reduced aggregation, and immunogenicity. Figure 1.12 illustrates NHS ester-spacer-iodoacetyl heterobifunctional cross-linkers whereby NHS ester reacts with primary amines at pH 7–9 to form stable amide bond while haloacetyl group reacts with thiols at pH > 7.5 to afford stable thioether linkage. Figure 1.13 shows cleavable heterobifunctional cross-linkers containing NHS ester-*spacer*-dipyridylsulfide. Disulfide bond in the spacer arm can be readily cleaved by reducing agents such as 10–50 mM DTT or tris-2-(carboxyethyl)phosphine (TCEP) at pH 8.

Red color indicates the amine-reactive group and blue color denotes thiol-reactive group.

1.2.7.3 Photoreactive Cross-linkers

Photoreactive reagents are chemically inert compounds that become reactive upon exposure to ultraviolet or visible light. Photoreactive cross-linking reagents are vital tools for determining proximity of two sites; for instance, they can reveal interactions among proteins, nucleic acids, and membranes in living cells. They usually possess a chemical reactive group as well as a photoreactive group. These cross-linkers are first chemically reacted with one molecule and then this modified molecule is conjugated to a second molecule in the presence of ultraviolet light. Depending on the reactive properties of the chemical and photoreactive groups, they can be used to couple like or unlike functional groups. The chemical reaction is normally carried out in subdued light with reaction vessels wrapped in foil. The photoactivation can be initiated with a bright camera flash or ultraviolet hand-held lamp about 1–2 inches above the reaction vessels. Historically, aryl azides (also known as phenylazides) have been the most popular photoreactive chemical group used in cross-linking and labeling reagents, mainly due to their

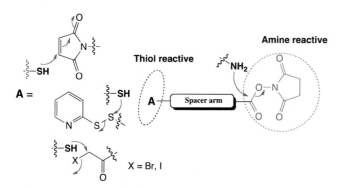

FIGURE 1.10 Schematic representation of heterobifunctional cross-linker reactive toward amines and thiols.

TABLE 1.6 Noncleavable heterobifunctional cross-linkers reactive (NHS ester-*spacer*-Maleimide) toward amine and thiol groups

Chemical Structures, Names, and Abbreviations	Characteristics	References
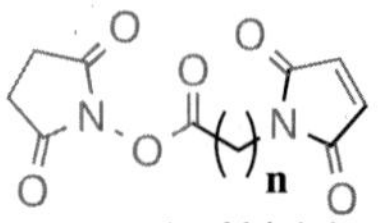 n = 1, *N*-(alpha-maleimidoacetoxy) succinimide ester (AMAS) n = 2, *N*-(beta-maleimidopropyloxy) succinimide ester (BMPS) n = 3, *N*-(gamma-maleimidobutyryloxy) succinimide ester (GMBS) n = 5, *N*-(-maleimidocaproyloxy) succinimide ester (EMCS)	– Noncleavable – Water insoluble – NHS ester reacts with amines at pH 7–9 – Maleimide reacts with thiols at pH 6.5–7.5 – Low potential to elicit immune response – Water-soluble Sulfo-GMBS, Sulfo-EMCS are also available. – Increased sphere of coupling for EMCS vs. GMBS	219–224
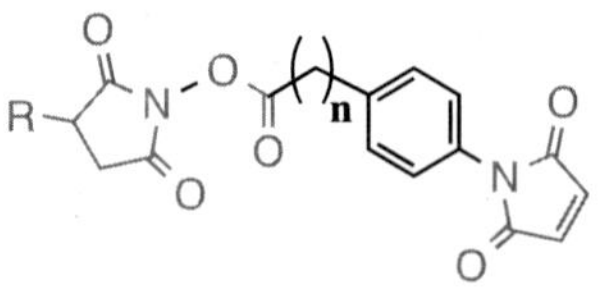 R =H, n = 0, *m*-Maleimidobenzoyl-*N*-hydroxysuccinimide ester (MBS) R =SO₃Na, n = 0, *m*-Maleimidobenzoyl-*N*-hydroxy sulfosuccinimide ester (Sulfo-MBS) R =H, n = 3, Succinimidyl 4-(*p*-maleimidophenyl) butyrate (SMPD) R =SO₃Na, n = 3, *m*-Maleimidobenzoyl-*N*-hydroxy sulfosuccinimide ester (Sulfo-SMPD)	**MBS:** – Noncleavable – Water insoluble – For enzyme immunoconjugates **SMPD:** – Extended chain analog of MBS limits steric effect – More stable conjugates than SPDP **Sulfo-MBS and Sulfo-SMPD:** – Non cleavable – Water soluble – Membrane impermeable	225–228
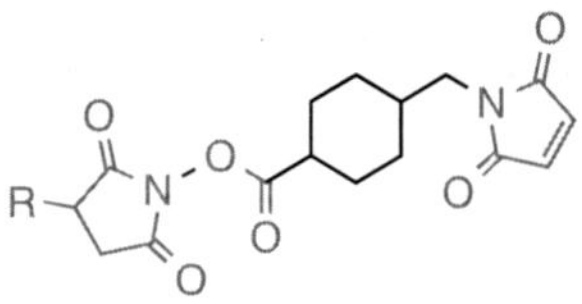 R =H, Succinimidyl 4-(*N*-maleimidomethyl) cyclohexane-1-carboxylate (SMCC) R =SO₃Na, Succinimidyl 4-(*N*-maleimidomethyl) cyclohexane-1-carboxylate (Sulfo-SMCC)	**SMCC:** – Noncleavable – Membrane permeable – NHS ester at pH 7–9 – Maleimide at pH 6.5–7.5 **Sulfo-SMCC** – Water soluble, membrane impermeable **Both SMCC and Sulfo-SMCC:** – Unusual extra stability due to cyclohexane ring – Ideal for enzyme labeling of antibodies	229–235

high reaction efficiencies, rapid kinetics, excellent storage stability, and ease of preparation (Figure 1.14).

The photochemistry of aryl azides is complex; however, when an aryl azide compound is exposed to UV light, it decomposes by releasing nitrogen and generates highly reactive and short-lived intermediate known as aryl nitrenes. These nitrenes can initiate addition reactions with double bonds or insertion into C–H and N–H sites [250] to form stable covalent adducts. They can also undergo rearrangement, such as ring expansion to the corresponding seven-membered ketenimine (dehydroazepine) intermediate, and the latter is reactive toward primary amines, thiols, and reactive hydrogen groups (Scheme 1.35) [251–253].

It was found that in order to increase the efficiency of addition/insertion reactions while suppressing ring expansion reaction, halogen atoms (F or Cl) should be introduced on the aromatic rings [255, 256]. Furthermore, any thiol-containing reducing agents (e.g., DTT or 2-mercaptoethanol) must be avoided in the sample solution during all steps before and during photoactivation. These reagents are known to reduce the azide functional group to an amine, and hence hinder photoactivation. Reactions can be performed in a variety of amine-free buffer conditions. If heterobifunctional photoreactive cross-linkers are used, then buffers compatible with the chemically reactive portion of the reagent are required. Currently, there are a number of photoreactive cross-linking reagents available for the coupling of proteins, peptides, nucleic acids, and other biomolecules. Figure 1.15 shows heterobifunctional NHS-based photoreactive cross-linkers, which are reactive toward

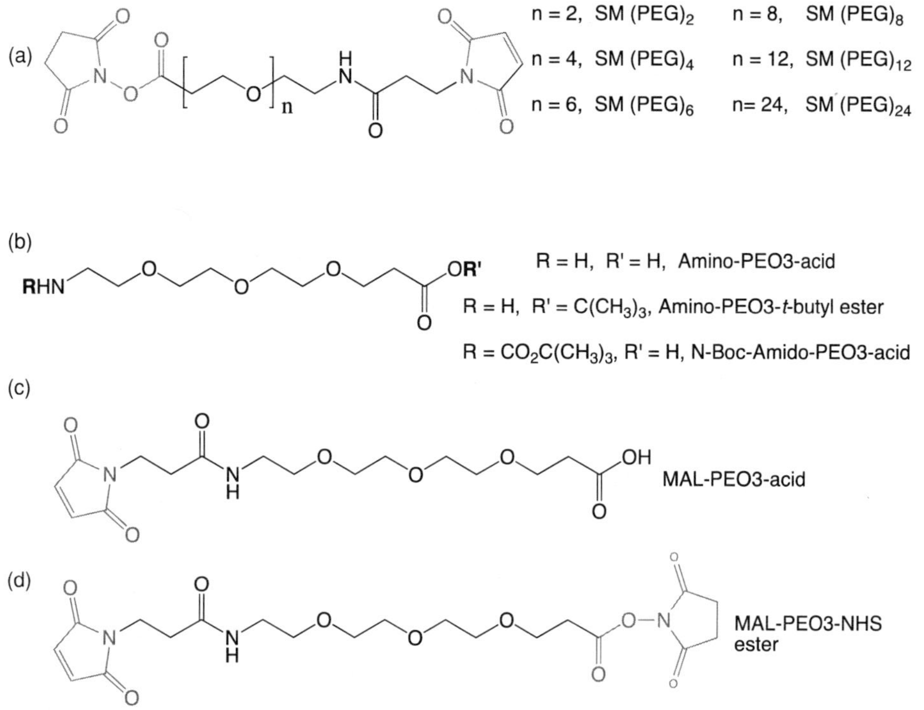

n = 2, SM (PEG)₂	n = 8, SM (PEG)₈

(a) ... n = 2, SM (PEG)$_2$ n = 8, SM (PEG)$_8$
n = 4, SM (PEG)$_4$ n = 12, SM (PEG)$_{12}$
n = 6, SM (PEG)$_6$ n= 24, SM (PEG)$_{24}$

(b) RHN ... OR'

R = H, R' = H, Amino-PEO3-acid

R = H, R' = C(CH$_3$)$_3$, Amino-PEO3-t-butyl ester

R = CO$_2$C(CH$_3$)$_3$, R' = H, N-Boc-Amido-PEO3-acid

(c) ... OH   MAL-PEO3-acid

(d) ... O–N   MAL-PEO3-NHS ester

FIGURE 1.11   General structures of PEG- or PEO-based heterobifunctional cross-linking reagents commercially available (a) Thermo Scientific Pierce (b), (c), (d) Cyanagen.

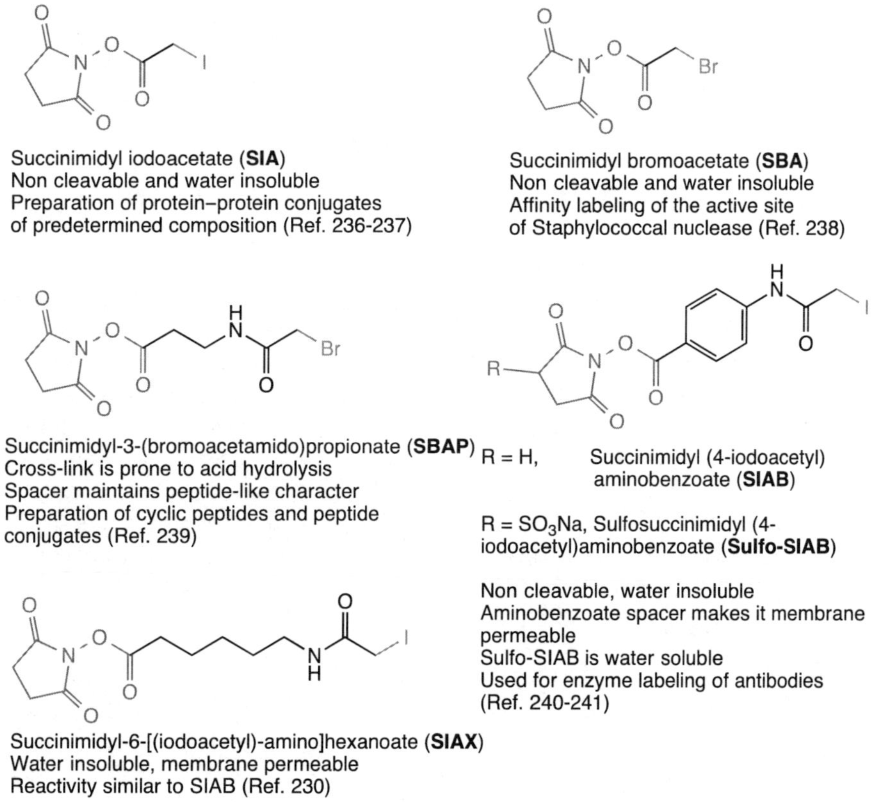

Succinimidyl iodoacetate (**SIA**)
Non cleavable and water insoluble
Preparation of protein–protein conjugates
of predetermined composition (Ref. 236-237)

Succinimidyl bromoacetate (**SBA**)
Non cleavable and water insoluble
Affinity labeling of the active site
of Staphylococcal nuclease (Ref. 238)

Succinimidyl-3-(bromoacetamido)propionate (**SBAP**)
Cross-link is prone to acid hydrolysis
Spacer maintains peptide-like character
Preparation of cyclic peptides and peptide
conjugates (Ref. 239)

R = H,   Succinimidyl (4-iodoacetyl)
                 aminobenzoate (**SIAB**)

R = SO$_3$Na, Sulfosuccinimidyl (4-
iodoacetyl)aminobenzoate (**Sulfo-SIAB**)

Non cleavable, water insoluble
Aminobenzoate spacer makes it membrane
permeable
Sulfo-SIAB is water soluble
Used for enzyme labeling of antibodies
(Ref. 240-241)

Succinimidyl-6-[(iodoacetyl)-amino]hexanoate (**SIAX**)
Water insoluble, membrane permeable
Reactivity similar to SIAB (Ref. 230)

FIGURE 1.12   NHS ester/iodoacetyl heterobifunctional cross-linking reagents.

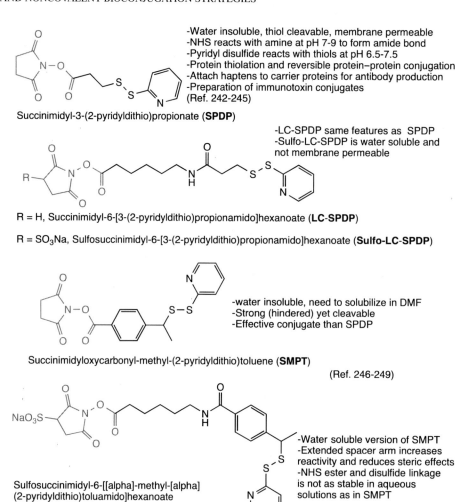

-Water insoluble, thiol cleavable, membrane permeable
-NHS reacts with amine at pH 7-9 to form amide bond
-Pyridyl disulfide reacts with thiols at pH 6.5-7.5
-Protein thiolation and reversible protein–protein conjugation
-Attach haptens to carrier proteins for antibody production
-Preparation of immunotoxin conjugates
(Ref. 242-245)

Succinimidyl-3-(2-pyridyldithio)propionate (**SPDP**)

-LC-SPDP same features as SPDP
-Sulfo-LC-SPDP is water soluble and not membrane permeable

R = H, Succinimidyl-6-[3-(2-pyridyldithio)propionamido]hexanoate (**LC-SPDP**)

R = SO₃Na, Sulfosuccinimidyl-6-[3-(2-pyridyldithio)propionamido]hexanoate (**Sulfo-LC-SPDP**)

-water insoluble, need to solubilize in DMF
-Strong (hindered) yet cleavable
-Effective conjugate than SPDP

Succinimidyloxycarbonyl-methyl-(2-pyridyldithio)toluene (**SMPT**)

(Ref. 246-249)

Sulfosuccinimidyl-6-[[alpha]-methyl-[alpha] (2-pyridyldithio)toluamido]hexanoate (**Sulfo-LC-SMPT**)

-Water soluble version of SMPT
-Extended spacer arm increases reactivity and reduces steric effects
-NHS ester and disulfide linkage is not as stable in aqueous solutions as in SMPT

FIGURE 1.13   Cleavable NHS ester/pyridyldisulfide heterobifunctional cross-linking reagents.

amine-functionalized biomolecules. The general reaction pathway (Scheme 1.36) involves, first, reaction of NHS ester of cross-linker with an amine-containing molecule to form a photosensitive intermediate via an amide linkage. Upon UV illumination, the photosensitive azide moiety will form the reactive nitrene intermediate, which can readily conjugate with other biomolecules containing amine, alkene, or active hydrogen groups as shown in Scheme 1.35.

Usually, presence of a negatively charged sulfonate group (R = SO₃Na) on the NHS ring will aid in the solubility of

the conjugate. The length of the spacer varies and could be short or long depending on the criteria of the experimental design. In some cases, the aromatic ring carries a hydroxyl group, which acts as an activating group (*ortho-* or *para-*directors) toward electrophilic aromatic substitution reactions. For example, iodination with radiolabeled-iodination reagents is a common reaction in order to facilitate the detection and/or purification of interacting proteins. Some of the photoreactive cross-linkers have a disulfide linkage, which render them cleavable under reducing conditions. The

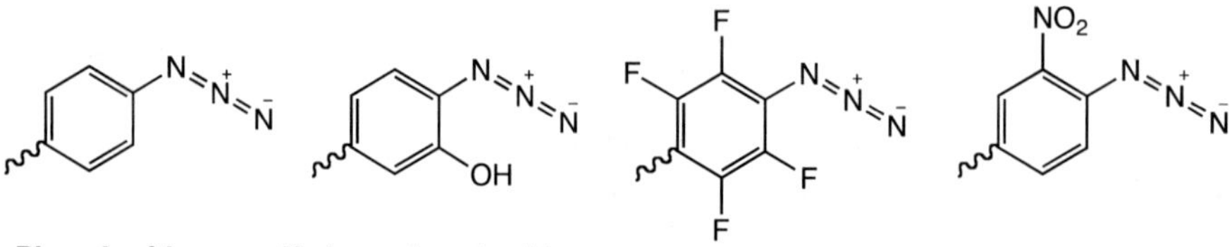

Phenyl azide   **Hydroxyphenyl azide**   **Tetrafluorophenyl azide**   **Nitrophenyl azide**

FIGURE 1.14   Aryl azide-reactive groups in photoreactive cross-linking reagents.

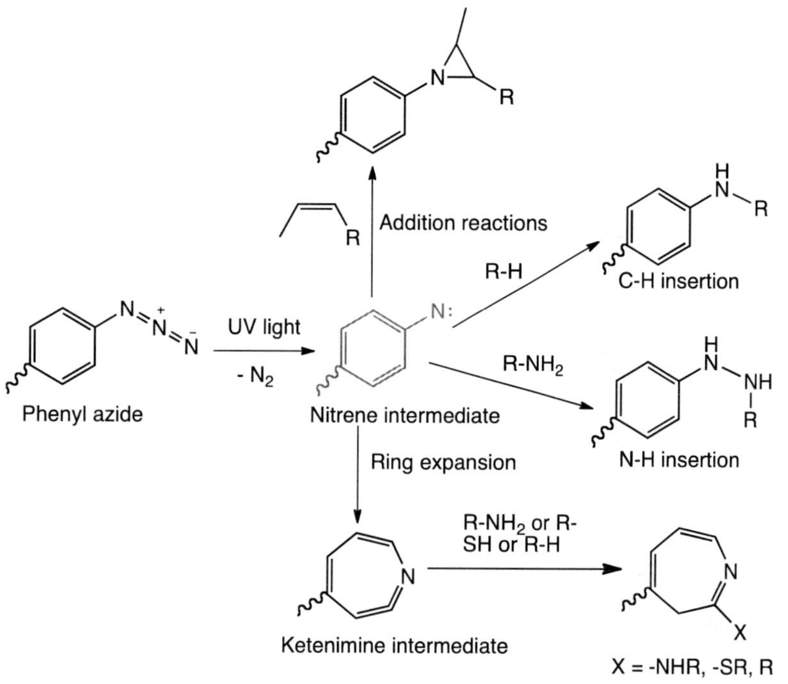

SCHEME 1.35   Reaction pathways of aryl nitrene intermediates. Adapted with permission from Reference [254], Copyright 2008, Elsevier Academic Press.

presence of a powerful deactivating group such as nitro ($-NO_2$) group on the aromatic ring will shift the optimal wavelength of photolysis to 320–350 nm and this helps in maintaining the biological integrity of proteins and nucleic acids. Fluorinated phenyl azides have also been used as photoreactive cross-linking reagents. Due to the ease of preparation of perfluorinated phenyl azide (PFPA) and its derivatives (Scheme 1.37), they have been extensively used to conjugate molecules possessing –CH, −NH and C=C bonds including polymers, organic materials, biomolecules, and carbon materials. The *p*-substituted PFPAs are attractive heterobifunctional cross-linking agents due to their two reactive centers, namely the fluorinated photoreactive phenylazide and the carboxyl functional group (R), which can be easily tailored through synthesis (Figure 1.16).

Another type of photoreactive cross-linking reagents is benzophenone derivatives. Upon UV activation, the benzophenone moiety generates a highly reactive triplet state ketone derivative, which can undergo insertion reactions with active hydrogen–carbon bonds even in the presence of solvent water and bulk nucleophiles. They can be repeatedly excited at <360 nm until they form covalent adducts, without loss of reactivity. They generally tend to give higher cross-linking yields than aryl azides photoreactive cross-linking reagents [275]. The key advantage of using benzophenone-based photoreactive reagents (Figure 1.17) is its lack of reactivity with water upon activation with light. The heterobifunctional nature of benzophenone-based photoreactive

cross-linking reagents usually consist of thiol-reactive groups (maleimides and iodo-acetamides) The maleimide/iodoacetamides chemically react with thiols and upon UV illumination, the photoreactive component reacts with nucleophiles or forms C–H insertion products.

Diazirines are a newer class of photo-activating chemical groups that are currently being used as cross-linking and photo-affinity labeling reagents [286–288]. Diazirines consist of a carbon bound to two nitrogen atoms, which are double-bonded to each other forming a cyclopropene-like ring. They are unstable in the presence of light and decay upon photolysis to generate highly reactive carbene species with evolution of nitrogen gas (Scheme 1.38). However, the diazirine moiety is known to have better photostability than phenyl azide groups, and can be easily and efficiently activated with long-wave UV light (330–370 nm). As a result, this has less potential to cause UV-induced protein damage or cellular damage. Diazirine photoreactive units can be easily incorporated into physiological ligands such as peptides, proteins, DNA, and sugars by chemoselective reactions [289].

The reactive carbene intermediates can afford covalent bonds via addition reactions with any amino acid side chain or peptide backbone at distances corresponding to the spacer arm lengths of the particular reagent. Thermo Scientific Pierce NHS–diazirine photoreactive cross-linkers have been developed to conjugate amine-containing molecules with nearly any other functional group via long-wave UV light activation. Succinimidyl–diazirine (SDA) reagents are a new

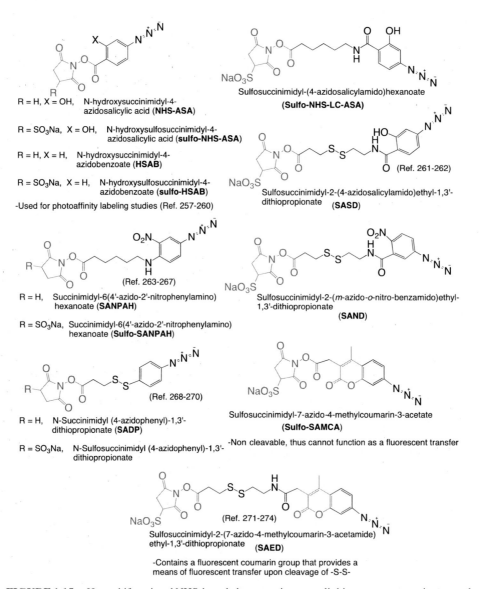

R = H, X = OH, N-hydroxysuccinimidyl-4-
azidosalicylic acid (**NHS-ASA**)

R = SO₃Na, X = OH, N-hydroxysulfosuccinimidyl-4-
azidosalicylic acid (**sulfo-NHS-ASA**)

R = H, X = H, N-hydroxysuccinimidyl-4-
azidobenzoate (**HSAB**)

R = SO₃Na, X = H, N-hydroxysulfosuccinimidyl-4-
azidobenzoate (**sulfo-HSAB**)

-Used for photoaffinity labeling studies (Ref. 257-260)

Sulfosuccinimidyl-(4-azidosalicylamido)hexanoate
(**Sulfo-NHS-LC-ASA**)

(Ref. 261-262)

Sulfosuccinimidyl-2-(4-azidosalicylamido)ethyl-1,3'-
dithiopropionate (**SASD**)

(Ref. 263-267)

R = H, Succinimidyl-6(4'-azido-2'-nitrophenylamino)
hexanoate (**SANPAH**)

R = SO₃Na, Succinimidyl-6(4'-azido-2'-nitrophenylamino)
hexanoate (**Sulfo-SANPAH**)

Sulfosuccinimidyl-2-(*m*-azido-*o*-nitro-benzamido)ethyl-
1,3'-dithiopropionate
(**SAND**)

(Ref. 268-270)

R = H, N-Succinimidyl (4-azidophenyl)-1,3'-
dithiopropionate (**SADP**)

R = SO₃Na, N-Sulfosuccinimidyl (4-azidophenyl)-1,3'-
dithiopropionate

Sulfosuccinimidyl-7-azido-4-methylcoumarin-3-acetate
(**Sulfo-SAMCA**)

-Non cleavable, thus cannot function as a fluorescent transfer

(Ref. 271-274)

Sulfosuccinimidyl-2-(7-azido-4-methylcoumarin-3-acetamide)
ethyl-1,3'-dithiopropionate (**SAED**)

-Contains a fluorescent coumarin group that provides a
means of fluorescent transfer upon cleavage of -S-S-

FIGURE 1.15 Heterobifunctional NHS-based photoreactive cross-linking reagents reactive toward amine-functionalized biomolecules.

class of heterobifunctional cross-linkers that combine amine-reactive chemistry with diazirine-based photochemistry for conjugating amine-containing molecules to other functional groups. Different types of SDA cross-linkers were prepared including compounds differing in spacer arm lengths, cleavable or noncleavable, and presence or absence of a charged group for membrane permeability (Figure 1.18) [290]. SDA cross-linkers are photo stable; hence reactions do not need to be carried in the dark. They have been mainly used to study protein structures and protein–protein interactions.

1.2.7.4 Other Cross-linking Reagents Carbodiimides, also commonly known as zero-length cross-linking reagents, have been used to conjugate carboxyl-containing biomolecules with amine or hydrazide-containing

biomolecules via amide or hydrazone linkages to form peptide–protein or ONT–protein bioconjugates. They allow direct conjugation without becoming part of the final cross-link between target molecules, that is, no spacer exists between the molecules being coupled. Among the carbodiimide cross-linkers (DCC or DIC, see Table 1.1, Section 1.2.1.2), the water-soluble EDC is the most popular with a number of applications such as for example, immobilization of peptides, proteins, and sugars, immunogen preparation, peptide–protein and oligomeric conjugates [291–294]. EDC reacts with carboxylic acid group and activates the carboxyl group to form an active *O*-acylisourea intermediate, which then couples to the amino group in the reaction mixture (see Scheme 1.3, Section 1.2.1.2). A soluble urea derivative by-product is released after displacement by the amine

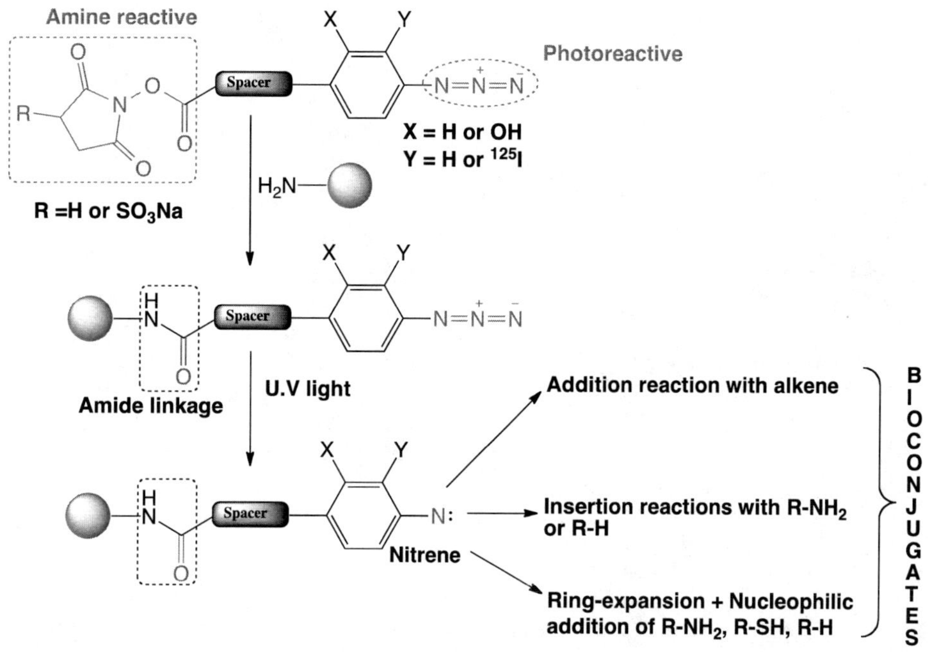

SCHEME 1.36 General reaction pathways for heterobifunctional NHS-based photoreactive cross-linking reagents.

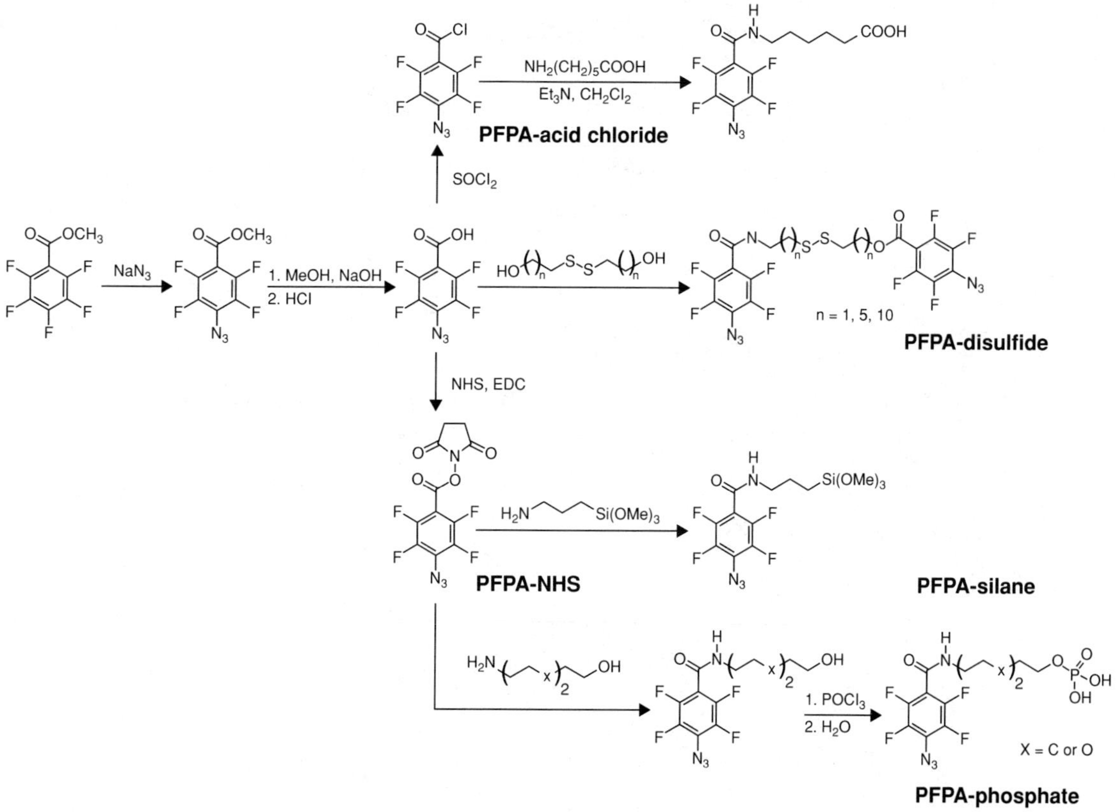

SCHEME 1.37 Synthetic scheme for PFPA and derivatives. Reprinted with permission from Reference 253, Copyright 2010, American Chemical Society.

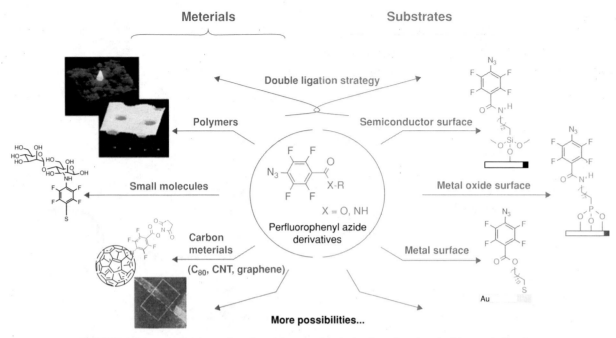

FIGURE 1.16   Applications of perfluorophenyl azide derivatives. Reprinted with permission from Reference 253, Copyright 2010, American Chemical Society. For a color version, see the color plate section.

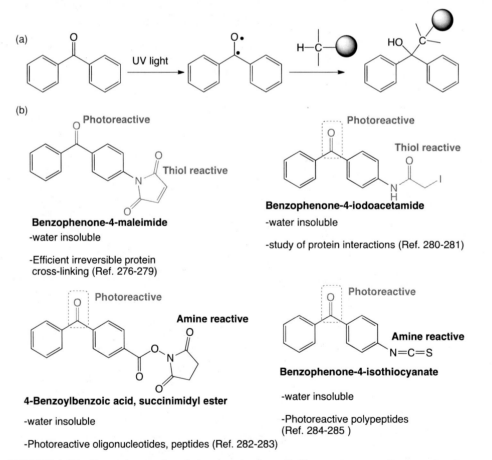

FIGURE 1.17   Benzophenone-based photoreactive cross-linking reagents reactive toward amines and thiols. For a color version, see the color plate section.

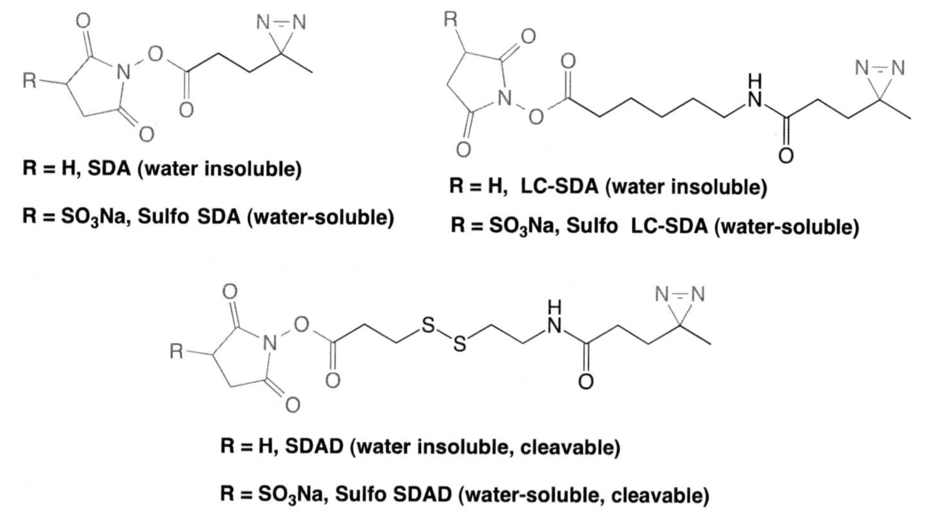

FIGURE 1.18   Heterobifunctional diazirine-based photoreactive cross-linking reagents reactive toward amines.

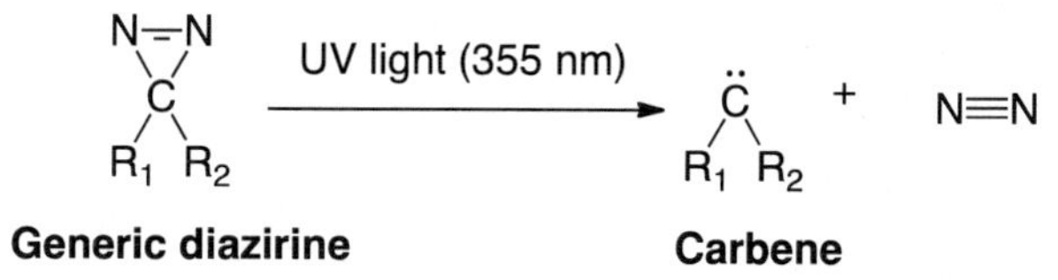

SCHEME 1.38   Decomposition of diazirine under UV light.

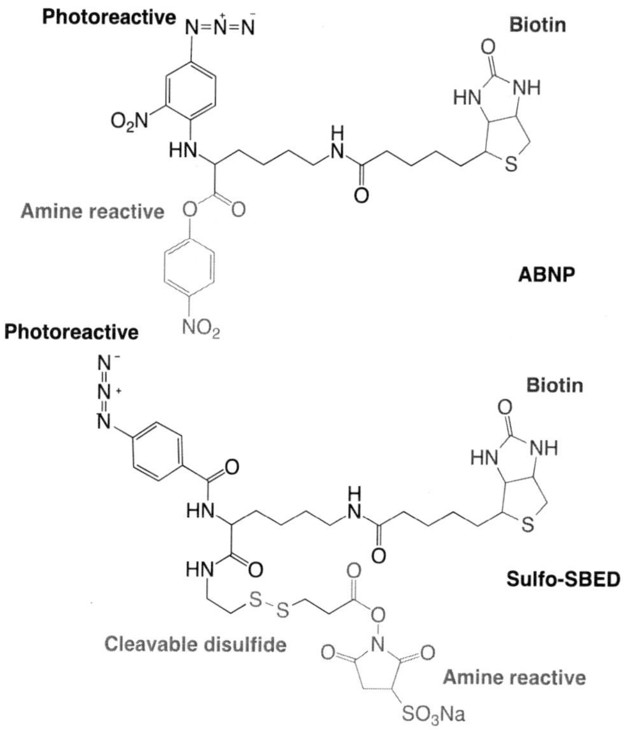

FIGURE 1.19   Chemical structures of ABNP and Sulfo-SBED trifunctional cross-linking reagents.

nucleophile. EDC cross-linking reactions must be carried out under conditions devoid of extraneous carboxyls and amines. Acidic (pH 4.5–5.5) MES buffer (4-morpholino-ethane-sulfonic acid) is most effective and *N*-hydroxysuccinimide (NHS) or its water-soluble analog (Sulfo-NHS) is usually included in EDC coupling protocols to improve efficiency or to create a more stable, amine-reactive intermediate.

Other than the most widely used bifunctional cross-linking reagents, trifunctional reagents have also been developed in which three distinct reactive groups are present. It combines the concept of heterobifunctional cross-linking and another different reactive group for specifically linking a third biological target. The third functional group can either incorporate an affinity tag [295] or another reactive group to cross link a third site. Biotin is the most commonly used affinity tag that facilitates the isolation of inserted cross-linked peptides after enzymatic digestion. Sometimes, this type of design (cleavable trifunctional cross-linking reagents) allows the transfer of a biotin tag to an interacting protein after cleavage. Thus, these trifunctional reagents are also known as label transfer reagents. Common examples of trifunctional cross-linkers are 4-azido-2-nitrophenylbiocytin-4-nitrophenyl ester (ABNP), sulfosuccinimidyl-2-[6-(biotinamido)-2-(p-azidobenzamido)hexanoamido]ethyl-1,39-dithiopropionate (Sulfo-SBED) [Thermo Scientific Pierce; Sufo-SBED technology is protected under US patent # 5,532, 379] (Figure 1.19) and methanethiosulfonate-azidotetrafluoro-

Biotin (Mts-Atf-Biotin) as well its longer chain version, Mts-Atf-LC-Biotin (Figure 1.20). ABNP consists of a nitrophenyl ester group (amine reactive at pH 7–9), phenyl azide group (photoreactive group), and a biotin handle. It is water insoluble but can be dissolved in DMF and has been used for studying hormone-binding site of the insulin receptor [296]. Sulfo-SBED is perhaps the most popular trifunctional cross-linker having a growing number

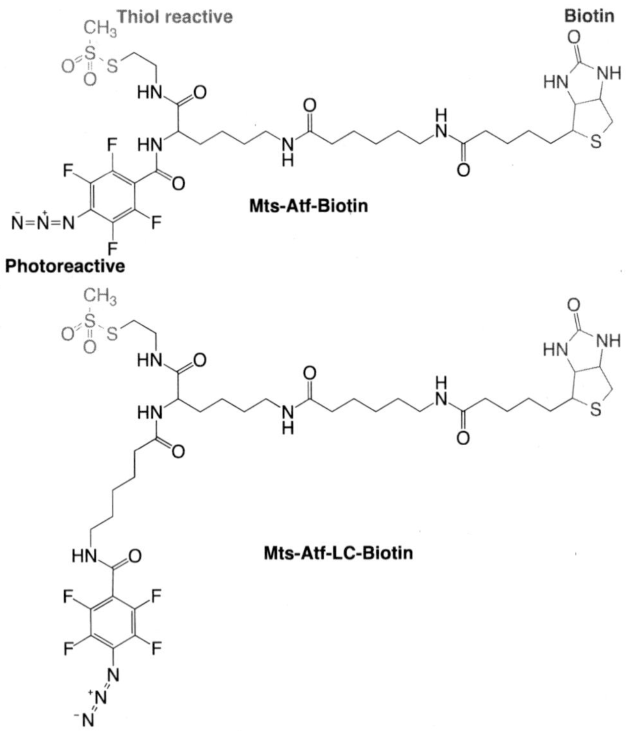

FIGURE 1.20   Chemical structures of Mts-Atf-Biotin and Mts-Atf-LC-Biotin trifunctional cross-linking reagents.

of protein interaction-based applications [295, 297–308]. Its trifunctional design is based on an amine-reactive sulfo-NHS ester, which is built off the α-carboxylate of the lysine core, a photoreactive phenyl azide group on the other side and a biotin handle connecting to the ε-amino group of lysine. The arm containing the sulfo-NHS ester is made up of a cleavable disulfide bond, which helps in the transfer of the biotin component to any captured protein. A typical protocol for performing a label transfer with SulfoSBED is given below [adapted from Thermo Scientific Pierce's Protocol]:

1. Add dissolved SulfoSBED reagent (a few microliters) to purified bait protein (0.5–1 mL) in phosphate buffered saline (PBS).

2. Incubate mixture (30–120 minutes) on ice or at room temperature in the dark.

3. Desalt or dialyze (in subdued light) to remove excess nonreacted Sulfo-SBED from the labeled bait protein.

4. Add labeled bait protein to cell lysate or other solution containing putative target protein.

5. When interaction complexes have formed, expose the solution to UV light (365 nm) for several minutes.

6. Analysis of products by one of several methods such as Western Blotting (cleave cross-links in DTT, separate proteins by SDS-PAGE, and detect biotinylated

bands) or by purification and mass spectrometry or sequencing (Affinity purify biotinylated proteins or peptide fragments following trypsin digestion and perform MS or sequencing to characterize the proteins involved).

Mts-Atf-Biotin and Mts-Atf-LC-Biotin are new trifunctional biotin-containing cross-linking reagents similar in design as Sulfo-SBED except they incorporate the benefits of the thiol-reactive methanethiosulfonate (Mts) group and the high yielding photoreactive tetrafluorophenyl-azide moiety (Thermo Scientific Pierce). In addition, a biotin tag is added on the other side of the arm. Mts-Atf-Biotin and Mts-Atf-LC-Biotin are both water insoluble and have to be dissolved in an organic solvent (DMF or DMSO). Their rapid reactions with thiols create disulfide linkages and when exposed to UV light, the photoreactive group activates to form covalent bonds to adjacent sites on the prey protein. Reducing the disulfide bond releases the bait protein and leaves the biotin label on the prey. It remains to see the applications of these new trifunctional cross-linking reagents in the study of protein interactions.

Recently, the synthesis of biotinylated lysine-reactive homobifunctional cross-linkers with varying lengths (BCCL1 and BCCL2) was reported [309]. The trifunctional design is based on the homobifunctional concept with a biotin handle in the middle of the linker, which was used for enriching cross-linked peptides. The synthesis of BCCL1 and BCCL2 is outlined in Scheme 1.39.

A representative protocol for the cross-linking of HIV-1 capsid protein (CA) with BCCLs is as follows [309]:

1. Add purified CA in 50 mM sodium phosphate and 100 mM NaCl, pH 8 to freshly prepared 0.1 M BCCL cross-linkers (1 mM final) in DMSO.

2. Incubate at room temperature for 30 seconds.

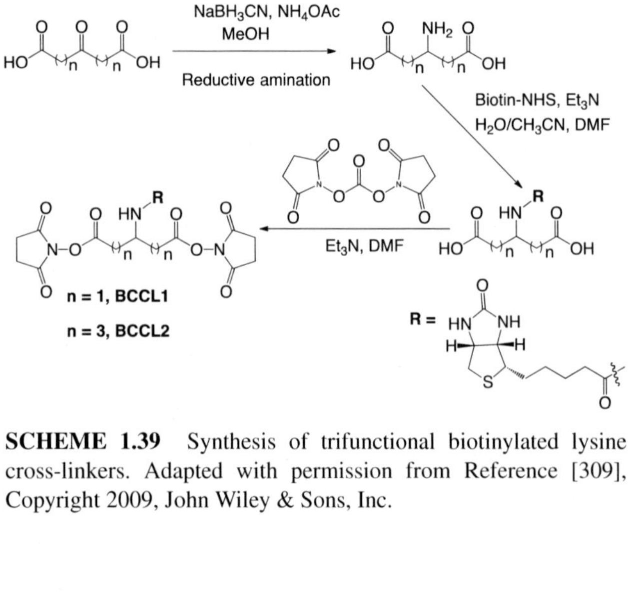

SCHEME 1.39   Synthesis of trifunctional biotinylated lysine cross-linkers. Adapted with permission from Reference [309], Copyright 2009, John Wiley & Sons, Inc.

3. Quench with 60 mM tris.

4. Remove free BCCLs by a Bio-Gel P-6 spin column (Micro Bio-spin P-6, Bio-Rad).

A new trifunctional traceless photocleavable cross-linker known as di-6-(3-succinimidyl carbonyloxymethyl-4-nitro-phenoxy)-hexanoic acid disulfide diethanol ester (SCNE) was also recently reported [310, 311]. The *N*-hydroxysuccinimidyl and the disulfide moieties allow conjugation of proteins to a gold-coated substrate. SCNE can be synthesized in three steps starting 2,2′-dihydroxyethyldisulfide (Scheme 1.40).

Upon irradiation at 330 nm, SCNE was found to retain the chemical structure and functionality of the cleaved proteins. This versatile cross-linker will find broad applications where photocleavage ensures clean and remote-controllable release of biological molecules from a substrate. For instance, it was applied in reversible phage particle immobilization, protein photoprinting, and guided protein delivery [311].

A representative protocol for the conjugation of homodimeric pilM protein (3HG9) with SCNE is as follows [311]:

Add 50 μL SCNE (5 mM in acetonitrile) to 300 μL of freshly prepared 3HG9 solution.

Keep the mixture at room temperature for 8 hours.

Quench with 250 mM Tris-HCl buffer (pH 8) to passivate any unreacted SCNE.

1.2.8   Bioorthogonal Reactions

The selective conjugation or derivatization of biomolecules has long been a significant barrier due to the vast array of functionality present in biological milieu. Many side reactions and nonspecific labeling are unavoidable. The vast complexity of the cellular systems renders the study of biomolecules in their native environments very challenging. While traditional bioconjugation strategies have a long-standing history in the *in vitro* modification of biomolecules, chemists started contemplating whether chemical reactions can be applied on or in living cells or whole organisms, the so-called bioorthogonal reactions. The first published use of the term "*bioorthogonal*" came from the laboratory of Carolyn R. Bertozzi in 2003 who subsequently pioneered this area [312]. Since its introduction, there has been a dramatic increase in the use of this term in the chemistry community, particularly, for the study of biomolecules such as glycans, nucleic acids, proteins, and lipids [313, 314].

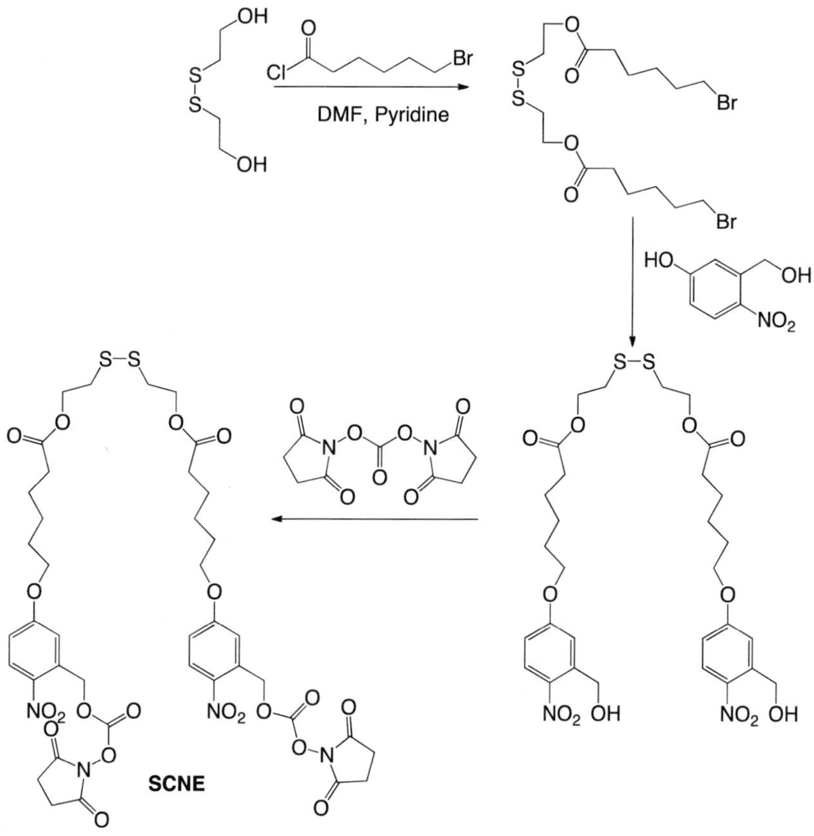

SCHEME 1.40   Synthetic route for trifunctional SCNE photocleavable cross-linker. Adapted with permission from Reference [311], supporting information. Copyright 2012, American Chemical Society.

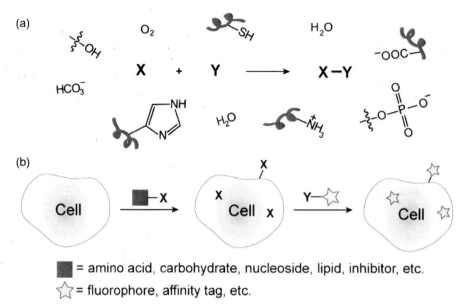

FIGURE 1.21 (A) A typical bioorthogonal reaction, (B) Experimental platform for biomolecule probing using bioorthogonal chemistry. Reprinted with permission from Reference 320, Copyright 2011, American Chemical Society.

Historically, the existence of the concept of bioorthogonality dated back in 1990s with great work by Rideout and coworkers [315] (selective condensation of hydrazine and aldehyde groups to assemble toxins from inactive prodrugs within live cells) and Tsien and coworkers [316] (first example of live cell protein labeling using bisarsenical dyes). These groundbreaking works laid the foundation for the ability to carry covalent reactions in cells among abiotic functional groups. Bioorthogonal reactions are chemical reactions that proceed inside of living systems without any interference with native biological processes [1(e), 317–319]. Overall, the reaction of two compounds (X and Y) bearing bioorthogonal functional groups occurs in the presence of a myriad of functionalities (amine, carboxyl, hydroxyl, thiol, disulfide, etc.) found in the biological living systems (Figure 1.21 (A)). Biorthogonal chemistry usually proceeds in two steps: first, a nonnative functional group (chemical reporter) is introduced in a biomolecule of interest; then, the modified biomolecule is labeled via a bioorthogonal chemical reaction [320].

Devising bioorthogonal reactions continues to be a big challenge for chemists and in general, for a reaction to be considered bioorthogonal, it must fulfill a number of requirements such as *selectivity* (avoid side reactions with biological compounds and provide good yields), *kinetics* (rapid reactions at low reagent concentrations), *chemical and biological inertness* (strong covalent linkage and remain inert to abundant biological nucleophiles, electrophiles, and redox-active metabolites), *reaction biocompatibility* (reactions proceed smoothly at physiological pH, aqueous environments, and temperature) and *toxicity* (reactions must produce nontoxic or any side products). A number of chemical ligation strategies have been developed that meet the criteria of bioorthogonality and the following sections describe them in detail. However, it is to be noted that condensation reactions of aldehyde/ketone with aminooxy or hydrazides (oxime/hydrazone ligations), are not strictly bioorthogonal, but can be effectively bioorthogonal if biological aldehydes/ketones are not native to a particular experimental matrix of interest [321]. Aldehydes and ketones have not been widely used for *in vivo* labeling of biomolecules due to competition with endogeneous aldehydes and ketones. However, since they are not present on cell surfaces, they can serve as unique chemical reporters [23, 322, 323].

1.2.8.1 Staudinger Ligation (Traceless and Nontraceless) In 1919, Staudinger and Meyer disclosed the reaction between an azide and a phosphine, which was termed the Staudinger reaction (Scheme 1.41) [324]. The reaction is considered to be a mild process for the reduction of azides to the corresponding amine in the presence of a phosphine (triphenylphosphine as soft nucleophiles). Mechanistically, the reaction proceeds via an aza-ylide (iminophosphorane) intermediate, which is formed upon loss of a nitrogen molecule. This intermediate is then hydrolyzed by water to afford the amine product and triphenylphosphine oxide as the by-product (Scheme 1.42(A)) [325].

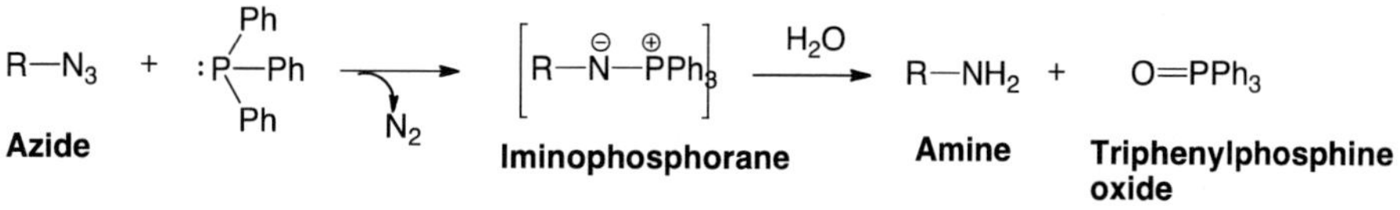

SCHEME 1.41   The classical Staudinger reaction.

In 2000, Saxon and Bertozzi realized the potential of this powerful reaction, which fulfilled the criteria for developing chemoselective bioorthogonal reaction [326]. The azide functionality is small in size, chemically inert, stable under physiological conditions, and extremely rare in biological systems. Furthermore, the other reactive component, phosphine, is naturally absent from living systems. Bertozzi's group mechanistically modified the classic Staudinger reaction by introduction of an intramolecular trap into the phosphine reagent to prevent the loss of the initial covalent linkage formed (intermediate **3**) due to hydrolysis (Scheme 1.42(B)). As such, an ester group *ortho* to the phosphorus atom on one of the aryl rings (**6**) was incorporated. This opened a new mode of reactivity whereby the nucleophilic nitrogen atom reacted with this electrophilic trap to form intermediate **8**. Hydrolysis of intermediate **8** afforded a stable ligated product (**9**), which carries the phosphine oxide within its structure [320]. The overall transformation was termed the Staudinger

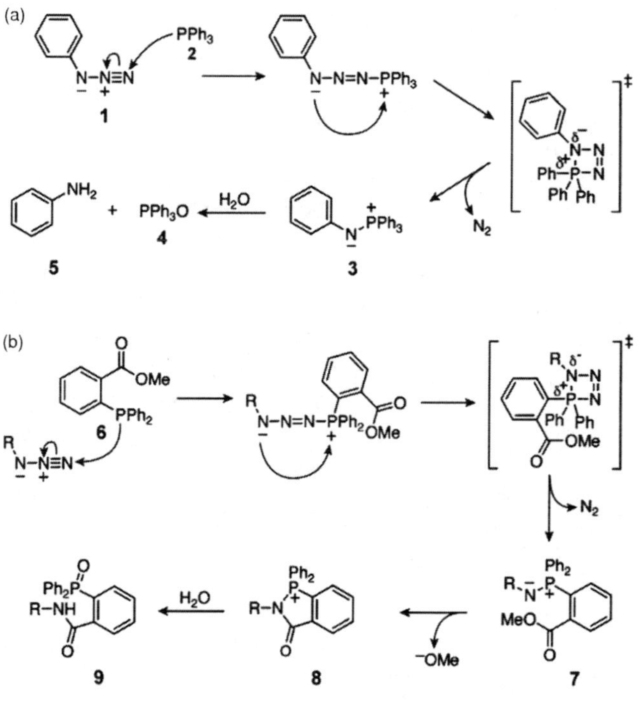

Ligation, which also commonly referred to as the *nontraceless Staudinger ligation*.

The introduction of the Staudinger ligation method essentially opens the field of bioorthogonal chemistry as it can be potentially translated to *in vivo* studies. For an extensive use of the Staudinger ligation as a bioorthogonal reaction, readers are directed to excellent reviews by Kohn and Breinbauer in 2004 [327], Schilling et al. [328] and van Berkel et al. [329] in 2011. Extensive mechanistic studies by Bertozzi et al. unravel the mechanism and the influence of parameters associated with the Staudinger ligation. It was found that basic and more nucleophilic phosphines tend to react faster; however aliphatic basic phosphines are prone to oxidation [330]. Shortly after the disclosure of the nontraceless Staudinger ligation, the so-called "*traceless Staudinger ligation*" was reported simultaneously by Bertozzi et al. [331] and Raines et al. [332]. In this modified version of the Staudinger ligation, the goal was to eliminate the triphenylphosphine oxide by-product via cleavage from the product after the ligation is completed leaving a native amide bond (Scheme 1.43). An important structural requirement in these compounds was the presence of two aromatic phosphine substituents to avoid excessive oxidation of the phosphine.

An interesting feature of the traceless Staudinger ligation developed by Raines and coworkers for peptide coupling is that it does not require the presence of a cysteine residue at the ligation site, unlike the standard NCL process. Since its introduction in 2000, the Staudinger ligation has been mainly used in the labeling of biomolecules (glycans [317, 326, 333, 334], DNA [335–337], proteins [338–340], and lipids [341, 342]) in their native environment. Other than its use as an efficient labeling agent, the Staudinger ligation has shown to be a useful synthetic method for the construction of polypeptides [332, 343, 344], microarrays [328, 329], functional biopolymers, glycopeptides, glycoproteins [329], and surface modifications (Figure 1.22) [345–347, 329].

The Staudinger ligation remains the reaction of choice for many bioconjugation applications; however, like any reaction, it is not without limitations. Steric hindrance imposed by groups close to the ligation site can significantly reduce the ligation yield. Phosphine reagents slowly undergo air oxidation in living systems. The major drawback is the relatively slow reaction kinetics and as such, the ligation requires high concentrations of a triarylphosphine (>250 mm). Extensive kinetic studies have shown that the rate-determining step is

SCHEME 1.42   Mechanism of the (A) classical Staudinger reaction and (B) Staudinger ligation. (Reprinted with permission from Reference 320, Copyright 2011, American Chemical Society.

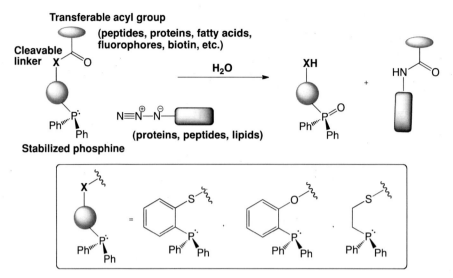

SCHEME 1.43 Traceless Staudinger ligation. Adapted from Reference 328 with permission from the Royal Society of Chemistry.

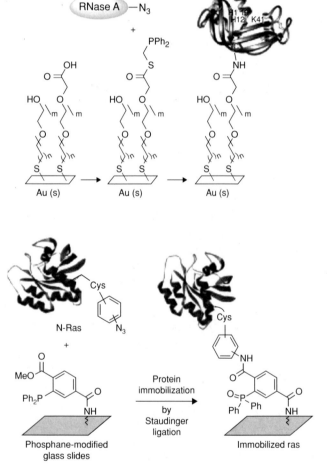

FIGURE 1.22 Surface modifications via Staudinger ligation. Reprinted with permission from Reference 329, Copyright 2011, John Wiley & Sons, Inc.

the initial nucleophilic attack of the phosphine on the azide [330]. Attempts to improve the kinetics of the Staudinger ligation by increasing the nucleophilicity of the phosphine reagents were unsuccessful due to rapid phosphine oxidation in air [1 (e), 348]. The sluggish kinetics (with second-order rate constants around 0.0020 M/s) of the Staudinger ligation has turned the attention of researches to investigate other modes of reactivity of azides, namely 1,3-dipolar azide–alkyne cycloaddition reactions.

1.2.8.2 Click Chemistry (Copper(I)-catalyzed Azide–Alkyne Cycloaddition)

The azide moiety can participate in a [3 + 2] cycloaddition process with alkenes and alkynes. In 1893, Michael was the first to report the thermal reaction of terminal or internal alkynes with organic azides to form 1,2,3-triazole derivatives [349] (Scheme 1.44). He proposed that the reaction proceeded by a concerted cycloaddition reaction, but not until decades later when Huisgen and coworkers described the reaction as a 1,3-dipolar cycloaddition [350]. Rolf Huisgen was among the first to understand and extend the scope of this organic reaction and the reaction was commonly referred as the azide–alkyne Huisgen 1,3-dipolar cycloaddition. However, the reaction has to be performed under high temperature or pressure due to its high activation energy barrier and as such, it has not been popular

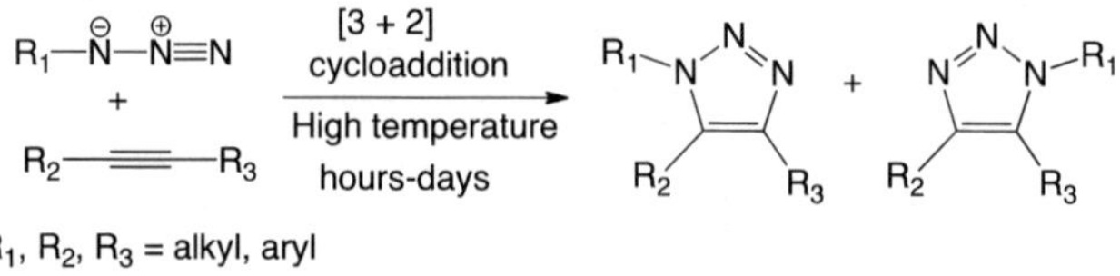

R_1, R_2, R_3 = alkyl, aryl

SCHEME 1.44 Classical azide-alkyne Huisgen 1,3-dipolar cycloaddition.

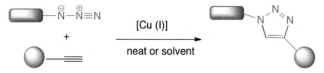

SCHEME 1.45 Copper (I)-catalyzed azide-alkyne cycloaddition (CuAAC) for conjugation of an azide moiety with a terminal alkyne.

in the synthetic organic community as well as its applications in living systems.

In 2002, Sharpless and Meldal laboratories independently showed that the rate of 1,3-dipolar cycloaddition reactions of azides with terminal alkynes could be dramatically increased ($\sim 10^6$ fold) in the presence of catalytic amount of copper (I) [351, 352] (Scheme 1.45). This notable variant of the Huisgen 1,3-dipolar cycloaddition proceeds at ambient temperatures and pressures in excellent yield. It can take in many protic and aprotic solvents, including water. However, it is recommended that acetonitrile be avoided as solvent due to the high coordinating ability of nitriles toward copper (I). Reaction is unaffected by most organic and inorganic functional groups; hence, no protecting group chemistry is required [353]. 1,4-disubstituted 1,2,3-triazoles are formed exclusively as compared to a mixture of 1,4- and 1,5-regioisomers in the thermal process (Scheme 1.44). The 1,2,3-triazole moiety possesses several attributes such as high chemical stability to reactive conditions (oxidation, reduction, and hydrolysis), strong dipole moment (4.8–5.6 Debye), and aromatic character. The copper-mediated Huisgen 1,3-dipolar cycloaddition is preferably termed as Copper(I)-catalyzed azide-alkyne Cycloaddition (CuAAC) [354].

Common commercial sources of copper(I) are cuprous bromide or iodide, however, the reaction works much better when generating Cu(I) *in situ* by using a mixture of copper(II) (e.g., copper(II) sulfate) and a reducing agent (e.g., sodium ascorbate). Addition of stabilizing ligands such as tris-(benzyltriazolylmethyl)amine (TBTA) greatly improve reaction rate and inhibits oxidative alkyne coupling [355]. CuAAC turns out to have all the attributes of "click" chemistry, a term that was coined by Sharpless and coworkers in 2001 to describe a set of reactions that covalently combine two components with high yielding and minimal by-products [356]. In fact, CuAAC is now simply called "click" chemistry. The simplicity, efficiency, and selectivity of CuAAC

have opened the door to bioorthogonal conjugation strategy [355, 357, 358]. The azide group is very small and is favorable for cell permeability. Azides are not present in cells and hence have no competing biological side reactions. However, they can be introduced into target molecules either metabolically or via chemical modifications. The other coupling partner, alkyne, though not very small still has the stability and orthogonality required for *in vivo* labeling. Furthermore, CuAAC reactions proceed at a much faster rate than the Staudinger ligation. Finn and coworkers were the first to disclose the use of CuAAC as a bioconjugation strategy for the attachment of dyes to cowpea mosaic virus [355]. Since then, CuAAC has been applied in a variety of fields, including proteomics applications [359]; immobilization of proteins, peptides, and carbohydrates [360, 361]; DNA functionalization [362]; small molecule libraries for drug discovery [363]; polymers and dendrimers chemistry [364]; and the selective labeling of biological molecules [365]. Despite its countless applications, CuAAC has a main limitation: the toxic nature of the copper catalyst [366]. This precludes many applications of CuAAC in the living biological systems [367, 368].

1.2.8.3 Copper-free Click Chemistry

As the cellular toxicity of the metal catalyst in copper-catalyzed click chemistry was a major concern, there was a quick need to develop a copper-free click chemistry version. The Bertozzi's laboratory found an alternative way to activate the alkyne functionality toward azide without use of any toxic transition metal [320, 369]. This wonderful approach relies on the ring strain effect and it was found that cyclooctynes react readily with azides at room temperature without the need for a catalyst (Scheme 1.46) [370]. This led to a new strategy called the strain-promoted [3 + 2] azide-alkyne cycloaddition or simply copper-free click chemistry, which can be used for covalent modification of biomolecules in living systems. The roots of this approach date back to pioneering works of Alder and Stein [371, 372] (rapid reaction of dicyclopentadiene with azides) and Wittig and Krebs [373] (the explosive reaction of cyclooctyne with phenylazide).

The first application of copper-free click chemistry involved the effective labeling of azides within cell-surface glycans using a biotin conjugate of cyclooctyne **1** (OCT) (Figure 1.23) [370]. No cytotoxic effects were observed, but

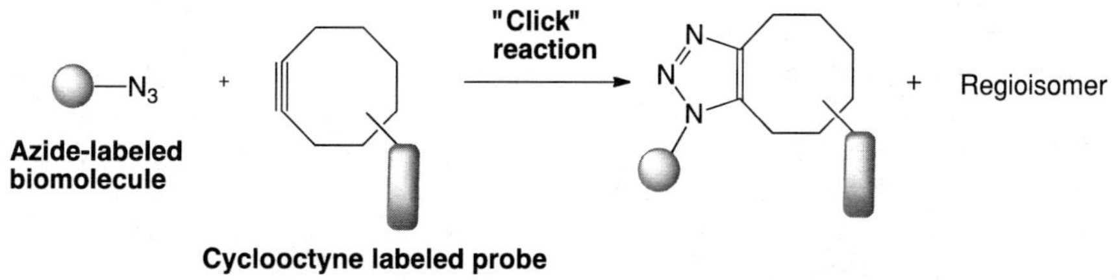

Azide-labeled biomolecule

Cyclooctyne labeled probe

SCHEME 1.46 Copper-free click chemistry for bioorthogonal conjugation of biomolecules.

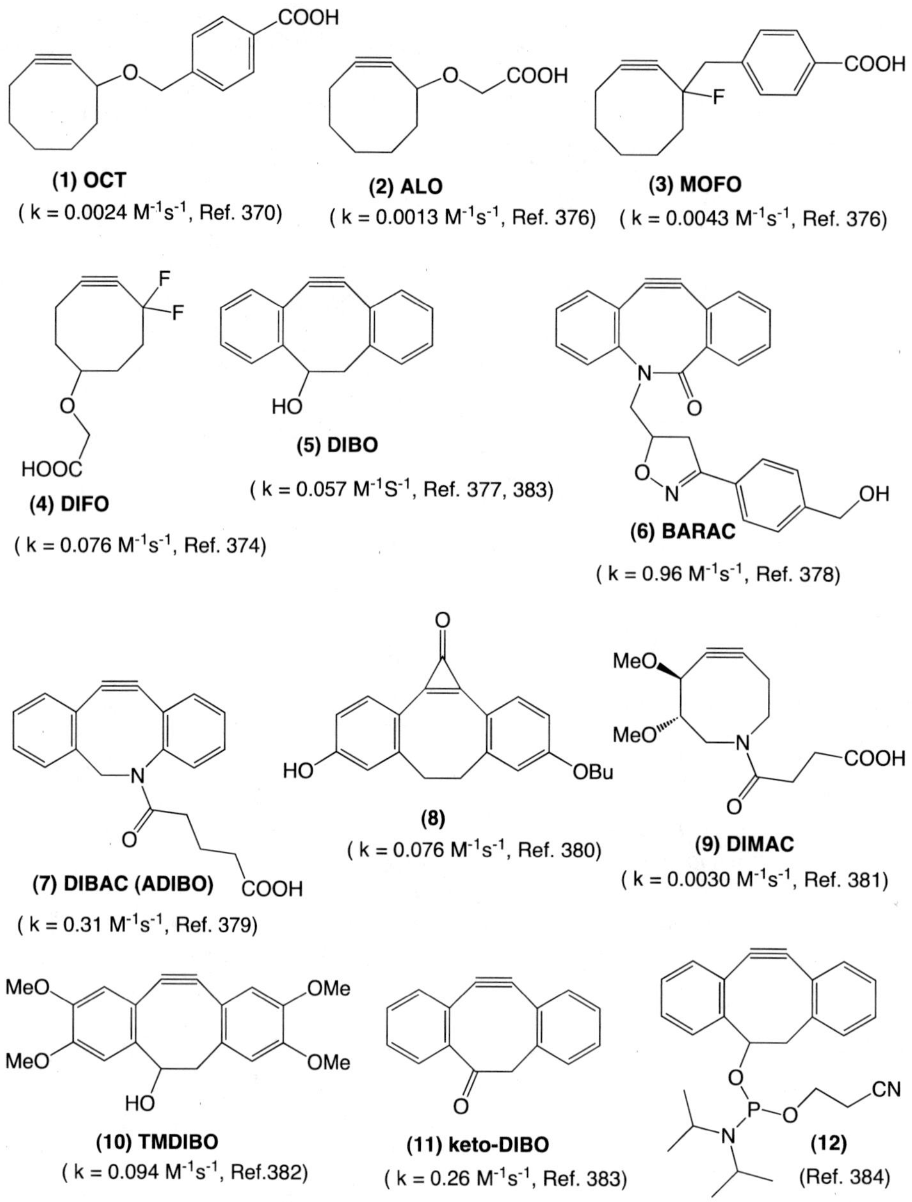

FIGURE 1.23 Cyclooctynes investigated and/or used for copper-free click chemistry. Adapted with permission from Reference 320, Copyright 2011, American Chemical Society.

the reaction rate was sluggish (slower than the Staudinger ligation and CuAAC) and there were solubility issues. As such, a number of cyclooctyne derivatives (**2-11**, Figure 1.23) were investigated in order to improve the kinetics of the cycloaddition [320]. Initial studies indicated that fluorinated cyclooctynes turned out to be good candidates. Difluorinated cyclooctyne **4** (DIFO) gave a surprisingly 60-fold rate enhancement ($k = 10^{-1}$ M/s) and was used for imaging azides labeled biomolecules in live cells [374a], mice [374b], and zebrafish embryos [375]. Boons and coworkers recently designed a highly strained cyclooctyne **5** (DIBO) (fusion of two aryl rings on cyclooctyne core), which was shown to

be nontoxic and exhibit similar reactivity as DIFO [377]. However, compared to DIFO reagents, the major advantages of DIBO are their synthetic accessibility and the possibility of rate enhancement by substituent effect on the benzene ring [1(f)]. Hydrophilic cyclooctyne **9** (DIMAC) was also designed to address solubility problems. Recent theoretical/computational studies have shed more light into the mechanistic understanding of how different parameters can influence the rate of cycloaddition of this powerful emerging copper-free click chemistry [385, 386].

Factors like increasing stability of cycloalkyne reagents, minimization of side reactions, and the fine tuning of

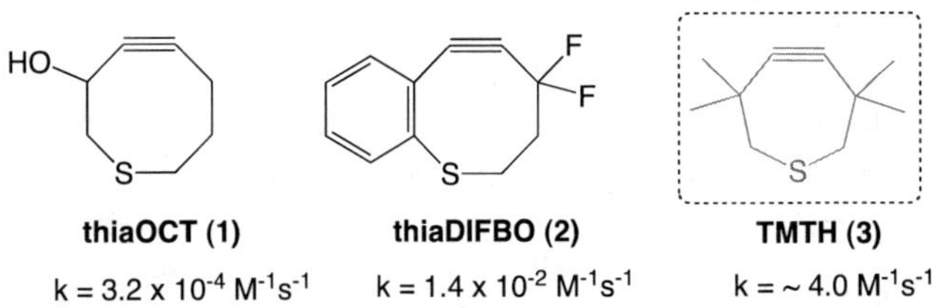

thiaOCT (1)
$k = 3.2 \times 10^{-4}$ M⁻¹s⁻¹

thiaDIFBO (2)
$k = 1.4 \times 10^{-2}$ M⁻¹s⁻¹

TMTH (3)
$k = \sim 4.0$ M⁻¹s⁻¹

FIGURE 1.24 Thiacycloalkynes for copper-free click chemistry.

reactivity have recently led to the engineering of new cycloalkynes for metal-free click chemistry. In 2011, a new dibenzocyclooctyne (DIBO) phosphoramidite (**12**, Figure 1.23) was synthesized and reacted with azides via copper-free, strain-promoted [3 + 2] azide–alkyne cycloaddition for the labeling of terminal DNA [384]. Early in 2012, Bertozzi and coworkers disclosed the investigation of a new promising class of reagents called thiacycloalkynes for copper-free click chemistry (Figure 1.24) [387]. Initial studies showed that the introduction of a sulfur atom into the cyclooctyne ring (**2-3**, Figure 1.24) greatly increased stability but unfortunately proceeded with low rate of cycloaddition. Further studies indicated that a combination of the stabilizing effect of endocyclic sulfur atom together with ring contraction identified a known thiacycloheptyne (**3**), 3,3,6,6-tetramethylthiacycloheptyne (TMTH) [388] with the fastest reactivity (second-order rate constant, $k = \sim 4.0$ M/s) for copper-free click chemistry to date.

Bertozzi and Houk et al. carried an extensive study on the effects of strain and electronics on the reactivity of biarylazacyclooctynones in copper-free click chemistry using both experimental and computational (density functional theory) methods [389]. In the same year, another group reported cyclononyne reagents, in particular, benzocyclononyne (**1**, Figure 1.25), which reacted spontaneously with azides [390]. Cyclononynes are more stable than the typical cyclooctynes. More reactive variants of benzocyclononyne platform for metal-free click chemistry are still under investigation.

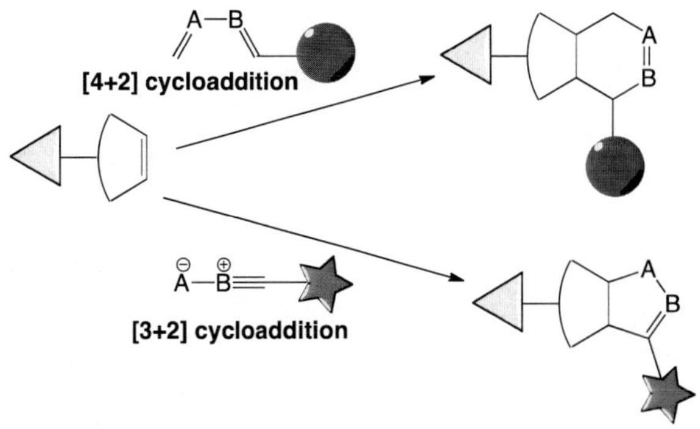

FIGURE 1.26 Covalent bioconjugation of strained alkenes via cycloaddition reactions.

Copper-free click chemistry has been mainly applied in areas such as chemical biology (molecular imaging and labeling, macromolecular noncovalent interactions) and materials science (nanomaterials functionalization, new materials design, initiation of gelation process) [369].

1.2.8.4 Bioconjugation Via Alkene Chemistry While strained alkynes have played a significant role in the development of powerful bioorthogonal reactions, strained alkenes have also established themselves as efficient tools in reagent-free covalent bioconjugations [391]. Most of the alkene reactions invoked the reactive double bond of the cyclic alkene (dipolarophiles) to participate in many cycloaddition reactions such as [4 + 2] and [3 + 2] cycloadditions (Figure 1.26). A landmark example is the classic Diels–Alder cycloaddition, which has become the most widely used synthetic tools.

1.2.8.4.1 Diels–Alder Cycloaddition A typical Diels–Alder reaction involves a [4 + 2] cycloaddition process between a diene and a dienophile (alkene) to generate a cyclohexene derivative. The reaction between an electron-rich diene (electron-donating groups (EDG)) and an electron-poor dienophile (electron-withdrawing group (EWG)) is termed as a *normal electron-demand Diels–Alder* reaction.

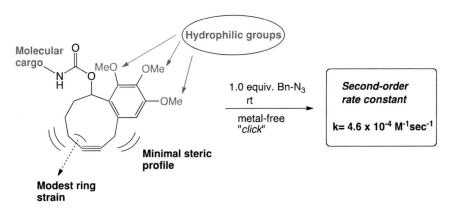

FIGURE 1.25 Engineering cyclononynes for copper-free click chemistry. Adapted with permission from Reference 390, Copyright 2012, American Chemical Society.

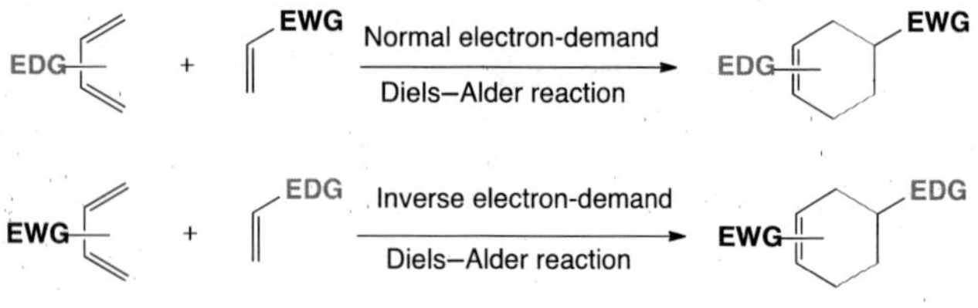

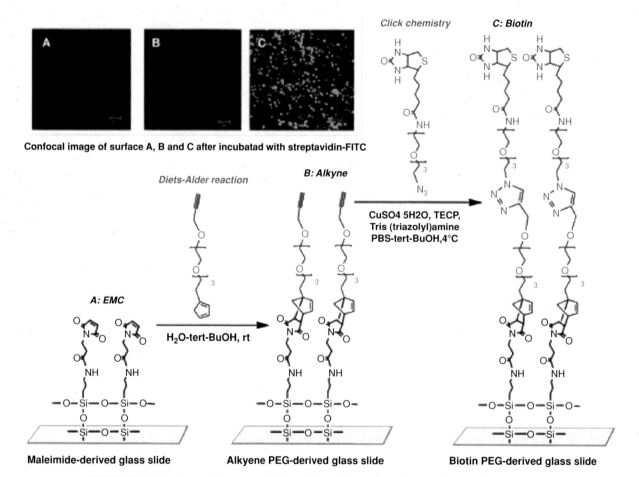

FIGURE 1.27   Normal or inverse electron-demand Diels–Alder reactions.

When the diene is electron poor and the dienophile is electron rich, then it is called an *inverse electron-demand Diels–Alder* cyclization (Figure 1.27).

While the Diels–Alder reaction remains one of the most useful carbon–carbon bond forming reactions to synthetic organic chemist for decades, it was not until in 2001 that the first report of the use of the normal electron-demand Diels–Alder was disclosed for covalent conjugation of biomolecules [392]. Breslow's groundbreaking discovery [393] that aqueous solvents increased the rate of Diels–Alder cycloaddition led to a renewed interest in this powerful cycloaddition

process for covalent biomolecule modifications [394–396]. The Diels–Alder reaction is highly selective, compatible with most functional groups found on proteins (except when furans as dienes and maleimides as dienophiles are used in the presence of proteins with unrestricted SH groups) and proceeds under very mild conditions. The 2,4-hexadiene moiety is pretty stable under physiological conditions and could be easily incorporated chemically into biomolecules from the commercially available precursor trans, trans-2,4-hexadienol [397]. The dienophiles are usually maleimide derivatives or probes, which are commercially available from many suppliers. The main concern is the maleimide dienophiles that can cross-react with biological nucleophiles. Nevertheless, the compatibility of the Diels–Alder reaction with biomolecules has led to its applications in the bioconjugation and/or immobilization of ONTs, carbohydrates, proteins, and other biomolecules [360, 392, 396, 398–401] as well as microarray development [402]. A sequential Diels–Alder and azide–alkyne cycloadditions was recently employed for the immobilization of carbohydrate and protein onto solid surfaces [360] (Scheme 1.47).

SCHEME 1.47   Diels–Alder cycloaddition followed by surface biotinylation via click chemistry. Reprinted with permission from Reference 360, Copyright 2006, American Chemical Society. For a color version, see the color plate section.

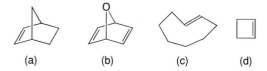

(a)　　　　**(b)**　　　　**(c)**　　　　**(d)**

FIGURE 1.28 Chemical structures of strained alkenes (i) norbonene, (ii) oxanorbornadiene, (iii) *trans*-cyclooctene, and (iv) cyclobutene.

Inverse electron-demand Diels–Alder cycloaddition has recently emerged as a versatile click-reaction concept for covalent conjugation of molecular fragments [406, 407].

1.2.8.4.2 Strained Alkenes in Bioconjugation Dipolar cycloaddition reactions usually require heating above 50°C due to the high activation energies. As such, this has limited their applications in the biological system; however, the use of strained alkene (ring strain effect) provides a solution to this problem. Typical examples of strained alkenes that are involved in cycloaddition reactions are norbonene, oxanorbornadiene, *trans*-cyclooctene, and cyclobutene (Figure 1.28).

Cycloaddition reaction of strained alkenes with dipoles has long been known to possess reactivity about hundred times more than unstrained alkenes [371, 372, 408], but has only been exploited recently as a tool for bioconjugation. Oxanorbornadiene was found to react with azide via a [3 + 2] cycloaddition to give an intermediate that underwent a spontaneous retro-Diels–Alder reaction to exclude a furan moiety and formed stable triazole regioisomeric products (Scheme 1.48(a)) [409]. The cycloaddition occurs preferentially at the electron-poor double bond but with a slow reaction rate ($k = 1.8 \times 10^{-4}$ M/s). However, instead of using an azide, tetrazine turns out to be a better candidate for cycloaddition reaction with strained alkenes [410]. For instance, an inverse electron-demand Diels–Alder reaction of *trans*-cyclooctene with dipyridyl tetrazine proceeded very rapidly in water ($k = 2.0 \times 10^3$ M/s) (Scheme 1.48(b)) [411]. After a [4 + 2] cycloaddition, the resulting intermediate loses a nitrogen molecule and subsequent isomerization leads to the final ligation product. *Trans*-cyclooctene is readily accessible either synthetically or photochemically [412] and is the most reactive alkene/tetrazine pair for cycloaddition. Norbonene reaction with tetrazine has also been investigated [413] (Scheme 1.48(c)) but does not proceed as rapidly as the *trans*-cyclooctene/tetrazine pair. Fused cyclobutene–norbonene system was found to react at a reasonably fast rate with high yield and purity [406] (Scheme 1.48(d)). Although the *trans*-cyclooctene/tetrazine cycloaddition has exceptionally high kinetics, it does suffer a limitation such as poor stability of bispyridyltetrazine in serum, PBS, and blood [391, 414]. Stability of the conjugate remains one of the crucial criteria of bioorthogonality in addition to fast kinetics.

1.2.8.4.3 Tetrazole Photo-click Chemistry *Rapid inducibility*, in addition to high bioorthogonality and fast reaction kinetics, is another critical attribute for a bioorthogonal reaction to be broadly useful in studying biological processes in living systems [1(f)]. In this regard, Lin's laboratory has developed a photoinducible bioorthogonal reaction also known as "photo-click chemistry" [415, 416]. The photo-click chemistry involves a photoinduced cycloreversion of tetrazole to liberate nitrogen. This generates a short-lived nitrile imine 1,3-dipole, which undergoes a 1,3-dipolar cycloaddition with dipolarophiles, such as alkenes to afford pyrazoline cycloadducts (Scheme 1.49).

Initial work made use of light with a wavelength of 302 nm to produce the nitrile–imine dipole, which can lead to photodamage to living systems. In order to circumvent this issue, the aryl groups on the tetrazole can be modified such that the reaction can proceed at a wavelength of 365 nm and without the use of an activated alkene [417]. Potential advantages of photo-click chemistry include mild reaction conditions (incident photons as the only reagents), easily accessible tetrazole reagents, spatiotemporal reaction control via focused UV light irradiation, and formation of fluorescent pyrazoline cycloadduct, which can help in monitoring reaction. The photoinduced cycloaddition has been used for the selective *in vivo* labeling of proteins with *Escherichia coli* cells overexpressing an alkene-containing Z-domain protein [418].

1.2.8.4.4 Quadricyclane Ligation The quadricyclane ligation is a fairly new bioorthogonal reaction developed by Bertozzi and coworkers, which joins the growing list of tools for selective covalent modification of biomolecules [419]. It utilizes a highly strained quadricyclane to undergo a [2 + 2 + 2] cycloaddition with π-systems. Bis(dithiobenzil)nickel(II) (**2**) was found to be the ideal reaction partner and to prevent light-induced reversion to norbornadiene, diethyldithiocarbamate (**5**) is added to chelate the nickel in the product (Scheme 1.50).

Quadricyclane is abiotic, unreactive with native biomolecules, relatively small, and highly strained (~80 kcal/mol). These attributes make it a promising bioorthogonal reagent. Reactions are enhanced by aqueous conditions with a second-order rate constant of 0.25 M/s. Method was applied for the selective labeling of proteins using a quadricyclane-modified bovine serum albumin (BSA), even in the presence of cell lysate (Scheme 1.51). It was also shown to be bioorthogonal to both copper-free and oxime chemistries during a one-pot reaction with differently functionalized protein substrates. It remains to see in the future the use of this new ligation method in living systems.

Protocol for conjugation of quadricyclane (**1**) with bis(dithiobenzil)nickel(II) (**2**) to form complex **3** [419]:

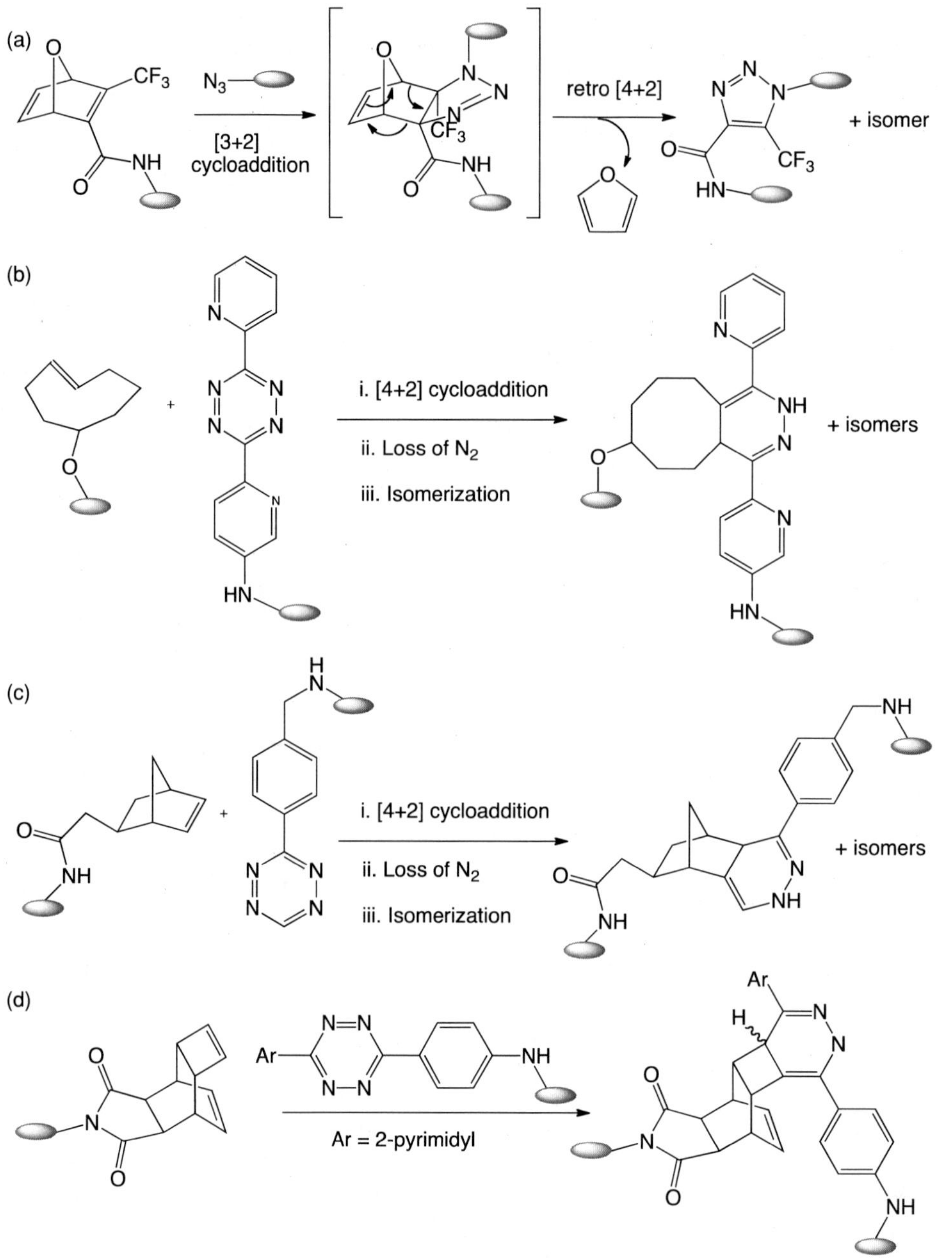

SCHEME 1.48 Bioconjugation via strained alkenes (a) oxanorbornadiene/azide pair, (b) *trans*-cyclooctene/tetrazine pair, (c) norbornene/tetrazine pair, and (d) cyclobutene-norbornene/tetrazine pair.

1. Combine 7-acetoxyquadricyclane (2.2 equiv.) with bis(dithiobenzil)nickel(II) (2.0 equiv.) in dichloromethane.

2. React for 5 days in the dark.

3. Evaporate reaction mixture and add methanol.

4. Place mixture in the fridge until a brown precipitate is formed.

5. Collect precipitate and wash with minimal amounts of methanol.

1.2.8.4.5 Alkyne–Nitrone Cycloaddition A modified version of the metal-free click chemistry employed nitrones as the 1,3-dipole rather than azides [420]. This 1,3-dipolar cycloaddition involves reaction between a nitrone and a cyclooctyne to form stable *N*-alkylated isoxazoline product (Scheme 1.52). The reaction rate is extremely fast with second-order rate constants ranging from 12 M/s to 32 M/s, depending on the substitution of the nitrone. Nitrones containing ester or amide α-substituents displayed much faster kinetics than similar reactions with azides. Although the

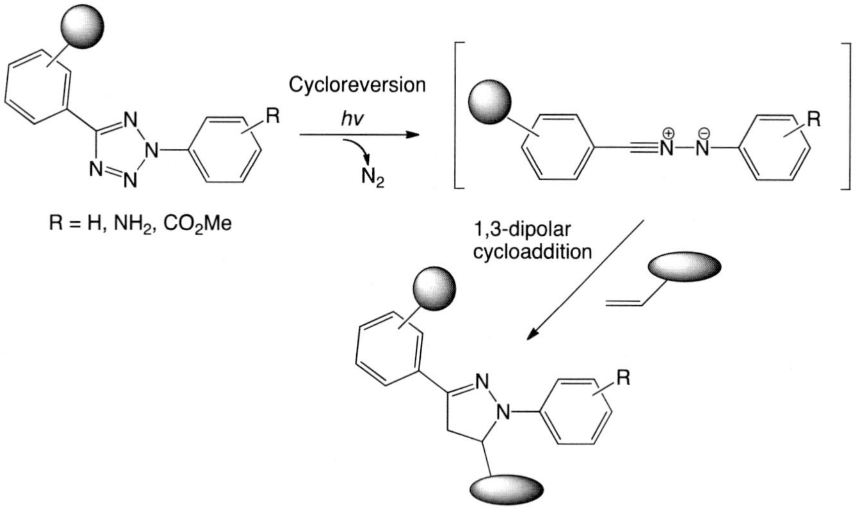

SCHEME 1.49   Bioconjugation via photo-click chemistry of tetrazoles and alkenes.

reaction proceeds with an exceptionally high reaction rate, attempts to incorporate the nitrone into biomolecules through metabolic labeling were unsuccessful. This new covalent bioconjugation method has mainly been applied for site-specific modification of peptides and proteins.

1.2.8.4.6   Isonitrile-click Reaction   Recently, a biocompatible isonitrile-click conjugation was reported, which proceeded in aqueous media via an initial [4 + 1] cycloaddition followed by a retro Diels–Alder elimination of nitrogen molecule [421] (Scheme 1.53). The reaction outcome depends on the nature of the isonitrile; if primary and secondary isonitriles are used, an imine is formed which can

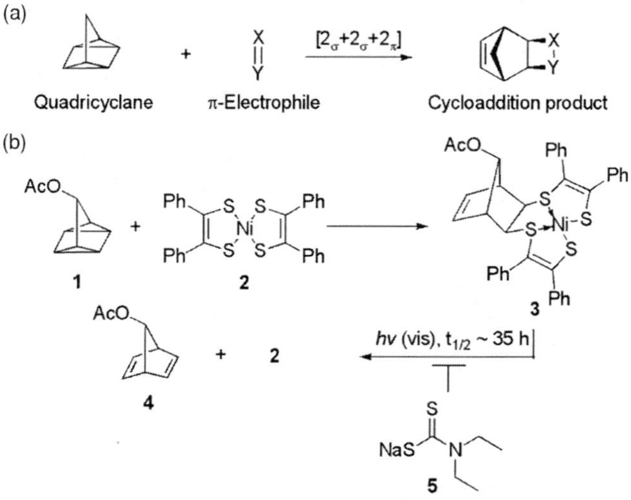

SCHEME 1.50   (A) Generic [2 + 2 + 2] cycloaddition of quadricyclane with π-systems, and (B) Quadricyclane ligation. Reprinted with permission from Reference 419, Copyright 2011, American Chemical Society.

easily be hydrolyzed; however, a stable adduct with tetrazine is formed if a tertiary isonitrile or isocyanopropanoate is used.

The isonitrile is an attractive bioorthogonal partner due to its small size, stability, nontoxicity, and absence from living systems. The main drawback at this stage is the slow reaction kinetics (second-order rate constants on the order of 10^{-2} M/s), however, it is expected to be a useful ligation to proteins and other biomolecules.

1.2.9   Bioconjugation Via Transition Metal-catalyzed/Mediated Reactions

Covalent linkages mediated by transition metals had a striking impact in the field of organic chemistry in the last few decades. New transformations that were previously difficult or impossible to create can now be achieved easily with the use of transition metal-catalyzed or mediated reactions. The rapid expansion in exploring the magical power of transition metal chemistry has eventually led to a Nobel Prize in Chemistry in 2010 shared by the three pioneers in this field, namely Professors R.F. Heck, E.I. Negishi, and A. Suzuki. As such, transition metal-catalyzed/mediated reactions have been recognized as a useful set of transformations that can expand the toolkit of bioconjugation. Initial investigations on the use of transition metal chemistry for bioconjugation has been mainly focused toward applications such as site-specific protein modification [422, 423]. These reactions can be easily tuned to proceed with high chemoselectivity and functional group tolerance under mild conditions by the judicious choice of metal centers, ligands, solvents, and additives. With the advent of transition metal-mediated reactions in water [424], their use for covalent bioconjugation has become even more possible. The Francis group has pioneered this area by modifying natural and unnatural residues of proteins using transition metals such as rhodium and palladium [422].

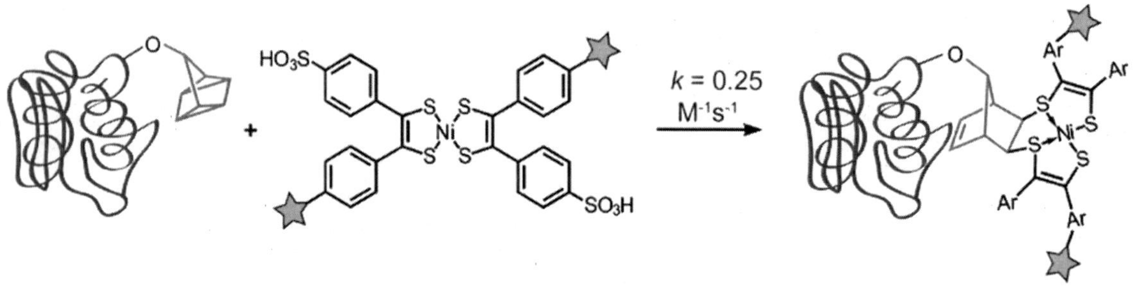

SCHEME 1.51 Quadricyclane ligation for labeling of proteins. Reprinted with permission from Reference 419, Copyright 2011, American Chemical Society. For a color version, see the color plate section.

Reactive rhodium carbenoid was used for selective modification of the indole moiety of tryptophan residues (Scheme 1.54) [425]. Effective modification was observed at low protein concentration, which also minimizes competitive O–H insertion. However, this method is limited by its low pH reaction medium that arises due to the addition of the additive, H$_2$NOH.HCl.

The same group also reported the use of palladium [108] and cerium [109] complexes for tyrosine modifications (refer to Section 1.2.5.3, Scheme 1.25(D, E)). Olefin cross-metathesis has emerged as a powerful technique for the conjugation of two terminal alkenes into an internal alkene (Scheme 1.55a). This method has eventually led to a Nobel Prize in Chemistry in 2005. Previously, the use of olefin cross-metathesis for bioconjugate preparation was limited due to its poor compatibility in aqueous solvents. However, a renewed interest in this method arises as a result of advances in catalyst design for aqueous olefin metathesis [423, 426]. As such, several water-soluble (pre)catalysts (Scheme 1.55c) have been engineered for olefin metathesis, in particular, for ring-closing metathesis (RCM) and ring-opening metathesis (ROM) (Scheme 1.55b) [427–430]. Studies have shown that allyl sulfides are privileged substrates in aqueous olefin cross-metathesis [431].

Ruthenium-catalyzed cross-metathesis is considered to be a bioorthogonal reaction due to its remarkable functional group tolerance, high selectivity, and the ease of alkene introduction into biomolecules. An early example of the application of aqueous ring-opening cross-metathesis involved the synthesis of carbohydrate inhibitors of cell agglutination [432]. Recent applications have been focused mainly on site-selective modification of protein surfaces [423, 431]. With the rapid development of aqueous olefin metathesis, the application to *in vivo* protein modification might be the next future challenge. Palladium-catalyzed cross-coupling reactions for carbon–carbon bond formation has been recognized to be among the most powerful and versatile organic transformations. Thus, palladium catalysis has shown to be useful in the area of bioconjugation. Typical palladium catalyzed cross-coupling reactions are the Heck reaction (cross-coupling between unsaturated halides and alkenes) [433], Suzuki–Miyaura reaction (cross-coupling between aryl halide or triflate (Ar–X, X = Br, I, OTf) and organoboron compounds (aryl or alkenyl)) [434], and Sonogashira reaction (cross-coupling between unsaturated halides and alkynes in the presence of copper catalysis) [435] (Scheme 1.56).

In 2006, Yokoyama and coworkers disclosed the first example of a Heck coupling for bioconjugation [436]. Iodophenylalanine moiety was installed on RAS protein, which was then selectively conjugated with a biotin-tethered alkene via the Heck reaction albeit in poor yield (2%) (Scheme 1.57(a)). A year later, the same group reported a site-specific biotinylation of RAS protein using the Sonogashira cross-coupling reaction to afford the desired conjugate in only 25% yield (Scheme 1.57(b)) [437]. Schultz and coworkers later showed Suzuki cross-coupling between a genetically coded *p*-boronophenylalanine moiety and a bodipyaryl iodide probe (Scheme 1.57(c)) [438]. The reaction proceeded at pH = 8.5 and temperature of 70°C in about 30% yield. In order to make these coupling reactions

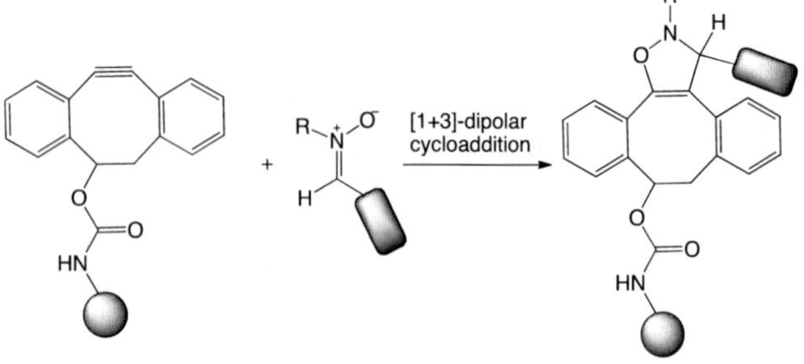

SCHEME 1.52 Alkyne–nitrone cycloaddition for peptide and protein modifications.

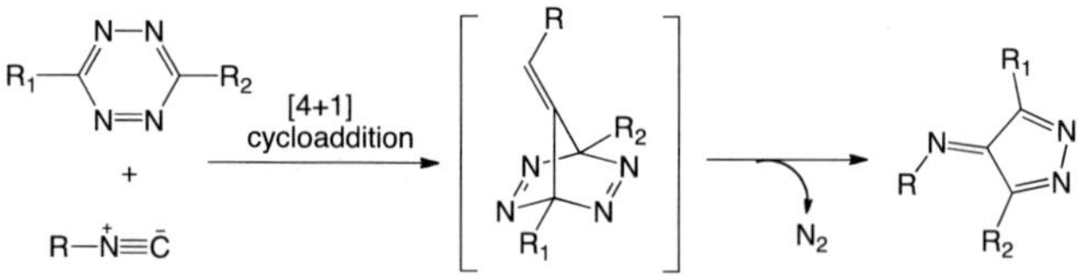

SCHEME 1.53 Isonitrile–tetrazine click ligation process.

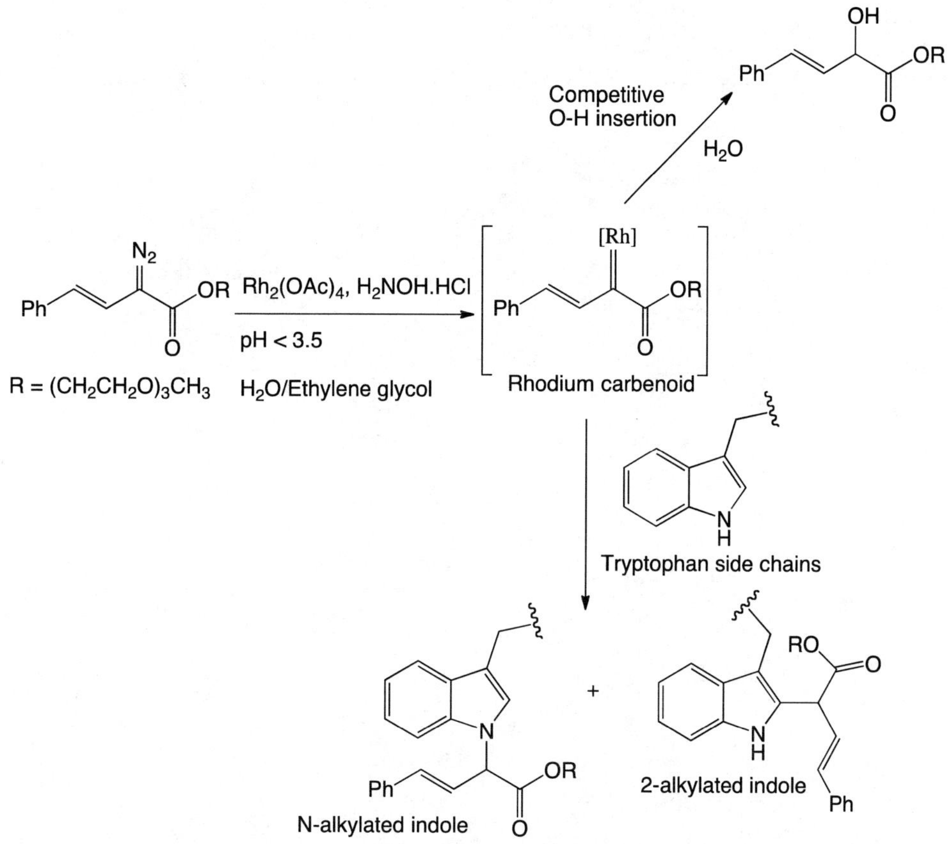

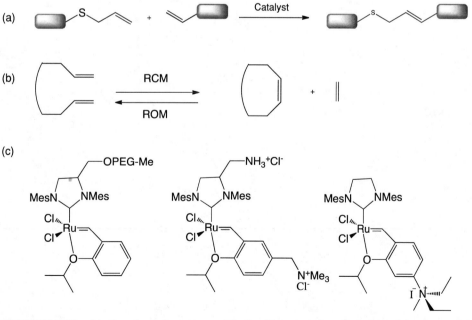

SCHEME 1.54   Modification of tryptophan residues with rhodium carbenoid.

SCHEME 1.55   (a) Generic olefin cross-metathesis, (b) typical ring-closing/opening metathesis, and (c) selected examples of ruthenium-based water-soluble catalysts.

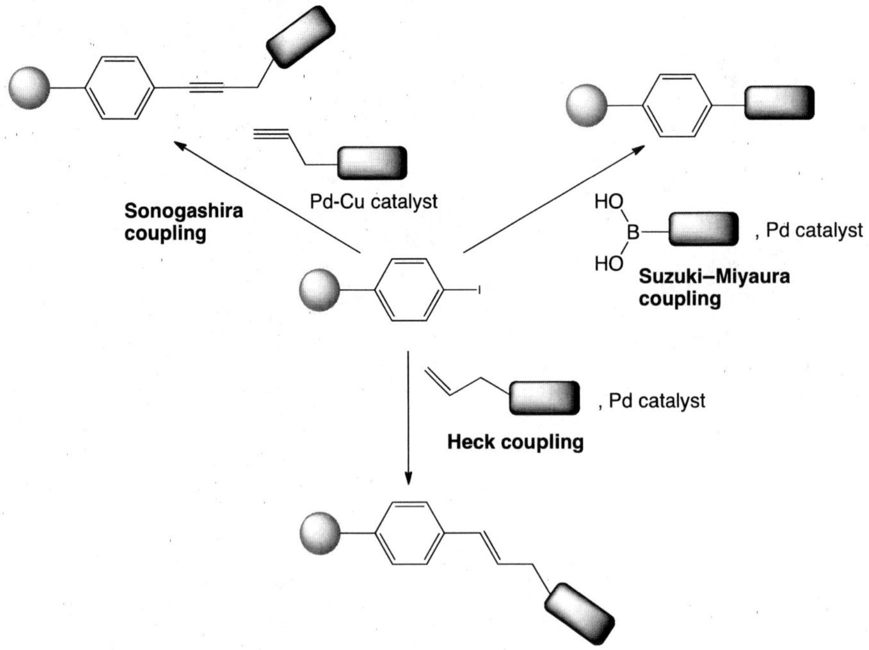

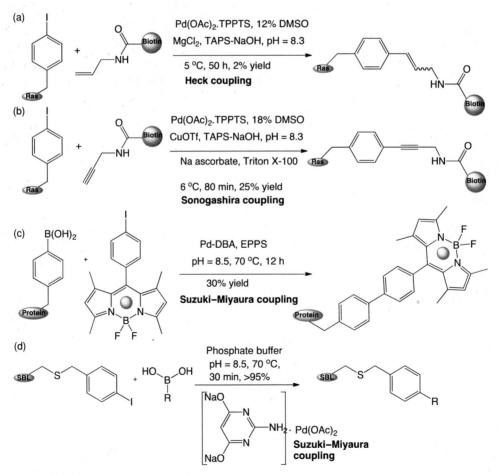

SCHEME 1.56   Bioconjugation via palladium-catalyzed cross-coupling reactions.

SCHEME 1.57   Protein modifications via palladium-catalyzed cross-coupling reactions Adapted from Reference 1f with permission from the Royal Society of Chemistry.

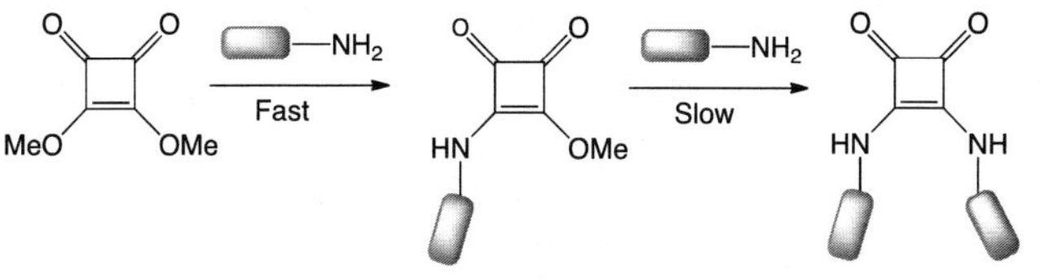

SCHEME 1.58   Bioconjugation via squaric acid chemistry.

more amenable to the biological system, Davis and coworkers recently demonstrated a Suzuki–Miyaura cross-coupling reaction with protein substrates using a water-soluble palladium catalyst (Scheme 1.57(d)) [439]. The reaction is carried out at 37°C at pH = 8 with almost >95% conversion in about 30–60 minutes.

While initial progress in transition metal-catalyzed reactions for bioconjugation looks promising, there are still a few concerns that need to be addressed. For instance, the toxicity and cell permeability of palladium catalyst for *in vivo* applications still need to be investigated. Furthermore, boronic acids are known to have strong affinity toward polysaccharides [440] and thus might not be bioorthogonal.

1.2.10   Other Covalent Bioconjugation Methods

1.2.10.1   *Squaric Acid Chemistry*   Squaric acid chemistry has been exploited for bioconjugation due to the special reactivity of squarates [441–443]. Bioconjugation takes place by reaction of amino groups with squarates via a nucelophilic conjugate addition followed by elimination to afford two vinylogous amide linkages (Scheme 1.58). The small size of the squarate and the slow kinetics of its reaction with a second amino group make this conjugation method attractive. Tietze et al. first reported the use of squaric acid chemistry in 1991 for the conjugation of *p*-aminophenyl glycosides to protein (BSA) using diethyl squarate [441]. It was used to couple oligosaccharideamines to carrier proteins [442]. Squaramides are remarkable four-membered ring systems and ideal units for bioconjugation and supramoloecular chemistry [444].

1.2.10.2   *Pyridoxal 5′phosphate (PLP)-mediated Bioconjugation*   PLP-mediated bioconjugation is a method for preparation of protein bioconjugates that achieves site specificity by targeting the N-terminus [445–447]. First, a protein is incubated with pyridoxal 5′phosphate (PLP) under mild, aqueous conditions, which oxidizes the N-terminus of the protein to a ketone or an aldehyde without modifying lysine side-chain amines (Scheme 1.59 (a)). This provides a unique functional group for further modification. Then, the ketone is conjugated to an alkoxyamine-bearing reagent via oxime formation (Scheme 1.59(b)) [448]. The proposed mechanism involves a Schiff base formation between the PLP aldehyde and the N-terminal amine followed by tautomerization and

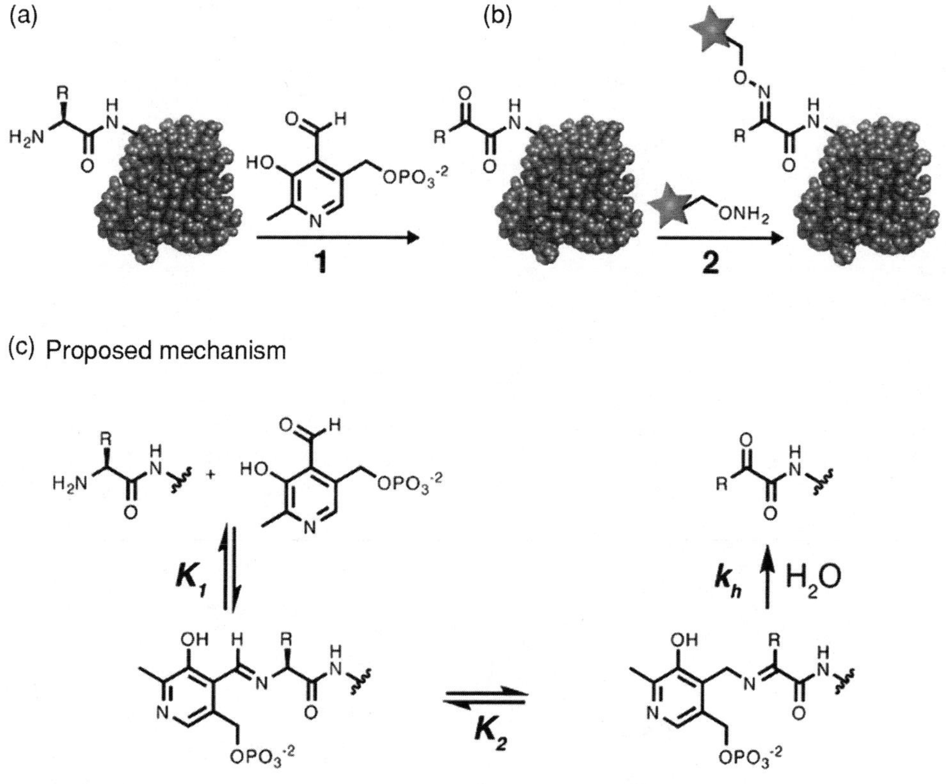

SCHEME 1.59   PCL-mediated bioconjugation and proposed mechanism. Reprinted with permission from Reference 448, Copyright 2010, American Chemical Society.

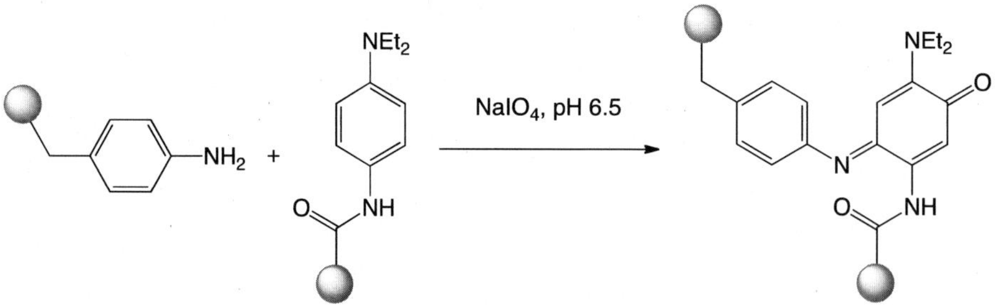

SCHEME 1.60 Bioconjugation via oxidative coupling between aniline side chains and *N,N*-diethyl-*N'*-acylphenylene diamine.

hydrolysis to give the keto-protein product (Scheme 1.59(c)). PLP-mediated bioconjugation has been used to modify proteins for applications in surface and polymer attachment [449] and light harvesting systems [450].

1.2.10.3 Oxidative Coupling of Anilines and Aminophenols

Recently, a new bioconjugation method was developed for the oxidative coupling of aniline with *N,N*-diethyl-*N'*-acylphenylene diamine in the presence of sodium periodate (Scheme 1.60) [451]. The reaction is chemoselective, efficient at low reactant concentrations and occurs within 30–60 minutes. It was used for the attachment of peptides to the surface of bacteriophage MS2.239. Bioconjugation via oxidative coupling to aniline was also applied to decorate the exterior surface of MS2 particles with zinc porphyrins for photocatalysis [452] and a DNA aptamer for cellular delivery [453].

A similar and highly efficient protein bioconjugation method was later disclosed, which involved the addition of anilines to *o*-aminophenols under the same oxidizing conditions. This new modified reaction takes place in aqueous buffer at pH 6.5 and can reach high conversion in 2–5 minutes to afford stable protein bioconjugates. Method is compatible with thiol modification chemistry. The rapid, chemoselective bioconjugation through oxidative coupling of aniline and aminophenol renders this method well suited for achieving protein modification in time-sensitive situations such as radiolabeling or for the quick assembly of multicomponent structures from building blocks at low concentrations [454].

The mechanism of this unprecedented oxidative coupling reaction is yet to be elucidated.

1.2.10.4 Diazonium Coupling

A diazonium coupling reaction is a well-known electrophilic aromatic substitution reaction in organic synthesis for the coupling of a phenol with a diazonium salt via a diazo or azo ($-N = N-$) linkage. This chemistry has been exploited mainly with proteins, whereby diazonium salts can selectively react with the phenol and imidazole groups of tyrosine and histidine, respectively. Diazonium coupling reaction is commonly used to label the *ortho* position of tyrosine residues (Scheme 1.61) [455–458].

Diazonium coupling method has also been used to introduce functional moieties that allowed subsequent bioconjugation to other biomolecules. For instance, tyrosine was used to introduce ketones on the surface of tobacco mosaic virus (TMV) particles for coupling to PEG through secondary oxime ligation [459] or alkyne-functionalized phenyl diazonium salts for subsequent CuAAC ligation with PEG, peptides, and dyes [321, 460].

1.2.10.5 Thiol-Ene and Thiol-Yne "Click" Reactions

The highly efficient reactions of thiols with reactive carbon–carbon double bonds (or commonly referred as "enes") were known since the early 1900s [461]. A renewed interest in these reactions emerged with two thiol reactions namely the thiol-ene free-radical reactions and the catalyzed Michael addition to electron-deficient carbon–carbon double bonds [462]. These reactions have recently attracted many researchers in other areas of syntheses as they possess many

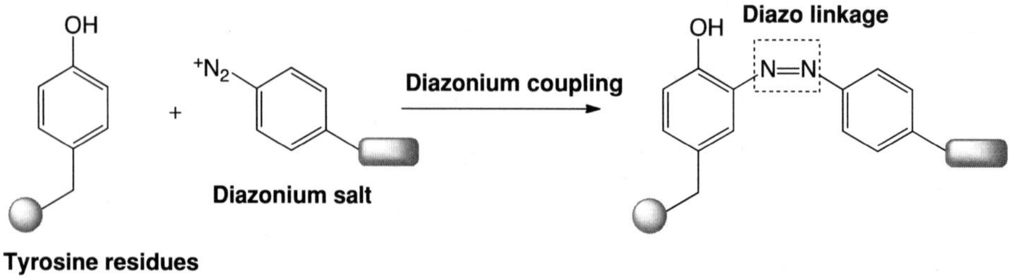

SCHEME 1.61 Diazonium coupling bioconjugation method.

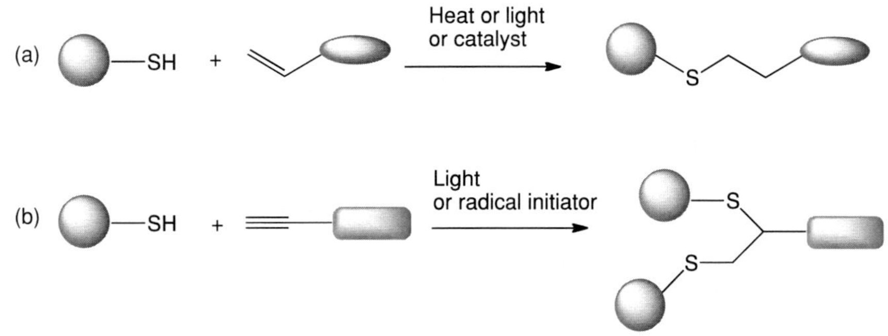

FIGURE 1.29   Covalent bioconjugation via (i) thiol-ene, (ii) thiol-yne click reactions.

of the attributes of click chemistry such as ease of implementation with no side products, mild reaction conditions, use of benign catalysts and solvents, insensitivity to water and oxygen, rapid reaction rates with high yield, and the ready availability of a wide range of alkenes and thiols. Maleimides are activated substrates for thiol-ene reactions (Figure 1.29 (i)) and their reactive C = C bonds render the reactions to occur extremely rapidly. As such, the high efficiency of these reactions has led to their applications as a convenient bioconjugation tool other than their routine applications in polymer and materials synthesis [463]. Radical thiol-yne coupling, that is, free-radical addition of two thiols to a terminal alkyne (Figure 1.29 (ii)) has recently emerged as one of the most appealing click chemistry procedures, which could eventually replace other popular click reactions such as copper-catalyzed azide–alkyne cycloaddition and thiol-ene coupling [464]. Early examples of thiol-yne coupling were independently reported in the 1930s by Finzi [465] and Kohler [466]. Being a metal-free process, thiol-yne coupling was eventually recognized in 2009 as a useful click reaction for biomedical applications [467]. It has shown to exhibit the same "click" features as its thiol-ene counterparts with an additional advantage of being able to create molecular complexity in a very straightforward manner [464, 468].

Recently, the dual modification of sugar alkynes through photoinduced thiol-yne coupling and thiol-ene coupling heterosequences was investigated [469], and the same strategy was extended for the glycosylation and fluorescent labeling of BSA (Scheme 1.62) [464, 470].

1.2.10.6 Metalloporphyrin-catalyzed Carbenoid Transfer for Protein Bioconjugation
Recently, a novel method was disclosed for the site-selective modification of the N-terminus of proteins via metal-mediated carbenoid N−H insertion, together with the modification, via metal-catalyzed alkene cyclopropanation, of a protein prefunctionalized with a styrene moiety [471]. The reaction takes place in the presence of a water-soluble metalloporphyrin ruthenium catalyst, [Ru^{II}(4-Glc-TPP)(CO)] (**1**). This ruthenium-catalyzed carbenoid transfer to alkene-tethered protein in aqueous media

could open up an entry to new bioconjugation reactions for protein modifications using metalloporphyrins as catalysts.

1.2.10.7 Bifunctional Silane Coupling Agents
Silane coupling agents are silicon-based molecules that act as an interface between an inorganic material (glass, metal, fillers, mineral, silica stone) and an organic material (polymers, coatings, adhesives, synthetic resins) via covalent linkages.

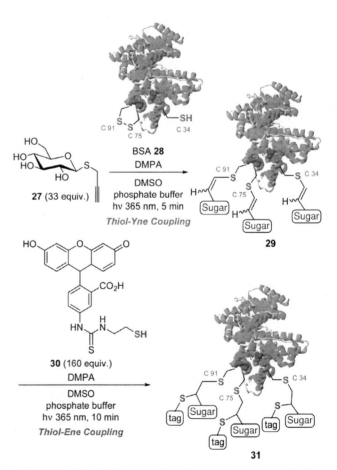

SCHEME 1.62   Thiol-yne and thiol-ene coupling strategies for glycosylation and fluorescent labeling of BSA. Reprinted from Reference 464 with permission from the Royal Society of Chemistry.

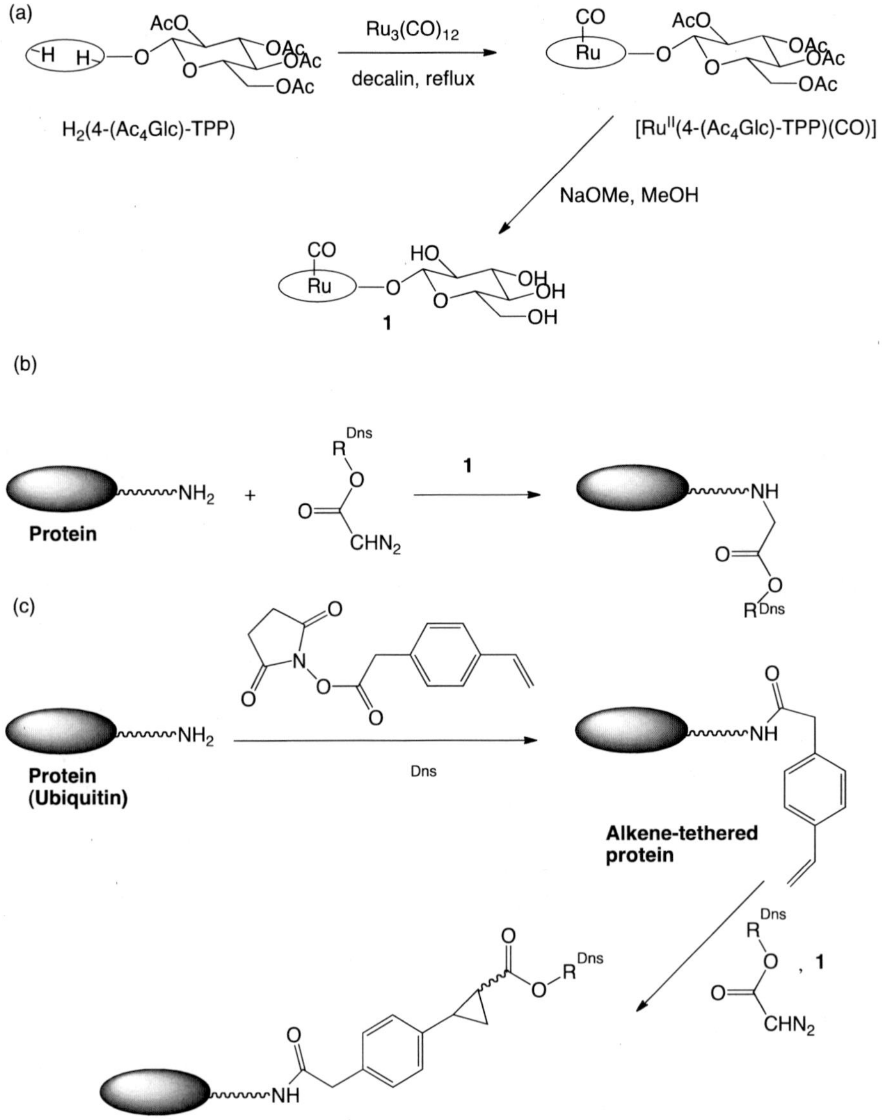

SCHEME 1.63 (a) Synthesis of water-soluble ruthenium glycosylated porphyrin [Ru^{II}(4-Glc-TPP)(CO)] (**1**), (b) modification of N-terminus of proteins, (c) modification of alkene-tethered protein through carbenoid transfer reactions mediated by [Ru^{II}(4-Glc-TPP)(CO)] (**1**) in aqueous media. Adapted with permission from Reference 471, Copyright 2010, American Chemical Society.

These organofunctional silanes are becoming increasingly important nowadays in many applications [472, 473]. A general representative structure of a silane consists of a central silicon atom carrying two different functional groups namely alkoxy and organofunctional groups. These groups exhibit different reactivities and can subsequently be used to react and couple with different materials such as inorganic surfaces and organic materials. Overall, the process allows the conjugation of an organic material and an inorganic material with the silane coupling agent as a mediator (Figure 1.30).

The alkoxy groups on the silicon will normally bind well to the metal hydroxyl groups on most inorganic surfaces to form stable Si–O–M bonds (M = Si, Fe, Al, etc.). The Si–OR bonds are readily hydrolyzed in the presence of water to form silanols (Si–OH), which subsequently generate oligomers via a partial condensation reaction. The silanol oligomers then hydrogen bond to the inorganic surface and after a drying process, strong chemical bonds are formed via a dehydration–condensation sequence (Scheme 1.64).

Many organofunctional silanes are commercially available (Shin-Etsu, Dow Corning, Gelest) and due to their bifunctional nature, they have been applied in the field of bioconjugate chemistry for the covalent linkage of proteins, ONTs, antibodies, or other biomolecules to inorganic substrates. The characteristics of the organofunctional groups can vary from hydrophilic to hydrophobic as well

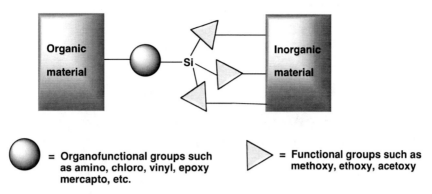

FIGURE 1.30   General structure of a bifunctional silane coupling agent.

as orgonoreactive and organophilic. Figure 1.31 displays some of the commonly used organofunctional silanes for bioconjugation of organic and inorganic materials. They carry a number of terminal reactive functional groups that open opportunities for typical conjugation reactions.

Inorganic surfaces modified with 3-acryloxy-propyltrichlorosilane (APTS) coupling agent will introduce terminal amino groups, which can be subsequently modified to reactive carboxylate or azide groups by reaction with succinic anhydride or NHS-PEG$_4$-azide, respectively. Presence of carboxylate groups will allow conjugation with amine-containing biomolecules via covalent amide linkages. Hydrophilic PEG spacer terminated with azido groups will open routes to click chemistry by reaction with terminal alkyne-containing biomolecules or Staudinger ligation by treatment with phosphine-derived biomolecules (Scheme 1.65). Carboxyethylsilanetriol has been used to introduce carboxylate groups to fluorescent silica nanoparticles for subsequent coupling with

antibodies for multiplexed bacteria monitoring [475]. *N*-(Trimethoxysilylpropyl)ethylenediamine triacetic acid (TMS–EDTA) possesses a metal-chelating functional group (EDTA) and once it is coated on a surface, it can be used for affinity separations (Figure 1.32).

Surface-containing reactive epoxy groups can react with thiol-, amine- or alcohol-containing biomolecules via ring-opening reactions of epoxide. This strategy has been exploited for preparation of high density PEG surface on glass slides for arrays applications [477] and in the activation of glass surfaces for antibody detection [481].

1.3   NONCOVALENT BIOCONJUGATION STRATEGIES

A noncovalent bond is a special type of chemical bond, typically between macromolecules, which does not involve the sharing of a pair of electrons. While most bioconjugation

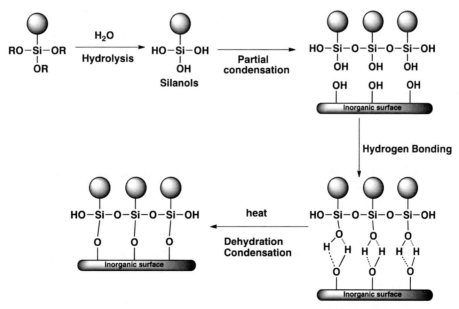

SCHEME 1.64   Chemical binding of silane coupling agent to an inorganic surface.

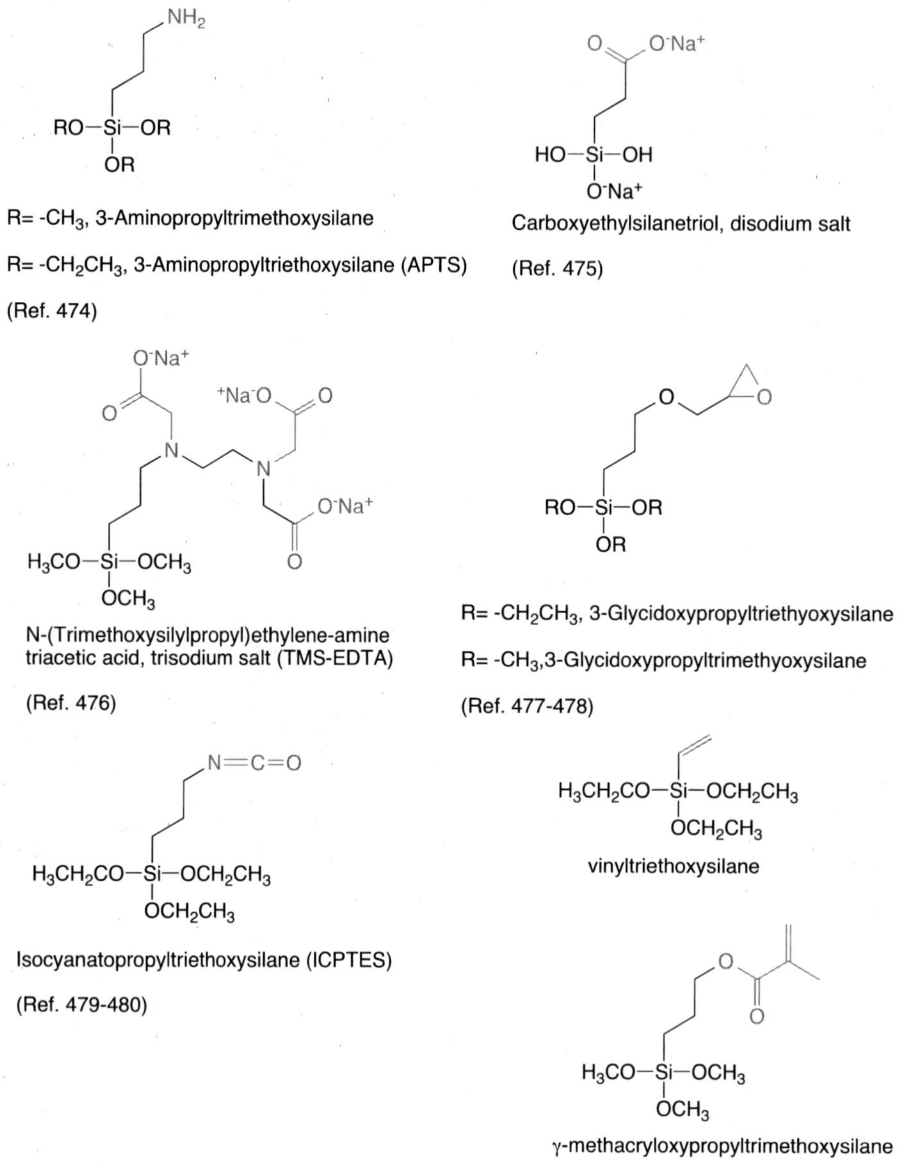
NH2
RO—Si—OR
OR
R= -CH3, 3-Aminopropyltrimethoxysilane
R= -CH2CH3, 3-Aminopropyltriethoxysilane (APTS)
(Ref. 474)
O O-Na+
HO—Si—OH
O-Na+
Carboxyethylsilanetriol, disodium salt
(Ref. 475)
O-Na+
+Na-O
O
N
N
O-Na+
H3CO—Si—OCH3
OCH3
N-(Trimethoxysilylpropyl)ethylene-amine triacetic acid, trisodium salt (TMS-EDTA)
(Ref. 476)
RO—Si—OR
OR
R= -CH2CH3, 3-Glycidoxypropyltriethyoxysilane
R= -CH3,3-Glycidoxypropyltrimethyoxysilane
(Ref. 477-478)
N=C=O
H3CH2CO—Si—OCH2CH3
OCH2CH3
Isocyanatopropyltriethoxysilane (ICPTES)
(Ref. 479-480)
H3CH2CO—Si—OCH2CH3
OCH2CH3
vinyltriethoxysilane
H3CO—Si—OCH3
OCH3
γ-methacryloxypropyltrimethoxysilane

FIGURE 1.31 Examples of bifunctional silane coupling agents with different reactive groups.

strategies are predominantly based on the covalent bond formation, which offers the advantage of a strong (100–400 KJ/mol) and stable linkage, noncovalent approaches rely heavily on physical interactions via the intermolecular noncovalent forces (Figure 1.33). Noncovalent forces are weak by nature and they must work together in order to have a significant effect (cooperative effect). Noncovalent interactions encompass affinity interactions, electrostatic, metal-mediated coordination, hydrophobic, and hydrogen bonding.

In the past few decades, advances in exploiting noncovalent interactions have led to the design of sophisticated biomolecules with complex architectures. Traditional covalent-based synthetic methods become increasingly tedious as macromolecular structures and functional materials continue to evolve with higher level of complexity and function. Noncovalent chemistries can overcome some of the drawbacks of covalent strategies such as lengthy synthesis and reagents' incompatibility. Noncovalent forces generally fall into three major classes: (i) weak interactions (0–15 kcal/mol), (ii) medium interactions (15–60 kcal/mol), and (iii) strong interactions (above 60 kcal/mol) [482]. The degree of the noncovalent interactions is dependent on the type of application of the bioconjugate.

1.3.1 Biotin-(Strept)Avidin System

Many supramolecular bioconjugation strategies employ biological recognition motifs, which are almost perfect guest–host systems. A classic example of such a system

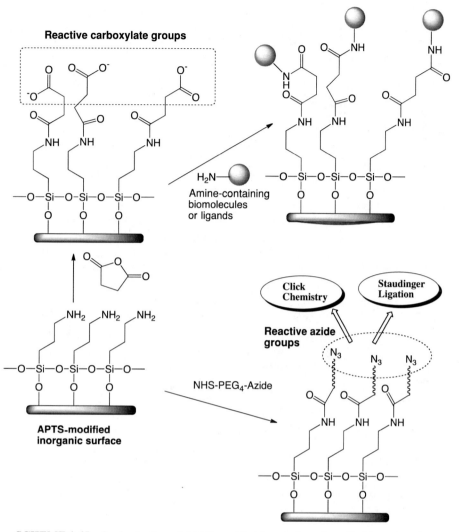

SCHEME 1.65 Derivatization of APTS-modified inorganic surfaces for bioconjugation.

is the biotin-(strept)avidin interaction. Biotin (5-(2-oxo-hexahydro-1*H*-thieno[3,4-d]imidazol-4-yl)pentanoic acid) (Figure 1.34), also known as vitamin H, B7, or co-enzyme R, is abundant in certain plant and animal tissues. It is a cofactor

in the metabolism of fatty acids and leucine and plays crucial role in a number of biological processes. The carboxylic acid group of biotin is the site of attachment of the molecule to amino-containing biomolecules via an amide linkage.

Due to its small size (244.3 Da) and good stability, biotin can be easily covalently conjugated with many proteins,

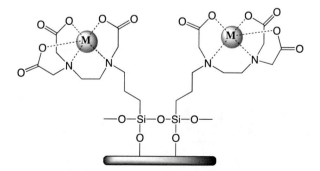

FIGURE 1.32 EDTA-modified inorganic substrate for complexation with metal ions.

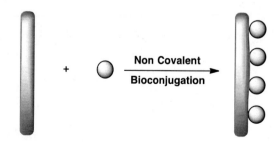

FIGURE 1.33 Schematic representation of noncovalent bioconjugation strategy.

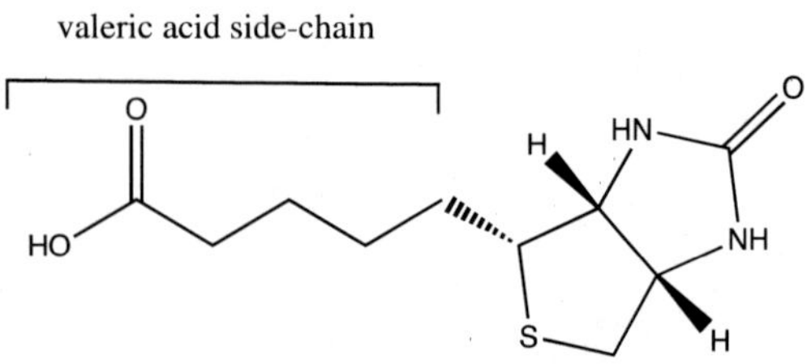

valeric acid side-chain

FIGURE 1.34 Chemical structure of biotin.

polymers, enzymes, and antibodies without any significant change in their biological activities (e.g., enzymic catalysis or antibody binding). Avidin is a glycoprotein (64 kDa) located in egg white and tissues of birds, reptiles, and amphibians that has high binding affinity and specificity with biotin. It is highly glycosylated with an isoelectric point of 10–10.5 and also highly water soluble. It is stable over a wide range of pH and temperature, and hence it retains its activity after chemical modifications. Streptavidin is another biotin-binding glycoprotein (~53 kDa), which is isolated from *Streptomyces avidinii*. Unlike avidin, streptavidin is nonglycosylated with an acidic isoelectric point of 5, thereby a lower water solubility and degree of nonspecific binding (especially lectin binding). A recombinant streptavidin with a near-neutral isoelectric point (6.8–7.5) and mass of 53 kDa is commercially available (ThermoScientific Pierce). The avidin–biotin and streptavidin–biotin complexes are the strongest, noncovalent, biological interaction known to date with $K_a = 10^{15}$ and 10^{13}/M, respectively, between a protein and a ligand [483]. Both avidin and biotin has the ability to bind noncovalently up to four biotin molecules rendering this interaction ideal for purification and detection strategies. They both have advantages and disadvantages: avidin has low cost production, high solubility but high isoelectric point, and high degree of nonspecific interactions and lectin binding; streptavidin has near-neutral isoelectric point and lower nonspecific binding, but a high cost of production. As such, deglycosylated avidin called NeutrAvidin is much more ideal biotin-binding reagent that overcomes the major drawbacks of avidin and streptavidin. NeutrAvidin (60 kDa) has a cost-effective production, commercially available (ThermoScientific), a near-neutral isoelectric point, high biotin-binding affinity, and absence of nonspecific binding issues.

Conjugates derived from strept(avidin)-biotin complex are highly stable even under harsh conditions with a slow kinetics of dissociation compared to the time scale of most of the experimental procedures. These excellent features have been instrumental in protein immobilization [484] and more recently for synthesis of protein–polymer bioconjugates [485].

Strept(avidin)-biotin method became a popular method for protein immobilization due to the widespread availability of supports (microtiter plates, microarray substrates, magnetic particles) coated with these proteins. However, the protein of interest must be labeled with biotin prior to immobilization. This can be achieved either enzymatically or chemically with a number of nonselective chemical biotinylation reagents, such as biotin NHS ester. The desire for site-specific biotinylation has led to the development of biotin ligase enzyme-catalyzed bioconjugation systems. For example, *E. coli* biotin ligase/synthetase (BirA) transfers endogenous biotin to a specific lysine side chain and has been used *in vitro* to label proteins with the 75 amino acid biotin carboxyl carrier protein (BCCP) [484] (Scheme 1.66).

A number of biotinylated initiators were designed for the synthesis of bioconjugates (Figure 1.35). Both atom-transfer radical polymerization (ATRP) and reversible addition–fragmentation technique (RAFT) were used for the synthesis of polymer-protein bioconjugates. Initiator **A** was employed for the preparation of biotinylated PNIPAAm polymers by ATRP, which was then conjugated with streptavidin at room temperature for 24 hours [486]. A "*grafting-from*" approach was used in the construction of streptavidin–polymer bioconjugate using biotinylated ATRP initiator **B** [487].

A simple route toward heterotelechelic polymers for bioconjugation of two different proteins (BSA and Streptavidin) was reported using RAFT agent **C** (Scheme 1.67) [488]. However, the biotinylated RAFT agent **D**-containing amide group was found to minimize loss of biotin end-groups upon retro Diels–Alder deprotection at high temperature.

Biotin-terminated gradient glycopolymers were prepared using RAFT agent **E** and subsequent *in situ* photochemical reduction generates gold nanoparticles [489]. The presence of biotin on the surface of the glyconanoparticles allows bioconjugation with streptavidin. Biotinylated PEG

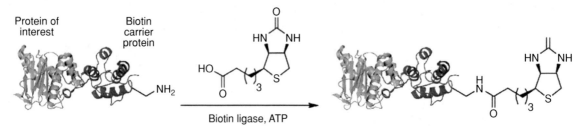

SCHEME 1.66 Protein ligation with protein of interest fused with BCCP. Reprinted with permission from Reference 484, Copyright 2009, American Chemical Society.

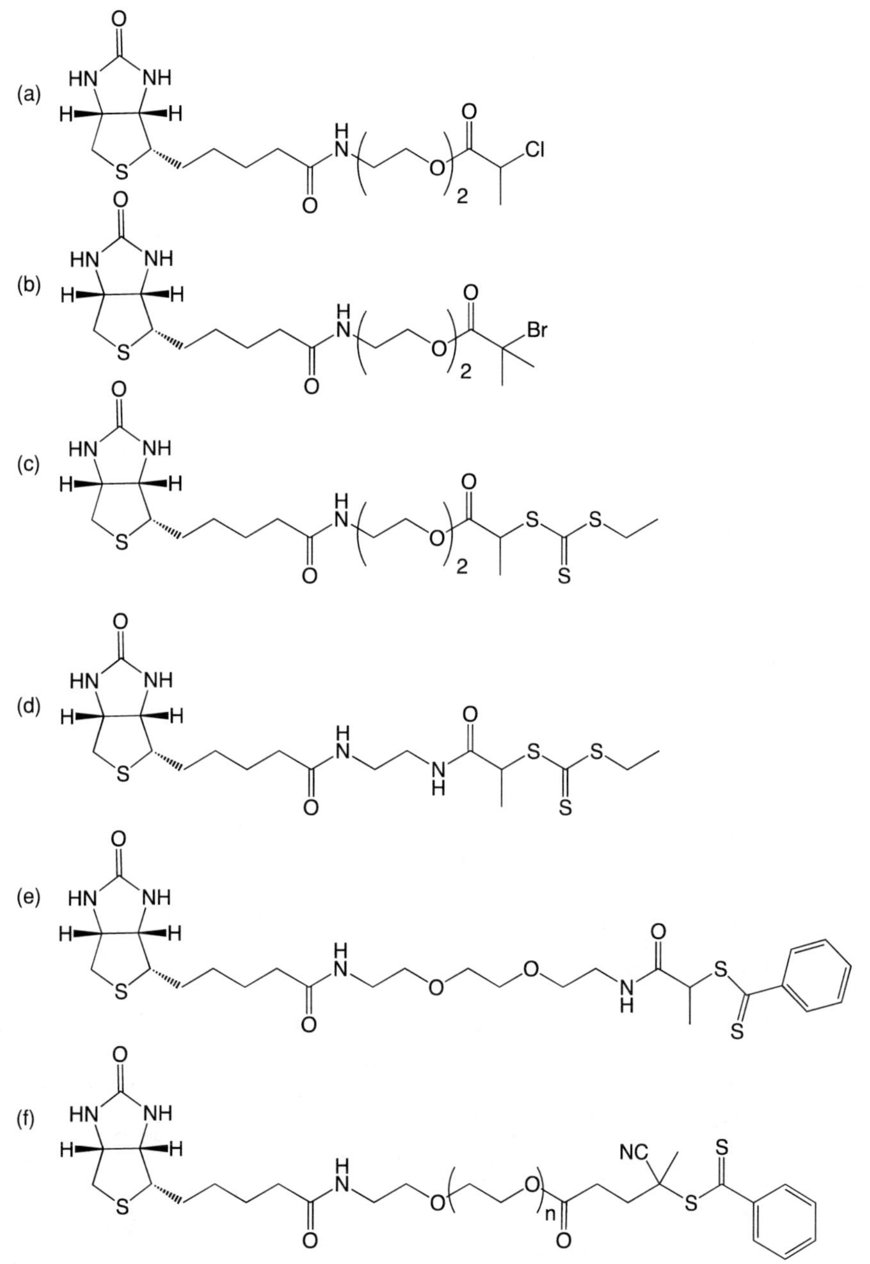

FIGURE 1.35   Biotinylated initiators: **A**, **B** as ATRP agents and **C**, **D**, **E**, **F** as RAFT agents.

macro-RAFT agent **F** was used to form a block copolymer that was able to self-assemble in methanol [490]. Recently, avidin–polymer conjugates were synthesized using a combination of ATRP and "click" chemistry between an alkynylated biotin and an azide-functionalized polymer (Scheme 1.68) [491].

In the biotin-(Strept)-avidin system, at least one of the participating component must be biotinylated, and luckily many biotinylation reagents [492] are now commercially available. They are easily used by simply following well-established procedures from the literature or the reagent manufacturers.

The use of this system has exploded in the last few decades, with a diverse range of applications, such as protein and nucleic acid detection system, affinity system, protein interactions, immunological assays, nucleic acid hybridization assays, DNA sequencing, etc. [493].

1.3.2   Electrostatic Interactions

Electrostatic interactions are generally defined as the attractive or repulsive forces between atoms and/or groups of atoms and/or molecules as a result of the presence

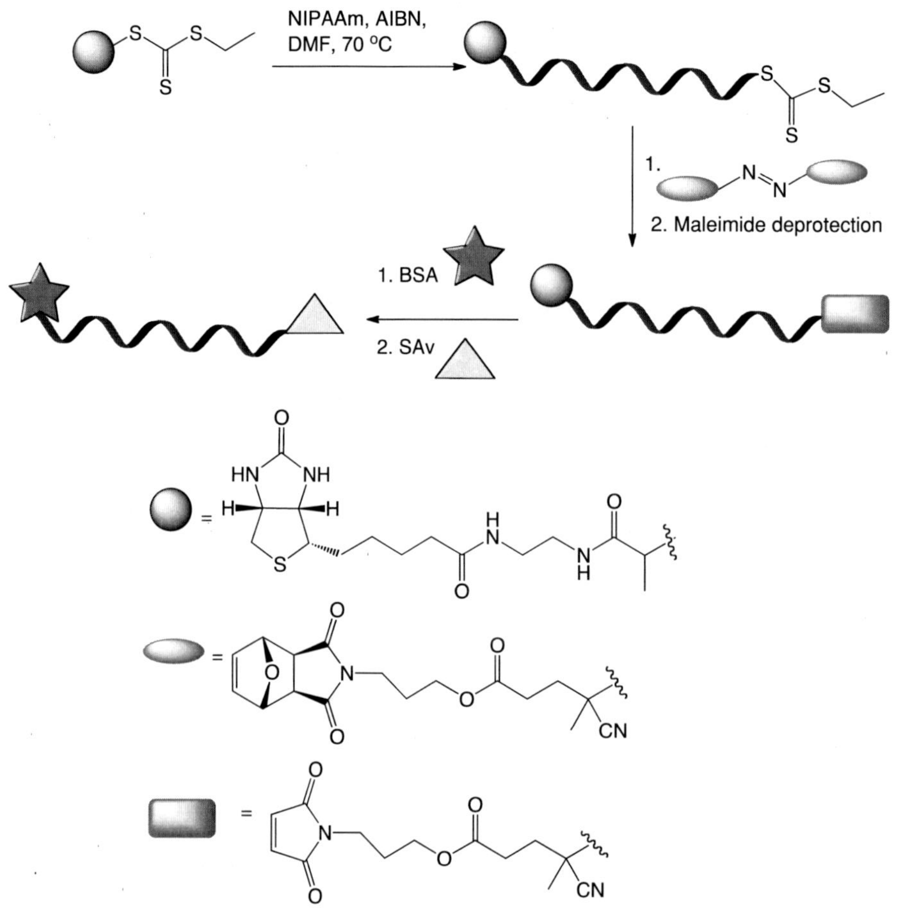

SCHEME 1.67 Synthesis of protein–heterodimer conjugates by RAFT method. Adapted from Reference 485 with permission from the Royal Society of Chemistry.

of ionized charged species. The ionic complexation of charged entities is an attractive route of noncovalent bioconjugation since many biological structures exhibit charged surfaces (Figure 1.36). For instance, the negatively charged backbone of DNA can be complexed electrostatically with various synthetic polycations or block copolymers possessing polycationic segments. The resulting complexations commonly known as polyplexes, have been extensively studied in the field of nonviral gene delivery [494]. Electrostatic interactions have been mainly useful for loading therapeutic genes or siRNA onto nanoparticles coated with cationic polyethylene imine (PEI) [495].

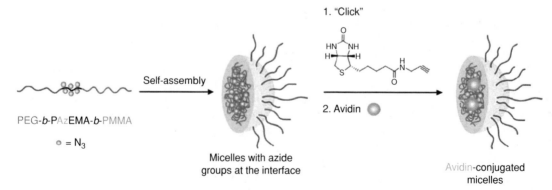

SCHEME 1.68 Bioconjugation of biotin to the interfaces of polymeric micelles and subsequent binding with avidin. Reprinted from Reference 485 with permission from the Royal Society of Chemistry. For a color version, see the color plate section.

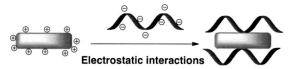

FIGURE 1.36 Electrostatic interaction between two oppositely charged biological entities.

Interaction between cationic polymer–nucleic acid complex and negatively charged cell membranes can enhance cell uptake and hence, transfection efficiency [496].

Bioconjugation of proteins to colloidal gold can be achieved noncovalently using a combination of electrostatic and hydrophobic interactions. Citrate-capped gold NPs are negatively charged due to the presence of a layer of negative citrate ions and thus positively charged amino groups of the antibody will be attracted to the gold surface [497]. When the protein is close enough for binding, the hydrophobic pockets of the protein will make contact and bind with the gold [498]. The bioconjugation process can generally be optimized by maintaining the pH of the antibody and the gold sol at or slightly above the isoelectric point of the antibody.

The covalent bioconjugation of antibodies to quantum dots (QDs) is a relatively tedious process in addition with the resulting low quantum yield and low quantity [499], which severely limit their utility. Thus, noncovalent bioconjugation of antibodies to QDs has been made possible on the basis of electrostatic interaction, for example, between the negatively charged capped QDs and an engineered adaptor protein having the immunoglobulin G (IgG)-binding domain of streptococcal protein G modified by a positively charged leucine zipper tag [500, 501]. Noncovalent nanoparticle–protein bioconjugates have also been prepared via complementary electrostatic interactions between the nanoparticle and the protein, which have enabled applications in sensing and delivery [502]. However, this approach is particularly useful when specificity is not required.

1.3.3 Metal-mediated Non-covalent Conjugation

Metal-mediated interactions could be a useful strategy for preparation of bioconjugates, in particular, controlled nanoparticle bioconjugates. These interactions are mainly via weak coordination (dative) bonding, whereby stability of conjugates is dictated by the equilibrium dissociation constants. The interesting properties of polyhistidine moiety (Figure 1.37) have been exploited for the synthesis of bioconjugates from inorganic nanoparticles. The polyhistidine motif is well known for binding with divalent metal cations (Co^{2+}, Cu^{2+}, Ni^{2+}, Zn^{2+}) immobilized on solid supports through carboxylated chelates. The imidazole rings of histidine interact with divalent metal ions as well as with noble metals such as gold and silver.

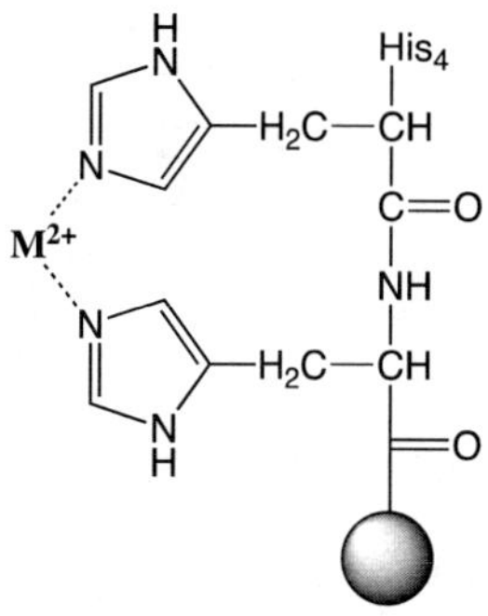

FIGURE 1.37 Polyhistidine motif for coordination with divalent metal ions.

The intrinsic properties of histidine motif include its small size, which does not disrupt native protein function and its bioorthogonality as it is not normally found in natural proteins. It also enables good control over biomolecular orientation via its single point of conjugation. Polyhistidine tags have been mainly employed for self-assembly with QDs through coordination with Zn^{2+} ions at the inorganic surface of CdSe/ZnS QDs [503, 504]. The method has been extended to the preparation of QD—ONT conjugates [505, 506]. The use of nickel(II)–nitrilotriacetic acid (Ni^{2+} −NTA) functionalized nanoparticles turns out to be a better approach for the self-assembly with polyhistidine tags rather than direct polyhistidine coordination (Figure 1.38). Polyhistidine tags are known to bind to Ni^{2+}–NTA with dissociation constants on the order of $K_d = 10^{-13}$ M [507]. QDs [508–510], single-walled carbon nanotubes [511], gold [512], silica [513, 514], and magnetic nanoparticles [515] have been modified with NTA for subsequent binding with polyhistidine tagged proteins.

Recently, the synthesis of well-defined NTA–end-functionalized polystyrenes prepared by ATRP and their bioconjugation with histidine-tagged GFPs (His$_6$-GFP) were disclosed [516]. It was demonstrated that site-specific and noncovalent bioconjugation occurred between polymer and protein. The reversible association of His$_6$-GFP with polystyrene spherical aggregates (with Ni^{2+}) was

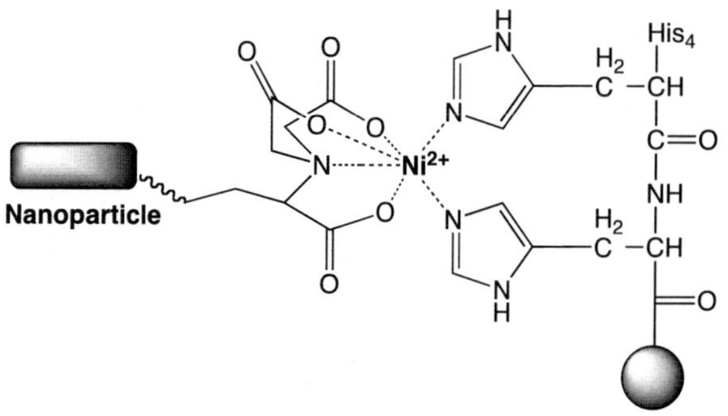

FIGURE 1.38 Conjugation of polyhistidine through coordination with Ni^{2+} ions with NTA-coated nanoparticles.

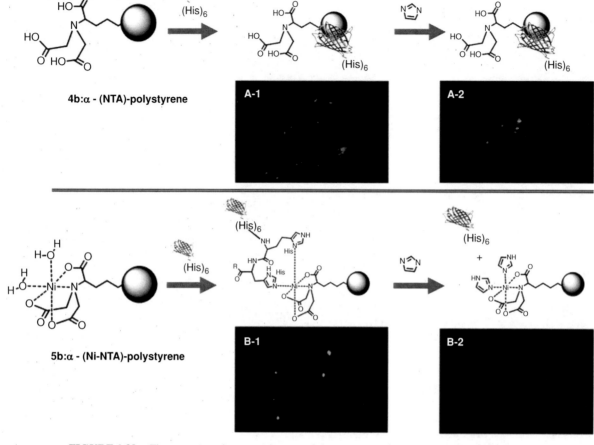

FIGURE 1.39   Fluorescence microscope images of the aggregates from the α-(NTA)-polystyrene: (A-1) nonspecific binding of His6–GFP with **4b**; (A-2) binding after rinse with excess imidazole; (B-1) specific binding of His6–GFP to NTA of **5b** in the presence of Ni; (B-2) binding after rinse with excess imidazole. Reprinted with permission from Reference 516, Copyright 2011, American Chemical Society.

successfully controlled with excess imidazole treatment and monitored by fluorescence microscope (Figure 1.39). Controlled bioconjugation of His-tagged protein with α-(Ni-NTA)-polystyrene may eventually have broad applications in protein purification and enzyme immobilization studies.

1.3.4   Hybridization Method

Recently, hybridization method (Figure 1.40) has emerged as a powerful strategy for the conjugation of nanoparticles with nucleic acid ligands (aptamers) to form nanoparticle–aptamer bioconjugates [517]. Aptamers are short nucleic acid molecules with 15–40 bases (single-stranded DNA or RNA ONTs) that bind with target molecules with high affinity [518]. The unique specificity of aptamers renders them suitable for use as targeting molecules and thus drug-encapsulated nanoparticle–aptamer bioconjugates can be used for the delivery of chemotherapeutics to primary and metastatic tumors [519].

Aptamer-targeted gold nanoparticles were developed using the hybridization method for use as molecular-specific contrast agents for reflectance imaging [517]. Conjugation of the aptamers to gold nanoparticles was accomplished using an extended aptamer design where the extension is complementary to an ONT sequence attached to the surface of gold nanoparticles (20 nm). The resulting aptamer–gold nanoparticle bioconjugate was used for the detection of prostate-specific membrane antigen (PSMA).

One major advantage of this approach is that the integrity and stability of the aptamers is easily preserved during the bioconjugation process since the aptamers are only introduced for a short period of time during the hybridization process. Furthermore, a smaller amount of aptamers is required for binding to gold nanoparticles through the short cheap capture ONTs. Same hybridization strategy was employed for

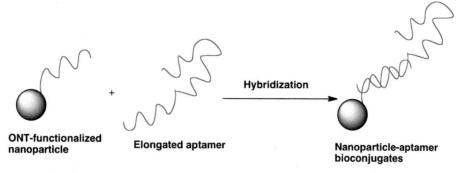

FIGURE 1.40 Hybridization method for nanoparticle-aptamer bioconjugates

conjugation of an anti-EGFR–aptamer onto gold nanoparticles [520] and an anti-PSMA–aptamer to iron-oxide nanoparticle surfaces [521].

1.4 CONCLUSIONS AND OUTLOOK

For many years, researchers have largely relied upon the traditional covalent and noncovalent bioconjugation chemistries for the preparation of bioconjugates for various applications. While the majority of bioconjugation reactions are mainly based on the formation of strong covalent bonds, noncovalent strategies have also been developed depending on the type of applications. Although traditional bioconjugation chemistries continue to be adequate for proof-of-concept studies, the engineering of bioconjugates for "*real*" biological applications still require much greater control than these chemistries can offer. In the past few years, novel applications of reactions drawn from knowledge of organic chemistry and biochemistry have surfaced as novel techniques for the preparation of the desired bioconjugates under physiological conditions and with highly desirable properties. The last decade has also seen the emergence of the powerful bioorthogonal chemistry and as well lay out by C.R. Bertozzi, bioorthogonal chemistry is like "bringing chemistry to life." The field of bioorthogonal conjugation chemistry is growing at a fast pace as it allows a wide range of biomolecules to be specifically labeled and probed in living cells and whole organisms. Bioorthogonal reactions have nowadays become essential tools for chemical biologists since their applications span over many areas. However, improvements in the kinetics and selectivities of some of the developed bioorthogonal reactions still need further investigations. Furthermore, the discovery of novel bioorthogonal reactions for applications in living systems will be a future challenge to synthetic chemists. Bioconjugation methods are active areas of research and the design of robust covalent or noncovalent conjugation reactions in order to expand the "toolbox" of bioconjugation will undoubtedly continue to be developed in the coming years.

REFERENCES

1. For previous books and reviews on bioconjugation techniques, see: (a) Hermanson TG.*Bioconjugate Techniques*, 2nd ed. Elsevier Academic Press; Amsterdam; Boston. 2008, 1-1323. (b) Niemeyer CM, *Bioconjugation Protocols: Strategies and Methods*, Humana Press Inc., Totowa, New Jersey 2004, 1-330. (c) Veronese FM, Morpurgo M. *IL Farmaco* 1999;54:497–516. (d) Kalia J, Raines RT *Curr Org Chem* 2010;14:138–147. (e) Farkas P, Bystricky S. *Chem Pap* 2010;64:683–695. (e) Sletten EM, Bertozzi CR. *Angew Chem Int Ed* 2009;48:6974–6998. (f) Lim RKV, Lin Q. *Chem Commun* 2010;46:1589–1600.

2. Moore JA, Reed DE. *Org Synth* 1973;5:351.

3. Shioiri T, Aoyama D, Mori S. *Org Synth* 1993;8:612.

4. DeMar Jr JC, Disher RM, Wensel TG. *Biophys J* 1992;61:A81.

5. Vale P, de M Sampayo MA. *Toxicon* 2002;40:979–987.

6. Suzuki M, Yamaguchi S, Iida T, Hashimoto I, Teranishi H, Mizoguchi M, Yano F, Todoroki Y, Watanabe N, Yokoyama M. *Plant Cell Physiol.* 2003;44:35–43.

7. García C, Pereira P, Valle L, Lagos N. *Biol Res* 2003;36:171–183.

8. Brekke OL, Sagen E, Bjerve KS. *J Lipid Res* 1997;38:1913–1922.

9. Nimura N, Kinoshita T, Yoshida T, Uetake A, Nakai C. *Anal Chem* 1988;60:2067–2070.

10. Tojo H, Ono T, Kuramitsu S, Kagamiyama H, Okamoto M. *J Biol Chem* 1988;263:5724–5731.

11. (a) Anderson GW, Paul R. *J Am Chem Soc* 1958;80:4423–4423. (b) Armstrong A. *Encyclopedia of Reagents for Organic Synthesis*, J. Wiley & Sons, New York, 2001.

12. Paul R, Anderson GW. *J Am Chem Soc* 1960;82:4596–4600.

13. Zhang X, Rodrigues J, Evans L, Hinkle B, Ballantyne L, Peña, M. *J Org Chem* 1997;62:6420–6023.

14. Sheehan JC, Hess GP. *J Am Chem Soc* 1955;77:1067–1068.

15. Sarantakis D, Teichman J, Lien EL, Fenichel RL. *Biochem Biophys Res Commun* 1976;73:336–342.

16. Radzicka A, Wolfenden R. *J Am Chem Soc* 1996;118:6105–6109.

17. (a) Updegrove TB, Correia JJ, Chen Y, Terry C, Wartell RM. *RNA* 2011;17:489–500. (b) Naue N, Fedorov R, Pich A, Manstein DJ, Curth US. *Nucleic Acids Res* 2011;39:1398–1407.

18. Grabarek Z, Gergely J. *Anal Biochem* 1990;185:131–135.

19. Carpino LA. *J Am Chem Soc* 1993;115:4397–4398.

20. Tulla-Puche J, Torres A, Calvo P, Royo M, Albericio F. *Bioconjugate Chem* 2008;19:1968–1971.

21. Zatsepin TS, Stetsenko DA, Gait MJ, Oretskaya TS. *Bioconjugate Chem* 2005;16:471–489.

22. (a) Chen I, Howarth M, Lin W, Ting AY. *Nat Methods* 2005;2:99–104. (b) Zhang Z, Smith BAC, Wang L, Brock A, Cho C, Schultz PG. *Biochemistry* 2003;42: 6735–6746.

23. Mahal LK, Yarema KJ, Bertozzi CR. *Science* 1997;276:1125–1128.

24. Baxter EW, Reitz AB. *Organic Reactions* 2002;1:59.

25. Tarasevich VA, Kozloz NG. *Russian Chemical Reviews* 1999;68:55–72.

26. (a) Abdel-Magid AF, Carson KG, Harris BD, Maryanoff CA, Shah RD. *J Org Chem* 1996;61:3849–3862. (b) Dalpathado DS, Jiang H, Kater MA, Desaire H. *Analytical and Bioanalytical Chemistry*, 2005;381:1130–1137.

27. (a) Vliegenthart JFG. *FEBS Lett* 2006;580:2945–2950. (b) Seeberger PH, Werz DB. *Nature*, 2007;446, 1046–1051. (c) Oppenheimer SB, Alvarez M, Nnoli J. *Acta Histochemica* 2008;110, 6–13. (d) Cipolla L, Peri F, Airoldi C. *Anticancer Agents Med Chem* 2008;8: 92–121.

28. Bovin NV. *Biochem Soc Symp* 2002;69:143–160.

29. (a) Mieszala M, Kogan G, Jennings HJ. *Carbohyd Res* 2003;338:167–175. (b) Pawlowski A, Källenius G, Svenson SB. *Vaccine* 1999;17:1474–1483.

30. Roy R, Katzenellenbogen E, Jennings HJ. *Can J Biochem Cell Bio* 1984;62:270–275.

31. Gildersleeve JC, Oyelaran O, Simpson TJ, Allred B. *Bioconjugate Chem* 2008;19:1485–1490.

32. Sander EG, Jencks WP. *J Am Chem Soc* 1968;90:6154–6162.

33. Jencks WP. *J Am Chem Soc* 1959;81:475–481.

34. Sayer JM, Peshkin M, Jencks WP. *J Am Chem Soc* 1973;95:4277–4287.

35. Zatsepin TS, Stetsenko DA, Arzumanov AA, Romanova EA, Gait MJ, Oretskaya TS. *Bioconjugate Chem* 2002;13:822–830.

36. (a) Melnyk O, Duburcq X, Olivier C, Urbes F, Auriault C, Gras-Masse H. *Bioconjugate Chem* 2002;13:713–720. (b) Duburcq X, Olivier C, Malingue F, Desmet R, Bouzidi A, Zhou F, Auriault C, Gras-Masse H, Melnyk O. *Bioconjugate Chem*, 2004;15: 307–316.

37. (a) Dirksen A, Hackeng TM, Dawson PE. *Angew Chem Int Ed* 2006;45:7581–7584. (b) Dirksen A, Dawson PE. *Bioconjugate Chem* 2008;19: 2543–2548.

38. Kohler JJ. *Chem Bio Chem* 2009;10:2147–2150.

39. Carrasco MR, Alvarado CI, Dashner ST, Wong AJ, Wong MA. *J Org Chem* 2010;75:5757–5759.

40. Dirksen A, Dirksen S, Hackeng TM, Dawson PE. *J Am Chem Soc* 2006;128:15602–15603.

41. Byeon JY, Limpoco FT, Bailey RC. *Langmuir* 2010;26:15430–15435.

42. (a) Mannich C. *Arch Pharm* 1917;255:261–276. (b) Blicke FF. *Organic Reactions* 1942;1:303–341.

43. Joshi NS, Whitaker LR, Francis MB. *J Am Chem Soc* 2004;126:15942–15943.

44. Romanini DW, Francis MB. *Bioconjugate Chem* 2008;19:153–157.

45. Xie J, Chen K, Lee HY, Xu C, Hsu AR, Peng S, Chen X, Sun S. *J Am Chem Soc* 2008;130:7542–7543.

46. Li GL, Kung KKY, Zou L, Chong, HC, Leung YC, Wong KH, Wong MK. *Chem Commun* 2012;48:3527–3529.

47. (a) Basavaiah D, Rao KV, Reddy RJ. *Chem Soc Rev* 2007;36:1581–1588. (b) Basavaiah D, Reddy BS, Badsara SS. *Chem Rev*, 2010;110:5447–5674.

48. (a) Gutsche AT, Parsons-Wingerter P, Chand D, Saltzman WM, Saltzman WM, Leong KW. *Biotech Bioeng* 1994;43:801. (b) Jagendorf AT, Patchornik A, Sela M. *Biochim Biophys Acta* 1963;78: 516–527.

49. Drumheller PD, Hubbell JA. Surface immobilization of adhesion ligands for investigations of cell-substrate interactions. In: Bronzino JD (ed.), *The Biomedical Engineering Handbook*, 2nd ed. Boca Raton, FL: CRC Press LLC; 2000.

50. Banks PR, Paquette DM. *Bioconjugate Chem* 1995;6:447–458.

51. Dong S, Roman M. *J Am Chem Soc* 2007;129:13810–13811.

52. Swoboda G, Hasselbach WZ. *Z Naturforsch C* 1985;40:863–875.

53. Gee KR, Archer EA, Kang HC. *Tetrahedron Lett* 1999;40:1471–1474.

54. Fischer MJ. *Methods Mol Biol* 2010;627:55–73.

55. (a) Hand ES, Jencks WP. *J Am Chem Soc* 1962;84:3505–3514. (b) Mattson G, Conklin E, Desai S, Nielander G, Savage D, Morgensen S. *Mol Biol Rep* 1993;17:167–183.

56. Seiler N. *Methods Biochem Anal* 1970;18:259–337.

57. Boyle RE. *J Org Chem* 1966;31:3880–3882.

58. Kondoh A, Makino K, Matsuda T. *J Appl Polym Sci* 1993;47:1983–1988.

59. Kobayashi H, Ikada Y. *Biomaterials* 1991;12:747–751.

60. Wachter E, Machleidt W, Hofner H, Otto J. *FEBS Lett* 1973;35:97–102.

61. Streeter HB, Rees DA. *J Cell Biol* 1987;105:507–515.

62. Singer II, Kawka DW, Scott S, Mumford RA, Lark MW. *J Cell Biol* 1987;104:573–584.

63. Nathan A, Bolikal D, Vyavahare N, Zalipsky S, Kohn J. *Macromolecules* 1992;25:4476–4484.

64. Zalipsky, Shmuel (Edison, NJ). Active carbonates of polyalkylene oxides for modification of polypeptides. United States, Enzon, Inc. (Piscataway, NJ) 5612460. 1997.

65. Elling L, Kula MR. *Biotech Appl Biochem* 1991;13:354–362.

66. Pope NM, Kulcinski DL, Hardwick A, Chang YA. *Bioconjugate Chem* 1993;4:166–171.

67. Maisano F, Gozzini L, de Haen C. *Bioconjugate Chem* 1992;3:212–217.

68. Lowe CR, Dean PDG. *Affinity Chromatography*, New York: Wiley; 1974. pp 228–229.

69. Curtius Th. *Ber Dtsch Chem Ges*. 1902;35:3226–3228.

70. Shimada K, Mitamura K. *J Chromatogr B* 1994;659:227–241.

71. Di Marco M, Shamsuddin S, Razak KA, Aziz AA, Devaux C, Borghi E, Levy L, Sadun C. *Int J Nanomedicine* 2010;5:37–49.

72. (a) Han JC, Han GY. *Anal Biochem* 1994;220:5–10. (b) Rhee SS, Burke DH. *Anal Biochem* 2004;325: 137–143.

73. Shafer DE, Inman JK, Lees A. *Anal Biochem* 2000;282:161–164.

74. Carlsson J, Drevin H, Axén R. *Biochem J* 1978;173:723–737.

75. Duncan RJ, Weston PD, Wrigglesworth R. *Anal Biochem* 1983;132:68–73.

76. Burns JA, Butler JC, Moran J, Whitesides GM. *J Org Chem* 1991;56:2648–2650.

77. Davis NJ, Flitsch SL. *Tetrahedron Lett* 1991;32:6793–6796.

78. Pawlowski A, Kallenius G, Svenson SB. *Vaccine* 1999;17:1474–1483.

79. Gurd FRN. *Methods Enzymol* 1972;25:424.

80. Aslam M, Dent A. *Bioconjugation: Protein Coupling Techniques for the Biomedical Sciences*. London: Macmillan Reference Ltd; 1998.

81. Nielsen ML, Vermeulen M, Bonaldi T, Cox J, Moroder L, Mann M. *Nat Methods* 2008;5:459–460.

82. Zou W, Abraham M, Gilbert M, Wakarchuk WW, Jennings HJ. *Glycoconj J* 1999;16:507–515.

83. Masuko T, Minami A, Iwasaki N, Majima T, Nishimura SI, Lee YC. *Biomacromolecules* 2005;6:880–884.

84. Clarke S. *Ageing Res Rev* 2003;2:263–285.

85. Thibaudeau K, Leger R, Huang XC, Robitaille M, Quraishi O, Soucy C, Bousquet-Gagnon N, van Wyk P, Paradis V, Castaigne JP, Bridon D. *Bioconjugate Chem* 2005;16:1000–1008.

86. Kalia J, Raines RT. *Bioorg Med Chem Lett* 2007;17:6286–6289.

87. Lin D, Saleh S, Liebler DC. *Chem Res Toxicol* 2008;21:2361–2369.

88. Sweeney JB. *Chem Soc Rev* 2002;31:247-258.

89. (a) Nakajima K, Oda H, Okawa K. *Bull Chem Soc Jpn* 1983;56:520–522. (b) Kogami Y, Okawa K. *Bull Chem Soc Jpn* 1987;60:2963–2965. (c) Wipf P, Uto Y. *J Org Chem* 2000;65:1037–1049.

90. Wakamiya T, Shimbo K, Shiba T, Nakajima K, Neya M, Okawa K. *Bull Chem Soc Jpn* 1982;55:3878–3881.

91. Galonić DP, Ide ND, van der Donk WA, Gin DY. *J Am Chem Soc* 2005;127:7359–7369.

92. Watanabe Y, Imai K. *Anal Chem* 1983;55:1786–1791.

93. Simard JR, Getlik M, Grütter C, Pawar V, Wulfert S, Rabiller M, Rauh D. *J Am Chem Soc* 2009;131:13286–13296.

94. (a) Toyo'oka T, Imai K. *Anal Chem* 1984;56:2461–2461. (b) Toyo'oka T, Miyano H, Imai K. *Peptide Chem* 1986;23:403.

95. Imai K, Uzu S, Kanda S. *Anal Chim Acta* 1994;290:3.

96. King P, Li Y, Kochoumian L. *Biochemistry* 1978;17:1499–1506.

97. (a) Durand A, Brown T. *Nucleosides Nucleotides Nucleic Acids* 2007;26:785–794. (b) Raouane M, Desmaeïe D., Urbinati G, Massaad-Massade L, Couvreur P. *Bioconjugate Chem.* 2012;23:1091–1104.

98. (a) Nathani R, Moody P, Smith MEB, Fitzmaurice RJ, Caddick S. *Chem Bio Chem* 2012;13:1283–1285. (b) Bernardes GJL., Chalker JM, Errey JC, Davis BG. *J Am Chem Soc*, 2008;130:5052–5053.

99. Chalker JM, Gunnoo SB, Boutureira O, Gerstberger SC, Fernán-dez-González M, Bernardes GJL, Griffin L, Hailu H, Schofield CJ, Davis BG. *Chem Sci* 2011;2:1666–1676.

100. Hearn MTW, Smith PK, Mallia AK, Hermanson GT. In: *Affinity Chromatography and Biological Recognition*, New York: Academic Press; 1983. pp 191–196.

101. Miron T, Wilchek M *Bioconjugate Chem.* 1993;4:568-569.

102. Pereira M, Lai EPC. *J Nanobiotechnology*. 2008;6:1–15.

103. (a) Burdine L, Gillette TG, Lin H-J, Kodadek T. *J Am Chem Soc* 2004;126:11442–11443. (b) Bo L, Burdine L, Kodadek T. *J Am Chem Soc.* 2006;128:15228–15235.

104. Ayyadurai N, Prabhu NS, Deepankumar K, Jang YJ, Chitrapriya N, Song E, Lee N, Kim SK, Kim B-G, Soundrarajan N, Lee S, Cha HJ, Budisa N, Yun H. *Bioconjugate Chem* 2011;22:551–555.

105. Trader DJ, Carlson EE. *Mol BioSyst* 2012;8:2484–2493.

106. Hooker JM, Kovacs EW, Francis MB. *J Am Chem Soc* 2004;126:3718–3719.

107. Ban H, Gavrilyuk J, Barbas CF. *J Am Chem Soc* 2010;132:1523–1525.

108. Tilley SD, Francis MB. *J Am Chem Soc* 2006;128:1080–1081.

109. Seim KL, Obermeyer AC, Francis MB. *J Am Chem Soc* 2011;133:16970–16976.

110. Wieland T, Bokelmann E, Bauer L, Lang HU, Lau H. *Liebigs Ann Chem* 1953;583:129–149.

111. Dawson PE, Muir TW, Clark-Lewis I, Kent SBH. *Science* 1994;266:776–778.

112. Johnson ECB, Kent SBH. *J Am Chem Soc* 2006;128:6640–6646.

113. Wang C, Guo QX, Fu Y. *Chem Asian J* 2011;6:1242–1251.

114. Dawson PE, Churchill MJ, Ghadiri MR, Kent SBH. *J Am Chem Soc* 1997;119:4325–4329.

115. Muir TW. *Annu Rev Biochem* 2003;72:249–289.

116. (a) Tam JP, Xu J, Eom KD. *Biopolymers* 2001;60:194–205. (b) Tam JP, Lu YA, Liu CF, Shao J. *Proc Natl Acad Sci USA* 1995;92:12485–12489.

117. (a) Dawson PE, Kent SBH. *Annu Rev Biochem* 2000;69:923–960. (b) Torbeev VY, Kent SB H. *Angew Chem Int Ed* 2007;46:1667.

118. Becker CF, Hunter CL, Seidel R, Kent SB, Goody RS, Engelhard M. *Proc Natl Acad Sci USA* 2003;100:5075–5080.

119. van Baal I, Malda H, Synowsky SA, van Dongen JL, Hackeng TM, Merkx M, Meijer EW. *Angew Chem Int Ed Engl* 2005;44:5052–5057.

120. Lutz JF, Borner HG. *Prog Polym Sci* 2008;33:1–39.

121. Reulen SW, Brusselaars WW, Langereis S, Mulder WJ, Breurken M, Merkx M. *Bioconjugate Chem* 2007;18:590–596.

122. Sewing A, Hilvert D. *Angew Chem Int Ed Engl* 2001;40:3395–3396.

123. Futaki S, Sogawa K, Maruyama J, Asahara T, Niwa M. *Tetrahedron Lett* 1997;38:6237–6240.

124. von Eggelkraut-Gottanka R, Klose A, Beck-Sickinger AG, Beyermann M. *Tetrahedron Lett* 2003;44:3551–3554.

125. Clippingdale AB, Barrow CJ, Wade JD. *J Pept Sci* 2000;6:225–234.

126. Li X, Kawakami T, Aimoto S. *Tetrahedron Lett* 1998;39:8669–8672.

127. Bu X, Xie G, Law CW, Guo Z. *Tetrahedron Lett* 2002;43:2419–2422.

128. Muir TW, Sondhi D, Cole PA. *Proc Natl Acad Sci USA* 1998;95:6705–6710.

129. Severinov K, Muir TW. *J Biol Chem* 1998;273:16205–16209.

130. Hofmann RM, Muir TW. *Curr Opin Biotechnol* 2002;13:297–303.

131. Schwarzer D, Cole PA. *Curr Opin Chem Biol* 2005;9:561–569.

132. Evans TC Jr, Xu M-Q. *Biopolymers* 2000;51:333–342.

133. Berrade L, Camarero JA. *Cell Mol Life Sci* 2009;66:3909–3922.

134. Karagöz GE, Sinnige T, Hsieh O, Rüdiger SG. *Protein Eng Des Sel* 2011;24:495–501.

135. Canne LE, Bark SJ, Kent SBH. *J Am Chem Soc* 1996;118:5891–5896.

136. Yan LZ, Dawson PE. *J Am Chem Soc* 2001;123:526–533.

137. Vila-Perelló M, Muir TW. *Cell* 2010;143:191–200.

138. Muralidharan V, Muir TW. *Nat Methods* 2006;3:429–438.

139. Scott CP, Abel-Santos E, Wall M, Wahnon DC, Benkovic S. *J Proc Natl Acad Sci USA* 1999;96:13638–13643.

140. Evans TC, Martin D, Kolly R, Panne D, Sun L, Ghosh I, Chen LX, Benner J, Liu XQ, Xu MQ. *J Biol Chem* 2000;275:9091–9094.

141. Giriat I, Muir TW. *J Am Chem Soc* 2003;125:7180–7181.

142. Ozawa T, Nogami S, Sato M, Ohya Y, Umezawa Y. *Anal Chem* 2000;72:5151–5157.

143. Wong SS. *Chemistry of Protein Conjugation and Cross-linking*, 2nd ed. CRC Press, Inc. 2009.

144. (a) Hemaprabha E. *Journal of Pharmaceutical and Scientific Innovation* 2012;1:22–26. (b) www.piercenet.com

145. Green NS, Reisler E, Houk KN. *Protein Sci* 2001;10:1293–1304.

146. Hartman FC, Wold F. *J Am Chem Soc* 1966;88:3890–3891.

147. Hunter MJ, Ludwig ML. *J Am Chem Soc* 1962;84:3491–3504.

148. Browne DT, Kent SBH. *Biochem Biophys Res Comm* 1975;67:126–132.

149. Ji TH. *Biochim Biophys Acta* 1979;559:39–69.

150. Wilbur DS. *Bioconjugate Chem* 1992;3:433–470.

151. Bragg PD, Hou C. *Arch Biochem Biophys* 1975;167:311–321.

152. Lomant AJ, Fairbanks G. *J Mol Biol* 1976;104:243–261.

153. Staros JV. *Biochemistry* 1982;21:3950–3955.

154. Anjaneyulu PSR, Staros JV. *Int J Pept Protein Res* 1987;30:117–124.

155. Dihazi GH, Sinz A. *Rapid Commun Mass Spectrom* 2003;17:2005–2014.

156. Hartman FC, Wold F. *Biochemistry* 1967;6:2439–2448.

157. Pearson KM, Pannell LK, Fales HM. *Rapid Commun Mass Spectrom* 2002;16:149–159.

158. Bauer PI, Buki KG, Hakam A, Kun E. *Biochem J* 1990;270:17–26.

159. O'Keeffe ET, Mordick T, Bell JE. *Biochemistry* 1980;19:4962–4966.

160. Schneider C, Newman RA, Sutherland DR, Asser U, Greaves MF. *J Biol Chem* 1982;257:10766–10769.

161. Sinha SK, Brew K. *J Biol Chem* 1981;256:4193–4204.

162. Wang D, Moore S. *Biochem* 1977;16:2937–2941.

163. Packman LC, Perhan RN. *Biochem* 1982;21:5171–5175.

164. Jerng HH, Kunjilwar K, Pfaffinger PJ. *J Physiol* 2005;568:767–788.

165. Wang K, Richards F. *J Biol Chem* 1974;249:8005–8018.

166. Waugh SM, DiBella EE, Pilch PF. *Biochemistry* 1989;28:3448–3455.

167. Hauser NC, Martinez R, Jacob A, Rupp S, Hoheisel JD, Matysiak S. *Nucleic Acids Res* 2006;34:5101–5111.

168. Chataway TK, Barritt GJ. *Mol Cell Biochem* 1995;145:111–120.

169. Donato R, Giambanco I, Aisa MC. *J Neurochem* 1989;53:566–571.

170. Ishmael FT, Shier VK, Ishmael SS, Bond JS. *J Biol Chem* 2005;280:13895–13901.

171. Koller D, Ittner LM, Muff R, Husmann K, Fischer JA, Born W. *J Biol Chem* 2004;279:20387–20391.

172. Law BK, Chytil A, Dumont N, Hamilton EG, Waltner-Law ME, Aakere ME, Covington C, Moses HL. *Mol Cell Biol* 2002;22:8184–8198.

173. Farries TC, Atkinson JP. *J Immunol* 1989;142:842–847.

174. Park LS, Friend D, Gillis S, Urdal DL. *J Biol Chem* 1986;261:205–210.

175. Predescu SA, Predescu DN, Palade GE. *Mol Biol Cell* 2001;12:1019–1033.

176. Lomant AJ, Fairbanks G. *J Mol Biol* 1976;104:243–261.

177. Xiang CC, Mezey E, Chen M, Key S, Ma L, Brownstein MJ. *Nucleic Acids Res.* 2004;32:185.

178. Grubor NM, Shinar R, Jankowiak R, Porter MD, Small GJ. *Biosens Bioelectron* 2004;19:547–556.

179. Staros JV. *Biochemistry* 1982;21:3950–3955.

180. Paine-Saunders S, Viviano BL, Economides AN, Saunders S. *J Biol Chem* 2002;277:2089–2096.

181. Schaefer LL, Babelova A, Kiss E, Hausser HJ, Baliova M, Krzyzankova M, Marsche G, Young MF, Mihalik D, Götte M, Malle E, Schaefer RM, Gröne HJ. *J Clin Invest* 2005;115:2223–2233.

182. Abdella PMSmith PK, Royer GP. *Biochem Biophysics Res Commun* 1979;87:734–742.

183. Browning J, Ribolini A. *J Immunol* 1989;143:1859–1867.

184. Sinz A. *J Mass Spectrom* 2003;38:1225–1237.

185. Bouizar Z, Fouchereau-Person M, Taboulet J, Moukhtar MS, Milhaud G. *Eur J Biochem* 1986;155:141–147.

186. Zarling DA, Watson A, Bach FH. *J Immunol* 1980;124:913–920.

187. Fenton RA, Brond L, Nielsen S, Praetorius J. *Am J Physiol Renal Physiol* 2007;293:F748–F760.

188. Edelhoch H, Katchalsk E, Maybury RH, Hughes Jr WL, Edsall JT. *J Am Chem Soc* 1953;75:5058.

189. Edsall JT, Maybury RH, Simpson RB, Straessle R. *J Am Chem Soc* 1954;76:3131.

190. Fasold H, Groschel-Stewart U, Turba F. *Biochem Z* 1963;337:425–430.

191. Gaffney BJ, Willingham GL, Schopp RS. *Biochemistry* 1983;22:881–892.

192. Bloxham DP, Cooper CK. *Biochemistry* 1982;21:1807–1812.

193. Mandy WJ, Rivers MM, Nisonoff A. *J Biol Chem* 1961;236:3221–3226.

194. Moore JF, Ward WH. *J Am Chem Soc* 1956;78:2414–2418.

195. Moroney JV, Warncke K, McCarthy RF. *J Bioenerg Biomembr* 1982;14:347–359.

196. Ozawa H. *J Biochem* (Tokyo) 1967;62:531–536.

197. Partis MD, Griffiths DG, Roberts GC, Beechey RB. *J Protein Chem* 1983;2:263–277.

198. Tawney PO, Snyder RH, Conger RP, Leibbrand KA, Stiteler CH, Williams AR. *J Org Chem* 1961;26:15–21.

199. Chen LL, Rosa JJ, Turner S, Pepinsky RB. *J Biol Chem* 1991;266:18237–18243.

200. Tona R, Haner R. *Mol BioSyst* 2005;1:93–98.

201. Schmidt CJ, Thomas TC, Levine MA, Neer EJ. *J Biol Chem* 1992;267:13807–13810.

202. Yi F, Denker BM, Neer EJ. *J Biol Chem* 1991;266:3900–3906.

203. Klem E, Kottner N. *Angew Makromol Chem* 1998;254:39–45.

204. Stalteri MA, Mather SJ. *Bioconj Chem* 1995;6:179–186.

205. Ahmed H. *Principles and Reactions of Protein Extraction, Purification, Characterization.* Boca Raton, FL: CRC Press; 2005, pp 254–255.

206. Swanton E, Holland A, High S, Woodman P. *Proc Natl Acad Sci USA* 2005;102:4342–4347.

207. Hermanson TG. *Bioconjugate Techniques*, 2nd ed. Academic Press, 2008, pp 268–269.

208. Allmer K, Hilborn J, Larsson PH, Hult A, Ranby B. *J Polym Sci: Part A: Polym Chem* 1990;28:173–183.

209. Brillhart KL, Ngo TT. *J Immunol Meth* 1991;144:19–25.

210. Schneerson R, Barrera O, Sutton A, Robbins JB. *J Exp Med* 1980;152:361–376.

211. Kubler-Kielb J, Coxon B, Schneerson R. *J Bacteriol* 2004;186:6891–6901.

212. Carmenate T, Canaan L, Alvarez A, Delgado M, Gonzalez S, Menendez T, Rodes L, Guillen G. *FEMS Immunol Med Mic* 2004;40:193–199.

213. O'Shannessy DJ, Wilchek M. *Anal Biochem* 1990;191:1–8.

214. Avrameas S. *Immunochemistry* 1969;6:43–52.

215. Avrameas S, Ternynck T. *Immunochemistry* 1969;6:53–66.

216. Avrameas S, Ternynck T. *Immunochemistry* 1971;8:1175.

217. Nimni EM, Cheung D, Strates B, Kodama M, Sheikh K. *J Biomed Mater Res* 1987;21:74–771.

218. Migneault I, Dartiguenave C, Bertrand MJ, Waldron KC. *BioTechniques* 2004;37:790–802.

219. May JM. *Biochemistry* 1989;28:1718–1725.

220. Sayre LM, Larson DL, Takemori AE, Portoghese PS. *J Med Chem* 1984;27:1325–1335.

221. McKenzie JA, Raison RL, Rivett EE. *J Protein Chem* 1988;7:581–592.

222. Kitagawa T. Enzyme labeling with N-hydroxysuccinimidyl ester of maleimide. In: Ishikawa E, Kawai T, Miyai K (eds), *Enzyme Immuno-assay*, Tokyo and New York: IGAKU-SHOIN; 1981. pp 81–89.

223. Chrisey LA, Lee GU, O'Ferrall CE. *Nucleic Acids Res.* 1996;24:3031–3039.

224. Fujiwara K, Matsumoto N, Yagisawa S, Tanimori H, Kitagawa T, Hirota M, Hiratani K, Fukushima K, Tomonaga A, Hara K, Yamamoto K. *J Immunol Meth* 1988;112:77–83.

225. Myers DE, Uckun FM, Swaim SE, Vallera DA. *J Immunol Meth* 1989;121:129–142.

226. Kitagawa T, Aikawa T. *J Biochem (Tokyo)* 1976;79:233–236.

227. Iwai K, Fukuoka S-I, Fushiki T, Kido K, Sengoku Y, Semba T. *Anal Biochem* 1988;171:277–282.

228. Teale JM, Kearney JR. *J Mol Cell Immunol* 1986;2:283–292.

229. Bieniarz C, Husain M, Barnes G, King CA, Welch CJ. *Bioconjugate Chem* 1996;7:88–95.

230. Brinkley MA. *Bioconjugate Chem* 1992;3:2–13.

231. Yoshitake S, Imagawa M, Ishikawa E, Niitsu Y, Urushizaki I, Nishiura M, Kanazawa R, Kurosaki H, Tachibana S, Nakazawa N, Ogawa H. *J Biochem* 1982;92:1413–1424.

232. Muller L, de Escauriaza MD, Lajoie P, Theis M, Jung M, Müller A, Burgard C, Greiner M, Snapp EL, Dudek J, Zimmermann R. *Mol Biol Cell* 2010;21:691–703.

233. Dickgreber N, Stoitzner P, Bai Y, Price KM, Farrand KJ, Manning K, Angel CE, Dunbar PR, Ronchese F, Fraser JD, Bäckström BT, Hermans IF. *J Immunol* 2009;182:1260–1269.

234. Kersemans V, Cornelissen B, Minden MD, Brandwein J, Reilly RM. *J Nucl Med* 2008;49:1546–1554.

235. Duval M, Posner MR, Cavacini LA. *J Virol* 2008;82:4671–4674.

236. Thorpe PE, Ross WC, Brown AN, Myers CD, Cumber AJ, Foxwell BM, Forrester JT. *Eur J Biochem* 1984;140:63–71.

237. Rector ES, Schwenk RJ, Tse KS, Sehon AH. *J Immunol Methods* 1978;24:321–336.

238. Cuatrecasas P, Wilchek M, Anfinsen CB. *J Biol Chem* 1969;244:4316–4329.

239. Inman JK, Highet PF, Kolodny N, Robey FA. *Bioconjugate Chem* 1991;2:458–463.

240. Cumber AJ, Forrester JA, Foxwell BMJ, Ross WCJ, Thorpe PW. *Methods Enzymol* 1985;112:207–225.

241. Hermanson GT. *Bioconjugate Techniques*. San Diego: Academic Press, 1996. pp 542, 553, 568.

242. Carlsson J, Drevin H, Axén R. *Biochem J* 1978;173:723–737.

243. Engstler M, Thilo L, Weise F, Grünfelder CG, Schwarz H, Boshart M, Overath P. *J Cell Sci* 2004;117:1105–1115.

244. Itoh Y, Cai K, Khorana HG. *Proc Natl Acad Sci USA* 2001;98:4883–4887.

245. Wang D, Li Q, Hudson W, Berven E, Uckun F, Kersey JH. *Bioconjugate Chem* 1997;8:878–884.

246. Ghetie V, Till MA, Ghetie MA, Tucker T, Porter J, Patzer EJ, Richardson JA, Uhr JW, Vitetta ES. *Bioconjugate Chem* 1990;1:24–31.

247. Na DH, Woo BH, Lee KC. *Bioconjugate Chem* 1999;10:306–310.

248. van Oosterhout YVJM, van Emst L, Schattenberg AV, Tax WJ, Ruiter DJ, Spits H, Nagengast FM, Masereeuw R, Evers S, de Witte T, Preijers FW. *Blood* 2000;95:3693–3701.

249. Warren HS, Matyal R, Allaire JE, Yarmush D, Loiselle P, Hellman J, Paton BG, Fink MP. *J Infec Dis* 2003;188:1382–1393.

250. Keana JFW, Cai SX. *J Org Chem* 1990;55:3640–3647.

251. Brunner J. *Annu Rev Biochem* 1993;62:483–514.

252. Schuster GB, Platz MS. *Adv Photochem* 1992;17:69–143.

253. Liu LH, Yan M. *Acc Chem Res* 2010;43:1434–1443.

254. Hermanson GT. *Bioconjugate Techniques*, 2nd ed. Elsevier Academic Press; Amsterdam, Boston. 2008. p. 303.

255. Platz MS. *Acc Chem Res* 1995;28:487–492.

256. Banks RE, Sparkes GR. *J Chem Soc, Perkin Trans* 1972;1:2964–2970.

257. Ji I, Ji TH. *Proc Natl Acad Sci USA* 1981;78:5465–5469.

258. van der Horst GTJ, Mancini GMS, Brossmer R, Rose U, Verheijen FW. *J Biol Chem* 1990;265:10801–10804.

259. Yeung CWT, Moule ML, Yip CC. *Biochemistry* 1980;19:2196–2203.

260. Cai H, Song C, Endoh I, Goyette J, Jessup W, Freedman SB, McNeil HP, Geczy CL. *J Immunol* 2007;178, 1852-1860.

261. Wollenweber HW, Morrison DC. *J Biol Chem* 1985;260:15068–15074.

262. Sorenson P, Farber NM, Krystal G. *J Biol Chem* 1986;261:9094–9097.

263. Ballmer-Hofer K, Schlup V, Burn P, Burger MM. *Anal Biochem* 1982;126:246–250.

264. Wood CL, O'Dorisio MS. *J Biol Chem* 1985;260:1243–1247.

265. Khatiwala CB, Peyton SR, Putnam AJ. *Am J Physiol Cell Physiol* 2006;290:C1640–C1650.

266. Sasuga Y, Tani T, Hayashi M, Yamakawa H, Ohara O, Harada Y. *Genome Res* 2006;16:132–139.

267. Marinković A, Mih JD, Park J-A, Liu F, Tschumperlin DJ. *Am J Physiol Lung Cell Mol Physiol* 2012;303:L169–L180.

268. Jung SM, Moroi M. *Biochim Biophys Acta* 1983;761:152–162.

269. Mann WA, Meyer N, Weber W, Meyer S, Greten H, Beisiegel U. *J Lipid Res* 1995;36:517–525.

270. Coscoy S, Lingueglia E, Lazdunski M, Barbry P. *J Biol Chem* 1998;273:8317–8322.

271. Thevenin BJ-M, Shahrokh Z, Williard RL, Fujimoto EK, Kang J-J, Ikemoto N, Shohet SB. *European Journal of Biochemistry* 1992;206:471–477.

272. Yano M, Okuda S, Oda T, Tokuhisa T, Tateishi H, Mochizuki M, Noma T, Doi M, Kobayashi S, Yamamoto T, Ikeda Y, Ohkusa T, Ikemoto N, Matsuzaki M. *Circulation* 2005;112:3633–3643.

273. Mochizuki M, Yano M, Oda T, Tateishi H, Kobayashi S, Yamamoto T, Ikeda Y, Ohkusa T, Ikemoto N, Matsuzaki M. *J Am Coll Cardiol* 2007;49:1722–1732.

274. Liu SH, Cheng HH, Huang SY, Yiu PC, Chang YC. *Mol Cell Proteomics* 2006;5:1019–1032.

275. Dormán G, Prestwich GD. *Biochemistry* 1994;33:5661–5673.

276. Tao T, Lamkin M, Scheiner CJ. *Arch Biochem Biophys* 1985;240:627–634.

277. Yang HC, Reedy MM, Burke CL, Strasburg GM. *Biochemistry* 1994;33:518–525.

278. Agarwal R, Rajasekharan KN, Burke M. *Arch Biochem Biophys* 1991;288:584–590.

279. Leszyk J, Collins JH, Leavis PC, Tao T. *Biochemistry*. 1987;26:7042–7047.

280. Lu RC, Wong A. *Biochemistry* 1989;28:4826–4829.

281. Ismaili N, Sha M, Gustafson EH, Konarska MM. *RNA* 2001;7:182–193.

282. Berens C, Courtoy PJ, Sonveaux EA. *Bioconjugate Chem* 1999;10:56–61.

283. Gariépy J, Schoolnik GK. *Proc Natl Acad Sci USA* 1986;83:483–487.

284. Koepf EK, Burtnick LD. *FEBS Lett* 1992;309:56–58.

285. Harnish DG, Leung WC, Rawls WE. *J Virol* 1981;38:840–848.

286. Sadakane Y. *Yakugaku Zasshi* 2007;127:1693–1699.

287. Das J. *Chem Rev* 2011;111:4405–4417.

288. Dubinsky L, Krom BP, Meijler MM. *Bioorg Med Chem* 2012;20:554–570.

289. Sadakane Y, Hatanaka Y. *Yakugaku Zasshi* 2008;128:1615–1622.

290. Gomes AF, Gozzo FC. (*J Mass Spectrom* 2010;45:892-899.

291. Ferreira JP, Sasisekharan R, Louie O, Langer R. *Eur J Biochem* 1994;223:611–616.

292. Wilkens S, Capaldi RA. *J Biol Chem* 1998;273:26645–26651.

293. Burgener M, Sanger M, Candrian U. *Bioconjugate Chem* 2000;11:749–754.

294. Aebersold R, Pipes GD, Wettenhall RE, Nika H, Hood LE. *Anal Biochem* 1990;187:56–65.

295. Trester-Zedlitz M, Kamada K, Burley SK, Fenyo D, Chait BT, Muir TW. *J Am Chem Soc* 2003;125:2416–2425.

296. Wedekind F, Baer-Pontzen K, Bala-Mohan S, Choli D, Zahn H, Brandenburg D. *Biol Chem Hoppe-Seyler* 1989;370:251–258.

297. Neely KE, Hassan AH, Brown CE, Howe L, Workman JL. *Mol Cell Biol* 2002;22:1615–1625.

298. Ishmael FT, Alley SC, Benkovic SJ. *J Biol Chem* 2002;277:20555–20562.

299. Alley SC, Ishmael FT, Jones AD, Benkovic SJ. *J Am Chem Soc* 2000;122:6126–6127.

300. Trotman LC, Mosberger N, Fornerod M, Stidwill RP, Greber UF. *Nat Cell Biol* 2001;3:1092–1100.

301. Horney MJ, Evangelista CA, Rosenzweig SA. *J Biol Chem* 2001;276:2880–2889.

302. Daum JR, Tugendreich S, Topper LM, Jorgensen PM, Hoog C, Hieter P, Gorbsky GJ. *Curr Biol* 2000;10:R850–R857. S1-S2.

303. Kleene R, Classen B, Zdzieblo J, Schrader M. *Biochemistry* 2000;39:9893–9900.

304. Minami Y, Kawasaki H, Minami M, Tanahashi N, Tanaka K, Yahara I. *J Biol Chem* 2000;275:9055–9061.

305. Sharma KK, Kumar RS, Kumar GS, Quinn PT. *J Biol Chem* 2000;275:3767–3771.

306. Ishmael FT, Trakselis MA, Benkovic SJ. *J Biol Chem* 2003;278:3145-3152.

307. Bower K, Djordjevic SP, Andronicos NM, Ranson M. *Infect Immuno* 2003;71:4823–4827.

308. Ilver D, Johansson P, Miller-Podraza H, Nyholm PG, Teneberg S, Karlsson KA. *Methods Enzymol* 2003;363:134–157.

309. Kang S, Mou L, Brouillette WJ, Prevelige Jr PE. *Rapid Commun Mass Spectrom* 2009;23:1719–1726.

310. Yan F. Novel photocleavable cross-linkers and their applications in biotechnologies [PhD dissertation]. Chicago (IL): Department of Biological, Chemical and Physical Sciences, Illinois Institute of Technology; 2006. Available from: http://proquest.umi.com/pqdweb?index=0&did=1172109241&SrchMode=2&sid=3&Fmt=2&VInst=PROD&VType=PQD&RQT=309&VName=.

311. Wang R, Yan F, Qiu D, Jeong J-S, Jin Q, Kim TY, Chen L. *Bioconjugate Chem* 2012;23:705–713.

312. Hang HC, Yu C, Kato DL, Bertozzi CR. *Proc Natl Acad Sci USA* 2003;100:14846–14851.

313. Plass T, Milles S, Koehler C, Schultz C, Lemke EA. *Angew Chem Int Ed* 2011;50:3878-3881.

314. Neef AB, Schultz C. *Angew Chem Int Ed* 2009;48:1498–1500.

315. Rideout D, Calogeropoulou T, Jaworski J, McCarthy M. *Biopolymers* 1990;29:247–262.

316. Griffin BA, Adams SR, Tsien RY. *Science* 1998;281:269–272.1

317. Prescher JA, Dube DH, Bertozzi CR. *Nature* 2004;430:873–877.

318. Prescher JA, Bertozzi CR. *Nat Chem Biol* 2005;1:13–21.

319. Bertozzi CR. *Acc Chem Res* 2011;44:651–653.

320. Sletten EM, Bertozzi CR. *Acc Chem Res* 2011;44:666–676.

321. Algar WR, Prasuhn DE, Stewart MH, Jennings TL, Blanco-Canosa JB, Dawson PE, Medintz IL. *Bioconjugate Chem* 2011;22:825–858.

322. Hang HC, Bertozzi CR. *J Am Chem Soc* 2001;123:1242–1243.

323. Zeng Y, Ramya TNC, Dirksen A, Dawson PE, Paulson JC. *Nat Methods* 2009;6:207–209.

324. Staudinger H, Meyer J. *Helv Chim Acta* 1919;2:635–646.

325. Gololobov YG, Kasukhin LF. *Tetrahedron* 1992;48:1353–1406.

326. Saxon E, Bertozzi CR. *Science* 2000;287:2007–2010.

327. Kohn M, Breinbauer R. *Angew Chem Int Edn Engl* 2004;43:3106–3116.

328. Schilling CI, Jung N, Biskup M, Schepers U, Bräse S. *Chem Soc Rev* 2011;40:4840–4871.

329. van Berkel SS, van Eldijk MB, van Hest JCM. *Angew Chem Int Ed* 2011;50:8806–8827.

330. Lin FL, Hoyt HM, van Halbeek H, Bergman RG, Bertozzi CR. *J Am Chem Soc* 2005;127:2686–2695.

331. Saxon E, Armstrong JI, Bertozzi CR. *Org Lett* 2000;2:2141--2143.

332. Nilsson BL, Kiessling LL, Raines RT. *Org Lett* 2000;2:1939–1941.

333. Jacobs CL, Goon S, Yarema KJ, Hinderlich S, Hang HC, Chai DH, Bertozzi CR. *Biochemistry* 2001;40:12864–12874.

334. Dube DH, Prescher JA, Quang CN, Bertozzi CR. *Proc Natl Acad Sci USA* 2006;103:4819–4824.

335. Weisbrod SH, Marx A. *Chem Commun* 2007;1828–1830.

336. Weisbrod SH, Baccaro A, Marx A. *Nucleic Acids Symp Ser* 2008;52:383–384.

337. Comstock LR,Rajski SR. *J Am Chem Soc* 2005;127:14136–14137.

338. Slavoff SA, Chen I, Choi YA, Ting AAY. *J Am Chem Soc* 2008;130:1160–1162.

339. Hosoya T, Hiramatsu T, Ikemoto T, Nakanishi M, Aoyama H, Hosoya A, Iwata T, Maruyama K, Endo M, Suzuki M. *Org Biomol Chem* 2004;2:637–641.

340. Verdoes M, Florea BI, Hillaert U, Willems LI, van der Linden WA, Sae-Heng M, Filippov DV, Kisselev AF, van der Marel GA, Overkleeft HS. *Chem Bio Chem* 2008;9:1735–1738.

341. Kho Y, Kim SC, Jiang C, Barma D, Kwon SW, Cheng JK, Jaunbergs J, Weinbaum C, Tamanoi F, Falck J, Zhao YM. *Proc Natl Acad Sci USA* 2004;101:12479–12484.

342. Nguyen UTT, Cramer J, Gomis J, Reents R, Gutierrez-Rodriguez M, Goody RS, Alexandrov K, Waldmann H. *Chem Bio Chem* 2007;8:408–423.

343. Nilsson BL, Kiessling LL, Raines RT. *Org Lett* 2001;3:9–12.

344. Soellner MB, Tam A, Raines RT. *J Org Chem* 2006;71:9824–9830.

345. Kalia J, Abbott NL, Raines RT. *Bioconjugate Chem* 2007;18:1064–1069.

346. Watzke A, Köhn M, Gutierrez-Rodriguez M, Wacker R, Schröder H, Breinbauer R, Kuhlmann J, Alexandrov K,

Niemeyer CM, Goody RS, Waldmann H. *Angew Chem* 2006;118:1436–1440.

347. Kohn M. *J Pept Sci* 2009;15:393–397.

348. Bertozzi CR. Unpublished results.

349. Michael A. *J Prakt Chem* (presently part of *Adv Synth Catal*) 1893;48:94–95.

350. Huisgen R. *Angew Chem Int Ed Engl* 1963;2:565–632.

351. Rostovtsev VV, Green LG, Fokin VV, Sharpless KB. *Angew Chem IntEd* 2002;41:2596–2599.

352. Tornoe CW, Christensen C, Meldal M. *J Org Chem* 2002;67:3057–3064.

353. Hein JE, Fokin VV. *Chem Soc Rev* 2010;39:1302–1315.

354. Meldal M, Tornøe CW. *Chem Rev* 2008;108:2952–3015.

355. Wang Q, Chan TR, Hilgraf R, Fokin VV, Sharpless KB, Finn MG. *J Am Chem Soc* 2003;125:3192–3193.

356. Kolb HC, Finn MG, Sharpless KB. *Angew Chem Int Ed* 2001;40:2004–2021.

357. Breinbauer R, Köhn M. *Chem Bio Chem* 2003;4:1147–1149.

358. New K, Brechbiel MW. *Cancer biotherapy and radiopharmaceuticals*. 2009;24:289–302.

359. Speers AE, Adam GC, Cravatt BF. *J Am Chem Soc* 2003;125:4686–4687.

360. Sun XL, Stabler CL, Cazalis CS, Chaikof EL. *Bioconjugate Chem* 2006;17:52–57.

361. Gauchet C, Labadie GR, Poulter CD. *J Am Chem Soc* 2006;128:9274–9275.

362. Gramlich PM, Wirges CT, Manetto A, Carell T. *Angew Chem Int Ed* 2008;47:8350–8358.

363. Kolb HC, Sharpless KB. *Today* 2003;8:1128–1137.

364. Lutz JF. *Angew Chem Int Ed* 2007;46:1018–1025.

365. Wu P, Fokin VV. *Aldrichim Acta* 2007;40:7–17.

366. Link AJ, Tirrell DA. *J Am Chem Soc* 2003;125:11164–11165.

367. Wolbers F, ter Braak P, Le Gac S, Luttge R, Andersson H, Vermes I, van den Berg A. *Electrophoresis* 2006;27:5073–5080.

368. Link AJ, Vink MKS, Tirrell DA. *J Am Chem Soc* 2004;126:10598–10602.

369. Baskin JM, Bertozzi CR. *Aldrichimica Acta* 2010;43:15–23.

370. Agard NJ, Prescher JA, Bertozzi CR. *J Am Chem Soc* 2004;126:15046–15047.

371. Alder K, Stein G. *Liebigs Ann Chem* 1933;501:1–48.

372. Alder K, Stein G. *Liebigs Ann Chem* 1931;485:211–222.

373. Wittig G, Krebs A. *Chem Ber* 1961;94:3260–3275.

374. (a) Baskin JM, Prescher JA, Laughlin ST, Agard NJ, Chang PV, Miller IA, Lo A, Codelli JA, Bertozzi CR. *Proc Natl Acad Sci USA* 2007;104:16793–16797. (b) Chang PV, Prescher JA, Sletten EM, Baskin JM, Miller IA, Agard NJ, Lo A, Bertozzi CR. *Proc Natl Acad Sci USA* 2010; 107:1821–1826.

375. Laughlin ST, Baskin JM, Amacher SL, Bertozzi CR. *Science* 2008;320:664–667.

376. Agard NJ, Baskin JM, Prescher JA, Lo A, Bertozzi CR. *ACS Chem Biol* 2006;1:644–648.

377. Ning XH, Guo J, Wolfert MA, Boons GJ. *Angew Chem Int Ed Engl* 2008;47:2253–2255.

378. Jewett JC, Sletten EM, Bertozzi CR. *J Am Chem Soc* 2010;132:3688–3690.

379. Debets MF, van Berkel SS, Schoffelen S, Rutjes FPJT, van Hest JCM, van Delft FL. *Chem Commun* 2010:97–99.

380. Poloukhtine AA, Mbua NE, Wolfert MA, Boons G-J, Popik VV. *J Am Chem Soc* 2009;131:15769–15776.

381. Sletten EM, Bertozzi CR. *Org Lett* 2008;10:3097–3099.

382. Stockmann H, Neves AA, Stairs S, Ireland-Zecchini H, Brindle KM, Leeper FJ. *Chem Sci* 2011;2:932–936.

383. Mbua NE, Guo J, Wolfert MA, Steet R, Boons G-J. *Chem Bio Chem* 2011;12:1912–1921.

384. Marks IS, Kang JS, Jones BT, Landmark KJ, Cleland AJ, Taton TA. *Bioconjugate Chem* 2011;22:1259–1263.

385. Ess DH, Jones GO, Houk KN. *Org Lett* 2008;10:1633–1636.

386. Chenoweth K, Chenoweth D, Goddard WAIII. *Org Biomol Chem* 2009;7:5255–5258.

387. de Almeida G, Sletten EM, Nakamura H, Palaniappan KK, Bertozzi CR. *Angew Chem Int Ed* 2012;51:2443–2447.

388. Banert K, Plefka O. *Angew Chem Int Ed Engl* 2011;50:6171–6174.

389. Gordon CG, Mackey JL, Jewett JC, Sletten EM, Houk KN, Bertozzi CR. *J Am Chem Soc* 2012;134:9199–9208.

390. Tummatorn J, Batsomboon P, Clark RJ, Alabugin IV, Dudley GB. *J Org Chem* 2012;77:2093–2097.

391. Debets MF, van Berkel SS, Dommerholt J, Dirks AJ, Rutjes FPJT, van Delft FL. *Acc Chem Res* 2011;44:805–815.

392. Hill KW, Taunton-Rigby J, Carter JD, Kropp E, Vagle K, Pieken W, McGee DPC, Husar GM, Leuck M, Anziano DJ, Sebesta DP. *J Org Chem* 2001;66:5352–5358.

393. Breslow R, Rideout DC. *J Am Chem Soc* 1980;102:7816–7817.

394. Garner PP. Diels-Alder reactions in aqueous media. In: Grieco PA (ed.), *Organic synthesis in water*. London: Blackie Academic and Professional; 1998. pp 1–46.

395. Kumar A. *Chem Rev* 2001;101:1–19.

396. Pozsgay V, Vieira NE, Yergey A. *Org Lett* 2002;4:3191–3194.

397. de Araujo AD, Palomo JM, Cramer J, Seitz O, Alexandrov K, Waldmann H. *Chem Eur J* 2006;12:6095–6109.

398. Husar GM, Anziano DJ, Leuck M, Sebesta DP. *Nucleic Acids Res* 2001;29:559–566.

399. Latham-Timmons HA, Wolter A, Roach JS, Giare R, Leuck M. *Nucleic Acids Res* 2003;31:1495–1497.

400. Berkin A, Coxon B, Pozsgay V. *Chem Eur J* 2002;8:4424–4433.

401. Tona R, Haner R. *Bioconjugate Chem* 2005;16:837–842.

402. Houseman BT, Mrksich M. *Chem Biol* 2002;9:443–454.

403. Houseman BT, Huh JH, Kron SJ, Mrksich M. *Nat Biotechnol* 2002;20:270–274.

404. Yeo W, Yousaf MN, Mrksich M. *J Am Chem Soc* 2003;125:14994–14995.

405. Dillmore WS, Yousaf MN, Mrksich M. *Langmuir* 2004;20:7223–7231.

406. Pipkorn R, Waldeck W, Didinger B, Koch M, Mueller G, Wiessler M, Braun K. *J Peptide Sci* 2009;15:235–241.

407. Wiessler M, Waldeck W, Kliem C, Pipkorn R, Braun K. *Int J Med Sci* 2010;7:19–28.

408. Huisgen R, Ooms PHJ, Mingin M, Allinger NL. *J Am Chem Soc* 1980;102:3951–3953.

409. van Berkel SS, Dirks AJ, Debets MF, van Delft FL, Cornelissen JJLM, Nolte RJM, Rutjes FPJT. *Chem Bio Chem* 2007;8:1504–1508.

410. Devaraj NK, Weissleder R. *Acc Chem Res* 2011;44:816–827.

411. Blackman ML, Royzen M, Fox JM. *J Am Chem Soc* 2008;130:13518–13519.

412. Royzen M, Yap GPA, Fox JM. *J Am Chem Soc* 2008;130:3760–3761.

413. Devarah NK, Weissleder R, Hilderbrand SA. *Bioconjugate Chem* 2008;19:2297–2299.

414. Rossin R, Verkerk PR, van den Bosch SM, Vulders RCM, Verel I, Lub J, Robillard MS. *Angew Chem Int Ed* 2010;49:3375–3378.

415. Song W, Wang Y, Qu J, Madden MM, Lin Q. *Angew Chem Int Ed* 2008;47:2832–2835.

416. Lim RKV, Lin Q. *Acc Chem Res* 2011;44:828–839.

417. Wang Y, Hu WJ, Song W, Lim RK, Lin Q. *Org Lett* 2008;10:3725–3728.

418. Song W, Wang Y, Lin Q. *J Am Chem Soc* 2008;130:9654–9655.

419. Sletten EM, Bertozzi CR. *J Am Chem Soc* 2011;133:17570–17573.

420. Ning X, Temming Rinske P, Dommerholt J, Guo J, Ania, Daniel B, Debets Marjoke F, Wolfert Margreet A, Boons G-J, van Delft Floris L. *Angew Chem Int Ed* 2010;49:3065–3068.

421. Stöckmann H, Neves AA, Stairs S, Brindle KM, Leepe FJ. *Org Biomol Chem* 2011;9:7303–7305.

422. Antos JM, Francis MB. *Curr Opin Chem Biol* 2006;10:253–262.

423. Lin YA, Chalker JM, Davis BG. *Chem Bio Chem* 2009;10:959–969.

424. Herrerias CI, Yao X, Li Z, Li C-J. *Chem Rev* 2007;107:2546–2562.

425. Antos JM, Francis MB. *J Am Chem Soc* 2004;126:10256–10257.

426. Binder JB, Raines RT. *Curr Opin Chem Biol* 2008;12:767–773.

427. Hong SH, Grubbs RH. *J Am Chem Soc* 2006;128:3508–3509.

428. Jordan JP, Grubbs RH. *Angew Chem Int Ed Engl* 2007;46:5152–5155.

429. Binder JB, Blank JJ, Raines RT. *Org Lett* 2007;9:4885–4888.

430. Gulajski L, Michrowska A, Naroznik J, Kaczmarska Z, Rupnicki L, Grela K. *Chem Sus Chem* 2008;1:103–109.

431. Lin YA, Chalker JM, Floyd N, Bernardes GJL, Davis BG. *J Am Chem Soc* 2008;130:9642–9643.

432. Mortell KH, Gingras M, Kiessling LL. *J Am Chem Soc* 1994;116:12053–12054.

433. Heck RF. *Acc Chem Res* 1979;12:146–151.

434. Miyaura N, Suzuki A. *Chem Rev* 1995;95:2457–2483.

435. Chinchilla R, Najera C. *Chem Rev* 2007;107:874--922.

436. Kodama K, Fukuzawa S, Nakayama H, Kigawa T, Sakamoto K, Yabuki T, Matsuda N, Shirouzu M, Takio K, Tachibana K, Yokoyama S. *Chem Bio Chem* 2006;7:134–139.

437. Kodama K, Fukuzawa S, Nakayama H, Sakamoto K, Kigawa T, Yabuki T, Matsuda N, Shirouzu M, Takio K, Yokoyama S, Tachibana K. *Chem Bio Chem* 2007;8:232--238.

438. Brustad E, Bushey ML, Lee JW, Groff D, Liu W, Schultz PG. *Angew Chem Int Ed* 2008;47:8220-8223.

439. Chalker JM, Wood CSC, Davis BG. *J Am Chem Soc* 2009;131:16346-16347.

440. James TD, Sandanayake KRAS, Shinkai S. *Angew Chem Int Ed Engl* 1996;35:1910–1922.

441. Tietze LF, Arlt M, Beller M, Glüsenkamp KH, Jähde E, Rajewsky MF. *Chem Ber* 1991;124:1215–1221.

442. Kamath VP, Diedrich P, Hindsgaul O. *Glycoconjugate J* 1996;13:315–319.

443. Carlson CB, Mowery P, Owen RM, Dykhuizen EC, Kiessling LL. *ACS Chem Biol* 2007;2:119–127.

444. Storer RI, Aciro C, Jones LH. *Chem Soc Rev* 2011;40:2330–2346.

445. Gilmore JM, Scheck RA, Esser-Kahn AP, Joshi NS, Francis MB. *Angew Chem, Int Ed* 2006;45:5307–5311.

446. Scheck RA, Francis MB. *ACS Chem Biol* 2007;2:247–251.

447. Scheck RA, Dedeo MT, Iavarone AT, Francis MB. *J Am Chem Soc* 2008;130:11762–11770.

448. Witus LS, Moore T, Thuronyi BW, Esser-Khan AP, Scheck RA, Iavarone AT, Francis MB. *J Am Chem Soc* 2010;132:16812–16817.

449. (a) Christman KL, Broyer RM, Tolstyka ZP, Maynard HD. *J Mater Chem* 2007;17:2021–2027. (b) Lempens EHM, Helms BA, Merkx M, Meijer EW. *Chem Bio Chem* 2009;10, 658–662. (c) Gao W, Liu W, Mackay JA, Zalutsky MR, Toone EJ, Chilkoti A. *Proc Natl Acad Sci USA* 2009;106, 15231–15236.

450. Dedeo MT, Duderstadt KE, Berger JM, Francis MB. *Nano Lett* 2010;10:181–186.

451. Carrico ZM, Romanini DW, Mehl RA, Francis MB. *Chem Commun* 2008;1205–1207.

452. Stephanopoulos N, Carrico ZM, Francis MB. *Angew Chem Int Ed* 2009;48:9498–9502.

453. Tong GJ, Hsiao SC, Carrico ZM, Francis MB. *J Am Chem Soc* 2009;131:11174–11178.

454. Behrens CR, Hooker JM, Obermeyer AC, Romanini DW, Katz EM, Francis MB. *J Am Chem Soc* 2011;133:16398--16401.

455. Datta A, Hooker JM, Botta M, Francis MB, Aime S, Raymond KN. *J Am Chem Soc* 2008;130:2546–2552.

456. Hooker JM, O'Neil JP, Romanini DW, Taylor SE, Francis MB. *Mol Imaging Biol* 2008;10:182–191.

457. Kovacs EW, Hooker JM, Romanini DW, Holder PG, Berry KE, Francis MB. *Bioconjugate Chem* 2007;18:1140–1147.

458. Hooker JM, Datta A, Botta M, Raymond KN, Francis MB. *Nano Lett* 2007;7:2207–2210.

459. Schlick TL, Ding Z, Kovacs EW, Francis MB. *J Am Chem Soc* 2005;127:3718–3723.

460. Bruckman MA, Kaur G, Lee LA, Xie F, Sepulveda J, Breitenkamp R, Zhang X, Joralemon M, Russell TP, Emrick T, Wang Q. *Chem Bio Chem* 2008;9:519–523.

461. Posner T. *Ber Dtsch Chem Ges* 1905;38:646–657.

462. Hoyle CE, Bowman C. *Angew Chem Int Ed* 2010;49:1540–1573.

463. Lowe AB. *Polym Chem* 2009;1:17–36.

464. Massi A, Nanni D. *Org Biomol Chem* 2012;10:3791–3807.

465. Finzi C, Venturini G, Sartinni L. *Gazz Chim Ital* 1930;60:798–811.

466. Kohler EP, Potter H. *J Am Chem Soc* 1935;57:1316–1321.

467. Chen G, Kumar J, Gregory A, Stenzel MH. *Chem Commun* 2009;6291–6293.

468. Lowe AB, Hoyle CE, Bowman CN. *J Mater Chem* 2010;20:4745–4750.

469. Minozzi M, Monesi A, Nanni D, Spagnolo P, Marchetti N, Massi A. *J Org Chem* 2011;76:450–459.

470. Conte ML, Staderini S, Marra A, Sanchez-Navarro M, Davis BG, Dondoni A. *Chem Commun* 2011;47:11086–11088.

471. Ho CM, Zhang JL, Zhou CY, Chan OY, Yan JJ, Zhang FY, Huang JS, Che CM. *J Am Chem Soc* 2010;132:1886–1894.

472. Plueddemann EP. *Silane Coupling Agents*, 2nd ed. New York: Plenum Press; 1991.

473. VanDerVoort P, Vansant EF. *J Liq Chrom Relat Tech* 1996;19:2723–2752.

474. Benters R, Niemeyer CM, Drutschmann D, Blohm D, Wöhrle D. *Nucl Acids Res* 2002;30:e10.

475. Wang ZV, Schraw TD, Kim J-Y, Khan T, Rajala MW, Follenzi A, Scherer PE. *Mol Cell Biol* 2007;27:3716–3731.

476. Posewitz MC, Tempst P. *Anal Chem* 1999;71:2883–2892.

477. Piehler J, Brecht A, Valiokas R, Liedberg B, Gauglitz G. *Biosens. Bioelectron* 2000;15:473–481.

478. FitzGerald SP, Lamont JV, McConnell RI, Benchikh EO. *Clin Chem* 2005;51:1165–1176.

479. Silva SS, Ferreira RAS, Fu L, Carlos LD, Mano JF, Reis RL, Rocha J. *J Mater Chem* 2005;15:3952-3961.

480. Boeva VI, Solovievb A, Silva CJR, Gomes MJM. *Solid State Sci* 2005;8:50–58.

481. Duan L, Wang Y, Li SS-C, Wan Z, Zhai J. *BMC Infect Dis* 2005;5:53.

482. Pollino JM, Weck M. *Chem Soc Rev* 2005;34:193–207.

483. Wilchek M, Bayer EA. *Anal Biochem* 2009;171:1–32.

484. Wong LS, Khan F, Micklefield J. *ChemRev* 109:4025–4053.

485. Le Droumaguet B, Nicolas J. *Polym Chem* 2010;1:563–598.

486. Bontempo D, Li RC, Ly T, Brubaker CE, Maynard HD. *Chem Commun* 2005:4702–4704.

487. Qi K, Ma Q, Remsen EE, Clark Jr CG, Wooley KL. *J Am Chem Soc* 2004;126:6599–6607.

488. Heredita KL, GroverGN, Tao L, Maynard HD. *Macromolecules* 2009;42:2360–2367.

489. Narain R, Housni A, Gody G, Boullanger P, Charreyre MT, Delair T. *Langmuir* 2007;23:12835–12841.

490. Jia Z, Wong L, Davis TP, Bulmus V. *Biomacromolecules* 2009;10:3253–3258.

491. Wang X, Liu L, Luo Y, Zhao H. *Langmuir* 2009;25:744–750.

492. Wilchek M, Bayer EA. *Methods Enzymol* 1990;184:123–138.

493. Diamandis EP, Christopoulos TK. *Clin Chem* 1991;37/5:625–636.

494. Lutz J-F. In: Niewohner J, Tannert C (eds), *Gene Therapy: Prospective Technology Assessment in its Societal Context*. Amsterdam: Elsevier; 2006. pp 57–72.

495. Yu MK, Park J, Jon S. *Theranostics* 2012;2:3–44.

496. Han S, Mahato RI, Sung YK, Kim SW. *Mol Ther* 2000;2:302–317.

497. Rayavarapu RG, Petersen W, Ungureanu C, Post JN, van Leeuwen TG, Manohar S. *Int J Biomed Imaging* 2007;2007:29817.

498. Hermanson GT. Preparation of colloidal-gold-labeled proteins. In: *Bioconjugate Techniques*. New York: Academic Press; 1996. pp 593–605.

499. Medintz IL, Uyeda HT, Goldman ER, Mattoussi H. *Nat Mater* 2005;4,:435–446.

500. Lao UL, Mulchandani A, Chen W. *J Am Chem Soc* 2006;128:14756–14757.

501. Medintz IL, Clapp AR, Mattoussi H, Goldman ER, Fisher B, Mauro JM. *Nat Mater* 2003;2:630–638.

502. Hong R, Fischer NO, Verma A, Goodman CM, Emrick TS, Rotello VM. *J Am Chem Soc* 2004;126:739–743.

503. Irrgang J, Ksienczyk J, Lapiene V, Niemeyer CM. *Chem Phys Chem* 2009;10:1483–1491.

504. Delehanty JB, Medintz IL, Pons T, Brunel FM, Dawson PE, Mattoussi H. *Bioconjugate Chem* 2006;17:920–927.

505. Medintz IL, Berti L, Pons T, Grimes AF, English DS, Alessandrini A, Facci P, Mattoussi H. *Nano Lett* 2007;7:1741–1748.

506. Berti L, D'Agostino PS, Boeneman K, Medintz IL. *Nano Res* 2009;2:121–129.

507. Kim MJ, Park HY, Kim J, Ryu J, Hong S, Han SJ, Song R. *Anal Biochem* 2008;379:124–126.

508. Gupta M, Caniard A, Touceda-Varela A, Campopiano DJ, Mareque-Rivas JC. *Bioconjugate Chem* 2008;19:1964–1967.

509. Bae PK, Kim KN, Lee SJ, Chang HJ, Lee CK, Park JK. *Biomaterials* 2009;30:836–842.

510. Susumu K, Medintz IL, Delehanty JB, Boeneman K, Mattoussi H. *J Phys Chem C* 2010;114:13526–13531.

511. Graff RA, Swanson TM, Strano MS. *Chem Mater* 2008;20:1824--1829.

512. Lee SK, Maye MM, Zhang YB, Gang O, van de Lelie D. *Langmuir* 2009;25:657–660.

513. Kim I, Park YH, Rey DA, Batt CA. *J Drug Targeting* 2008;16:716–722.

514. Kim SH, Jeyakumar M, Katzenellenbogen JA. *J Am Chem Soc* 2007;129:13254–13264.

515. Xie HY, Zhen R, Wang B, Feng YJ, Chen P, Hao J. *J Phys Chem C* 2010;114:4825–4830.

516. Cho HY, Kadir MA, Kim BS, Han HS, Nagasundarapandian S, Kim Y-R, Ko SB, Lee S-G, Paik H-J. *Macromolecules* 2011;44:4672–4680.

517. Javier DJ, Nitin N, Levy M, Ellington A, Richards-Kortum R. *Bioconjugate Chem* 2008;19:1309–1312.

518. Barbas AS, Mi J, Clary BM, White RR. *Future Oncol* 2010;6:1117–1126.

519. Farokhzad OC, Karp JM, Langer R. *Expert Opin Drug Deliv* 2006;3:311–324.

520. Li N, Larson T, Nguyen HH, Sokolov KV, Ellington AD. *Chem Commun* 2010;46:392–394.

521. Yu MK, Kim D, Lee IH, So JS, Jeong YY, Jon S. *Small* 2011;7:2241–2249.

SECTION II

POLYMER BIOCONJUGATES

2

BIOCONJUGATES BASED ON POLY(ETHYLENE GLYCOL)S AND POLYGLYCEROLS

Rajesh A. Shenoi, Fang Gao, Muhammad Imran ul-Haq, and Jayachandran N. Kizhakkedathu

Centre for Blood Research and the Department of Pathology and Laboratory Medicine
Department of Chemistry, University of British Columbia, Vancouver BC, Canada

2.1 INTRODUCTION

The covalent attachment of bioactive molecules such as drugs, proteins, and peptides to polymers, collectively referred to as polymer therapeutics, has attracted significant attention in modern pharmaceutical technology [1–5]. There are several polymer-based conjugates in clinical practice or under clinical trials. Improvement in the therapeutic index of drugs is desired for the treatment of cancer, inflammatory diseases, and infection. One potential way to achieve this is by the covalent or noncovalent attachment of drugs to polymeric systems which results in reduced renal clearance, prolonged circulation in the blood stream, and improved tissue distribution at the disease sites. In addition, the toxicities of the drugs are significantly reduced. The covalent attachment of proteins to polymeric carriers can lead to more favorable profile in terms of decreased immunogenicity, antigenicity, increased body residence time, and stability of this class of therapeutic agents. Among the different polymeric systems, poly(ethylene glycol) (PEG) has been the most widely explored and currently there are several PEG-based therapeutics available in the market [6–11]. Although PEG has been used extensively, there are several drawbacks associated with PEG and some recent reports have provided evidence for anti-PEG antibodies in healthy individuals and in patients treated with PEG-based therapeutic agents. Also, PEG is not multifunctional and possess high intrinsic viscosity. In the context of alternatives to PEG, one of the potential candidates that have recently been explored is polyglycerols, both

hyperbranched polyglycerol (HPG) and linear polyglycerol (LPG) [12–15]. HPG has a dendritic structure and possess all the characteristic features of PEG, in addition being equally or more biocompatible, nontoxic and multifunctional. In this chapter, various bioconjugates based on PEG and polyglycerols will be discussed.

2.2 POLYETHYLENE GLYCOL-BASED BIOCONJUGATES

Polyethylene glycol (Figure 2.1) is the most widely used polymer in the field of polymer drug conjugates. It is a polyether diol synthesized by the anionic ring-opening polymerization of ethylene oxide and could be obtained in a wide range of molecular weights with low polydispersity index. When anhydrous methanol or methoxy ethoxy ethanols is used as initiators for the polymerization, monofunctional methoxy polyethylene glycols (mPEGs) are obtained. The amphiphilic nature of PEG results in excellent solubility in both aqueous and organic solvents and hence it is much easier to perform further chemical modifications of the polymer. PEG is also nontoxic and has the lowest level of protein or cellular adsorption of any known polymer [16]. It has been shown that grafting PEG to surfaces reduces the protein adsorption and this property could also be used for preventing bacterial surface growth. The unique features of PEG are due to its highly hydrated polyether backbone that can form hydrogen bonds with water and results in large exclusion

Chemistry of Bioconjugates: Synthesis, Characterization, and Biomedical Applications, First Edition. Edited by Ravin Narain.
© 2014 John Wiley & Sons, Inc. Published 2014 by John Wiley & Sons, Inc.

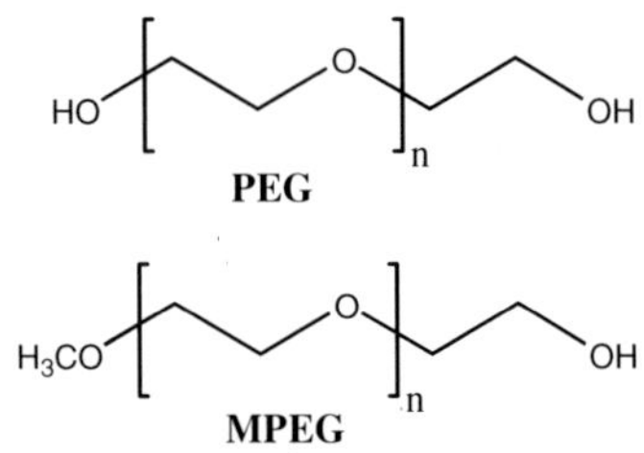

FIGURE 2.1 Structures of polyethylene glycol (PEG) and methoxy polyethylene glycol (mPEG).

volume [17]. The conjugation of drugs or drug carriers to PEG most often dramatically changes their pharmacokinetic properties by increasing their bioavailability and efficacy. Hence the majority of the drug conjugates in the market or in advanced clinical trials are based on PEG-based systems and in fact it is the only polymer-based stealth drug delivery system available in the market now. In the following sections we will cover the bioconjugates based on PEG.

2.2.1 PEG-protein Conjugates

Although proteins have been recognized as therapeutics for several years, their short circulation half-lives and relatively low stability require higher doses to maintain therapeutic efficacy which could possibly lead to adverse immune responses. As an approach toward improving the therapeutic index of protein drugs, Abuchowksi, Davis, and coworkers in 1977 reported the first covalent conjugation of bovine serum albumin and bovine liver catalase to mPEG [18]. The PEGylated albumin was found to be nonimmunogenic and the circulation half-life of the PEGylated bovine liver catalase was increased from 12 hours to 48 hours while retaining a large portion of the biological activity. The improved properties of these conjugates were attributed to the protective shell of mPEG around the protein that minimizes the interaction with immune cells and proteolytic inactivation.

Several PEGylated protein therapeutics have now been well established and clinically approved (Table 2.1). The first PEGylated protein to enter the market was ADAGEN® which is PEG-conjugated bovine adenosine deaminase

(ADA) commercialized by Enzon, Inc. for the treatment of severe combined immunodeficiency syndrome (SCID) [19, 20]. ADA has a half-life of only 30 minutes and induces immunogenic reactions that limit its therapeutic potential. PEGylation increased the circulation half-life of the protein by a factor of 6.4 and prevented its interaction with antibodies. ADAGEN has been used as an alternative to bone marrow transplantation and enzyme replacement by gene therapy [21]. There are also reports on the long-term improvement in the immune responses of patients treated with ADAGEN [21]. The PEGylated L-asparaginase conjugate, ONCASPAR®, has also been developed by Enzon for the treatment of acute lymphoblastic leukemia (ALL) and the drug was approved by FDA in 1994. The PEGylated enzyme exhibited three times higher plasma half-life, slower renal clearance, and reduced hypersensitivity (only 8% of the patients developed hypersensitivity) compared to the native enzyme [22, 23]. This requires dosing of PEG-L-asparaginase only once every 2 weeks compared to two to three times a week dosing for the native enzyme. Recently, PEG-asparaginase has been approved by FDA for the first-line treatment of children with ALL [24].

In the early 2000s, two of the PEG-α-interferon conjugates, PEG-INTRON® (PEG-IFN-α-2b, Schering), and PEGASYS® (PEG-IFN-α-2a, Roche) [25–27] were approved for the treatment of hepatitis C. IFN-α-2a and IFN-α-2b have similar biological activity, but differ in only a single amino acid residue and different molecular weight PEG with different linkers were used for their conjugation. PEG-INTRON used the succinimidyl carbonate derivative of PEG-12000 for conjugating the protein and possesses only a single chain of PEG per IFN. The plasma circulation half-life of PEG-INTRON is about eight times that of the native IFN-α-2b and requires only once weekly subcutaneous dosing instead of the thrice weekly dosing for the unmodified protein. However, the conjugate was hydrolytically unstable in plasma and when stored at ambient temperatures. A phase I/II clinical study using PEG-INTRON in patients with a variety of advanced solid tumors demonstrated that a 1 year dose of 6 μg/kg per week was well tolerated and the safety of the PEGylated construct was comparable to that of the non-PEGylated IFN-α-2b [28]. There are several clinical trials

TABLE 2.1 Clinically Approved PEG-protein Conjugates

Product Name	Biomolecule conjugated	Clinical use
Adagen	Adenosine deaminase asparaginase	SCID
Oncaspar	Asparaginase	Acute lymphocytic leukemia
PEGIntron	Interferon α-2b	Hepatitis C
Pegasys	Interferon α-2a	Hepatitis C
Neulasta	rhGCSF	Chemotherapy-induced neutropenia
Cimzia	anti-TNF Fab	Rheumatoid arthritis, Crohn's disease
Puricase	Uricase	Gout
Mircera	Erythropoietin	Anemia associated with chronic kidney disease

in progress using PEG-INTRON for other indications such as cancers, multiple sclerosis, and HIV/AIDS. The objective response rate was found to be 14% among 44 previously untreated renal cell carcinoma patients with a median survival of 13.2 months. PEGASYS uses a branched PEG of higher molecular weight (40,000 Da) for the conjugation to lysine residues *via* an amide bond and has a single strand of a branched PEG per IFN-α molecule. This conjugate has higher *in vitro* and *in vivo* stability than PEGINTRON, but retains only 7% of the activity of the native protein [27]. A 4- and 13-week toxicology study of PEGASYS in monkeys showed no PEG-related toxicities. PEGASYS was found to be cleared mainly *via* the liver and the metabolic products were excreted through the kidney [29]. Approximately 61–80% bioavailability of PEGASYS was observed up on subcutaneous administration in humans. The clinically approved dose for PEGASYS is 2.7–3.6 μg/kg per week for a 50 kg person [30]. Both PEG-INTRON and PEGASYS showed similar antiviral activities when used in combination therapy of hepatitis C with ribavirin, but there are no reports on direct comparative studies. Another PEG-protein conjugate that has been approved by FDA for the treatment of chemotherapy-induced neutropenia is NEULASTA® (Amgen) which is a PEG-modified form of granulocyte-colony stimulating factor (G-CSF). In NEULASTA a single chain of PEG-20000 is covalently attached to the N-terminal amino group of the methionyl residue of G-CSF [31]. The molecular weight of the conjugate is more than twice that of the native G-CSF which significantly reduces the renal clearance, prolongs the circulation time, and stimulates the proliferation and differentiation of neutrophils [32]. Clinical trials have shown that PEG-G-CSF requires less frequent administration compared to the native protein; a single dose of 100 μg/kg on day 2 of chemotherapy was equivalent to 14 daily doses of 5 μg/kg of the unmodified G-CSF.

Mircera® is another PEGylated product marketed by Hofmann-La Roche that has been approved in the United States and Europe in 2007 for the treatment of anemia associated with chronic renal failure (CRF) [33]. Mircera is made by the conjugation of mPEG butanoic acid with the N-terminal amino group of ϵ-amino group of any lysine present in the epoetin β. Due to its extended half-life and different receptor-binding activity, Mircera allows continuous stimulation of erythropoiesis. Mircera is usually administered at a dose of 0.6 μg/kg once every 2 weeks. The safety and efficacy of Mircera in hemodialysis and peritoneal dialysis patients has been reported [34–36].

Pegloticase (Krystexxa®) is a PEGylated mammalian urate oxidase (uricase) developed by Savient Pharmaceuticals and approved by FDA in 2010 for the treatment of severe, treatment-refractory chronic gout. Pegloticase is a tetrameric peptide composed of four identical chains of 300 amino acids each and nine of the 30 lysine residues in each chain are PEGylated. The elimination half-life of uricase was increased from 8 hours to 10–12 days upon PEGylation and this also decreases the immunogenicity of the foreign uricase protein. Pegloticase prevents inflammation and pain resulting from the formation of urate crystals in the plasma and was found to be more effective than other standard treatments for reducing gout tophi [37–40]. Several clinical trials reported the production of antibodies when pegloticase was injected subcutaneously or intravenously, but these were found to be due to PEG and not because of uricase. Moreover, subcutaneous injections of pegloticase results in slow absorption and causes transient local pain and allergic reactions, but these effects were not observed for intravenous injections. However, multidose trials with intravenous administration have also reported several infusion reactions. Hence the long-term safety of pegloticase needs to be established.

In addition to the described PEGylated proteins that are available in the market, there are several products that are under clinical investigation. Conjugation of recombinant human interleukin-2 (IL-2) to PEG was found to enhance the solubility and markedly prolong the circulation half-life, while maintaining the *in vitro* and *in vivo* activity of IL-2 in murine Meth A sarcoma model [41]. Patients in second remission of acute myelogenous leukemia (AML) were treated with PEG-IL-2 which prolonged the second remission duration in AML [42]. There are several clinical trials reported for the use of PEG-IL-2 for the treatment of metastatic renal cell carcinoma, melanoma, and basal cell carcinoma (BCC) [43, 44].

Pegvisomant is a PEGylated growth hormone receptor antagonist used for the treatment of acromegaly which is a chronic debilitating disorder resulting from excessive secretion of growth hormone and increase in the production of insulin-like growth factor I (IGF-I). Trainer et al. conducted a 12-week clinical study for testing the efficacy and tolerability of pegvisomant in patients with acromegaly [45]. Preliminary results indicated that pegvisomant was well tolerated without any adverse effects and there was reduction in the serum IGF-I concentrations in the patients [46, 47].

PEG-modified recombinant staphylokinase is another PEGylated protein that has been developed for the treatment of acute myocardial infarction (AMI). Staphylokinase (SakSTAR) is a 136-amino acid profibrinolytic bacterial protein that has the potential for coronary artery recanalization and has significantly high fibrin selectivity. But its relatively short plasma half-life and the formation of neutralizing antibodies upon administration limits its therapeutic use. The PEGylated SakSTAR was made by the site-directed substitution of selected amino acids with cysteine and derivatization with linear PEG molecules containing thiol-specific reactive groups [48]. Bolus administration of the PEGylated SakSTAR in experimental animals resulted in the reduction of plasma clearances by 2.5–20-fold and pulmonary clot lysis was observed at doses inversely related to their clearance [49]. In AMI patients, bolus intravenous injection of

PEGylated SakSTAR prolonged the plasma half-life by 4–40-fold for different variants compared to the unmodified SakSTAR. One of the variants, SY-161-P5 modified with PEG-5000 was found to be a promising fibrin-selective agent for single-bolus coronary thrombolysis [50].

Enhancing the molecular size of the metalloprotein hemoglobin (Hb) by PEGylation was explored as an approach to reduce the vasoactivity of acellular Hb by preventing its extravasation into the interstitial spaces. Some of the earlier work by Winslow and coworkers showed that PEG conjugation to the surface amino groups of human and bovine hemoglobin results in increased viscosity and osmotic pressure that can be used as an efficient way to modulate vasoactivity of Hb [51–53]. However, this approach used PEG conjugation to the surface amino groups of the protein through isopeptidyl linkage, resulting in the loss of the net surface positive charge (nonconservative PEGylation). Conservative PEGylation in which the PEGylation does not alter the surface charge of Hb was developed by Manjula and coworkers in order to study the role of PEGylation in modulating the vasoactivity of Hb. They developed a new maleimide-based protocol for the site-specific conjugation of PEG to Hb at the Cys-93 (β) [54]. A series of PEGylated Hbs carrying two copies of PEG per tetramer were synthesized using maleimidophenyl PEG of varying molecular weights (5, 10, and 20 kDa). PEGylation resulted in increase in the hydrodynamic volume, molecular radius, viscosity, and osmotic pressure of Hb, and correlated directly with the PEG molecular weight. However, there was no direct correlation between the vasoactivity, colligative properties, and the PEG molecular weight. This indicated that the ability to modulate the vasoactivity was decreased beyond certain PEG chain length, even if there is an increase in the colligative properties. There also appeared to be a relation between the surface coverage of Hb by PEG and the vasoactivity, although the surface coverage was not proportional to the PEG chain length. In another study the authors showed that the enhanced molecular size and solution properties achieved by conjugation of multiple copies of low molecular weight PEG chains to Hb is more effective in decreasing the vasoactivity than those achieved by the conjugation of smaller number of high molecular weight PEG chains [55]. Thus the conjugate, $(SP-PEG5K)_6$-Hb that contains six copies of the 5 kDa PEG chains per Hb exhibited significantly reduced vasoconstriction-mediated response than $(SP-PEG20K)_2$-Hb with two copies of 20 kDa PEG chains. The effect of different conjugation chemistry on the functional and solution properties of the PEG-Hb conjugates was reported [56]. The hexaPEGylated Hb, $(propyl-PEG5K)_6$-Hb obtained by reductive alkylation showed higher colloidal osmotic pressure than $(SP-PEG5K)_6$-Hb that used thiol-maleimide chemistry. But the functional properties of both the conjugates were comparable. This revealed that the functional properties of PEG-Hb conjugates are independent of the site and chemistry of the conjugation, while their solution properties depend on these factors.

The toxicity and hemodynamic functions of the PEGylated human hemoglobin MP4 (PEG-Hb containing several 5 kDa PEG per Hb) used as a hemoglobin-based oxygen carrier (HBOC) was evaluated in rhesus monkeys [58]. MP4 was administered by exchange transfusion to rhesus monkeys at a dose of 21 mL/kg (approximately 30% of the blood volume) and the blood samples were analyzed for clinical chemistry and hematology at 3, 7, and 13 days after dosing. There was no significant toxicity imparted by the conjugate, with only modest, transient elevation of aspartate aminotransferase (AST), alanine aminotransferase (ALT), and lactate dehydrogenase (LDH) on day 3. Histopathological analysis revealed vacuolation in the liver, renal tubules, and macrophages in the spleen, bone marrow, and lymph nodes at higher doses of MP4. The results demonstrated the safety of MP4 when administered as an exchange transfusion of 30% of estimated blood volume.

Very recently, the PEGylation of coagulation factors has been developed as a promising strategy for the treatment of hemophilic patients. Hemophilia is a bleeding disorder caused by the deficiency of coagulation factors, factor VIII (Hemophilia A) and factor IX (Hemophilia B). At present the bleeding events associated with hemophilia are treated by frequent infusions (up to three times a week) of the recombinant coagulation factors [59]. By modification of the coagulation factors with PEG, their half-life can be prolonged thereby requiring less frequent infusions which will improve the quality of life of hemophilic patients.

Recombinant factor VIIa (rFVIIa) has proven to be a safe and effective treatment for hemophilic patients with inhibitors. Novo Nordisk developed a long-acting glycoPEGylated rFVIIa (N7-GP) by selective and reproducible PEGylation of N-glycan in rFVIIa [60]. N7-GP retained its catalytic activity in *in vitro* tests and interacted efficiently with tissue factor (TF), factor X (FX), and the plasma inhibitors tissue factor plasma inhibitor (TFPI) and antithrombin (AT) [61]. The circulation half-life of N7-GP in mice was 25 hours, sixfold higher than that of rFVIIa. When intravenously administered in mice at a dose of 10 mg/kg, the plasma FVIIa activity of N7-GP was significantly higher than that of rFVIIa after 90 minutes. In hemophilic mice treated with 20 mg/kg of N7-GP, the hemostatic effect was sustained for 24 hours and the bleeding was prevented at 6 hours post dosing [62]. The safety and pharmacokinetics assessment of N7-GP in humans demonstrated that the FVIIa activity was measurable up to at least 72 hours and the circulation half-life was 15 hours [63]. The pharmacokinetics was dose dependent and no thromboembolic complications were observed. The long circulating nature of N7-GP may find potential in the treatment of hemophilic patients with inhibitors.

Bleeding events in Hemophilia A are currently treated with intravenous infusion of recombinant or plasma-derived factor VIII (rFVIII) 2–4 times a week. The frequent infusions are required due to the short circulating half-life of FVIII (12–14 hours) in patients, but this affects their quality of life. In addition the interaction of FVIIIa with von Willebrand factor (VWF) is necessary to maintain its stability in circulation. In patients with type 3 von Willebrand disease (VWD), the half-life of FVIII is further reduced to about 3 hours. In an attempt toward prolonging the half-life of FVIII without reduction in its activity, researchers at Bayer Health Care developed a series of PEGylated rFVIII [64]. These molecules were conceived by the introduction of single cysteine residues on the surface of B-domain depleted FVIII (BDD-FVIII) and the site-specific PEGylation of the cysteine with PEG-maleimide. Different molecular weight PEGs (12–60 kDa) were conjugated to sites spanning all the five domains of rFVII. Several of these conjugates retained their coagulation activity and VWF binding *in vitro*, depending on the site of PEGylation. In Hemophilia A mice and rabbits, the PEGylated FVIII variants exhibited improved pharmacokinetic profiles compared to BDD-FVIII. For example, in both animal species, the terminal half-life of FVIII was extended to 9.8–13.6 hours for the conjugates (depending on the PEG molecular weight and the PEGylation site) compared to 5.9 hours for BDD-FVIII. The 60 kDa PEG-conjugated FVIII variants were found to be the best in terms of extended half-life and they were also efficacious in treating acute tail bleeding in hemophilic mice. The extended PK of the PEGylated FVIII variants also resulted in prolonged prophylactic efficacy of FVIII in protection from venous bleeding in hemophilic mice. A recently completed phase I study in humans using 60 kDa PEG-rFVIII (Bay 94-2027) demonstrated its tolerance and efficacy without any adverse events. A phase II/III clinical study using this conjugate is under progress. Baxter and Nektar developed a long-acting PEGylated rFVIII (BAX855). A recently conducted toxicology and preclinical study in rats and cynomolgus monkeys demonstrated the safety of intravenous administration of BAX855 without any serious systemic effects [65].

Replacement therapy with plasma-derived or recombinant factor IX concentrates has been the mainstay for Hemophilia B treatment. The relatively short half-life of FIX (18–24 hours) requires frequent infusions in patients that pose several challenges. In order to overcome these problems, a longer-acting recombinant FIX (N9-GP) was developed by the enzymatic glycoconjugation of 40 kDa PEG to the native N-glycans of FIX which are attached to the activation peptide [66]. The glycoPEGylated rFIX was converted to native FIXa by the cleavage of the peptide by the physiologic activators. The PEGylation did not perturb rFIX as evidenced by the retention of its specific activity (73–100%) in plasma and blood-based assays. N9-GP was shown to be as efficacious as rFIX in stopping acute bleeding in Hemophilia B mice. Considerable improvement in the pharmacokinetic profile of N9-GP was observed in several animal models; the half-life

of N9-GP was found to be 67 hours in Hemophilia B mice, 113 hours in Hemophilia B dogs and 76 hours in mini pigs compared to the 16 hours for rFIX. The safety and pharmacokinetics of N9-GP was evaluated in 16 Hemophilia B patients by single intravenous administration [67]. The conjugate was well tolerated in 15 patients, while one patient developed hypersensitivity reaction soon after administration and was withdrawn from the study. None of the patients developed inhibitors. The half-life of N9-GP was 93 hours, which is fivefold higher than plasma-derived or recombinant FIX. The results indicate the potential of the glycoPEGylated rFIX as an effective treatment of bleeding in Hemophilia B patients using single and reduced doses.

2.2.2 PEG-peptide Conjugates

In contrast to several PEG-protein conjugates that are clinically available, the area of peptide PEGylation is still in its infancy. There are only a few PEG-peptide conjugates being studied and these were found not to be clinically superior. However, there is still much interest in the development of PEGylated peptides.

There are currently several clinical trials involving recombinant hirudin (r-hirudin) for use as an antithrombotic agent. However, the rapid elimination of r-hirudin from circulation (terminal $t_{1/2}$ of 50–100 minutes) demands continuous infusions or multiple daily injections for maintaining the therapeutic efficiency. Conjugation of r-hirudin to two molecules of PEG-5000 afforded PEG-hirudin with significant prolongation of $t_{1/2}$, permitting once daily subcutaneous administration. A dose-dependent anticoagulant activity was observed upon intravenous injection of varying doses (0.03–0.3 mg/kg) of PEG-hirudin. PEG-hirudin was very well tolerated without immunoallergic side effects which makes it a promising compound for antithrombotic therapy [68]. In patients on maintenance hemodialysis, a single dose of PEG-hirudin was well tolerated and maintained a residual anticoagulant effect in the intervals between hemodialysis sessions [69].

Salmon calcitonin (sCT) is a therapeutic polypeptide that can be used for the treatment of hypercalcemia and osteoporosis. Mono- and di-PEGylated sCT were prepared by covalently linking with succinimidyl carbonate of mPEG [70]. The stability of PEG-sCT was significantly improved in the rat liver homogenate compared to native sCT, while retaining its biological activity. The mono-PEG-sCT showed prolonged elimination half-life (189.1 minutes versus 59.8 minutes for the native sCT) while the di-PEG-sCT exhibited significantly reduced renal clearance. The extent of urinary excretion of the PEG-conjugated sCTs was higher than that of the unmodified sCT and reduced accumulation of the PEGylated sCTs was observed in the tissues [71]. The reduced systemic clearance and prolonged elimination half-life impart greater therapeutic potential for the PEGylated sCTs. Ryan et al. recently reported the conjugation of sCT to a novel comb-shaped end-functionalized poly((PEG)methyl ether methacrylate) via cysteine-1 to generate conjugates of different molecular weights. The conjugates retained 85% of the bioactivity of the peptide *in vitro* and the proteolytic stability was significantly increased [72]. Enhanced potency and efficacy were observed for the conjugates when compared to sCT when intravenously administered in rats, with the high molecular weight conjugates showing superior properties [73].

The peptides, RGD, and EILDV, that are active fragments of the cell adhesion protein fibronectin, were conjugated to a bifunctional amino acid type PEG (aaPEG) of molecular weight 10,000 Da. The hybrid, RGD-aaPEG-EILDV, containing 0.43 mmol each of RGD and EILDV was found to exhibit antiadhesive effect, while a mixture of the same molar concentrations of the peptides did not demonstrate this effect [74].

Although PEGylation has several beneficial effects on the therapeutic potential of proteins and peptides, the PEGylation most often is nonspecific that results in a mixture of products which often requires costly purification steps. Site-specific PEGylation on the other hand, results in chemically well-defined PEGylated products with optimized properties and was considered to be an effective way to improve the bioactivity of these molecules. Peptides possess more specific active or metabolic sites than proteins and hence they were studied as ideal targets for site-specific PEGylation in an attempt toward increasing their therapeutic potential. Site-specific PEGylation of sCT was reported by Youn et al. by the covalent attachment of PEG to the Lys(18) amine of the peptide as an attempt toward improving the intrapulmonary delivery of sCT. Eighty percent of the bioactivity of the peptide was retained in the conjugate, at the same time prolonging the circulation half-life and increasing the proteolytic stability. The pulmonary stabilities and pharmacokinetic properties of the conjugates were considerably higher than that of sCT [75]. Kim et al. recently reported the site-specific PEGylation of Exendin-4 (Ex4-Cys), a potential therapeutic peptide that acts as a GLP-1 receptor agonist for the treatment of type 2 diabetes mellitus [76]. Due to its short *in vivo* half-life, the commercial Ex4-Cys requires twice daily injections in diabetic patients. Moreover, exendin-4 may induce immunogenic response in humans due to its nonmammalian origin and unique C-terminal sequence. In an attempt to minimize these effects, C-terminal-specific PEGylated Ex4-Cys (C40-tPEG-Ex4-Cys) was prepared using cysteine and amine residue-specific coupling of Ex4-Cys with activated trimeric PEG of different molecular weights (23 kDa and 50 kDa). The receptor-binding affinity of C40-PEG5K-Ex4-Cys was about 3.5-fold higher than that of the N-terminal PEGylated Ex4-Cys and the trimeric conjugate C40-tPEG23K-Ex4-Cys exhibited much higher receptor binding than the branched PEG conjugate, C40-brPEG20K-Ex4-Cys. Prolonged blood circulation was observed with the high

molecular weight PEG conjugate, C40-tPEG50K-Ex4-Cys and the hypoglycemic duration was 8.22 times longer than that of the native Ex4-Cys. All these results suggest the therapeutic potential of the PEGylated exendin-4 as an effective long-acting GLP-1 receptor agonist for the treatment of type 2 diabetes mellitus. Similar results were also reported by Gao et al. [77].

PEGylation of insulin has been studied by several groups as an approach to overcome its unfavorable properties such as chemical and biological instability, inherent immunogenicity and antigenicity, and rapid clearance from systemic circulation. Site-specific covalent attachment of low molecular weight PEG (750 and 2000 Da) to insulin *via* the N-terminal (PheB1) amino group and penultimate C-terminal ε-amino (LysB29) group was studied by Hinds and Kim [78]. When administered intravenously in Sprague Dawley rats, the conjugates with PEG-750 exhibited equivalent biological activity to that of Humulin (regular insulin formulation), whereas those with PEG-2000 showed lower activities probably due to larger steric hindrance of the larger polymer on insulin–receptor interaction. The PEG-insulin conjugates showed significant reduction in the humoral and cellular immunogenicity, antigenicity, and allergenicity compared to insulin in both the A/J (low insulin responders) and C57BL/10 (high insulin responders) models. The interaction of insulin with IgG and IgE antibodies was also reduced substantially upon conjugation with PEG. The biological activity and pharmacokinetics of the conjugates were investigated in a dog model by both intravenous and subcutaneous administration. The blood glucose lowering abilities of the conjugates were similar to that of insulin in both the routes of administration and a measurable increase in the circulation half-life of the conjugates relative to insulin was observed. Calceti et al. developed an oral PEG-insulin system by the reaction of mono and di-tert-butyl carbonate insulin derivatives with a 750 Da PEG [79]. The PEG conjugates showed enhanced stability toward proteases *in vitro* compared to native insulin. The insulin permeation behavior across the intestinal mucosa was not significantly altered in the conjugates. One of the PEG conjugate was formulated into mucoadhesive tablets with thiolated polyacrylic acid-cysteine. When orally administered in diabetic mice, a 40% reduction in the glucose levels was observed after 3 hours and the biological activity was maintained for up to 30 hours. Shechter et al. designed a hydrolyzable PEGylated insulin by conjugating the amino acid residues with a sulfhydryl containing 40 kDa PEG via a heterobifunctional linker, 9-hydroxymethyl-7-(amino-3-maleimidopropinate)-fluorene-N-hydroxysuccinimide (MAL-Fmoc-Su). The conjugate underwent hydrolysis under the physiological conditions and released the biologically active insulin in a controlled manner, with a hydrolysis half-life of 30 hours. A single subcutaneous administration of the conjugate in diabetic rodents showed 4- to 7-fold prolonged glucose lowering effects than similar doses of native insulin [80].

2.2.3 PEG-antibody Conjugates

PEGylation has also been extended for the modification of antibodies, both single chain antibodies (SCAs) and monoclonal antibodies (mAbs) for improving the solubility and *in vivo* circulation half-life. Lee et al. studied the conjugation of the single chain antibody CC49/218 sFv with PEG of different molecular weights (2–20 kDa) and architectures (linear and branched) [81]. The conjugates with PEG:sFv ratio of 1:1 and 2:1 were found to exhibit sFv affinity values within twofold of the unmodified sFv protein. It was also observed that PEG conjugation to carboxylic acid moieties using PEG-hydrazide chemistry resulted in significant activity retention of the bioconjugates. Prolonged half-lives were observed in mice with all the PEGylated conjugates and conjugation with higher molecular weight PEG was found to be more effective in extending the serum half-life of the conjugates than multiple PEGylation with low molecular weight PEG. The conjugation of a high molecular weight PEG (40 kDa) to two types of mAbs N12 and L26 that are specific to the ErbB2 (HER2) oncoprotein was reported by Hurwitz et al. [82]. These antibodies suppress the growth of tumor overexpressing ErbB2. N-hydroxysuccinimide-activated branched PEG for conjugation through the amino groups of the protein was used for binding the whole antibody or its monomeric Fab' fragment. The binding affinity and tumor inhibiting activity of the antibodies was found to be retained in the conjugates when tested *in vitro* against human gastric carcinoma N87 cells. The antitumor activity of the conjugates was also demonstrated *in vivo* and the same pattern of tumor development was observed during the first few weeks after administration. The most striking observation was the significant and complete inhibition of a second tumor implanted into the same mice during the later stage by the PEGylated conjugates compared to the results in mice injected with the unmodified antibody.

In an effort toward a more clinically acceptable therapeutic, the research group at Genentech studied the conjugation of PEG to the ε-amino groups of the F(ab')$_2$ form of a humanized anti-interleukin IL-8 antibody [83]. The objective of the study was to modulate the *in vivo* clearance rate with minimal loss of bioactivity by changing the hydrodynamic size of the PEGylated antibody fragments which in turn was achieved by using different molecular weight PEG molecules. The conjugates were prepared by the reaction of N-hydroxysuccinimide-activated PEG through the primary amines of the anti-IL-8 F(ab')$_2$ fragment. It was observed that the conjugation of two 40 kDa branched PEG to the antibody resulted in a serum half-life of 48 hours when compared to the 8.5 hours for the parent F(ab')$_2$. This study demonstrated that desired pharmacokinetics could be achieved by the conjugation of high molecular weight PEG to one or two sites of the antibody fragments.

Raffler and coworkers reported a novel and potentially useful application of PEGylated monoclonal antibody (mAb) to an antibody-directed enzyme prodrug therapy (ADEPT) system. In this work, the F(ab')$_2$ fragment of the anti-TAG-72 antibody, B72.3 was conjugated to β-glucuronidase that was modified with mPEG-5000 [84]. When tested in LS174T xenografts (human colon adenocarcinoma cells), the conjugate B72.3-βG-PEG localized to a peak concentration within 48 hours of administration. However, the enzyme activity persisted in plasma such that the administration of the prodrug has to be delayed for at least 4 days in order to avoid the systemic prodrug and toxicity effects. At this time, the level of the conjugate in the tumor decreased to 36% of the peak levels. When AGP3, an IgM mAb against mPEG, was intravenously administered, the clearance of the conjugate from serum was accelerated and the tumor to blood ratio increased from 3.9 hours to 29.6 hours without significant decrease in the accumulation of the conjugate in the tumor. Thus the treatment of mice bearing human colon adenocarcinoma xenografts with B72.3-βG-PEG conjugate followed 48 hours later with AGP3 and a glucuronidase prodrug of p-hydroxyaniline mustard significantly delayed tumor growth with minimal toxicity compared to the control conjugate and conventional chemotherapy.

Certolizumab Pegol (Cimzia, UCB) is a PEGylated humanized Fab' fragment of an antitumor necrosis factor-α mAb that has a high affinity for TNF-α. Certolizumab pegol has demonstrated a long-lasting effect on reducing the progression of joint damage and improving the physical function in rheumatoid arthritis (RA) patients [85–87]. It has been approved by FDA in 2009 as monotherapy as well as in combination therapy with disease-modifying antirheumatic drugs (DMARDs) for the treatment of moderate to severe RA in adults. In Europe and Canada, certolizumab pegol in combination with methotrexate (MTX) is being approved for patients with inadequate response to DMARDs. Certolizumab pegol has also been approved in the United States and Switzerland for reducing the signs and symptoms of Crohn's disease (CD) and maintaining the clinical response in patients with moderate to severe active disease where the conventional therapies are inadequate. Unlike the other TNF-α inhibitors, certolizumab does not contain an Fc region and hence does not cause complement-dependent cytotoxicity, antibody-dependent cellular toxicity, and apoptosis. Other than RA and CD, clinical trials have demonstrated the potential of certolizumab pegol across a range of other clinical and radiographic outcomes.

2.2.4 PEGylation of Cells and Tissues

Recently PEGylation has also been applied for the modification of cells and tissues to create new drug delivery systems, to reduce the immunogenicity, and for the development of nonimmunogenic cells for safer transplantations. Several groups have reported the chemical modification of red blood cell (RBC) membranes with PEG as a strategy toward universal donor cells [88–91]. Scott and coworkers studied the covalent conjugation of different mammalian RBCs with mPEG (molecular weight 5 kDa) via cyanuric chloride coupling reaction [90]. The PEG modification did not show any detrimental effect on the RBC structure and functions such as lysis, morphology, osmotic fragility, hemoglobin oxidation state, hemostasis, oxygen binding, and cellular deformability. Upon PEGylation, human RBCs lost their ABO blood group sensitivity and a considerable reduction in the antiblood antibody binding was observed. The PEG-modified sheep RBCs were not effectively phagocytosed by human monocytes (white blood cells) in contrast to untreated sheep RBCs that were readily cleared. Normal survival (50 days) was observed *in vivo* with the PEGylated mouse RBCs with no allosensitization even after repeated infusions [91]. When the PEG-derivatized sheep RBCs were intraperitoneally transfused in mice, a 360-fold improvement in the survival was observed when compared to the unmodified sheep RBCs [90]. Nacharaju et al. used a thiolation-mediated maleimide chemistry to modify the RBCs by covalently linking maleimidophenyl PEG (Mal-Phe-PEG) of varying molecular weights 5 kDa and 20 kDa [92]. The Mal-Phe-PEG (5 kDa)-derivatized RBCs masked most antigens of the Rhesus system (C, c, E, e, D) and a combination of the 5 kDa and 10 kDa Mal-Phe-PEG-derivatized cells was required to mask the A and B antigens. By this modification, Group A Rh(D) + and Group B Rh(D) + RBCs were transformed into Group O Rh (D)-RBCs which makes it suitable as universal RBCs for transfusion.

The covalent modification of cell surfaces with PEG was also shown to alter the immune recognition of foreign tissues (such as major histocompatibility complex, MHC class II) thereby preventing the necessary cell–cell interactions between T-cells and the antigen-presenting cell (APC) molecules [93]. This in turn will likely prevent the initial adhesion and costimulatory events necessary for immune recognition and response. Such PEGylated blood products may find potential application in blood transfusion in immunosuppressive and chronically transfused patients.

Pancreatic islet transplantation has been considered as one of the most promising strategies for curing type 1 diabetes mellitus, since it can provide strict control over the blood sugar levels [94–96]. However, pancreatic islets are recognized and easily rejected by the host immune system. Hence patients undergoing islet transplantation requires simultaneous treatment with immunosuppressive medications which in turn cause severe adverse effects. PEGylation of pancreatic islets has been studied as a potential strategy for the immunoprotection of transplanted islets without the use of immunosuppressants [97]. In a report by Panza et al., rat pancreatic islets were derivatized with 5 kDa PEG-isocyanate by reaction with the surface amino groups of the islets. PEGylation did not alter the cell viability and similar insulin responses were observed with the PEGylated islets compared

to the unmodified islets in glucose perifusion assay [98]. Xie et al. presented the modification of adult porcine islets using monosuccinimidyl and disuccinimidyl PEG of varying molecular weight [99]. All the PEG derivatives showed significant cytoprotection against the cytotoxic effects on porcine islets by human serum *in vitro*. The cytoprotection was also demonstrated *in vivo* by intraportal transplantation in streptozotocin-induced immunocompromised diabetic mice. Both the PEG molecular weight and concentration of the derivatives were found to have a significant effect on the cytoprotection. The disuccinimidyl PEG derivative with porcine islet conjugated at one end and human albumin at the other end demonstrated better shielding than the monosuccinimidyl derivative. Lee et al. showed that PEG-modified rat pancreatic islets exhibited normal functioning *in vivo* for longer periods of time than the unmodified islets (14 days versus 5 days) [100]. They also demonstrated the use of PEGylated islets in combination with a low dose of the immunosuppressant drug cyclosporine as an effective therapy for the protection of transplanted islets for at least 1 year, in contrast to the unmodified islets that were eliminated within 2 weeks of transplantation [101, 102]. In order to increase the amount of PEG conjugated on the islet surface, multiple PEGylation was explored as a potential strategy. In this method, the islets were subjected to repetitive conjugation reactions with PEG-succinimidyl propionic acid (PEG-SPA) [103]. The viability and functionality of the islets were found to be unaltered upon triple PEGylation. When transplanted into diabetic patients, these islets survived for more than 100 days without the use of immunosuppressant in 43% of the patients, while the unmodified islets were rejected by the immune system within 1 week [103].

2.2.5 PEG Conjugates of Oligonucleotides, Aptamers, and siRNAs

During the past two decades oligonucleotides, mainly antisense oligonucleotides and aptamers, have been explored as potential drugs due to their extremely high selectivity in target recognition [104, 105]. However, their use in human therapy was limited by their low *in vivo* stability toward exo- and endonucleases (present in plasma and inside cells) and inability to reach effective concentrations at cellular targets. In addition, they are rapidly excreted from the body due to their relatively small size and they possess low cellular uptake due to their anionic nature. Hence conjugation of oligonucleotides to PEG has been investigated to improve their stability and cellular uptake. Chemical conjugation of PEG (both linear and branched) at the 3′ and 5′-terminus of ODNs was found to significantly improve their enzymatic stability, binding affinity, and *in vivo* retention time for the conjugates in mice when compared to the corresponding unmodified and phosphorothioate ODNs [106–111]. Veronese and coworkers tested the conjugates of an anti-HIV ODN with high molecular weight linear and branched PEG for their activity as substrate toward the enzyme RNase H [110]. The conjugates exhibited higher enzymatic stability and did not impede the formation of the regular hybrid duplex with the target RNA sequence. The hydrolysis of RNA by the enzyme at the same site and with the same extent of cleavage as the native sequence was observed with the PEGylated conjugates. The PEGylation of triplex-forming oligonucleotides (TFOs) was studied by Rapozzi et al. [111]. TFOs can silence gene transcription by site-specific binding to the transcription initiation sites and have the potential for gene-targeting therapies especially for cancer. In this work, a 13-mer AG motif oligonucleotide was covalently attached to a 9 kDa PEG and the conjugate PEG-ODN(13) was found to exhibit superior cellular uptake and biological properties to those of the free ODN(13). The PEG-ODN(13) formed a stable triple helix with a natural polypurine–polypyrimidine target of a human oncogene and downregulated the gene transcription at 65 $\pm$ 5% with respect to the control. In addition, PEG-ODN(13) was more resistant against S1 and fetal bovine serum nucleases than the free ODN and less inclined to self-associate into multistrand structures in solution.

Pegaptanib sodium (Macugen) is a PEGylated antivascular endothelial growth factor (anti-VEGF) aptamer that has been approved by the FDA in 2004 for the treatment of age-related macular degeneration (AMD) of retina. Macugen consists of 28 nucleotides in length that terminates in a pentamino linker to which a branched 40 kDa PEG is covalently attached. It specifically binds to VEGF-165, a protein that plays a critical role in angiogenesis and increased permeability responsible for vision loss associated with AMD. The safety and efficacy of Macugen has been demonstrated in several clinical trials [112–116].

RNA interference (RNAi) has been studied as a promising platform for the treatment of gene-related diseases such as cancer and several viral infections [117]. Small interfering RNAs (siRNAs) that are 21–23 nucleotides in length are investigated as one of the best established RNAi mediators [118, 119]. However, the therapeutic potential of siRNAs is often limited by their rapid clearance, immunogenicity, and nonspecific distribution. Hence PEGylation of siRNAs has been studied by several groups in order to improve their therapeutic efficiency. Covalent conjugation of PEG to siRNA is often achieved by using either disulfide or acid-labile bonds that are cleaved at the reducing condition inside the cytoplasm or acidic environment in the endosomal/lysosomal compartments of the cell respectively. Park and coworkers conjugated VEGF-siRNA to PEG *via* disulfide linkage and observed greater enzymatic stability for the conjugates than naked VEGF-siRNA [120]. Complexation of PEG-VEGF siRNA with polyethyleneimine (PEI) resulted in polyelectrolyte complex micelles (PEC) with the siRNA/PEI forming the inner core and the PEG forming the surrounding shell layer. Superior VEGF gene-silencing effect was observed with the PEC in prostate carcinoma cells (PC-3) than the VEGF siRNA/PEI complexes even in the presence of serum. There was significant inhibition of VEGF expression after intravenous and intratumoral administration of the PEC micelles in animal models and no detectable inflammatory response was observed in mice. *In vivo* optical imaging also revealed enhanced accumulation of the PEC micelles in tumor regions [121]. Improved intracellular delivery and better gene-silencing efficiency by PECs obtained by the conjugation of PEGylated siRNAs with the fusogenic peptide KALA and luteinizing hormone-releasing hormone (LHRH) peptide analog has also been reported by the same group [122, 123]. In another work, GFP-siRNA was conjugated to a 6-arm PEG and functionalized with a cell penetrating peptide HPh1 to enhance the cellular uptake. The PEGylated siRNA formed PECs with KALA and showed enhanced gene-silencing efficiency in MDA-MB-435 cells in the presence of serum [124]. PEGylated polyplexes of poly (L-lysine) and lactosylated PEG-siRNA containing acid-labile β-thiopropionate linkage were reported by Kataoka and coworkers as an efficient siRNA delivery system [125]. The active siRNA molecules were released inside the cells by the cleavage of the β-thiopropionate linkage in the endosomal compartment. This PEC micellar system exhibited 100-fold higher gene-silencing activity compared to the free conjugate. The clustering of lactose moieties on the periphery of these PEGylated polyplexes act as targeting ligands for asialoglycoprotein receptor expressing human hepatocarcinoma cell lines (HuH-7) and showed appreciable growth inhibition of the HuH-7 spheroids for up to 21 days [126]. The FITC-tagged conjugate in the lac-PEGylated polyplexes exhibited enhanced penetration into the HuH-7 cells when compared with that of the cationic lipoplexes; this effect was due to the presence of the lactosylated PEG layer. HuH-7 cells treated with the PEGylated polyplexes also showed cellular apoptosis with programmed cell death [126].

In another study, siRNA was conjugated to a combtype PEG, pyridyl disulfide-functionalized poly(PEG methyl ether acrylate) (pPEGA) that demonstrated significantly higher *in vitro* serum stability and nuclease resistance when compared to unmodified and thiol-modified siRNAs [127]. The siRNA-p(PEGA) conjugate when complexed with the cationic fusogenic peptide KALA inhibited green fluorescent protein (GFP) expression that was much higher than the KALA complex of the unmodified siRNA and the protein inhibition was comparable to that of the unmodified siRNA-lipofectamine complex [128].

Dicer substrate small interfering RNAs (DsiRNAs), 25–30 nucleotide long siRNAs have been studied recently as more potent RNA designs in triggering RNAi in mammalian cells, where they are cleaved by the enzyme Dicer into 21–23 nucleotide long siRNAs. Like siRNAs, the DsiRNAs also suffer from unfavorable pharmacokinetics such as nonspecific distribution and rapid renal clearance. Hence the conjugation of DsiRNA to PEG was explored as a strategy in order to improve their therapeutic potential. In this study, the 3'-sense or 5'-antisense thiol-modified blunt-ended DsiRNAs targeting enhanced GFP (eGFP) gene were conjugated to PEG-acrylate of different molecular weights *via* a physiologically stable thioether bond [129]. Conjugates with 2 kDa PEG (both the 3'-sense and 5'-antisense strand conjugated) and 10 kDa PEG (3'-sense strand conjugated) were efficiently cleaved by recombinant Human Dicer resulting in the release of the 21-mer siRNA. Human neuroblastoma (SH-EP) cells treated with the 2 kDa and 10 kDa PEG conjugated to the 3'-sense strand of DsiRNAs exhibited gene-silencing activity at both mRNA and protein levels. PEG-conjugated DsiRNAs showed reduced immunogenicity compared to the unmodified DsiRNA.

2.2.6 PEG-drug Conjugates (PEG prodrugs)

Design of prodrugs for the optimized drug delivery has been of significant attention in the area of drug research. In prodrug design, the drug is first converted into a biologically inactive form by attachment to a carrier molecule through a chemical bond that can undergo enzymatic or hydrolytic cleavage in the body and thereby release the active drug [130–133]. Thus the prodrugs normally possess improved delivery efficiencies over the parent drug molecule. It is also important to control the breakdown of the prodrugs in the body as rapid cleavage can lead to toxicity while too slow release reduces the drug efficacy. Conjugation of drugs to polymeric systems is one of the methods used for the prodrug strategy. The conjugation of drugs to PEG has many advantages over other polymeric systems because of the enhanced solubility of the PEG-drug conjugates over the parent drug that may result in more effective drug delivery. Conjugation of drugs to PEG also increases their circulation half-life which could be beneficial in the case of water-soluble drugs where rapid elimination is problematic. It is also possible to achieve site-specific delivery of the drug to solid tumors using the PEG prodrug strategy.

Most of the PEG prodrug design is based on hydrolyzable or enzymatically cleavable bonds such as esters, carbonates, carbamates, and hydrazones. Prodrugs based on esters are the most commonly employed as they are the easiest to synthesize chemically.

One of the earliest studies in PEG-drug conjugates involved the use of low molecular weight (5 kDa) PEG for the conjugation of paclitaxel [134]. Paclitaxel is an extremely effective anticancer diterpene species, but has poor water solubility (<0.01 mg/mL). Highly water-soluble PEG-ester of paclitaxel was prepared by the reaction of PEG-carboxylic acid with the 2'-hydroxyl group of paclitaxel (Figure 2.2). The electron withdrawing α-alkoxy substituent of the PEG-acid was found to be more effective in the prodrug design as it helps in the rapid hydrolysis of the ester bond and releases

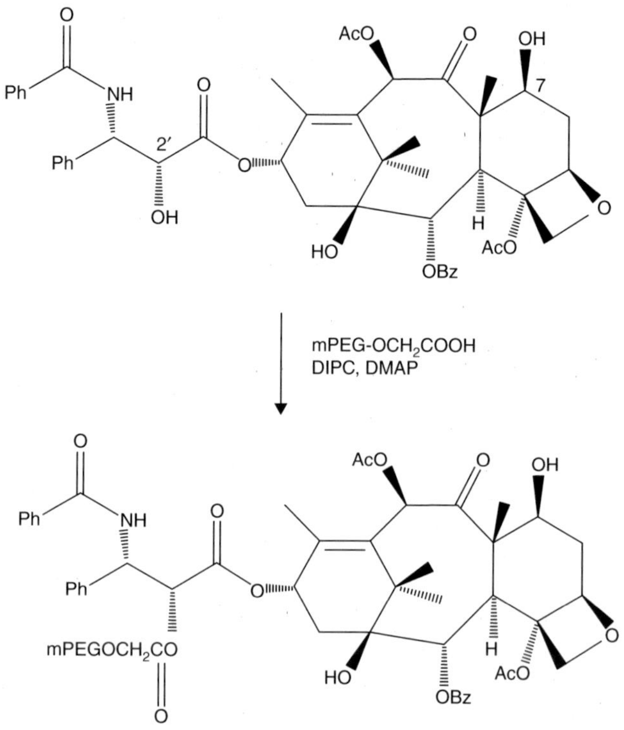

FIGURE 2.2 PEG-paclitaxel prodrug with ester bond. Reprinted with permission from Reference 135, Copyright® 1996, American Chemical Society.

the drug more effectively. The PEG-paclitaxel conjugates functioned as efficient prodrugs and the hydrolysis $t_{1/2}$ was 5.5 hours *in vitro* (PBS at pH 7.4) and 1 hour in rat plasma. When tested in murine leukemia cell lines P388 and L1210, the conjugate showed IC50 values that were comparable to the commercial TAXOL formulations. However, there was some discrepancy in the *in vitro* and *in vivo* efficacy for this low molecular weight PEG-paclitaxel conjugate as no acute toxicity was observed in mice at a dose of 50 mg/kg while profound toxicity was observed for TAXOL at the same dose. This lowered toxicity may be due to the faster excretion of the low molecular weight prodrug prior to its timed breakdown [135].

It became evident from the initial studies that the plasma circulation half-life of the prodrug has to be higher than its rate of hydrolysis for achieving potency equivalent or greater than that of the parent drug. Upon conjugation to PEG, the drug is transformed into a more hydrophilic form that has a tendency for rapid renal excretion. One way to prevent this rapid clearance was to design longer circulating drugs by increasing the molecular weight of the PEG used for conjugation. In addition, enhanced tumor accumulation could be achieved by such large macromolecules that show prolonged circulation time [136]. Greenwald et al. synthesized prodrugs of paclitaxel with different molecular weight PEG (5, 20, and 40 kDa) and observed that the drug delivery and efficacy was

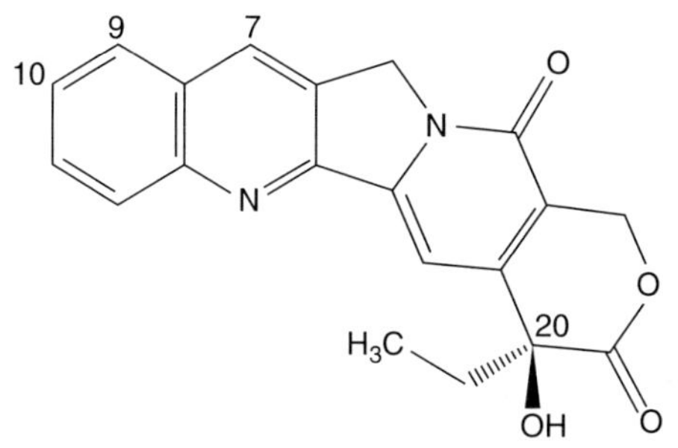

20(s)-Camptothecin (1)

FIGURE 2.3 Structure of camptothecin (CPT). Reprinted with permission from Reference 137, Copyright® 1996, American Chemical Society.

significantly influenced by the PEG molecular weight. Thus, murine acute lethality studies revealed that while the 5 kDa PEG-paclitaxel was not lethal at a dose of 10 μmol/mouse, the 40 kDa PEG-paclitaxel exhibited lethality at the same dose. The 20 kDa PEG-paclitaxel showed 50% lethality at 10 μmol/mouse, but was not lethal at 5 μmol/mouse. The *in vivo* efficacy study in P388/0 murine leukemia model revealed that the 40 kDa PEG-conjugated prodrug was equivalent to the TAXOL formulation [135].

Camptothecin (CPT), an extract from the Chinese tree *Camptotheca acuminata*, exhibits significant antitumor activity in mice with lung, ovarian, breast, pancreas, and stomach cancers. The structure of CPT consists of a lactone ring and a tertiary alcohol at the 20-position (20-OH group) both of which are essential for its anticancer activity (Figure 2.3). However, CPT is virtually insoluble in water that limits its use and its conversion into salts of amino acid esters has been studied as an approach to increase the water solubility. A PEG prodrug strategy was applied to CPT to obtain a nonionic solubilized form by conjugating a 40 kDa PEG at the 20-position of CPT [137]. The PEG(40K)-CPT exhibited a water solubility of 2.5 mg/mL that is considerably higher than that of the native CPT (0.0025 mg/mL). The hydrolysis of PEG(40K)-CPT occurred at much slower rates in PBS buffer and rat and human plasma when compared to the salt form of CPT. Thus the $t_{1/2}$ for PEG(40K)-CPT was respectively 27 hours, 2 hours, and 1 hour in PBS buffer, rat plasma, and human plasma when compared to the corresponding values of 3.5 hours, 0.5 hours, and 0.25 hours for CPT-20-glycinate TFA salt. When administered in P388 cells-treated mice, PEG(40K)-CPT showed increased life expectancies (ILS) of 200% and a cure rate of 80% at a dose of 16 mg/kg with no acute toxicity. These values were markedly higher when compared to the 70% cure rate for the native CPT and no cure for the salt form at the same dose. Pharmacokinetic studies by intravenous administration of PEG(40K)-CPT in CD-1 mice showed that the conjugate circulates sufficiently to release the drug over a substantial period of time. Significant reduction (60%) in tumor burden without significant weight loss was observed when colorectal carcinoma (HT-29) xenografts were treated with the PEG(40K)-CPT. This study revealed that PEG modification of CPT at the 20-position stabilizes the active lactone ring under physiological conditions and also results in relatively fast hydrolysis of the ester group thereby regenerating the free 20-OH group necessary for activity [138, 139]. This in turn results in enhanced bioactivity and more effective drug delivery for the prodrug.

PEG prodrug of another anticancer agent podophyllotoxin, an aryl tetralin lactone, was also studied. The anticancer activity of podophyllotoxin results from its binding to tubulin during mitosis thereby preventing the formation of microtubules. The drug is insoluble in water and was formulated in alcohol/water or propylene glycol for early *in vivo* experiments. However, the unpredictable systemic behavior arising from the water insolubility of podophyllotoxin generated interest in its modification to more soluble versions. Water-soluble PEG derivatives with ester, carbonate, and carbamate bonds were synthesized and tested for their *in vivo* oncolytic activity [140]. The prodrugs were found to be only slightly more efficacious than the native drug in solid lung tumor model (A549); those with faster release showed greater antitumor activity while slower release rates resulted in higher toxicity. This study also revealed that the conjugation of anticancer drugs to high molecular weight PEG does not always result in enhanced bioactivity.

The introduction of spacer groups between PEG and the drug molecule in prodrug design was considered as another strategy toward achieving enhanced activities and drug efficacies. There are reports on the application of this method for paclitaxel and camptothecin using heterobifunctional amino acid moieties as the spacer group [141, 142]. Amides, amines, and carbamates when present as linker groups accelerate the hydrolysis rates due to neighboring group participation [143–147]. The effect of various spacer groups on the rate of hydrolysis was studied for the PEG-spacer-CPT prodrugs [142]. A range of hydrolysis $t_{1/2}$ from 0.2 hours to 102 hours was observed in PBS buffer (pH 7.4) for the different spacer groups (Table 2.2). Although the variation in $t_{1/2}$ was less extreme in rat plasma, there were clear differences between the spacer groups. The observed difference may be due to the different extent of steric hindrance offered by the α-substituent. Among the different spacer groups studied for CPT, only glycine resulted in water-soluble prodrug with potency similar to the simple ester, with prolonged circulation $t_{1/2}$ and lowered *in vivo* toxicity [148]. In some cases, the hydrolysis of the ester bond between amino acid and CPT by esterases or pH may be favored by certain spacer groups. The amide bond between PEG and the amino acid could also be cleaved by the exopeptidases or proteinases in the tumor to release the amino acid-CPT that will be subsequently cleaved by esterases to release the free drug.

TABLE 2.2   *In Vitro* and *In Vivo* Characteristics of Spaced Prodrugs of Camptothecin

X-PEG	IC$_{40}$ (nM) P38S/0	t$_{1/2}$(h)[a]		P388 *in vivo*[b]		
		PBS pH 7.4R	at plasma	Mean time to death (day)[c]	%ILS[d]	Survivors on day 40
Control	–	–	–	13.0	–	0/10
Camptothecin	7	–		38.0*	192	7/10
(structure: —O—CH$_2$CH$_2$—O—PEG)	15	27	2	38.0*	192	9/10
(structure: —O—CH$_2$—C(=O)—NH—CH$_2$CH$_2$—PEG)	16	5.5	0.8	17.4**	34	4/10
(structure: —O—CH$_2$—C(=O)—N(CH$_3$)—CH$_2$CH$_2$—PEG)	21	27	3	31.6*,**	143	6/10
(structure: —O—C(=O)—N(CH$_3$)—CH$_2$CH$_2$—PEG)	18	28	5	23.4	80	0/10
(structure: —CH$_2$—C(=O)—NH—CH$_2$—PEG)	12	40	6	35.0*	169	8/10
(structure: —CH$_2$—C(=O)—N(CH$_3$)—CH$_2$—PEG)	15	97	10	19.3*,**	48	0/10
(structure: —CH$_2$—NH—CH$_2$CH$_2$—PEG)	24	12	3	30.6*	135	0/10
(structure: —CH$_2$—N(CH$_3$)—CH$_2$CH$_2$—PEG)	42	102	24	21.4*,**	65	0/10

Source: Reprinted with permission from Reference 7, Copyright® 2003, Elsevier.

All experiments were done in duplicate: standard deviation of measurements = ± 10%.

[a]These results more appropriately represent the half-lives of disappearance of the transport form.

[b]*In vivo* efficacy study of the water-soluble camptothecin derivatives using the P388/O murine leukemia model. Camptothecin or prodrug derivatives were given in equivalent dose of camptothecin (total dose of 16 mg/kg) daily [intraperitoneal × 5], 24 h following an injection of P388/O cells into the abdominal cavity with survival monitored for 40 days.

[c]Kaplan–Meier estimates with survivors censored.

[d]Increased life span (% ILS) is (T/C−1) × 100.

*Significant ($p < 0.001$) compared to control (untreated).

**Significant ($p < 0.001$) compared to camptothecin.

The spaced prodrug strategy was also applied for paclitaxel using heterobifunctional spacer groups that were used for the CPT systems. Thus the introduction of a glycine spacer between PEG and paclitaxel resulted in lower toxicity and enhanced *in vivo* antitumor activity compared to the native drug and the simple paclitaxel-PEG ester [141]. In another study, Dosio and coworkers conjugated 2 kDa and 5 kDa PEGs to human serum albumin-paclitaxel (HSA-PCT) resulting in stable PEG-HSA-PCT thioamide conjugates [149]. The objective was to further increase the $t_{1/2}$ of the HSA-PCT conjugate and reduce its uptake by the macrophages. These PEG-grafted conjugates exhibited higher cytotoxicity (similar to the HSA-PCT conjugate), efficient cell binding, and internalization. The prodrug was also efficiently cleaved inside the cell releasing the free drug. When administered intravenously in mice, the PEG-grafted conjugates showed reduced total clearance compared to the free drug and the HSA-PCT and the uptake by the liver and spleen was also significantly reduced.

Site-specific PEGylation of the Ara-C (cytosine arabinose, 1-(β-D-arabinofuranosyl)cytosine), a pyrimidine nucleoside analog used for the treatment of acute and chronic leukemias (ALL, AML, CML) has been reported [150]. The short plasma half-life of Ara-C requires frequent infusions for therapeutic effect, which in turn leads to severe adverse effects. Acyl-N^4-amino derivatives were prepared by using thiazolidine thiones. The conjugates showed *in vitro* hydrolysis half-life ranging from 1 to 3 days, depending on the spacer group employed. Some of the prodrugs exhibited superior antitumor activity than Ara-C in LX1 lung tumor model. However, the drug loading for these prodrugs was relatively low (about 1% Ara-C w/w) that limited the administration of higher doses necessary for better therapeutic effect. In order to achieve higher drug loading, the conjugation of Ara-C to dendritic PEG was employed [151]. The octamer-loaded PEG Ara-C conjugates were much more effective in several *in vivo* cancer models, LX-1 (solid lung tumor), PANC-1 (human pancreatic cancer), and P388/0 (murine leukemia).

Three of the PEG-drug conjugates from the camptothecin family, camptothecin itself, irinotecan (camptothecin-11), and SN-38 (7-ethyl-10-hydroxy-camptothecin) are presently under clinical investigation. Enzon Pharmaceuticals, Inc. developed Pegamotecan (PEG-camptothecin) by the conjugation of two CPT molecules to a glycine-bifunctionalized PEG of molecular weight 40 kDa. The PEG:CPT ratio in Pegamotecan was 60:1 that corresponds to an active drug loading of only 1.7 wt% [152]. In a phase I clinical trial in patients with various solid tumors, the maximum tolerated dose of the conjugate was determined to be 7 g/m^2 when intravenously administered for 1 hour every 3 weeks. At this dose, the maximum concentration of the free drug was estimated to be 0.5 $\mu g/mL$ which was much lower than that observed for sodium camptothecinate (30–60 $\mu g/mL$). Anticancer activity

in the form of partial tumor responses and prolonged stable disease was observed.

NKTR-102, a covalent conjugate of the drug irinotecan with 4-arm PEG was developed by Nektar Therapeutics for the treatment of various solid tumors [153, 154]. It possesses a unique pharmacokinetics providing a continuous concentration of the active drug with lower peak concentrations. In preclinical models, a 300-fold increase in the tumor concentration was achieved compared to the free drug. A phase I clinical trial was conducted to assess the safety, pharmacokinetics, and antitumor activity of NKTR-102 in patients with refractory solid tumors. The half-life of the active drug metabolite (SN-38) was extended to 50 days for NKTR-102 compared to 2 days for irinotecan and a 1.2- to 6.5-fold higher exposure of SN-38 was observed. The maximum tolerated dose was found to be 144 mg/m^2 with manageable toxicity. Forty to fifty-six percent reduction in tumor size was observed for patients with advanced solid tumors such as breast, ovarian, cervical, colorectal, and nonsmall cell lung cancers. A phase III study in patients with locally recurrent and metastatic breast cancer and phase II study in those with solid tumor malignancies (breast, ovarian, and colorectal) are currently under investigation.

Enzon Pharmaceuticals have also undertaken clinical trials with EZN-2208 which is a PEGylated conjugate of SN-38, an active metabolite of irinotecan (Figure 2.4). SN-38 has 100- to 1000-fold higher cytotoxic activity *in vitro* compared to irinotecan. In EZN-2208, four SN-38 molecules are linked to a 4-arm PEG through glycine spacers and this resulted in increased drug loading of about 3.7 wt% (versus 1.7 wt% for pegamotecan) [155]. A 1000-fold increase in the water solubility was observed upon conjugation of SN-38 to PEG. EZN-2208 showed 10- to 245-fold higher potency *in vitro* than irinotecan in human tumor cell lines. In mouse xenograft models of breast (MX-1), pancreatic (MiaPaCa-2), and colon carcinoma (HT-29), treatment with either single dose or multiple injections of EZN-2208 was more efficacious than irinotecan [156]. Pharmacokinetic and biodistribution studies in mice revealed prolonged blood circulation times for the conjugate and a 207-fold higher exposure to free SN-38 was observed compared to irinotecan. The tumor to plasma concentration ratio of EZN-2208 or the released SN-38 also increased over time. In a phase I clinical trial, 39 patients with various advanced solid tumors were given intravenous dose between 1.25 mg/m^2 and 25 mg/m^2 of EZN-2208 once every 21 days [157]. The maximum tolerated dose (MTD) was found to be 16.5 mg/m^2 and 10 mg/m^2 with and without granulocyte-colony stimulating factor (G-CSF). Pharmacokinetic studies showed a terminal half-life of 19.4 $\pm$ 3.4 hours for the conjugate. Partial tumor responses were observed in 18% of heavily treated patients, including those who failed prior irinotecan therapy. The company is pursuing a phase II study for metastatic breast and colorectal cancers and a phase I/II study in pediatric cancer patients.

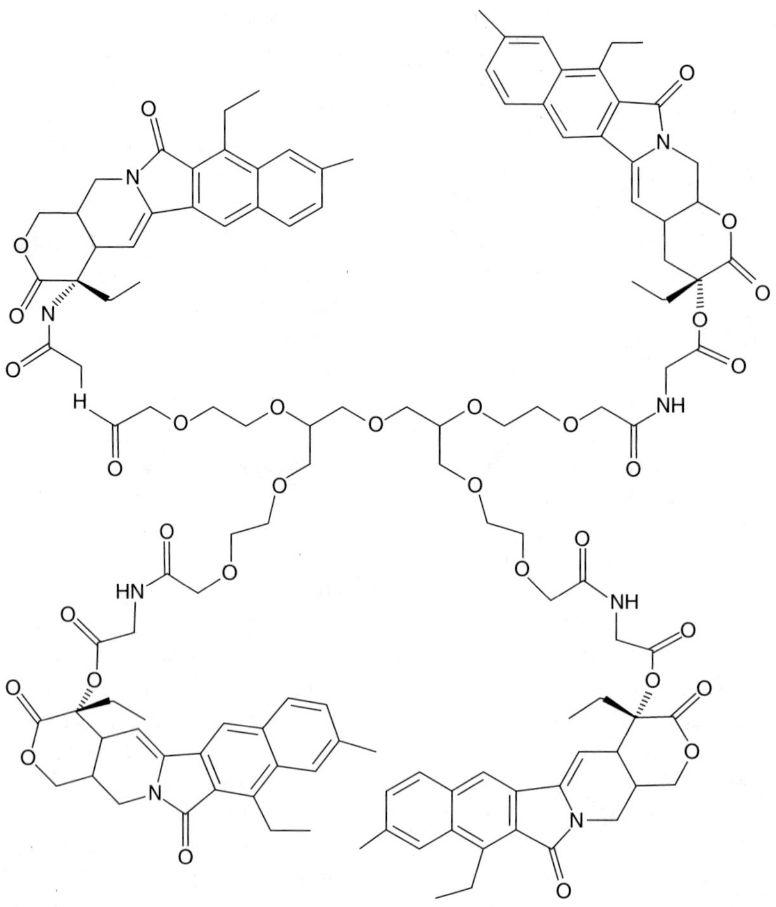

FIGURE 2.4 Structure of EZN-2208. Reprinted with permission from Reference 155, Copyright® 2008, American Chemical Society.

Besides camptothecin and paclitaxel that have been widely explored for produg design, the PEGylation of other drugs such as methotrexate (MTX), doxorubicin (Dox), and dexamethasone has generated academic interest. Rodriguez et al. synthesized conjugates of Dox with high molecular weight PEG (20 kDa and 70 kDa) with an acid-sensitive hydrazone linkage between PEG and the drug [158]. The conjugates were obtained by the coupling of Dox-maleimide derivatives with a stable amide or acid-sensitive hydrazone groups to PEG *via* thiopropionic acid spacer (Figure 2.5). The PEG-Dox prodrug was designed for the cleavage of the hydrazone linkage under the acidic environment of the tumor cell after uptake by endocytosis. *In vitro* antitumor activity was observed with the acid-sensitive PEG-Dox conjugates against human bladder carcinoma (BXF24) and lung cancer (LXFL 529L) cell lines, while the conjugates with the stable amide bond between PEG and drug did not exhibit any activity. It was also revealed by fluorescence microscopy studies that the acid-labile PEG-Dox conjugates were primary located in the cytoplasm, in contrast to the free drug which mainly accumulates in the cell nucleus. The PEG (20 kDa)-Dox exhibited lower embryo toxicity than the free drug, but

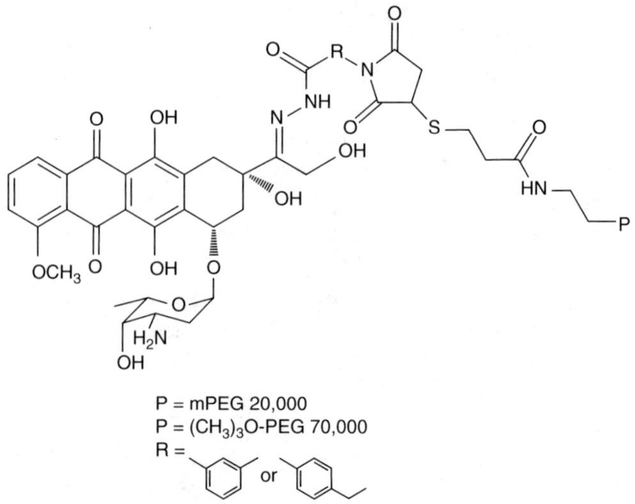

FIGURE 2.5 PEG-Doxorubicin prodrug with acid-sensitive hydrazone linker. Reprinted with permission from Reference 7, Copyright® 2003, Elsevier.

no antiangiogenic effects were observed. In another work, Veronese, Duncan and coworkers reported a series of PEG-Dox conjugates synthesized using different molecular weight (5–20 kDa) linear and branched PEGs with different peptidyl linkers (GFLG, GLFG, GLG, GGRR, RGLG) between the polymer and the drug [159]. The authors were successful in obtaining drug loading of 2.7 to 8 wt% with the conjugates that corresponds to 2% of the free drugs. The release of Dox from the conjugates with GFLG linker was found to be the same, irrespective of the PEG molecular weight or architecture, with 30% release observed after 5 hours. The *in vitro* degradation of the conjugates by lysosomal enzymes showed different degradation rates depending on the peptide linker, the conjugates with GLFG degraded faster (with 57% of the drug released in 5 hours) while the other linkers showed very slow degradation (16% release after 5 hours). *In vitro* toxicity studies in B16F10 melanoma cells showed all the conjugates to be greater than tenfold less toxic than the free drug. *In vivo* studies in mice-bearing B16F10 tumors with ^{125}I-labeled conjugates showed prolonged plasma clearance and enhanced tumor accumulation. However, there was no effect of the PEG molecular weight on biodistribution that was attributed to the formation of aggregates of PEG-DOX conjugates in solution. The conjugate linear PEG (5 kDa)-GFLG-Dox showed enhanced tumor activity in mice-bearing subcutaneous B16F10 (melanoma) and L1210 (lymphocytic leukemia) tumors at a dose of 5 mg/kg and 10 mg/kg Dox equivalent respectively. Antitumor activity against B16F10 tumor was also observed with the branched PEG (10 kDa)-GLFG-Dox conjugate at a dose of 5 mg/kg. Recently a PEG-Dox conjugate for the targeted anticancer therapy was reported by Polyak et al. by the conjugation of an RGD peptidomimetic on one end of a linear PEG and Dox at the other end [160]. Acid-sensitive hydrazone linker was used as the spacer between PEG and Dox. The conjugate, Dox-PEG-c(RGDfK)$_2$ showed active targeting of endothelial and tumor cells overexpressing $\alpha_v\beta_3$ integrin. The conjugate exhibited similar cytotoxic effects as that of the free drug in U87-MG human glioblastoma cells. Dox-resistant M109 murine lung carcinoma cells treated with the conjugate showed inhibition of cell proliferation at lower IC50 values than the free drug. *In vivo* studies revealed preferential accumulation of the conjugate in mCherry-labeled DA3 murine mammary tumors.

The conjugation of MTX to PEG was investigated by Kratz and coworkers using amino group containing PEG of varying molecular weights (750–40,000 Da) [161]. The conjugates showed higher IC$_{50}$ values compared to the free MTX in *in vitro* studies using human tumor cell lines. In contrast, *in vivo* studies in mesothelioma (MSTO-211H) xenograft models showed higher antitumor activities for the conjugates than the free MTX. The efficacy was dependent on the molecular weight of the PEG used for the conjugation, the conjugate with 5 kDa PEG showed lower activity than free MTX at equivalent doses, the 20 kDa PEG conjugate showed comparable activity as that of the free MTX at a dose of 40 mg/kg while the conjugate with 40 kDa PEG showed superior activity at a dose of 20 mg/kg. In another work, Kohler et al. conjugated MTX to a PEG immobilized on a magnetic nanoparticle surface. The PEGylated nanoparticles showed enhanced cellular uptake by 9L glioma cells than the control nanoparticles and showed higher cytotoxicity than the free MTX [162]. Transmission electron microscopy (TEM) revealed internalization of the conjugate into the cellular cytoplasm of 9L glioma cells and the particles were retained there for 144 hours. The prolonged retention of the PEGylated nanoparticle conjugate makes it an ideal platform for imaging of tumor cells during an extended therapeutic course.

Corticosteroids such as dexamethasone are commonly used for the treatment of inflammatory diseases such as rheumatoid arthritis, psoriasis, and inflammatory bowel diseases. Although corticosteroids are considered as potent drugs, their large volume distribution in the body causes severe side effects such as osteoporosis and hypertension. Funk et al. developed a prodrug system which could minimize these side effects by selectively targeting the sites of inflammation [163]. In their work, a corticosteroid was conjugated to a 22 kDa PEG *via* an acid-labile hydrazone linker that can undergo lysosomal degradation to release two new derivatives of dexamethasone. Confocal microscopy studies showed efficient lysosomal uptake and subsequent degradation of the PEG conjugate in human mammary carcinoma (MDA-kb2) and murine lymphoma (308) cell lines. The efficacy of the released dexamethasone derivatives was also demonstrated in the study. In an earlier study, Zacchigna et al. linked dexamethasone to a PEG-amine through a succinate linker and performed drug release and bioavailability studies in rabbits [164]. A 50% release of the drug in 15 hours was observed in plasma. In rabbits, the area under the concentration (AUC) for the PEGylated conjugate was twice as that of the parent drug after oral administration. A novel linear multifunctional PEG-dexamethasone conjugate employing click chemistry was reported by Liu et al. for the treatment of rheumatoid arthritis. Acid-labile hydrazone linker was used to conjugate the drug to a click-PEG. When tested in an adjuvant-induced arthritis (AA) rat model, a single injection of the conjugate resulted in sustained (>15 days) amelioration of joint inflammation, in contrast to the temporal resolution of arthritis by the free dexamethasone [165]. The superior anti-inflammatory and disease-modifying effects of the PEGylated conjugate over the free drug was also revealed by histological and bone mineral density evaluation.

2.2.7   PEGylation of Viruses

Over the past two decades recombinant viruses have been used as vectors in gene therapy due to their high efficiency at inducing transgene expressions in cellular targets. However, their clinical use is limited by significant toxicity, immunogenicity, and rapid clearance from circulation. Hence covalent modification of proteins contained in the virus coat with PEG has been explored to improve the biological and physicochemical properties of the viruses. Adeno-associated viruses (AAVs) are one of the least toxic vectors employed in the gene transfection due to their ability to produce long-term transgene expression. Hence most of the efforts were directed toward the PEGylation of AAV. This was achieved by the conjugation of PEG activated by cyanuric chloride (CC-PEG), tresyl chloride (TM-PEG), and succinimidyl succinate (SS-PEG) to the ε-amino groups of the lysine residues on the virus capsid proteins [166, 167]. The transduction efficiency of the viruses *in vitro* was not compromised up on PEGylation and the viruses were also protected from serum neutralization. *In vivo* studies showed that the PEGylated viruses increased the transduction efficiency in the lung and liver by a factor of 3 and 5 respectively [168, 169]. The plasma half-life of the virus was significantly improved upon PEGylation and was also dependent on the degree of modification. Thus AAV conjugated to 5 kDa PEG at degrees of modification of 90% and 100% showed half-life of 6.4 minutes and 22 minutes respectively, compared to 2 minutes for the unmodified virus in mice [170]. The systemic circulation of the virus was also prolonged by the use of higher molecular weight PEG even at lower degrees of modification [171, 172]. There was also reduction in the clearance rate of the PEGylated viruses in mice and nonhuman primates. PEGylation significantly reduced the virus-induced hepatotoxicity and cytotoxicity in mice and nonhuman primates [173, 174]. It has also been reported that conjugation of virus to PEG completely prevented the abnormalities associated with blood coagulation such as thrombocytopenia (reduced platelet counts) that are often encountered by administration of an unmodified virus [175]. There are several reports on the reduction in the immunogenicity of the virus upon PEGylation that was attributed to the PEG preventing uptake and processing of the virus by antigen-presenting cells [175].

The PEGylation of other viruses such as lentivirus [176], retrovirus [177], baculovirus [178], and virus-like particles influenza virosomes [179] has also been reported. In most cases, the *in vivo* transduction efficiencies were improved upon PEG modification. The area of virus PEGylation has been recently reviewed by Wongnan and Coyle [180].

2.3 LIMITATIONS OF PEG CONJUGATES

In spite of the potential advantages of PEG over other polymeric systems for drug delivery that resulted in the commercialization of several PEG conjugates, there are certain limitations to the use of PEG for human therapy. It was assumed in several animal models that PEG is nonimmunogenic by itself which may be due to the rapid renal clearance of the unconjugated polymer. However, strong anti-PEG immune response was observed with some of the PEGylated conjugates such as pegloticase (PEG-uricase) and PEG-ovalbumin [181]. Richter et al. reported the formation of antibodies for the conjugate of ovalbumin with mPEG in rabbits, although the antibody formation was strongly dependent on the degree of PEGylation of the protein [182]. A clinical study in patients treated for acute lymphoblastic leukemia with PEG-asparaginase showed the presence of anti-PEG that resulted in rapid renal clearance of the conjugate and limited its therapeutic efficacy. In patients with severe chronic gout treated with pegloticase, anti-PEG was detected in 89% of the patients resulting in loss of efficacy and increased risk of subsequent infusion reactions. High prevalence of anti-PEG was also observed in patients with hepatitis C. One interesting observation was the detection of anti-PEG in over 25% of healthy blood donors with no prior exposure to PEG conjugates, compared to the 0.2% occurrence observed 20 years ago by Richter and Akerblom [183]. This increase may be due to the improved detection limits of PEG antibodies over the years as well as to greater exposure to PEG and PEG-containing products in pharmaceuticals, cosmetics, and processed foods. Further studies are required to confirm if the observed anti-PEG are due to PEG itself or due to the complex environment of the conjugated proteins. All the currently approved PEGylated therapeutic products are based on mPEG. In a recent work, Sherman et al. immunized rabbits with mPEG and hydroxyl PEG conjugates of human serum albumin, human interferon-α, and porcine uricase and observed that the affinities of the antibodies raised against mPEG were significantly higher than those against hydroxyl PEG for all the three protein conjugates [184]. This study revealed that the anti-PEG with high affinity for the methoxy group was responsible for the reduction in the efficacy of mPEG conjugates and that the use of hydroxyl PEG instead of mPEG may overcome the undesirable therapeutic effects in clinical use.

Another area of concern is the lack of knowledge regarding the toxicity and excretion of PEG, and the fate of PEG and the PEGylated drug delivery systems at the cellular level after *in vivo* administration. The nonbiodegradable nature of PEG prefers the use of low molecular weight PEGs; however, those with molar mass below 400 Da were shown to be toxic in humans and the toxicity decreases with increase in the molar mass [181]. On the other hand, PEGs of higher molar mass may limit its renal clearance and complete excretion from the body. It appears that PEG of molar mass less than 2 kDa are easily excreted via urine, while those with higher molecular weight undergo slow clearance through the liver. The use of multiarm and branched biodegradable PEGs will overcome some of these uncertainties as they will form easily excretable low molecular weight forms after cleavage in the body.

There are several other drawbacks for PEG-based systems such as complement activation that leads to hypersensitivity reactions and the phenomenon called ABC effect (accelerated blood clearance) that affects its pharmacokinetic behavior [15]. Besides, PEG chains are susceptible to degradation under stress due to its ether structure and can result in the formation of hydroperoxides. The toxicity of the side products such as 1,4-dioxane, or the residual ethylene oxide and formaldehyde during the synthesis of PEG also necessitates the use of pharmaceutical grade PEG for biomedical applications. Another important disadvantage of PEG is their high intrinsic viscosity which prevents its use in high dose formulations [15].

Several other hydrophilic polymers have been explored as alternatives to PEG for use in therapeutics. The following sections discuss the conjugates based on one such class of polymer, polyglycerols.

2.4 POLYGLYCEROL-BASED CONJUGATES

In 1999, Sunder and coworkers reported the first controlled synthesis of HPG by the anionic ring-opening multibranching polymerization (ROMBP) of glycidol (Figure 2.6) [185]. Since then, several approaches have been developed for the synthesis of HPGs of controlled molecular weight and polydispersity index [12, 13]. HPGs are characterized by a highly flexible aliphatic polyether backbone, compact structure, and the presence of multiple hydroxyl groups, high water solubility, and excellent biocompatibility. It has been reported that HPGs in the molecular weight range, 4.2–670 kDa exhibits similar or even better biocompatibility profile than PEG by various analyses [14, 186, 187]. In fact, HPGs outperformed

PEG in some of the blood compatibility assays such as complement activation, platelet activation, red blood cell aggregation and morphology, plasma protein precipitation, and cytotoxicity. HPGs are well tolerated in mice up to a maximum injected dose of 1 g/kg and there was no molecular weight-dependent toxicities observed. It has also been demonstrated that the *in vivo* circulation half-life of HPGs could be tuned by changing the molecular weight. These characteristics are being currently explored in the biomedical field for the conjugation of bioactive molecules such as drugs, proteins, peptides, and carbohydrates. HPG is advantageous over PEG in this context since its multifunctional nature allows the covalent attachment of multiple bioactive molecules, imaging agents, and targeting moieties onto a single polymer molecule making it a novel drug delivery platform. The following sections outline the bioconjugates based on HPG.

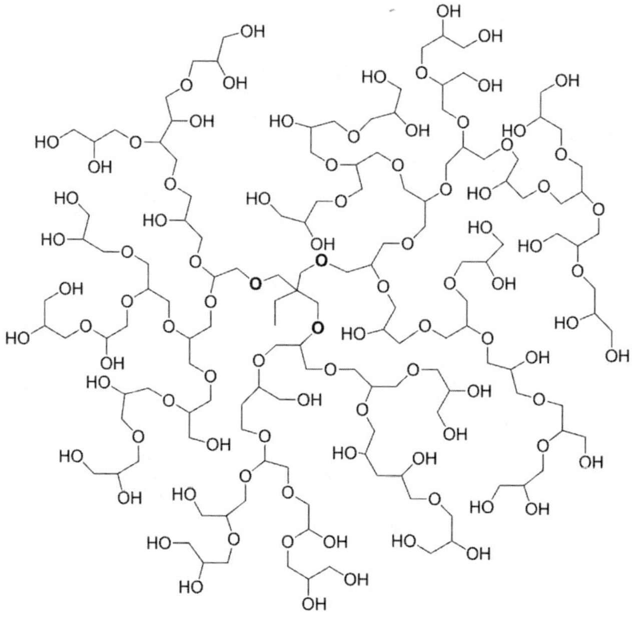

2.4.1 HPG-drug Conjugates

The first prodrug design using HPG was reported by Kannan and coworkers by the conjugation of ibuprofen to a 6 kDa HPG via an ester linkage [188]. A multiple functional system with 67% w/w of the drug conjugated to HPG was obtained. The HPG-drug conjugate did not release the drug up to 72 hours in methanol indicating the stability of the ester bond during this period. The intracellular delivery and localization of FITC-labeled conjugate was studied in A549 lung epithelial cells by fluorescence microscopy. The conjugate was found to rapidly deliver the drug inside the cells after distribution into the cytosol. The conjugate showed considerably faster anti-inflammatory activity than the free drug as investigated by prostaglandin inhibition.

Haag and coworkers reported the conjugation of maleimide-bearing Dox and MTX to a thiolated HPG that also incorporated a self-immolative p-aminobenzyloxy carbonyl spacer coupled to dipeptide Phe-Lys or the tripeptide D-Ala-Phe-Lys as the protease substrate. Both the prodrugs showed an effective release of the drug by the cleavage of the HPG-Phe-Lys-Dox and HPG-D-Ala-Phe-Lys-MTX by the enzyme cathepsin B that is overexpressed in several solid tumors [189]. The activity of the drugs was retained in the conjugates as revealed by the cytotoxicity studies in human tumor cell lines. In another report, the same group prepared a series of conjugates of HPG 10 kDa with Dox via the acid-sensitive hydrazone linker [15]. The conjugates were stable at pH 7.4 and showed only a marginal release of the drug, while acid-triggered release of the drug was observed at pH 5.0. The conjugates showed a 2- to 10-fold lower cytotoxicity than the free drugs in the human tumor cell lines AsPC1 LN (pancreatic carcinoma) and MDA-MB-231 LN (mammary carcinoma). Excellent antitumor effects were observed with the conjugates with complete tumor remission up to day 30 without significant changes in the body weight, even after

FIGURE 2.6 Structure of hyperbranched polyglycerol (HPG).

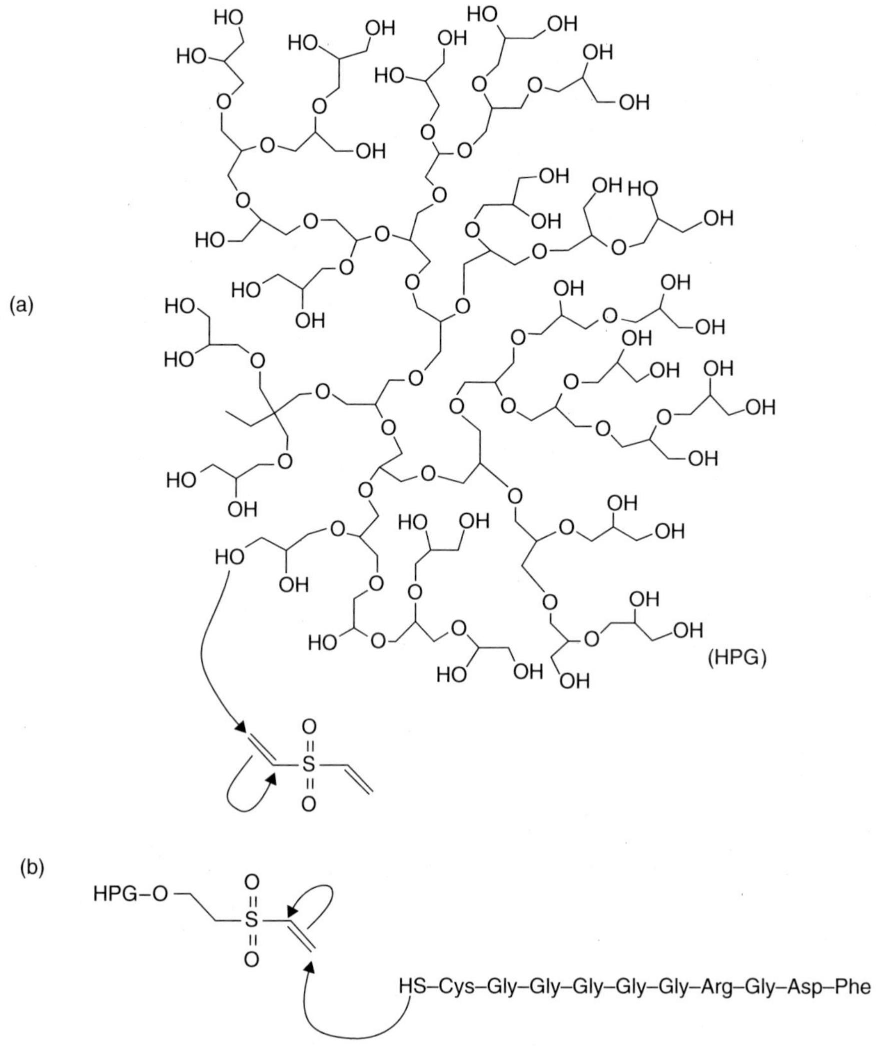

FIGURE 2.7   Synthesis of HPG-RGD conjugates. Reprinted with permission from Reference 190, Copyright® 2008, American Chemical Society.

administration of threefold higher dose than the maximum tolerated dose of free Dox.

2.4.2   HPG-peptide and Protein Conjugates

The conjugation of RGD peptide to a high molecular weight HPG was reported by Gyongyossy Issa and coworkers as a model for a new class of antithrombotics [190]. RGD is a tripeptide of arginine, glycine, and aspartic acid, which binds to platelet integrin receptor GPIIbIIIa thereby disrupting platelet-fibrinogen binding and platelet cross-linking during thrombus generation. However, its high IC$_{50}$ values and low *in vivo* residence times limit the use of RGD for clinical applications. In order to address these issues, RGD was conjugated to HPG via divinyl sulfone linker (Figure 2.7). A 2- to 3-fold increase in the platelet inhibitory activity of RGD was observed upon conjugation to the polymer and the

effectiveness of the conjugates depended on the molecular weight of the HPG and the number of RGD molecules per polymer. The multivalent inhibition of platelet aggregation by the conjugate decreased the IC$_{50}$ values of RGD in an inverse linear manner depending on the number of RGD per HPG molecule. The conjugates did not cause platelet activation by degranulation and did not increase fibrinogen binding to the resting platelets.

Recently Klok and coworkers reported squaric acid-mediated synthesis and biological activity of a library of polyglycerol-protein conjugates [191]. The study was directed toward the feasibility of squaric acid diethyl ester-mediated coupling as an amine-selective, hydroxyl-tolerant, and hydrolysis-insensitive approach for the synthesis of side chain-functionalized hydroxyl containing polymer–protein conjugates. A diverse library of BSA and lysozyme conjugates was prepared using different molecular weights and different architectures of polyglycerol to study their effect on the biological activity of the protein. No obvious structure–activity relationship on biological activity was observed for the low molecular weight polyglycerol–lysozyme conjugates. However, the conjugates prepared with higher molecular weight polyglycerols exhibited activities that were strongly dependent on the polymer architecture.

2.4.3 HPG Glycoconjugates

Multivalent carbohydrate interactions are believed to play important roles in many biological processes such as cell binding and biological recognition which make them as attractive therapeutic agents. However, the development of carbohydrate-based therapeutic agents was limited by the low binding affinity of monovalent carbohydrate ligands to proteins and receptors. The enhancement of binding affinities could be achieved by presenting the carbohydrates in a multivalent fashion by the use of linear and dendritic polymers. HPG, due to its highly flexible structure, can potentially change its conformation to make the maximum number of ligands available for binding without much entropic penalty and hence has been evaluated by several groups as a platform for the synthesis of multivalent glycoarchitectures.

Lectins are multivalent carbohydrate-binding proteins that have received significanace recently. Haag and workers reported the use of multivalent glycoarchitectures based on HPG for the inhibition of L- and P-selectins, a class of receptors that displays selective adhesion [192]. The glycoconjugates were obtained in high yields by the reaction of alkyne-functionalized HPGs with galactose-azide. The selectin inhibition activity of the glycoconjugates based on HPG-galactose and HPG-sulfated galactose was compared using surface plasmon resonance (SPR). All the HPG conjugates showed significant enhancement of selectin inhibition due to multivalent effect. Binding affinities in the nanomolar range were observed for L-selectin with both the HPG conjugates and the IC_{50} values higher for the HPG-sulfated galactose conjugate when compared to HPG-galactose. However, the binding to E-selectin was hindered by the sulfated glycoconjugate.

Our group reported the synthesis of multivalent conjugates of mannose (Figure 2.8) with HPGs of different molecular weights (100–500 kDa), sizes, and mannose densities (22–303 per HPG molecule) [193]. The HPG-mannose conjugates inhibited concanavalin A (Con A, a lectin that binds mannose-containing molecules) induced hemagglutination of fresh human red blood cells. The number of mannose units on the polymer and the size of the conjugates were found to affect the binding interaction between Con A and the HPG-mannose conjugates. The relative potency of the conjugates increased with increase in mannose density and the size of

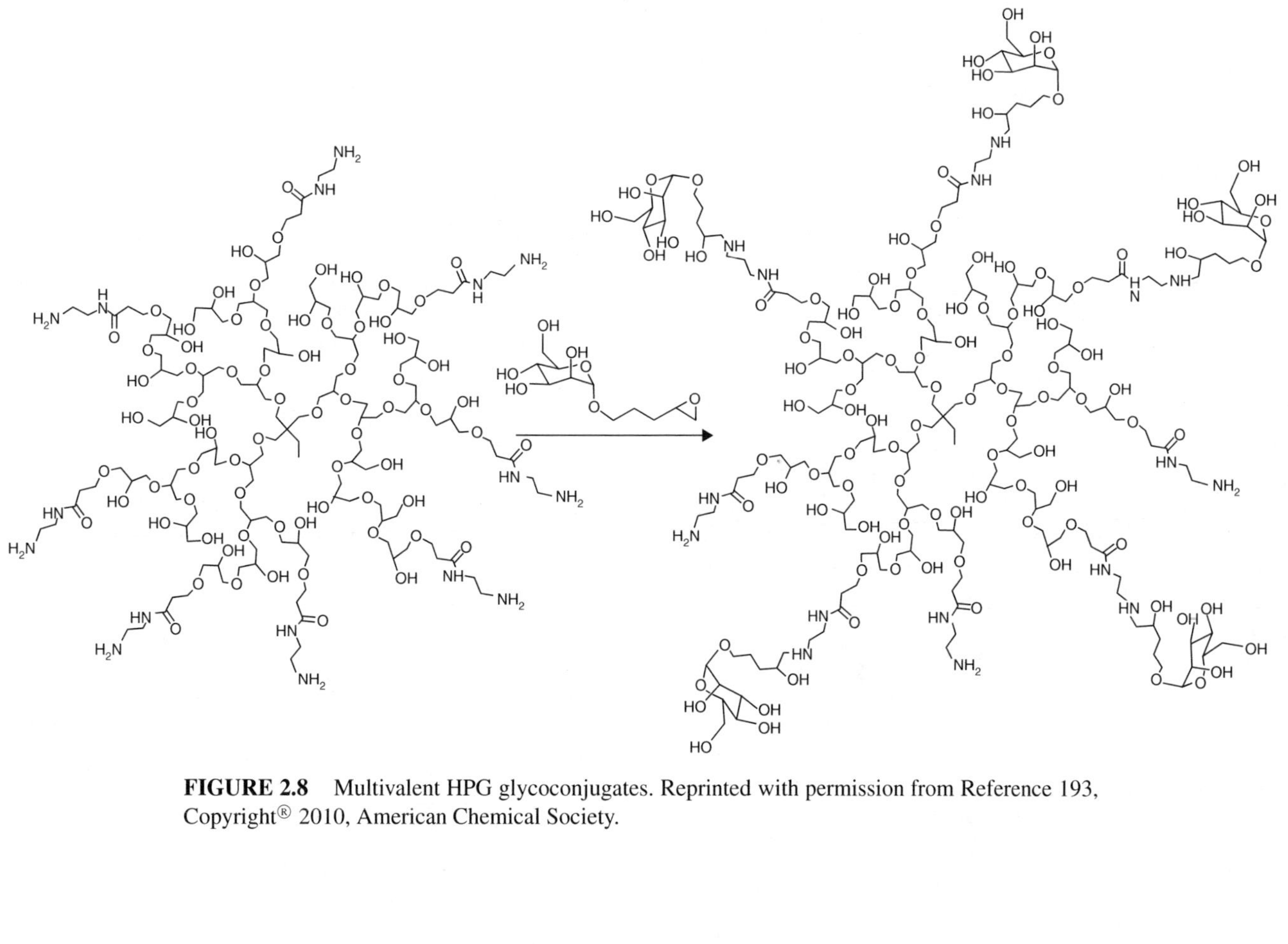

FIGURE 2.8   Multivalent HPG glycoconjugates. Reprinted with permission from Reference 193, Copyright® 2010, American Chemical Society.

the conjugates. The high molecular weight HPG-mannose conjugate showed a 40,000-fold enhancement in the potency relative to the methyl α-D-mannopyranoside and an increase in the relative activity per sugar unit. Isothermal titration calorimetry (ITC) experiments revealed a 50-fold increase in the binding affinity for Con A relative to methyl α-D-mannopyranoside suggesting multivalent effect of the mannose presented on the HPG scaffold. This enhanced binding was attributed to the favorable entropic contributions due to multiple bridging of Con A to high molecular weight HPG-mannose conjugates along with single interactions. Some of the conjugates also exhibited positive cooperative binding to Con A unlike methyl α-D-mannopyranoside or other small multivalent glycoclusters. The multivalent effect of HPG-mannose conjugates was also demonstrated by a later work by Papp et al. using SPR [194].

A new class of multivalent glycoarchitecture was developed by Haag and coworkers as powerful inhibitors of influenza A virus activity [195]. In this work, sialic acid (SA) was conjugated to HPG nanogels of different particle sizes (1–100 nm). The polymer particle size was varied in order to match the size of the influenza A virus that are typically 100 nm in diameter. The inhibitory activity of the HPG-SA conjugates showed dramatic increase with increase in the particle size. At comparable sugar concentrations, the conjugate with a particle size of 50 nm was 7000 times more effective towards inhibition than that with a size of 3 nm. Optimized larger-sized conjugates showed inhibition of viral infection up to 80%. This study emphasized the importance of matching the particle sizes and ligand densities to improve the multivalent effect necessary for biological interactions.

2.4.4   HPG-Red Blood Cell Conjugates for Antigen Protection

Rossi et al. reported the first example of covalent attachment of multifunctional polymers to red blood cell (RBC) surface using HPG. HPGs were first functionalized with succcinimidyl succinate (SS) groups that were further reacted with the primary amine groups present on RBC surface proteins to form stable covalent linkages [196, 197]. The SS-HPG concentration and the polymer molecular weight were proportional to cell surface coverage and the grafting density was also increased with increase in the number of SS groups on the HPG and the grafting time. The HPG-RBC conjugates exhibited excellent blood compatibility and cell viability *in vitro* tests such as cell lysis, hemoglobin oxidation, lipid peroxidation, serum stability, and osmotic fragility. The efficient cell-surface grafting of HPG was also studied by the interaction of a highly reactive fluorescently labeled monoclonal anti-D antibody with the cell surface antigens. Flow cytometry analysis revealed that the antigenicity of the RBC was significantly reduced upon HPG grafting [196]. It was also demonstrated using tritium labeled HPG-RBCs that a large portion (50%) of the HPG-grafted RBCs were functional and remained in circulation for more than 50 days. The HPG-RBCs did not undergo immediate complement-mediated lysis while in circulation as evidenced by the low radioactivity in the plasma on day 1 [198]. The HPG-grafted RBCs were also taken up by the reticuloendothelial system (RES) macrophages of the liver and spleen similar to the normal RBC clearance mechanism. There was no accumulation of the modified RBCs in other organs and no abnormalities were observed in necropsy [198]. All these results suggest the potential application of the HPG-grafted RBCs in drug delivery and in the development of antigen camouflaged RBCs for transfusion.

There is not much information available in the literature on the effect of topology of the grafted polymer on cell surface antigens camouflage and protection of cell surface proteins. This was the objective of a recent work in our group where HPG and a linear PEG with similar hydrodynamic radius were conjugated to RBC surfaces and their effect on cell surface antigens, protection of the self-marker protein CD47, and complement mediated lysis was investigated [199]. The HPG-grafted RBCs showed higher electrophoretic mobility than the PEG-grafted RBCs which indicates clear difference in the structure of the polymer exclusion layer formed on the cell surface. Both the polymer-grafted RBCs triggered complement activation at higher concentrations, with HPG-grafted RBCs showing higher activation than the PEG-grafted ones, possibly due to the large number of hydroxyl groups in HPG. The HPG-grafted RBCs exhibited significantly higher levels of CD47 accessibility than the PEG-grafted RBCs, while the latter provided better shielding and protection to ABO and other minor antigens from antibody recognition. This work further highlighted the potential of the polymer-grafted RBCs in transfusion medicine as universal red blood donor cells.

Although the use of polyglycerols in the field of macromolecular therapeutics is still in its infancy compared to PEG and most of the efforts were for the development of prodrugs, the polymer has the potential for the synthesis of other bioconjugates as well, and significant efforts are currently undergoing in this direction. In this context, HPG may have additional advantages over PEG due to its better biocompatibility profiles and the large number of hydroxyl groups present in the dendritic structure allows the incorporation of multiple bioactive molecules onto a single polymer molecule.

2.5   CONCLUSIONS

Polyethylene glycol, due to its hydrophilicity, decreased interaction with blood components, and high biocompatibility has been the polymer of choice for bioconjugation strategies in the field of polymer therapeutics for the last three decades. Since the first report of protein conjugation to

PEG in 1977, PEGylation has shown tremendous progress as demonstrated by the clinical approval of several PEGylated therapeutics. This includes PEG-based proteins, antibodies, and most recently an aptamer. In addition, several PEG-based drugs and coagulation proteins are under clinical investigation. Though PEGylation was initially designed for anticancer therapy, it has also shown potential in the treatment of other diseases such as hepatitis, rheumatoid arthritis, age-related macular degeneration, and hemophilia. Another significant development has been the design of branched and multiarm PEG for bioconjugation as another step toward achieving improved therapeutic effects. However, recent scientific results also gave insights into some of the limitations of PEG, most notably the generation of anti-PEG antibodies in humans and its limited functionality. A wide range of other synthetic polymers are currently being studied as alternatives to PEG, of which the polyglycerols deserve special mention due to their excellent biocompatibility profiles and multifunctional nature. The preliminary results obtained with some of the polyglycerol-based bioconjugates show considerable promise for this polymer as a potential alternative to PEG.

ACKNOWLEDGMENTS

This research was funded by the Canadian Institutes of Health Research (CIHR), Natural Sciences and Engineering Research Council (NSERC) of Canada, Canadian Blood Services (CBS), and Health Canada. The infrastructure facility is supported by the Canada foundation for Innovation (CFI) and the Michael Smith Foundation for Health Research (MSFHR). J.N. Kizhakkedathu is a recipient of MSFHR career investigator scholar award.

REFERENCES

1. Duncan R. *Nat Rev Drug Discov* 2003;2:347–360.
2. Haag R, Kratz F. *Angew Chem Int Ed* 2006;45:1198–1215.
3. Liu SR, Maheswari RKL, Kiick KL. *Macromolecules* 2009; 42:3–13.
4. Duncan R. *Curr Opin Biotechnol* 2011;22:492–501.
5. Duncan RMJ, Vincent MJ. *Adv Drug Deliv Rev, press* 2013;65:60–70.
6. Veronese FM. *Biomaterials* 2001;22:405–417.
7. Greenwald RB, Choe YH, McGuire J, Conover CD. *Adv Drug Deliv Rev* 2003;55:217–250.
8. Veronese FM, Pasut G. *Drug Discov Today* 2005;10:1451–1458.
9. Milton Harris J, Chess RB. *Nat Rev Drug Discov* 2003;2:214–221.
10. Knop K, Hoogenboom R, Fischer D, Schubert U. *Angew Chem Int Ed* 2010;49:6288–6308.
11. Banerjee SS, Aher N, Patil R, Khandare J. *J Drug Deliv* 2012;2012:1–17.
12. Wilms W, Striba S, Frey H. *Acc Chem Res* 2010;43:129–141.
13. Calderon M, Quadir MA, Sharma SK, Haag R. *Adv Mater* 2010;22:190–218.
14. Kainthan RK, Janzen J, Levin E, Devine DV, Brooks DE. *Biomacromolecules* 2006;7:703–709.
15. Imran ul-Haq MI, Lai BFL, Chapanian R, Kizhakkedathu JN. *Biomaterials* 2012;33:9135–9147.
16. Hooftman G, Herman S, Schacht E. *J Bioact Compat Polymers* 1996;11:135–159.
17. Vandegriff KD, McCarthy M, Rohlfs RJ, Winslow RM. *Biophys Chem* 1997;69:23–30.
18. Abuchowski A, McCoy TJR, Palczuk NC, Van Es T, Davis FF. *J Biol Chem* 1977;252:3582–3586.
19. David S, Abuchowski A, Park YK, Davis FF. *Clin Exp Immunol* 1981;46:649–652.
20. Nucci ML, Shorr R, Abuchowski A. *Adv Drug Deliv Rev* 1991;6:133–151.
21. Hershfield. MS. *Semin Hematol* 1998;35:291–298.
22. Graham ML. *Adv Drug Deliv Rev* 2003;55:1293–1302.
23. Fu CH, Sakamoto KM. *Exp Opin Pharmacother* 2007;8:1977–84.
24. Anne Dinndorf PA, Gootenberg J, Cohen MH, Keegan P, Pazdur R. *Oncologist* 2007;12:991–998.
25. Wang Y, Youngster S, Grace M, Bausch J, Bordens R, Wyss DF. *Adv Drug Deliv Rev* 2002;54:547–570.
26. Reddy KR, Modi MW, Pedder S. *Adv Drug Deliv Rev* 2002;54:571–586.
27. Foster GR. *Aliment Pharmacol Ther* 2004;20:825–830.
28. Bukowski R, Ernstoff MC, Gore ME, Nemunaitis JJ, Amato R, Gupta SK, Tendler CL. *J Clin Oncol* 2002;20:3841–3849.
29. Modi MW, Fulton JS, Buchann DK, Wrigh TL, Moore DJ. *Hepatol* 2000;32:371A.
30. EMA EPAR. Available at http://www.ema.europa.eu/ema/index.jsp?curl=/pages/medicines/landing/epar_search.jsp&mid=WC0b01ac058001d124. Last accessed: 23 September 2013.
31. Holes FA, O'Shaughessy JA, Vukelja S, Jones SE, Shogan J, Savin M, Glaspy J, Moore M, Meza L, Wiznitzen I, Neumann TA, Hill LR, Liang BD. *J Clin Oncol* 2002;20:727–731.
32. Molineux G. *Curr Pharm Des* 2004;10:1235–1244.
33. FDA. Available at http://www.accessdata.fda.gov/scripts/cder/drugsatfda/. Last accessed: 20 September 2013.
34. Locatelli F, Mann JF, Aldigier JC, Guajardo D Sanz, Schmidt R, Van Vlem B, Sulowicz W, Dougherty FC, Beyer U. *Clin Nephrol* 2010;73:94–103.
35. Fliser D, Kleophas W, Dellana F, Winkler RE, Backs W, Kraatz U, Fassbinder W, Wizemann V, Strack G. *Curr Med Res Opin* 2010;26:1083–1089.
36. Cano F, Alarcon C, Azocar M, Lizama C, Lillo AM, Delucchi A, Gonzalez M, Arellano P, Delgado I, Droguett MT. *Pediatr Nephrol* 2011;26:1303–1310.
37. Sundy JS, Ganson NJ, Kelly SJ, Scarlett EL, Rehrig CD, Huang W, Hershfield MS. *Arthritis Rheum* 2007;56:1021–1028.

38. Biggers K, Scheinfild N. *Curr Opin Investig Drugs* 2008; 9:422–429.

39. Sundy JS, Baraf HSB, Yood RA, Edwards NL, Gutierrez-Urena SR, Treadwell EL, Vazquez-Mellado J, White WB, Lipsky PE, Horowitz Z, Huang W, Maroli AN, Waltrip RW, Hamburger SA, Becker MA. *J Am Med Assoc* 2011;306:711–720.

40. George RL, Sundy JS. *Drugs Today* 2012;48:441–449.

41. Katre NV, Knauff MJ, Laird WJ. *Proc Natl Acad Sci USA* 1987;84:1487–1491.

42. Yang JC, Topalian SL, Schwartzentruber DJ, Parkinson DR, Marincola FM, Weber JS, Seipp CA, White DE, Rosenberg SA. *Cancer* 1995;76:687–694.

43. Wiernik PH, Dutcher JP, Todd M, Caliendo G, Benson L. *Am J Hematol* 1994;47:41–44.

44. Kaplan B, Moy RL. *Dermatol Surg* 2000;26:1037–1040.

45. Trainer PJ, Drake WM, Katznelson L, Freda PU, Herman-Bonert V, van der Lely AJ, Dimaraki EV, Stewart PM, Friend KE, Vance ML, Besser GM, Scarlett JA, Thorner MO, Parkinson C, Klibanski A, Powell JS, Barkan AL, Sheppard MC, Malsonado M, Rose DR, Clemmons DR, Johannsson G, Bengtsson BA, Stavrou S, Kleinberg DL, Cook DM, Phillips LS, Bidlingmaier M, Strasburger CJ, Hackett S, Zib K, Bennett WF, Davis RJ. *N Engl J Med* 2000;342:1171–1177.

46. Schreiber I, Buchfelder M, Droste M, Forssman K, Mann K, Saller B, Strasburger CJ. *Eur J Endocr* 2007;156:75–82.

47. Van der Lely AJ, Biller BMK, Brue T, Buchfelder M, Ghigo E, Gomez R, Hey-Hadavi J, Lundgren F, Rajicic N, Strasburger CJ, Webb SM, Koltowska-Haggstrom M. *J Clin Endocrinol Metab* 2012;97:1589–1597.

48. Collen D, Sinnarve P, Demarsin, e, Moreau H, Maeyer MD, Jespers L, Laroche Y, Van de Werf F. *Circulation* 2000;102:1766–1772.

49. Vanwetswinkel S, Plaisance S, Zhi-yong Z, Vanlinthout I, Brepoels K, Lasters I, Collen D, Jespers L. *Blood* 2000;95:936–942.

50. Moons L, Valinthout I, Roelants I, Moreadith R, Collen D, Rapold HJ. *Toxicol Pathol* 2001;29:285–291.

51. Rohlfs RJ, Bruner E, Chiu A, Goonzales A, Gonzales ML, Magde MD, Vandegriff KD, Winslow RM. *J Biol Chem* 1998;273:12128–12134.

52. Winslow RM, Gonzales A, Gonzales ML, Magde MD, cCarthy MD, Rohlfs RJ, Vandegriff KD. *J Appl Physiol* 1998;85:993–1003.

53. Vandegriff KD, McCarthy M, Rohlfs RJ, Winslow RM. *Biophys Chem* 1997;69:23–30.

54. Manjula BN, Tsai A, Upadhya R, Perumalsay K, Smith PK, Malavalli A, Vandegriff K, Winslow RM, Intaglietta M, Prabhakaran M, Friedmand JM, Acharya AS. *Bioconjug Chem* 2003;14:464–472.

55. Manjula BN, Tsai AG, Intaglietta M, Tsai CH, Ho C, Smith PK, Perumalsaruy K, Kanika ND, Friedman JM, Acharya SA. *Protein J* 2005;24:133–146.

56. Acharya SA, Friedman JM, Manjula BN, Intaglietta M, Tsai AG, Winslow RM, Malavalli A, Vandegriff K, Smith PK. *Artif Cells Blood Substit Immobil Biotechnol* 2005;33:239–255.

57. Hu T, Prabhakaran M, Acharya SA, Manjula BN. *Biochem J* 2005;392:555–564.

58. Young MA, Malavalli A, Winslow N, Vandegriff KD, Winslow RM. *Transl Res* 2007;149:333–342.

59. Blanchette VS, Manco-Johnson M, Santagostino E, Ljung R. *Haemophilia* 2004;10:97–104.

60. Stennicke HR, Ostergaard H, Bayer RJ, Kalo MS, Kinealy K, Holm PK, Sorensen BB, Zopf D, Bjorn SE. *Thromb Haemost* 2008;100:920–928.

61. Ghosh S, Sen P, Pendurthi UR, Rao LV. *J Thromb Haemost* 2008;6:1525–1533.

62. Holmberg H, Elm T, Karpf D, Tranholm M, Bjorn SE, Stennicke H, Ezban M. *J Thromb Haemost* 2011;9:1070–1072.

63. Moss J, Rosholm A, Lauren A. *J Thromb Haemost* 2011; 9:1368–1374.

64. Mei B, Pan C, Jiang H, Tjandra H, Strauss J, Chen J, Gu J, Subramanyam B, Fournel MA, Pierce GF, Murphy JE. *Blood* 2010;116:270–279.

65. Dietrich B, Spatzeneger M, Stidl R, Wolfsegger M, Ehrlich HJ, Scheiflinger F, Schwartz HP, Muchitsch E. *Blood* 2011;118:21.

66. Ostergaard H, Bjelke JR, Hansen L, Petersen LC, Pedersen AA, Elm T, Møller F, Hermit MB, Holm PK, Krogh TN, et.al. *Blood* 2011;118:2333–2341.

67. Negrier C, Knobe K, Tiede A, Giangrande P, Moss J. *Blood* 2012;118:2695–2701.

68. Esslinger HU, Hass S, Maurer R, Lassman A, Dubbers K, Muller-Peltzer H. *Thromb Haemost* 1997;77:911–919.

69. Poschel KA, Bucha E, Esslinger HU, Ulbricht. K, Norter-Sheuser P, Stein G, Nowak G. *Kidney Int* 2004;65:666–674.

70. Lee KC, Tak KK, Park MO, Lee JT, Woo BH, Yoo SD, Lee HS, Deluca PP. *Pharm Dev Technol* 1999;4:269–275.

71. Yoo SD, Jun H, Shim BS, Lee HS, Park MO, Deluca PD, Lee KC. *Chem Pharm Bull* 2000;48:1921–1924.

72. Ryan SM, Wang X, Mantovani G, Sayers CT, Haddleton DM, Brayden DJ. *J Control Release* 2009;135:51–59.

73. Ryan SM, Frias JM, Wang X, Sayers CT, Haddleton DM, Brayden DJ. *J Control Release* 2010;149:126–132.

74. Maeda M, Izuno Y, Kawasaki K, Kaneda Y, Mu Y, Tsutsumi Y, Lem KW, Mayumi T. *Biochem Biophys Res Commun* 1997;241:595–598.

75. Youn YS, Kwon MJ, Na DH, Chae SY, Lee S, Lee KC. *J Control Release* 2008;125:68–75.

76. Kim TH, Jiang HH, Lim SM, Youn YS, Choi KY, Lee S, Chen X, Byun Y, Lee KC. *Bioconjug Chem* 2012;23:2214–2220.

77. Gao M, Jin Y, Tong Y, Tian H, Gao X, Yao W. *J Pharmacol* 2012;64:1646–1653.

78. Hinds KD, Kim SW. *Adv Drug Deliv Rev* 2002;54:505–530.

79. Calceti P, Salmaso S, Walker G, Bernkop-Schnurch A. *Eur J Pharm Sci* 2004;22:315–323.

80. Shechter Y, Mironchik M., Rubinraut S, Tsubery H, Sasson K, Marcus Y, Fridkin M. *Eur J Pharm Biopharm* 2008;70: 19–28.

81. Lee LS, Conover C, Shi C, Whitlow M, Filpula D. *Bioconjug Chem* 1999;10:973–981.

82. Hurwitz E, Klapper LN, Wilchek M, Yarden Y, Sela M. *Cancer Immunol Immunother* 2000;49:226–234.

83. Koumenis IL, Shahrokh Z, Leong S, Hsei V, Deforge L, Zapata G. *Int J Pharm* 2000;198:83–95.

84. Cheng T-L, Chen BM, Chern JW, Wu MF. *Bioconjug Chem* 2000;11,258–266.

85. Horton S, Walsh C, Emery P. *Expert Opin Biol Ther* 2012;12:235–249.

86. Mease PJ. *Rheumatology* 2011;50:261–270.

87. Vavricka SR Schoepfer AM, Bansky G, Binek J, Felley C, Geyer M, Manz M, Rogler G, de Saussure P, Sauter B, Scharl M, Seibold F, Straumann A, Michetti P; Swiss IBDnet. *Inflamm Bowel Dis* 2011;17:1530–1539.

88. Hortin GL, Huang ST. *Blood* 1996;88(S1):181.

89. Armstrong JK, Heiselman HJ, Fisher TC. *Am J Hematol* 1997;56:26–28.

90. Scott MD, Murad KL, Koumpouras F, Talbot M, Eaton JW. *Proc Natl Acad Sci USA* 1997;94:7566–7571.

91. Murad KL, Mahany KL, Kuypers FA, Brunara C, Eaton JW, Scott MD. *Blood* 1999;93:2121–2127.

92. Nacharaju P, Boctor FN, Manjula BN, Acharya SA. *Transfusion* 2005;45:374–383.

93. Murad KL, Gosselin EJ, Eaton JW, Scott MD. *Blood* 1999;94:2135–2141.

94. Ricordi C, Strom TB. *Nat Rev Immunol* 2004;4:259–268.

95. Shapiro AMJ, Nanji SA, Lakey JR. *Immunol Rev* 2003; 196:219–236.

96. Robertson RP. *N Engl J Med* 2004;350:694–705.

97. Lee DY, Byun Y. *Biotechnol Bioprocess Eng* 2010;15:76–85.

98. Panza JL, Wagner WR, Rilo HLR, Rao RH, Beckman EJ, Russell AJ. *Biomaterials* 2000;21:1155–1164.

99. Xie D, Smyth CA, Eckstein C, bilbao G, Mays J, Eckhoff DE, Contreras JL. *Biomaterials* 2005;26:403–412.

100. Lee DY, Yang K, Lee S, Chae SY, Kim KW, Lee MK, Han DJ, Byun Y. *J Biomed Mater Res* 2002;62:372–377.

101. Lee DY, Nam JH, Byun Y. *Biomaterials* 2007;28:1957–1966.

102. Lee DY, Park SJ, Nam JH, Byun Y. *Tissue Eng* 2006;12:615–623.

103. Lee DY, Park SJ, Lee S, Nam JH, Byun Y. *Tissue Eng* 2007;13:2133–2141.

104. Zhao X, Pan F, Holt CM, Lewis AL, Lu. JR. *Exp Opin Drug Deliv* 2009;6:673–686.

105. Irenson CR, Kelland LR. *Mol Cancer Ther* 2006;5:2957–2962.

106. Kawaguchi T, Asakawa H, Tashiro Y, Juni K, Sueishi T. *Biol Pharm Bull* 1995;18:474–476.

107. Bonora GM, Ivanoa E, Zarytova V, Burcovich B, Veronese FM. *Bioconjug Chem* 1997;8:793–797.

108. Jaschke A, Furste JP, Nordhoff E, Hillenkamp F, Cech D, Erdmann VA. *Nucleic Acids Res* 1994;22:4810–4817.

109. Burcovich B, Veronese FM, Zarytova V, Bonora GM. *Nucleos Nucleot* 1998;17:1567–1570.

110. Vorobjev PE, Zarytova VF, Bonora GM. *Nucleos Nucleot* 1999;18:2745–2750.

111. Rapozzi V, Cogoi S, Spessotto P, Risso A, Bonora GM, Quadrifoglio F, Xodo LE. *Biochemistry* 2002;41, 502–510.

112. Ng EWM, Shima DT, Calias P, Emmett T, Cunningham Jr, Guyer DR, Adamis AP. *Nat Rev Drug Discov* 2006;5:123–132.

113. Gragoudas ES, Adamis AP, Cunningham ET Jr, Feinsod M, Guyer DR; VEGF Inhibition Study in Ocular Neovascularization Clinical Trial Group. *N Engl J Med* 2004;351:2805–2816.

114. VEGF Inhibition Study in Ocular Neovascularization (V.I.S.I.O.N.) Clinical Trial Group; Chakravarthy U, Adamis AP, Cunningham ET Jr, Goldbaum M, Guyer DR, Katz B, Patel M. *Ophthalmology* 2006;113:1508–1521.

115. VEGF Inhibition Study in Ocular Neovascularization (V.I.S.I.O.N.) Clinical Trial Group; D'Amico DJ, Masonson HN, Patel M, Adamis AP, Cunningham ET Jr, Guyer DR, Katz B. *Ophthalmology* 2006;113:992–1001.

116. Singerman LJ, Masonson H, Patel M, Adamis AP, Buggage R, Cunningham E, Goldbaum M, Katz B, Guyer D. *Br J Opthalmol* 2008;92:1606–1611.

117. Tuschl T. *Chembiochem* 2001;2:239–245.

118. Whitehead KA, Langer R, Anderson DJ. *Nat Rev Drug Discov* 2009;8:129–138.

119. Jeong JH, Mok H, Oh YK, Park TG. *Bioconjug Chem* 2009;20:5–14.

120. Kim SH, Jeong JH, Lee SH, Kim SW, Park TG. *J Control Release* 2006;116:123–129.

121. Kim SH, Jeong JH, Lee SH, Kim SW, Park TG. *J Control Release* 2008;129:107–116.

122. Lee SH, Kim SH, Park TG. *Biochem Biophys Res Commun* 2007;357:511–516.

123. Kim SH, Jeong JH, Lee SH, Kim SW, Park TG. *Bioconjug Chem* 2008;19:2156–2162.

124. Choi SW, Lee SH, Mok H, Park TG. *Biotechnol Prog* 2010;26:57–63.

125. Oishi M, Nagasaki Y, Itaka K, Nishiyama N, Kataoka K. *J Am Chem Soc* 2005;127:1624–1625.

126. Oishi M, Nagasaki Y, Nishiyama N, Itaka K, Takagi M, Shimamoto A, Furuichi Y, Kataoka K. *ChemMed Chem* 2007;2:1290–1297.

127. Heredia KL, Nguyen TH, Chang CW, Bulmus V, Davis TP, Maynard HD. *Chem Commun* 2008;28:3245–3247.

128. Gunasekaran K, Nguyen TH, Maynard HD, Davis TP, Bulmus V. *Macromol Rapid Commun* 2011;32:654–659.

129. Kow SC, McCarroll J, valade D, Boyer C, Dwarte T, Davis TP, Kavallaris M, Bulmus V. *Biomacromolecules* 2011;12:4301–4310.

130. Rautio R, Kumpulainen H, Himbach T, Oliyai R, Oh D, Jarvinen T, Savolainen J. *Nat Rev Drug Discov* 2008;7:255–270.

131. Huttunen KM, Raunio H, Rautio J. *Pharmacol Rev* 2011;63:750–771.

132. Bundgaard H. *Drug Future* 1991;16:443–458.

133. Sinhababu AK, Thakker DR. *Adv Drug Deliv Rev* 1996;19:241–273.

134. Greenwald RB, Pendri A, Bolikal D, Gilbert CW. *Bioorg Med Chem Lett* 1994;4:2465–2470.

135. Greenwald RB, Gilbert CW, Pendri A, Conover CD, Xia J, Martinez A. *J Med Chem* 1996;39:424–431.

136. Noguchi Y, Wun J, Duncan R, Strohalm R, Ulbrich K, Akaike T, Maeda H. *Jpn J Cancer Res* 1998;89:307–614.

137. Greenwald RB, Pendri A, Conover C, Gilbert C, Yang R, Xia J. *J Med Chem* 1996;39:1938–1940.

138. Conover CD, Pendri A, Lee C, Gilbert CW, Shum KL, Greenwald RB. *Anticancer Res* 1997;17:3361–3368.

139. Zhao H, Lee C, Sai P, Choe YH, Boro M, Pendri A, Guan S, Greenwald RB. *J Org Chem* 2000;65:4601–4606.

140. Greenwald RB, Conover CD, Pendri A, Choe YH, Martinez A, Wu D, Guan S, Yao Z, Shum KL. *J Control Release* 1999;61:281–294.

141. Pendri A, Conover CD, Greenwald RB. *Anticancer Drug Des* 1998;13:387–395.

142. Greenwald RB, Pendri A, Conover CD, Lee C, Choe YH, Gilbert C, Martinez A, Xia J, Wu D, Hsue M. *Bioorg Med Chem* 1998;6:551–562.

143. Bernhard SA, Berger A, Carter JH, Katchalski E, Sela M, Shalitin Y. *J Am Chem Soc* 1962;84:2421–2434.

144. Tadayoni BM, Friden PM, Walus LR, Musso GF. *Bioconjug Chem* 1993;4:139–145.

145. Gogate US, Repta AJ, Alexander J. *Int J Pharmaceut* 1987;40:235–248.

146. Fife TH, DeMark BR. *J Am Chem Soc* 1976;98:6978–6982.

147. Saari WS, Schwering JE, Lyle PA, Smith SJ, Englehardt EL. *J Med Chem* 1990;33:97–101.

148. Conover CD, Greenwald RB, Pendri A, Gilbert CW, Shum KL. *Cancer Chemother Pharmacol* 1998;42:407–414.

149. Dosio F, Arpicco S, Brusa P, Stella B, Cattel L. *J Control Release* 2001;76:107–117.

150. Choe YH, Conover CD, Wu D, Royzen M, Greenwald RB. *J Control Release* 2002;79:41–53.

151. Choe YH, Conover CD, Wu D, Royzen M, Gervacio Y, Borowski V, Mehlig M, Greenwald RB. *J Control Release* 2002;79:55–70.

152. Scott LC, Yao JC, Benson AB, Thomas AL, Falk S, Mena RR, Picus J, Wright J, Mulcahy MF, Ajani JA, Evans TR. *Cancer Chemother Pharmacol* 2009;63:363–370.

153. Antonian L, Burton K, Goodin R, Eldon MA. *Eur J Cancer* 2007;5(Suppl):115.

154. http://ir.nektar.com.

155. Zhao H, Rubio B, Sapra P, Wu D, Reddy P, Sai P, Martinez A, Gao Y, Lozanguiez Y, Longley C, Greenberger LM, Horak ID. *Bioconjug Chem* 2008;19:849–859.

156. Sapra P, Zhao H, Mehlig M, Malaby J, Kraft P, Longley C, Greenberger LM, Horak ID. *Clinical Cancer Res* 2008;14:1888–1896.

157. Kurzock R, Goel S, Wheler J, Hong D, Fu S, Rezai K, Morgan-Linnell SK, Urien S, Manis S, Chaudhary I, Ghalib MH, Buchbinder A, Lokiec F, Mulcahy M. *Cancer* 2012;118(24):6144–6151.

158. Rodrigues PCA, Beyer U, Schumacher P, Roth T, Fiebig HH, Unger C, Messori L, Orioli P, Paper EH, Mulhaupt R, Kraft F. *Bioorg Med Chem* 1999;7:2517–2524.

159. Veronese FM, Schiavon O, Pascut G, Mendichi R, Andersson L, Tsirk A, Ford J, Wu G, Kneller S, Davies J, Duncan R. *Bioconjug Chem* 2005;16:775–784.

160. Polyak D, Ryppa C, Eldar-Boock A, Ofek P, Many A, Licha K, Kratz F, Satchi-Fainaro R. *Polym Adv Tech* 2011;22:103–113.

161. Riebeseel K, Biedermann E, Loser R, Breiter N, Hanselmann R, Mulhaupt R, Unger C, Kratz F. *Bioconjug Chem* 2002;13:773–785.

162. Kohler N, Sun C, Fichtenholtz A, Gunn J, Fang C, Zhang M. *Small* 2006;2:785–792.

163. Funk D, Schrenk H, Frei E. *J Drug Target* 2011;19:434–445.

164. Zacchigna M, Cateni F, Di LG, Voinovich D, Perissutti B, Drioli S, Bonora GM. *J Drug Deliv Sci Tech* 2008;18:155–159.

165. Liu X, Quan L, Tian J, Laquer FC, Ciborowski P, Wang D. *Biomacromolecules* 2010;11:2621–2628.

166. Le HT, Yu QC, Wilson JM, Croyle MA. *J Control Release* 2005;108:161–177.

167. Lee GK, Maheshri N, Kaspar B, Schaffer DV. *Biotechnol Bioeng* 2005;92:24–34.

168. Croyle MA, Chirmule N, Zhang Y. *J Virol* 2001;75:4792–4801.

169. Croyle MA, Chirmule N, Zhang Y, Wilson JM. *Hum Gene Ther* 2002;13:1887–1900.

170. Gao JQ, Eto Y, Yoshioka Y, Sekiguchi F, Kurachi S, Morishige T, Yao X, Watanabe H, Asavatanabodee R, Sakurai F, Mizuguchi H, Okada Y, Mukai Y, Tsutsumi Y, Mayumi T, Okada N, Nakagawa S. *J Control Release* 2007;122:102–110.

171. Yao X, Yoshioka Y, Morishige T, Eto Y, Watanabe H, Okada Y, Mizuguchi H, Mukai Y, Okada N, Nakagawa S. *Gene Ther* 2009;16(12):1395–1404.

172. Hofherr SE, Shashkova EV, Weaver EA, Khare R, Barry MA. *Mol Ther* 2008;16:1276–1282.

173. Alemany R, Suzuki K, Curiel DT. *J Gen Virol* 2000;81:2605–2609.

174. Wonganan P, Croyle MA. *Viruses* 2010 2:468–502.

175. De Geest B, Snoeys J, Van Linthout S, Lievens J, Collen D. *Hum Gene Ther* 2005;16:1439–1451.

176. Croyle MA, Callahan SM, Auricchio A, Schumer G, Linse KD, Wilson JM, Brunner LJ, Kobinger GP. *J Virol* 2004;78:912–921.

177. Katakura H, Harada A, Kataoka K, Furusho M, Tanaka F, Wada H, Ikenaka K. *J Gene Med* 2004;6:471–477.

178. Kim YK, Park IK, Jiang HL, Choi JY, Je YH, Jin H, Kim HW., Cho MH., Cho CS. *J Biotechnol* 2006;125:104–109.

179. Khoshnejad M, Young PR, Toth I, Minchin RF. *Curr Med Chem* 2007;14:3152–3156.

180. Wonganan P, Coyle MA. *Viruses* 2010;2:468–502.

181. Garay RP, El-Gewely R, Armstrong JK, Garratty G, Richette P. *Exp Opin, Drug Deliv* 2012;9:1319–1323.

182. Richter AW, Akerblom E. *Int Arch Allergy Appl Immunol* 1983;70:124–131.

183. Richter AW, Akerblom E. *Int Arch Allergy Appl Immunol* 1984;74:36–39.

184. Sherman MR, Williams LD, Sobczyk MA, Michaels SJ, Saifer MGP. *Bioconjug Chem* 2012;23:485–499.

185. Sunder A, Hanselmann R, Frey H, Mulhaupt R. *Macromolecules* 1999;32:4240–4246.

186. Kainthan RK, Hester SR, Levin E, Devine DV, Brooks DE. *Biomaterials* 2007;28:4581.

187. Kainthan RK, Brooks DE. *Biomaterials* 2007;28:4779.

188. Kolhe P, Khandare J, Pillai O, Kannan S, Lich-Lai M, Kannan R. *Pharm Res* 2004;21:2185–2195.

189. Calderon M, Graeser R, Kratz F, Haag R. *Bioorg Med Chem Lett* 2009;14:3725.

190. Zhang JG, Krajden OB, Kainthan RK, Kizhakkedathu JN, Constantinescu I, Brooks DE, Gyongyossy-Issa MIC. *Bioconjug Chem* 2008;19:1241–1247.

191. Wurm F, Dingels C, Frey H, Klok H. *Biomacromolecules* 2012;13:1161–1171.

192. Papp I, Dernedde J, Enders S, Haag R. *Chem Commun* 2008:5851.

193. Kizhakkedathu JN, Creagh AL, Shenoi RA, Rossi NAA, Brooks DE, Chan T, Lam J, Dandepally SR, Haynes CA. *Biomacromolecules* 2010;11:2567–2575.

194. Papp I, Dernedde J, Enders S, Riese SB, Shiao TC, Roy R, Haag R. *Chembiochem* 2011;12:1075–1083.

195. Papp I, Sieden C, Sisson AL, Kostka J, Bottcher C, Ludwig K, Herrmann A, Haag R. *Chembiochem* 2011;12:887–895.

196. Rossi NAA, Constantinescu I, Kainthan R, Brooks DE, Scott MD, Kizhakkedathu JN. *Biomaterials* 2010;31:4167–4178.

197. Rossi NAA, Constantinescu I, Brooks DE, Scott MD, Kizhakkedathu JN. *J Am Chem Soc* 2010;132:3423–3430.

198. Chapanian R, Constantinescu I, Brooks DE, Scott MD, Kizhakkedathu JN. *Biomaterials* 2012;33:3047–3057.

199. Chapanian R, Constantinescu I, Rossi NAA, Medvedev N, Brooks DE, Scott MD, Kizhakkedathu JN. *Biomaterials* 2012;33:7871–7883.

3

SYNTHETIC POLYMER BIOCONJUGATE SYSTEMS

Marya Ahmed and Ravin Narain
Department of Chemical and Materials Engineering, Alberta Glycomics Centre, University of Alberta, Edmonton, AB, Canada

3.1 INTRODUCTION

The combination of synthetic polymers with biomolecules is an appealing strategy to produce macromolecules with distinctive properties [1–6]. The macromolecules obtained by the covalent conjugation of natural molecules such as proteins, peptides, nucleic acids, carbohydrates, and lipids with synthetic polymers are known as *"biohybrids,"* polymer bioconjugates, or *"macromolecular chimeras"* [1, 5]. The construction of these polymer bioconjugates has been an important topic of research in past decades, in an effort to comprehend and utilize the complex biological properties of relatively simple natural molecules. Moreover, these conjugation approaches have proven to be useful in providing physiological stability, solubility, and therapeutic potential to biomolecules [1–6]. Figure 3.1 represents a comprehensive list of biomolecules and synthetic polymers, which have been studied for bioconjugation approaches [1].

This chapter will focus on the bioconjugation of amino acids/peptides/proteins, carbohydrates, nucleic acids, and bioactive molecules such as drugs and contrast agents with synthetic polymers. The polymers synthesized by radial polymerization and living radical polymerization (LRP) will be used to produce these biohybrids, using a variety of bioconjugation techniques. Typical conjugation of biomolecules with synthetic polymers is done either by *"grafting to"* or *"grafting from"* approach. In "grafting from" approach, biomolecule-functionalized monomers are polymerized to produce synthetic bioconjugates. However, in *"grafting to"* approach biomolecules are immobilized by reactive coupling reactions. Both techniques have their own benefits and limitations. For example, *"grafting from"* approach may cause the denaturation of biomolecules due

to the stringent polymerization conditions. Similarly, the problems associated with *"grafting to"* approach are the need of excess functional polymers to overcome steric stabilization, separation of unreacted polymers and proteins, and the presence of significantly low concentration of functional groups. Due to the high tolerance of reversible addition fragmentation chain transfer (RAFT) and atom transfer radical polymerization (ATRP) techniques toward reactive functional groups, protein biohybrids are also prepared by *"grafting from"* approach. In *"grafting from"* approach, the proteins are immobilized with RAFT agent (in the case of RAFT polymerization) or initiator (in the case of ATRP) and polymerization is performed *in situ* directly on the surface of protein [6]. In addition, creative combination of available polymerization approaches and covalent and noncovalent conjugation reactions have produced biomolecular chimeras with controlled properties. Hence, site-specific, stimuli-responsive, and well-defined (linear, stars, hyperbranched, brushes) biomolecule hybrids have been produced.

3.2 PEPTIDE OR PROTEIN BIOCONJUGATION TECHNIQUES

The conjugation of synthetic polymers with peptides or proteins is one of the well-explored fields of bioconjugates [6–21]. These biohybrids fall into three main categories, based on the construction principle of proteins [1].

1. Conjugation with single amino acid
2. Conjugation with peptide chain
3. Conjugation with proteins

Chemistry of Bioconjugates: Synthesis, Characterization, and Biomedical Applications, First Edition. Edited by Ravin Narain.
© 2014 John Wiley & Sons, Inc. Published 2014 by John Wiley & Sons, Inc.

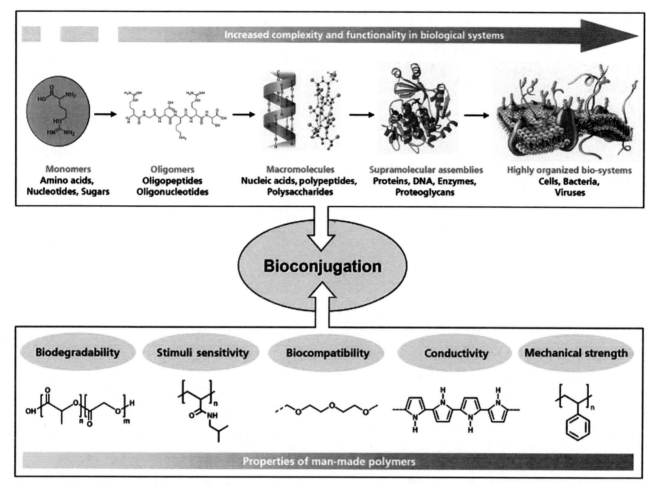

FIGURE 3.1 Schematics of possible bioconjugates [1]. For a color version, see the color plate section.

3.2.1 Conjugation with Amino Acid

The covalent attachment of single amino acid per polymer chain produces biomolecules with synthetic polymer backbone and pendent amino acid moieties. These biomolecular chimeras represent the lowest level of complexity and functionality of the class [1]. The nonionic block copolymers of polybutadiene-block-poly(ethylene oxide) (PBD-*b*-PEO) were grafted with cysteine and cysteine-containing oligopeptide to synthesize self-assembling peptide–hybrid amphiphiles [21]. The amino acid was grafted on the hydrophobic back bone of PBD through free radical addition reaction. The addition reaction of cysteine on polymer backbone was performed in tetrahydrofuran (THF) using azoisobutyronitrile (AIBN) as a radical source in inert atmosphere. The molar concentration of $[C=C]_0/[\text{amino acid}]_0/[\text{AIBN}]_0$ was 1:10:0.33, where $C=C$ represents the butadiene block of polymer chain. The mixture was allowed to react for 24 hours at 70°C and the product was analyzed by nuclear magnetic resonance (NMR) spectroscopy and size exclusion chromatography (SEC) [21].

3.2.2 Conjugation with Peptide Chain

Polymer–peptide bioconjugation is an interesting example of a relatively facile approach to produce bioinspired macromolecules. In contrast to carbohydrates and nucleic acids, peptide's modification offer several advantages such as resistance to hydrolysis, presence of modification sites, and ability to interact with complex biological systems [2]. Peptide–polymer conjugates were prepared *via* consecutive ester–amide/thiol–ene postpolymerization modification of poly(pentafluorophenyl methacrylate) (pPFMA). Pentafluorophenyl methacrylate was polymerized *via* RAFT in the presence of AIBN and chain transfer agent (CTA) to yield well-defined polymers. pPFMA were modified with 1 mol equivalent of allylamine or a mixture of allylamine and 2-hydroxypropylamine (1:4 molar ratio). The modification of RAFT-polymerized pPFMA with synthetic peptide is shown in Scheme 3.1 [10].

The degree of modification of polymers was determined by ^{1}H-NMR. The resultant poly(allyl methacrylamide-*co*-hydroxypropyl methacrylamide) copolymers were

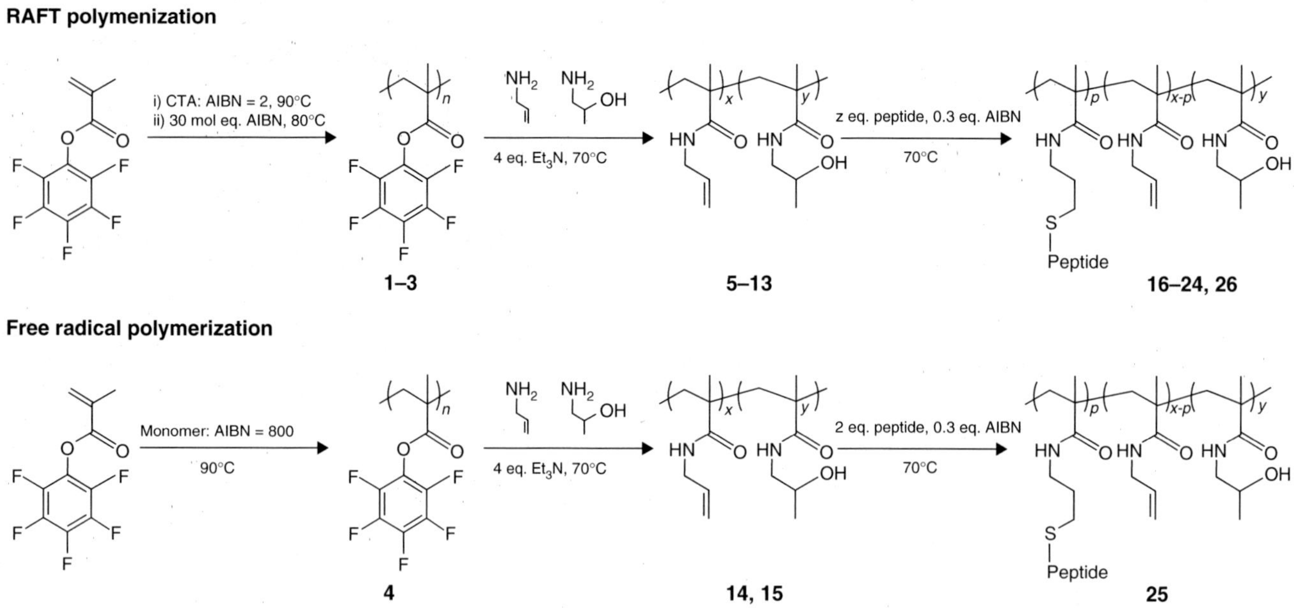

RAFT polymenization

Free radical polymerization

SCHEME 3.1 Synthesis of p(PFMA)–peptide conjugates *via* RAFT and free radical polymerization [10].

modified with thiol (SH)-containing peptide *via* thiol–ene coupling reaction. The modification was done by adding 1.6–3 mol equivalent of peptide and 1 mol equivalent of poly(allyl methacrylamide-*co*-hydroxypropyl methacrylamide) copolymers in DMF in inert atmosphere in the presence of AIBN at 70°C for 24 hours. The mixture was precipitated in ether and was dissolved in Milli-Q water. The postpolymerization modification was confirmed by ^{1}H-NMR [10].

pHEMA was conjugated with polyhistidine (pHis) by grafting to approach. pHis was obtained by ring-opening polymerization (ROP) of benzyl-histidine-*N*-carboxyanhydride (Bn-His-NCA). Bn-His-NCA was first synthesized from Boc-L-His(Bn)-OH. Boc-L-His(Bn)-OH was dissolved in anhydrous 1,4-dioxane, followed by the addition of PCl$_5$ at room temperature (r.t). The clear solution was filtered through a glass filter and was precipitated in diethyl ether to obtain Bn-His-NCA crystals. For the copolymerization of pHEMA with pHis, the macroinitiator of pHEMA was first synthesized using a multistep procedure. pHEMA was synthesized by ATRP to obtain pHEMA-Br, and was modified at bromide end to produce amine-containing pHEMA. pHEMA-Br was reacted with slight molar excess of phthalimide (PI) potassium salt in refluxing DMF for 8 hours. The solvent was removed under vacuum, washed, and dried to obtain pHEMA-PI. pHEMA-PI and KOH (molar ratio of 1:10 respectively) were heat reflexed in *t*-butanol for 12 hours. pHEMA-NH$_2$ was obtained by removing the solvent and acidifying the solid, followed by its extraction in dichloromethane (DCM). The macroinitiator was then obtained by the conversion

of pHEMA-NH$_2$ into pHEMA-N-TMS-NH$_2$ in anhydrous DMF. pHEMA-NH$_2$ and *N,O*-bis(trimethylsilyl)acetamide, (BSA) (in excess) were reacted in DMF at r.t for 48 hours. The product was precipitated in hexane and was dried under vacuum to obtain pHEMA-N-TMS-NH$_2$. pHEMA-N-TMS-NH$_2$ was used for the ROP of Bn-His-NCA. The reaction was stirred for 48 hours at r.t. The solvent was concentrated and the copolymer of p(HEMA)-*b*-p(Bn-His) was precipitated in diethyl ether. The copolymers were deprotected in trifluoroacetic acid (TFA), in the presence of 33 wt% HBr in acetic acid. The reaction was stirred for 2 hours at 0°C. p(HEMA)-*b*-p(His) was purified in diethyl ether. The step-by-step reaction procedure is shown in Scheme 3.2. The copolymers were found to be useful for producing pH-responsive micelles bearing drug-loading abilities [22].

pH-responsive conjugates of poly(vinylpyridine) (PVP) and glutamate-based polypeptide were prepared, by the combination of ROP, RAFT polymerization approach, and thiazolidine linkage. Poly(γ-benzyl-L-glutamate) (BLG) was prepared *via* ROP of γ-benzyl-L-glutamate NCA. BLG-NCA was polymerized in DMF in the presence of benzylamine, as initiator, for 3 days at 0°C. The polypeptide was precipitated in diethyl ether and was dried under vacuum. Cysteine-modified PBLG were then prepared in a stepwise manner. Pentafluorophenyl (Pfp)-activated ester, Fmoc-LCys(Acm)-OPfP was reacted with PBLG in the presence of hydroxybenzotriazole (HOBt) in distilled DMF in the presence of *N,N*-diisopropylethylamine (DiPEA). The molar ratios of ester:peptide:HOBT were 1:0.5:1, respectively. The mixture was stirred under N$_2$ at 35°C for 2 hours and the solution was precipitated in diethyl ether. The deprotection

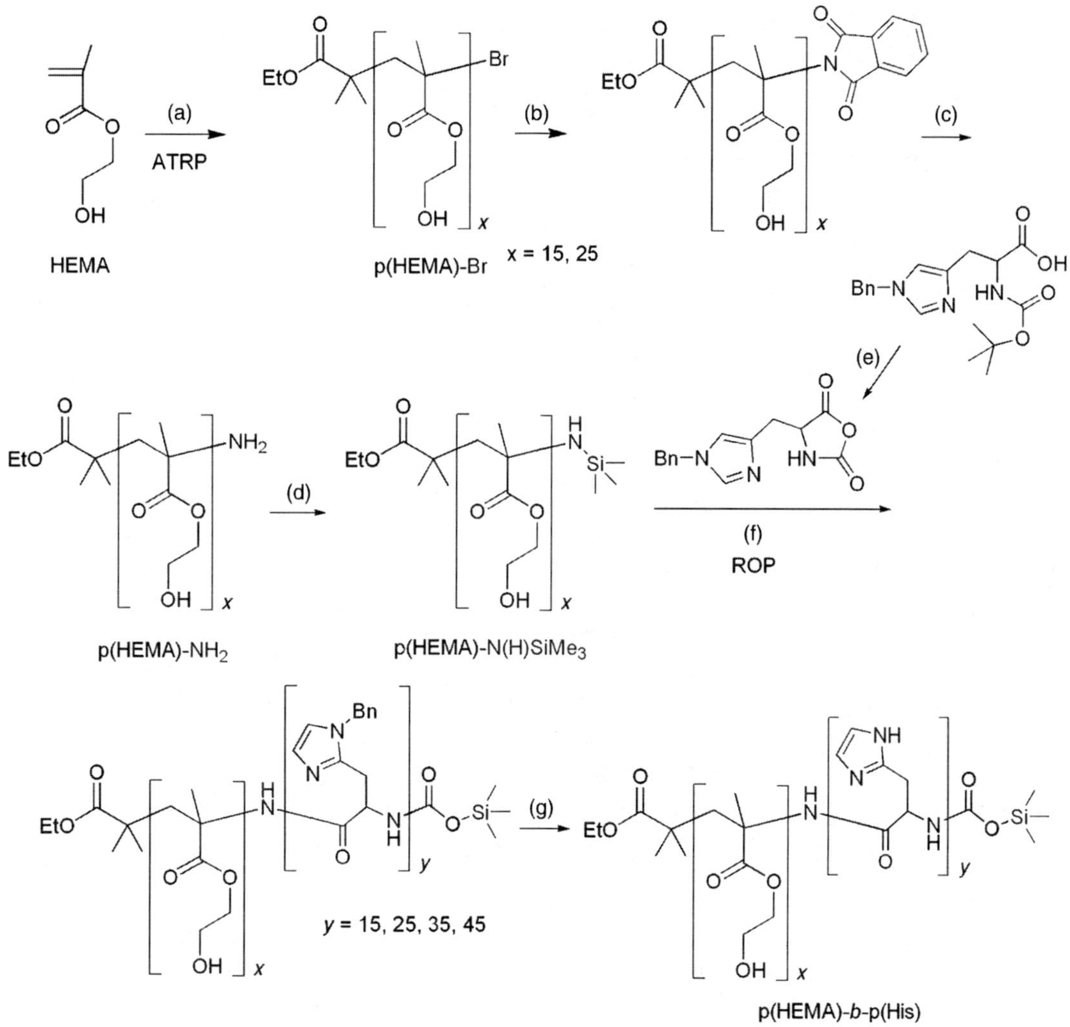

SCHEME 3.2 Synthesis of p(HEMA)-b-p(Lys) by combination of ATRP and ROP approaches. (i) ethyl 2-bromoisobutyrate, Cu(I)Br, bpy; (ii) phthalimide potassium salt, DMF, (iii) KOH, t-BuOH, (iv) N,O-bis(trimethylsilyl)acetamide, DMF, (v) PCl$_5$, 1,4-dioxane, (vi) DMF, and (vii) HBr/CH$_3$COOH, TFA; 0°C [22].

of primary amine was done by reacting PBLGFmoc-L-Cys(Acm) with piperidine in DMF in the presence of catalytic amount of 1,8-diazobicyclo[5.4.0]undec-7-ene (DBU). The mixture was stirred for 2 hours at r.t. The solution was concentrated and was precipitated in diethyl ether. Thiol deprotection was done by the removal of Acm group. PBLG-L-Cys(Acm) and benzyl alcohol (excess) were added in TFA in the presence of a catalytic amount of p-toluenesulfonic acid (p-TSA). The mixture was added to a solution of anisole (10-fold molar excess to peptide) in DMSO and TFA at 50°C. The reaction was allowed to occur for 5 hours, and the mixture was precipitated in diethyl ether. PBLG-(Cys-(s-s)-Cys)-PBLG was confirmed by NMR, SEC, and MALDI-ToF MS. N-vinylpyrrolidone (NVP) was synthesized by RAFT and was modified postpolymerization, to obtain PVP-OH. The conversion of PVP-OH to PVP-CHO (PVP-aldehyde)

was obtained by heating the solid at 120°C in vacuum. The polymer was analyzed by ^{1}H-NMR and ^{13}C-NMR. The conjugation of PVP-CHO with PBLG-(Cys-(s-s)-Cys)-PBLG was done by at equimolar ratios in DMF in the presence of excess DTT. The mixture was stirred in inert atmosphere for 24 hours at 37°C, as shown in Scheme 3.3. The product was purified by dialysis and was freeze-dried. The block copolymers were characterized by NMR and SEC [23].

ROP of NCA from macroinitiators is a well-studied strategy to produce synthetic peptide-based polymers. However, the polypeptides produced by this method are limited to one amino acid. The combination of solid-phase peptide synthesis and LRP techniques (NMRP and ATRP) has been studied to synthesize complex bioactive peptide-based bioconjugates [23]. GRGDS peptide has been studied for cell adhesion properties. GRGDS-pHEMA conjugates were prepared

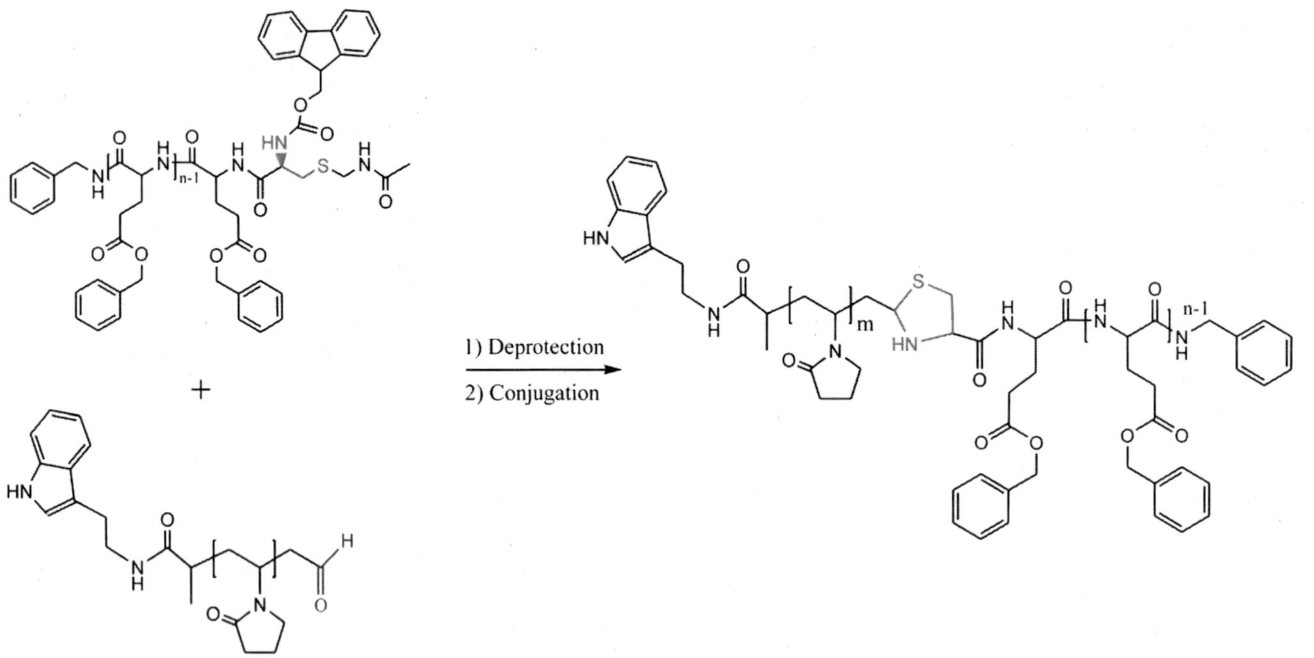

SCHEME 3.3 Bioconjugation of PVP with pGlu-polypeptide *via* thiazolidine linkage [23].

by the combination of solid-phase peptide synthesis and ATRP approach. The peptide was synthesized using Fmoc-protected amino acids and coupling reaction was performed in the presence of *N,N′*-diisopropylcarbodiimide (DIC) and HOBt. The synthesis was performed on Fmoc-protected Wang resins. The peptide initiator was synthesized by reacting resin-conjugated peptide (0.6 mmol/g of NH₂ functionality) in THF with DiPEA (2.7 mmol) followed by the dropwise addition of 2-bromopropionyl bromide (2.7 mmol). The solution was stirred overnight. The resin-bound initiator was washed withDCM, acetone, and large amounts of deionized (DI) water. The product was obtained by vacuum filtration. GRGD-pHEMA conjugates were then prepared by ATRP. The beads were isolated from peptide conjugates by dissolving them in a mixture of water, TFA, and triisobutylsilane (TIS) (Scheme 3.4). The solution was stirred for 1.5 hours, filtered, and was washed three times with TFA. The resulting solution was freeze-dried and was redissolved in methyl ethyl ketone (MEK)/1-propanol (70:30) mixture. The peptide–polymer conjugate was precipitated into hexane [24].

Click chemistry introduced by Sharpless et al. in 2001 has gained a lot of attention as a bioconjugation tool due to its high efficacy, regioselectivity, tolerance to different functionalities, mild reaction conditions, fast reaction rates, and facile product purification [25]. TAT peptide was conjugated with polystyrene (PS) *via* a combination of ATRP and click chemistry approaches. TAT-oligopeptide GGYGRKKRRQRRRG was synthesized using solid-phase support synthesis using *N*-methyl-2-pyrrolidone (NMP) as a solvent. Fmoc-protected amino acid derivatives were reacted by sequential coupling in 2-(1H-benzotriazole-1-yl)-1,1,3,3-tetramethyluronium

hexafluorophosphate (HBTU) and DiPEA. PS-(2-chlorotrityl chloride) resin was loaded with 9H-fluoren-9-ylmethyl *N*-(2-hydroxyethyl) carbamate (Fmoc-*N*-(2-aminoethanol)) followed by the addition of Fmoc-protected amino acid derivatives to obtain resin-bound peptides. The peptide was deprotected followed by the amidation upon the treatment with 20 equivalent excess of 4-pentynoic acid. The alkyne-functionalized TAT peptide was released from the solid support upon treatment with 2 vol% TFA inDCM, followed by the addition of dioxane. The mixture was concentrated in vacuum and was precipitated in diethyl ether. Azide-functionalized PS was obtained by the modification of bromine end-functionalized PS, obtained from ATRP approach. Sodium azide (NaN₃) and PS were added in DMF at 1:1 molar ratio and mixture was stirred at r.t for 3 hours. PS-N₃ was precipitated in methanol and was filtered and dried. The azide content of the polymer was determined by ¹H-NMR. Click reaction was performed by reacting PS-N₃ with alkyne-modified peptide at 1.5:1 molar ratios, respectively in the presence of 2-2′-bipyridyl and copper bromide (CuBr) in NMP solution, in inert atmosphere (Scheme 3.5). The reaction was stirred at r.t for 24 hours. The copper catalyst was removed using silica column chromatography [20].

3.2.3 Conjugation with Proteins

Polymer–protein hybrids (PPHs) are interesting candidates of research because of the improved pharmacokinetics and enhanced physical and proteolytic stability of the resulting biohybrids. The modification of proteins with

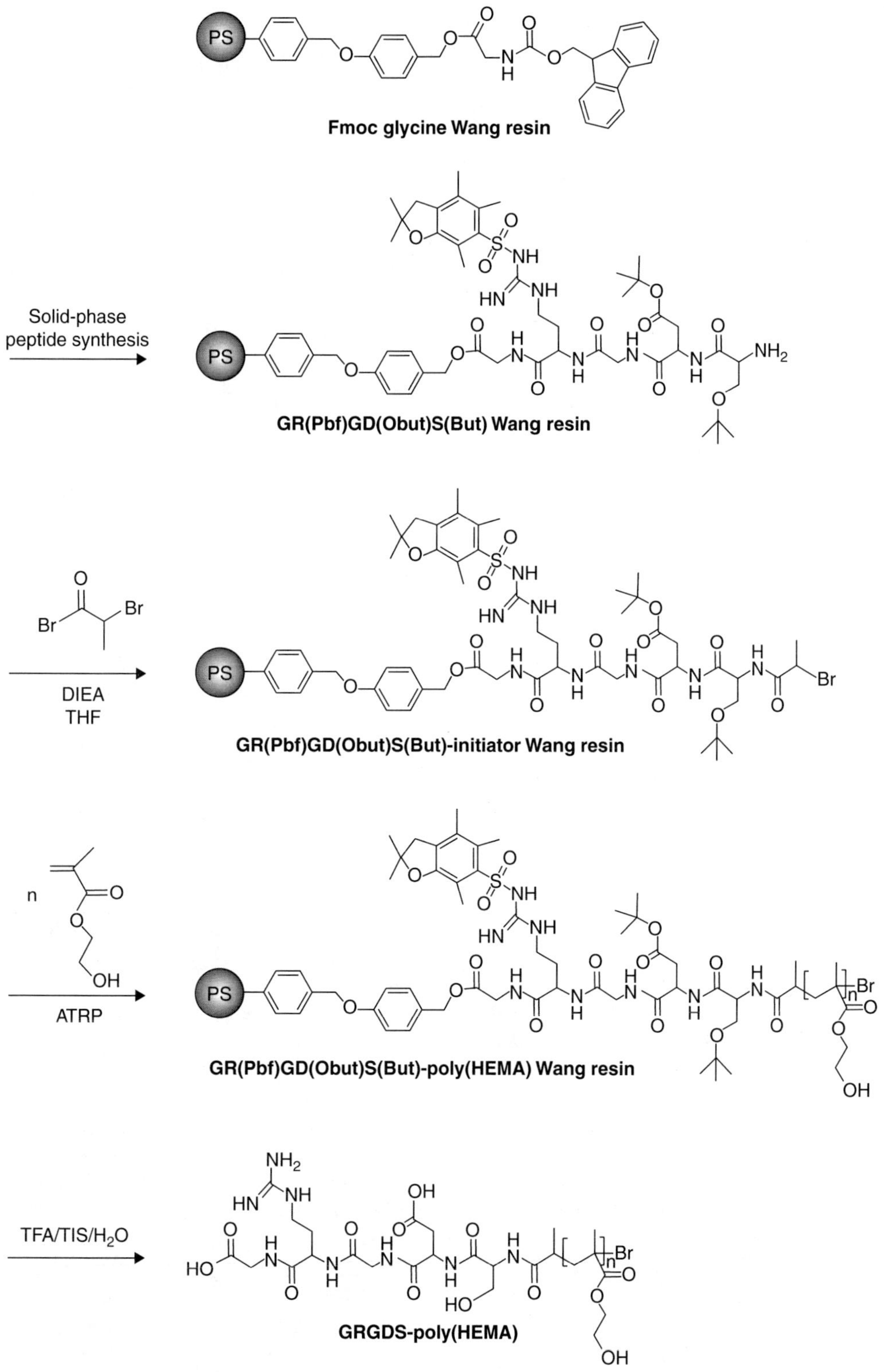

Fmoc glycine Wang resin

Solid-phase peptide synthesis →

GR(Pbf)GD(Obut)S(But) Wang resin

Br $\stackrel{O}{\diagdown}$ Br, DIEA THF →

GR(Pbf)GD(Obut)S(But)-initiator Wang resin

n, ATRP →

GR(Pbf)GD(Obut)S(But)-poly(HEMA) Wang resin

TFA/TIS/H$_2$O →

GRGDS-poly(HEMA)

SCHEME 3.4   Solid-support synthesis of peptide–polymer conjugates [24].

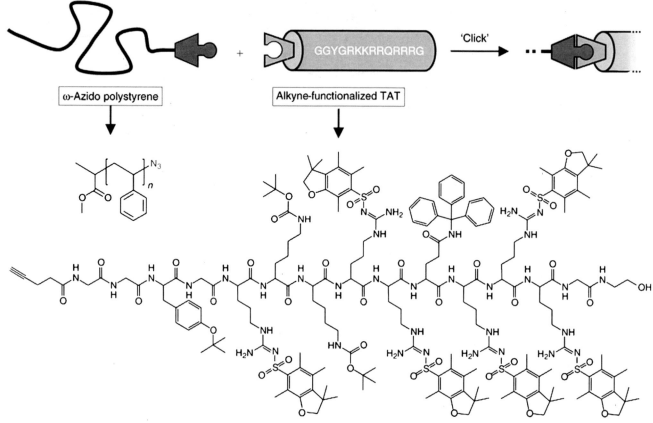

SCHEME 3.5   Click chemistry approach for the coupling of alkyne-functionalized TAT peptide with azide-modified PS [20].

polymers can occur through *"grafting to approach"* or *"grafting from"* approach [26]. The conjugation of synthetic end-functionalized polymers with functional groups of proteins is a facile approach to produce well-defined bioconjugates. The modification of proteins with synthetic polymers is challenging, as the chemical modification can alter the structural and functional integrity of these biomolecules, rendering them inactive. [8] A combination of RAFT polymerization approach and click chemistry has been employed as tools to prepare stimuli-responsive protein–polymer conjugates. *N*-isopropylacrylamide (NIPAM) monomer was polymerized in 1,4-dioxane in the presence of 2-dodecylsulfanylthiocarbonylsulfanyl-2-methylpropionic acid 3-azidopropyl ester (CTA), and AIBN at 60°C in inert atmosphere to obtain well-defined telechelic polymers bearing azide functional groups, as shown in Scheme 3.6.

The resulting polymer was analyzed using gel permeation chromatography (GPC) for molecular weight and polydispersity index. Native bovine serum albumin (BSA) and reduced BSA (BSA$_{red}$) were used for their conjugation with thermoresponsive polymer. To increase the number of conjugation sites of native BSA, BSA was dissolved in phosphate buffer saline (PBS) followed by addition of, ethylenediaminete-

traacetic acid (EDTA) and tris(2-carboxyethyl)phosphine hydrochloride (TCEP) in a dropwise manner. The mixture was stirred for 4 hours and the solution was dialyzed to obtain pure BSA$_{red}$. BSA and BSA$_{red}$ (1 equivalent) were activated with propargyl amine (20 equivalent) (60 equivalent for BSA$_{red}$) in PBS at room temperature for 24 hours. The alkyne-functionalized BSA was purified by dialysis and was lyophilized. The alkyne-functionalized BSA was conjugated to PNIPAM-N$_3$ *via* click chemistry. The catalyst solution containing copper sulfate (CuSO$_4$) and ascorbic acid at a molar ratio of 1:5 was degased in PBS-EDTA solution. BSA–alkyne and PNIPAM-N$_3$ were dissolved in PBS at a molar ratio of 1:1.2, respectively. For BSA$_{red}$, the polymer to BSA ratio was 3.6:1. The catalyst solution was added to protein–polymer mixture and the reaction was allowed at r.t for 24 hours. The mixture was dialyzed and lyophilized to obtain PNIPAM-BSA conjugates. The bioconjugates were analyzed for free thiol content using Ellman's assay. The size and molecular weights of bioconjugates were determined by GPC and DLS as a function of temperature to study their thermosensitive behavior [8].

In another study, thermoresponsive polymer–protein bioconjugates were prepared using a combination of RAFT polymerization approach and EDC/NHS chemistry [23].

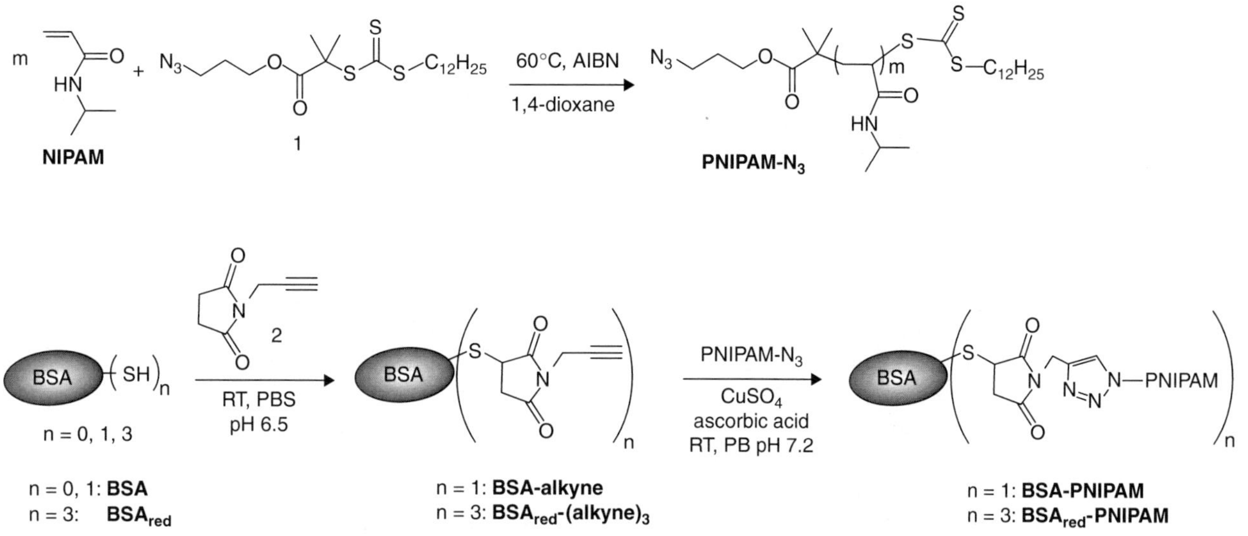

SCHEME 3.6   Synthesis of PNIPAM-BSA conjugates by a combination of RAFT polymerization and click chemistry [8].

N-Hydroxysuccinimidyl (NHS) ester-modified CTAs have been prepared to produce well-controlled polymers bearing NHS functionalities, which can then specifically react with amines of proteins to produce well-defined bioconjugates. To obtain NHS-modified CTA, S-butyl-S′-(α,α′-dimethyl-α″-acetic acid)-trithiocarbonate (CTA) was reacted with NHS at 1:1 molar ratio in DCM in inert atmosphere

at 4°C for 30 minutes. An equimolar amount of *N,N′*-dicyclohexylcarbodiimide (DCC) was dissolved in DCM and was added to the reaction mixture under N_2. The reaction was incubated for 2 hours at 4°C and was then stirred at r.t for 24 hours. The modification of CTA with NHS and bioconjugation with lysozyme is depicted in Scheme 3.7 [26].

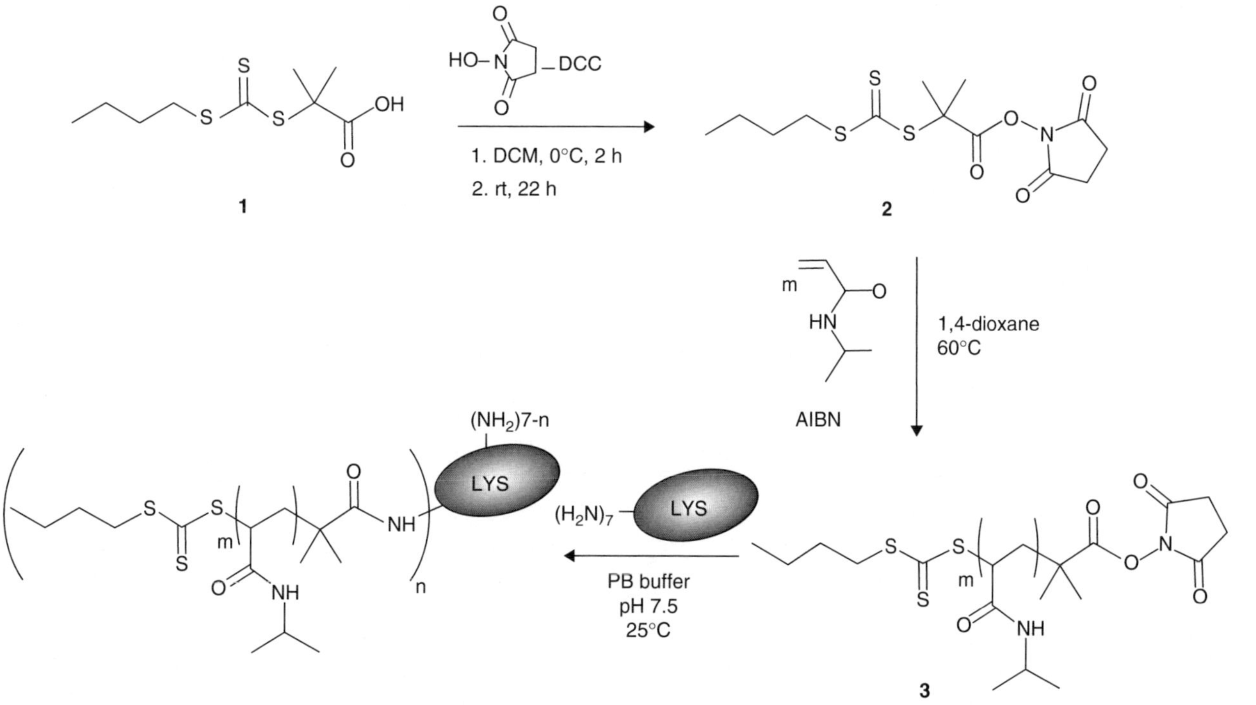

SCHEME 3.7   Synthesis of NIPAM-Lys conjugates by a combination of RAFT polymerization and EDC/NHS chemistry [26].

The product was purified using column chromatography and was recrystallized and dried under vacuum. NIPAM was polymerized using NHS-modified CTA in the presence of AIBN in 1,4-dioxane, in inert atmosphere. The polymer was precipitated in ether and the product was dried under vacuum to obtain succinimidyl-terminated-PNIPAM (NHS-PNIPAM). Hen egg lysozyme was reacted with NHS-PNIPAM at a molar ratio of 2:1, respectively in PBS for 20 hours at r.t. The progress of reaction was recorded by sodium deodecyl sulfate polyacrylamide gel electrophoresis (SDS-PAGE). The bioconjugates were heated at 40°C, and centrifuged to remove excess protein. The white precipitates were collected and were redispersed in DI water. The mixture was dialyzed and was lyophilized to obtain pure bioconjugates [26].

2-Methacryloxy ethyl phosphorylcholine (MPC) is a biocompatible zwitterionic monomer, which has been used for a variety of biological applications. The modifications of enzyme with polymers are shown to improve their physiological stability of proteins. The *"grafting to"* approach was used to modify papain with PMPC *via* EDC/NHS chemistry. The carboxyl group containing MPC was synthesized by free radical polymerization approach. 4-(*N,N*-diethyldithiocarbamoylmethyl)-benzoic acid (BDC) was reacted with MPC by UV radiation in inert atmosphere in THF/ethanol mixture at r.t for 3 hours. The polymer was precipitated in chloroform and was dried in vacuum. The polymer was analyzed by GPC and ^{1}H-NMR. The conjugation of PMPC-COOH with amine groups of papain was done by activating carboxyl group with NHS, as discussed before. The enzymatic activity of native papain and PMPC-papain was determined using benzoyl-L-arginine-ethylester (BAEE), as a substrate. BAEE and native papain/PMPC-papain were dissolved in PBS (pH = 6.1, 0.1 M) in the presence of 1 mM EDTA and 5 mM L-cysteine. The rate of reaction was studied by UV absorbance at 258 nm at 40°C. [27]

A variety of noncovalent bioconjugation techniques has been applied to produce protein–polymer bioconjugates. Biotin–avidin interactions, ConA–carbohydrate interactions, and metal–ligand interactions are some of the noncovalent approaches used to prepare these biohybrids. Polymer–protein bioconjugates have also been prepared using cofactor reconstitution approach. The first step in the synthesis of protein–polymer bioconjugates, using cofactor reconstitution approach is the synthesis of cofactor-modified polymers. PNIPAM was prepared by the ATRP approach and was modified at chain terminal end by porphyrin moiety using click reaction, as shown in Scheme 3.8 [28].

Alkynyl-PNIPAM was prepared using propargyl 2-chloropropionate, as ATRP initiator. Alkynyl-PNIPAM was reacted with azide-modified protoporphyrin IX-Zinc (PPIXZn) at 1:1 molar ratio, in the presence of *N,N,N′,N″,N″*-pentamethyldiethylenetriamine (PMDETA) and CuBr in DMF. The incorporation of Zn(OAc)$_2$ in PPIX solution ensures the integrity of PPIX during click reaction. The click reaction was performed in inert atmosphere, for 24 hours at r.t to yield PNIPAM–PPIXZn. Apomyoglobin was prepared from myoglobin and was reconstituted with PNIPAM–PPIXZn at a molar ratio of 1: 2, respectively in aqueous phosphate buffer upon gentle shaking for 24 hours. The conjugation reaction was confirmed by SDS-PAGE. [28]

Biotin/avidin interaction are one of the most strong noncovalent and specific interactions with $K_d = 10^{-15}$ M, which have been used to produce macromolecules. As compared to covalent approaches, noncovalent functionalization is desirable, as it maintains protein's native structure, and relevant biological activity. The copolymers of varying architectures, such as star block copolymers, heteroatom-star polymers and star polymers are synthesized owing to the special interactions between biotin and avidin. Alkynyl-terminated PNIPAM was synthesized by ATRP by polymerizing NIPAM and using propargyl 2-chloropropionate and CuCl/Me$_6$TREN, as initiator and catalyst, respectively. Biotin–azide was synthesized by dissolving biotin in DMF, followed by the addition of 1,1′-carbonyldiimidazole (CDI). The mixture was stirred at r.t for 3 hours. 1-azido-3-aminopropane was dissolved in DMF and was slowly added in biotin solution at a molar ratio of 3:1 of azide to biotin, respectively. The reaction was stirred for 12 hours, the solvent was removed under vacuum and biotin-azide was recrystallized in butanol/acetic acid/water mixture to obtain pure solid. The biotinylated polymers are then synthesized by reacting alkyne-PNIPAM with biotin-azide *via* click chemistry, using *N,N,N′,N″,N″*-pentamethyldiethylenetriamine (PMEDTA) and CuBr as catalyst. The incubation of biotinylated polymers with avidin produced diverse biohybrids as shown in Figure 3.2 [12].

In another study, avidin-conjugated thermoresponsive biomolecules were prepared by free radical NHS chemistry. The copolymers of NIPAM and acrylamide (Am) containing carboxyl end-groups were prepared, by free radical polymerization approach, in the presence of 3-mercaptopropanoic acid, as CTA (Scheme 3.9). The copolymers were then activated with NHS ester and were reacted with avidin in 0.2 M borate buffer of pH = 8 at 1:2 protein:polymer ratio. The solution was stirred for 2 hours at r.t and was dialyzed and purified using gel filtration FPLC with superose 6 column. The eluted components were detected with UV at 282 nm to detect the protein and with iodine to detect polymer component. The number of polymer chains bound to protein was quantified by NH$_2$ analysis using TNBS test [7].

ATRP has been used largely for the synthesis of well-defined polymers. However the use of ATRP for the synthesis of bioconjugates has always been questioned because of the potential toxicity of Cu metal for biological applications. The variations in ATRP approach such as AGET–ATRP are found to be useful for the synthesis of biohybrids while avoiding the possible complications of metal catalysts [12]. ATRP

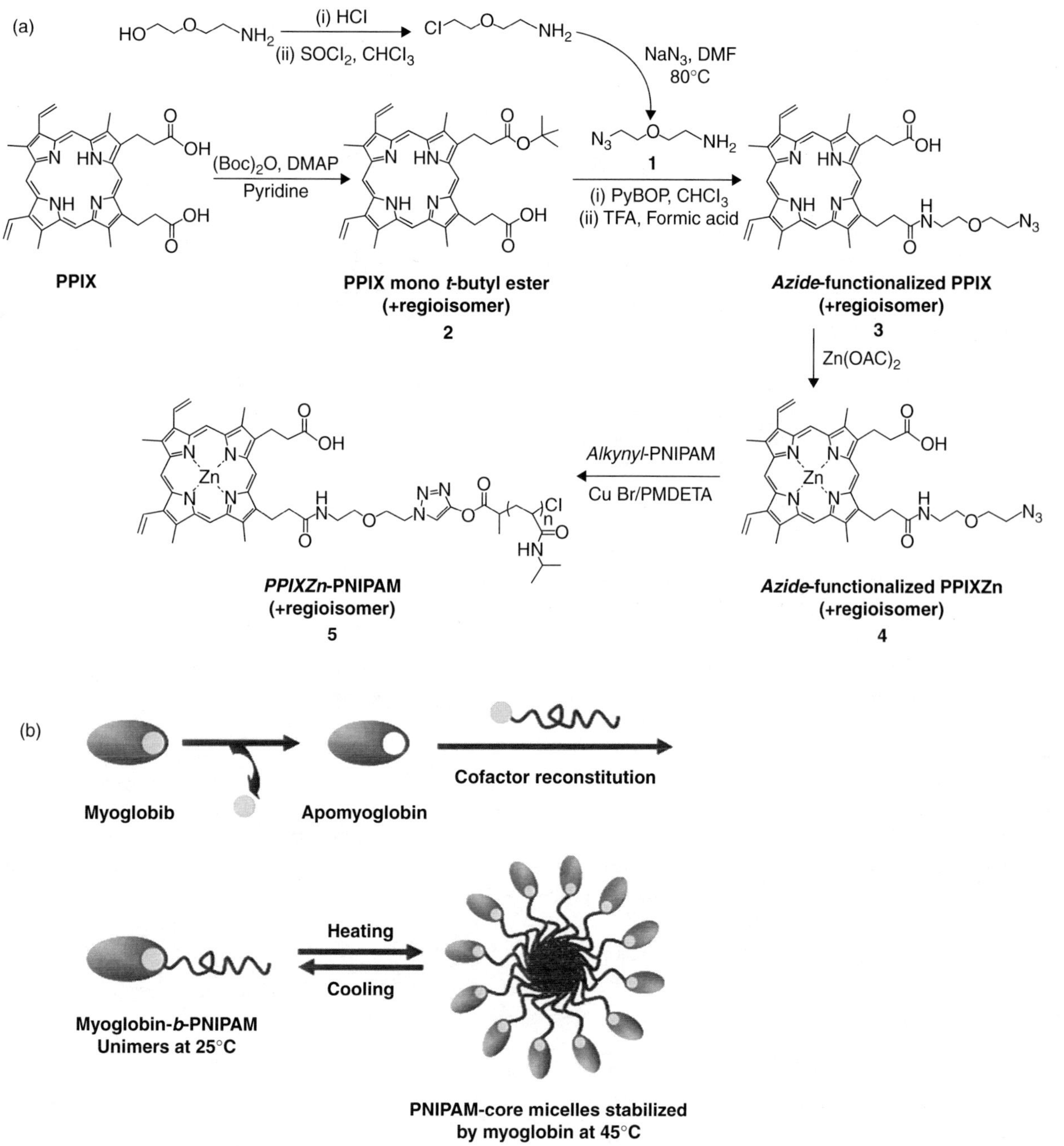

SCHEME 3.8 Synthesis of PNIPAM-b-myoglobin conjugates using cofactor reconstitution approach [28].

and AGET-ATRP approaches were used to produce PPHs *via* grafting from approach in PBS at 30°C (Scheme 3.10). BSA was modified with ester-based initiator, bromoisobutyrate (iBBr)) to obtain BSA-O-iBBr. MALDI-ToF analysis was performed to confirm the conjugation of BSA with initiator, and it was determined that there were 20 initiating sites per BSA [BSA-O-(iBBr)$_{30}$]. Thermoresponsive polymer OEOMA was polymerized by ATRP in the presence of CuX, CuX$_2$, and ligand [29].

3.3 CARBOHYDRATE BIOCONJUGATION TECHNIQUES

The synthesis of neoglycopolymers has gained a lot of attention in research, due to their ability to produce *"cluster glycoside effect,"* upon carbohydrate–carbohydrate and carbohydrate–protein interactions. The synthesis of glycopolymers with well-defined structures (block, statistical, stars, hyperbranch, and dendrimers) has been made

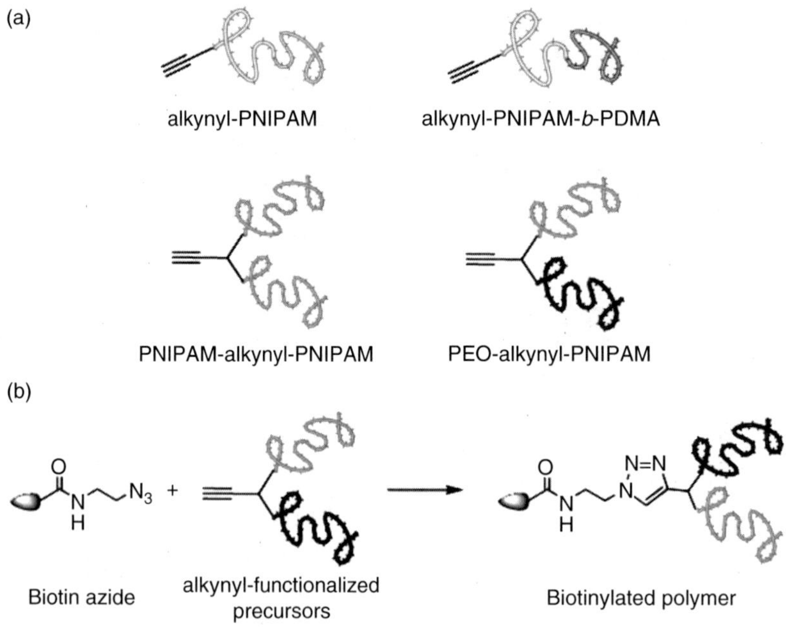

(a)

alkynyl-PNIPAM　　　　　alkynyl-PNIPAM-*b*-PDMA

PNIPAM-alkynyl-PNIPAM　　　　　PEO-alkynyl-PNIPAM

(b)

Biotin azide　　alkynyl-functionalized precursors　　　Biotinylated polymer

FIGURE 3.2　The schematics of synthesis of alkyne-modified polymer and azide-modified biotin and their conjugation by click reaction [12].

possible by LRP techniques. The number of reports showing the synthesis of well-defined glycopolymers (carbohydrate–polymer conjugates) is still limited as compared to protein bioconjugates, possibly due to the inherent difficulties in the synthesis of glycomonomers, the incompatibility of functional groups with polymerization techniques, and requirement of protected group chemistries. The grafting of sugar monomers on polymer backbone *via* click chemistry has been used as a strategy to produce sugar bioconjugates. Alkyne-containing monomer (trimethylsilyl methacrylate) was synthesized by the reaction of 3-(trimethylsilyl)propyn-1-ol

with methacryloyl chloride. Trimethylsilyl methacrylate was copolymerized with methylmethacrylate in the presence of *O*-benzyl-α-bromoester, as initiator and CuIBr/*N*-(*n*-ethyl)-2-pyridylmethanimine, as catalyst, using ATRP approach. The copolymers produced were deprotected in tetrabutyl ammonium fluoride (TBAF) and acetic acid solution to obtain alkyne-containing polymers. The sugar monomers were modified with azide group. Azide-modified monosaccharides, namely 2,3,4,6-tetra-*O*-acetyl-β-D-glucopyranosyl azide, 2,3,4,6-tetra-*O*-acetyl-β-D-galactopyranosyl azide, and methyl α-D-6-azido-6-deoxymannopyranoside were

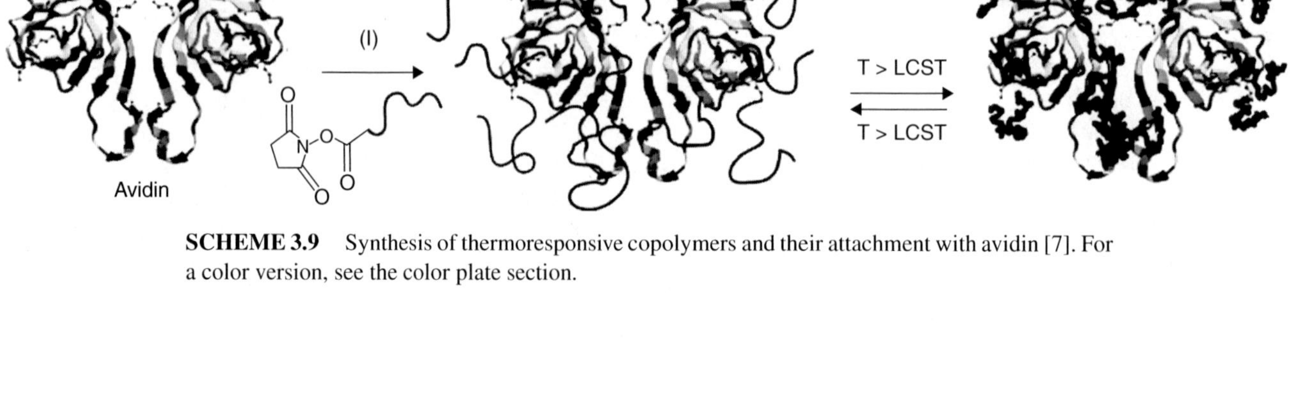

Poly(NIPAm-co-AM-NHS (I))

Avidin

(I)

T > LCST

SCHEME 3.9　Synthesis of thermoresponsive copolymers and their attachment with avidin [7]. For a color version, see the color plate section.

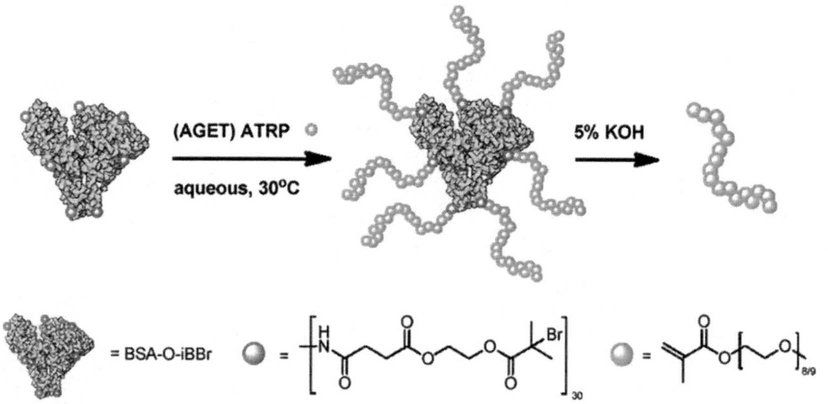

SCHEME 3.10 Synthesis of PPH-BSA hybrids prepared by ATRP [29]. For a color version, see the color plate section.

conjugated with alkyne-modified polymers in the presence of [(PPh$_3$)$_3$CuBr] as the catalyst and DIPEA. [30]

Trehalose was attached to styrene monomer *via* 4,6-acetal linkage. 4-vinylbenzaldehyde diethyl acetal was synthesized by wittig reaction and was reacted with trehalose to obtain regioselective polymerizable trehalose monomer, also called 4,6-O-(4-vinylbenzylidene)-α,α-trehalose. 4-vinylbenzaldehyde diethyl acetal was synthesized by dissolving methyltriphenylphosphonium bromide in THF. The reaction mixture was cooled to -78°C and was stirred for 20 minutes. n-Butyllithium (n-BuLi) was slowly added to the reaction mixture (over a time of 20 minutes) and the solution was stirred for 30 minutes at -78°C. The mixture was warmed to 25°C and was stirred for another 10 minutes. The orange color solution obtained was cooled to -78°C and the mixture of terephthaldehyde monodiethyl acetal in THF (0.8 mol equivalent to methyltriphenylphosphonium-bromide) was added dropwise over a period of 1 hour. The flask was warmed to 0°C and was stirred for 3 hours. Then the flask was warmed to 25°C and was stirred for another hour. The reaction was quenched by NaHCO$_3$ and the organic layer was collected and dried with MgSO$_4$. The product was purified by silica gel column chromatography and the product in liquid form was obtained. Trehalose was dissolved in DMF and the mixture was warmed at 100°C. 4-Vinylbenzaldehyde diethyl acetal (1 equivalent) and p-toluenesulfonic acid (p-TsOH, 0.12 equivalent) were added and the solution was stirred at 100°C for 3 hours (Scheme 3.11). The solvent was removed to obtain a white solid which was further purified by HPLC. The pure product was analyzed by HNMR, CNMR, HMQC, and COSYNMR to confirm the regioselectivity of the reaction. The trehalose-based monomer was polymerized *via* RFAT to obtain well-defined synthetic glycopolymers. These glycopolymers were further conjugated with hen egg lysozyme. For this purpose hen egg lysozyme was constituted with a thiol group upon treatment with N-succinimidyl-S-acetylthiopropionate (SATP). Hen egg lysozyme was dissolved in phosphate buffer in the presence of EDTA (50 mM

PB, 1 mM EDTA) at 10 mg/mL concentration. 10 μL SATP (16 mg/mL, 65 mM DMF) was added in protein solution and the mixture was cooled to 4°C for 4 hours. The resulting modified protein was purified by centriprep-ultrapurification, and was deprotected with 0.5M hydroxylamine solution. The final product was repurified and SH content was determined by Ellman's assay. For the bioconjugation with glycopolymer, thiol-modified lysozyme (70 μL, 10 mg/mL in PBS) was added to LoBind centrifuge tube. Trehalose polymer synthesized was added to the tube (15 equivalent) and the final reaction volume was diluted to 500 μL. The reaction was stored at 4°C for 3 hours. The resultant mixture was purified by FPLC and fractions were concentrated by centriprep-ultrafiltration. The fractions were stored at 4°C for their analysis by SDS-PAGE. The effect of heat and lyophilization on protein native structure was studied [31].

The facile synthesis of methacrylate and methacrylamide-based glycomonomers and their polymerization *via* RAFT and ATRP has produced well-defined glycopolymers of linear and branched architectures [32, 33]. The synthesis of linear glycopolymer is depicted in Scheme 3.12 [32].

In another approach, Hawker and group has prepared well-defined dendrons modified with glycomonomers *via* click chemistry approach [34]. Hawker and colleagues have prepared monosaccharide-functionalized hydrophilic and hydrophobic dendrons using click chemistry. Recently, carbohydrate-based bioconjugates were synthesized by RAFT approach, without any need for postmodification [35].

Graft copolymers bearing amylose side chains were prepared from amylose-substituted styrene macromonomer (vinylbenzyl amylose amide, VAA). An amylose chain was connected to vinylbenzyl group *via* gluconamide linkage, catalyzed by phosphorylase enzyme. Vinylbenzyl maltopentaose amide (VMA) was synthesized by dissolving maltopentaose in distilled water. Iodine was dissolved in methanol and solution was added to maltopentaose solution followed by the addition of 4% KOH methanolic solution dropwise at 40°C for 15 minutes. The mixture was poured into iced water

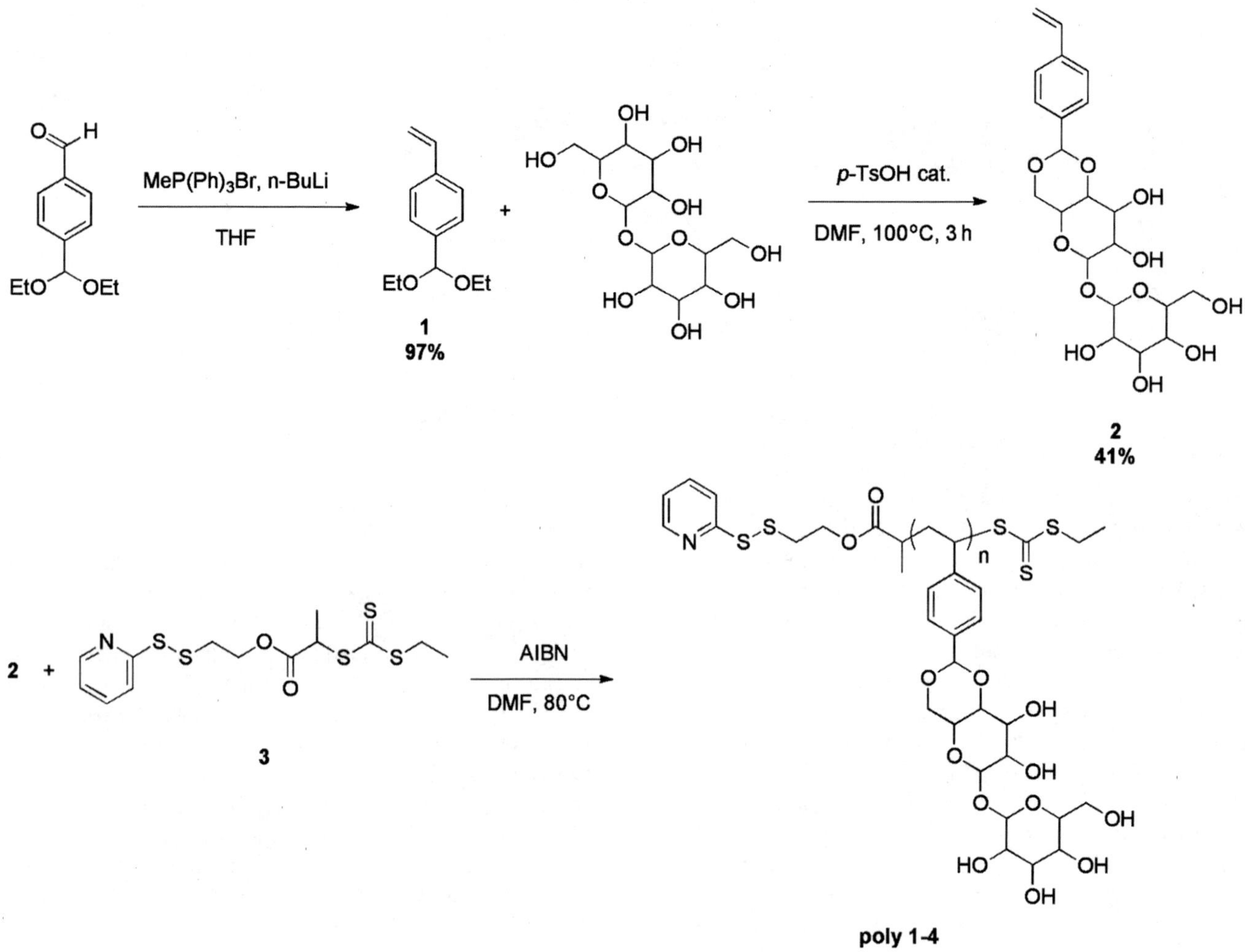

SCHEME 3.11   Synthesis of trehalose-based glycopolymers *via* RAFT [31].

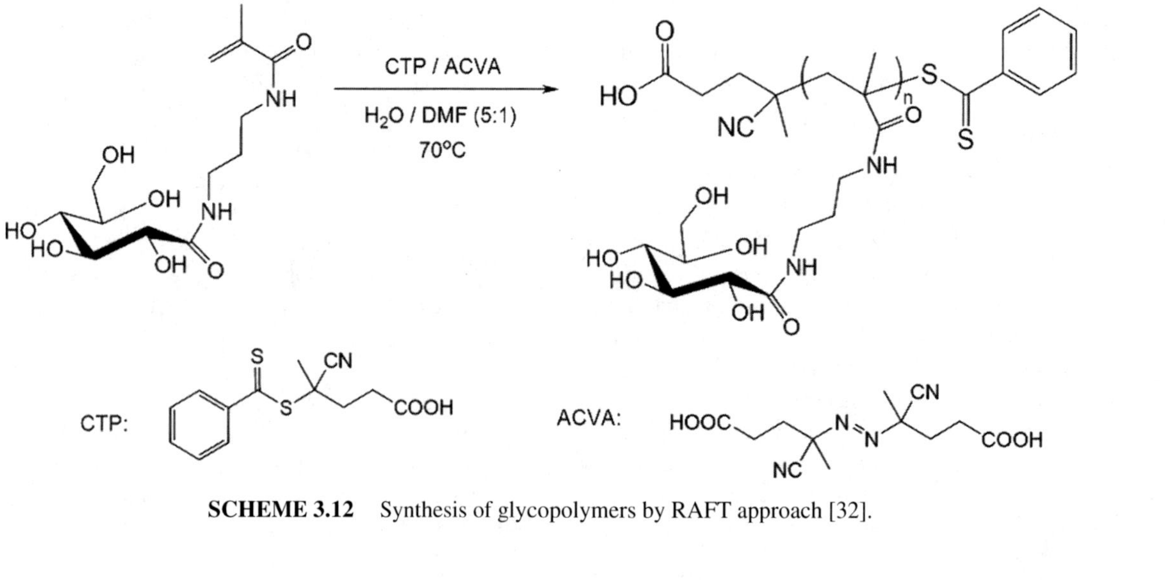

SCHEME 3.12   Synthesis of glycopolymers by RAFT approach [32].

and the precipitates were collected and dissolved in water. The solution was treated with activated carbon, was filtered, and freeze-dried to yield maltopentaose lactone. The lactone was then dissolved in ethylene glycol solution, followed by the addition of *p*-vinylbenzylamine and the mixture was then heated at 70°C for 6 hours. The final product was precipitated in acetone and was dried under reduced pressure at 60°C. The unreactive product was removed by washing the crude product with water through preparative high performance chromatography system equipped with YMC ODS column and RI detector, using a mixture of water:acetonitrile as eluent. To synthesize VAA, VMA (1.1 mmol) and potassium α-D-glucose-1-phosphate dehydrate (38 mmol) were dissolved in a mixture of 0.1M maleic acid buffer (pH = 7.5, 363 mL) and DMSO (120 mL) and the solution was heated at 45°C. Potato α-glucan phosphorylase (80 units) was added and the mixture was incubated at 45°C for 10 hours. The mixture was then inactivated by heating at 95°C for 5 minutes, and was immediately cooled to 40°C. The filtrate was poured into ethanol and the supernatant was removed by centrifugation. The precipitates were washed with ethanol and diethyl ether and dried in vacuum at 70°C to get pure VAA monomer. VAA and VMA monomers were polymerized by free radical polymerization to obtain glycoconjugates [36].

Glycocylindrical brushes were prepared *via* ATRP, using a bulky sugar carrying methacrylate monomer, (3-*O*-methacryloyl-1,2:5,6-di-*O*-isopropylidene-α-D-gluco furanose) (MAIGlc). MAIGlc was synthesized by the reaction of 1,2:5,6-di-*O*-isopropylidene-D-glucofuranose and methacrylic anhydride in pyridine and monomer was purified by vacuum distillation. MAIGlc was polymerized *via* ATRP using polyinitiator, poly(2-(2-bromoisobutyryloxy)ethylmethacrylate) (pBIEM-II) in ethyl acetate. Hexamethyltriethylenetetramine (HMTETA) was then added and the flask was put in oil bath at 60°C for 25 minutes. The polymer was precipitated in petroleum ether and the solid was dried under vacuum. The polymer obtained was deprotected in acidic solution and was dialyzed to obtain poly(3-*O*-methacryloyl-α,β-D-glucopyranose) (PMAGlc) [37].

The linear glycopolymers bearing pedant D-glucaric and D-gluconic moieties were also prepared. For this reason, novel styrene-modified glycomonomers, *N*-(*p*-vinylbenzyl)-1-D-glucaramide (VB-1-GlcaH) and *N*-(*p*-vinylbenzyl)-D-gluconamide (VB-1-Glco), were synthesized. *p*-Vinylbenzylamine (2 equivalent) was added to D-glucaro-1,4-lactone in methanol and the mixture was stirred at 50°C for 2 hours. The mixture was treated with acetic anhydride and was extracted with chloroform. The chloroform-insoluble portion was recrystallized from ethanol to provide a glycopolymer. The glycomonomers obtained were polymerized in DMSO using AIBN as an initiator [38].

Polymethacrylates containing pendent maltoheptaose (PMA-SM7) units were prepared by radical polymerization

of acetylated 1-*O*-methacryloyl maltoheptaoside (MA-AcM7). The methacrylate-based monomer synthesis is as follows. The acetylated maltoheptaoside (1 mmol) was peracetylated in dry THF in the presence of benzylamine (1.5 mmol) after stirring for 22 hours at r.t. THF was evaporated and the reaction mixture was dissolved in chloroform. The chloroform layer was extracted and was washed with 1N HCl, 5% NaHCO₃ solution, and water several times. The residue was passed through silica gel chromatogram and was eluted with hexane:ethyl acetate mixture (1:2) to obtain pure peracetylated maltoheptaoside. To obtain MA-AcM7, methacrloyl was added to peracetylated maltoheptaoside (1:1 molar ratio) in methylene chloride in the presence of triethylamine at 0°C. The mixture was stirred overnight at r.t. Chloroform was then added to the reaction mixture and the chloroform layer was washed with 5% NaHCO₃ and water several times and was evaporated. The product was purified by silica gel column chromatography and by freeze-drying from benzene to obtain pure peracetylated 1-*O*-methacryloyl maltoheptaoside (MA-AcM7). MA-AcM7 was polymerized by free radical polymerization and polymers were deprotected with sodium methoxide (NaOMe) in THF after stirring for 3 days at r.t to get poly(MA-M7). The sulfated analogs of glycopolymers were also prepared by mixing poly(MA-M7) and piperidine-*N*-sulfonic acid in DMF. The solution was stirred at 85°C for 1.5 hours. The reaction was cooled to room temperature and was neutralized by NaHCO₃. The sulfated polymer was dialyzed to obtain a pure product [39].

Comb-shaped peptide–glycopolymer conjugates containing acid-labile β-thiopropionate linkage were prepared by a combination of RAFT polymerization and thiol–ene click chemistry. *O*-isopropylidene-D-glucofuranose (MAIpGlcP) monomer was polymerized *via* RAFT to obtain PMAIpGlc macro CTA. The glycopolymer was blocked with HEMA to obtain pMAIpGlc-*b*-pHEMA. The removal of RAFT-CTA (thiocarbonylthio moiety) was done in DMF, in the presence of AIBN under inert atmosphere. The mixture was heated at 80°C for 3 hours and polymer was precipitated in petroleum ether and was dried under vacuum. The polymer was modified with acrylate moiety. The copolymer (16 μmol) was dissolved in THF (4.4 mL) and pyridine (26 μL) was added in inert atmosphere. The solution was cooled to 0°C followed by the dropwise addition of acryloyl chloride (26 μL) in THF (2.2 mL). The reaction was stirred at 0°C for 2 hours and at r.t for 24 hours. The product was dialyzed and was lyophilized. The deprotection of isopropylidene groups of MAIpGlc was done before the conjugation of copolymer with glutathione (GSH). The copolymer was dissolved in 80% formic acid and was stirred for 48 hours at r.t. The water was added and the solution was stirred for another 3 hours. The solution was then dialyzed and freeze-dried to obtain pure copolymer pMAGlc-*b*-pHEMA. pMAGlc-*b*-pHEMA (4.3 μmol) was dissolved in DMF, and a solution of reduced GSH (86 μmol) and pyridine (86 μmol) in water were added (Scheme 3.13).

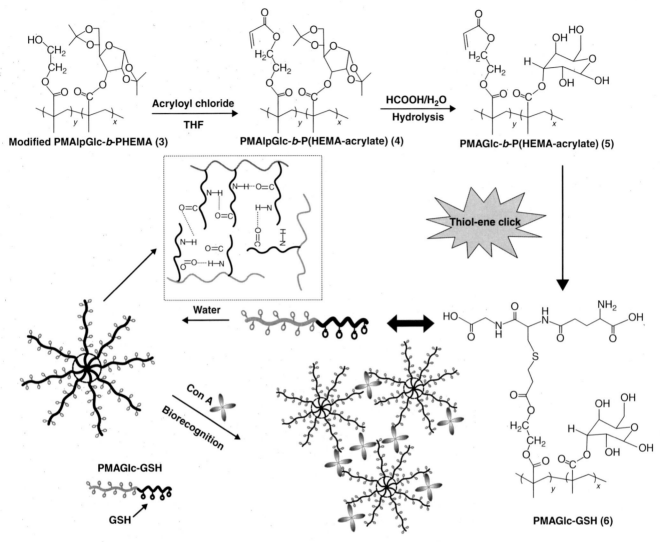

SCHEME 3.13 The synthesis of block-glycopolymers *via* RAFT and their conjugation with GSH by thiol–ene click chemistry [40].

The reaction was done in inert atmosphere for 72 hours at r.t. The solution was dialyzed and freeze-dried to obtain pure conjugates. The release of GSH from glycopolymer–peptide conjugates was studied overtime in sodium acetate/acetic acid buffer solution (pH = 5) at 37°C [40].

3.4 CONJUGATION WITH NUCLEIC ACID

The synthesis of DNA polymer hybrids is an interesting field, which has been studied extensively to produce novel materials with combined biological and physiochemical properties of both DNA and synthetic polymers. The electrostatic interactions between anionic DNA and cationic polymers are the most common type of bioconjugation approach, used for gene delivery applications. A variety of cationic polymers, peptide-modified cationic polymers, and cationic

glyopolymers have been synthesized and are studied for their bioconjugation with DNA and gene delivery applications [41–43]. The *block*, *statistical* and *"block-statistical"* cationic glycopolymers are prepared *via* RAFT. Synthesis of *"block-statistical"* polymers is shown in Scheme 3.14 [42].

These cationic glycopolymers are incubated with *gWiz*-β-galactosidase plasmid in saline solution at varying polymer/plasmid w/w ratios for 30 minutes at room temperature, as shown in Figure 3.3. The formation of polyplexes is confirmed using agarose gel electrophoresis, dynamic light scattering (DLS), and zeta potential instruments [42].

Synthetic nucleotides are an important part of genomic research. They are widely used as primers, in polymerase chain reactions and for therapeutic applications. The introduction of functional group at 3′ or 5′ termini provides a facile way to conjugate these oligonucleotides with polymers and other desired functional moieties. The introduction of

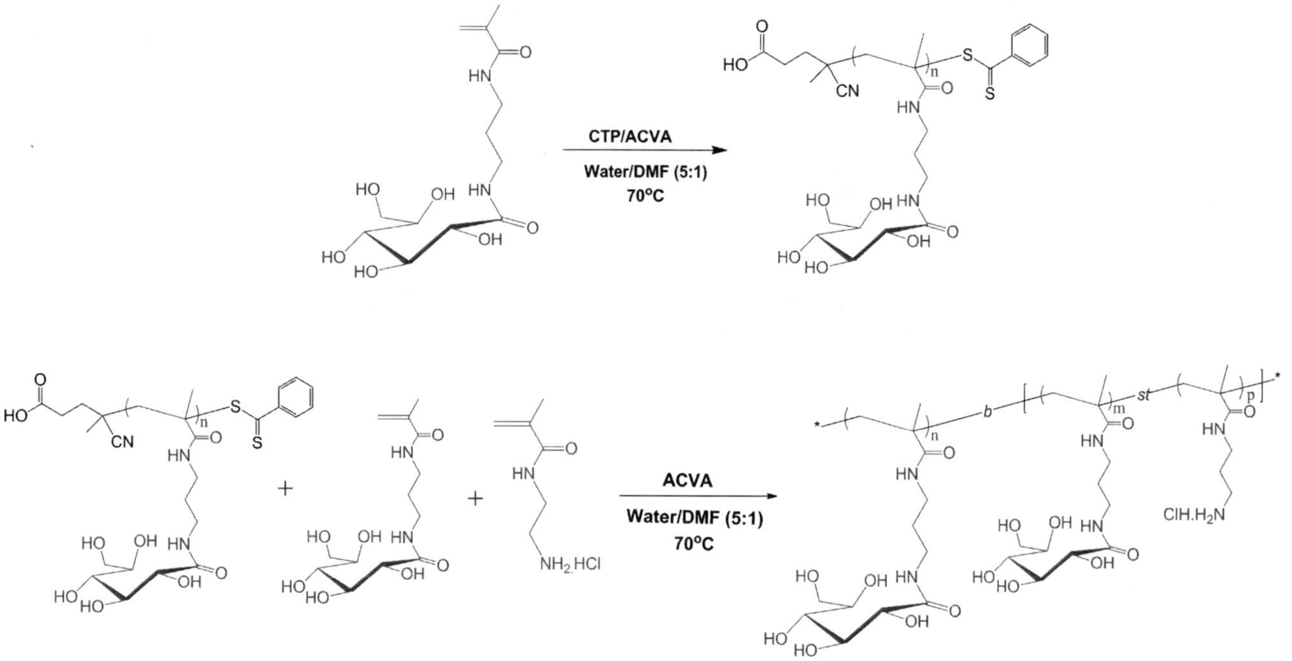

SCHEME 3.14 Synthesis of *"block-statistical"* cationic glycopolymers by RAFT [42].

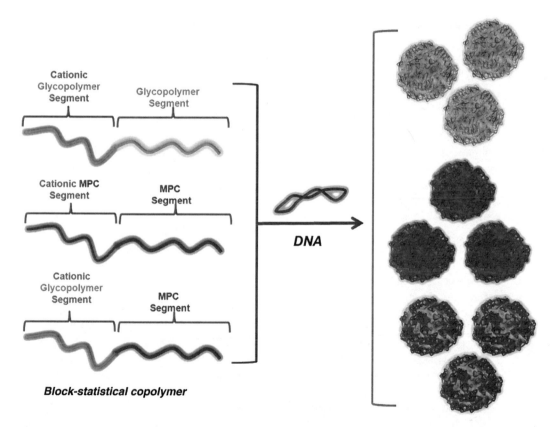

Block-statistical copolymer

Block-statistical copolymer based polyplexes

FIGURE 3.3 Formulation of polyplexes using *"block-statistical"* cationic glycopolymers [42]. For a color version, see the color plate section.

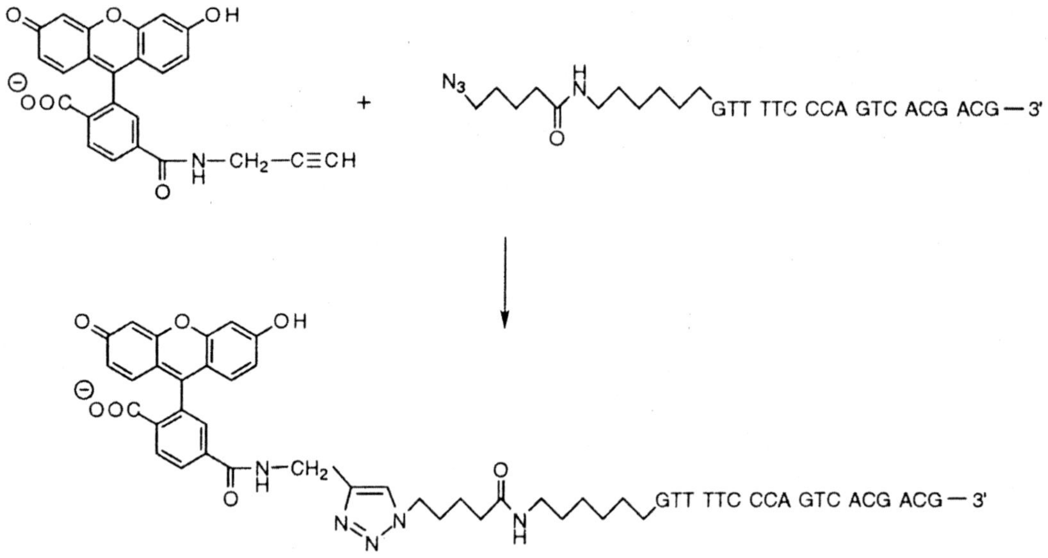

SCHEME 3.15   1,3-dipolar cycloaddition of 6-carboxyfluorescein (Fam) and azide-labeled DNA [45].

functionalities in DNA can be achieved by using phosphoramidite reagent in solid-phase synthesis. Azide-functionalized DNA was synthesized by incorporating azide moiety at 5′ end of oligonucleotide. Succinimidyl-5-azidovalerate was first synthesized. 1-(3-Dimethylaminopropyl)-3-ethylcarbodiimide hydrochloride (EDC) and NHS were added to the solution of 5-azido valeric acid at equimolar ratios in CH_2Cl_2 and the mixture was stirred for 7 hours at r.t. The organic layer was washed with water and brine, and was dried on Na_2SO_4. CH_2Cl_2 was evaporated to yield succinimidyl-5-azidovalerate, as a pale yellow liquid. To incorporate azide moiety in nucleotides, amino-modified nucleotides (10 nmol) were incubated with succinimidyl-5-azidovalerate (10 μmol) in 0.25M $Na_2CO_3/NaHCO_3$ buffer (pH = 9) (40 μL) and 12 μL DMSO for 12 hours at r.t. The excess succinimidyl-5-azidovalerate was removed by SEC using a PD-10 column. The azide-labeled DNA was purified by purification cartridge and was quantified by UV absorbance. The azide-labeled DNA was conjugated with fluorescent probes *via* click chemistry. 6-Carboxyfluorescein propargylamine was synthesized by reacting 3.4 μL of propargylamine (0.05 mmol) with 500 μL of carboxyfluorescein NHS ester (0.023 mmol) in DMF and 0.1M $NaHCO_3$ (100 μL) solution. The mixture was stirred for 5 hours and the solvent was removed under vacuum. The crude product was purified by silica gel TLC plate. The fluorescent DNA was prepared by reacting azide-modified DNA with alkyne-modified fluorescent probe *via* click chemistry [44].

In another study fluorescent DNA was prepared by Staudinger ligation reaction, which allows chemoselective coupling reaction between azide and triarylphosphines of two compounds. Azide-modified DNA was prepared as described above. Fam-tethered triphenylphosphine was then prepared from 3-diphenylphosphino-4-methoxycarbonylbenzoic acid. 3-Diphenylphosphino-4-methoxycarbonylbenzoic acid was activated with DCC/NHS and the resultant 3-diphenyl phosphino-4-methoxycarbonylbenzoate was reacted with 5-(aminoacetamido)fluorescein at a ratio of 1:0.8 in DMF and inert atmosphere to obtain 5-[(N-(3′ diphenyl phosphinyl-4′-methoxycarbonyl) phenylcarbonyl)amino acetamido]fluorescein. Fam-labeled DNA was obtained by adding DMF solution (30 μL, 0.5 μmol) of fluorescein-tethered triphenylphosphine in a 100 μL solution of azido-DNA (2.4 nmol) in $Na_2CO_3/NaHCO_3$ buffer (pH = 9) (Scheme 3.15). The tube was incubated at r.t for 12 hours and DNA was purified by PD-10 column [45].

The alkyne-modified oligodinucleotides (ODNs) were also prepared, to incorporate functional moieties to DNA. The modified uridine nucleosides were prepared from their corresponding phosphoramidites. Modified nucleoside (**3**) (4.26 mmol) was added to CuI (0.852 mmol) and $PdCl_2(PPh_3)_2$ (0.426 mmol) in DMF. The mixture was degased and degased *N,N*-diisopropylethyl amine (21.3 mmol) was added. The reaction was stirred at r.t for 10 minutes. A degased solution of trimethylsilyl-1,7-octadiyne (5.54 mmol) was dissolved in DMF and was added to the reaction mixture slowly over a period of 1 hour. The mixture was then stirred overnight at r.t. The mixture was concentrated and was diluted with ethyl acetate. The organic layer was then washed with water and brine and was dried over $MgSO_4$. The mixture was concentrated and was purified by flash column chromatography to obtain alkyne-modified oligonucleosides (**4**), as shown in Scheme 3.16.

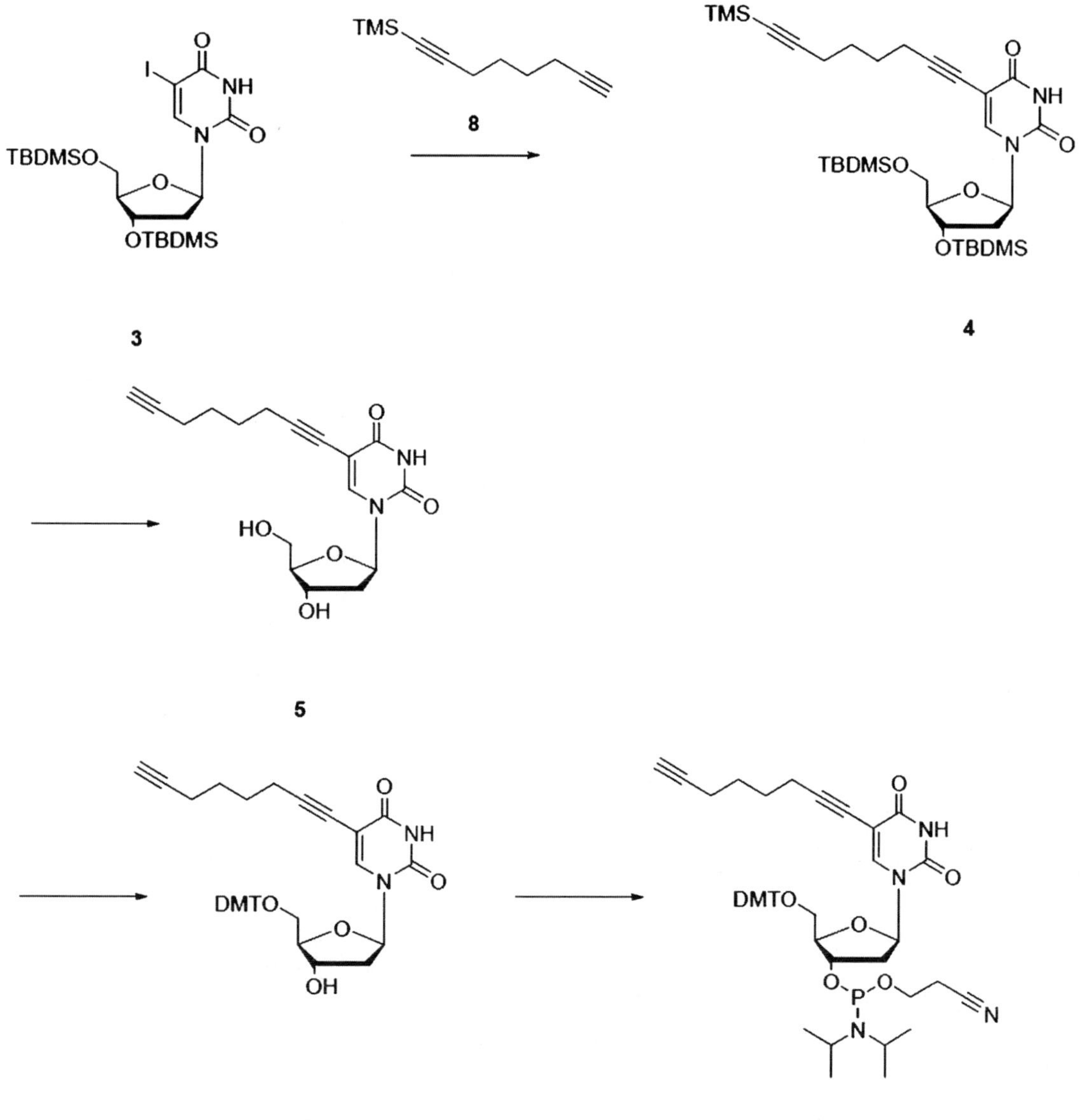

SCHEME 3.16   Synthesis of alkynated nucleosides and nucleotides [47].

The deprotection of (**3**) was performed in THF to obtain (**4**). (**4**) (3.4 mmol) was added in THF and the solution was cooled to 0°C. The mixture was then added to the TBAF solution in THF (1 M, 10.9 mmol) in inert atmosphere. The mixture was stirred for 3 hours and was quenched with glacial acetic acid. The mixture was concentrated and was purified with flash column chromatography to obtain pure alkyne-modified nucleoside (**5**). The alkynated nucleosides were reacted with azide-containing dyes to produce fluorescent probes *via* click chemistry. The introduction of another alkyne group at phosphate moiety of nucleotides was obtained in a two-step reaction. (**5**) (4.97 mmol) was reacted with 4,4′-dimethoxytriphenylmethyl chloride (5.5 mmol) in dry pyridine at 0°C and the mixture was stirred overnight in inert atmosphere. The mixture was concentrated and was

diluted with ethyl acetate. The organic layer was washed with brine and water and was dried over MgSO$_4$. The crude product was purified by flash column chromatography to obtain (**6**). The solution of (**6**) (0.32 mmol) in DCM at 0°C was added to triisopropyl tetrazolide (0.4 mmol) followed by the addition of 2-cyanoethyl tetraisopropyl phosphoramidite (0.97 mmol) in inert atmosphere. The mixture was stirred at r.t for 4 hours. The mixture was concentrated and was purified by flash column chromatography to obtain (**7**). These alkynated nucleosides and nucleotides were incorporated to synthesize 16-mer nucleotides *via* solid-phase DNA synthesis or by PCR reaction [46, 47].

A typical PCR reaction is as follows. 0.5 μg of PCR template, 0.3 mM of each forward and reverse primers, 1.25 U polymerase, and 10X polymerase buffer with magnesium,

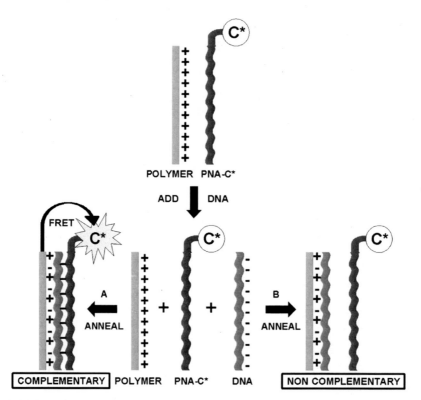

FIGURE 3.4   A representation of CP-PNA reporter conjugates to detect ssDNA [49]. For a color version, see the color plate section.

200 μM of unmodified dNTPs and 200 μM of dUTPs were used. The reaction was performed on Eppendorf Master Cycle Personal. The hot start (2 minutes, 94°C) was used for amplification, followed by 10 cycles (15 seconds at 94°C, 30 seconds at 53°C, 45 seconds at 72°C), 25 cycles (15 seconds at 94°C, 30 seconds at 56°C, 45 seconds at 72°C) and the mixture was finally incubated for 2 minutes at 72°C. PCR products were analyzed by agarose gel electrophoresis and were purified by PCR purification kit [47].

DNA metallization is an interesting approach to introduce conductivity properties to DNA nanostructures. The alkynated nucleotides prepared above were reacted with azide–galactose-containing protective aldehydes. The reaction of bioconjugates with Ag ions was studied by Tollen's reaction [47].

Poly(acrylic acid) (pAA) and 20-mer ssDNA hybrids were prepared to produce highly sensitive probes for ssDNA detection. The DNA–polymer conjugates were then modified with fluorescent probe 3-(2-pyridyldithio)propionyl hydrazide (pDPH) for detection and analysis. 3′-amino group of ssDNA was first modified to increase the chain length. The reaction of ssDNA with N-(ε-trifluorocoacetylcaproyloxy) succinimide ester (TFCS) was done at pH 7 for 1 hour, at 4°C. Trifluoroacetyl functional group was removed by changing the pH to 8 and modified ssDNA was purified by ultrafiltration. ssDNA (0.1 mM) was then reacted with pAA (20 mM) in the presence of excess EDC in 0.1M phosphate buffer at

pH = 8 containing 1M NaCl for 3 days. pAA–ssDNA were then modified with pDPH (in DMSO solution) in 0.1M PBS (pH = 8), containing 0.15M NaCl. The molar ratio of carboxyl group of pAA and hydrazide group of pDPH was 10:1. The reaction was performed at r.t for 24 hours. The conjugates were purified by ultracentrifugation and freeze-drying [48].

Conjugated polymers (CPs) are highly sensitive optical reporters which are characterized by their delocalized electronic structure. The conjugation of CPs with peptide nucleic acid (PNA) is found to produce efficient bioconjugates for DNA detection. PNAs are negatively charged DNA like polymers, where phosphates of DNA are replaced with peptide–mimetic neutral linkages. The conjugates of CP–PNA were prepared by electrostatic interactions. The water-soluble CP namely, poly(9,9-bis(6-N,N,N-trimethylammonium)-hexyl)-fluorenephenylene) was conjugated to PA and was used for the detection of DNA as shown in Figure 3.4 [49].

30-mer ODN, containing antisense sequence for ribosomal binding site for mRNA of EGFP protein was conjugated to PNIPAM. 3′ and 5′-methacryloly-modified antisense ODNs were prepared and were reacted to PNIPAM in 10 mM tris-HCl (pH = 8). Then 100 μL of aqueous ammonium persulfate (13 mM) and 40 μL of N,N,N′,N″-tetramethylethylenediamine (2.15M) was then added and the mixture was incubated for 1 hour at r.t, under N₂. ODN-PNIPAM conjugates were purified and tested for their temperature responsiveness [50].

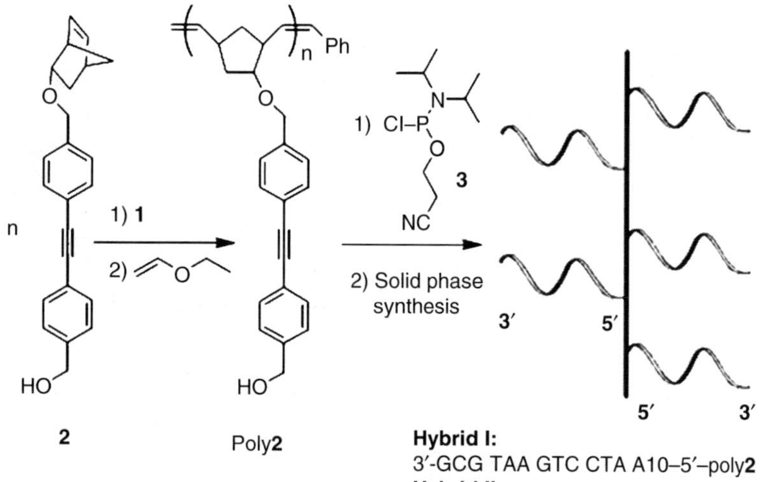

SCHEME 3.17   Synthesis of polymer–DNA hybrids on CPG-modified solid support [53].

DNA–polymer amphiphiles were also prepared by solid-phase DNA synthesis. A hydrophobic PS-based polymer was reacted with alcohol-terminated ODN. The coupling of polymer with 5′-OH group of nucleotide strand bound to CPG is done for 3 hours by a "syringe synthesis" method. The unbound polymer is removed by washing CPG with CH_2Cl_2 and DMF. The conjugates are deprotected and cleaved to obtain pure polymer–DNA conjugates. [51]

Poly(indole) (PIn) and DNA conjugates were prepared in the form of nanowires bearing electric and sensing properties. PIn-DNA nanowires were synthesized in the presence of $FeCl_3$ as antioxidant. Indole (5 μL, 6.24 mM) and λ-DNA (20 μL, 0.5 μg/μL) was mixed in the presence of $MgCl_2$ (0.5 mM) and $FeCl_3$ (5 μL, 1 mM) was added dropwise. The solution was reacted for 1 hour at r.t to allow the synthesis of nanowires. The nanowires produced were analyzed by probe microscopy, FTIR, and XPS [52].

Norbene-based, well-defined polymers were prepared by ROMP and were used for DNA conjugation by post-polymerization modification. The polymers produced were reacted with chlorophosphoramidite and derivatives were reacted with CPG-supported DNA using syringe technique. The DNA is deprotected and hybrids are cleaved from solid support in aqueous ammonia at 60°C, as shown in Scheme 3.17 [53].

Covalent conjugation of PNA with PHEA was obtained. PNA was synthesized by solid-phase synthesis method and was modified with ATRP-initiator, by reacting amine of PNA with 4-(1-bromoethyl)benzoic acid. PNA-initiator was used to synthesize PHEA–PNA conjugates *via* ATRP technique, as shown in Scheme 3.18 [54].

3.5   CONJUGATION WITH DRUGS

Polymer–drug conjugates are ideal carriers to enhance the circulation, efficacy, and stability of drugs *in vivo*. Polyaspartamide, polyamidoamine, poly-2-hydroxypropyl methacrylate are some of the synthetic polymers, studied as drug carriers.

Hyperbranched polyol and ibuprofen conjugates were prepared in anhydrous DMF; polymer was added in excess (40%

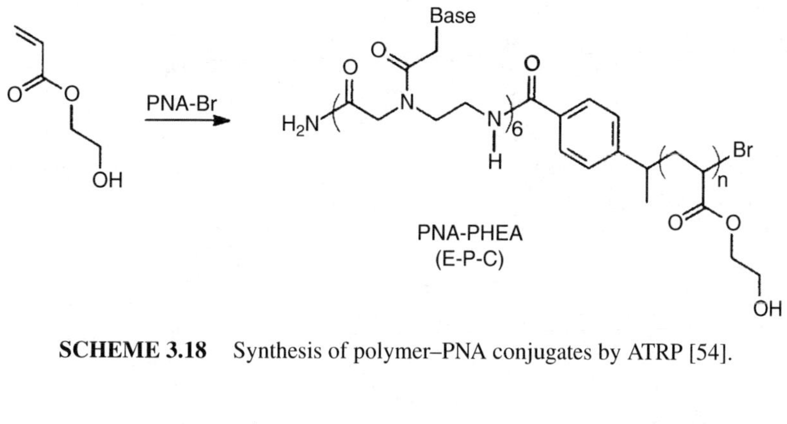

SCHEME 3.18   Synthesis of polymer–PNA conjugates by ATRP [54].

molar excess of OH groups). DCC was added as a coupling agent and the reaction was stirred for 3 days at r.t. The solution was filtered to remove the by-product, dicyclohexyl urea (DCU) formed during the reaction. The filtrate was dialyzed to remove free DCC and ibuprofen. The degree of conjugation was determined by ^{1}H-NMR [55].

O-(chloracetyl-carbamoyl) fumagillol (TNP-470) is a potent angiogenesis inhibitor, and is an emerging anticancer agent. HPMA copolymer-based TNP-470 conjugates were prepared. The random copolymer of HPMA containing 10 mol% tetrapeptide (Gly-Phe-Leu-Gly-ethylenediamine) was obtained and was conjugated to TNP-470 by nucleophilic displacement of the terminal chlorine. For the conjugation reaction, HPMA copolymer and TNP-470 at 100 mg/mL concentration in DMF were prepared separately. The solutions were combined and were incubated at 4°C in dark for 12 hours. DMF was then evaporated, the mixture was dissolved in water, and was dialyzed and freeze-dried to obtain pure conjugates [56].

Paclitaxel (Taxol) was conjugated of poly($_L$-glutamic acid) (PG) by DCC-mediated coupling reaction between COOH groups of PG and OH groups of taxol. PG (75 mg) was dissolved in 1.5 mL of DMF, followed by 20 mg of Taxol, 15 mg of DCC, and trace amount of dimethylaminopyridine. The mixture was incubated at r.t overnight and was precipitated by pouring the mixture in chloroform. The precipitates were dissolved into NaHCO$_3$ to obtain sodium salt of conjugates. The aqueous solution of conjugates was dialyzed in water to obtain pure PG-TXL conjugates [57].

Poly(*N*-(2-hydroxyethyl)-$_L$-glutamine (PHEG) was conjugated to antibiotic mitomycin C (MMC), using peptidyl spacers. PHEG was prepared by aminolysis of PBLG. PBLG (4.57 mmol) and 2-hydroxypyridine (22.8 mmol) were dissolved in DMF. 2-Aminoethanol was added dropwise over a period of 1 hour. The reaction was stirred at 40°C for 24 hours and PHEG was precipitated in diethyl ether/ethanol mixture (4:1 v/v). PHEG was then activated by 4-nitrophenyl chloroformate. Briefly, PHEG (1.45 mmol), and 4-dimethylaminopyridine (0.13 mmol) were dissolved in NMP/pyridine solution (4:1 v/v). Chloroformate (0.87 mmol) was then added at 0°C. The reaction was done at 0°C for 4 hours. The reaction mixture was precipitated in diethyl ether/ethanol mixture. Then MMC–peptide conjugates were prepared and Fmoc-protected peptides were derivatized into ester form. Fmoc-protected peptide (0.35 mmol) and pentafluorophenol (0.425 mmol) were dissolved in dry THF. The mixture was cooled to 0°C and DCC (0.355 mmol) was added. The reaction was stirred for 2 hours at 0°C and was incubated overnight at r.t. The by-product was removed by filtration and peptide conjugate was precipitated in 1/1 mixture of diethyl ether/hexane. MMC was then conjugated with Fmoc-modified oligopeptide–pentafluorophenyl ester. MMC and peptide were dissolved in 5 mL DMF and 0.1 mL of pyridine at 1:1 molar ratio.

The mixture was incubated in the dark for 48 hours at r.t. The solvent was evaporated and MMC–peptide conjugates were purified by column chromatography. MMC–peptide conjugates were deprotected using triethylamine. The amine-containing MMC–peptide and activated PHEG were reacted in NMP/pyridine solution (4/1, v/v) for 48 hours in the dark. The mixture was precipitated in diethylether/ethanol (2/1, v/v) to obtain pure MMC-PHEG conjugates [58].

Methotrexate–PAMAM conjugates were prepared by EDC/NHS chemistry. Acetamide carboxyl or hydroxypropyl capped PAMAM dendrimers were first prepared. PAMAM dendrimer (1.7 μmol) was dissolved in DMSO and 0.1M sodium phosphate buffer (pH = 9). Acetic anhydride or glycidol (833 μmol) were added dropwise and the mixture was stirred for 6 hours. The pH of the solution was adjusted to 9 and the mixture was stirred in the dark overnight. This procedure was repeated four times and on the fifth day, the mixture was filtered and the filtrate was dialyzed to obtain pure acetamide or hydroxypropyl capped PAMAM dendrimers. Carboxyl-functionalized dendrimer was prepared by the reaction of succinic anhydride (179 μmol) with PAMAM (1.2 μmol) in DMSO at r.t for 18 hours, in the dark in inert atmosphere. Methotrexate was activated by adding methotrexate (9.9 × 10^{-7} mol) and 1-[3-(dimethylamino) propyl-3-ethylcarbodiimide hydrochloride (1.4 × 10^{-5} mol) in DMSO. The mixture was reacted at r.t in inert atmosphere for 1 hour. The activated methotrexate was added dropwise to capped PAMAM dendrimer (9.9 × 10^{-8} mol) in water. The reaction was stirred at r.t for 3 days, and the solution was dialyzed and lyophilized to obtain pure conjugates [59].

pH-sensitive PHPMA-DOX conjugates were prepared by hydrazone linkage. PHPMA derivatives containing reactive side chain hydrazide group were prepared by hydrazinolysis of 4-nitrophenyl ester group of polymer precursor with hydrazine monohydrate, as shown in Scheme 3.19.

Polymer–DOX conjugates were prepared by the reaction of polymer precursor II with DOX-HCl in the dark at r.t. Polymer precursor II was dissolved at 10 wt% concentration in anhydrous methanol and DOX-HCl at 60 wt% concentration was as added in polymer solution, while stirring. A drop of acetic acid was added and the reaction was stirred for 48 hours. The polymer conjugates were purified by multipleGPC [60].

In another study, PHPMA-DOX conjugates were prepared by pH-sensitive hydrazone bond. However, in the study statistical copolymers of HPMA with *N-N′*(6-methacrylamidohexanoyl)hydrazine were synthesized by radical polymerization and were used for the conjugation of DOX, as described above [61].

5-aminosalicylic acid (ASA) is a drug for colonic infection, which is largely unabsorbed by colon. ASA–dendrimer conjugates are prepared and are shown to improve the efficacy of ASA by several folds. PAMAM dendrimer were modified with *p*-amino benzoic acid (PBA), as shown in

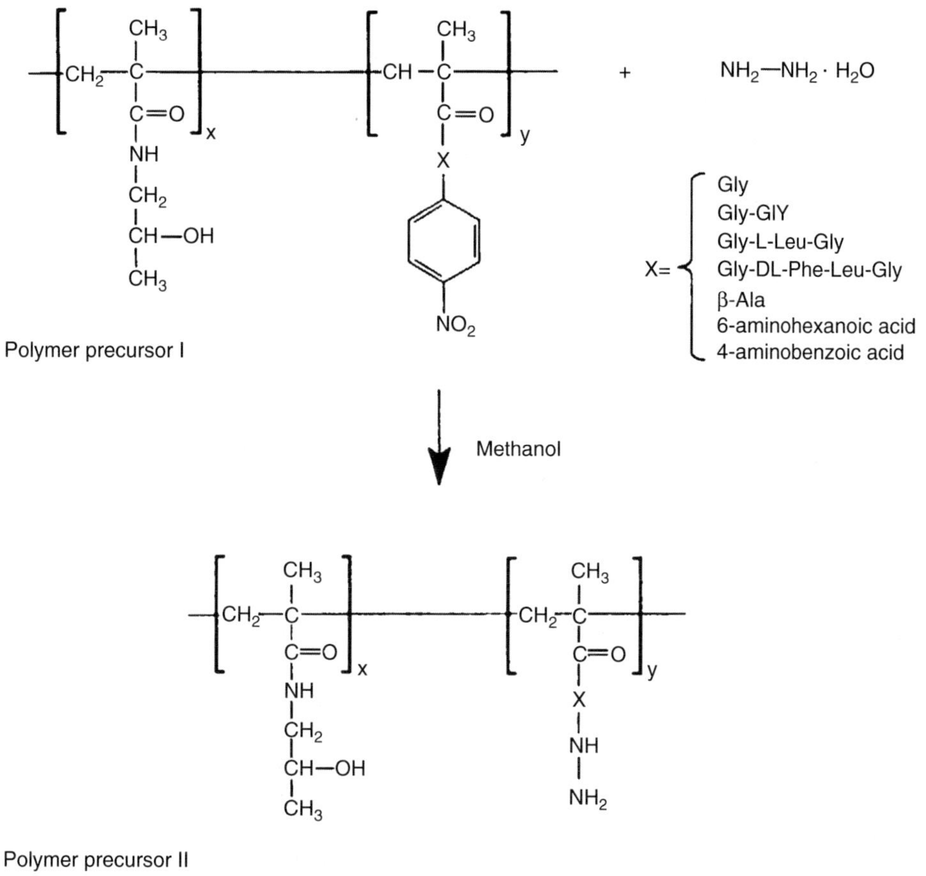

SCHEME 3.19   Synthesis of HPMA-DOX conjugates [60].

Scheme 3.20. PAMAM-PBA were then conjugated to ASA to yield PAMAM-PBA-ASA. PAMAM-PBA were dissolved in a solution of NaNO₂ (6.4 mmol) and the mixture was stirred on ice bath for 15 minutes, followed by acidification with HCl. The solution was stirred for 5 minutes. ASA solution (6.4 mmol) in 0.5M NaOH was prepared and was cooled at ice bath for 15 minutes. Diazonium solution was then added to ASA solution and pH was adjusted to 10. The mixture was stirred for 30 minutes and the solution was dialyzed and lyophilized to obtain PAMAM-PBA-ASA conjugates [62].

Methylprednisolone (MP), a corticosteroid 6α-methyl-11β,17α,21-trihydroxy-1,4-pregnadiene-3,20-dione, was conjugated to PAMAM dendrimers by using DCC chemistry as describe above [63].

Platinum drugs are effective anticancer treatment, which are in clinical trials now. Cisplatin [cisdichlorodiamine platinum(II) (CDDP)] is one of the most studied platinum-based anticancer drug. In order to improve the efficacy and to reduce the side effects of cisplatin, its conjugation with nontoxic polymers have been proposed. Degradable thermoresponsive PEGMA and 1,1-di-tert-butyl 3-(2-(methacryloyloxy)ethyl) butane-1,1,3-tricarboxylate (MAETC)-based statistical copolymer were prepared by RAFT. The conjugation

of polymers with cisplatin was done in the presence of AgNO₃; cisplatin:AgNO₃ molar ratio was 1:2. Cisplatin was first reacted with AgNO₃ to make complexes in water. The precipitates formed by AgCl were removed and the polymer solution was added to the filtrate. The mixture was reacted at 37°C for 12 hours with gentle shaking to obtain CDPP–polymer conjugates. (Figure 3.5) [64].

In another study, therapeutic index of platinum-based drugs was improved by producing polymer-linked DACH (diaminocyclohexyl)–platinum conjugates. Poly(HPMA)–camptothecin conjugates (p(HPMA)-CPT) were prepared in a three-step reaction. The copolymers of HPMA were first synthesized by radical precipitation of HPMA with methylacryloyl-glycyl-p-nitrophenyl ester (molar ratio of 9:1). The content of p-nitrophenyl ester incorporated into the polymer was determined by UV-absorbance. CPT was modified with (tert-butylcarboxy)-phenylalanyl-leucyl-glycyl p-nitrophenyl ester to obtain 20-O-(phenylalanyl-leucyl-glycyl)-camptothecin. The modification of copolymer of HPMA with CPT peptide was done by aminolysis in DMSO in inert atmosphere for 15 hours (Scheme 3.21). The residual p-nitrophenol groups were quenched by propanolamine. The polymer conjugates were precipitated in acetone [65].

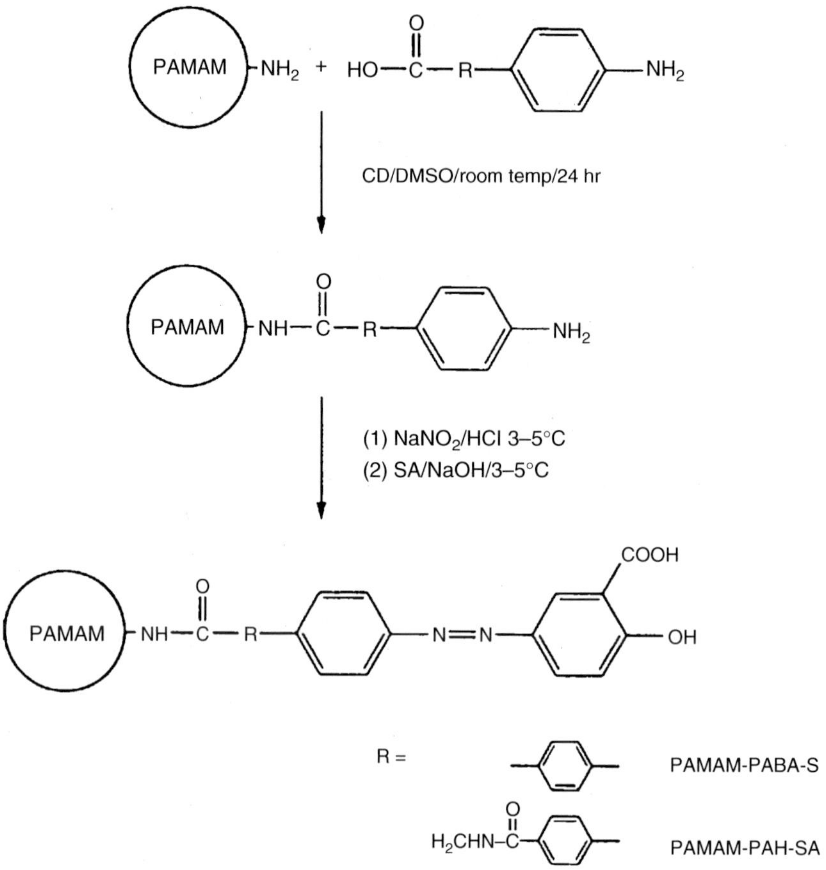

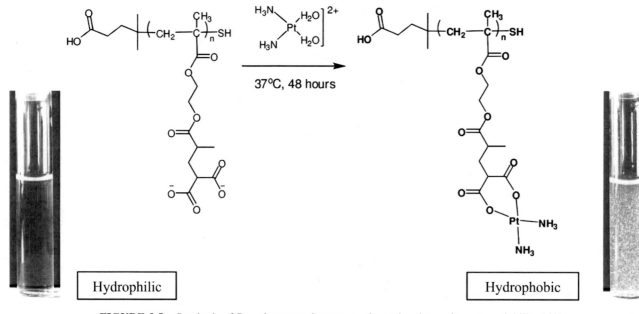

SCHEME 3.20   Depiction of PAMAM-SA conjugates [62].

FIGURE 3.5   Synthesis of Pt–polymer conjugates, as shown by change in water solubility [64]. For a color version, see the color plate section.

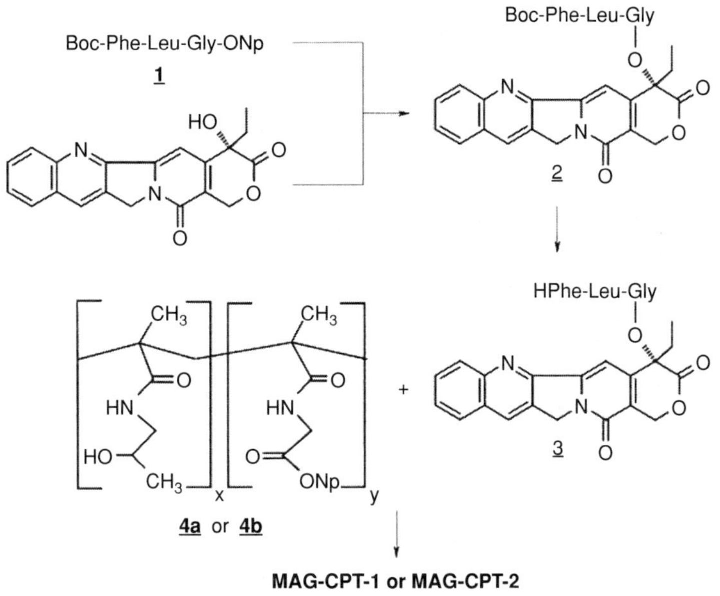

SCHEME 3.21   Synthesis of PHPMA–camptothecin conjugates, using a linker peptide [65].

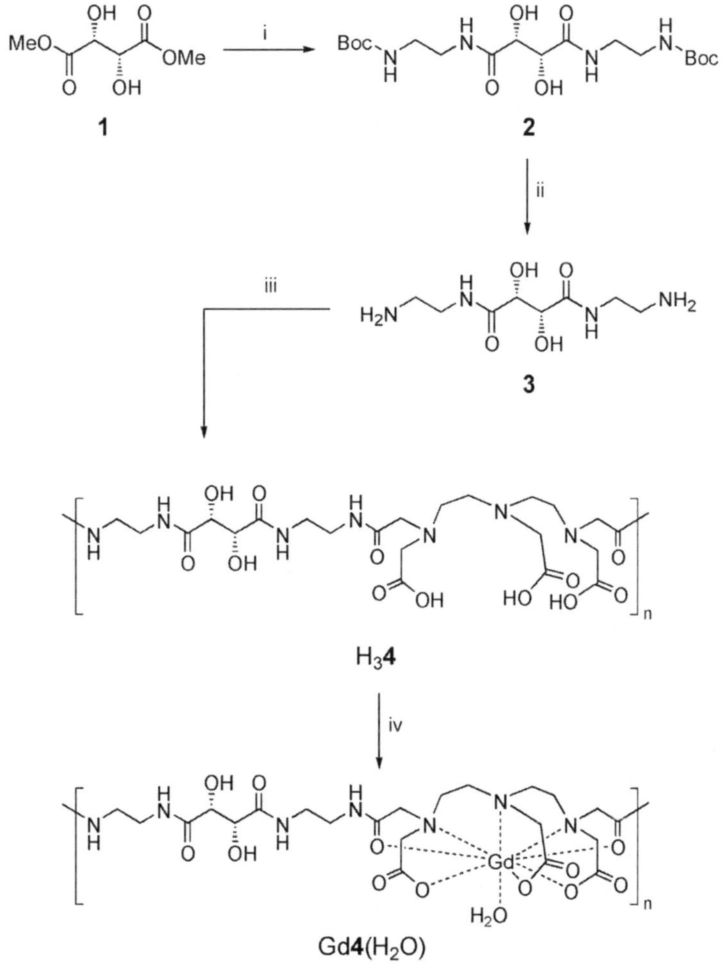

SCHEME 3.22   Synthesis of Gd-conjugated glycopolymer for MRI [66].

3.6   CONJUGATION WITH CONTRAST AGENTS

3.6.1   Polymers for Magnetic Resonance Imaging

Magnetic resonance imaging (MRI) is a leading imaging technique in medicine today. Gadolinium(III)-like contrast agents containing paramagnetic metal ions are used to improve the sensitivity of MRI technique. The modification of Gd(III) with different chelators is one approach to improve the image enhancing properties of Gd. A number of chelators such as DTPA, DOTA, and HOPO have been studied for this purpose. The modification of Gd-complexes with polymers, improves the solubility of the former. The synthetic glycopolymer-modified Gd-DTPA-complexes were prepared as shown in Scheme 3.22. In a stepwise manner carbohydrate–diamine monomer (**3**) was synthesized from tartaric acid (**1**). The compound (**3**) was polymerized by condensing (**3**) with diethylenetriaminepentaacetic acid–bisanhydride (DTPA-BA) to obtain polymeric precursor (**H₃4**). **H₃4** was reacted with $GdCl_3.6H_2O/EuCl_3.6H_2O$ at 1:1 molar ratio in 0.1M NaOH. The pH of the solution was adjusted to 6 and the solution was stirred at r.t for 1 hour to obtain **Gd4H₂O**. The pure Gd(III)-complex or Eu(III)-complexes were obtained by dialysis and freeze-drying [66].

pLL and random copolymers of p(Lys–Glu) (4:1) were conjugated with Gd contrast agent in the presence of different chelators, such as DOTA and DTPA. The anhydride of DTPA was added to PLL solution in 100 mM $NaHCO_3$ buffer in ice bath. The reaction was allowed to occur for 16 hours at r.t. The mixture was concentrated and was purified by dialysis [67].

3.6.2   Polymers for Optical Imaging

"Grafting from" technique was also used to produce fluorescent protein–polymer bioconjugates. Materials containing fluorescent properties are interesting as they can provide imaging capabilities *in vitro* and *in vivo*. Hostasol methacrylate monomer and rhodamine B-based monomer (fluorescent probes) were copolymerized with PEGMA or DMAEMA *via* ATRP, as shown in Scheme 3.23. All synthetic steps were monitored by SDS-PAGE [17]. The native BSA was isolated from modified BSA by HPLC technique. BSA-macroinitiator was used to polymerize PEGMA or DMAEMA and hostasol or rhodamine-based fluorescent monomers. The bioconjugates were purified by dialysis and lyophilization. Similarly, lysozyme–NHS-macroinitiator was used to produce fluorescent bioconjugates to confirm the versatility of the approach [17].

In another study, fluorescent thermoresponsive polymer PEGMA was synthesized by ATRP and was grafted on keratin fibers (hair), bearing free thiol and amine

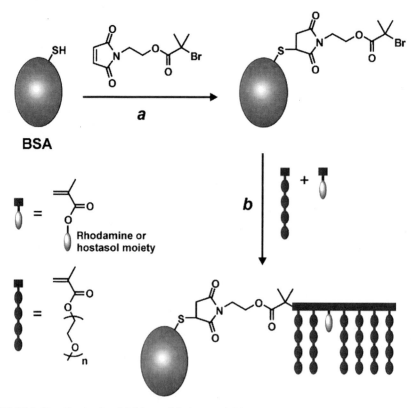

SCHEME 3.23   Synthesis of BSA-modified macroinitiator and LRP of PEGMA and fluorescent monomer [17].

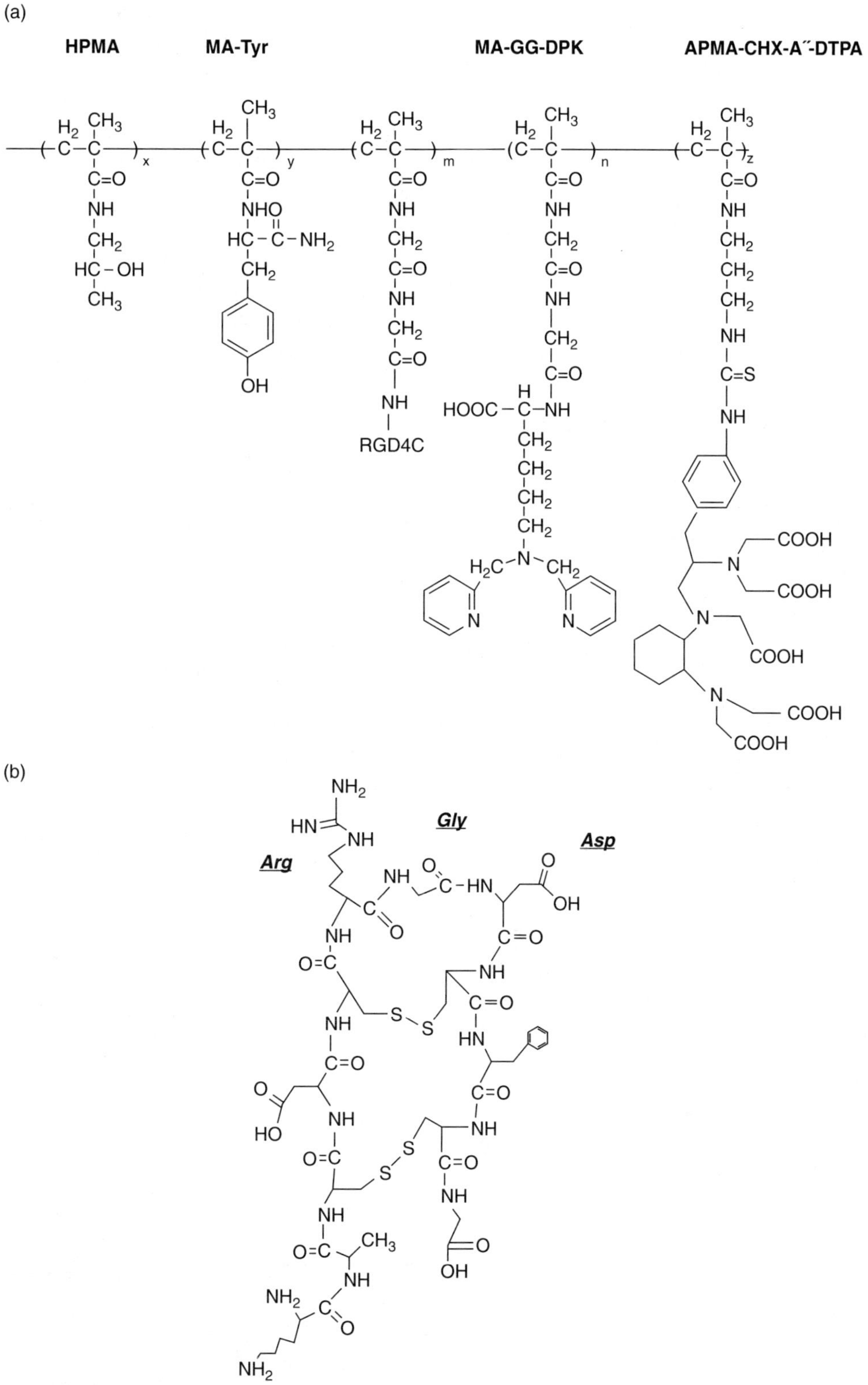

(a)

(b)

SCHEME 3.24   HPMA copolymers, containing RGD4C peptide and ^{99m}Tc, and ^{90}Y chelators (A), chemical structure of RGD4C (B) [69].

functionalities on the surface. Hostasol methacrylate (HMA) monomer was copolymerized with PEGMA by ATRP using *N*-hydroxy-succinimide-2-bromopropionate as initiator. The reaction was done at 50°C in toluene for 48 hours. A range of fluorescently labeled polymers of PEGMA-*co*-HMA were thus obtained. The bioconjugation reaction was performed on untreated hair in anhydrous DMSO with 5 wt% triethylamine in the presence of fluorescent polymers for 70 hours at r.t. The hairs were intensely rinsed with water to remove excess polymers and were air dried. The bioconjugation was confirmed using confocal microscopy [68].

3.6.3 Polymers for Nuclear imaging

HPMA-based copolymers containing $\alpha_v\beta_3$ integrin-targeting peptide (RGD4C) and chelators of ^{99m}Tc and ^{90}Y were prepared and were studied for the effectiveness of β-radiotherapy in tumor vasculature, as shown in Scheme 3.24. The copolymers of HPMA-containing chelator side chains of ^{99m}Tc and ^{90}Y were prepared by free radical polymerization. RGD4C conjugates of polymers were prepared by aminolysis of *p*-nitrophenyl ester. ^{99m}Tc radiolabeling of peptide–polymer conjugate was then obtained by coupling of ^{99m}Tc tricarbonyl to lysine moiety (*N*-ω-bis(2-pyridylmethyl)-L-lysine, DPK) of the chelator. Sodium pertechnetate (NaTcO$_4$) was boiled with sodium tartarate, sodium borate, sodium boranocarbonate, and sodium bicarbonate for 20 minutes to obtain ^{99m}Tc-tricarbonyl [^{99m}Tc(H$_2$O)$_3$(CO)$_3$]. 120 μL of 1N HCl was then added to decompose residual boranocarbonate and neutralize the pH to 6–7. The polymer–peptide mixture in saline was then added to ^{99m}Tc-containing solution and the mixture was heated at 75°C for 30 minutes. The labeled conjugates were purified by Sephadex G-25 column.

^{90}Y was labeled with polymer–peptide conjugate via *trans*-(*S,S*)-cyclohexane-1,2-diamine-*N,N-N′,N′,N″N″*-pentaacetic acid (CHX-A″-DTPA) side chain. HCl solution of ^{90}Y was purged with nitrogen and pH of the solution was adjusted to 5 with 100 μL of 1M acetate buffer. The polymer–peptide solution was added to ^{90}Y solution and the mixture was incubated at 22°C for 30 minutes. The reaction was quenched by 0.1M EDTA and was stirred for 10 minutes at 22°C. The labeled conjugates were purified by Sephadex G-25 column [69].

3.7 CONCLUSION AND PERSPECTIVE

The progress in the field of polymer chemistry has provided valuable tools for the synthesis of well-defined polymers. The tailor-made polymers of specific shape, predetermined molecular weights, narrow polydispersities, and presence of functional groups for bioconjugation have resulted in well-organized "macromolecular chimeras." In this chapter,

a combination of polymerization and bioconjugation approaches used to produce "biohybrids" are discussed in detail. Polymer–protein hybrids are extensively studied bioconjugates to date. The polymer–amino acid, polymer–peptide and polymer–protein "biohybrids" are produced either by covalent or noncovalent conjugation approach. The polymers produced by free radical polymerization, ATRP and RAFT polymerization techniques, are conjugated to biomolecules by either "grafting to" or "grafting from" approach. The synthetic carbohydrate-based bioconjugates are now a focus of research, due to the availability of facile polymerization approaches to yield well-defined "glycoconjugates." The synthetic methacrylate or methacrylamide-based sugar monomers are produced and are polymerized by RAFT or ATRP to yield linear or branched polymers. The oligosaccharide-based monomers are also polymerized to obtain well-defined glycopolymers. These glycopolymers are further conjugated with proteins, lectin, and DNA to produce complex biomolecules. ODNs, ssDNA, DNA, and PNA-based bioconjugates are obtained using various methods. The electrostatic interactions between cationic polymers and anionic NA is a well-studied approach to produce biomolecules, which are further utilized for various biological applications such as gene delivery, and siRNA delivery. In addition, synthetic alkyne-modified oligonucleosides and oligonucleotides are synthesized and are conjugated to a variety of molecules such as fluorescent probes. The synthesis of polymer–drug conjugates is interesting, as this bioconjugation strategy increases the therapeutic efficacy, specificity, and half-life of the drug *in vivo*. A variety of polymer–drug conjugates are prepared and are studied for their therapeutic efficacies *in vitro* and *in vivo*. The polymeric vectors are also conjugated with imaging agents such as MRI agents, fluorescent probes, and nuclear imaging agents. These fluorescent probes have enabled the tracking of biomolecules *in vitro* as well as *in vivo*.

REFERENCES

1. Lutz JF, Borner HG. *Prog Polym Sci* 2008;33:1–39.
2. Borner HG. *Macromol Chem Phys* 2007;208:124–130.
3. Summerlin B, Van Hest JM. *Polym Chem* 2011;2:1427.
4. Van Hest JM. *J Macromol Sci Part C: Polymer Rev* 2007;47:63–92.
5. Le Durmanguet B, Vellonia K. *Macromol Rapid Commun* 2008;29:1073–1089.
6. Canalle LA, Lowik DWPM, Van Hest JM. *Chem Soc Rev* 2010;39:329–353.
7. Salmaso S, Bersani S, Pennadam SS, Alexander C, Caliceti P. *Intl J Pharma* 2007;340:20–28.
8. Li M, De P, Gondi SR, Sumerlin BS. *Macromol Rapid Commun* 2008;29:1172–1176.

9. Jin J, Wu D, Sun P, Liu L, Zhao H. *Macromolecules* 2011;44:2016–2024.

10. Danial M, Root MJ, Klok HA. *Biomacromolecules* 2012;13:1438–1447.

11. Klok HA. *Macromolecules* 2009;42:7990–8000.

12. Wan X, Zhang G, Ge Z, Narain R, Liu S. *Chem Asian J* 2011;6:2835–2845.

13. Canalle LA, Van der Knaap M, Overhand M, Van Hest JCM. *Macromol Rapid Commun* 2011;32:203–208.

14. Kostiainen MA, Szilavy GR, Lehtinen J, Smith DK, Linder MB, Urtti A, Ikkala O. *ACS Nano* 2007;1:103–113.

15. Thordarson P, Le Droumaguet B, Velonia K. *Appl Microbiol Biotechnol* 2006;73:243–245.

16. Hoffman AS. *Clinical Chem* 2000;46:1478–1486.

17. Nicolas J, Miguel VS, Mantovani G, Haddleton DM. *Chem Commun* 2006:4697–4699.

18. Nicolas J, Mantovani G, Haddleton DM. *Macromol Rapid Commun* 2007;28:1083–1111.

19. Hoffman AS, Stayton PS. *Prog Polym Sci* 2007;32:922–932.

20. Lutz JF, Borner HG, Weichanhan K. *Aust J Chem* 2007;60:410–413.

21. Geng Y, Discher DE, Justynska J, Schlaad H. *Angew Chem Int Ed* 2006;45:7578–7581.

22. Johnson RP, Jeong YI, Choi E, Chung CW, Kang DH, Oh SO, Kim I. *Adv Funct Mater* 2012;22:1058–1068.

23. Jacobs J, Pound-Lana G, Klumperman B. *Polym Chem* 2012;3:2551–2560.

24. Mei Y, Beers KL, Byrd HCM, VanderHart DL, Washburn NR. *J Am Chem Soc* 2004;126:3472–3476.

25. Lutz JF. *Angew Chem Int Ed* 2007;46:1018–1025.

26. Li H, Bapat AP, Li M, Sumerlin BS. *Polym Chem* 2011;2:323–327.

27. Miyamoto D, Watanabe J, Ishihara K. *Biomaterials* 2004;25:71–76.

28. Wan X, Liu X. *Macromol Rapid Commun* 2010;31:2070–2076.

29. Averick S, Simakova A, Park S, Konkolewicz D, Magenau AJD, Mehl RA, Matyjaszewski K. *ACS Macro Lett* 2012;1:6–10.

30. Ladmiral V, Mantovani G, Clarkson GJ, Cauet S, Irwin JL, Haddleton DM. *J Am Chem Soc* 2006;128:4823–4830.

31. Mancini RJ, Lee J, Maynard HD. *J Am Chem Soc* 2012;134:8474–8479.

32. Deng Z, Ahmed M, Narain R. *J. Polym Sci Part A: Polym Chem* 2009;47:614–627.

33. Ahmed M, Narain R. *Biomaterials* 2012, 33, 3990–4001. ASAP article.

34. Wu P, Malkoch M, Hunt JN, Vestberg R, Kaltgrad E, Finn MG, Fokin VV. *Chem Commun* 2005:5775–5777.

35. Glassner M, Delaittre G, Kaupp M, Blinco JP, Barner-Kowollik C. *J Am Chem Soc* 2012;134:7274–7277.

36. Kobayashi K, Kamiya S, Enomoto N. *Macromolecules* 2006;29:8670–8676.

37. Muthukrishnan S, Zhang M, Burkhardt M, Drechsler M, Mori H, Muller AHE. *Macromolecules* 2005;38:7926–7934.

38. Hashimoto K, Siato H, Ohsawa R. *J Polym Sci Part A: Polym Chem* 2006;44:4895–4903.

39. Yoshida T, Akasaka T, Choi Y, Hattori K, Yu B, Mimura T, Kaneko Y, Nakashima H, Aragaki E, Premanathan M, Yamamoto N, Uryu T. *J Polym Sci Part A: Polym Chem* 1999;37:789–800.

40. Wang X, Liu L, Luo Y, Shi H, Li J, Zhao H. *Macrom ol Biosci* 2012, 11, 1575–1582. ASAP article.

41. Synatschke VC, Schallon A, Jérôme V, Freitag R, Müller AHE. *Biomacromolecules* 2011;12:4247–4255.

42. Ahmed M, Jawanda M, Ishihara M, Narain R. *Biomaterials* 2012, 33, 7858–7870. ASAP article.

43. Uchida H, Miyata K, Oba M, Ishii T, Suma T, Itaka K. *J Am Chem Soc* 2011:133:15524–15532.

44. Seo TS, Li Z, Ruparel H, Ju J. *J Org Chem* 2003;68:609–612.

45. Wang CC-Y, Seo TS, Li Z, Ruparel H, Ju J. *Bioconjugate Chem* 2003;14:697–701.

46. Gierlich J, Burley GA, Gramlich PME, Hammond DM, Carell T. *Org Lett* 2006;8:3639–3642.

47. Burley GA, Gierlich J, Mofid MR, Nir H, Tal S, Eichen Y, Carell T. *J Am Chem Soc* 2006;128:1398–1399.

48. Taira S, Yokoyama K. *Biotechnol Bioeng* 2004;88:35–41.

49. Gaylord BS, Heeger AJ, Bazan GC. *Proc Natl Acad Sci USA* 2002;99:10954–10957.

50. Murata M, Kaku W, Anada T, Sato Y, Kano T, Maeda M, Katayama Y. Novel DNA/Polymer Conjugate for Intelligent Antisense Reagent with Improved Nuclease Resistance. *Bioorg Med Chem Lett* 2003;13:3967–3970.

51. Li Z, Zhang Y, Fullhart P, Mirkin CA. *Nano Lett* 2004;4:1055–1058.

52. Hassanien R, Al-Hinai M, Al-Said SAF, Little R, Siller L, Wright NG, Houlton A, Horrocks BR. *ACS Nano* 2010;4:2149–2159.

53. Watson KJ, Park SJ, Im JH, Nguyen ST, Mirkin CA. *J Am Chem Soc* 2001;123:5592–5593.

54. Wang Y, Armitage BA, Berry GC. *Macromolecules* 2005;38:5846–5848.

55. Kolhe P, Khandare J, Pillai O, Kannan J, Lieh-Lai M, Kannan, R. *Pharm Res* 2004;21:2185–2195.

56. Satchi-Finaro R, Puder M, Davies JW, Tran HT, Sampson DA, Greene AK, Corfas G, Folkman J. *Nature Med* 2004;10:255–261.

57. Li C, Yu DF, Newman RA, Carbal F, Stephens CL, Hunter N, Milas L, Wallace S. *Cancer Res* 1998;58:2404–2409.

58. De Marre A, Soyez H, Schacht E. *J. Control Release* 1994;32:129–137.

59. Quintana A, Raczka E, Piehler L, Lee I, Myc A, Majoros I, Patri AK, Thomas T, Mule J Jr, Baker JR. *Pharm Res* 2002;19:1310–1316.

60. Etrych T, Chytil P, Jelinkova M, Rihova B, Ulbrich K. *Macromol Biosci* 2002;2:43–52.

61. Kovar L, Strohalm J, Chytil P, Mrkvan T, Kovar M, Hovorka O, Ulbrich K, Rihova B. *Bioconjugate Chem* 2007;18:894–902.

62. Wiwattanapatapee R, Lomlim M, Saramunee K. *J Control Release* 2003;88:1–9.

63. Khandare J, Kolhe P, Pillai O, Kannan S, Lieh-Lai M, Kannan MR. *Bioconjugate Chem* 2005;16:330–337.

64. Huynh VT, De Souza PL, Martina H, Stenzel MH. *Macromolecules* 2011;44:7888–7900.

65. Caiolfa VR, Zamai M, Fiorino A, Frigerio E, Pellizoni C, d' Argy R, Ghiglieri A, Castelli MG, Farao M, Pesenti E, Gigli M, Angelucci F, Suarato A. *J Control Release* 2000;65:105–119.

66. Lucas RL, Benjamin M, Reineke TM. *Bioconjugate Chem* 2008;19:24–27.

67. Uzgiris EE, Cline H, Moasser B, Grimmond B, Amaratunga M, Smith JF, Goddard G. *Biomacromolecules* 2004;5:54–61.

68. Nicolas J, Khoshdel E, Haddleton DM. *Chem Commun* 2007:1722–1724.

69. Mitra A, Nan A, Papadimitriou JC, Ghandehari H, Line BR. *Nucl Med Biol* 2006;33:43–52.

4

NATURAL POLYMER BIOCONJUGATE SYSTEMS

LICHAO LIU, XIUWEI PAN, AND WEIAN ZHANG

Shanghai Key Laboratory of Functional Materials Chemistry, East China University of Science and Technology, Shanghai, PR China

4.1 INTRODUCTION

The field of covalent polymer-based bioconjugation is fast expanding and its potentials have already been proved by the recent progress in biomedical and pharmaceutical research, especially including polymer–drug conjugates [1, 2], polymer–peptide/protein conjugates [3–5] using controlled/living radical polymerization, click chemistry, and Staudinger ligation. Polymers are widely found in nature, namely natural polymers, such as collagen, chitin, silk fibroin, and cellulose. Naturally derived polymers, as their name implies, are derived from natural sources such as animals, plants, trees, and bacteria. Natural polymers can be divided into protein-origin polymers and polysaccharidic polymers. The extensive use of natural polymeric materials has been considered vital in bioconjugation technology and methodology, because they possess unique physical–chemical properties such as outstanding biodegradability, biocompatibility, nontoxicity, similarity with extracellular matrix (ECM), intrinsic cellular interactions, and typically good biocharacteristics compared with synthetic polymers.

Over the last few decades, impressive progress has been recorded in terms of developing new materials or refining existing material composition or microstructure in order to obtain better performance for a variety of biomedical applications with the advancement of biotechnology. Naturally derived materials can also be easily engineered and are amenable to ligand conjugation, cross-linking, and other appropriate surface modification to provide an ideal candidate for a range of applications, in particular as gene/drug delivery systems. Leong et al. [6] highlighted the past and present research on various applications of natural polymers as gene carriers and tissue engineering scaffolds for gene delivery in the field of nonviral gene therapy and regenerative medicine. Although synthetic polymeric and viral carriers still dominate the field of gene delivery, natural polymers have unique, intrinsic properties that render them appealing to serve as a targeted gene carrier and a tissue engineering scaffold with improvement in transfection efficiency as well as prolonging the residence time once delivered *in vivo*. Malafaya et al. [7] reviewed the biocharacteristics of a wide range of natural-origin polymers with special focus on proteins and polysaccharides that render them as one of the most attractive options to be used in the combination of tissue engineering field and drug delivery, aiming at different clinically relevant applications. They verified that combining both applications into a single material was very challenging though. Fu et al. [8] also summarized the contribution of natural materials and natural materials-based protein delivery systems to regenerative medicine research, with emphasis on the roles of natural materials in the methodology of the application of multifunctional vehicles for cell and growth factor delivery in skin regeneration research. The desirable properties of naturally derived materials render them attractive for applications in the dual role as a scaffold material and as a vehicle for active biomolecules.

Natural-origin polymers have received extensive interest in the bioconjugation field in recent years, even though they present some drawbacks, namely the difficulties in controlling the variability from batch to batch, mechanical properties or limited processability. Many efforts have been made to chemically modify them or conjugate them with bioactive molecules for biomedical applications. The properties and applications of natural polymers are reviewed in the following details according to their nature and origins.

Chemistry of Bioconjugates: Synthesis, Characterization, and Biomedical Applications, First Edition. Edited by Ravin Narain.
© 2014 John Wiley & Sons, Inc. Published 2014 by John Wiley & Sons, Inc.

4.2   NATURAL POLYMER SYSTEMS

4.2.1   Protein-origin Polymer Bioconjugates

4.2.1.1   Introduction
Protein-origin polymers, which are a group of natural polymers composed of natural amino acid sequences linked by amide (or peptide) bonds, have recently emerged as a promising new class of bioconjugated materials, including collagen, gelatin, silk protein, and other proteins such as fibrin and elastin. For example, wheat proteins (WPs), as one of the most interesting plant proteins, have demonstrated a series of polymer properties such as high tensile strength, good viscoelastic properties, and excellent gas barrier performance when used in packaging, coating, and biomedical applications.

There has been enormous progress in the design and engineering of natural protein-based functional nanomaterials over the past few decades in the field of biomedicine, bioengineering, pharmaceutical and therapeutical science due to their low inherent toxicity, outstanding biocompatibility, and biodegradability. These polymers are easily absorbed into the microenvironment by biological and natural physical degradation processes, which can be used as effective delivery carriers of drugs, gene, and imaging agents. Protein-based therapeutics represents a powerful series of clinically approved drugs for the prevention and treatment of various diseases. Precise and specific structural properties can be obtained via chemical modification or covalent conjugation to yield exactly functional features of the polymer, such as hydrophobicity, secondary structures and biorecognizable motifs, to improve, for example, cellular and tissue interactions and distribution as well as bioconjugation and interaction. However, as a member of natural-origin polymers, protein-based polymers have the common disadvantages of natural polymers, including limitations involved in statistical characterization of conventional polymer synthetic techniques, namely possible batch variation. One interesting strategy to overcome this issue is the recombinant protein technologies which realize the monodispersity and precisely defined properties of polymers as well as the predictable placement of cross-linking groups; binding moieties at specific sites along the polypeptide chain or their programmable degradation rates makes them very attractive and useful for drug delivery and tissue engineering [7]. These protein-based polymers such as collagen, gelatin, and silk, and their applications are described in more detail in the following sections.

4.2.1.2   Collagen
Collagen is the most abundant protein found in animals, the most abundant extracellular protein in vertebrates and the predominant component of all connective tissues such as basement membranes, tendons, ligaments, cartilage, bones and the skin. More than 90% of the extracellular protein in tendons and bones, and more than 50% of the skin consists of collagen [8]. Collagen is a highly cross-linked material which is usually insoluble in water. At molecular level, each chain of collagen contains 1000 amino acids. Also, collagen has a complex hierarchical conformation composed of a triple left-handed, rod-like helix held by weaker bonds, namely hydrogen bonds, dipole–dipole bonds, ionic bonds, van der Waals interactions. The triple helix structure contains three basic amino acids: glycine, proline, and hydroxyproline, which generally consists of two identical chains (α1) and an additional chain that differs slightly in its chemical composition (α2). The pattern is glycine, proline, and X, with X being any amino acid. Specific amino acids can cause specific functions for collagen [9].

Collagen has potential as a biomaterial compared with other natural polymers due to its abundance, superior biocompatibility, high porosity, facility for combination with other materials, high mechanical strength, hydrophilicity, weak antigenicity, absorbability in the body, etc. Much of the effort and energy in the past was focused on biomimetic mineralization, chemical modification and cross-linking of collagen for a variety of applications such as cartilage tissue engineering or skin replacement. For example, Landis et al. [10] elaborated collagen potentially can be utilized as a scaffold for biomimetic mineralization of vertebrate tissues, and provided details concerning collagen–mineral interaction and its implications with respect to designing biomineralizing collagen that will be functionally competent in its biological, chemical, and biomechanical properties.

Simultaneously, collagen has been widely applied in fields including drug delivery and tissue engineering. Kono et al. [11] synthesized a collagen-mimic peptide-attached dendrimer to induce a collagen-like thermally reversible triple helix as a thermosensitive drug carrier and a potential cellular matrix for controllable drug release (Figure 4.1). Friessc et al. [12] highlighted some of the recent developments and applications of collagen as a biomaterial in drug delivery systems for antibiotics, especially gentamicin mostly in trauma, orthopedics, soft tissue infection, dirty abdominal surgery, and wound healing. Nair et al. [13] also summarized information available on collagen dosage forms for drug delivery systems, as collagen shields in ophthalmology, sponges for burns/wounds, mini-pellets and tablets for protein delivery, gel formulation in combination with liposomes for sustained drug delivery, as controlling material for transdermal delivery, and as nanoparticles for basic matrices for cell culture systems.

During the last decades, collagen and its derivatives have been extensively applied as a nonviral vehicle in gene delivery and transfection. Scherer et al. [14] prepared equine collagen type I sponges loaded by a lyophilization method with naked DNA, polyethylenimine (PEI)-DNA, 1,2-dioleoyl-3-trimethylammoniumpropane (DOTAP)/cholesterol–DNA and copolymer-protected PEI-DNA for superiorly mediating sustained gene delivery *in vitro* and local transfection *in vivo* as compared to naked DNA-loaded sponges due to their

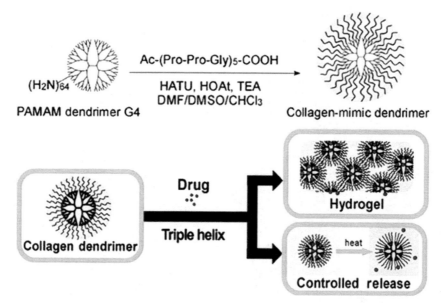

FIGURE 4.1   A collagen-mimic dendrimer capable of controlled release. From Reference 11.

stabilizing and opsonization-inhibiting properties. Spector et al. [15] prepared a gene-supplemented type II collagen-glycosamino-glycan (GSCG) scaffold as a nonviral insulin-like growth factor (IGF)-1 gene delivery with carbodiimide cross-linking of the plasmid to the scaffold resulting in prolonged release. By incorporating a lipid transfection reagent Gene Porter (GP) in conjunction with the plasmid DNA, a local, elevated and prolonged overexpression of IGF-1 by cells results in enhanced biosynthesis and chondrogenesis.

4.2.1.3 Gelatin
Gelatin, the principal protein component of white fibrous connective tissues such as tendons, bones, and the skin, is a water-soluble natural protein derived from collagen, and is commonly used for pharmaceutical and medical applications due to its biodegradability and biocompatibility in physiological environments. However, unlike the triple helix structure of collagen, gelatin is composed of single-strand molecules. Gelatin is a denatured, biodegradable protein obtained by acid and alkaline processing of collagen [16]. Two different types of gelatin can be prepared depending on the method in which collagen is pretreated prior to the extraction process. Moreover, this processing method affects the electrical nature of collagen, producing a variety of gelatin samples with different isoelectric point (IEP) values. The alkaline process targets the amide groups of asparagine and glutamine and hydrolyzes them into carboxyl groups, thus converting many of these residues to aspartate and glutamate. In contrast, acidic pretreatment does little to affect the amide groups present [17]. As the alkaline-processed gelatin possesses a greater proportion of carboxyl groups, rendering it negatively charged and lowering its IEP compared to acid-processed gelatin which possesses an IEP

similar to collagen. As a result, the gelatin processed with an alkaline pretreatment is electrically different in nature from acid-processed gelatin.

Gelatin is extensively used for industrial, pharmaceutical, and medical purposes and the biosafety has been proven through its clinical use as biomaterials and drugs due to its various physicochemical natures effective in the controlled release as a carrier. Mikos et al. [17] modified the isoelectric point of gelatin during its extraction from collagen to yield either a negatively charged acidic gelatin, or a positively charged basic gelatin at physiological pH as a gelatin-based controlled-release carrier system to achieve sustained release, which can be used in tissue engineering, drug delivery, therapeutic angiogenesis, and gene therapy applications. The studies showed that the cross-linking density of gelatin hydrogels affect their degradation rate *in vivo* and the release rate of bioactive molecules released by enzymatic degradation of the drug carrier. Chen et al. [18] successfully synthesized an amphiphilic gelatin macromolecule capable of self-assembling to form micelle-like nanospheres for entrapping hydrophobic therapeutic molecules by grafting hydrophobic hexanoyl anhydrides to the amino groups of primitive gelatin with controlled biodegradability and rapid cellular internalization for intracellularly based nanotherapy.

Recently, local therapeutic gene delivery systems have been paid much attention in the cartilage natural healing process. Gelatin carriers can release plasmid DNA/siRNA extracellularly and intracellularly. Zhang et al. [19] designed gene-activated matrix (GAM) consisting of plasmid DNA and biodegradable chitosan–gelatin complex as three-dimensional scaffolds for cartilage regeneration by studying cell proliferation, adherence, and the protein expression. Plasmid DNA was entrapped in the scaffolds encoding transforming growth factor-b1 (TGF-b1), which was proposed as a promoter to increase the chondrocyte-specific ECM synthesis and therefore enhanced cartilage tissue regeneration. Milam et al. [20] investigated uncross-linked gelatin as a cost-effective vehicle for temporarily encapsulating, then releasing active, short oligonucleotide stands. Adsorption of poly(allylamine hydrochloride) and poly(acrylic acid) on either gelatin blocks or gelatin microspheres (GMS) effectively hinders DNA release at room temperature, then permits release of active DNA once warmed to 37°C, which can be used to control release of DNA in gene therapy.

Gelatin also can be utilized as a matrix of new, desirable biocomposites for various tissue-engineering applications. Ramakrishna et al. [21] fabricated random and aligned poly(ε-caprolactone) (PCL)/gelatin biocomposite scaffolds by varying the ratios of PCL and gelatin concentrations. They demonstrated that the properties of nanofibrous scaffolds were strongly influenced by the concentration of gelatin in the biocomposite. PCL/gelatin 70:30 nanofibers were selected for cell culture study as they provided outstanding mechanical and biodegradation properties. It was found that PCL/gelatin 70:30 enhanced the nerve differentiation and proliferation compared to PCL nanofibrous scaffolds (Figure 4.2).

4.2.1.4 Silk Silks are fibrous proteins with remarkable mechanical properties produced by some lepidoptera larvae such as silkworms, spiders, scorpions, mites, and caddisflies. From a materials science perspective, silk represents the strongest and toughest natural fibers known (Table 4.1; UTS, ultimate tensile strength) for the outstanding performance of their components. Due to the limitless opportunities for functionalization, processing and biological integration of silk, silks have been successfully investigated as suture biomaterials for centuries [22]

Silks are protein-based polymers produced naturally; thus, in the early days, silks had partial application as medical sutures because of their excellent mechanical properties, environmental stability, biocompatibility, and biodegradability. Beyond their traditional use, the extra performance of silks have been excavated including their versatility, favorable characteristics, and potential for processing in aqueous solution under ambient conditions, thus silk-based materials have potential applications in biomedical fields such as drug delivery systems and scaffolds for tissue engineering. For example, Kaplan et al. [23] creatively prepared silk fibroin microspheres consisting of physically cross-linked β-sheet structure and about 2 μm in diameter, using lipid vesicles as templates to efficiently load protein drugs horseradish peroxidase (HRP) in active form for controlled enzyme-release profiles.

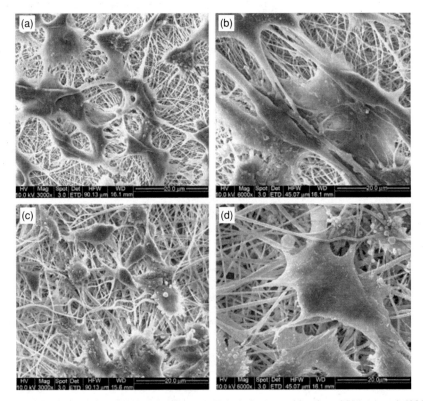

FIGURE 4.2 Morphology of C17.2 cells on PCL/gelatin [magnification: 3000 (a) and 6000 (b)] and PCL [magnification: 3000 (c) and 6000 (d)] after 6 days of cell culture. From Reference 20.

TABLE 4.1   Comparison of Mechanical Properties of Biodegradable Polymeric Materials

Source of Biomaterial	Modulus (GPa)	UTS (MPa)	Strain (%) at Break
Bombyx mori silk (with sericin)	5–12	500	19
B. mori silk (without sericin)	15–17	610–690	4–16
B. mori silk	10	740	20
Nephila clavipes silk	11–13	875–972	17–18
Collagen	0.0018–0.046	0.9–7.4	24–68
Cross-linked collagen	0.4–0.8	47–72	12–16
Polylactic acid	1.2–3.0	28–50	2–6

Source: Adapted from Reference 22.

Silk proteins can self-assemble into mechanically robust material structures that are also biodegradable and noncytotoxic, suggesting their utility for gene delivery. In this respect, silk-based materials can be designed with selective features owing to the controllable secondary structure which affects their solubility, biocompatibility, and biodegradability. For instance, Kaplan et al. [24] prepared recombinant silks modified to contain polylysine sequences which were utilized to form globular complexes with pDNA, and also prepared silk films containing the pDNA complexes on their surface which were used to directly transfer the pDNA complexes to HEK cells (Figure 4.3). The results demonstrate the potential and feasibility of bioengineered silk proteins as a new family of highly tailored gene delivery systems. Furthermore, they prepared and studied a novel nanoscale genetically engineered silk-based ionic complex containing both poly(L-lysine) domains to interact with pDNA and tumorhoming peptides (THPs) to bind to specific tumor cells for target-specific pDNA delivery [25], which are globular and approximately 150–250 nm in diameter. Moreover, the experimental results suggested that the bioengineered silk delivery systems containing functional peptides, such as THP, can serve as a versatile and useful new platform for nonviral gene delivery.

4.2.1.5 Other Protein-origin Polymer Bioconjugates
There are other very practical and attractive protein-based polymers such as elastin and fibrin that have been applied to some extent in bioconjugation applications. Here, just a general overview has been given in this section due to their narrow applications (but potential) as carriers of drug/gene delivery, or other biomaterials.

Fibrin is a fibrous, nonglobular protein formed from fibrinogen by the protease thrombin, and then polymerized to form a "mesh" that forms a hemostatic plug or clot (in conjunction with platelets) over a wound site. As such, fibrin can be autologously harvested from the patient, providing an immunocompatible carrier for delivery of active biomolecules, especially cells. Fibrin has an innate ability to induce improved cellular interaction and subsequent scaffold remodeling; therefore, fibrin is being increasingly used as a scaffold for delivery of cells, drugs, plasmids, and, more recently, gene vectors. Pandit et al. [26] developed a series of fibrin scaffolds, hydrogels, or microspheres as carriers for gene delivery in the past few years, making a great contribution to natural-origin gene carrier systems. For example, they fabricated a fibrin scaffold containing adenovirus encoding endothelial nitric oxide synthase (AdeNOS) comparing with AdeNOS alone, fibrin alone, and no treatment, in an

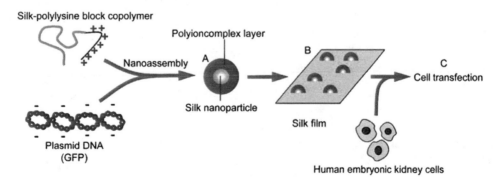

FIGURE 4.3   A schematic presentation of the strategy used in this study. pDNA complex formation with silk-polylysine block copolymer (a). Preparation of a silk film containing pDNA complex (b). Cell transfection using the silk film containing pDNA complex (c). From Reference 24.

alloxan rabbit ear ulcer model to verify that fibrin scaffold provides an enhanced method of gene transfer of adenovirus encoding eNOS, resulting in augmented nitric oxide production and improved wound healing. As such, they prepared a fibrin–lipoplex system as a nonviral gene delivery carrier which showed higher transfection efficiency for lipoplexes when compared to lipoplexes alone, suggesting that the prepared fibrin–lipoplex system is suitable for extended release of lipoplexes for topical gene delivery applications [27]. Furthermore, to design dynamic extended-release systems with spatiotemporal control in gene delivery, they fabricated fibrin microspheres of around one μm size encapsulating lipoplexes for extending gene delivery capacity of a fibrin scaffold without any degradation of DNA by modified preheated oil emulsion method [28].

Elastin is an ECM protein abundant in organs where elasticity is of major importance, providing the elastic recoil properties and resilience essential for the proper functioning of tissues which are subject to repetitive distension and physical stress. Self-assembling elastin-based materials are an emerging and promising topic within the elastin biomaterials field, which may result in microtubes and nanotubes. Elastin-based materials exhibit interesting biological, biomechanical, biochemical, and biophysical properties in the field of tissue engineering and beyond [29].

Elastin-like polymers (ELPs), derived from elastin, have a repeating sequence originating in the sequences found in the mammalian elastic protein, elastin. Thus, some of the main characteristics of these ELPs are derived from the natural protein they are based on. ELPs have found application in the drug delivery field and in genetic engineering due to their ability to be synthesized with a precise molecular weight (MW) and low polydispersity, their potential biocompatibility and controlled degradation. Furgeson et al. [30] successfully constructed a new breed of ELP-based diblock copolymers featuring polyglutamic/aspartic acids for facile drug conjugation and thermo-targeted chemotherapy of hyperthermic tumor margins.

4.2.2 Polysaccharidic Polymer Bioconjugates

4.2.2.1 Introduction
Polysaccharides are a class of biopolymers that consist of simple sugar monomers (monosaccharides) linked together by *O*-glycosidic bonds that endow polysaccharides with the capacity to form both linear and branched polymers [7]. The diversification of their monosaccharide composition, linkage shapes, patterns, chain shapes, and MW dictates their various physical properties, including solubility, gelation, and surface interfacial properties. Polysaccharides, derived from renewable resources that can be found in abundance and are widely available, are able to select properties according to their monosaccharides [31]. Natural polysaccharides, due to their outstanding nontoxicity, good biocompatibility, and biodegradability, have been the recipients of increasing attention in the biomedical field as a carrier in targeted drug and gene delivery systems as well as scaffold materials in tissue engineering applications as described in more detail in the following sections, for example, chitosan, cellulose, dextran, cycloamylose, pullulan, and pectin.

4.2.2.2 Chitin and Chitosan
Chitin and chitosan naturally abundant polysaccharides of major importance, have been widely used in the last decades in various applications, especially in biomedical, pharmaceutical, and drug delivery research. Chitin, the supporting material of crustaceans, insects, etc., is widely known to consist of 2-acetamido-2-deoxy-β-D-glucose through a β(1→4) linkage first identified in 1884 (Figure 4.4). In consideration of the amount of chitin produced annually in the world, it is the second most abundant natural polymer after cellulose. Chitosan, which is soluble in acidic aqueous media, is a cationic polymer obtained from the *N*-deacetylated derivative of chitin and is used in many applications that follow from its unique character [7, 32]. The chemical structure of chitosan is shown in Figure 4.4. The positive attributes of excellent biocompatibility and superb biodegradability with ecological safety and low toxicity, with versatile biological activities such as antimicrobial activity and low immunogenicity, have provided abundant opportunities for further development [33].

Chitin is a natural polymer obtained particularly from the exoskeleton of crustaceans (crabs, shrimp, etc.), cuticles of insects, and cell walls of fungi. In industrial processing, chitin is extracted by deproteination with alkali, demineralization with acid, and decolorization to obtain a colorless product. Chitosan is obtained by partial deacetylation under alkaline condition at 120°C for 1–3 hours, which is the most important chitin derivative in terms of applications. The degree of deacetylation (DD) of typical commercial chitosan is usually between 70% and 95%, and the MW between 10 and 1000 kDa [34]. The properties, biodegradability and biological role of chitosan, are dependent on DD, viscosity, and MW. Chitosan, the fully or partially deacetylated form of chitin, is readily soluble in dilute acidic solutions below pH 6.0 due to the large quantities of amino groups on its chain, while the parent chitin is highly hydrophobic and is insoluble in most organic solvents [35]. In addition, chitosan molecule can be tailored to specific chemical modifications, because it has both active amine and hydroxyl groups in its backbone. Chitosan is also a bioadhesive material. It has attracted attention as an excellent mucoadhesive in its swollen state, and as a natural bioadhesive polymer that can adhere to hard and soft tissues. Chitosan is currently sold in the United States as a dietary supplement to aid weight loss and lower cholesterol and is approved as a food additive in Japan, Italy, and Finland.

During the past few decades, a number of chitosans and chitosan-derived bioconjugated nanosystems have been

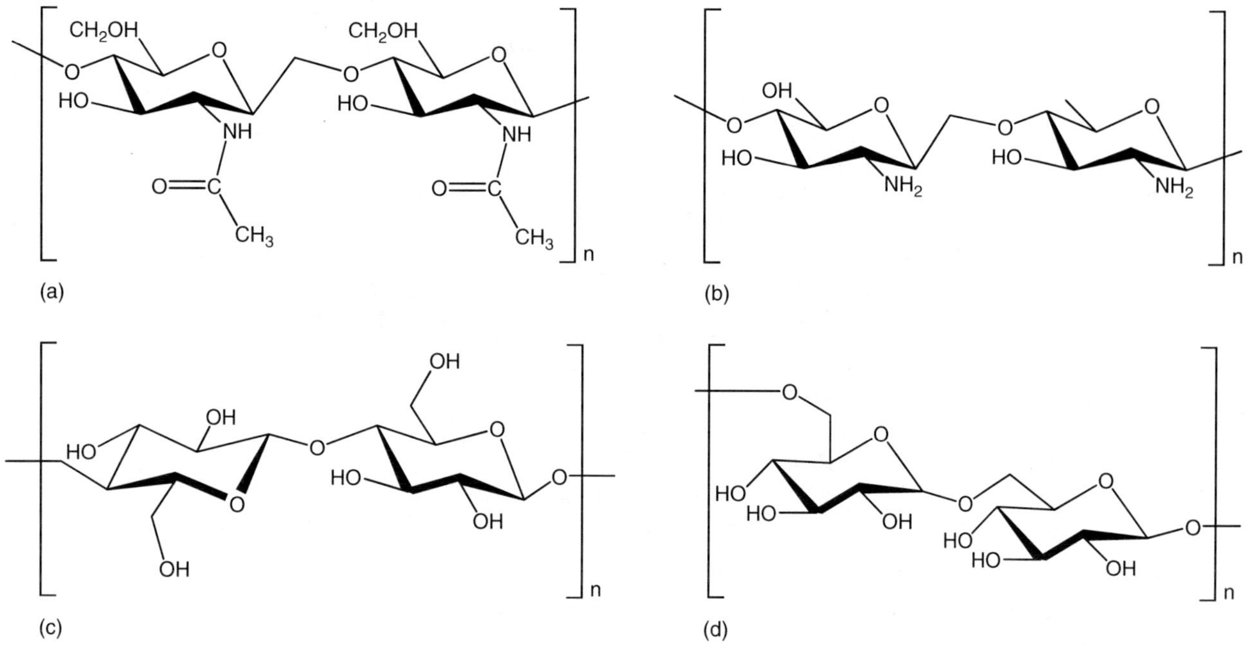

FIGURE 4.4   Chemical structure of (a) chitin, (b) chitosan, (c) cellulose, and (d) dextran.

investigated for the mucosal delivery of drugs, peptides, proteins, gene, and vaccines, including unmodified chitosans with different MWs and degrees of quaternized, deacetylated chitosans, PEGylated chitosan, and chitosans bearing specific ligands [36]. Furthermore, the bioconjugates of chitosan and its derivatives can be used for organ-specific drug delivery, such as colon-specific, gastrointestinal, oral, and ophthalmic delivery systems [37]. A pH-responsive nanoparticle delivery system, consisting of chitosan and poly(γ-glutamic acid) for oral delivery of insulin, has been studied to provide an alternative way for oral protein drugs [38]. The orally administered chitosan nanoparticle, as a nontoxic, soft-tissue-compatible cationic polysaccharide, can adhere to and infiltrate the mucosa of the tight junctions between contiguous epithelial cells to protect insulin from enzymatic degradation and enhance their absorption in the small intestine. In recent years, chitosan-based nanotherapeutics has offered novel strategies for various cancer therapies in the field of oncology, used as drug-conjugated and photosensitizer-conjugated chitosan nanoparticles [36]. Tumor-targeting glycol chitosan nanoparticles, with physically loaded and chemically conjugated photosensitizers, have been studied in photodynamic therapy (PDT) in cancer treatment. These comparative studies showed that hydrophobic photosensitizer-conjugated chitosan nanoparticles have excellent targeting properties, prolonged circulation time, and efficient accumulation in the tumor compared with drug-loaded nanoparticles [39].

In the past several years, gene therapy has received remarkable attention due to its great potential in the replacement of dysfunctional gene and treatment of acquired

diseases. There have been many different approaches applied to gene delivery. Polysaccharides and other cationic polymers have been extensively studied as nonviral DNA/siRNA carriers for gene delivery and therapy. Chitosan is a nontoxic alternative to other cationic polymers and it forms a platform for chitosan-based gene delivery system [37, 40, 41]. Chitosan has positive charges under slightly acidic conditions, which allows the efficient electrostatic between chitosan and DNA/siRNA to be strong enough that the chitosan–DNA complex does not dissociate until it has entered the cell (Figure 4.5). The DD, the charge ratio between chitosan and DNA, the pH, the MW of chitosan, and the N/P ratio can influence the capacity of chitosan binding to cells and therefore the transfection effectiveness [42, 43]. Nanoparticles were prepared from antibody-conjugated chitosan and siRNA by a coacervation method useful for siRNA delivery, specifically targeting T cells for the treatment of various infectious diseases [44] (Figure 4.6). To improve transfection efficiency of nonviral vectors, biotinylated chitosan was applied to the complex with DNA in different N/P ratios and different avidin-coated surfaces. These results showed that bioconjugation is accessible to improve and control gene delivery pathway of chitosan [45].

Conjugated to additional materials such as Fe_3O_4, photosensitizer, or quantum dots (QDs), chitosan and its derivatives provide an attractive approach for advanced bioimaging application, because it has both active amine and hydroxyl groups in its backbone [46]. Besides, polymer nanoparticles based on chitosan are known to accumulate *in vivo* by the so-called enhanced permeation and retention (EPR) effect, resulting in efficient passive targeting in tumor tissues,

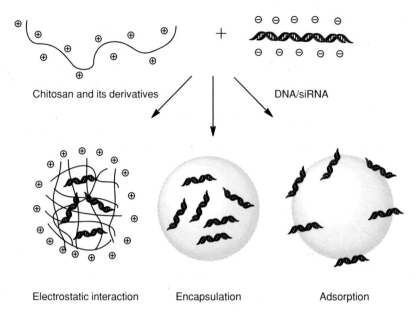

FIGURE 4.5   Preparation of chitosan-based DNA/siRNA nanoparticles based on different mechanisms.

which can also be used in bioimaging application. Jeong et al. [47] prepared iron oxide nanoparticles (ION)-loaded Cy5.5 dye and oleic acid-conjugated chitosan (ION-Cy5.5-oleyl-chitosan) nanoparticles based on dual probe for near-infrared fluorescence (NIRF) optical and MR imaging *in vivo*. This nanoparticle may be used as an effective imaging probe for tumor diagnosis and detection.

4.2.2.3 Cellulose   As a chemical raw material, cellulose constitutes the most ubiquitous and abundant natural polymers on earth and an almost inexhaustible renewable resource

for the increasing demand for environmental friendly and biocompatible products. Cellulose is a natural polymer obtained from plants, trees, bacteria, and some animals (tunicates) via the condensation polymerization of glucose. Basically, the cellulose is a syndiotactic semicrystalline homopolymer composed of β-1,4-linked glucopyranose units, with polymer chains associated with intramolecular and intermolecular hydrogen bonds with a large number of free reactive hydroxyl groups (Figure 4.4). The illustration of the polymeric structure of cellulose can be traced to 1920 and the pioneering work of Staudinger. It has found applications in

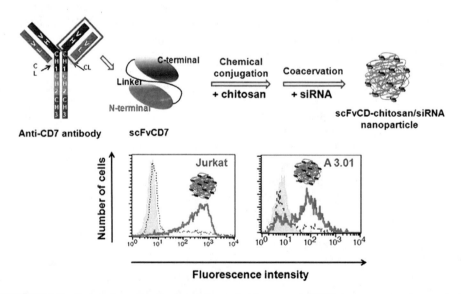

FIGURE 4.6   Preparation of antibody-conjugated chitosan and siRNA by a coacervation method for siRNA delivery. From Reference 44.

areas as diverse as protein, gene, and drug delivery systems, bioimaging probes and tissue engineering, and so on, which has attracted great attention due to its physicochemical properties. Currently, novel forms of cellulose are extensively studied, termed cellulose crystallites, nanocrystals, whiskers, nanofibrils, and nanofibers, to meet the increasing demand for renewable and sustainable resources that are biodegradable, nonpetroleum based, carbon neutral, and have low environmental, animal/human health and safety risks [48–52].

Extensive studies have been performed on polymer nanoparticle delivery systems for drugs, proteins, and enzymes in the field of biomedical research for the effective delivery and release of the targeted carrier molecules, over the past few decades, due to their excellent stability in aqueous environments, prolonged blood circulation times, and their ability to solubilize hydrophobic molecules, as well as their higher payload capacity. As promising candidates for targeted delivery of therapeutics, cellulose and its derivatives have been extensively studied as an effective carrier system with modification or conjugation with drug, peptide, protein, etc. For example, Shoichet et al. [53] investigated an injectable, fast-gelling blend of hyaluronan (HA) and methyl cellulose (MC) which was chemically modified with peptide as a drug delivery matrix to achieve tunable and prolonged release of any bioactive protein. De Smedt et al. [54] prepared electrospun cellulose acetate phthalate (CAP) fibers loaded with antiviral drugs Rhodamine 6G against HIV-1. The CAP fiber can immediately dissolve in semen and rapidly release entrapped fluorophore in a couple of seconds to prevent HIV-1spread during sexual intercourse. Also, cellulose nanowhisker-based drug delivery systems have attracted attention in the biomedical field due to their outstanding hydrophilicity, biocompatibility, biodegradability, and high surface area. Recently, Ragauskas et al. [55] synthesized a novel drug delivery system consisting of syringyl alcohol linker and cellulose nanowhiskers by employing a series of oxidation, reductive-amination and esterification reactions for the conjugation of amine-containing drugs and biologically active compounds.

Nowadays, exploitation of an efficient method to introduce a therapeutic gene into target cells *in vivo* is a key issue in gene therapy because naked DNA is very susceptible to degradation by nucleases in the ECM and in the cells. Recently, cationic polysaccharides (such as cationic cellulose, dextran, and chitosan) are investigated as appropriate pharmaceutical carriers for nonviral gene delivery and gene transfection purposes, because they are natural, renewable, nontoxic, biodegradable, and excellent biocompatible materials compared with polyethylenimine (PEI)-based gene delivery systems. Zhou et al. [56] synthesized homogeneous quaternized cellulose derivatives by reacting cellulose with 3-chloro-2-hydroxypropyltrimethylammonium   chloride (CHPTAC) in NaOH/urea aqueous solutions for the first time as a promising nonviral gene carrier. And then, they

studied the influence of MW and degree of substitution (DS) of quaternized cellulose on the efficiency of gene transfection as gene carriers. The results proved that quaternized cellulose derivatives could bind DNA efficiently [57]. Xu et al. [58] synthesized well-defined comb-shaped copolymers composed of biocompatible hydroxypropyl cellulose backbones and short cationic poly(2-(dimethylamino)ethyl methacrylate) (PDMAEMA) side chains via Atom transfer radical polymerization (ATRP), which were further quaternized with quaternary ammonium groups. The quaternized copolymers show stronger ability to complex with pDNA than nonquaternized copolymers to form nanoparticle complexes as gene carriers.

Also, cellulose and its derivatives can be used as carriers for targeted delivery of distinct therapeutic and diagnostic agents for *in vivo* imaging via covalent conjugation. Daub et al. [59] studied the chemistry of the metal-free and zincated prophyrin-appended cellulose at C(6)-*O* with a helical structure to use of cellulose as a stiff backbone and the porphyrin chromophore as a photo- and electrosensitive unit by optical and electrochemical methods. The helical architecture of the prophyrin-appended cellulose arrangement provides precursor systems for nanoscale materials with optoelectronic potential. Roman et al. [60] exploited a simple method to covalently attach fluorescent fluorescein-5′-isothiocyanate (FITC) moieties to the surface of cellulose nanocrystals for bioimaging application. Then, they studied the interaction of fluorescently labeled cellulose nanocrystals with cells, the uptake, and the biodistribution *in vivo*.

4.2.2.4 Dextran

Detran is a highly water-soluble branched bacterial polysaccharide composed of glucose units mainly linked by α-1,6-glycosidic linkages. The degree of branching depends on its origin. Dextran possesses many reactive hydroxyl functional groups present in each monomer unit, which makes it amenable to chemical modification and conjugation with macromolecules via ATRP, Reversible Addition-Fragmentation chain Transfer (RAFT), and Ring-opening polymerization (ROP). Consequently, dextran is an attractive candidate employed in biomedical applications, owing to its biodegradability, remarkable degree of biocompatibility, nonimmunogenicity, ease of modification, wide availability, and improved transfection efficiency with reduced toxicity. It also has been used as a two-dimensional (2D) or porous soft tissue engineering scaffold for vascular tissue engineering.

Over the past few decades, a series of dextran-conjugated nanoparticles and microcapsules have been prepared as drug/gene delivery vehicles on account that dextran is a highly water-soluble nonelectrolyte polymer and presents no tendency to self-assemble in its native form. Zhong et al. [61] synthesized disulfide-linked dextran-b-poly(ϵ-caprolactone) diblock copolymer (Dex-SS-PCL) reduction-responsive biodegradable nanomicelles by thiol–disulfide

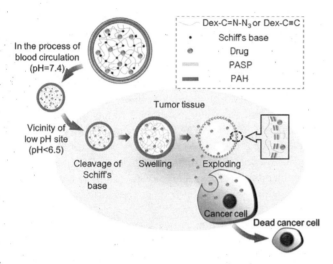

FIGURE 4.7   A schematic diagram of the structure of a dextran-based microcapsule and tumor-triggered exploding of the microcapsule. From Reference 62.

exchange reaction which can be used for efficient intracellular release of doxorubicin (DOX) *in vitro* and inside cells. Zhang et al. [62] designed and prepared a novel, intelligent "active defense" system consisting of a biodegradable dextran microgel core cross-linked by a Schiff's base and a surrounding layer formed by Layer-by-Layer (LbL) assembly, effectively loaded with macromolecular model drug, dex-FITC and antineoplastic drug, DOX. The results verified that the entrapped drugs could be explosively released, triggered by the acidic environment of tumor tissues, to achieve the desirable therapy effect (Figure 4.7).

Recently, a series of dextran-based cationic polymers, especially dextran-spermine, were prepared for efficient gene delivery. Xu et al. [63] developed a simple method to prepare well-defined comb-shaped copolymers Dextran-g-P(DMAEMA) (DPDs) composed of biocompatible dextran

backbones and cationic PDMAEMA side chains via ATRP for highly efficient nonviral gene delivery. Liu et al. [64] prepared a lauric acid-modified dextran–agmatine bioconjugate (Dex-L-Agm) as a gene delivery vector by the nucleophilic reaction between the tosyl of tosylate dextran and the primary amine of agmatine. The incorporated laurate groups enhanced the complexation of the gene delivery vector with DNA due to the hydrophobically cooperative binding effect.

The ability to combine different imaging modalities with dexran has become an exciting area of biomedical applications, including cell labeling, magnetic resonance imaging (MRI), targeted molecular imaging, molecular diagnostics, and therapy. Kai et al. [65] successfully synthesized dextran-based chemiluminescent compounds containing luminol (or isoluminol) and biotin as CL-labeling macromolecular probes for the sensitive CL imaging of a cytochrome P450 (CYP) protein on a poly(vinylidene difluoride) (PVDF) membrane (Figure 4.8). Weissleder et al. [66] used a superparamagnetic iron oxide nanoparticle with a cross-linked dextran coating as a versatile platform for conjugation to targeting ligands with wide applications in biologic discovery, molecular imaging, diagnostic analyte detection, and therapeutic decision making and monitoring. The primary amines distributed throughout the nanoparticle provide chances for conjugate ligands bearing various functional groups that can label cancer cells, peptides, and small molecules for efficient targeting and signal amplification (Figure 4.9).

4.2.2.5   Other Polysaccharidic Polymer Bioconjugates
Other polysaccharides have also been found in abundance, are widely available and inexpensive, and have a variety of structures and properties due to their similarity with biological macromolecules. Many attempts have been made to produce smart naturally derived systems conjugated with drug moiety to form a prodrug for site-specific drug delivery, and

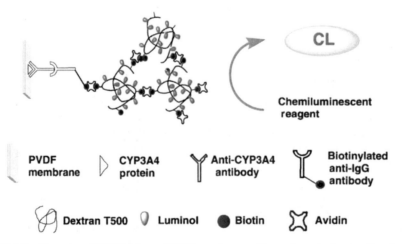

FIGURE 4.8   Detection of CYP3A4 on a PVDF membrane with a polymeric dextran-based chemiluminescent probe. From Reference 65.

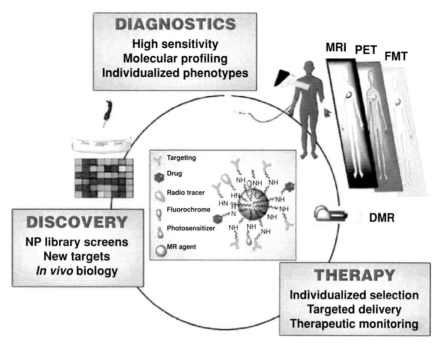

FIGURE 4.9   A superparamagnetic ION with a cross-linked dextran coating. From Reference 66. For a color version, see the color plate section.

conjugated with pDNA for efficient gene delivery and transfection in gene engineering, owing to their similarity with the ECM.

Hyaluronan (HYA) is an extracellular and cell-surface-associated linear polysaccharide found in all tissues and body fluids of vertebrates as well as in some bacteria, consisting of alternating disaccharide units of α-1,4-D-glucuronic acid and β-1,3-N-acetyl-D-glucosamine, linked by $\beta(1\rightarrow3)$ bonds and exists *in vivo* as a polyanion. It plays key roles in the structure and organization of the ECM, regulation of cell adhesion, morphogenesis and modulation of inflammation. In recent years, hyaluronan has been extensively studied for drug delivery, gene delivery, tissue engineering in the form of matrices, micro/nanoparticles, and hydrogels [67]. For example, Alonso et al. [68] prepared a new nanocarrier system composed of two main components, hyaluronic acid (HA) and chitosan, with different proportions of polysaccharides and different MWs to associate with significant amounts of pDNA, to enter the cells and very efficiently deliver the associated pDNA.

Alginate, composed of alternating blocks of 1–4-linked α-L-guluronic (G units) and β-D-mannuronic acid (M units) residues which vary in amount and sequential distribution along the polymer chain depending on the source of the alginate, is a family of water-soluble linear unbranched polysaccharides extracted from brown seaweed and some soil bacteria. Alginate-based materials are pH-sensitive and could be potential candidates in smart delivery system. Also alginate possesses excellent biocompatibility, bioadhesiveness, and mild gelation conditions, which may aid in its utility as a potential delivery vehicle for protein, nucleic acid, and cell encapsulation [7, 69].

Pectins, predominantly composed of α-(1–4)-linked D-galacturonic acid residues interrupted by 1,2-linked L-rhamnose residues, are nonstarch, linear polysaccharides extracted from the primary cell walls of terrestrial plants. Pectin has a few hundred to about one thousand building blocks per molecule, with an average MW of about 50,000 to about 180,000 g/mol. Pectin remains as an aggregate of macromolecules in acidic environments and tends to dissociate and expand at neutral solution pH. It can be used for protein and polypeptide drugs, especially in colon-specific delivery systems. Liu et al. [70] summarized pectin-based systems for colon-specific drug delivery via oral route and discussed their advantages, limitations, and possible future developments in pectin-based formulations with particular emphasis in the field of colon-specific drug delivery.

4.3 CONCLUSION AND FUTURE DIRECTIONS

To date, a great number of attempts have been made to design and select an ideal biomaterial in the development of bioconjugation technology. It is essential to unite multidisciplinary researchers, including materials scientists, biologists, clinicians, engineers, and pharmacologists, to continue to work together to meet this challenge. Natural-based polymeric bioconjugates are under intensive investigation with expectedly encouraging outcomes for different biomedical applications. Ground-breaking advances in natural polymeric bioconjugates have proposed novel candidates used as carriers for drug/gene delivery systems, scaffolds for tissue engineering, and cargos for bioimaging agents *in vivo* and *in vitro* briefly summarized in this part, which can potentially revolutionize current medical practice.

Various precisely tunable structures and properties of natural polymeric systems have been designed and synthesized in recent years with special focus on creating more multifunctional materials in biomedical applications. Although natural-origin bioconjugates present some drawbacks, namely the difficulties in controlling the variability from batch to batch, their unique properties, for instance, excellent biodegradability, biocompatibility, nontoxicity, similarity with the ECM, and typically good biocharacteristics, make them promising and attractive for biomedical and biotechnological applications. Therefore, the field has room for improvement. Many efforts are continuing to be put in to prepare different well-defined formulations based on natural polymers with adequate chemical modifications, conjugations, or surface treatments.

ACKNOWLEDGMENT

This work was financially supported by the National Natural Science Foundation of China (Nos. 21074035 and 51173044).

REFERENCES

1. Khandare J, Minko T. *Prog Polym Sci* 2006;31:359–397.
2. Pasut G, Veronese FM. *Prog Polym Sci* 2007;32:933–961.
3. Heredia KL, Maynard HD. *Org Biomol Chem* 2007;5:45–53.
4. Gauthier MA, Klok H-A.*Chem Commun* 2008;23:2591–2611.
5. Canalle LA, Lowik DWPM, van Hest JCM. *Chem Soc Rev* 2010;39:329–353.
6. Dang JM, Leong KW. *Adv Drug Deliv Rev* 2006;58:487–499.
7. Malafaya PB, Silva GA, Reis RL. *Adv Drug Deliv Rev* 2007;59:207–233.
8. Sionkowska A. *Prog Polym Sci* 2011;36:1254–1276.
9. Paul S. *J Arch Sci* 2011;38:3358–3372.
10. Landis WJ, Silver FH, Freeman JW. *J Mater Chem* 2006;16:1495–1503.
11. Kojima C, Tsumura S, Harada A, Kono K. *J Am Chem Soc* 2009;131:6052–6053.
12. Ruszczak Z, Friess W. *Adv Drug Deliv Rev* 2003;55:1679–1698.
13. Nair R, Sevukarajan M, Mohammed Badivaddin T, Kumar CKA. *JITPS* 2010;7:288–304.
14. Scherer F, Schillinger U, Putz U, Stemberger A, Plank C. *J Gene Med* 2002;4:634–643.
15. Capito RM, Spector M. *Gene Ther* 2007;14:721–732.
16. Tabata Y, Ikada Y. *Adv Drug Deliv Rev* 1998;31:287–301.
17. Young S, Wong M, Tabata Y, Mikos AG. *J Control Release* 2005;109:256–274.
18. Li W-M, Liu D-M, Chen S-Y.*J Mater Chem* 2011;21:12381–12388.
19. Guo T, Zhao J, Chang J, Ding Z, Hong H, Chen J, Zhang J. *Biomaterials* 2006;27:1095–1103.
20. Hardin JO, Milam VT. *Soft Matter* 2011;7:2674–2681.
21. Ghasemi-Mobarakeh L, Prabhakaran MP, Morshed M, Nasr-Esfahani M-H, Ramakrishna S. *Biomaterials* 2008;29:4532–4539.
22. Tao H, Kaplan DL, Omenetto FG. *Adv Mater* 2012;24:2824–2837.
23. Wang X, Wenk E, Matsumoto A, Meinel L, Li C, Kaplan DL. *J ControlRelease* 2007;117:360–370.
24. Numata K, Subramanian B, Currie HA, Kaplan DL. *Biomaterials* 2009;30:5775–5784.
25. Numata K, Reagan MR, Goldstein RH, Rosenblatt M, Kaplan DL. *Bioconjug Chem* 2011;22:1605–1610.
26. Breen AM, Dockery P, O'Brien T, Pandit AS. *Biomaterials* 2008;29:3143–3151.
27. Kulkarni M, Breen A, Greiser U, O'Brien T, Pandit A. *Biomacromolecules* 2009;10:1650–1654.
28. Kulkarni MM, Greiser U, O'Brien T, Pandit A. *Mol Pharm* 2011;8:439–446.
29. Daamen WF, Veerkamp JH, van Hest JC, van Kuppevelt TH. *Biomaterials* 2007;28:4378–4398.
30. Rodriguez-Cabello JC, Reguera J, Girotti A, Alonso M, Testera AM. *Prog Polym Sci* 2005;30:1119–1145.
31. Hovgaard L, Brondsted H. *Crit Rev Ther Drug Carrier Syst* 1996;13:185–223.
32. Mourya VK, Inamdar NN.*React Funct Polym* 2008;68:1013–1051.
33. Pillai CKS, Paul W, Sharma CP.*Prog Polym Sci* 2009;34:641–678.
34. George M, Abraham TE. *J Control Release* 2006;114:1–14.
35. Dash M, Chiellini F, Ottenbrite RM, Chiellini E. *Prog Polym Sci* 2011;36:981–1014.
36. Lakshmanan V-K, Snima KS, Bumgardner JD, Nair SV, Jayakumar R. *Adv Polym Sci* 2011;243:55–92.
37. Kumar MNVR, Muzzarelli RAA, Muzzarelli C, Sashiwa H, Domb AJ. *Chem Rev* 2004;104:6017–6084.
38. Sung H-W, Sonaje K, Liao Z-X, Hsu L-W, Chuang E-Y. *Acc Chem Res* 2012;45:619–629.
39. Lee SJ, Koo H, Jeong H, Huh MS, Choi Y, Jeong SY, Byun Y, Choi K, Kim K, Kwon IC. *J Control Release* 2011;152:21–29.
40. Jayakumar R, Chennazhi KP, Muzzarelli RAA, Tamura H, Nair SV, Selvamurugan N. *Carbohydr Polym* 2010;79:1–8.
41. Mao S, Sun W, Kissel T. *Adv Drug Deliv Rev* 2010;62:12–27.
42. Riva R, Ragelle H, des Rieux A, Duhem N, Jérôme C, Préat V. *Adv Polym Sci* 2011;244:19–44.
43. Kim T-H, Jiang H-L, Jere D, Park I-K, Cho M-H, Nah J-W, Choi Y-J, Akaike T, Cho C-S. *Prog Polym Sci* 2007;32:726–753.

44. Lee J, Yun K-S, Choi CS, Shin S-H, Ban H-S, Rhim T, Lee SK, Lee KY. *Bioconjug Chem* 2012;23(6):1174–1180.

45. Hu W-W, Syu W-J, Chen W-Y, Ruaan R-C, Cheng Y-C, Chien C-C, Li C, Chung C-A, Tsao C-W. *Bioconjug Chem* 2012;23(8):1587–1599.

46. Agrawal P, Strijkers GJ, Nicolay K. *Adv Drug Deliv Rev* 2010;62:42–58.

47. Lee C-M, Jang DR, Kim J, Cheong S-J, Kim E-M, Jeong M-H, Kim S-H, Kim DW, Lim ST, Sohn M-H, Jeong YY, Jeong H-J. *Bioconjug Chem* 2011;22(2):186–192.

48. Klemm D, Kramer F, Moritz S, Lindström T, Ankerfors M, Gray D, Dorris A. *Angew Chem Int Ed* 2011;50:5438–5466.

49. Eichhorn SJ. *Soft Matter* 2011;7:303–315.

50. Moon RJ, Martini A, Nairn J, Simonsenf J, Youngblood J. *Chem Soc Rev* 40:3941–3994.

51. Habibi Y, Lucia LA, Rojas OJ.*Chem Rev* 2010;110:3479–3500.

52. Roy D, Semsarilar M, Guthrie JT, Perrier S. *Chem Soc Rev* 2009;38:2046–2064.

53. Vulic K, Shoichet MS. *J Am Chem Soc* 2012;134:882–885.

54. Huang C, Soenen SJ, van Gulck E, Vanham G, Rejman J, Van Calenbergh S, Vervaet C, Coenye T, Verstraelen H, Temmerman M, Demeester J, De Smedt SC. *Biomaterials* 2012;33:962–969.

55. Dash R, Ragauskas AJ. *RSC Adv* 2012;2:3403–3409.

56. Song Y, Sun Y, Zhang X, Zhou J, Zhang L. *Biomacromolecules* 2008;9:2259–2264.

57. Song Y, Wang H, Zeng X, Sun Y, Zhang X, Zhou J, Zhang L. *Bioconjug Chem* 2010;21:1271–1279.

58. Xu FJ, Ping Y, Ma J, Tang GP, Yang WT, Li J, Kang ET, Neoh KG. *Bioconjug Chem* 2009;20:1449–1458.

59. Redl FX, Lutz M, Daub J. *Chem Eur J* 2001;7:5350–5358.

60. Dong S, Roman M. *J Am Chem Soc* 2007;129:13810–13811.

61. Sun H, Guo B, Li X, Cheng R, Meng F, Liu H, Zhong Z. *Biomacromolecules* 2010;11:848–854.

62. Zhang J, Xu X-D, Liu Y, Liu C-W, Chen X-H, Li C, Zhuo R-X, Zhang X-Z. *Adv Funct Mater* 2012;22:1704–1710.

63. Wang ZH, Li WB, Ma J, Tang GP, Yang WT, Xu FJ. *Macromolecules* 2011;44:230–239.

64. Yang J, Liu Y, Wang H, Liu L, Wang W, Wang C, Wang Q, Liu W. *Biomaterials* 2012;33:604–613.

65. Zhang H, Smanmoo C, Kabashima T, Lu J, Kai M. *Angew Chem Int Ed* 2007;46:8226–8229.

66. Tassa C, Shaw SY, Weissleder R. *Acc Chem Res* 2011;44(10):842–852.

67. Fraser JR, Laurent TC, Laurent UB. *J Intern Med* 1997;242:27–33.

68. de la Fuente M, Seijo B, Alonso MJ. *Nanotechnology* 2008;19(7):075105.

69. Rowley JA, Madlambayan G, Mooney DJ. *Biomaterials* 1999;20:45–53.

70. Liu LS, Fishman ML, Kost J, Hicks KB. *Biomaterials* 2003;24:3333–3343.

5

DENDRIMER BIOCONJUGATES: SYNTHESIS AND APPLICATIONS

ALI NAZEMI AND ELIZABETH R. GILLIES

Department of Chemistry and Department of Chemical and Biochemical Engineering, The University of Western Ontario, London, Canada

5.1 INTRODUCTION—DENDRIMERS FOR BIOCONJUGATE CHEMISTRY

Branching is frequently observed in nature, where the multiplicity of terminal functionalities leads to enhanced properties and functions. For example, neurons exhibit highly branched "dendritic" structures, allowing them to make many excitatory synaptic contacts, while highly branched structures within the lungs allow for the efficient transfer of oxygen into the blood for delivery to tissues. In the late 1970s and the 1980s, a new class of highly branched, synthetic macromolecules was introduced. In 1978, Vögtle and coworkers developed an iterative method for the synthesis of low molecular weight (MW) branched amines, which were termed "cascade" molecules [1]. In 1984–1985, Tomalia and coworkers reported conditions that were less prone to side reactions and were more suitable for repetitive growth [2, 3]. Inspired by the naturally occurring structures, these molecules were termed dendrimers, based on the Greek word *dendron*, meaning tree.

In comparison with hyperbranched polymers, which possess an irregular architecture and contain incomplete branching points due to their noniterative synthesis (Figure 5.1a), the stepwise synthesis of dendrimers results in a regularly branched structure (Figure 5.1b). Dendrimers comprise three structural regions: (a) a core; (b) layers of branching repeat units comprising the backbone, where each layer typically results from one stage of growth and is termed a "generation"; and (c) end groups on the peripheral layer. It is these end groups that are generally of greatest interest in the context of bioconjugation chemistry. Alternatively, when dendrimers are prepared from a monovalent core moiety (focal point), a wedge-like structure typically called a "dendron" results (Figure 5.1c). In this case, bioconjugation to the peripheral and/or the focal point can be pursued.

Due to their iterative syntheses and highly branched architectures, dendrimers and dendrons possess several properties that are unique relative to traditional polymers. For example, while most syntheses of linear and hyperbranched polymers lead to a range of molecules differing in MWs, the iterative syntheses of dendrimers, with a focus on driving reactions to completion and purifying intermediate molecules, leads to molecules with a single or very narrow range of MWs. Furthermore, while linear or hyperbranched polymers can theoretically be grown infinitely, the growth of dendrimers is mathematically limited. This is due to the exponential increase in the number of monomer units with each generation, while the volume available for these units increases with the cube of the dendrimer radius. At a certain generation, sterics, known as de Gennes dense packing [4], limit regular growth and if growth is continued structural flaws will result. This also results in higher generation dendrimers exhibiting globular conformations. Finally, one of the most important differences in the context of bioconjugate chemistry is that while linear polymers have only two end groups, dendrimers have an exponentially increasing number of end groups. This results in the properties of dendrimers being dominated by these end groups at high generations, and also provides many sites for the conjugation of functional moieties.

The iterative synthesis of dendrimers can generally be categorized into two strategies, the divergent approach and

Chemistry of Bioconjugates: Synthesis, Characterization, and Biomedical Applications, First Edition. Edited by Ravin Narain.
© 2014 John Wiley & Sons, Inc. Published 2014 by John Wiley & Sons, Inc.

(a)   (b)   (c)

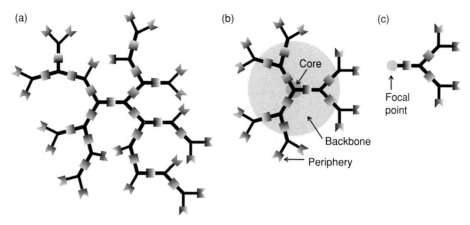

FIGURE 5.1 Schematic representation of (a) a hyperbranched polymer, (b) a dendrimer, and (c) a dendron.

the convergent approach. In the divergent approach [3, 5–8], the dendrimer is grown outward from the core by the repetition of coupling and activation steps. This approach is the preferred one for the large-scale preparation of dendrimers because the quantity of dendrimer sample increases with each generation and the removal of excess reagents by techniques such as precipitation, distillation, or ultrafiltration is facilitated by their differences in mass. However, the exponentially increasing number of coupling reactions required for each subsequent generation means that the number of side reactions or incomplete couplings also increases, ultimately leading to incomplete branching and flawed structures that are nearly impossible to separate from the target molecule. In the convergent approach [9], growth initiates from what will become the dendrimer periphery and progresses toward the core. When the desired generation is reached, the resulting "dendrons" are coupled to a core molecule. As this approach only involves a small number of coupling reactions at each generation, the molecules that result from incomplete couplings can often be separated from the desired molecules as they are sufficiently different in structure. This affords dendrimers with higher structural homogeneity and monodispersity than the divergent approach. Nevertheless, the couplings become increasingly challenging due to steric hindrance as the dendrons approach higher generations. Furthermore, although the MW increases with each generation, the excesses of dendrons used in the couplings, incomplete couplings, and losses associated with the purification generally result in a decrease in the overall mass of material at each step, making this approach less attractive on a large or industrial scale.

Over the past few decades, tremendous progress has been made in the optimization of dendrimer syntheses and a diverse array of backbones are now readily accessible. Some of the more commonly used backbones include the poly(amido amine) (PAMAM) "Starburst" (Figure 5.2a), poly(propylene imine) (PPI) (Figure 5.2c), polyester (PE)

dendrimers based on 2,2-bis(methylol)propionic acid (Figure 5.2b), and poly(L-lysine) (PLL) (Figure 5.2d). Many of these dendrimers are now available from commercial suppliers. However, whether prepared by a convergent or divergent approach, these molecules are still much more costly than conventional polymers. This means that over the longer term, the applications of dendrimers will almost certainly be limited to high value-added products. One area that meets this criterion is biomedical materials, where the cost of a material is less important than its performance. In addition, very well defined materials are typically required by regulatory agencies to approve their use in the human body. For this, the structural homogeneity provided to dendrimers as a result of their iterative syntheses is a distinct advantage over other classes of synthetic macromolecules. As a result, the biomedical applications of dendrimers are some of the most widely investigated. This chapter will explore how bioconjugation chemistry can be used to covalently attach biologically relevant molecules to dendrimers. It will explore how the specific conjugation chemistries are determined based on the application, the chemical functionalities available on the molecules of interest and those on the dendrimer's focal point or periphery. First, the conjugation of drug molecules will be discussed, followed by carbohydrates, imaging agents, oligonucleotides, and peptides/proteins. The impact of the conjugation chemistry on the biological properties of the resulting molecules will be illustrated through selected examples.

5.2 DENDRIMER–DRUG CONJUGATES

5.2.1 Motivation for the Development of Dendrimer–Drug Conjugates

Over the past several decades, many advances have been made in the development of therapeutics to treat human diseases. However, many current drugs and new drug candidates

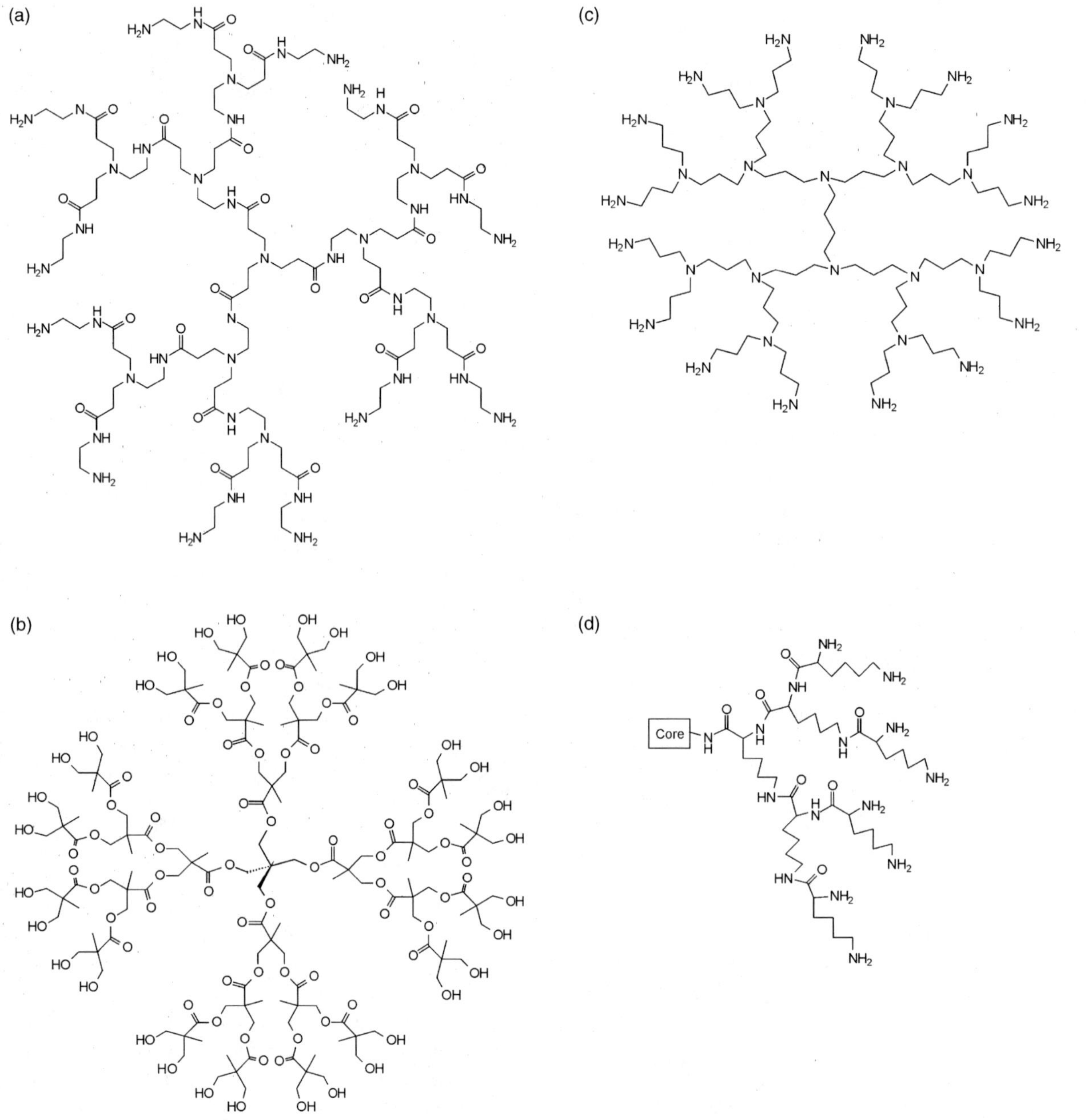

(a)

(c)

(b)

(d)

FIGURE 5.2   Readily accessible dendrimer backbones: (a) PAMAM, (b) PE, (c) PPI, and (d) PLL.

still suffer from significant limitations. For example, the low aqueous solubilities of hydrophobic drugs are major obstacles for their administration. One of the ways to overcome solubility problems is the use of excipients. However, this can result in undesirable side effects, such as when Cremophor EL or ethanol are used for the solubilization of paclitaxel (TAX) [10]. An additional challenge encountered is the rapid elimination of drug molecules from the blood stream, which limits their therapeutic efficacy and increases the required dose [11]. Moreover, many drugs exhibit a lack of specificity for their therapeutic target. For example, many anticancer drugs not only attack cancer cells but also kill noncancerous and healthy cells, causing severe side effects. These challenges have motivated significant interest in the development of drug delivery systems, where the incorporation of a drug into a polymeric system can enhance its solubility, prolong its circulation time, and enhance its specificity for its target. Among the currently studied drug delivery systems, dendrimers have emerged as an attractive class of materials, mainly because of their well-defined structures. In addition, they possess many peripheral groups for drug conjugation, and their nanoscale sizes can lead to enhanced blood circulation times and selective accumulation in tumors via the enhanced permeability and retention effect [12–14].

As reviewed recently in the context of chemotherapeutics [12, 15], drugs can be incorporated into dendrimers either by covalent conjugation to the periphery or by noncovalent encapsulation within the backbone of the dendrimer. Both classes of delivery systems have been demonstrated to be more effective than the free drug in certain laboratory studies and each approach is associated with its own advantages and disadvantages [11]. However, control over the drug:dendrimer ratio in a noncovalent system can present challenges, and noncovalently incorporated drugs are often released too rapidly under physiological conditions. This has limited their *in vivo* efficacy thus far [14]. With the use of optimized chemical reactions, the covalent attachment of drugs to dendrimers benefits from superior control over the ratio of drug:dendrimer in the resulting conjugate. Moreover, the problem of the burst release observed for physically entrapped drugs upon injection can be mitigated to a great extent by covalently attaching drugs to dendrimers.

In selecting the appropriate bioconjugation chemistry for dendrimer–drug conjugates, there are some important considerations. In order to achieve a controlled release of the drug from a dendrimer–drug bioconjugate, the linker stability under various physiological conditions is crucial. The lability of a given linker in a specific microenvironment plays an important role in the specificity and the rate of drug release [12]. For example, to obtain selective and controlled release of drug in cancerous tissues or within the endosomes and lysosomes of cells, which are known to be more acidic than healthy tissues, an acid-labile linkage such as an ester or hydrazone can prove effective [16, 17]. In addition, the released drug derivative needs to be identical to, or as active as the original drug in order to be effective [18]. In the remainder of this section, through the selection of doxorubicin (DOX), methotrexate (MTX), TAX, *N*-acetyl-*L*-cysteine (NAC), and antiviral sulfonic acid moieties as example therapeutics, the conjugation chemistry will be described and the merits of the different conjugation chemistries for the different drug molecules will be discussed.

5.2.2 Dendrimer–DOX Conjugates

DOX is a chemotherapeutic commonly used in the treatment of hematological malignancies, various carcinomas, soft tissue sarcomas, and a wide range of other types of cancers. It is an anthracycline antibiotic and the mechanism by which it operates involves the intercalation of DNA. Because of its fast cellular binding and trafficking to the nucleus, it has a short half-life in plasma and very wide biodistribution which in turn requires high doses to be administered. These doses can cause adverse effects such as nausea, vomiting, neutropenia, heart arrhythmias, and complete hair loss. Because of these limitations, delivery systems that allow DOX to be

delivered more effectively, with fewer side effects have been actively sought. Indeed a liposomal carrier termed DOXIL® was approved in the United States in 1995 for intravenous infusion. As shown in Figure 5.3, there are several functional groups on DOX that can be used for its conjugation to dendrimers. The sugar amino group can be used to form amide and carbamate linkages while the ketone of DOX can be used to form acid-labile hydrazone linkages.

A wide variety of dendrimer backbones including PAMAM [19, 20], PE [16, 21, 22], PLL [23], and poly(L-glutamic acid) (PGA) [24] have been used for DOX conjugation and delivery via the acid-labile hydrazone linkage. In order to construct the hydrazone linkage between DOX and the dendrimer, the periphery of the dendrimer needs to be prefunctionalized with hydrazide moieties. Szoka, Fréchet, and coworkers have shown that PE dendrimers and dendrons with hydroxyl peripheral groups can be converted to hydrazide-terminated analogs in three steps by first activating the hydroxyl groups using 4-nitrophenyl chloroformate, reaction of these activated carbonates with *tert*-butyl carbazate, and then deprotection of the *t*-butyloxycarbonyl (Boc) groups by treatment with TFA (Figure 5.4a). Treatment of these dendrimers with the HCl salt of DOX in methanol (MeOH) at 60°C resulted in the formation of the dendrimer–DOX bioconjugates with hydrazone linkage in relatively high yields [16]. Interestingly, it was later found that this carboxyhydrazide was unsuitable for the conjugation of DOX, as an intramolecular cyclization of the hydroxyl of DOX on the carboxyhydrazide released a modified and much less active form of the drug [25]. Therefore, the conjugation strategy used in subsequent work avoided this problematic carboxyhydrazide linkage. For example, the use of a β-alanine spacer provided DOX-functionalized bow-tie dendrimer–PEG hybrids that cured mice-bearing C-26 colon carcinomas (Figure 5.4b) [26]. A slightly different approach involved PEGylated PE dendrimers with trifunctional amino acids such as serine, tyrosine, aspartic acid, glycine, and phenylalanine on their peripheries (Figure 5.4c). In a mouse model, these dendrimers were found to be as effective as DOXIL with less drug released in healthy tissues [21].

Gu and coworkers have employed a simple method for hydrazide introduction to carboxylic acid-terminated PGA dendrimers with a polyhedral oligomeric silsesquioxane nanocubic core [24]. In this method, *tert*-butyl carbazate was directly reacted with carboxylic acid termini of the dendrimer under EDC/1-hydroxybenzotriazole (HOBt) coupling conditions. Boc deprotection and subsequent reaction with DOX gave the desired dendrimer–drug conjugate. Biotin was also used as a targeting group in this study and these materials showed better antitumor efficacy against HeLa cell lines when compared to the conjugate without biotin. In a study by Porter and coworkers, PEGylated PLL dendrimers with outer generation of L-lysine or succinimyldipropyldiamine (SPN), were functionalized with hydrazides by

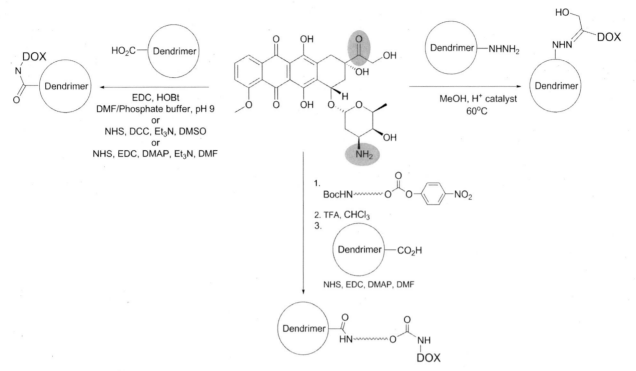

FIGURE 5.3 Amide, carbamate, and hydrazine conjugation strategies for DOX. EDC, 1-ethyl-3-(3-dimethylaminopropyl) carbodiimide; HOBt, 1-hydroxybenzotriazole; DCC, *N,N'*-dicyclohexylcarbodiimide; NHS, *N*-hydroxysuccinimidyl; DMAP, 4-dimethylaminopyridine; TFA, trifluoroacetic acid; DMF, *N,N*-Dimethylformamide; DMSO, dimethyl sulfoxide.

first reacting the dendrimer's peripheral amines with 4-(hydrazinosulfonyl)benzoic acid. DOX was then conjugated in acidic MeOH. Both of these materials showed uptake into tumors. However, DOX conjugates prepared from the dendrimers containing the L-lysine groups were demonstrated to have a lower accumulation in nontarget organs, such as the liver and were degraded more effectively *in vivo* compared to PLL-SPN-DOX conjugates [23].

To prepare dendrimer–DOX conjugates containing the more stable amide bond, Kono, Shieh, and coworkers have reacted PAMAM dendrimers bearing carboxylic acid peripheral groups directly with the amino group on DOX by either EDC- or DCC-mediated coupling reactions [19, 20]. Kono and coworkers prepared PAMAM-DOX conjugates via both amide and hydrazone linkages and observed that the conjugates containing hydrazone linkages exhibited seven times higher cytotoxicity to HeLa cancer cells than the conjugates containing amides. This result highlights the importance of the more labile hydrazone linkage for the treatment of the cancerous cells. However, both conjugates showed lower cytotoxicity to the same cell line when compared to free DOX [20]. In recent work by Baker and coworkers, DOX was introduced to the periphery of carboxylic acid-functionalized PAMAM dendrimers by means of a photocleavable spacer. In this approach, DOX was first protected with the photocleavable *o*-nitrobenzyl group through a carbamate

linkage with its amino group. Next, an amine group on the terminus of the spacer was used for the conjugation of DOX along with folic acid (FA) as a targeting group, via amide bond formation. They showed that these conjugates were cytotoxic to KB cells only upon exposure to UV light, which liberated the DOX from the conjugates [27].

5.2.3 Dendrimer–MTX Conjugates

MTX is an antimetabolite and antifolate drug that has been used in the treatment of cancer and autoimmune diseases. MTX treats cancer by inhibiting dihydrofolate reductase (DHFR), which catalyzes the conversion of dihydrofolate to the active tetrahydrofolate. However, MTX shows dosage-dependent hepatotoxicity, and cannot be prescribed to patients with liver disorders. Therefore, scientists have been actively exploring the use of vehicles for MTX delivery. Similar to DOX, there are several functional groups on MTX that can be used for conjugation. The presence of α- and β-carboxylic acid groups on the glutamic acid portion of the molecule provides the opportunity for amide and ester bond formation. In addition, amine groups at 2 and 4 positions in the pteridine heterocycle can be used for amide bond formation with carboxylic acid-functionalized dendrimers (Figure 5.5).

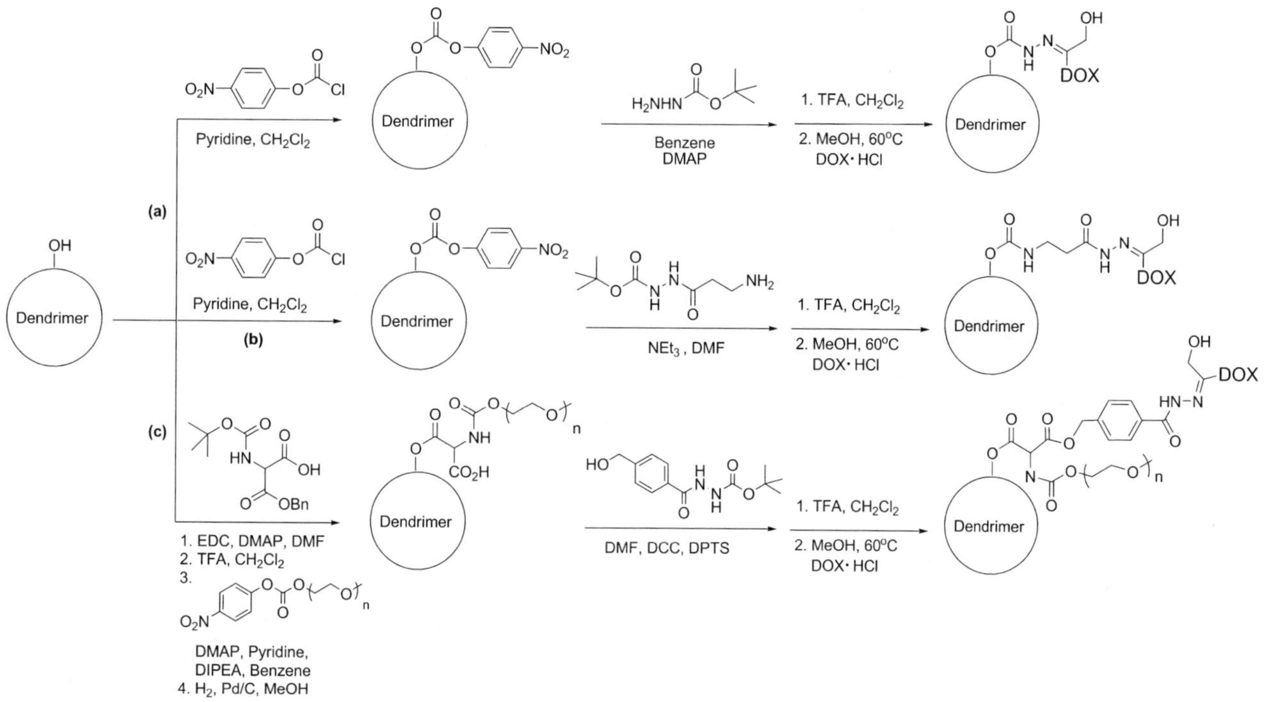

FIGURE 5.4   Approaches for the conjugation of DOX to the peripheral hydroxyl groups of polyester dendrimers: (a) a problematic approach where cyclization of the DOX on the carboxyhydrazide results in the release of modified, inactive drug; (b) use of a β-alanine spacer avoids this undesired cyclization; (c) conjugation to carboxylic acid-functionalized dendrimers. DIPEA, *N,N*-diisopropylethylamine; DPTS, 4,4-dimethylaminopyridinium *p*-toluenesulfonate.

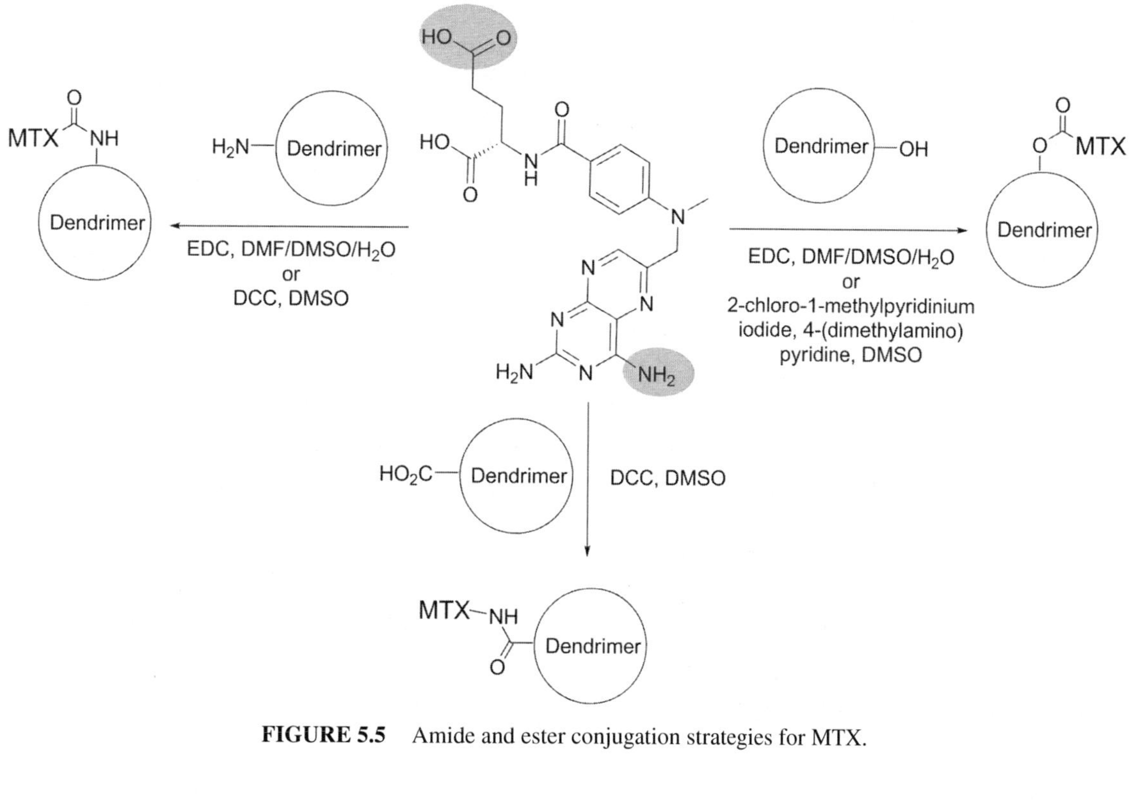

FIGURE 5.5   Amide and ester conjugation strategies for MTX.

Hydroxyl-terminated PAMAM dendrimers have been widely used for MTX conjugation [28–32]. These dendrimers were first synthesized through the reaction of amine-terminated PAMAM dendrimers with glycidol, and then the resulting hydroxyls were used in EDC-mediated coupling reactions with MTX [28–30]. Selectivity for the β-carboxylic acid was observed due to decreased steric hindrance relative to the α-position. Baker and coworkers prepared multifunctional PAMAM-fluorescein (FITC)-FA-MTX bioconjugates that provided increased anticancer potencies in mice-bearing KB tumors, relative to the free drug. The same group has also reported a more efficient synthesis of these constructs in a one pot synthesis manner using the Mukaiyama reagent (2-chloro-1-methylpyridinium iodide) and DMAP as coupling agents [31]. In a more recent study, a polyvalent saccharide-functionalized PAMAM dendrimer-bearing MTX via an ester linkage was developed as a potential anticancer drug [32]. D-glucoheptono-1,4-lactone was conjugated to amine-terminated PAMAM dendrimer via a stable amine bond. The resulting dendrimer presented approximately 192 hydroxyl groups on its surface which were then used to attach MTX and FITC via ester linkages using the Mukaiyama reagent. This construct, which is much simpler than the aforementioned PAMAM-FITC-FA-MTX, was shown to be as cytotoxic as PAMAM-FITC-FA-MTX against folate receptor-expressing KB cells.

Amide linkages have also been explored in PAMAM-MTX conjugates. The conjugation approach can be simple, involving a reaction of the carboxylic acid in MTX and the amines on the PAMAM periphery. Applying this conjugation method, PAMAM-FITC-FA-MTX and PAMAM-FA-MTX bioconjugates have been synthesized [20, 28, 33]. However, compared to free MTX, lower cytotoxicity was observed against KB cells, likely due to the slower rate of intracellular drug release resulting from the amide linkage [33].

Kannan and coworkers have conjugated MTX to PAMAM dendrimers via two different approaches [17]. In one approach, the peripheral amine groups of the PAMAM dendrimers were conjugated to one of the carboxylic acid groups of MTX by DCC-mediated coupling to form an amide linkage. In the second approach, carboxylic acid-functionalized PAMAM dendrimers were conjugated to MTX via one of its pteridine amino groups. Their rationale for the second approach was based on the literature, in which the presence of free carboxylic acid groups, mainly the α-carboxylic acid, is known to be essential for the cytotoxicity of MTX conjugates [34, 35]. The drug loading achieved in the first approach was higher than that for the later approach, presumably due to the steric effects on the amine sites of MTX. On the other hand, it was found that the conjugate prepared by the second approach exhibited much higher cytotoxicity in MTX-resistant cells when compared to free MTX. Conjugates prepared by the first approach showed only slightly increased cytotoxicity relative to free drug against the same cell lines.

Porter and coworkers have reported a series of PEGylated PLL dendrimers functionalized with MTX via either a stable amide linkage or a hexapeptide linkage in which α-carboxylic acid group in MTX was either capped with tert-butoxide (OtBu) group or left free [23, 36]. Interestingly, they observed that OtBu-capped MTX dendrimers exhibited higher cytotoxicity against HT1080 cells than those of the uncapped MTX conjugates. These capped MTX dendrimers, unlike the uncapped constructs, also significantly reduced tumor growth in mice-bearing HT1080 tumors. The authors concluded that capping the α-carboxylic acid group in MTX might be an effective way of increasing their circulation time and reducing their liver accumulation while preserving antitumor efficacy.

5.2.4 Dendrimer–TAX Conjugates

TAX is an anticancer drug that operates by inhibition of mitosis. The major limitation of TAX is its very poor water solubility, which has resulted in the use of solubilizing agents such as Cremophor EL and ethanol. However, the use of such excipients has been shown to cause undesirable side effects upon injection [10]. Therefore, the development of drug delivery systems that can enhance TAX's solubility is of significant interest in order to minimize these side effects. The molecule only contains free hydroxyl groups that can be used for bioconjugation, typically via ester formation. In addition, via esterification, different spacers have been conjugated to TAX to introduce other functional groups for further modifications. As shown in Figure 5.6, due to steric hindrance at the other positions, the hydroxyl on the 2′ position can generally be reacted selectively.

PAMAM dendrimers were the first backbones used in TAX conjugation [17, 37]. Minko and coworkers first introduced a hemisuccinate spacer to the 2′ hydroxyl position through an ester linkage. In the next step, the carboxylic acid functional group on the new TAX derivative was used for conjugation to dendrimer surface hyroxyl groups via either an EDC-mediated coupling or preactivation of the acid by formation of an NHS ester. In vitro studies in A2780 human ovarian carcinoma cells showed a 10-fold enhancement in cytotoxicity of the PAMAM-TAX conjugate compared to the free drug [17]. Baker and coworkers reported similar cytotoxicity for their PAMAM-FA-TAX conjugate against KB cells, and noted decreased cytotoxicity when testing against a cell line lacking folate receptors [37].

In a series of studies by Simanek and coworkers, PEGylated triazine-based dendrimers were used for the conjugation and delivery of 12 molecules of TAX [36, 38, 39]. Briefly, TAX was first acylated with glutaric anhydride to introduce a carboxylic acid. An NHS ester of this acid

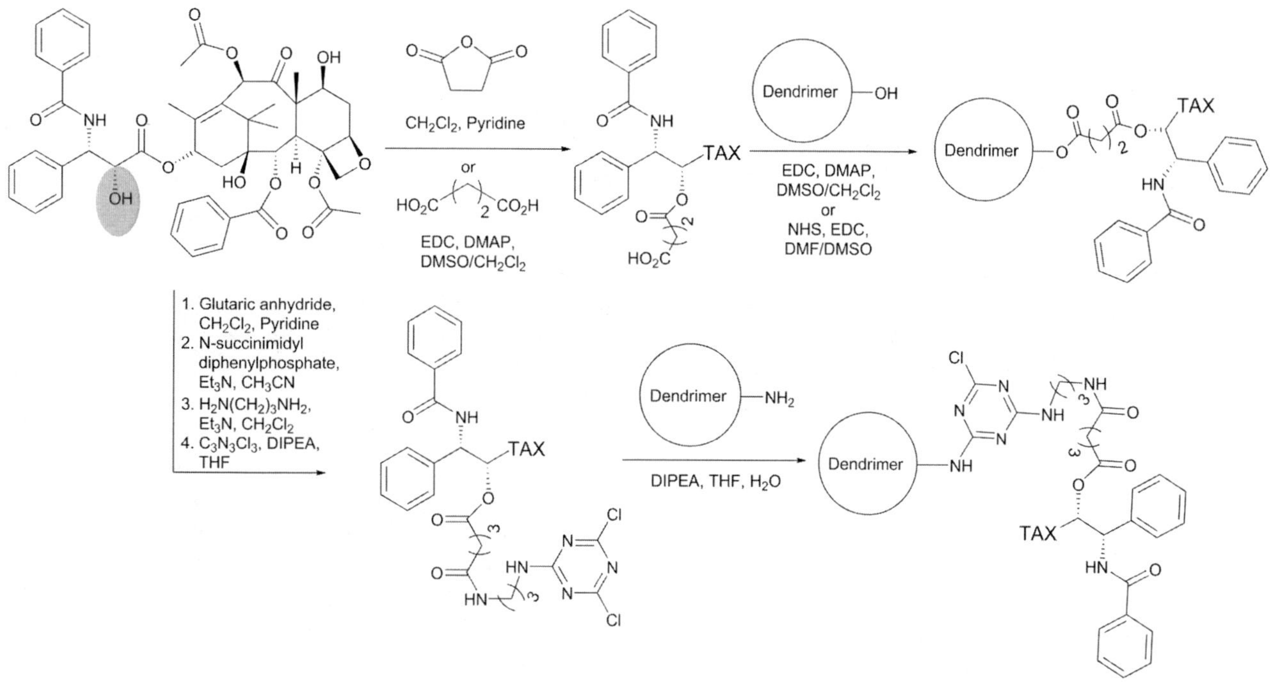

FIGURE 5.6 Linker-based conjugation strategies for TAX. THF, tetrahydrofuran.

was then reacted with 1,3-diaminopropane to afford an amine functionality. This spacer was chosen to induce high flexibility and to reduce steric hindrance. Reaction of this derivative with cyanuric chloride yielded a dichlorotriazine derivative that was conjugated to the dendrimer's peripheral amine groups in the presence of DIPEA. Finally, PEGylation was performed to introduce solubility. Studies of these

conjugates in mice showed they were well-tolerated at doses 200% of the maximum tolerated dose of free drug, but unfortunately they exhibited relatively low efficacy [36, 39].

Self-immolative dendrimers have also been reported for TAX conjugation. In a report by de Groot and coworkers, first and second generation self-immolative dendrimers based on a 2-(4-aminobenzylidene)propane-1,3-diol branching

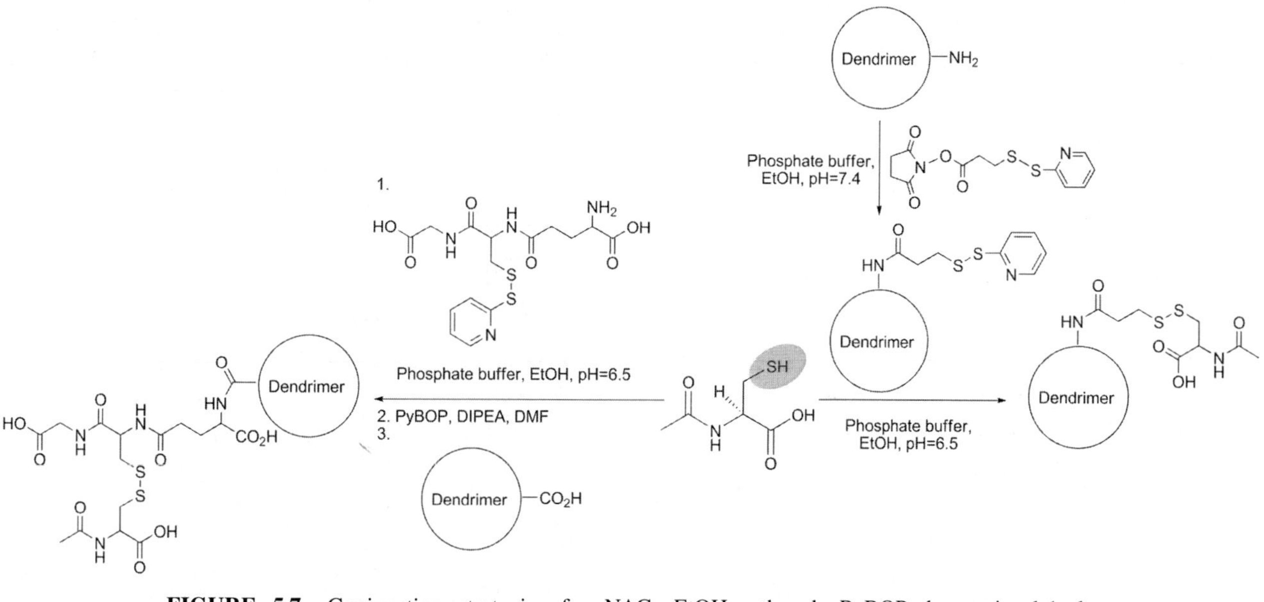

FIGURE 5.7 Conjugation strategies for NAC. EtOH, ethanol; PyBOP, benzotriazol-1-yl-oxytripyrrolidinophosphonium hexafluorophosphate.

monomer and a nitro group to mask the aniline focal point in the oxidized form [40]. Two or four TAX molecules were conjugated to the first and second generation dendrimers, respectively, via carbonate linkages. TAX was released upon reduction of the nitro group with zinc metal in acetic acid, and subsequent fragmentation of the self-immolative dendrimer backbone. More recently, a first generation dendrimer fragmenting by the 1,6-elimination-decarboxylation cascade was used to conjugate TAX to N-(2-hydroxypropyl)-methacrylamide-based polymer to obtain a copolymer–drug conjugate, with an AB_3 self-immolative dendritic linker, via a focal point oligopeptide spacer sensitive to cathepsin B enzyme [41]. Cleavage of this peptide spacer initiated dendrimer disassembly, resulting in the release of three equivalents of TAX. This new polymer–drug conjugate exhibited enhanced cytotoxicity to murine prostate adenocarcinoma (TRAMP-C2) cells in comparison with the classic monomeric polymer–drug conjugate.

5.2.5 Dendrimer–NAC Conjugates

NAC is an antioxidant and anti-inflammatory drug used in the treatment of neuroinflammation, colon cancer, detoxification of heavy metals from the body, and human immunodeficiency virus (HIV) infection. The free sulfhydryl group in the drug is capable of oxidation and formation of disulfide bonds with plasma proteins [42]. For these reasons, NAC has poor bioavailability and blood stability, which results in a requirement for high and repeated dosing of the drug. However, high doses of NAC can be cytotoxic and lead to side effects such as high blood pressure [43]. On this basis, the construction of delivery systems that can enhance its bioavailability and stability is of significant interest. The two functional handles on NAC that can be used for conjugation are a thiol and a carboxylic acid group. The thiol can be used

to form thioether or disulfide linkages while the carboxylic acid can form either ester or amide linkages (Figure 5.7).

PAMAM dendrimers having amine peripheral groups and those having carboxylic acid peripheral groups have both been studied for NAC conjugation via disulfide linkages [44–46]. For the conjugation of NAC to amine-terminated dendrimers, an N-succinimidyl 3-(2-pyridyldithio)propionate linker was first appended to the dendrimer surface through an amide bond. The thiopyridyl disulfide was then used for reaction with the thiol of NAC by a thiol–disulfide exchange mechanism to install NAC on the dendrimer periphery via a disulfide linkage. In order to conjugate NAC to the surface of anionic PAMAM dendrimers, NAC was first functionalized with glutathione (GSH) peptide via disulfide bond formation, and then the GS-S-NAC intermediate was conjugated to the dendrimer periphery through its amine group to form an amide linkage. The release of NAC from both dendrimer conjugates relied on the disulfide bond cleavage, which was dependent on intracellular glutathione levels. *In vitro* studies in activated microglial cells showed that both conjugates were significantly more cytotoxic, even at drug levels 16 times lower than that the free drug.

5.2.6 Dendrimer–Sulfonic Acids

Though not strictly a conjugate between a dendrimer and free, active drug molecules, the development of SPL7013, the active ingredient in VivaGel®, is a noteworthy example due to its recent successes in clinical trials. As shown in Figure 5.8, SPL7013 comprises a fifth generation PLL dendron to which aromatic sulfonic acid moieties have been conjugated through an amide linkage. Binding of the sulfonic acid moieties on the periphery of the dendron to receptors on the surface of the virus prevents the virus from attaching to human cells. VivaGel has been shown to inactivate and

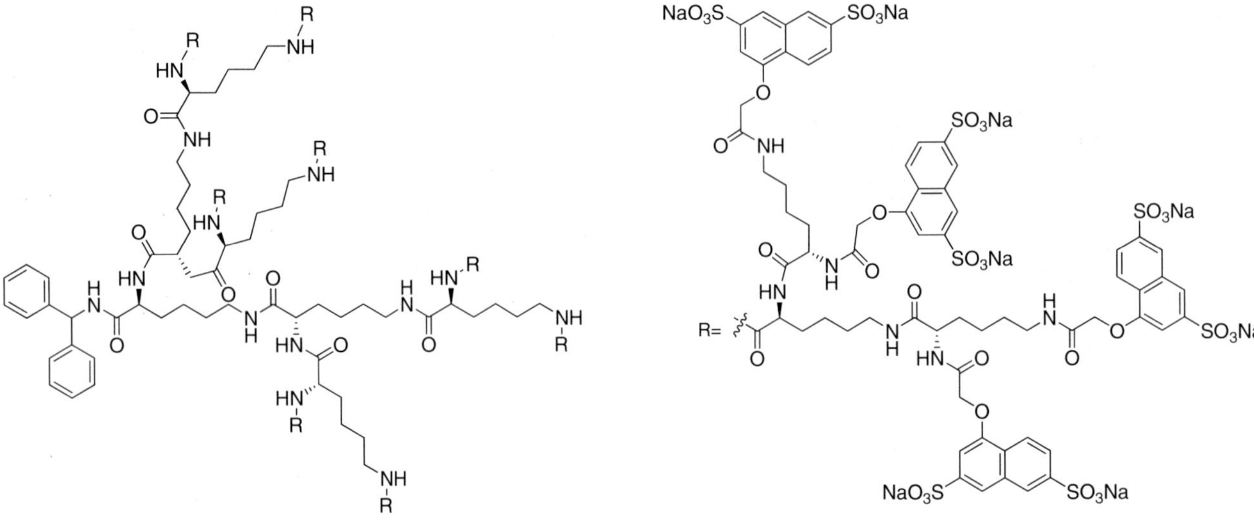

FIGURE 5.8   Chemical structure of SPL7013.

inhibit HIV and the genital herpes virus and will be sold as a condom coating. In addition, recent Phase 2 clinical studies have demonstrated VivaGel also has efficacy for the treatment of bacterial vaginosis (BV), and a Phase 3 trial is currently underway for this application. It is hoped that VivaGel will overcome some of the limitations of the current BV treatments, such as low cure rates and adverse systemic effects. For example, as the VivaGel treatment is topical, it is free from systemic effects. It should be noted that the activity of this therapeutic relies on the multivalent display of sulfonic acids and therefore the cleavage of these moieties from the dendrimer backbone is not desired.

5.3 DENDRIMER–CARBOHYDRATE CONJUGATES

5.3.1 Motivation for the Development of Dendrimer–Carbohydrate Conjugates

Carbohydrates are the most abundant group of natural products found on the earth. Aside from their important roles in supplying energy to cells and structural support to plants, carbohydrates are implicated in a vast array of biological processes. These include hormonal activities, fertilization, embryogenesis, neural development, and many other cellular processes such as cell–cell recognition, cell proliferation, cellular transport, viral infection, bacterial adhesion, and tumor cell metastasis [47]. Thus, it is not surprising that the efficient synthesis of saccharides and their incorporation into various systems to obtain specific biological effects has attracted much attention in the past few decades. One such example is the tremendous effort that has been devoted to enhancing the multivalency of carbohydrates by incorporating them onto multivalent architectures such as dendrimers. It is known that multivalent interactions are prevalent in biology, such as in the adhesion of viruses and bacteria to cell surfaces and in the binding of cells to other cells. Many of these processes involve the interactions of carbohydrates with protein receptors called lectins. While the interactions of individual carbohydrate ligands with lectins are often weak, multivalency provides a means of significantly increasing the strength of the interaction.

A wide variety of nanoscale materials, such as linear and hyperbranched polymers, nanoparticles, and polymer assemblies, can be used as backbones to present carbohydrates in a multivalent manner but the well-defined nature of the dendrimer backbone provides advantages. For example, the number of carbohydrate ligands present on a given molecule can be precisely determined, allowing advancements in the fundamental understanding of carbohydrate–lectin interactions. In addition, the product monodispersity and reproducibility in its synthesis is advantageous for the development of a clinical therapeutic. A wide range of saccharides including mannose, galactose, glucose, lactose, maltose,

xylose, N-acetylneuraminic acid (AcNA) (sialic acid), and other oligosaccharides have been conjugated to various dendrimer peripheries via different linkages such as amide, hydrazide, amine, thioether, thiourea, and triazole linkages. Unlike dendrimer–drug conjugates, in which fine-tuned lability of the linkage is essential for controlled release of the drug, carbohydrates typically do not need to be released from the dendrimer periphery in order to exhibit activity. Thus, although the linkages can have modest effects on the binding affinities of multivalent carbohydrates, the choice of linkage is determined primarily by synthetic requirements. In this section, we will discuss the conjugation of mannose, galactose, and AcNA to the peripheries of various dendrimer backbones and will briefly introduce the biological properties of the resulting bioconjugates. This discussion provides representative examples of the different conjugation chemistries that can be for coupling of dendrimers with carbohydrates.

5.3.2 Dendrimer–Mannose Conjugates

Mannose is a monomeric sugar that is a C-2 epimer of glucose. In its cyclic form it possesses a pyranose or furanose ring. It has been shown that terminal mannoside residues can interact with receptors found on macrophages [48], hepatic sinusoidal cells [49], and different invading pathogens [50], thus motivating interest in the development of various multivalent derivatives of mannose. To prepare a dendrimer–mannose bioconjugate, it is necessary to introduce to mannose molecules reactive functional groups that are complimentary to those on dendrimer's periphery. This has been accomplished using standard synthetic methods in carbohydrate chemistry that will not be reviewed here, but as shown in Figure 5.9, many different mannose derivatives including amines, carboxylic acids, azides, alkynes, isothiocyanates, and thiols have been prepared and can be conjugated to the peripheries of dendrimers.

Amine-functionalized mannose can be used as a starting point for various bioconjugation reactions. For example, Rojo and coworkers have conjugated these derivatives to the surfaces of carboxylic acid-functionalized Boltorn dendrimers under conventional HOBt/DIC coupling conditions [51, 52]. They have shown that these glycodendrimers can efficiently bind to the lectin *Lens culinaris* and that the dendritic backbone supports the necessary multivalency for lectin cluster formation [51]. Moreover, they have reported that these glycodendrimers with 32 mannose molecules can bind with submicromolar affinity to dendritic cell-specific ICAM-3 grabbing nonintegrin (DC-SIGN), which is involved in the initial steps of numerous infection processes, and they inhibit DC-SIGN binding to HIV glycoprotein gp120 with IC_{50} values in micromolar range. Thus, these materials can potentially serve as antiviral drugs [52]. Using the same amine-functionalized mannose molecule, Seeberger and coworkers have prepared

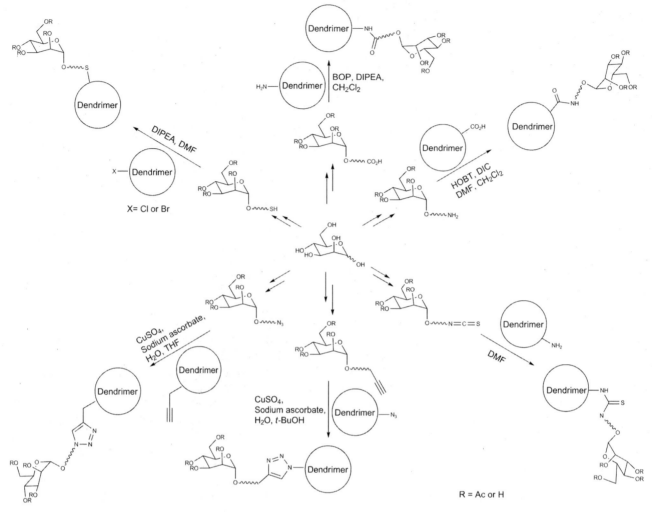

FIGURE 5.9   Conjugation strategies for mannose. These reactions can involve the unprotected (R = H) or protected (R = Ac) mannose derivatives. In the latter case, deprotection is necessary for following conjugation. t-BuOH, *tert*-butanol; DIC, *N,N'*-diisopropylcarbodiimide.

a series of tris(bipyridine)ruthenium dendrimers having 2–18 mannose molecules on their peripheries [39, 53]. They have shown that these glycodendrimers along with galactose-based glycodendrimers can be used in a digital, single-operation analytical method to study glycodendrimer–lectin interactions. Owing to the fluorescence and electron transfer character of the tris(bipyridine)ruthenium core, these materials were able to respond to change in pH, electron transfer agents, and different lectins, such as Concanavalin A (Con A), *Galantus nivalis* agglutinin, and asialoglycoprotein, with the output being the relative change in fluorescence quantum yield of the glycodendrimer [39].

There are also many examples involving the preparation of isothiocyanate derivatives of α-D-mannopyranoside and their coupling to dendrimers having peripheral amine groups. For example, protected *p*-aminophenyl mannopyranoside was reacted with thiophosgene to give the desired

protected *p*-isothiocyanatophenyl mannopyranoside in good yields [54]. This isothiocyanate derivative was reacted with the dendrimer's peripheral amines in 0.2 M borate buffer at pH 9, and then the acetate-protecting groups were removed under basic conditions. PAMAM dendrimers having 32 peripheral mannose molecules were synthesized by this approach and shown to bind to phytohemmaglutinins from Con A and *Pisum sativum* with 400-fold higher affinity than the reference monomeric mannose molecule [54].

In a different synthetic approach, Lindhorst and coworkers have used unprotected mannose derivatives *p*-isothiocyanatophenyl, α-D-mannopyranoside, and 2-isothiocyanatoethyl α-D-mannopyranoside, for direct coupling to tris(2-aminoethyl)amine and PAMAM–dendrimer peripheries [55, 56]. By applying this method, the final deprotection of mannose acetyl groups on the dendrimers can be avoided as the final yield-lowering step. Bogdan,

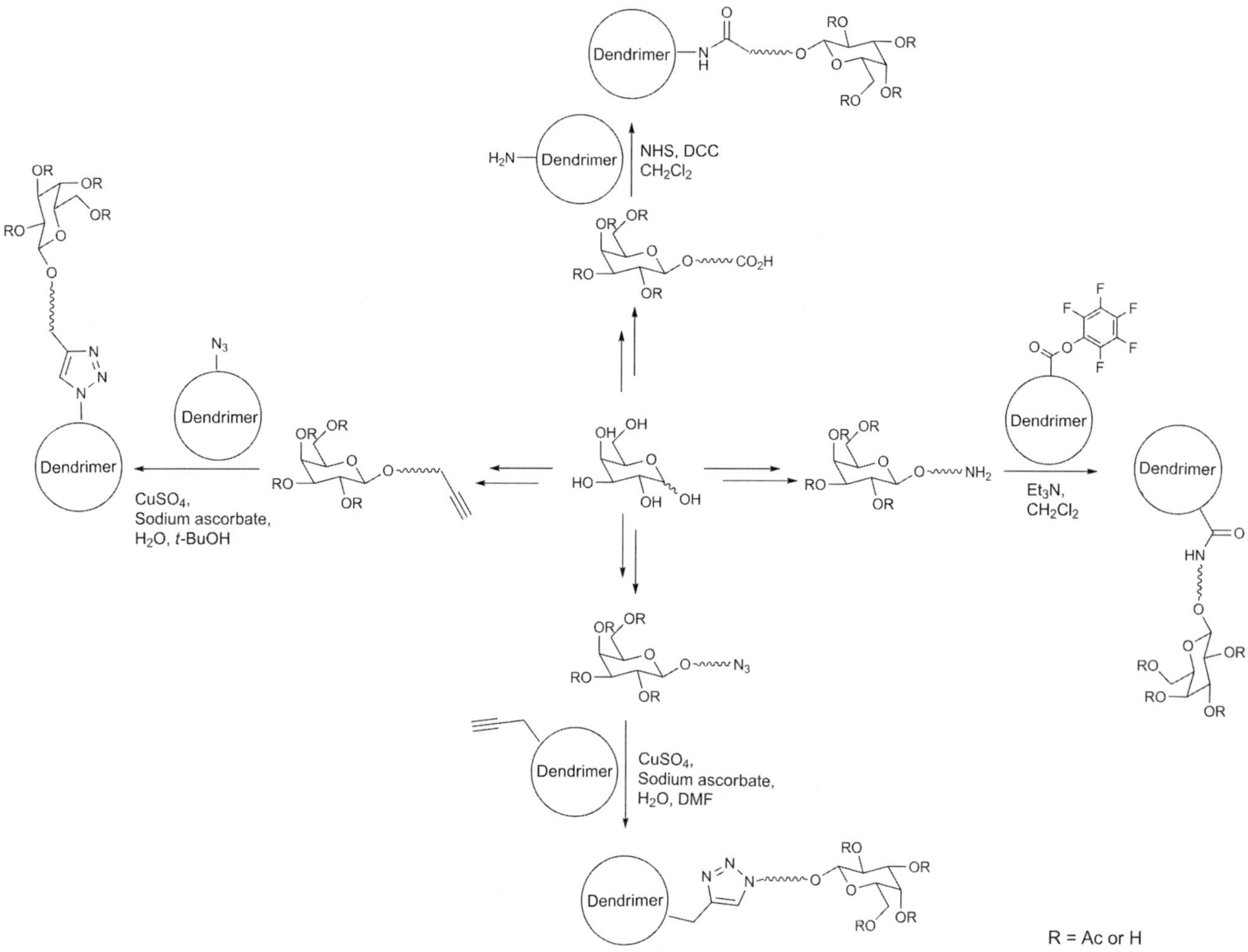

FIGURE 5.10   Conjugation strategies for galactose. These reactions can involve the unprotected (R = H) or protected (R = Ac) mannose derivatives. In the latter case, deprotection is necessary following conjugation.

Capobianco, and coworkers have also taken advantage of this strategy for the postfunctionalization of gold nanoparticles and upconverting lanthanide-doped nanoparticles coated with PAMAM dendrimers [57, 58]. These materials were used to study Con A-carbohydrate interactions via surface energy transfer process mechanisms. Cloninger and coworkers have prepared mannose-functionalized PAMAM dendrimers ranging from the first to sixth generation and their binding affinities to Con A were shown to depend on the sizes and multivalencies of the bioconjugates [59]. Moreover, they were able to systematically control the degree of lectin clustering and overall activity by varying the mannose/hydroxyl ratios on the surfaces of third to sixth generation dendrimers [60]. However, in contrast to the results obtained with Con A, the same mannose-functionalized PAMAM dendrimers were shown to bind to *P. sativum* (pea lectin) in a monovalent fashion [61]. This result indicates that specific glycodendrimers

may be good therapeutic agents for some lectins but at the same time inappropriate for other ones.

Owing to its high yield, steric and functional group tolerance, the copper(I)-catalyzed alkyne–azide click reaction has also been widely used for the functionalization of dendrimer peripheries with carbohydrates. To achieve this, two general approaches have been reported: (1) alkyne functionalization of the dendrimer periphery and its subsequent reaction with azide-functionalized mannose [39, 62–67] or (2) azide functionalization of dendrimer periphery and its subsequent reaction with alkyne-decorated mannose [68, 69]. The widely used catalytic system for the cycloaddition reaction is CuSO$_4$/sodium ascorbate, while different reaction solvents have been investigated. For example, Riguera and coworkers have used a H$_2$O/t-BuOH as the solvent for the reaction of azide-terminated gallic acid–triethylene glycol dendrimers with alkyne-functionalized mannose derivatives

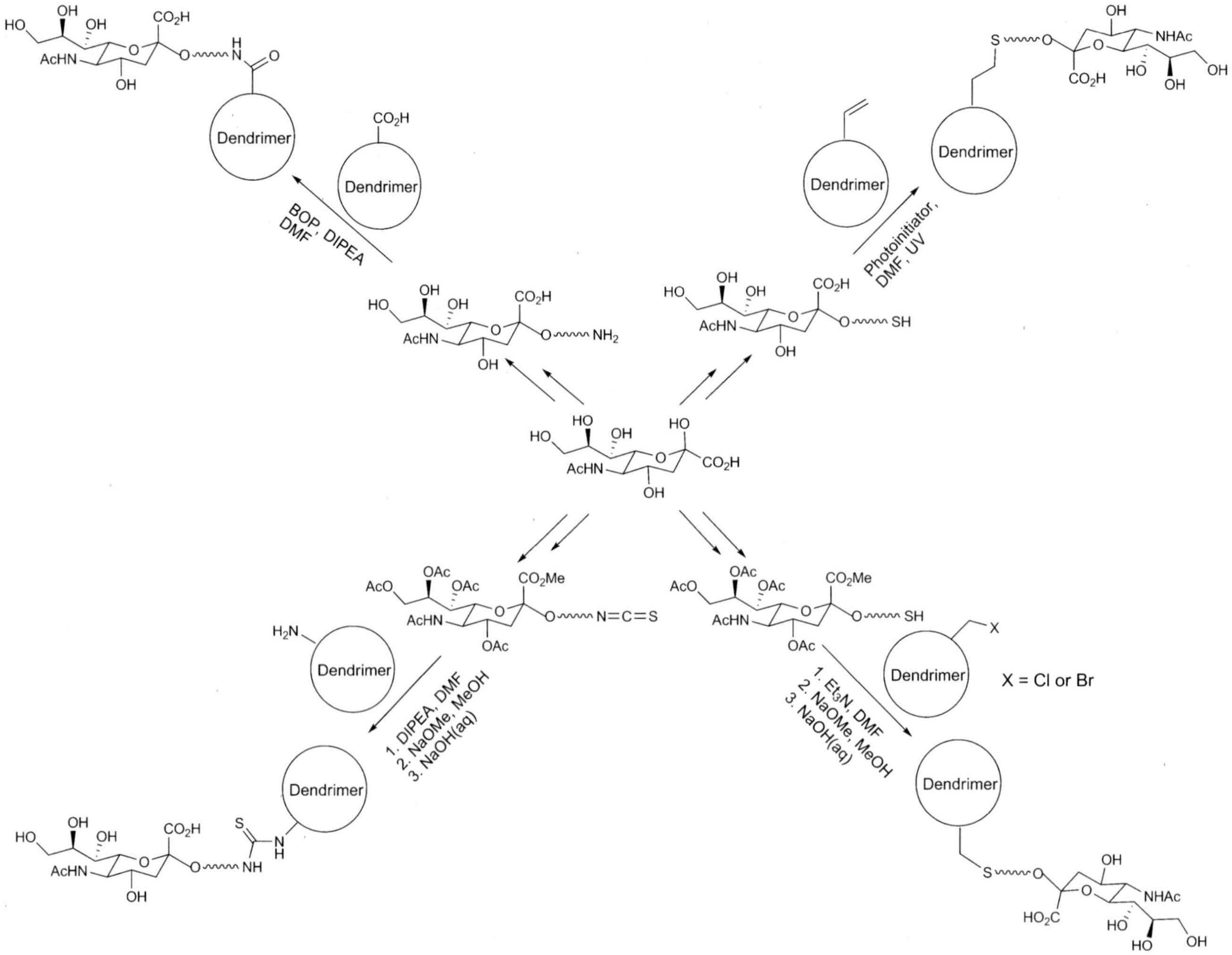

FIGURE 5.11 Conjugation strategies for AcNA.

in good yields, and have shown that aggregation of these conjugates with Con A increases with the generation of the conjugates [68, 69]. Pieters and coworkers have taken advantage of the click reaction under microwave conditions, which effectively reduced the reaction time, using a mixture of H$_2$O/MeOH as their solvent system to yield the desired glycodendrimer systems based on 3,5-di-(2-aminoethoxy)-benzoic acid repeating unit [62, 63]. Bertozzi and coworkers have reacted alkyne-functionalized PE dendrons with azide-functionalized mannose derivatives using THF/H$_2$O as the reaction medium [66, 67]. The focal points of these glyco-dendrons were functionalized with pyrene, which was used to bind to single-walled carbon nanotube (SWCN) and boron nitride nanotube (BNNT) surfaces through π–π interactions. It was shown that these glycodendrons could function as homogeneous bioactive coatings, decreasing the cytotoxicity of SWCNs and allowing the BNNTs to interact with proteins and cells.

Unlike the amide, thiourea, and triazole linkages discussed above, there are a limited number of examples

involving the conjugation of mannose to dendrimers using a thioether linkage. Roy and coworkers have prepared a thiol derivative of α-D-mannopyranoside having acetate-protecting groups on the sugar's hydroxyl groups [70]. Reaction of the thiol with N-chloroacetylated polypeptide-based dendrimers, followed by removal of the acetate-protecting groups yielded the desired dendrimer–mannose conjugates. On the other hand, Melnyk and coworkers have prepared an unprotected thiol derivative of α-D-mannopyranoside and reacted it with N-chloroacetylated PLL dendrimers affording the target dendrimer–mannose conjugate in relatively low yield [71]. Terunuma and coworkers have used carbosilane dendrimers as platforms to conjugate mannose derivatives to their surfaces through thioether linkages [72]. The conjugation was accomplished by treating a mixture of thioacetic acid-functionalized α-D-mannopyranoside and bromide-functionalized dendrimer with sodium methoxide (NaOMe)/MeOH in DMF followed by the addition of acetic anhydride/pyridine. Fully deprotected dendrimers were obtained by treatment of these dendrimers with

NaOMe/MeOH, then 0.1 M NaOH solution. By performing isothermal titration microcalorimetry these dendrimers were shown to bind to Con A more strongly than free mannose or mannobiose.

More recently, Luo and coworkers have reported a general method for the conjugation of underivatized carbohydrates to dendrimer peripheries. In this approach, the periphery of PAMAM dendrimers was first modified with hydrazide moieties. Next, by taking advantage of the reactivity between carbohydrates containing reducing hemiacetals and peripheral hydrazides, carbohydrates such as mannose, glucose, galactose, N-acetylglucosamine, and lactose were conjugated to the PAMAM dendrimer surfaces via a hydrazone linkage [73]. It should be noted that this conjugation approach leads to the coupling of a mixture of cyclic and noncyclic versions of the sugars.

5.3.3   Dendrimer–Galactose Conjugates

Galactose is a monomeric carbohydrate that is a C-4 epimer of glucose. It can exist in both open-chain and cyclic forms. Like mannose, in its cyclic form it exists as pyranose and furanose rings (Figure 5.10). Galactose is also involved in many biological and disease processes. For example, sulfated multivalent galactose derivatives have been shown to be potentially effective inhibitors for HIV-1 infections [74]. Moreover, galactose has been used as a targeting agent to deliver the anticancer drug DOX to hepatocytes and induce apoptosis *in vitro* [39]. Cholera toxin is the major lethal agent associated with cholera and it has been shown that multivalent galactose ligands can serve as mimics of the naturally occurring ganglioside GM1 oligosaccharide, binding to and mitigating the effects of this toxin [75].

Meijer, Stoddart, and coworkers prepared a carboxylic acid-functionalized galactose-bearing acetate-protecting groups on the sugar hydroxyls [76]. Following activation with NHS, coupling reactions with the peripheral amine groups on PPI dendrimers were performed. Deprotection under basic conditions gave the target glycodendrimers in good yields. Taking advantage of this synthetic method, Schengrund and coworkers were able to synthesize a series of sulfated and nonsulfated galactopyranosyl-derivatized PPI dendrimers from the first to fifth generations. Evaluation of these glycodendrimers for the inhibition of infection of U373-MAGI-CCR5 cells by HIV Ba-L revealed that sulfated glycodendrimers were better inhibitors than nonsulfated dendrimers, but still not as effective as dextran sulfate [74].

Seeberger and coworkers have reacted amine-functionalized galactose and mannose derivatives with dendrons based on N-(tris[(2-cyanoethoxy)methyl]methyl amine) having peripheral pentafluorophenol ester moieties [39, 77, 78]. The functionalization of the focal points of these glycodendrons with different functionalities have resulted in different applications of these materials [39, 77, 78]. For example, by installing an azide functional group at the focal

point of the glycodendrons and subsequently performing a click cycloaddition reaction with an alkyne-decorated β-cyclodextrin, it was possible to target the asialoglycoprotein receptor (ASGPR), which is a C-type lectin and is suitable for liver-specific delivery [39]. It was also shown that these materials were selectively taken up by HepG2 cells in comparison with β-cyclodextrins functionalized with mannose functionalities [39].

Copper(I)-catalyzed alkyne–azide click reactions have also been used to synthesize dendrimer–galactose conjugates via a triazole linkage. The azide-functionalized [39, 63, 75, 79] or alkyne-functionalized [80] galactose can be used. Using an azide derivative, Pieters and coworkers have taken advantage of the microwave-assisted reaction to perform conjugation reactions with three different alkyne-functionalized dendrons [79]. Using enzyme-linked immunosorbent assay (ELISA) experiments, they showed that the galactose–dendron conjugate containing a long spacer, based on a combination of PEG and an alkyl chain, was as effective as the natural ganglioside GM1 oligosaccharide in terms of inhibiting cholera toxin binding [75]. In a study by Roy and coworkers, alkyne-functionalized galactoside and/or fucoside were conjugated to the surfaces of azide-functionalized dendrimers based on 3,5-di(2-azidoethoxy)benzoic acid to obtain homo- and heterobifunctional glycodendrimers [80]. Using both PA-IL and PA-IIL lectins from *Pseudomonas aeruginosa*, it was shown that the heteroglycodendrimer containing four fucosides and four galactosides was able to recognize binding sites on both PA-IL and PA-IIL.

As previously described for mannose, a general method for the conjugation of underivatized galactose onto PAMAM dendrimers with hydrazide peripheral groups has also been reported, providing dendrimers with hydrazine linkages to the galactose [81]. Cellular uptake experiments using dye-labeled glycodendrimers with HepG2 cells showed that galactose-, lactose- and N-acetylglucosamine–PAMAM bioconjugates exhibited enhanced uptake compared to glucose–PAMAM and mannose–PAMAM dendrimers [73].

5.3.4   Dendrimer–AcNA Conjugates

AcNA is the most abundant sialic acid found in mammalian cells. This negatively charged molecule is found in complicated glycans on mucin and in glycoproteins that are embedded in cell membrane. It is known that all types of influenza viruses interact with AcNA residues on the host cell surface through their trimeric lectin hemagglutinin (HA), and this is followed by endocytosis of the virus into the cell [82]. Monovalent AcNA can inhibit this interaction at millimolar concentrations, but there is significant interest in the development of multivalent AcNA derivatives in order to obtain higher binding affinity. Different strategies for the functionalization of dendrimers with various derivatives of AcNA are shown in Figure 5.11.

Thus far, the formation of a thioether is the most commonly reported approach for the conjugation of AcNA to the peripheries of dendrimers [83–90]. To construct these conjugates, a thiol must be installed on either sialic acid or on the dendrimer periphery. Because unprotected thiols tend to dimerize readily, a process that is facilitated on the dense peripheries of dendrimers, the introduction of the thiol to AcNA has been a more viable approach. For example, Roy and coworkers have prepared a protected 2-α-thioacetyl–AcNA and reacted it with three types of N-chloroacetylated dendrimers, including polypeptide [83], PAMAM [84, 85], and gallic acid–oligoethylene glycol dendrimers [86]. The acetyl groups on the sugars were then removed under basic conditions to yield the unprotected dendrimers. An enzyme-linked lectin inhibition assay using human α_1-acid glycoprotein as the coating antigen and horseradish peroxidase-labeled LFA for detection purposes was performed. Their results showed that the globular dendrimer with a valency of 12 exhibited a 182-fold increase in inhibitory potency compared to the reference monomeric AcNA [85].

To investigate different spacers between AcNA and dendrimers, Matsuoka and coworkers prepared a library of brominated carbosilane dendrimers with different types of spacers on their peripheries [88, 89]. The thiol-functionalized AcNA derivative was similar to that employed by Roy and coworkers but with an additional five-carbon aliphatic spacer between the sugar and the thiol. Introduction of this molecule onto different brominated carbosilane dendrimers with either normal, ether-elongated, or amide-elongated peripheral groups was accomplished by initially treating the thioacetic acid-functionalized sugar and bromide-terminated dendrimer mixture with NaOMe/MeOH in DMF followed by addition of acetic anhydride (Ac$_2$O)/pyridine. Fully deprotected dendrimers were obtained by treatment of these dendrimers with NaOMe/MeOH and 0.1 M sodium hydroxide (NaOH) solution. Biological evaluations of these glycodendrimers showed that all of the ether- and amide-elongated compounds had inhibitory activities for the influenza sialidases in the millimolar range. Surprisingly, the glycodendrimers having normal aliphatic linkages did not exhibit any activities except for a dendrimer with a valency of 12 [89].

Hawker and coworkers recently reported the glycosylation of a fourth generation dendrimer via a free-radical thiol–ene coupling reaction between thiol-functionalized carbohydrates including AcNA, mannose, glucose, and lactose and a dendrimer having peripheral alkene moieties [90]. This reaction results in the formation of a thioether linkage in high yield.

McReynolds and coworkers have investigated amide linkages between AcNA and PAMAM dendrimers [90]. They have constructed their bioconjugates with or without spacers between the dendrimer and AcNA. In the case without any spacers, commercially available AcNA was directly conjugated to the amine peripheral groups of the dendrimers using

BOP. Alternatively, to minimize steric congestion between AcNA and the dendrimer, a bifunctional spacer molecule was first conjugated to the carboxylic acid functionality of AcNA. After deprotection of the other terminus of the spacer, which resulted in the formation a free amine, it was coupled to the periphery of acid-functionalized PAMAM dendrimers. Subsequent sulfation of the conjugates was accomplished by reacting the obtained dendrimers with an SO$_3$–pyridine complex. When evaluated for inhibition of HIV-1 infection, the sulfated AcNA-PAMAM glycodendrimer-bearing 16 AcNA moieties with 11 sulfate groups was found to inhibit all four HIV-1 strains tested in the low micromolar range.

Lastly, thiourea conjugates of AcNA and dendrimers have also been prepared. An acetate-protected p-isothiocyanatophenyl derivative of AcNA was prepared and was coupled to the peripheral amines of PAMAM dendrimers to give the protected AcNA dendrimers in high yields (71–100%) [91, 92]. Complete deprotection was accomplished by sequential ester hydrolysis in first NaOMe/MeOH followed by 50 mM NaOH solution to hydrolyze the acetyl followed by methyl ester groups. By performing a competitive enzyme-linked lectin assay, it was demonstrated that these dendrimers exhibited a substantial 210-fold increase in the inhibitory activity compared to monomeric AcNA [91].

5.4 DENDRIMER CONJUGATES WITH IMAGING AGENTS

5.4.1 Motivation for the Development of Dendrimer-Imaging Agent Conjugates

With rapid developments in imaging technology, along with an increased focus on the early detection of diseases and the monitoring of treatment effects, there has been great interest in the development of new contrast agents for various imaging modalities including magnetic resonance imaging (MRI), X-ray computed tomography (CT), single photon emission computed tomography (SPECT), positron emission tomography (PET), and optical imaging. These contrast agents aid in distinguishing between normal and diseased tissues through their localization at specific sites *in vivo*. Among the new contrast agents under development, nanosized agents based on materials such as linear polymers, organic and inorganic nanoparticles, proteins, and dendrimers have received particular attention in recent years. When the size and chemical functionalities of these agents are optimized, they can exhibit significantly longer *in vivo* circulation times than small molecule analogs. This enables new applications such as vascular imaging and the targeting of specific disease sites, such as tumors, to be explored. In addition, nanosized agents enable the conjugation of multiple contrast agent molecules to a nanomaterial, enhancing the contrast on a per molecule or per particle basis. Furthermore, this same attribute can

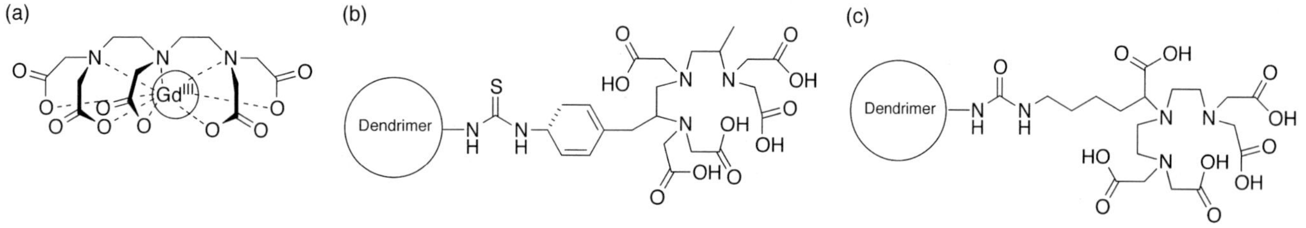

FIGURE 5.12 Chemical structures of (a) the clinical agent Gd(III)-DTPA (Magnevist) and dendrimer conjugates of DTPA derivatives containing (b) an aromatic isothiocyanate and (c) a more flexible aliphatic isocyanate.

allow the conjugation of both contrast agents and targeting moieties or multiple contrast agents for different imaging modalities to the same system providing enhanced, multifunctional properties.

Among the various nanomaterials available, as described in Section 5.2.1, the well-defined chemical structures of dendrimers provide a significant advantage in terms of reproducibility in the synthesis and resulting properties of the agents, allowing well-characterized materials to be prepared to the satisfaction of regulatory agents. Below, various examples involving the conjugation of imaging contrast agents for the different imaging modalities will be described.

5.4.2 Dendrimer Conjugates for Magnetic Resonance Imaging

MRI is a prominent noninvasive imaging modality due to its excellent spatial resolution, soft tissue contrast, and the absence of harmful ionizing radiation in its application. Despite its high levels of soft tissue contrast, contrast agents based on small molecule chelates of Gd(III) are frequently employed in clinical MRI scans to aid in the differentiation between healthy and diseased tissues [93–95]. These agents, which act by altering the relaxation times of the protons in nearby water molecules, have enabled significant advancements in MRI over the last couple of decades. However, the low contrast efficiency (i.e., low relaxivity), fast renal excretion, and low specificity of these agents result in a requirement for high doses. This can be problematic for patients with chronic renal disease [96]. It can also limit their applicability in molecular imaging applications, where target receptors are present only at low concentrations [97]. Dendrimer-based MRI contrast agents have been intensively investigated over the past couple of decades for several reasons [98, 99]. First, they allow for the attachment of multiple MRI labels to a single scaffold, greatly increasing the molecular relaxivity and allowing a single targeting moiety to carry multiple labels. In addition, their size can be well-controlled by tuning both their core and generation, allowing their biodistribution properties to be tuned. Finally, because of the nanoscale dimensions of the dendrimer and steric hindrance at the periphery, the molecular tumbling rate of the conjugated Gd(III) chelates

is significantly slowed, resulting in an increase in τ_R, the rotational correlation. This can result in substantial enhancements in the relaxivity (r_1) of the contrast agents, as described below and as predicted by Solomon–Bloembergen–Morgan theory, which is described in detail elsewhere [93, 100].

An important class of clinically used small molecule Gd(III) chelates is based on the ligand diethylenetriaminepentaacetic acid (DTPA) (Figure 5.12a). Seminal work by Lauterbur and coworkers in the area of dendrimer MRI contrast agents involved the conjugation of a DTPA derivative containing an aromatic isothiocyanate to various generations of PAMAM dendrimers having peripheral amine groups (Figure 5.12b) [101]. Gd(III) was introduced in the final synthetic step. On a per ion basis, the relaxivity (r_1) of this agent was found to be 34 mM/s, about sixfold greater than that of the clinical agent Gd(III)-DTPA (Magnevist®). This result was attributed to the slower tumbling rate of the chelates at the dendrimer periphery. Subsequently, a series of PPI dendrimer–DTPA conjugates up to the fifth generation were conjugated by Kobayashi and coworkers using the same linker chemistry and increasing relaxivity was observed with increasing generation, up to 29 mM/s [102]. In other work, Meijer and coworkers used a different linker to conjugate the DTPA derivative (Figure 5.12c) [103]. This resulted in a less significant increase in r_1, up to a maximum of 20 mM/s for the fifth generation dendrimer. In this case, the flexibility of the linker likely allowed for relatively high local mobility of the chelates at the dendrimer periphery, illustrating the importance of the conjugation chemistry.

Another major class of clinical Gd(III) chelates is based on the ligand 1,4,7,10-tetraazacyclododecane-1,4,7,10-tetraacetic acid (DOTA) (Figure 5.13a). This chelate forms Gd(III) complexes that are kinetically and thermodynamically more stable than those formed with DTPA. Like DTPA, DOTA derivatives have also been conjugated to PAMAM dendrimers from the second to tenth generations. Isothiocyanate linkages have commonly been used to conjugate these ligands to the peripheral amine groups of the dendrimer (Figure 5.13b). It was found that the ionic (per Gd(III)) r_1 values for these dendrimers plateaued at 36 mM/s due to slow water exchange with the chelates, an important consideration for systems with long τ_R [104, 105]. A derivative

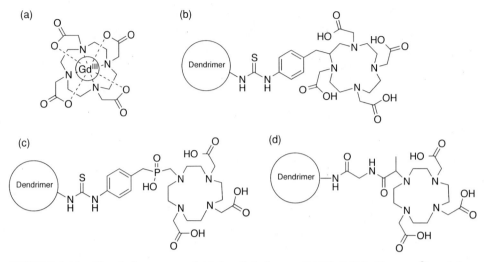

FIGURE 5.13   Chemical structures of (a) the clinical agent Gd(III)-DOTA (Dotarem®) and dendrimer conjugates of DOTA derivatives containing (b) an aromatic isothiocyanate linker; (c) a phosphinic acid linker; (d) an amino acid-based linker.

containing a phosphinic acid moiety in the linkage was also investigated with PAMAM dendrimers, resulting in good relaxivity values due to steric crowding and the formation of a secondary hydration sphere by the bulky phosphinate group (Figure 5.13c) [106, 107]. Researchers at Schering AG (Berlin, Germany) have developed Gadomer-17, bearing 24 DOTA derivatives attached to a lysine-based dendrimer backbone via amide linkages (Figure 5.13d) [108]. In a different approach, DOTA has also been incorporated at the core of polyglycerol dendrimers, where the motion of the Gd(III) should be coupled to the motion of the whole dendrimer [109]. The dendritic arms were conjugated to the chelates via amide linkages. This indeed resulted in a remarkably high relaxivity of 39 mM/s, though the rate of water exchange was slowest for the largest dendrimer limiting further gains in r_1.

As described above, for the highest generation dendrimers, when τ_R is increased by substantially slowing the tumbling rates of the Gd(III) chelates, the rate of water exchange can be a limiting factor in achieving higher relaxivity values. Chelates with faster water exchange rates are desired. In addition, the coordination of multiple water molecules can also increase r_1. However, due to the toxicity of unchelated Gd(III), it is also critical to maintain the stability of the complexes. These aspects have been addressed by Raymond and coworkers through the development of a new class of ligands based on hydroxypyridinone (HOPO) [110]. The Gd(III) chelates of these ligands bind two water molecules yet exhibit high stability due to their oxygen donor atoms and the high oxophilicity of the Gd(III) center. In addition they possess rapid, near-optimal water exchange rates. This results in enhanced relaxivities about twofold higher than the DTPA or DOTA chelates.

HOPO derivatives have also been incorporated into dendrimer systems. Initially, HOPO was conjugated to the

focal point of a dendron based on aspartic acid and tris(hydroxymethyl)aminomethane via the formation of a rigid amide linkage between an aromatic carboxylate of the ligand and the amine of the focal point aspartic acid (Figure 5.14a) [111]. HOPO derivatives have also been conjugated to the peripheries of PLL and esteramide (EA) dendrimers that also bear solubilizing PEG chains [112]. In this case, EDC-mediated amide bond formation was performed using the peripheral carboxylic acids of the dendrimers and the amines of the HOPO chelates containing precomplexed Gd(III). This precomplexation was argued to have prevented the nonspecific binding of Gd(III) to the dendrimer backbone, which might occur if a subsequent Gd(III) chelation step were performed. The effect of the linker was investigated and it was found that the shorter ethylene diamine spacer (Figure 5.14b) provided an r_1 of 38 mM/s with the EA dendrimer, whereas a more flexible diethylene triamine spacer resulted in a lower r_1 of 32 mM/s with the same dendrimer. In addition, the PLL dendrimer backbones resulted in lower relaxivity, perhaps due to increased hydrogen bonding between the dendrimer backbone and the water coordination sites or due to a shorter τ_R of the conjugated chelates. Overall, these results again demonstrate the importance of the bioconjugation chemistry, with shorter, more rigid spacers leading to the highest relaxivity values for Gd(III) contrast agents.

Several of the above dendrimer MRI contrast agents have been investigated *in vivo* [98, 99]. In general, it has been found that low generation dendrimers such as the third and fourth generations exhibit rapid renal clearance, while higher generations remain in the bloodstream for longer periods, making them useful for the visualization of vasculature. The highest generation dendrimers such as the eighth and ninth generations tend to accumulate in the liver but they have

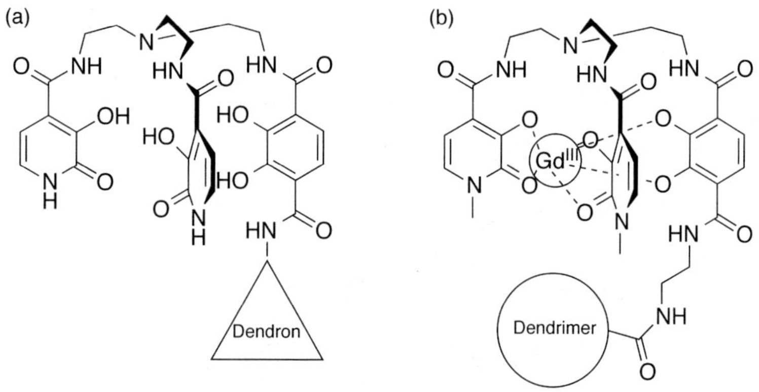

FIGURE 5.14 Dendrimer conjugates of HOPO derivatives using (a) a rigid amide linkage between the carboxylic acid on the ligand and the dendron's focal point amine; (b) an ethylene diamine spacer between the ligand and the dendrimer's peripheral carboxylic acids.

also been useful for MR lymphangiography. Exploiting the multivalent peripheries of dendrimers, moieties for targeting specific tissues *in vivo* have been conjugated along side the Gd(III) chelates. Many of these examples involve protein and peptides, which will be covered later in Section 5.6. However, a noteworthy example is a fourth generation dendrimer with DTPA and FA conjugated to the periphery, which enabled the selective labeling of ovarian cancer tumors overexpressing the folate receptor [113, 114].

5.4.3 Dendrimer Conjugates for X-ray Imaging

X-ray CT is a powerful noninvasive imaging method that can be used to obtain high resolution 3D images through the reconstruction of 1D or 2D projections corresponding to the attenuation of X-rays through tissues [115]. While the technology primarily lends itself toward the visualization of bones due to their high densities relative to soft tissues, soft tissues can be imaged using contrast agents based on heavy atoms. Iodinated small molecules are commonly used as CT contrast agents; however, these agents possess limitations, such as their rapid distribution into the extracellular space and rapid clearance from the body [116]. The development of a blood pool contrast agent with both higher intravascular concentrations and longer retention times is of significant interest both for vascular imaging applications and for the targeted imaging of disease sites such as tumors. In order to obtain these properties, various nanomaterials including dendrimers have been investigated [117]. There are currently several examples involving the encapsulation of metal nanoparticles such as gold and silver, which are known to effectively attenuate X-ray radiation, resulting in relatively high contrast [32, 73, 118–120]. However, as these do not involve true bioconjugation reactions, the focus here will be on the linkages of X-ray contrast agents to the peripheries of dendrimers.

In the first example of a dendritic CT contrast agent, Brechbiel and coworkers functionalized iodopanoic acid with

a tertiary amine side chain for solubility enhancement, and then conjugated it to a fourth generation PAMAM dendrimer via activation with NHS and subsequent reaction with the dendrimer's peripheral amines (Figure 5.15a) [121]. It was found that 37 of the dendrimer's 64 peripheral amines were functionalized, resulting in a molecule that was 33% iodine by mass. This high iodine content is essential for these agents as relatively high concentrations of agent are required for contrast in X-ray CT. In subsequent work, Brasch and coworkers prepared PEG-PLL hybrids by the growth of PLL dendrons of generations 3–5 from both termini of linear PEO with MWs ranging from 3 kDa to 12 kDa [122]. The peripheral amines on the dendrons were reacted with an NHS-ester-activated derivative of iobitridol (Figure 5.15b). For generation 4 or less, it was found that quantitative yields could be obtained in a single coupling or with one resubjection. However, for the fifth generation PEG-PLL the yields were much lower and the use of a β-alanyl spacer on the dendrimer periphery was required to obtain >90% functionalization. One agent based on 12 kDa PEG and a fourth generation dendron provided strong and persistent vascular enhancement in a rat model. In a different approach, it was found that a PAMAM dendrimer functionalized with Gd(III)-DTPA via a thiourea linkage, was capable of acting as a dual CT-MR agent for monitoring convection-enhanced drug delivery in the brain [123]. This can be attributed to the CT attenuation provided by the heavy Gd(III) ion.

5.4.4 Dendrimer Conjugates for Nuclear Imaging

Nuclear imaging employs radioactive substances for the diagnosis of disease. For example, SPECT requires the injection of a γ-emitting radionuclide into the bloodstream of a patient. A gamma camera acquires multiple 2D images from different angles and a computer is used to apply a reconstruction algorithm to yield a 3D image. In PET the radioisotope emits a positron that annihilates with a nearby

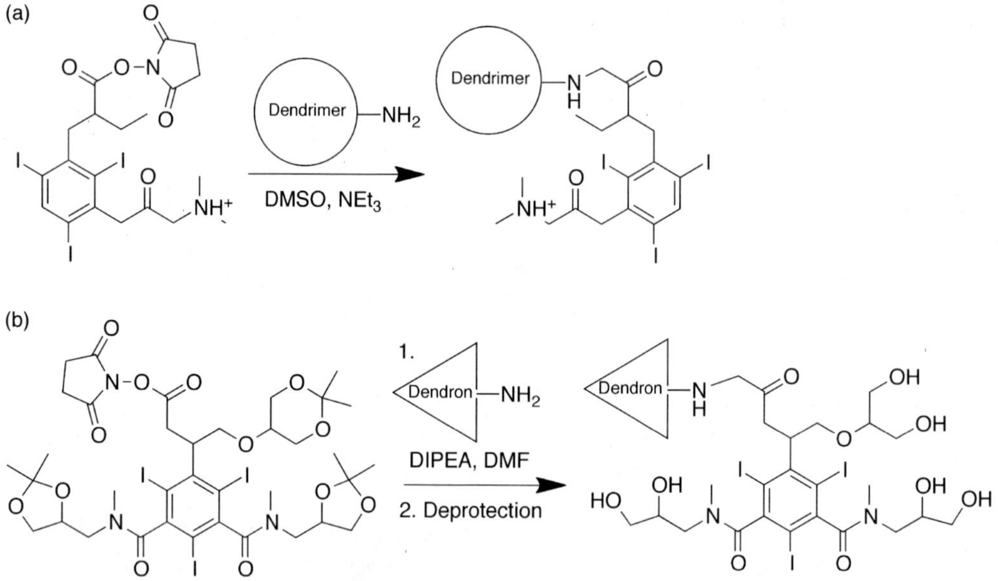

FIGURE 5.15 Dendrimer–CT agent conjugates synthesized by reaction of (a) the dendrimer's peripheral amines with an NHS-ester derivative of iodopanoic acid; (b) the dendron's peripheral amines with a protected NHS-ester derivative of iobitridol, followed by deprotection.

electron to emit two γ-photons in opposite directions. These two photons are detected coincidentally, allowing for increased localization information, resulting in higher resolution images than in SPECT. Occasionally, the radioisotope used in nuclear imaging is a free metal ion; however, in most cases it is chelated to a ligand that can bind to specific sites *in vivo*. In this context, there has been significant interest in the development of new diagnostic radiopharmaceuticals with enhanced properties, binding, and multifunctional capabilities [124]. Here, dendrimers again offer potential advantages but providing multiple sites for the conjugation of targeting moieties and radioisotopes to enhance targeting and/or the binding ratio of reporters to target [125]. In addition, the small size of dendrimers in comparison with other macromolecular and nanoparticulate systems allows them to exhibit rapid diffusion across blood capillaries, favorable biodistribution, and urinary elimination, which is critical for a radiopharmaceutical.

^{99m}Technetium (^{99m}Tc) is the most commonly used γ-emitting radionuclide in nuclear medicine [126]. There are currently several examples involving the conjugation of ^{99m}Tc chelates to dendritic systems. Seminal work in this area involved the development by Mukhtar and coworkers of first and second generation dendrimers comprising porphyrin cores and peripheral iminodiacetic acid groups (Figure 5.16a) [127]. These dendrimers were synthesized divergently and in this case the chelating capabilities of the dendrimer arose from the properties of the branching monomer rather than through the specific conjugation of ligands to the dendrimer periphery. They exhibited enhanced accumulation in

C_6-glioma tumors relative to surrounding tissues in Wistar rats. Adronov, Valliant, and coworkers have incorporated a single high affinity tridentate bis(pyridyl)amine ligand at the focal point of fifth, sixth, and seventh generation PE dendrons (Figure 5.16b) [128]. It was argued that this would minimize the impact of the ^{99m}Tc on the interactions of the dendrimer periphery with the biological environment. The ligand was introduced to the dendrimer focal point following its deprotection at the end of the synthesis and ^{99m}Tc$^+$ was chelated as the final step in a rapid microwave reaction. In SPECT biodistribution studies all generations were rapidly and cleanly eliminated via the kidneys, a promising result for the future development of targeted versions of the agents.

Shen and coworkers have conjugated ^{99m}Tc-DTPA chelates to the periphery of fifth generation PAMAM dendrimers via thiourea linkages in a similar manner to those described in the section on MRI contrast agents [129]. In addition, they have conjugated FA-targeting moieties via amide bond formation, resulting in preferential accumulation in tumors [130]. In subsequent work, they found that conjugation of the FA through a PEG spacer provided enhanced localization in tumors. This was attributed to the flexibility of the spacer, as well as the capacity of PEG to prevent nonspecific interactions. Recently, Tarek and coworkers prepared a dual SPECT/X-ray CT agent by conjugating 2,3,5-triiodobenzoic acid (TIBA), PEG, and a DTPA derivative to the periphery of a fourth generation PAMAM dendrimer [131]. The TIBA and PEG were conjugated via their NHS-ester derivatives while the DTPA was conjugated via an isothiocyanate derivative.

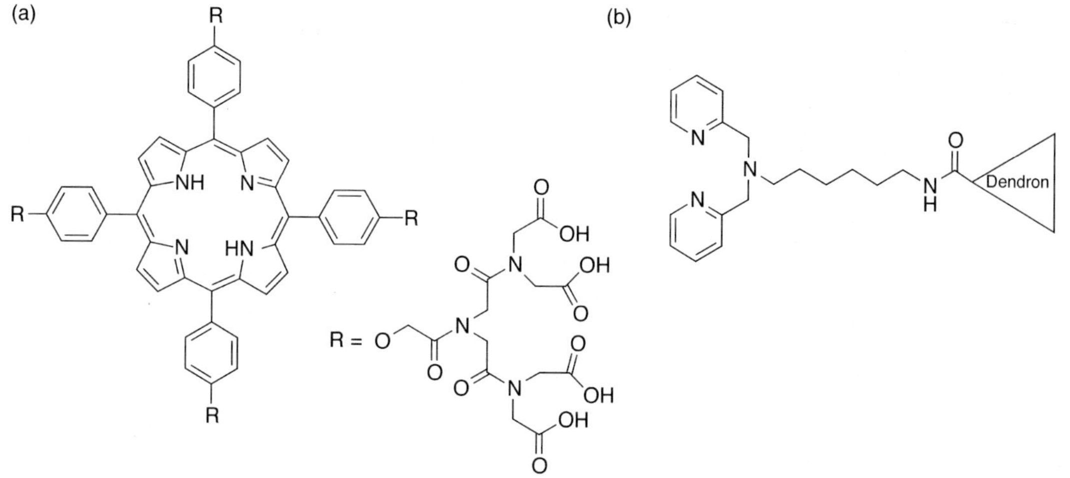

FIGURE 5.16 Dendrimer conjugates with ligands for radionuclides: (a) a dendrimer with a porphyrin core and peripheral iminodiacetic acid groups; (b) a PE dendron with a bis(pyridyl)amine ligand at the focal point.

[125]I-labeled dendrimers have primarily been investigated with the aim of determining the biodistribution properties of the dendrimers for applications in drug delivery and other biomedical applications. For example, Duncan and coworkers labeled various generations of cationic and anionic PAMAM dendrimers using the Bolton–Hunter reagent, which is commonly used to label proteins [132]. They found that all the dendrimers were rapidly cleared from circulation, with much of the dose accumulated in the liver. The anionic dendrimers circulated longer, but still accumulated in the liver to a great extent. Fréchet and coworkers investigated the biodistribution of a series of bow tie-shaped dendrimers with different MWs of PEG on one side of the bow tie and hydroxyls on the other side [133]. These hydroxyls were used to introduce a phenol for ^{125}I labeling. It was found that there were significant architectural effects on the biodistribution properties with the higher MW, more branched dendrimers exhibiting the longest circulation times. However, it was also found that when the PEG side of the bow tie was not highly branched, the dendrimers rapidly localized in the liver. This result was surprising and was attributed to the ^{125}I-labeled phenols on these molecules being accessible. It was thought that these labels directed the molecules to the liver, illustrating that the choice of conjugate and its linkage method can play a critical role in the results and one must be careful in interpreting the biodistribution properties as inherent to the system itself and not an artifact of the labeling method.

[68]Ga-labeled dendrimers are of significant interest for PET. The first example was reported recently by Fukase and coworkers [134]. Their dendrons were composed of a polylysine backbone with glycoclusters conjugated to their surfaces by a histidine-accelerated Cu(I)-mediated Huisgen

cycloaddition reaction. A DOTA derivative was introduced to the focal point by a 6π-azaelectrocyclization. Dendrons bearing 4, 8, or 16 glycoclusters on their peripheries were prepared and their biodistribution properties were studied using PET imaging in mice. The two smaller dendrons were rapidly excreted by the kidneys, while the largest dendron remained in the body longer and was excreted through the kidneys and the gallbladder. The fraction that was detected in the body after 4 hours, was primarily located in the liver. A number of radiolabeled dendrimers targeting $\alpha_v\beta_3$ integrins have also been developed, but will be described below in the section on peptide conjugates.

5.4.5 Dendrimer–Fluorophore Conjugates for Optical Imaging

Optical imaging has routinely been used for decades in *in vitro* studies to track the binding, cell uptake, and intracellular localization of various labeled or inherently fluorescent materials. More recently, it has gained popularity as an *in vivo* imaging modality, particularly for small animal imaging and has been suggested as a promising approach for guiding surgical resections in an operative setting [135]. Many of the early examples involving dendrimer–fluorophore conjugates were aimed at using the fluorophore to elucidate the properties of dendritic drug delivery systems. For example, the anticancer drug DOX is inherently fluorescent, and this property has been used to track the localization of the drug within cells as well as *in vivo* after extraction from organs [16]. In the case of the nonfluorescent drug MTX, the PAMAM dendrimer conjugates were labeled with FITC. This fluorescent labeling was performed by the reaction of the dendrimer's peripheral amines with FITC isothiocyanate to form a thiourea linkage.

In one example, 3.6% of the dendrimer's peripheral groups were functionalized with FITC, 4.5% with MTX, and 3.6% with FA-targeting moieties [136]. There have been a number of examples where the attributes of dendrimers have been exploited for the development of dendrimer-based optical imaging agents. Several examples of these dendrimers are described below.

The most common approach in dendritic optical imaging agents has involved the conjugation of chromophores to the core of the dendrimer or focal point of a dendron. The unique advantages of dendrimers in this area are exemplified by the development of phosphorescent dendritic probes for O_2 imaging by Vinogradov and coworkers. The imaging and quantification of O_2 *in vivo* is highly relevant to understanding various biological processes such as neuronal activation and diseases including diabetic retinopathy and cancer. Metalloporphyrins had previously been extensively used for the measurement of O_2 in various systems as O_2 is well known to quench their phosphorescence, allowing for applications such as *in vivo* phosphorescence lifetime imaging [137, 138]. However, these molecules tended to aggregate in aqueous solution and bound to various biomolecules *in vivo*, altering their sensitivity to O_2 and making quantitative measurements impossible. The encapsulation of the metalloporphyrin inside a dendritic "cage" was proposed as a means of imparting water solubility, preventing binding to biomolecules, and tuning the accessibility of the core chomophore to O_2 (Figure 5.17).

Early work involved the growth of polyglutamic acid dendrons from the metalloporphyrin core beginning with the coupling of the glutamic acid amine to carboxylic acid moieties on the metalloporphyrin [139]. It was later

found that arylglycine dendrimers were advantageous from a synthetic and photophysical standpoint, proving optimal O_2 quenching rate constants [140, 141]. These dendrons having focal point amines could be conjugated to the metalloporphyrin core using various coupling agents using a convergent approach. Subsequent attachment of PEG to the peripheries of these dendrimers provided further protection from binding to biomolecules, allowing these molecules to be used for *in vivo* O_2 imaging. A number of variations on these structures have been developed by the same group. For example, the use of a porphyrin core with more extended conjugation allows for O_2 imaging in the near-infrared (NIR) [140, 141]. NIR imaging is of particular interest for biomedical applications because light of wavelengths from approximately 700 nm to 900 nm can penetrate more deeply into tissue than the surrounding wavelengths due to decreased absorption from natural chromophores. In addition, the conjugation of two-photon absorbing chromophores to the dendrimer periphery allowed for excitation of these chromophores in the NIR with subsequent Förster resonance energy transfer (FRET) to the metalloporphyrin, followed by phosphorescent emission [142, 143]. Two-photon O_2 imaging offers several advantages over the linear method, such as reduced photodamage and improved depth resolution. These agents have successfully been applied for the high resolution measurement of O_2 *in vivo* in rat brains [144, 145].

In several other cases, the conjugation of fluorescent probes to the focal point of dendrons has reduced the self-quenching of fluorescence due to the site isolation provided by the bulky dendritic structure. For example, Weck and coworkers have conjugated NIR cyanine dye molecules to Newkome-type polyamide dendrons to provide a single dye molecule per dendron [146]. These conjugation reactions were performed using a variety of methods including copper-catalyzed azide–alkyne cycloaddition, strain-promoted azide–alkyne cycloaddition, peptide coupling reaction, or a direct radical nucleophilic substitution ($S_{RN}1$) reaction, and various linkers were explored for the focal point. It was found that the conjugates were obtained in good yields and that all compounds exhibited fluorescence in the NIR region. However, the formation of aggregates was highly dependent on the length of the linker and on the conjugation chemistry.

Fréchet and coworkers used a NIR cypate core surrounded by a PEG-functionalized PE dendrimer to monitor the *in vivo* biodistribution and biodegradation of the dendrimer [147]. In this work, the fluorescence intensity of the dye was relatively insensitive to encapsulation within the dendrimer, but the fluorescence lifetime of the dye was very sensitive to its surrounding environment, particularly in biological media. It was found that while the fluorescence lifetime of the free dye remained relatively unchanged *in vivo*, the dendritic probe underwent a change in the fluorescence lifetime over a period of a week. It was hypothesized that upon dendrimer

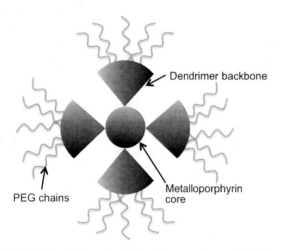

FIGURE 5.17   Schematic of a dendrimer conjugate for O_2 imaging. The metalloporphyrin core provides O_2-sensitive phosphorescence, while the dendrimer backbone controls O_2 access to the core and prevents aggregation. Peripheral PEG chains provide additional aqueous solubility.

degradation, the dye underwent binding to proteins, resulting in a change in fluorescence lifetime.

Another interesting system took advantage of the dendrimer's multiple conjugation sites [148]. First, the peripheral amines of a bow tie-shaped PAMAM dendrimer were conjugated to an NHS-ester derivative of DTPA to provide MRI contrast. After deprotection of the chelate and complexation of Gd(III), a disulfide linkage at the focal point of the bow tie was reduced and the resulting thiol was conjugated to biotin via a PEG linker. The biotin was introduced in order to induce strong binding to the avidin, a glycoprotein capable of targeting lectins overexpressed on ovarian cancer cells. The avidin was also labeled with rhodamine green, enabling simultaneous delivery of the optical probe and MRI probe to the cancer cells in mice.

Fewer dendritic systems for optical imaging have been developed containing peripheral chromophores due to the propensity of organic fluorophores to aggregate and self-quench when placed in close proximity on a dendrimer in water. For example, Mier and coworkers investigated the attachment of a large number of different fluorophores including sulforhodamine B, rhodamine 101, FITC, NIR 820, NIR 797, dansyl chloride, 4-chloro-7-nitro-1,2,3-benzoxadiazole (NBD) chloride, coumarin-343 to PAMAM dendrons of generations 0–2, bearing 1, 2, or 4 peripheral amines [149]. Only a limited number of these dye molecules could be coupled in high yields, and of these derivatives, all but the dansyl derivative exhibited significant self-quenching such that the fluorescence intensities of the higher generations were lower than those of the lower generation dendrons despite having more dye molecules. Larger generation dendrons, having up to 128 dansyl moieties per dendron, were synthesized and then a thiol group at the dendron focal point was deprotected and reacted with an antibody that had been functionalized to bear maleimide (MAL) groups. It was found that 54% of the immunoreactivity could be retained upon labeling with a dendron-bearing 16 dansyl groups, in comparison with complete loss of antibody integrity for the conjugate with 16 isolated dansyl moieties. This demonstrated the advantage of the dendritic approach for enhancing the signal amplification of dye-labeled antibodies.

Fréchet and coworkers have taken advantage of the self-quenching phenomenon to develop pH-sensitive probes for NIR fluorescence lifetime and intensity imaging [21]. Using the heterobifunctionality of a 4-acetylphenylalanine spacer introduced at the dendrimer periphery, eight PEG chains were conjugated via carbamate linkages to provide enhanced *in vivo* circulation times and stealthy properties, and eight NIR cypate dyes were conjugated via acid-sensitive hydrazone linkages. The close proximity of the dyes in the intact structure resulted in the formation of H-type homoaggregates which suppressed the fluorescence lifetime and intensity of the dye. In acidic environments, the NIR dyes were released, resulting in a sixfold increase in fluorescence.

5.5 DENDRIMER–DNA CONJUGATES

5.5.1 Motivation for the Development of Dendrimer–DNA Conjugates

Deoxyribonucleic acid (DNA) is a polymer composed of a sequence of nucleotide repeat units that provide a sugar–phosphate backbone and pendant nucleobases. It plays a critical role in biological systems by storing, through its specific sequence of nucleotides, the genetic information used in the development and function of living organisms. DNA generally exists as a double helix comprising two DNA strands held together by hydrogen bonds and $\pi–\pi$ stacking between the complementary nucleobases, with adenine binding specifically to thymine and guanine binding specifically to cytosine. During cell division, DNA is replicated by DNA polymerase to provide a copy of the entire DNA sequence for the daughter cells. In the process of transcription, the code contained within the DNA sequence is transcribed into related ribonucleic acid (RNA) sequences, which are subsequently translated into the sequences of amino acids in proteins. In recent years there has been substantial interest in DNA and related structures for both diagnostic and therapeutic applications. For example, in gene therapy, DNA is delivered to host cells to treat a disease through the replacement of a mutated gene, the correction of a mutation, or the introduction of a therapeutic protein [150]. In gene silencing, the expression of proteins is switched off through the delivery of RNAi, which prevents the translation of the complementary mRNA into protein [151]. In addition, many biosensors employed in the fields of molecular medicine, forensics, and food safety exploit DNA microarrays consisting of probe oligonucleotides immobilized on surfaces [152].

There are numerous examples involving combinations of dendrimers and DNA [153]. An area of intense interest has been the use of dendrimers as vectors for the delivery of nucleic acids into cells for gene delivery and gene silencing [154–156]. However these systems, referred to as "dendriplexes", generally involve noncovalent binding of the nucleic acids and the dendrimer and thus will not be discussed further in this review. There are also examples where the dendrimer backbone is composed of DNA [157, 158]. To highlight the conjugation of approaches to combining the unique topologies or dendrimers and DNA, we will describe here selected examples involving the covalent attachment of DNA to the focal points or peripheries of dendrimers with the goal of developing new complex, 3D architectures, as well as for the development of improved biosensors.

5.5.2 New Molecular Architectures Through Dendrimer–DNA Conjugates

As described above, dendrimers, particularly those of high generations exhibit globular structures, while DNA possesses

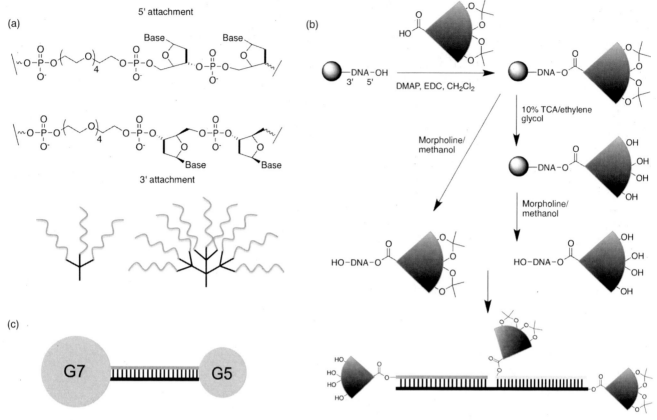

FIGURE 5.18   Assembly of new molecular architectures from (a) attachment of ODN sequences to phosphate-based dendrons on solid support via the 5′ or 3′ end and example schematics of the resulting dendrimer–DNA conjugate (dendron backbone shown in black and oligonucleotides shown in grey); (b) assembly of dendrons bearing different peripheral moieties via complementary focal point DNA strands; (c) assembly of dendrons of different generations through hybridization of complementary DNA strands on their peripheries.

a double helical structure [159]. Many of the early examples of DNA–dendrimer conjugates involved the use of DNA's double helical structure to provide new materials with complex architectures that would be challenging to access by other synthetic approaches. For example, although they can only be viewed as first generation dendrons, von Kiedrowski and coworkers prepared a phosphoramidite derivative of Newkome's dendrimer synthon-bearing dimethoxytrityl-protecting groups on its three hydroxyl groups [160]. Attachment to a controlled pore glass (CPG) solid support, followed by the standard phosphoramidite protocol for oligonucleotide (ODN) synthesis led to trisoligonucleotides. The assembly of two trisoligonucleotides having complementary sequences was investigated and topologies termed "nanoacetylenes" and "nanocyclobutadienes" were prepared through the hybridization of complementary strands. In later work by the same group, trisoligonucleotides bearing three different sequences were used as templates for artificial self-replication, where new trisoligonucleotides were covalently assembled via hydrazine formation following the binding

of the individual oligonucleotides to the complementary sequence on the template [161].

Southern and coworkers prepared branched phosphate-based dendrons on solid support using phosphoramidite chemistry [162]. After reaching the desired degree of branching, a penta(ethylene glycol) linker was introduced also via phosphoramidite chemistry and then decanucleotide sequences were grown from the peripheries of the dendrons using 3′- or 5′-nucleoside phosphoramidites to provide 3′- or 5′-attached oligonucleotides respectively (Figure 5.18a). Dendrons with 2, 3, 6, 9, or 27 decanucleotides were prepared in this manner. The assembly of various combinations of complementary structures, as well as the branched structures with surface-bound ODNs was investigated. It was found that it was possible to prepare a vast array of different nanostructures, and in general the stabilities of dendrimer-based assemblies were higher than those of the linear duplexes.

Fréchet and coworkers conjugated PE dendrons with peripheral acetonide-protecting groups to the 5′-OH of CPG-bound oligonucleotides using EDC [163]. Different peripheral groups on the dendron were generated by either leaving the acetonide groups intact or removing them using 3% trichloroacetic acid in ethylene glycol. While the ammonium hydroxide conditions usually used for deprotection and removal of DNA from the solid support led to hydrolysis of the PE dendrons, it was possible to perform this cleavage with morpholine/MeOH, leaving the esters intact. Subsequent assembly of ODN–dendron hybrids led to nanoassemblies bearing dendrons with different peripheral groups.

Tomalia and coworkers were also able to assemble dendrons with different sizes and peripheral functionalities [164]. They prepared PAMAM dendrons with focal point thiols through the reduction of cystamine core dendrimers and reacted the thiols with the MAL moiety of the heterobifunctional linker sulfosuccinimidyl 4-[*N*-maleimidomethyl] cyclohexane-1-carboxylate (sulfo-SMCC). Next, an amine-functionalized ODN was added, resulting in an amide linkage. Assembly of complementary dendron–ODN hybrids led to bow tie-shaped nanostructures. Baker and coworkers conjugated complementary ODNs that were 34, 50, or 66 base pairs in length to the peripheries of fifth and seventh generation PAMAM dendrimers via EDC/imidazole chemistry [165]. In order to prevent excessive electrostatic interactions between the negatively charged ODNs and the cationic dendrimer, 90% of the dendrimer's peripheral amine groups were first acetylated. It was estimated that one or two ODNs per dendrimer were conjugated. The assembly of fifth and seventh generation dendrimers bearing complementary sequences was studied and it was found that the sizes measured for the assemblies were similar to those predicted based on the ODN lengths and the target asymmetric structures comprising the different generations of dendrimers were visualized by atomic force microscopy (AFM) (Figure 5.18c).

In recent work, increasingly complex nanostructures have been generated from dendrimer–DNA hybrids. For example, Liu and coworkers have conjugated a second generation poly(aryl ether) dendron-bearing water solubilizing oligo(ethylene glycol) moieties on its periphery and a focal point NHS ester to an ODN that was modified to include a 5′-amino group and a 3′-biotin group [166]. Binding of the biotin group on the dendron–DNA hybrid to the four possible sites on the protein streptavidin produced a supramolecular complex composed of the protein and four dendron–DNA hybrids. Subsequently, it was possible to add a single-stranded DNA sequence that was complimentary to that in the above assemblies. Owing to its specific sequence, the resulting duplex underwent a pH-dependent conformational change that was able to induce a size change in the assembled complex. The same group has also prepared a second generation poly(benzyl ether) dendron-bearing peripheral dichlorobenzene moieties and a focal point phosphoramidite [167]. This phosphoramidite was reacted with the 5′-hydroxyl of DNA attached to the CPG support. After oxidation, deprotection, cleavage, and purification steps, the assembly of the dendron–DNA hybrid was investigated. It was found to form nanofibers that were tens of microns in length.

5.5.3 Dendrimer–DNA Conjugates for Biosensors

DNA microarrays and microchips serve as biosensors to provide rapid and reliable genetic information [152]. A typical device consists of nucleic acids immobilized at discrete positions on a surface (the probe), and the analyte which contains a complex mixture of fluorescently labeled nucleic acids including the target. The hybridization between the probe and the target is generally quantified based on fluorescence at high resolution. The ongoing goal in biosensor development is to improve the sensitivity and reliability of the device. In this regard, dendrimers can provide advantages in two ways [168]. When connected to the target the multivalent nature of their peripheries can be exploited to amplify the signal for detection. Alternatively, they can be connected to the surface to support the probe, providing enhanced device stability and improved hybridization.

In early work by Southern and coworkers, phosphate dendrimers based on pentaerythritol and ethylene glycol were grown from the 5′ end of an ODN on a CPG support using phosphoramidite chemistry [169]. At the second generation, 5-mers of thymidylates were grown from each of the nine peripheral hydroxyls either with or without a penta(ethylene glycol) spacer. These thymidylates were then tagged with ^{32}P-phosphates by reaction with polynucleotide kinase in the presence of [γ-^{32}P]ATP. It was found that the incorporation of the radiolabel was eightfold better than for a conventional ODN in the case of the spacer, but only twofold better in the absence of the spacer, a result that was attributed to tight clustering and steric hindrance at the dendron's periphery in the absence of the spacer. When used as probes in oligonucleotide arrays, the dendron–DNA hybrid containing the spacer provided a 7.5-fold increase in signal, indicating that the dendron component did not interfere with the binding of the DNA. In another example by Striebel et al., phosphate dendrimers were grown from CPG-immobilized ODNs using phosphoramidate chemistry, and after introduction of amines to the dendron peripheries they were labeled with the NHS-ester derivative of Cy3 [170]. Various dendron generations bearing one to nine dyes per dendron were investigated and it was found that the resulting DNA–dendron hybrids hybridized specifically to probe DNA on the microarrays resulting in up to 30-fold enhancement in fluorescence relative to a single fluorophore.

Unlike the above examples, the use of dendrimers as supports for DNA probe immobilization have generally involved reactions between surface-bound dendrimers and soluble DNA. In the first example by Wöhrle and coworkers, a fourth

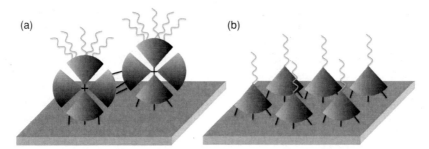

FIGURE 5.19 Dendrimer–DNA conjugates for biosensors: (a) use of surface-bound dendrimers can increase the loading of DNA strands on the surface while multiple bonds to the surface and cross-linking between dendrimers can provide enhanced surface stability; (b) use of dendrons can control the spacing of DNA strands on the surface, providing enhanced binding affinities.

generation PAMAM dendrimer was grafted onto glass surfaces that were pretreated with 3-aminopropyltriethoxysilane and either disuccinimidylglutarate (DSG) or phenylenediisocyanate (PDITC) (Figure 5.19a) [171]. After the grafting, DSG or PDITC was again used to intermolecularly cross-link the dendrimers, providing surface stability and at the same time generating reactive groups for the reaction with amino-derivatized ODNs. The best results were obtained with PDITC, where high surface homogeneity and stability were observed. As many of the dendrimer's peripheral amine groups were consumed in the cross-linking reaction, these surfaces had similar loadings of ODNs as classical surfaces. However, it was also possible to provide increased immobilization of ODNs by avoiding the cross-linking reaction and instead reacting the dendrimer's amines with glutaric anhydride, DCC, and NHS [172]. Amine-functionalized ODNs were then immobilized by reaction with the NHS esters. This resulted in a 10-fold increase in signal intensity upon binding of the target, and the surfaces were still quite stable.

Majoral and coworkers have used different immobilization chemistry involving the reaction of phosphorous-containing dendrimers bearing peripheral aldehydes with amine-functionalized surfaces [173, 174]. Amine-functionalized ODNs could then be directly reacted with the remaining aldehyde groups without the need for a linker. Upon hybridization of the target DNA, a 1000-fold increase in fluorescence intensity was obtained relative to simple aldehyde-functionalized surfaces. Another conjugation approach was recently reported by Norton and coworkers [175]. The amine periphery of a third generation PAMAM dendrimer was functionalized with ~1 MAL per dendron using sulfo-SMCC and then the remaining amines were reacted with excess N-succinimidyl-S-acetylthiopropionate (SATP) to provide protected thiols. A thiol-modified DNA was then reacted with the dendrimer. The purified dendrimer–DNA conjugate could be stored for months without deprotection, and then the thiols could be deprotected and the multithiolated dendrimers introduced to a gold surface. The

multiple gold–thiol bonds between the dendrimer and surface provided greatly enhanced stability, even in buffer at 95°C, in comparison with single thiol–gold linkages which exhibit limited stability.

Dendron–DNA hybrids have also been used to control the spacing of DNA probes on the surface (Figure 5.19b) [176]. In this work, peripheral carboxylic acid groups on the dendron were conjugated to hydroxyl groups on the surface by EDC- or DCC-mediated esterifications. After deprotection of the dendron's focal point amines, these amines were activated with di(N-succinimidyl)carbonate (DSC) and then reacted with ODNs having 5′-amino groups. The ample spacing between ODNs on the surface provided high hybridization yields, and the resulting microarrays had high sensitivity and selectivity [177].

5.6 DENDRIMER–PEPTIDE AND DENDRIMER–PROTEIN CONJUGATES

5.6.1 Motivation for the Development of Dendrimer–Peptide and Dendrimer–Protein Conjugates

Peptides and proteins control a variety of very important physiological and biochemical functions in life such as metabolism, pain, reproduction, and immune response, and influence cell–cell communication upon interaction with receptors. In receptor-mediated signal transduction, they are also involved as neuromodulators, neurotransmitters, and hormones. Moreover, synthetic peptides serve as antigens, enzyme substrates, and enzyme inhibitors to influence signaling pathways in research. In the field of drug development, peptide therapies have found great potential in various areas such as growth control, blood pressure management, neurotransmission, satiety, addiction, pain, digestion, reproduction, and others. Covalently constructed protein- and peptide-based conjugates are considered as potential synthetic vaccines and target antigens. Thanks to the targeting potential of many proteins and peptides, peptide-functionalized drug delivery vehicles and imaging contrast agents can exhibit improved biodistribution and hence increased efficacy and decreased nonspecific toxicity [178]. On this basis, multifunctional materials containing therapeutic/targeting [46, 179], imaging/targeting [180, 181], and therapeutic/imaging/targeting agents [182, 183] have been developed for biomedical applications.

As described in the above sections, among the various nanomaterials, the well-defined and multivalent chemical structures of dendrimers provide significant advantages in terms of reproducibility of the synthesis and resulting properties of the agents, allowing well-characterized materials to be prepared to the satisfaction of regulatory agencies. Thus, dendrimers have been widely used as novel platforms for the conjugation of peptides and proteins along with other therapeutics and imaging agents [184–186]. Generally, depending on the types of dendrimer core and dendrimer peripheral groups, one can divide peptide dendrimers into three categories [184]. The first group consists of conventional dendrimers (comprised of organic groups or unnatural amino acids) with their peripheries functionalized with peptide or proteins. In the second group, natural amino acids construct the dendrimer branching units and the termini of these amino acids form the surface groups of the dendrimers [187, 188]. The third type of peptide dendrimers comprise amino acid-based branching units as well as peptide chains on their peripheries. This family of peptide dendrimers is mostly known as multiple antigen peptides (MAPs) [189]. In order to highlight the bioconjugation chemistries, attention here will be focused on peptide dendrimers of the first type and examples involving amide, thioether, disulfide, oxime, phosphoramidate, and triazole linkages will be described below.

5.6.2 Dendrimer–Protein Conjugates

An amide bond between the dendrimer and the peptide/protein has been the most extensively employed linkage. In this context, it is worth introducing a powerful bioconjugation strategy termed "native chemical ligation" (NCL), which can be employed specifically in the case of peptide

and protein bioconjugates. As described in the other sections of this chapter, the formation of an amide bond by conventional chemical synthesis methods involves the activation of the carboxylic acid group of one of the components followed by subsequent reaction with the amine group of the other component. When applied to the synthesis of proteins and peptides, the ligation of two peptides to produce proteins or larger peptides, or attachment of proteins and peptides to other biomaterials, such as dendrimers, to construct dendrimer–peptide/protein conjugates, this method generally requires a protecting-group strategy for the other functionalities in peptide components. However when applied to the reactions with large peptides and proteins, the aforementioned conventional approach usually requires an exhaustive protection strategy, which in turn imparts limitations such as low solubility, low coupling efficiency, and the possibility of racemization [190]. To overcome these limitations, scientists have employed approaches to form hydrazone, thioether, and disulfide linkages between two peptides as well as between peptides/proteins and dendrimers instead of an amide bond [191–195]. Although these methods result in high conjugation efficiencies due to the orthogonality of the requisite functional groups with those on the polypeptide, their nonnatural placement within the polypeptide backbone can result in conformational distortion.

To address the above limitations, researchers have developed chemical reactions that ligate unprotected peptides to each other or to other biomaterials such as dendrimers. These include reactions such as NCL [196], the Staudinger reaction [197], and the thiaproline reaction [190]. Among these reactions, NCL has been used more widely than the other two for dendrimer–peptide/protein construction via an amide linkage. For this reason, it is worth mentioning the mechanism of this reaction briefly. As shown in Figure 5.20, first the thiolate anion of an N-terminal cysteine in one peptide attacks the carbonyl group of a C-terminal thioester in another peptide to produce the intermediate thioester ligation product, which is not observable as an isolable product. In the second step, this intermediate undergoes a spontaneous intramolecular rearrangement to form an amide bond at the ligation site [196]. In the context of this chapter, it is noteworthy

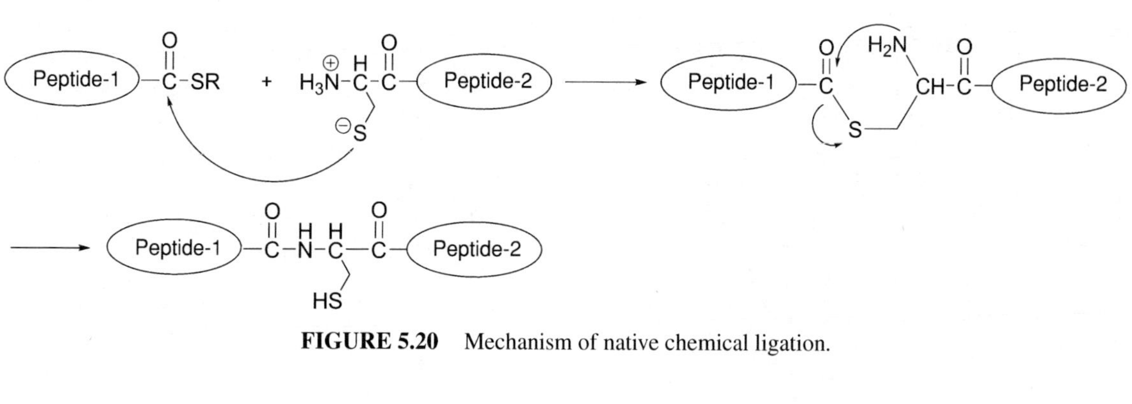

FIGURE 5.20 Mechanism of native chemical ligation.

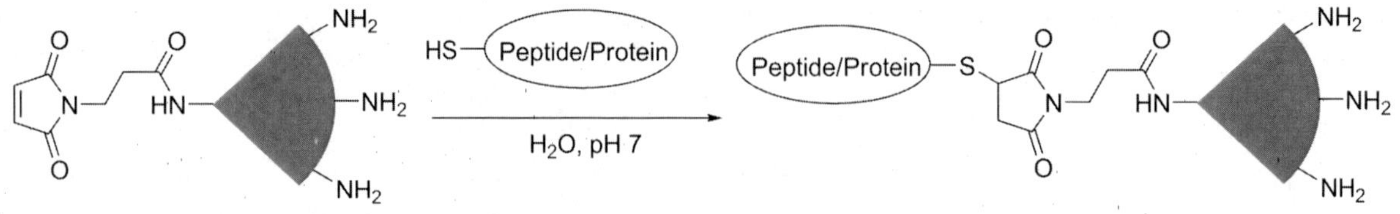

FIGURE 5.21   Maleimide–thiol 1,4-addition reaction for the conjugation of proteins/peptides to the dendron focal point.

that this reaction is not only limited to the ligation of two peptide components. It has also been used to attach peptides and proteins to dendrimer surfaces [198, 199]. Along with its advantages, NCL also possesses some limitations. For instance, NCL is restricted to reaction at an amino-terminal cysteine moiety. Moreover, cysteine has been shown to be the most reactive residue toward oxidation to disulfide bonds by oxygen and other electrophiles, an undesired side reaction [200]. Despite these limitations, NCL has been a powerful tool in constructing dendrimer–peptide/protein conjugates.

In the first example of a dendrimer–protein conjugate prepared via the NCL method, Meijer and coworkers functionalized the peripheries of first through third generation PPI dendrimers with cysteine residues via first activation of trityl-protected cysteine with disuccinimidylcarbonate followed by its coupling to the dendrimer's peripheral amines [198]. Subsequent deprotection of the cysteine residues by TFA in the presence of 2.5% triisopropylsilane and 2.5% water afforded the desired cysteine-functionalized PPI dendrimers. It was then shown that these dendrimers could be functionalized with either one or up to four green fluorescent proteins (GFPs) via amide linkage upon incubation of thioester-functionalized protein with dendrimer samples in 0.1 M sodium phosphate solution containing 2% (v/v) thiophenol and 2% (v/v) benzyl mercaptan at pH 7–7.5 for 20 hours at 20°C. They also demonstrated that this powerful method could be applied to the introduction of oligopeptides. Furthermore, a tetravalent cysteine-functionalized dendron was reacted with thioester-functionalized collagen-binding protein CNA35 and its nonbinding variant of CNA35-Y175K via an amide linkage by NCL method [199]. This dendron was shown to be as efficient as CNA35 micelles and liposomes in terms of collagen-binding properties and possessed even higher stability.

In addition to the NCL method, conventional amide bond-forming reactions have also been used to prepare dendrimer–protein conjugates. Kannan and coworkers have studied the performance of the relatively large therapeutic protein streptokinase, upon its conjugation to a third generation PAMAM dendrimer [201]. To prepare these conjugates, carobxylic acid-functionalized PAMAM dendrimers were first activated with EDC–methiodide and sulfo-NHS (s-NHS) as coupling reagents and were then reacted with an aqueous solution of streptokinase in 0.1 M borate buffer at pH 8.5 with different

PAMAM:streptokinase ratios of 1:1, 1:5, 1:10, and 1:20. It was found that bioconjugates with a 1:1 molar ratio showed that highest enzymatic activity retention compared to the other conjugates prepared here and also other macromolecular conjugates of streptokinase such as dextran and PEG.

Rather than conjugating the protein to the dendrimer periphery, Kostiainen, Smith, and coworkers have introduced protein molecules to the focal point of Newkome-type polyamine dendrons via a thioether linkage applying MAL–thiol 1,4-conjugation addition chemistry [202]. To construct these conjugates, dendrons functionalized with N-maleimide groups at their focal points were reacted with a free thiol group on bovine serum albumin (BSA) or a genetically engineered cysteine mutant of class II hydrophobin (HFBI) via 1,4-conjugation addition (Figure 5.21). The spermine surface groups of the dendrons were used to bind DNA with high affinity. It was shown that both BSA-G2 and HFBI-G2 conjugates bind to DNA with extremely high affinity. Moreover, the protein portion of the conjugates was used to promote their adhesion to hydrophobic surfaces and also their self-assembly at air–water interfaces. The preliminary studies showed that these conjugates are biocompatible and can enhance gene transfection in vitro.

The same group has recently extended the scope of the work to include optically degradable dendrons by incorporating a photo-labile o-nitrobenzyl linkage between the hydrophobic core and polyamine periphery of dendrons [203]. Conjugation of HFBI protein or a single-chain Fragment variable (scFv) antibody to the N-MAL focal point through the same mechanism as described above resulted in the formation of dendron–protein conjugates. These conjugates were shown to efficiently complex DNA, and subsequent exposure to UV light triggered the release of DNA and protein in a timescale of 3 minutes.

Very recently, Hartly and coworkers developed a method to construct dendrimer–protein/peptide conjugates via an oxime linkage [204]. In this method, PAMAM dendrimers were modified with pyruvate groups at their peripheries. These moieties were then reacted with aminoxyacetyl-functionalized model peptides (Leu-Tyr-Arg-Ala-Gly and the measles virus hemagglutinin-derived peptide MVHA49-72) and a recombinant protein such as insulin to form an oxime linkage (Figure 5.22). Moreover, by applying an orthogonal chemical conjugation method, the same group

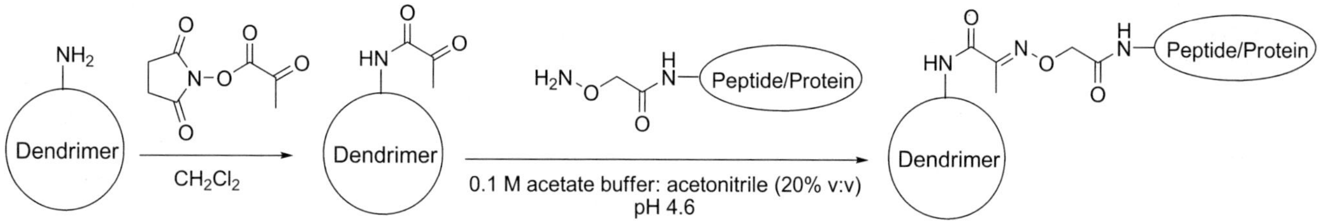

FIGURE 5.22 Functionalization of dendrimers with peptide/proteins via an oxime linkage.

was able to prepare PAMAM heterodendrimers functionalized with both peptide (Leu-Tyr-Arg-Ala-Gly) and mannose.

Shieh and coworkers have constructed PAMAM–saporin conjugates via a disulfide linkage [194]. Saporin is a protein that is known to be an efficient catalyst for the *in vitro* depurination of a specific adenine moiety in ribosomal RNAs. Here, to conjugate saporin to PAMAM dendrimers via a disulfide linkage, the dendrimer and saporin were reacted with *N*-succinimidyl 3-(2-pyridyldithio) propionate (SPDP) which installed a 2-pyridyldithio group on both reactants (Figure 5.23). PAMAM-SPDP was then treated with dithiothreitol (DTT) and subsequently with saporin-SPDP, resulting in the formation of a disulfide bond between the dendrimer and saporin. It was shown that cellular uptake and cytotoxicity of saporin was enhanced after its conjugation to dendrimer compared to free saporin.

5.6.3 Dendrimer–Peptide Conjugates

Depending on their application, dendrimer–peptide conjugates can be divided into two general categories that will be discussed below. First, the conjugation of limited numbers of peptides for targeting applications will be described. This will be followed by a discussion of dendrimers that can serve as model proteins through the conjugation of peptides to all or most of the peripheral groups.

5.6.3.1 Dendrimers Functionalized with a Limited Number of Peptides for Targeting
Integrins are heterodimeric cell surface receptors that bind to ligands with an exposed arginine–glycine–aspartate (RGD) sequence. They have been implicated in physiological processes such as thrombosis and inflammation, and are also involved in cancer. As a result, peptides containing the RGD sequence have found applications as inhibitors of apoptosis, angiogenesis, and tumor

formation as well as in biomaterials where they can provide targeted delivery of drugs and imaging agents.

In a series of studies by Baker and coworkers, PAMAM-RGD conjugates were synthesized by activating a carboxylic acid group on a cyclic-RGD (cRGD) peptide derivative using EDC in the presence of HOBt, followed by addition of the activated peptide to an amine-terminated PAMAM dendrimer to form an amide linkage [205–208]. Applying this method, FITC-labeled PAMAM–RGD dendrimers with and without entrapped gold nanoparticles were prepared and *in vitro* studies showed that these materials were efficiently taken up by cell expressing $\alpha_v\beta_3$ [205, 206]. It was also shown that when PAMAM-RGD conjugates were cultured with human dermal microvessel endothelial cells (HDMECs), with human vascular endothelial cells (HUVECs), or odontoblast-like MDPC-23 cells their targeting behavior was both time and dose dependent [207]. Moreover, these conjugates were shown to bind to dental pulp cells and regulate their differentiation, resulting in increases the odontogenic properties of these cells [208].

In a slightly different approach, Balogh and coworkers first activated the peripheral carboxylic acid groups on PAMAM dendrimers using EDC/HOBt coupling reagents, then allowed to react the activated dendrimer with an amine group on a cRGD peptide to form an amide linkage. Applying this method, biotinylated PAMAM-cRGD conjugates were prepared and were shown to be nontoxic within the relevant physiological concentrations. They were also shown to bind to $\alpha_v\beta_3$ integrins much more strongly than RGD peptide alone, presumably due to multivalent effect of the dendrimer [209].

Apart from the RGD conjugates described above, several other peptide–dendrimer conjugates have also been developed for targeted delivery. For example, Kong and coworkers have synthesized a nonsmall cell lung cancer-targeting

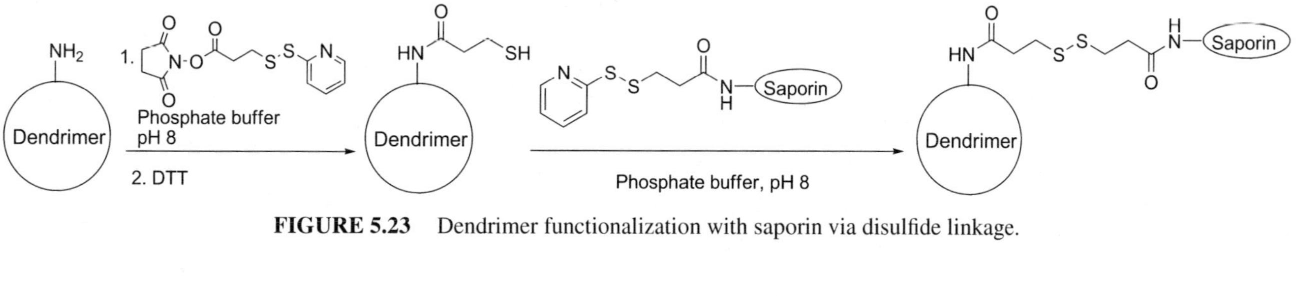

FIGURE 5.23 Dendrimer functionalization with saporin via disulfide linkage.

peptide (LCTP) with the sequence RCPLSHSLICY, and conjugated it to the periphery of a FITC-labeled PAMAM dendrimer through an EDC-mediated amide bond formation to obtain FITC-labeled PAMAM-LCTP conjugates. *In vivo* studies in athymic mice with lung cancer xenografts showed that these materials effectively accumulate in lung cancer cells and tumors [210].

In another study by Balian, Tomas, and coworkers, to target the delivery of exogenous genes to mesenchymal stem cells (MSCs), a family of gene delivery vehicles based on PAMAM dendrimers functionalized with either a low affinity MSC-binding (LAB) peptide, with the sequence of NSMIAHNKTRMHGGGSC, or high affinity MSCs-binding (HAB) peptide, with the sequence of SGHQLLLNKMP-NGGGSC, was synthesized employing disulfide linkages. PAMAM dendrimers were first reacted with SPDP to functionalize the dendrimers with 2-pyridyldithio moieties. In the next step, an excess amount of LAB or HAB peptides functionalized with cysteine dissolved in acetic acid (10% v/v) were reacted with these dendrimers in phosphate buffer at pH 8. These materials were shown to exhibit low toxicity, and to bind and deliver compact plasmid DNA to MSCs via a receptor-mediated gene delivery pathway. Moreover, their transfection efficiency was shown to be higher than that of the native dendrimers [195].

In another attempt to condense plasmid DNA and deliver it to cells expressing the glucose-regulating protein-78 kDa (GRP-78), Langer, Hammond, and coworkers have conjugated a peptide ligand with the sequence of WIFPWIQL, capable of selectively targeting GRP-78, to the focal point of PAMAM dendrons via a PEG spacer [179]. In their approach, the peptide was first modified with a bifunctional PEG spacer through an amide linker. Next, the MAL functional group on the other terminus of the PEG spacer was reacted with a free thiol functional group at the focal point of the dendron to afford the peptide–dendrimer conjugate via a thioether linkage. These conjugates showed better pDNA transfection efficiency in the cells expressing GRP-78 when compared to commercially available branched polyethylenimines.

More recently, Meijer and coworkers used the aforementioned aniline-catalyzed oxime formation strategy [204] to synthesize dendrimer–peptide conjugates with two different tumor-homing peptides [199]. The peptides used in this study were a linear peptide with the sequence of CREKA and a cyclic peptide with the sequence of CGNKRTRGC. It was shown that these conjugates were promising candidates for drug delivery and imaging purposes due to their tumor-targeting capabilities.

5.6.3.2 Fully Peptide-Functionalized Dendrimers as Model Proteins

In order to gain insight into the interactions involved in the protein-folding process and to develop biomaterials based on proteins, intense research has been devoted to the design and synthesis of model proteins. In recent years,

one of the approaches to design such model proteins has been through the use of rigid templates functionalized with polypeptides [211, 212]. Among the various rigid templates used for this purpose, dendrimers have attracted much attention because of their highly controllable sizes and valencies, as well as the versatility accessible in their surface chemistry.

In an early study by Niwa and coworkers, graft polymerization was used to polymerize γ-benzyl-ʟ-glutamate *N*-carboxy anhydride (BLG-NCA) from the peripheries of amine-terminated PAMAM dendrimers to obtain the corresponding peptide-functionalized PAMAM dendrimers [213]. It was found that the 3D PAMAM dendrimers effectively enhanced the helicity of the oligopeptides in a such a way that in lower pH ranges this conjugate exhibited an α-helix conformation with approximately 100% helicity. Moreover, upon interaction with α-amino acids, they were shown to preferentially bind to ᴅ-isomers rather than to ʟ-isomers [214]. Mihara and coworkers have reported α-helical peptide dendrimers prepared by the ligation reaction between perchloroacetylated PAMAM dendrimers and peptides containing cysteine via a thioether linkage [192, 215]. The peptides in these studies comprised 20 amino acids and were designed to adopt an amphiphilic α-helical structure. The resulting dendrimer–peptide conjugates were used to coordinate to Fe^{III}- or Zn^{II}-mesoporphyrin IX to mimic the natural light harvesting antennae in photosynthetic bacteria. It was shown that these conjugates served as effective artificial photosynthetic systems for the harvesting of light.

With the aim of developing collagen mimics, Kinberger and coworkers reacted a PAMAM dendrimer-bearing peripheral carboxylic acid moieties with the amine groups of peptides having the sequence of Gly-Pro-Nleu [216]. The reaction was performed using *N,N,N′,N′*-tetramethyl-*O*-(7-azabenzotriazol-1-yl)uronium hexafluorophosphate (HATU) and 1-hydroxy-7-azabenzotriazole (HOAt) as coupling reagents to afford an amide linkage. It was shown that these collagen mimetic dendrimers were able to complex biologically important Cu^{2+} and Ni^{2+} which is an important feature of collagen proteins. In an elegant study by Kono and coworkers, collagen peptides (Pro-Pro-Gly)₅, were grafted to the surfaces of the PAMAM dendrimers under the same HATU/HOAt coupling conditions resulting in the dendritic peptides exhibiting a collagen-like triple helix structure. Interestingly, in contrast to natural collagen, this helix was shown to be thermally reversible and was used for controlled release of a model drug rose bengal (RB) at different temperatures [217]. Moreover, a comprehensive study by same group using different generations of PAMAM dendrimers and binding ratios revealed that the formation of higher ordered structures is not influenced by the generation of dendrimers, but rather the binding ratios of polypeptide to dendrimer [218]. To further enhance the triple helicity of these conjugates, longer collagen model peptides (Pro-Pro-Gly)₁₀ were synthesized and conjugated to the PAMAM

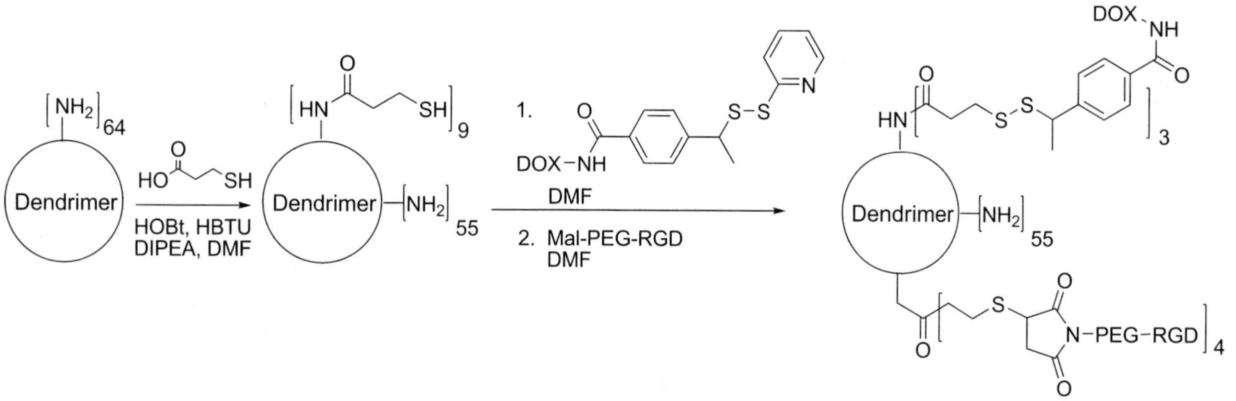

FIGURE 5.24   Preparation of dendrimer–DOX–RGD conjugates.

dendrimer surface using 4-(4,6-dimethoxy-1,3,5-triazin-2-yl)-4-methylmorpholinium chloride (DMT-MM) as a coupling reagent in aqueous solution at pH 7 to afford an amide linkage [219]. The reason why DMT-MM was used as a coupling reagent rather than HATU/HOAt, is that unlike (Pro-Pro-Gly)$_5$ which is soluble in organic solvents, (Pro-Pro-Gly)$_{10}$ is only soluble in water. DMT-MM is an effective water-soluble coupling reagent. It was shown that the thermal stability of the latter peptide-functionalized dendrimers was higher than the former dendrimer as a result of the longer peptide chains. The authors also studied the self-assembly behavior of these collagen mimetics and found that they undergo self-assembly to form spherical and rod-like morphologies [219].

In a series of studies by Mullen and coworkers, rigid and shape persistent polyphenylene dendrimers have been developed and used as scaffolds for polypeptide grafting by three different methods [193, 220–222]. In the first method, the amine peripheral groups of the dendrimers were used as initiators for polymerization of ε-benzyloxycarbonyl-L-lysine N-carboxyanhydride, which following deprotection of the ε-amines resulted in the grafting of PLL chains to the dendrimer surface. The second method involved the activation of the C-terminal carboxyl group of protected PLL followed by its coupling to the dendrimer's peripheral amines using HOBt/HBTU and subsequent deprotection. Lastly, the peripheries of the polyphenylene dendrimers were first decorated with MAL moieties, which were then reacted with sulfhydryl groups of the cysteine-functionalized unprotected peptides, via a Michael addition mechanism, to form thioether linkages [193, 220]. In each case, it was shown that the conjugated peptides were capable of forming α-helical secondary structures at pH 9.9 as well as in the presence of more than 75% 2,2,2-trifluoroethanol. Using techniques such as X-ray, solid-state NMR, calorimetry, and dielectric spectroscopy it was found that the self-assembly was strongly dependent on the poly-L-lysine chain length while the type of peptide secondary structure is governed

by packing limitations dictated by the polyphenylene core [221].

5.6.4 Multifunctional Dendrimer–Peptide Conjugates for Targeted Drug Delivery and Targeted Imaging

The development of multifunctional dendrimer–peptide–drug and dendrimer–peptide–contrast agent conjugates has also been an actively explored area. For example, Lu and coworkers reported the synthesis and *in vitro* study of dendrimer–DOX–RGD conjugates [223]. In this study, the dendrimer comprised a PLL backbone and a cubic silsesquinoxane core. As shown in Figure 5.24, the surface of this dendrimer was first decorated with 3-mercaptopropionic acid and then reacted with a DOX derivative that was prepared by the reaction of DOX with 4-succinimidyloxycarbonyl-a-methyl-a-[2-pyridyldithio]toluene (SMPT) to install the 2-pyridyldithio moiety. The resulting dendrimer–DOX conjugate contained a degradable disulfide linkage. The remaining free thiol groups on the dendrimer conjugate were then reacted with the MAL group on a MAL-PEG-RGD derivative to obtain the final dendrimer–DOX–RGD conjugate. It was shown that this multifunctional dendrimer conjugate had a higher cytotoxicity compared to free DOX in glioblastoma U87 cells. Moreover, upon complexation with siRNA, it was easily internalized by the same cell lines.

In a slightly different approach by Jiang and coworkers, dendrimer–PEG–RGD conjugates were first synthesized by reacting amine-terminated PAMAM dendrimers with a mixture of MAL-PEG-NHS and RGD peptide in 0.05 M borate buffer at pH 9.2 [32]. DOX was first converted to its succinic anhydride derivative, which was then activated to the corresponding NHS ester. The activated DOX derivative was then reacted with amine groups on the surface of the PAMAM-RGD conjugate to construct the target multifunctional PAMAM-RGD-DOX dendrimer. *In vitro* studies showed that the RGD peptide effectively increased the cytotoxicity of these materials in C6 cells. In addition, *in vivo* studies revealed that the conjugates had substantially longer half-lives and exhibited higher accumulation in brain tumor cells than normal brain tissues. In another study by the same group, PAMAM dendrimers were first functionalized with HAIYPRH (T7) peptide, which can target the transferrin receptor [58]. To accomplish this, PAMAM dendrimers were first reacted with a bifunctional NHS-PEG-MAL derivative to decorate their surfaces with MAL groups, which were then used in a reaction with a free thiol group on the peptide to construct PAMAM-PEG-T7 conjugates. Unlike the other aforementioned examples, DOX was encapsulated in the dendrimer via noncovalent interactions. It was shown that cellular uptake of DOX was enhanced when encapsulated within these peptide-functionalized dendrimers. Moreover, *in vivo* studies showed that tumor growth in mice was substantially inhibited when treated with PAMAM-PEG-T7/DOX compared to the treatment with PAMAM-PEG/DOX.

In addition to DOX, other anticancer drugs have also been conjugated to dendrimer–peptide conjugates and delivered in a targeted manner. For instance, Otto, Schluter, and coworkers have synthesized a multifunctional prototype dendrimer that is decorated with a pentapeptide H-Gly-Arg-Phe-Gly-OH, containing a cathepsin B cleavage site and a ligand for Pt^{2+} complexation at one terminus, as well as a fluorescent dansyl dye for intracellular visualization [224]. The peptide was conjugated to the dendrimer periphery via an amide linkage under HATU/DIPEA coupling conditions. It was shown that the resulting multifunctional materials were successfully

internalized by human HeLa cancer cells and were clustered next to the cell nucleus 18 hours after incubation.

In addition to drugs, various imaging contrast agents have also been linked to dendrimer–peptide conjugates in order to selectively image the tissues of interest. For example, Liskamp and coworkers reported an efficient microwave-assisted click reaction method to synthesize dendrimer–peptide conjugates by reacting alkyne-decorated dendrimers with azide-functionalized peptides under microwave condition using sodium ascorbate and copper(II) sulfate as the catalyst system. Applying this method, small and large peptides as well as cRGD peptides were successfully conjugated to the dendrimer's periphery [225]. The same group then extended their work to prepare DOTA-conjugated RGD peptide–dendrimer conjugates, applying the same microwave-assisted click reaction method. In this study, DOTA was installed at the focal point of the dendrimer, while RGD peptides were introduced to the periphery of the dendrimer via a triazole linkage to yield monomeric, dimeric, and tetrameric peptide conjugates [226]. *In vivo* studies showed that multivalent DOTA-conjugated ^{111}In-labeled peptide dendrimers had a higher cellular uptake in the $\alpha_v\beta_3$ integrin-expressing tumors.

In an approach reported by Meijer, Hackeng, and coworkers, NCL was used to synthesize nonsymmetric polylysine dendrimers functionalized with arginine–glycine–asparagine–serine (RGDS) peptide and DTPA (Figure 5.25) [227]. In their strategy, one wedge of the final dendrimer was

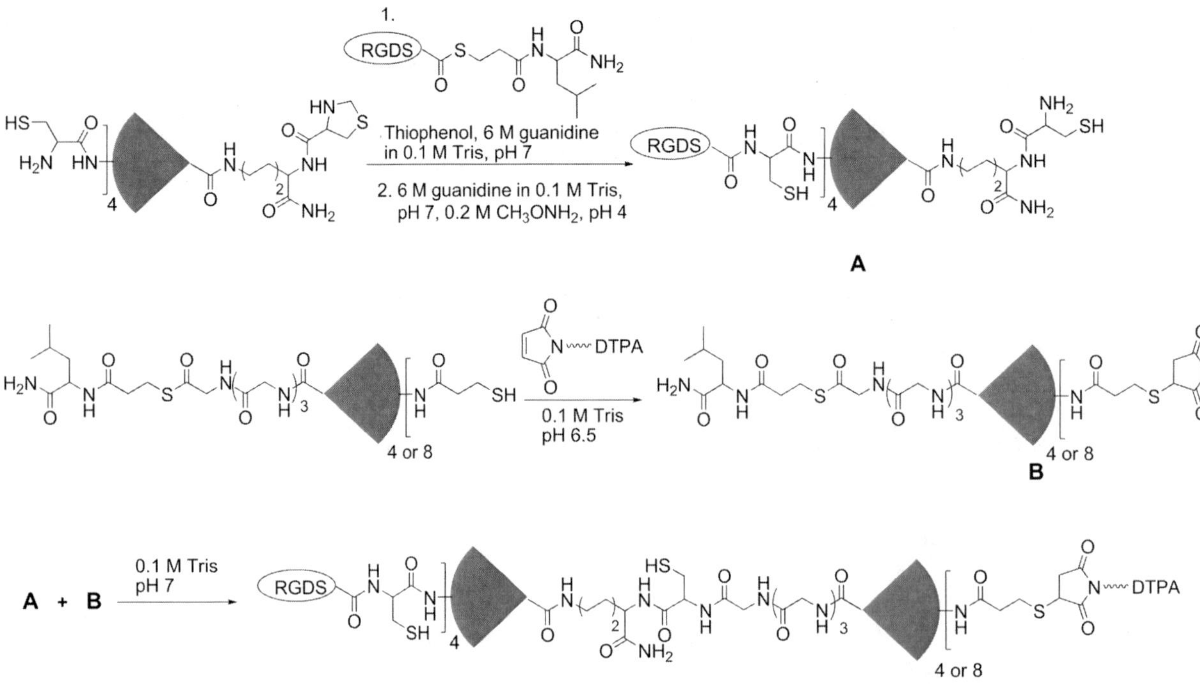

FIGURE 5.25 Synthesis of nonsymmetric dendrimers functionalized with targeting and imaging agents.

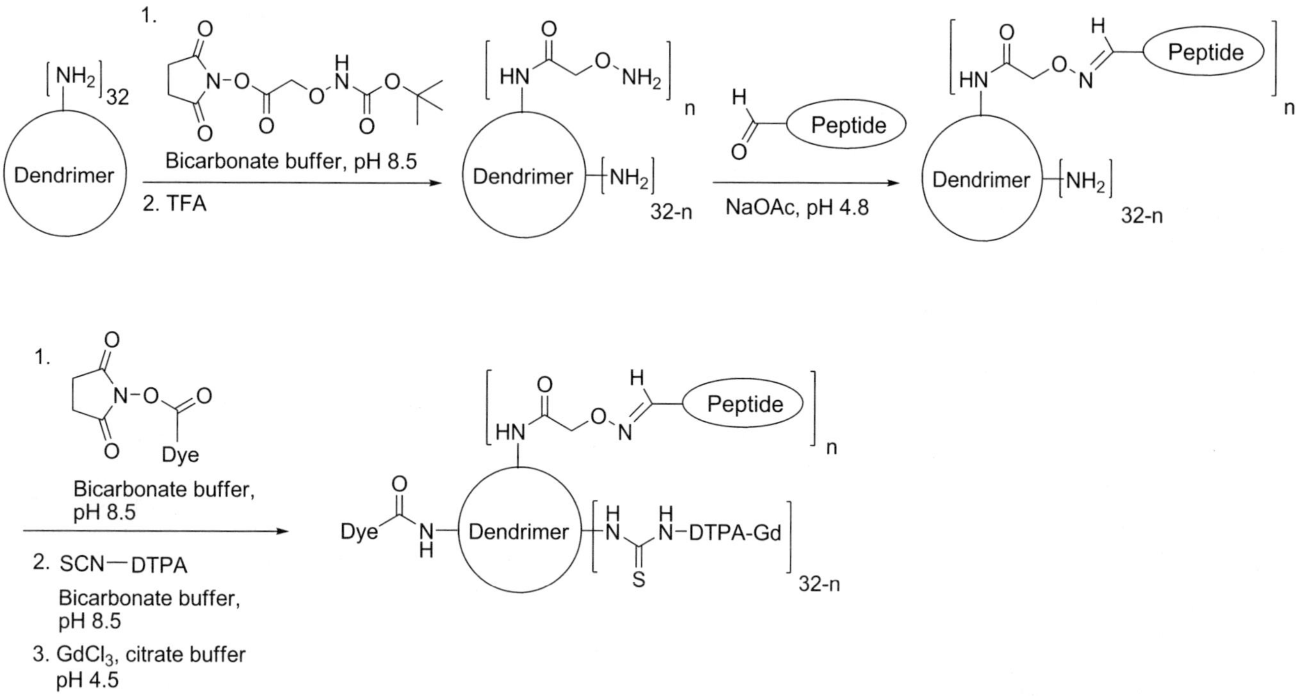

FIGURE 5.26 Synthesis of multifunctional dendrimers having targeting, imaging, and optical agents.

functionalized with DTPA through the reaction between the dendron's peripheral thiols and a MAL derivative of DTPA. Conjugation of the RGDS peptide to the second wedge was accomplished by the reaction between cysteine peripheral groups of the dendron and the *C*-terminal thioester of RGDS via NCL. At this stage, the thioproline residue at the focal point of the RGDS-functionalized wedge was converted to cysteine using methoxylamine and was subsequently conjugated to the thioester focal point of the DTPA containing wedge via NCL.

In a report by Brechbiel and coworkers, PAMAM dendrimers were used as scaffolds for the conjugation of an RGD cyclopeptide as a targeting agent, DTPA-Gd(III) as MRI contrast agent, and a fluorescent dye for optical imaging (Figure 5.26) [228]. In this strategy, the periphery of the dendrimer was first decorated with aminooxy groups, which were then reacted with an aldehyde-functionalized RGD peptide to form the PAMAM-RGD conjugate via an oxime linkage. In the next step, an NHS-activated Alexa Fluor 594 was reacted with the peripheral amine groups of PAMAM-RGD in bicarbonate buffer at pH 8.5 to construct dye-labeled PAMAM-RGD dendrimer via an amide linkage. The dendrimer was subsequently reacted with an excess amount of isothiocyanate-functionalized DTPA derivative in the same buffer solution to obtain dye-labeled PAMAM-RGD-DTPA conjugate. In the last step, Gd(III) was chelated by DTPA moieties in citrate buffer at pH 4.5 to give the final multifunctional dendrimer. Biological studies revealed that, the

bioconjugates without any Gd(III) selectively bound to $\alpha_v\beta_3$ integrin-expressing cells. However, upon the incorporation of Gd(III), reduced selectively in binding to cells was observed.

Multifunctional agents of even higher complexity have also been reported. For example, using a PAMAM backbone, Wei, Li, and coworkers conjugated rhodamine as a fluorescent probe in the visible range, Cy5.5 as an NIR probe, DOTA-Gd(III) as MRI contrast agent, RGDyK as a tumor vasculature-targeting peptide, and Angiopep-2 as a BBB-permeable peptide [122]. In this nanoprobe, RGDyK was conjugated to the dendrimer's surface through a PEG spacer via the Michael addition of the dendrimer's amine groups to the MAL group on PEG spacer while Angiopep-2 peptide was installed by first partially decorating the dendrimer with 2-pyridyldithio moieties followed by their reaction with the free thiol of the peptide to afford a disulfide linkage. *In vivo* studies revealed that this construct was able to efficiently cross an intact BBB in normal mice and it was possible to visualize orthotropic brain tumor xenografts with high sensitivity.

5.7 CONCLUSIONS AND FUTURE PROSPECTS

As described above, dendrimers exhibit many unique and advantageous properties that have motivated the preparation of a diverse range of bioconjugates over the last couple of decades. These advantages include the low polydispersities of dendrimers owing to their stepwise synthesis, their nanoscale and tunable dimensions, multiple functional groups at the focal point and periphery that are available for bioconjugation, and their globular architectures among others. For example, their well-defined structures and tunable biodistribution properties have been attractive properties for the development of both drug delivery vehicles and imaging agents based on dendrimers. Their tunable multivalency has made dendrimers ideal scaffolds for the development of carbohydrate conjugates for applications such as therapeutics and vaccines, and for the development of multifunctional systems. The high density of branching has proven to provide enhancements in the relaxivities of Gd(III)-based MRI contrast agents, and also to provide effective site isolation of chromophores for optical imaging applications. Their globular architecture has allowed for new nanomaterials with unique and well-defined architectures to be prepared through combinations with DNA and polypeptides.

This chapter has highlighted the diversity of conjugation strategies that can be used to covalently attach functional molecules including drugs, carbohydrates, imaging agents, DNA, and peptides/proteins to dendrimers and dendrons. Through this discussion, the different roles of the linkers in these different bioconjugates have emerged. For example, in the case of drug molecules, the choice of the linker is critical to the function of the bioconjugate. The linker must be selected to provide release of the drug molecule in its active form, ideally selectively at the therapeutic target. For this, esters, as well as the more stimuli-responsive hydrazone and disulfide linkages have proven to be promising. On the other hand, in dendrimer–carbohydrate conjugates the linker choice is generally dictated more by the synthetic challenges associated with the complete functionalization of dendrimer peripheries with these ligands, and has been found to play a lesser role in the biological function. For MRI contrast agents, more rigid linkers combined with higher generation dendrimers have been most effective in enhancing the properties of these agents. In the case of DNA and proteins the bioconjugation strategies have generally been dictated by the chemical functionalities available for conjugation, as well as the compatibility of the chemistry with these complex biomolecules.

In general, there has been a trend in recent years toward increasingly complex structures that can perform multiple functions. For example, dendrimer–drug conjugates with targeting groups, dendrimer–imaging agent conjugates with targeting groups, and dendrimers conjugated to multiple different imaging agents, or multiple different peptides have been prepared. On the one hand, this work elegantly illustrates the versatility and multifunctionality of the dendrimer scaffold. On the other hand, it does raise concerns with respect to product homogeneity. Following the recent failure of a PAMAM-MTX-FA conjugate in clinical trials, Banaszak Holl and coworkers performed studies to explore the structural homogeneity of dendrimer bioconjugates [229]. They demonstrated that when a statistical functionalization of the dendrimer periphery is performed, the distribution of resulting dendrimer conjugates is much more heterogeneous that commonly assumed. In addition, the mean ratio of dendrimer:conjugate, which is generally obtained from characterization techniques such as nuclear magnetic resonance or ultraviolet–visible spectroscopy, is insufficient to characterize the materials as batches with the same mean dendrimer:conjugate ratio can possess very different distributions. This distribution can be very sensitive to the synthetic history of the molecules and to slight changes in reaction conditions, making batch-to-batch reproducibility very challenging. When multiple species are conjugated to the dendrimer periphery, the situation becomes even more complicated and only a very small fraction of the resulting material may in fact possess the required number of each species to exhibit the desired biological effect. In biological experiments, this raises significant questions regarding the identity of the biological components.

This concern must be addressed especially in the case of dendrimers if the desired structural homogeneities, one of the key advantages of dendrimers, is to be preserved. One option is to work with relatively simple systems in which the entire dendrimer periphery is functionalized with the same molecule. The antiviral dendrimer SPL7013 is an example of such a structure, where optimized chemistry can provide very well defined materials. Where multiple moieties must be conjugated to the dendrimer it can be done in a controlled manner, avoiding statistical distributions. For example, in this chapter there were numerous examples where different species were attached to the core and to the periphery in a controlled manner. Examples were also highlighted where multifunctional units at the dendrimer periphery were used to conjugate different species such as PEO and drug or PEO and imaging agent in a controlled manner. However, when the number of different moieties becomes too large, this can become exceedingly complicated. Therefore, in the future, it will be critical to continue to exploit the unique and promising properties of dendrimers, using versatile and optimized chemistry to prepare functional yet well-defined materials in an efficient manner.

REFERENCES

1. Buhleier E, Wehner W, Vögtle F. *Synthesis* 1978;2:155–158.
2. Tomalia DA, Dewald J, Hall M, Martin S, Smith P. *Prepr 1st SPSJ Int Polym Conf Soc Polym Sci Jpn (Kyoto)* 1984:65.
3. Tomalia DA, Baker H, Dewald J, Hall M, Kallos G, Martin S, Roeck J, Ryder J, Smith P. *Polym J* 1985;17:117–132.
4. De Gennes PG, Hervet H. *J Phys Lett* 1983;44:351–360.
5. Tomalia DA, Baker H, Dewald J, Hall M, Kallos G, Martin S, Roeck J, Ryder J, Smith P. *Macromolecules* 1986;19:2466–2468.

6. de Bradander-van den Berg EMM, Meijer EW. *Angew Chem Int Ed* 1993;32:1308–1311.

7. Newkome GR, Yao Z, Baker GR, Gupta VK. *J Org Chem* 1985;50:2003–2004.

8. Launay N, Caminade AM, Lahana R, Majoral JP. *Angew Chem Int Ed* 1994;33:1589–1592.

9. Hawker CJ, Fréchet JMJ. *J Am Chem Soc* 1990;112:7638–7647.

10. Terwogt JMM, Nuijen B, Huinink WWT, Beijnen JH. *Cancer Treat Rev* 1997;23:87–95.

11. Reddy LH. *J Pharm Pharmacol* 2005;57:1231–1242.

12. El Kazzouli S, Mignani S, Bousmina M, Majoral JP. *New J Chem* 2012;36:227–240.

13. Boas U, Heegaard PMH. *Chem Soc Rev* 2004;33:43–63.

14. Patri AK, Kukowska-Latallo JF, Baker JR. *Adv Drug Delivery Rev* 2005;57:2203–2214.

15. Cheng YY, Xu ZH, Ma ML, Xu TW. *J Pharm Sci* 2008;97:123–143.

16. Padilla De Jesus OL, Ihre HR, Gagne L, Fréchet JMJ, Szoka FC. *Bioconjugate Chem* 2002;13:453–461.

17. Gurdag S, Khandare J, Stapels S, Matherly LH, Kannan RM. *Bioconjugate Chem* 2006;17:275–283.

18. Chau Y, Dang NM, Tan FE, Langer R. *J Pharm Sci* 2006;95:542–551.

19. Agashe HB, Babbar AK, Jain S, Sharma RK, Mishra AK, Asthana A, Garg M, Dutta T, Jain NK. *Nanomed-Nanotechnol* 2007;3:120–127.

20. Kono K, Kojima C, Hayashi N, Nishisaka E, Kiura K, Wataral S, Harada A. *Biomaterials* 2008;29:1664–1675.

21. Almutairi A, Guillaudeu SJ, Berezin MY, Achilefu S, Fréchet JMJ. *J Am Chem Soc* 2008;130:444–445.

22. van der Poll DG, Kieler-Ferguson HM, Floyd WC, Guillaudeu SJ, Jerger K, Szoka FC, Fréchet JM. *Bioconjugate Chem* 2010;21:764–773.

23. Kaminskas LM, Kelly BD, McLeod VM, Sberna G, Boyd BJ, Owen DJ, Porter CJH. *Mol Pharm* 2011;8:338–349.

24. Yuan H, Luo K, Lai YS, Pu YJ, He B, Wang G, Wu Y, Gu ZW. *Mol Pharm* 2010;7:953–962.

25. Lee CC, Cramer AT, Szoka FC, Fréchet JMJ. *Bioconjugate Chem* 2006;17:1364–1368.

26. Lee CC, Gillies ER, Fox ME, Guillaudeu SJ, Fréchet JMJ, Dy EE, Szoka FC. *Proc Natl Acad Sci USA* 2006;103:16649–16654.

27. Choi SK, Thomas T, Li MH, Kotlyar A, Desai A, Baker JR. *Chem Commun* 2010;46:2632–2634.

28. Kukowska-Latallo JF, Candido KA, Cao ZY, Nigavekar SS, Majoros IJ, Thomas TP, Balogh LP, Khan MK, Baker JR. *Cancer Res* 2005;65:5317–5324.

29. Han SQ, Yoshida T, Uryu T. *Carbohydr Polym* 2007;69:436–444.

30. van Haandel L, Stobaugh JF. *Anal Bioanal Chem* 2010;397:1841–1852.

31. Jiang YY, Tang GT, Zhang LH, Kong SY, Zhu SJ, Pei YY. *J Drug Targeting* 2010;18:389–403.

32. Guo R, Wang H, Peng C, Shen MW, Zheng LF, Zhang GX, Shi XY. *J Mater Chem* 2011;21:5120–5127.

33. Quintana A, Raczka E, Piehler L, Lee I, Myc A, Majoros I, Patri AK, Thomas T, Mule J, Baker JR. *Pharm Res* 2002;19:1310–1316.

34. Pignatello R, Guccione S, Forte S, Di Giacomo C, Sorrenti V, Vicari L, Barretta GU, Balzano F, Puglisi G. *Bioorg Med Chem* 2004;12:2951–2964.

35. Warlick CA, Sweeney CL, McIvor RS. *Biochem Pharmacol* 2000;59:141–151.

36. Kaminskas LM, Kelly BD, McLeod VM, Boyd BJ, Krippner GY, Williams ED, Porter CJH. *Mol Pharm* 2009;6:1190–1204.

37. Majoros IJ, Myc A, Thomas T, Mehta CB, Baker JR. *Biomacromolecules* 2006;7:572–579.

38. Lim JD, Simanek EE. *Org Lett* 2008;10:201–204.

39. Bernardes GJL, Kikkeri R, Maglinao M, Laurino P, Collot M, Hong SY, Lepenies B, Seeberger PH. *Org Biomol Chem* 2010;8:4987–4996.

40. de Groot FMH, Albrecht C, Koekkoek R, Beusker PH, Scheeren HW. *Angew Chem Int Ed* 2003;42:4490–4494.

41. Erez R, Segal E, Miller K, Satchi-Fainaro R, Shabat D. *Bioorg Med Chem* 2009;17:4327–4335.

42. Harada D, Naito S, Otagiri M. *Pharm Res* 2002;19:1648–1654.

43. Palmer LA, Doctor A, Chhabra P, Sheram ML, Laubach VE, Karlinsey MZ, Forbes MS, Macdonald T, Gaston B. *J Clin Invest* 2007;117:2592–2601.

44. Kurtoglu YE, Navath RS, Wang B, Kannan S, Romero R, Kannan RM. *Biomaterials* 2009;30:2112–2121.

45. Kurtoglu YE, Mishra MK, Kannan S, Kannan RM. *Int J Pharm* 2010;384:189–194.

46. Kim J, Yoon HJ, Kim S, Wang K, Ishii T, Kim YR, Jang WD. *J Mater Chem* 2009;19:4627–4631.

47. Nicolaou KC, Mitchell HJ. *Angew Chem Int Ed* 2001;40:1576–1624.

48. Taylor ME, Drickamer K. *J Biol Chem* 1993;268:399–404.

49. Ashwell G, Harford J. *Annu Rev Biochem* 1982;51:531–554.

50. Sharon N, Lis H. *Science* 1989;246:227–234.

51. Arce E, Nieto PM, Diaz V, Castro RG, Bernad A, Rojo J. *Bioconjugate Chem* 2003;14:817–823.

52. Tabarani G, Reina JJ, Ebel C, Vives C, Lortat-Jacob H, Rojo J, Fieschi F. *FEBS Lett* 2006;580:2402–2408.

53. Kikkeri R, Garcia-Rubio I, Seeberger PH. *Chem Commun* 2009;235–237.

54. Page D, Roy R. *Bioconjugate Chem* 1997;8:714–723.

55. Kieburg C, Lindhorst TK. *Tetrahedron Lett* 1997;38:3885–3888.

56. Woller EK, Cloninger MJ. *Biomacromolecules* 2001;2:1052–1054.

57. Bogdan N, Roy R, Morin M. *RSC Adv* 2012;2:985–991.

58. Bogdan N, Vetrone F, Roy R, Capobianco JA. *J Mater Chem* 2010;20:7543–7550.

59. Woller EK, Cloninger MJ. *Org Lett* 2002;4:7–10.

60. Woller EK, Walter ED, Morgan JR, Singel DJ, Cloninger MJ. *J Am Chem Soc* 2003;125:8820–8826.

61. Schlick KH, Udelhoven RA, Strohniever GC, Cloninger MJ. *Mol Pharm* 2005;2:295–301.

62. Branderhorst HM, Ruijtenbeek R, Liskamp RMJ, Pieters RJ. *Chembiochem* 2008;9:1836–1844.

63. Kikkeri R, Grunstein D, Seeberger PH. *J Am Chem Soc* 2010;132:10230–10232.

64. Xu JT, Boyer C, Bulmus V, Davis TP. *J Polym Sci, Part A: Polym Chem* 2009;47:4302–4313.

65. Chabre YM, Brisebois PP, Abbassi L, Kerr SC, Fahy JV, Marcotte I, Roy R. *J Org Chem* 2011;76:724–727.

66. Wu P, Chen X, Hu N, Tam UC, Blixt O, Zettl A, Bertozzi CR. *Angew Chem Int Ed* 2008;47:5022–5025.

67. Chen X, Wu P, Rousseas M, Okawa D, Gartner Z, Zettl A, Bertozzi CR. *J Am Chem Soc* 2009;131:890–891.

68. Fernandez-Megia E, Correa J, Riguera R. *Biomacromolecules* 2006;7:3104–3111.

69. Munoz EM, Correa J, Fernandez-Megia E, Riguera R. *J Am Chem Soc* 2009;131:17765–17767.

70. Page D, Aravind S, Roy R. *Chem Commun* 1996;1913–1914.

71. Grandjean C, Rommens C, Gras-Masse H, Melnyk O. *Tetrahedron Lett* 1999;40:7235–7238.

72. Mori T, Hatano K, Matsuoka K, Esumi Y, Toone EJ, Terunuma D. *Tetrahedron* 2005;61:2751–2760.

73. Liu H, Wang H, Guo R, Cao X, Zhao J, Luo Y, Shen M, Zhang G, Shi X. *Polym Chem* 2010;1:1677–1683.

74. Kensinger RD, Catalone BJ, Krebs FC, Wigdahl B, Schengrund CL. *Antimicrob Agents Chemother* 2004;48:1614–1623.

75. Branderhorst HM, Liskamp RMJ, Visser GM, Pieters RJ. *Chem Commun* 2007;5043–5045.

76. Ashton PR, Boyd SE, Brown CL, Nepogodiev SA, Meijer EW, Peerlings HWI, Stoddart JF. *Chem Eur J* 1997;3:974–984.

77. Kikkeri R, Hossain LH, Seeberger PH. *Chem Commun* 2008;2127–2129.

78. Laurino P, Kikkeri R, Azzouz N, Seeberger PH. *Nano Lett* 2011;11:73–78.

79. Appeldoorn CCM, Joosten JAF, el Maate FA, Dobrindt U, Hacker J, Liskamp RMJ, Khan AS, Pieters RJ. *Tetrahedron-Asymmetr* 2005;16:361–372.

80. Deguise I, Lagnoux D, Roy R. *New J Chem* 2007;31:1321–1331.

81. Liu XP, Liu J, Luo Y. *Polym Chem* 2012;3:310–313.

82. Mammen M, Choi SK, Whitesides GM. *Angew Chem Int Ed* 1998;37:2755–2794.

83. Roy R, Zanini D, Meunier SJ, Romanowska A. *J Chem Soc, Chem Commun* 1993;1869–1872.

84. Page D, Zanini D, Roy R. *Bioorg Med Chem* 1996;4:1949–1961.

85. Zanini D, Roy R. *J Am Chem Soc* 1997;119:2088–2095.

86. Meunier SJ, Wu QQ, Wang SN, Roy R. *Can J Chem* 1997;75:1472–1482.

87. Matsuoka K, Kurosawa H, Esumi Y, Terunuma D, Kuzuhara H. *Carbohydr Res* 2000;329:765–772.

88. Sakamoto JI, Koyama T, Miyamoto D, Yingsakmongkon S, Hidari K, Jampangern W, Suzuki T, Suzuki Y, Esumi Y, Hatano K, Terunuma D, Matsuoka K. *Bioorg Med Chem Lett* 2007;17:717–721.

89. Sakamoto JI, Koyama T, Miyamoto D, Yingsakmongkon S, Hidari K, Jampangern W, Suzuki T, Suzuki Y, Esumi Y, Nakamura T, Hatano K, Terunuma D, Matsuoka K. *Bioorg Med Chem* 2009;17:5451–5464.

90. Clayton R, Hardman J, LaBranche CC, McReynolds KD. *Bioconjugate Chem* 2011;22:2186–2197.

91. Zanini D, Roy R. *J Org Chem* 1998;63:3486–3491.

92. Reuter JD, Myc A, Hayes MM, Gan ZH, Roy R, Qin DJ, Yin R, Piehler LT, Esfand R, Tomalia DA, Baker JR. *Bioconjugate Chem* 1999;10:271–278.

93. Caravan P, Ellison JJ, McMurry TJ, Lauffer RB. *Chem Rev* 1999;99:2293–2352.

94. Merbach AE, Toth E. *The Chemistry of Contrast Agents in Medical Magnetic Resonance Imaging*. Chichester, UK: John Wiley & Sons, Inc.; 2001.

95. Aime S, Botta M, Terreno E. *Adv Inorg Chem* 2005;57:173–237.

96. Prince MR, Zhang HL, Roditi GH, Leiner T, Kucharczyk W. *J Magn Reson Imaging* 2009;30:1298–1308.

97. Terreno E, Delli Castelli D, Viale A, Aime S. *Chem Rev* 2010;110:3019–3042.

98. Villaraza AJL, Bumb A, Brechbiel MW. *Chem Rev* 2010;110:2921–2959.

99. Langereis S, Dirkson A, Hackeng TM, van Genderen MHP, Meijer EW. *New J Chem* 2007;31:1152–1160.

100. Caravan P. *Chem Soc Rev* 2006;35:512–523.

101. Wiener EC, Brechbiel MW, Brothers H, Magin RL, Gansow OA, Tomalia DA, Lauterbur PC. *Magn Res Med* 1994;31:1–8.

102. Kobayashi H, Kawamoto S, Jo SK, Bryant HL Jr, Brechbiel MW, Star RA. *Bioconjugate Chem* 2003;14:388–394.

103. Langereis S, de Lussanet QG, van Genderen MHP, Backes WH, Meijer EW. *Macromolecules* 2004;37:3084–3091.

104. Bryant LH, Brechbiel MW, Wu CC, Bulte JWM, Herynek V, Frank JA. *J Magn Reson Imaging* 1999;9:348–352.

105. Toth E, Pubanz D, Vauthey S, Helm L, Merbach AE. *Chem Eur J* 1996;2:1607–1615.

106. Rudovsky J, Hermann P, Botta M, Aime S, Lukes I. *Chem Commun* 2005;2390–2392.

107. Rudovsky J, Botta M, Hermann P, Hardcastle KI, Lukes I, Aime S. *Bioconjugate Chem* 2006;17:975–987.

108. Dong Q, Hurst DR, Weinmann HJ, Chenevert TL, Londy FJ, Prince MR. *Invest Radiol* 1998;33:699–708.

109. Fulton DA, O'Halloran M, Parker D, Senanayake K, Botta M, Aime S. *Chem Commun* 2005;474–476.

110. Raymond KN, Pierre VC. *Bioconjugate Chem* 2005;16:3–8.

111. Pierre VC, Botta M, Raymond KN. *J Am Chem Soc* 2005;127:504–505.

112. Floyd WC, Klemm PJ, Smiles DE, Kohlgruber AC, Pierre VC, Mynar JL, Fréchet JMJ, Raymond KN. *J Am Chem Soc* 2011;133:2390–2393.

113. Wiener EC, Konda S, Shadron A, Brechbiel M, Gansow O. *Invest Radiol* 1997;32:748–754.

114. Konda SD, Wang S, Brechbiel M, Wiener EC. *Invest Radiol* 2002;37:199–204.

115. Hounsfield G. *Br J Radiol* 1995;68:166–172.

116. Bourin M, Jolliet P, Ballereau F. *Clin Pharmacokinet* 1997;32:180–193.

117. Hallouard F, Anton N, Choquet P, Constantinesco A, Vandamme T. *Biomaterials* 2010;31:6249–6268.

118. Kojima C, Umeda Y, Ogawa M, Harada A, Magata Y, Kono K. *Nanotechnology* 2010;21:245104.

119. Peng C, Zheng L, Chen Q, Shen M, Guo R, Wang H, Cao X, Zhang G, Shi X. *Biomaterials* 2012;33:1107–1119.

120. Wang H, Zheng L, Peng C, Guo R, Shen M, Shi X, Zhang G. *Biomaterials* 2011;32:2979–2988.

121. Yordanov AT, Lodder AL, Woller EK, Cloninger MJ, Patronas N, Milenic D, Brechbiel MW. *Nano Lett* 2002;2:595–599.

122. Fu Y, Nitecki DE, Maltby D, Simon GH, Berejnoi K, Raatschen H-J, Yeh BM, Shames DM, Brasch RC. *Bioconjugate Chem* 2006;17:1043–1056.

123. Regino CAS, Walbridge S, Bernardo M, Wong KJ, Johnson D, Lonser R, Oldfield EH, Choyke PL, Brechbiel MW. *Contrast Media Mol Imaging* 2008;3:2–8.

124. Hamoudeh M, Kamleh MA, Diab R, Fessi H. *Adv Drug Delivery Rev* 2008;60:1329.

125. Ghobril C, Lamanna G, Kueny-Stotz M, Garofalo A, Billotey C, Felder-Flesch D. *New J Chem* 2012;36:310–323.

126. Abram U, Alberto R. *J Brazil Chem Soc* 2006;17:1486–1500.

127. Subbarayan M, Shetty SJ, Srivastava TS, Noronha OPD, Samuel AM, Mukhtar H. *Biochem Biophys Res Commun* 2001;281:32–36.

128. Parrott MC, Benhabbour SR, Saab C, Lemon JA, Parker S, Valliant JF, Adronov A. *J Am Chem Soc* 2009;131:2906–2916.

129. Zhang Y, Sun Y, Xu X, Zhu H, Huang L, Zhang X, Qi Y, Shen YM. *Bioorg Med Chem Lett* 2010;20:927.

130. Zhang Y, Sun Y, Xu X, Zhang X, Zhu H, Huang L, Qi Y, Shen YM. *J Med Chem* 2010;53.

131. Criscione JM, Dobrucki LW, Zhuang ZW, Papademetris X, Simons M, Sinusas AJ, Fahmy TM. *Bioconjugate Chem* 2011;22:1784–1792.

132. Malik N, Wiwattanapatapee R, Klopsch R, Lorenz K, Frey H, Weener JW, Meijer EW, Paulus W, Duncan R. *J Controlled Release* 2000;65:133–148.

133. Gillies ER, Dy E, Fréchet JMJ, Szoka FC. *Mol Pharm* 2005;2:129–138.

134. Tanaka K, Siwu ERO, Minami K, Hasagawa K, Nozaki S, Kamayama Y, Koyama K, Weihu CC, Paulson JC, Yasuyoshi W, Fukase K. *Angew Chem Int Ed* 2010;49:8195.

135. Lowik CWGM, Kaijzel E, Que I, Vahrmeijer A, Kuppen P, Mieog J, Van de Velde C. *Eur J Cancer* 1990;45(Suppl 1):391–393.

136. Thomas TP, Majoros IJ, Kotlyar A, Kukowska-Latallo JF, Bielinska A, Myc A, Baker JR. *J Med Chem* 2005;48:3729–3735.

137. Vinogradov SA, Lo L-W, Jenkins WT, Evans SM, Koch C, Wilson DF. *Biophys J* 1996;70:1609.

138. Rumsey WL, Vanderkooi JM, Wilson DF. *Science* 1988;241:1649.

139. Vinogradov SA, Lo L-W, Wilson DF. *Chem Eur J* 1999;5:1338–1347.

140. Lebedev AY, Cheprakov AV, Sakadzic S, Boas DA, Wilson DF, Vinogradov SA. *ACS Appl Mater Interfaces* 2009;1:1292–1304.

141. Esipova TV, Karagodov A, Miller J, Wilson DF, Busch TM, Vinogradov SA. *Anal Chem* 2011;83:8756–8765.

142. Brinas RP, Troxler T, Hochstrasser RM, Vinogradov SA. *J Am Chem Soc* 2005;127:11851–11862.

143. Finikova OS, Lebedev AY, Aprelev A, Troxler T, Gao F, Garnacho C, Muro S, Hochstrasser RM, Vinogradov SA. *Chem Phys Chem* 2008;9:1673–1679.

144. Sakadzic S, Roussakis E, Yaseen MA, Mandeville ET, Srinivasan VJ, Arai K, Ruvinskaya S, Devor A, Lo E-H, Vinogradov SA, Boas DA. *Nat Methods* 2010;7:755–759.

145. Lecoq J, Parpaleix A, Roussakis E, Ducros M, Houssen YG, Vinogradov SA, Charpak S. *Nat Med* 2011;17:893–898.

146. Ornelas C, Lodescar R, Durandin A, Canary JW, Pennell R, Liebes LF, Weck M. *Chem Eur J* 2011;17:3619–3629.

147. Almutairi A, Akers WJ, Berezin MY, Achilefu S, Fréchet JMJ. *Mol Pharm* 2008;5:1103–1110.

148. Xu H, Regino CAS, Koyama Y, Hama Y, Gunn AJ, Bernardo M, Kobayashi H, Choyke PL, Brechbiel MW. *Bioconjugate Chem* 2007;18:1474–1482.

149. Wängler C, Moldenhauer G, Saffrich R, Knapp E-M, Beijer B, Schnölzer M, Wängler B, Eisenhut M, Haberkorn U, Mier W. *Chem Eur J* 2008;14:8116–8130.

150. Giacca M. *Gene Therapy*. Milan: Springer-Verlag Italia S.r.l.; 2010.

151. Sioud M. *siRNA and Mirna Gene Silencing*. New York: Humana Press; 2011.

152. Fox KR, Brown T (eds). *DNA Conjugates and Sensors*. RSC Publishing; 2012.

153. Caminade AM, Turrin C-O, Majoral JP. *Chem Eur J* 2008;14:7422–7432.

154. Dufès C, Uchegbu IF, Schätzlein AG. *Adv Drug Delivery Rev* 2005;57:2177–2202.

155. Cho YN, Brumbach JS, Ramalingam M, Halder ZS. *J Bionanoscience* 2011;5:1–17.

156. Ainalem M-L, Nylander T. *Soft Matter* 2011;7:4577–4594.

157. Hudson RHE, Damha MJ. *J Am Chem Soc* 1993;115:2119–2124.

158. Hudson RHE, Robidoux S, Damha MJ. *Tetrahedron Lett* 1998;39:1299–1302.

159. Watson JD, Crick FHC. *Nature* 1953;171:737–738.

160. Scheffler M, Dorenbeck A, Jordan S, Wüstefeld M, von Kiedrowski G. *Angew Chem Int Ed* 1999;38:3311–3315.

161. Eckardt LH, Naumann K, Pankau WM, Rein M, Schweitzer M, Windhab N, von Kiedrowski G. *Nature* 2002;420:286–286.

162. Shchepinov MS, Mir KU, Elder JK, Frank-Kamenetskii MD, Southern EM. *Nucleic Acids Res* 1999;27:3035–3041.

163. Goh SL, Francis MB, Fréchet JMJ. *Chem Commun* 2002;2954–2955.

164. DeMattei CR, Huang B, Tomalia DA. *Nano Lett* 2004;4:771–777.

165. Choi Y, Mecke A, Orr BG, Banaszak Holl MM, Baker JRJ. *Nano Lett* 2004;4:391–397.

166. Chen P, Sun Y, Liu H, Xu L, Fan Q, Liu D. *Soft Matter* 2010;6:2143–2145.

167. Wang L, Feng Y, Sun Y, Li Z, Yang Z, He YM, Fan QH, Liu D. *Soft Matter* 2011;7:7187–7190.

168. Caminade AM, Padié C, Laurent R, Maraval A, Majoral JP. *Sensors* 2006;6:901–914.

169. Shchepinov MS, Udalova IA, Bridgman AJ, Southern EM. *Nucleic Acids Res.* 1997;25:4447–4454.

170. Striebel HM, Birch-Hirschfeld E, Egerer R, Földes-Papp Z, Tilz GP, Stelzner A. *Exp Mol Pathol* 2004;77:89–97.

171. Benters R, Niemeyer CM, Wöhrle D. *Chem Bio Chem* 2001;2:686–694.

172. Benters R, Niemeyer CM, Drutschmann D, Blohm B, Wöhrle D. *Nucleic Acids Res* 2002;30:e10-11–e10-17.

173. Le Berre V, Trévisol E, Dagkessamanskaia A, Sokol S, Caminade AM, Majoral JP, Meunier B, François J. *Nucleic Acids Res* 2003;31:e88-81–e-88-88.

174. Trévisol E, Leberre-Anton V, Leclaire J, Pratviel G, Caminade AM, Majoral JP, François J, Meunier B. *New J Chem* 2003;27:1713–1719.

175. Day RS, Fiegland LR, Vint ES, Shen W, Morris JR, Norton ML. *Langmuir* 2011;27:12434–12442.

176. Hong BJ, Oh SJ, Youn TO, Kwon SH, Park JW. *Langmuir* 2005;21:4257–4261.

177. Oh SJ, Ju J, Kim BC, Ko E, Hong BJ, Park JG, Park JW, Choi KY. *Nucleic Acids Res* 2005;33:e90-91–e90-98.

178. Zhang X-X, Eden HS, Chen X. *J Controlled Release* 2012;159:2–13.

179. Wood KC, Azarin SM, Arap W, Pasqualini R, Langer R, Hammond PT. *Bioconjugate Chem* 2008;19:403–405.

180. Veiseh O, Sun C, Gunn J, Kohler N, Gabikian P, Lee D, Bhattarai N, Ellenbogen R, Sze R, Hallahan A, Olson J, Zhang MQ. *Nano Lett* 2005;5:1003–1008.

181. Uchida M, Flenniken ML, Allen M, Willits DA, Crowley BE, Brumfield S, Willis AF, Jackiw L, Jutila M, Young MJ, Douglas T. *J Am Chem Soc* 2006;128:16626–16633.

182. Rowe MD, Thamm DH, Kraft SL, Boyes SG. *Biomacromolecules* 2009;10:983–993.

183. Gianella A, Jarzyna PA, Mani V, Ramachandran S, Calcagno C, Tang J, Kann B, Dijk WJR, Thijssen VL, Griffioen AW, Storm G, Fayad ZA, Mulder WJM. *ACS Nano* 2011;5:4422–4433.

184. Sadler K, Tam JP. *Reviews in Molecular Biotechnology* 2002;90:195–229.

185. Reymond JL, Darbre T. *Org Biomol Chem* 2012;10:1483–1492.

186. Darbre T, Reymond JL. *Acc Chem Res* 2006;39:925–934.

187. Chow HF, Mong TKK, Chan YH, Cheng CHK. *Tetrahedron* 2003;59:3815–3820.

188. Hartwig S, Nguyen MM, Hecht S. *Polym Chem* 2010;1:69–71.

189. Tam JP. *Proc Natl Acad Sci USA* 1988;85:5409–5413.

190. Liu CF, Tam JP. *Proc Natl Acad Sci USA* 1994;91:6584–6588.

191. Fischer-Durand N, Salmain M, Rudolf B, Dai L, Juge L, Guerineau V, Laprevote O, Vessieres A, Jaouen G. *Anal Biochem* 2010;407:211–219.

192. Sakamoto M, Ueno A, Mihara H. *Chem Eur J* 2001;7:2449–2458.

193. Herrmann A, Mihov G, Vandermeulen GWM, Klok HA, Mullen K. *Tetrahedron* 2003;59:3925–3935.

194. Lai PS, Pai CL, Peng CL, Shieh MJ, Berg K, Lou PJ. *J Biomed Mater Res A* 2008;87A:147–155.

195. Santos JL, Pandita D, Rodrigues J, Pego AP, Granja PL, Balian G, Tomas H. *Mol Pharm* 2010;7:763–774.

196. Dawson PE, Muir TW, Clarklewis I, Kent SBH. *Science* 1994;266:776–779.

197. Saxon E, Armstrong JI, Bertozzi CR. *Org Lett* 2000;2:2141–2143.

198. van Baal I, Malda H, Synowsky SA, van Dongen JLJ, Hackeng TM, Merkx M, Meijer EW. *Angew Chem Int Ed* 2005;44:5052–5057.

199. Breurken M, Lempens EHM, Temming RP, Helms BA, Meijer EW, Merkx M. *Bioorg Med Chem* 2011;19:1062–1071.

200. Nilsson BL, Kiessling LL, Raines RT. *Org Lett* 2000;2:1939–1941.

201. Wang XT, Inapagolla R, Kannan S, Lieh-Lai M, Kannan RM. *Bioconjugate Chem* 2007;18:791–799.

202. Kostiainen MA, Szilvay GR, Lehtinen J, Smith DK, Linder MB, Urtti A, Ikkala O. *ACS Nano* 2007;1:103–113.

203. Kostiainen MA, Kotimaa J, Laukkanen ML, Pavan GM. *Chem Eur J* 2010;16:6912–6918.

204. Gaertner HF, Cerini F, Kamath A, Rochat AF, Siegrist CA, Menin L, Hartley O. *Bioconjugate Chem* 2011;22:1103–1114.

205. Shukla R, Thomas TP, Peters J, Kotlyar A, Myc A, Baker JJR. *Chem Commun* 2005;5739–5741.

206. Shukla R, Hill E, Shi XY, Kim J, Muniz MC, Sun K, Baker JR. *Soft Matter* 2008;4:2160–2163.

207. Hill E, Shukla R, Park SS, Baker JR. *Bioconjugate Chem* 2007;18:1756–1762.

208. Kim JK, Shukla R, Casagrande L, Sedgley C, Nor JE, Baker JR, Hill EE. *J Dent Res* 2010;89:1433–1438.

209. Lesniak WG, Kariapper MST, Nair BM, Tan W, Hutson A, Balogh LP, Khan MK. *Bioconjugate Chem* 2007;18:1148–1154.

210. Liu JF, Liu JJ, Chu LP, Wang YM, Duan YJ, Feng LN, Yang CH, Wang L, Kong DL. *Int J Nanomed* 2011;6:59–69.

211. Wallimann P, Kennedy RJ, Kemp DS. *Angew Chem Int Ed* 1999;38:1290–1292.

212. Wallimann P, Kennedy RJ, Miller JS, Shalongo W, Kemp DS. *J Am Chem Soc* 2003;125:1203–1220.

213. Higashi N, Koga T, Niwa N, Niwa M. *Chem Commun* 2000;361–362.

214. Higashi N, Koga T, Niwa M. *Chembiochem* 2002;3:448–454.

215. Sakamoto M, Ueno A, Mihara H. *Chem Commun* 2000;1741–1742.

216. Kinberger GA, Taulane JP, Goodman M. *Tetrahedron* 2006;62:5280–5286.

217. Kojima C, Tsumura S, Harada A, Kono K. *J Am Chem Soc* 2009;131:6052–6053.

218. Suehiro T, Kojima C, Tsumura S, Harada A, Kono K. *Biopolymers* 2010;93:640–648.

219. Suehiro T, Tada T, Waku T, Tanaka N, Hongo C, Yamamoto S, Nakahira A, Kojima C. *Biopolymers* 2011;95:270–277.

220. Mihov G, Grebel-Koehler D, Lubbert A, Vandermeulen GWM, Herrmann A, Klok HA, Mullen K. *Bioconjugate Chem* 2005;16:283–293.

221. Mondeshki M, Mihov G, Graf R, Spiess HW, Mullen K, Papadopoulos P, Gitsas A, Floudas G. *Macromolecules* 2006;39:9605–9613.

222. Koynov K, Mihov G, Mondeshki M, Moon C, Spiess HW, Mullen K, Butt HJ, Floudas G. *Biomacromolecules* 2007;8:1745–1750.

223. Kaneshiro TL, Lu ZR. *Biomaterials* 2009;30:5660–5666.

224. Fuchs S, Otto H, Jehle S, Henklein P, Schluter AD. *Chem Commun* 2005;1830–1832.

225. Joosten JAF, Tholen NTH, El Maate FA, Brouwer AJ, van Esse GW, Rijkers DTS, Liskamp RMJ, Pieters RJ. *Eur J Org Chem* 2005;3182–3185.

226. Dijkgraaf I, Rijnders AY, Soede A, Dechesne AC, van Esse GW, Brouwer AJ, Corstens FHM, Boerman OC, Rijkers DTS, Liskamp RMJ. *Org Biomol Chem* 2007;5:935–944.

227. Dirksen A, Meijer EW, Adriaens W, Hackeng TM. *Chem Commun* 2006;1667–1669.

228. Boswell CA, Eck PK, Regino CAS, Bernardo M, Wong KJ, Milenic DE, Choyke PL, Brechbiel MW. *Mol Pharm* 2008;5:527–539.

229. Mullen DG, Banaszak Holl MM. *Acc Chem Res* 2011;44:1135–1145.

SECTION III

ORGANIC NANOPARTICLES BASED BIOCONJUGATES

6

BIOCONJUGATION STRATEGIES: LIPIDS, LIPOSOMES, POLYMERSOMES, AND MICROBUBBLES

ANIRBAN SEN GUPTA AND HORST A. VON RECUM
Department of Biomedical Engineering, Case Western Reserve University, Cleveland, OH, USA

6.1 INTRODUCTION

Lipids and lipid-based delivery vehicles, namely liposomes, micelles, and micro/nano bubbles have become the most researched and clinically applied area in the field of nanomedicine. Phospholipids are well known for their biocompatibility because many lipids are natural components of the biological cell membrane. Because lipids have a polar hydrophilic head and a nonpolar hydrophobic tail, in aqueous environment they can spontaneously undergo thermodynamically driven supramolecular self-assembly into micelles, planar bilayers, and closed bilayer vesicles (liposomes). Almost 50 years ago Sir Alec Bangham had reported on this pioneering observation of phospholipid assembly [1, 2], and since then liposomes and micelles have evolved from an object of biophysical research, to synthetic models of cell membrane, to one of the most clinically approved pharmaceutical nanocarrier of choice. Currently there are about ten liposomal drug formulations clinically approved for human use (Table 6.1), and several others are in advanced phase clinical trials. Almost all of the current clinical formulations use liposomes that are only meant for solubilization of hydrophobic drugs or encapsulation of hydrophilic/amphiphilic drugs and keeping them in circulation for longer time (sterically stabilized or "stealth" liposomes) to enhance drug bioavailability *in vivo*. However, in preclinical studies the technology has further evolved into conjugating a variety of ligand motifs onto the lipid molecules and lipid-based assemblies to promote targeting and binding to specific cell-surface receptors and disease site antigens, with a vision to enhance the selectivity and efficacy of drug delivery. Similar

to the lipid assembly systems, a similar strategy can be utilized with amphiphilic block-copolymer molecules to develop polymeric vesicles known as polymersomes [3]. Polymersomes are defined as bilayer polymeric vesicles made from amphiphilic diblock [3], triblock [4, 5], or multiarm [6, 7] copolymers. Similar to liposomes, polymersomes can range in size from hundreds of nanometers to greater than 50 μm [3].

Hydrophilic/hydrophobic block copolymers have long been known to self-assemble into micro- and nano-sized structures (such as micelles). Vessels with an aqueous core and polymeric bilayer membrane have been empirically shown to spontaneously occur when using polymers with ratios of hydrophilic block to total polymer mass of $\leq 35 \pm 10$ [8]. Polymers with more than 45% ratio generally form micelles, while polymers with less than 25% generally form inverted structures. The following sections will discuss the various bioconjugation strategies that have been developed in modifying the surface of such lipid- and polymer-based vehicles.

6.2 LIPIDS AND LIPOSOMES

Bioconjugation of lipids to proteins, peptides, small molecule agents, antibodies and other ligand motifs, oligosaccharides, and polymers have become exceedingly important in formulation and delivery of therapeutic agents. Drug delivery vehicles like liposomes and micelles can be made long circulating by surface conjugation of hydrophilic polymers like polyethylene glycol (PEG) [9–16]. This strategy has resulted

Chemistry of Bioconjugates: Synthesis, Characterization, and Biomedical Applications, First Edition. Edited by Ravin Narain.
© 2014 John Wiley & Sons, Inc. Published 2014 by John Wiley & Sons, Inc.

TABLE 6.1   Clinical Liposomal Formulations

Trade Name	Application	Drug description	Comany
DuanoXome	Kaposi's sarcoma	Duanorubicin in liposome	Nexstar/Gilead
Doxil	Kaposi's sarcoma	Doxorubicin in liposome	Sequus Pharma/Ortho Biotech
Myocet	Metastatic breast cancer	Doxorubicin in liposome	Elan/Zeneus
Depocyt	Lymphomatous meningitis	Cytarabine in liposome	SkyePharma/Pacira
Abelcet	Fungal infections	Amphotericin B in liposome	Enzon
Ambisome	Fungal infections	Amphotericin B in liposome	Nexstar/Gilead/Fujisawa
Epaxal	Hepatitis A	IRIV vaccine in liposome	Berna Biotech
Inflexal V	Influenza	IRIV vaccine in liposome	Berna Biotech
DepoDur	Postsurgical analgesia	Morphine in liposome	SkyePharma/Pacira
Visudyne	Macular degeneration PDT	Verteporfin in liposome	QLT/Novartis

in a major advancement in drug delivery technology where the PEG modification of the liposomal and micellar vehicles enables resistance to opsonization and macrophagic uptake, thereby resulting in longer plasma residence times and allowing passive accumulation of the drug-loaded vehicles into tumors via enhanced permeation and retention (EPR) effect [10, 12]. Besides PEG, several other hydrophilic polymers like hydroxypropyl methacrylamide systems, polyacrylamides, poly(vinyl pyrrolidone), poly(acryloyl morpholine), poly(2-methyl-2-oxazoline), poly(2-ethyl-2-oxazoline), phosphatidyl polyglycerols, polyvinyl alcohol, poly(glutamic acid) [17, 18], etc. have been tested as circulation time-enhancing conjugates to lipids and lipid-based vehicles, but with limited success. PEGylated lipids and lipid-based vehicles have emerged as the most successful formulation components clinically, have come to be known as the "stealth" nanoparticle design (because of their ability to resist opsonization and avoid macrophagic recognition) and have been FDA approved for a variety of therapeutic applications [19]. For PEG conjugation to lipids, the most popular strategy is to carry out reactions of active esters of PEG carboxylic acids as acylating agents with lipids or preformed lipid-based vehicles bearing primary amines, to form stable amides [20–23]. PEG derivatives like PEG-dichlorotriazine, PEG-tresylate, PEG-succinimidyl carbonate, PEG-benzotriazole carbonate, PEG-p-nitrophenyl carbonate, PEG-trichlorophenyl carbonate, PEG-carbonylimidazole and PEG-succinimidyl succinate, have all been used for this strategy. Scheme 6.1 shows some examples of these reactions using distearoylphosphatidyl ethanolamine (DSPE) as the amine-bearing lipid. An alternative strategy is to react thiol (-SH) bearing lipids to PEG-maleimide, PEG-vinylsulfone, PEG-iodoacetamide, or PEG-orthopyridyl disulfide, to form disulfide, thioether, or thioester conjugates [22]. Scheme 6.2 shows chemical schemes of such reactions using a thiol-terminated lipid.

Lipid–polymer conjugates like lipid-PEG can undergo self-assembly at concentrations greater than or equal to critical aggregation concentration to form micelles and liposomes, with the PEG molecules exposed principally on the exofacial surface. Although such self-assembled drug delivery systems have shown significant efficacy in encapsulating a variety of drugs, protecting them from plasma deactivation renal clearance and macrophagic uptake, improving their bioavailability and enhancing their passive (EPR-mediated) uptake in tumors, such passive processes mostly increase the accumulation of these formulations in the tissue stromal space (e.g., in tumor) but may not further facilitate the therapeutic delivery selectively to the diseased cell. Therefore, to further achieve intracellular delivery, the strategy of "active targeting" has been employed where the lipid components or preformed lipidic vehicles are further modified with cell receptor-specific ligands. The ligand categories that have been studied the most are antibodies (and their fragments), proteins, peptides, oligosaccharides, and vitamins [24–34].

Various strategies for conjugating antibodies and antibody fragments to lipids or lipid-formed vehicles (e.g., liposomes) have been studied [24–27]. Monoclonal antibody (MAb)-decorated liposomes selectively targeted toward cell-surface antigens (receptors) have come to be known as immunoliposomes. Based on the type of heavy chain-dependent class of antibody, most MAb-based immunoliposome fabrication methods have involved the use of IgG and sometimes IgM. The use of IgG and IgM in immunoliposomes can lead to immunogenic risks [35, 36]. Consequently significant research has been focused on the use of smaller functional antigen-binding fragments of antibodies. For example, treatment of antibodies with proteolytic enzymes can produce FAb [37] and Fab' [38] fragments, that have reduced immunogenicity issues due to absence of effector functions of the Fc chain [39]. Another strategy has been to generate chimeric antibodies that can avoid problems associated with species differences in the conserved regions of the antibody [40]. Bioconjugation of all these antibodies and antibody variants to lipids or preformed lipidic vehicles like liposomes, generally fall under four categories based upon the particular chemical functionality of the antibody being utilized, namely, amine mediated, carbohydrate mediated, disulfide mediated, and noncovalent mechanism mediated.

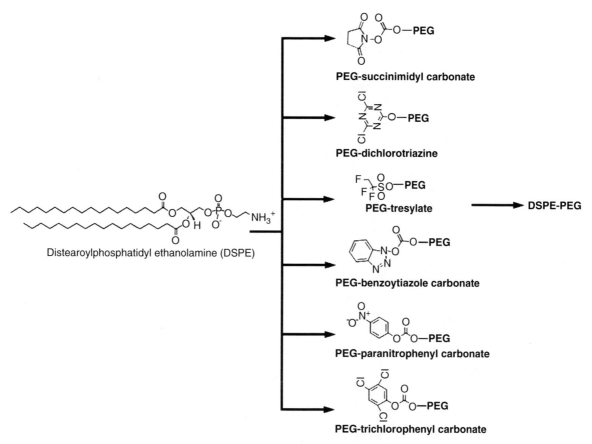

SCHEME 6.1 lipid-PEG via amide linkage

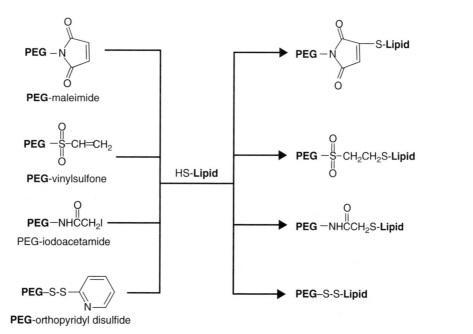

SCHEME 6.2 lipid-PEG via sulfhydryl-based reactions

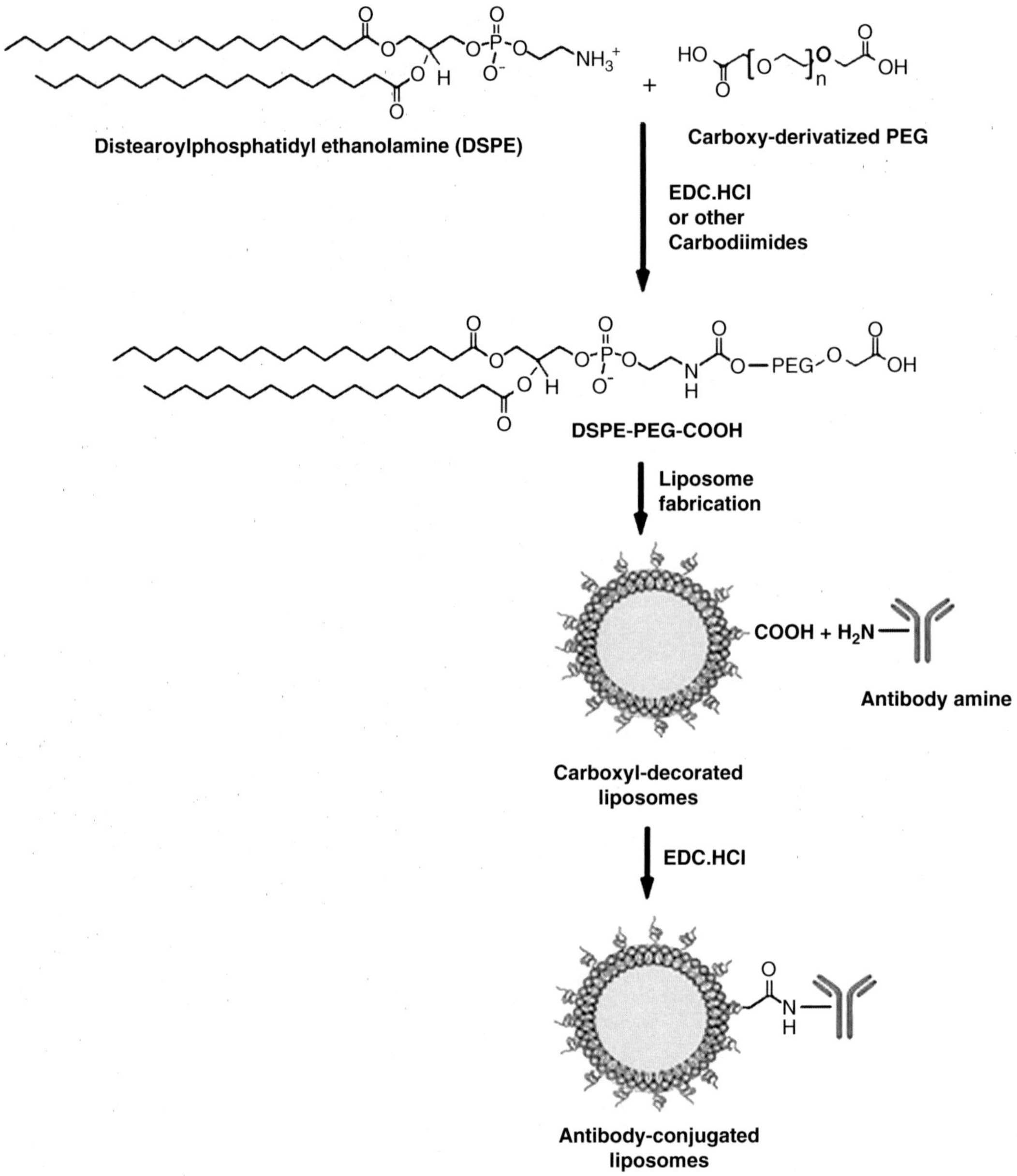

SCHEME 6.3   liposome-antibody conjugate via amide

In amine-based reactions, the most studied strategy is the conjugation of amine groups in the antibody or antibody fragment to exofacial carboxyl groups of preformed liposomes in the presence of carbodiimide-based cross-linking agents like 1-ethyl-3-(3′-dimethylaminopropyl)carbodiimide hydrochloride (EDC-HCl), as shown in Scheme 6.3 [41–43]. Using this strategy for antibodies may be problematic as EDC can end up cross-linking the antibodies themselves via reaction between mutual amine and carboxyl groups. Other strategies include activation of carboxyl-terminated lipids with N-hydroxysuccinimide (NHS) to form reactive lipid esters that can then conjugate to amine-containing antibodies and antibody fragments via reductive amidation [44, 45].

In carbohydrate-mediated reaction, the most popular strategy involves the oxidation of the carbohydrate groups on antibodies with galactose oxidase or sodium periodate to generate aldehyde groups, that can be further utilized to conjugate to amine-bearing lipids (or liposomes) via Schiff base reaction and hydrazide-bearing lipids (or liposomes) via hydrazine reaction [46–52]. The same reaction can also be utilized in an opposite scheme where carbohydrate-containing lipids can be used to preform liposomes, the vicinal hydroxyl groups on the carbohydrate domain of the lipid can be oxidized to form aldehydes and subsequent Schiff base reaction can be carried out with the primary amine groups of an antibody or antibody fragment. A representative schematic of these possibilities is

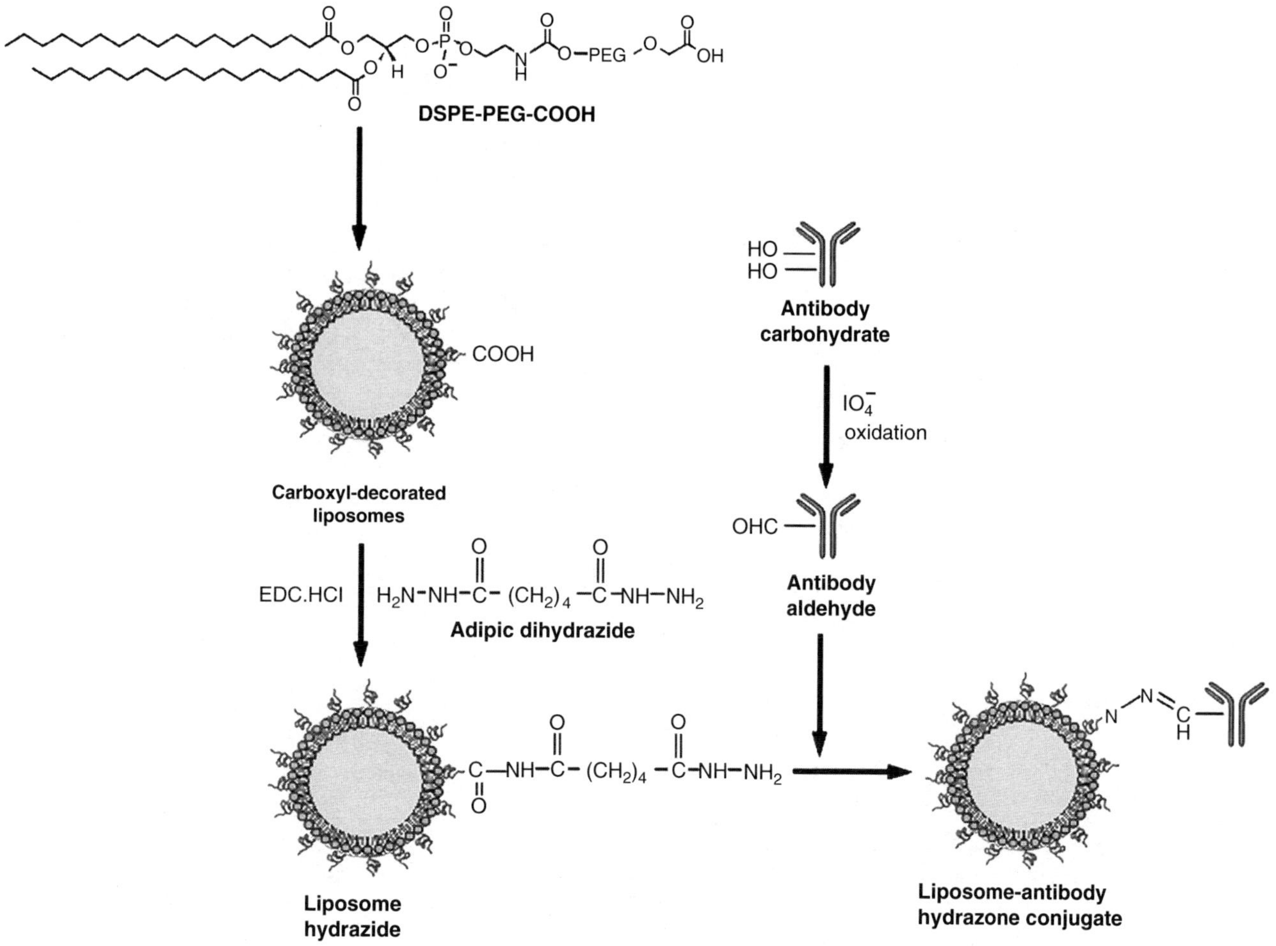

SCHEME 6.4   liposome-antibody conjugate via hydrazone

shown in Scheme 6.4. Such carbohydrate-based approaches may allow circumventing the antigen-binding site damage issues known to be associated with amine and disulfide-based approaches for conjugating antibodies [53]. Regarding disulfide-based bioconjugation of antibodies to lipids and lipidic vehicles, the most used strategy is the maleimide–thiol reaction to form thioether linkage. For example, treatment of antibodies or antibody fragments with reducing agents, such as dithiothreitol (DTT) or 2-mercaptoethylamine, leads to cleaving of the disulfide bonds in the antibody to form thiol (sulfhydryl) groups. These sulfhydryl groups can be further used to conjugate to maleimide-derivatized lipids or preformed liposomes [54–56]. Alternatively, antibodies can be thiolated with cross-linkers like 2-iminothiolane (also known as Traut's reagent), 3-(2-pyridyldithio) propionic acid-N-hydroxysuccinimide ester (SPDP), S-acetylthioglycolic acid N-hydroxysuccinimide ester (SATA), succinimidyl acetylthiopropionate (SATP), etc. [43, 57–59], followed by deprotection with agents like DTT (e.g., for SPDP reaction) or hydroxylamine (e.g., for SATA reaction) and resultant thiol groups

can be conjugated to maleimide-derivatized lipids or liposomes via thioether linkage [47, 60, 61] (e.g., Scheme 6.5). The noncovalent strategies for conjugating antibodies and antibody fragments to lipids and liposomes have included affinity-based interactions like the avidin–biotin complex and charge-based interactions like that between negatively charged phospholipid and a polycation. Avidin, a protein derived from avians and amphibians, has significant affinity for biotin, a cofactor in multiple eukaryotic biological processes. The avidin–biotin complex is one of the strongest noncovalent interactions ($K_d = 10^{-15}$M) between a protein and a ligand, and is resistant to pH, temperature, organic solvents, and other denaturing agents. Hence this interaction has been extensively utilized in bioconjugation strategies, for example, using avidin-modified preformed liposomes to interact with biotinylated antibodies [62, 63] or biotinylated liposomes to react with avidin-modified antibodies and other ligands [64]. Regarding charge-based methods, the electrostatic interactions of cationic poly(amino acids) like poly-L-lysine with negatively charged lipids or liposomal assemblies

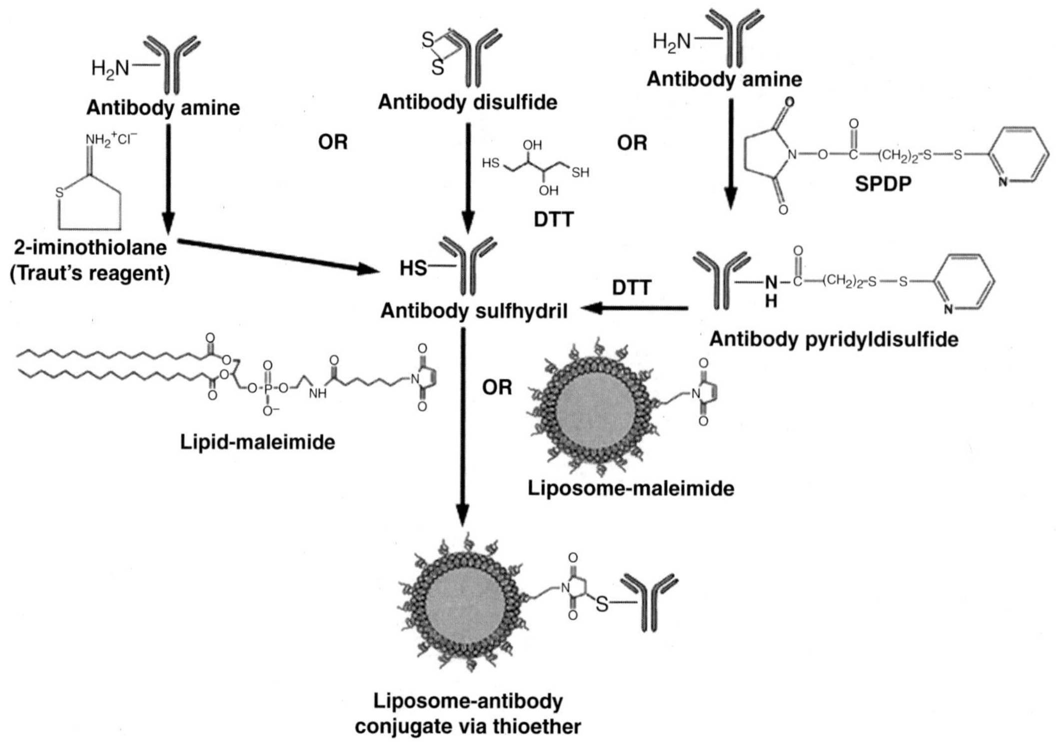

SCHEME 6.5 liposome-antibody conjugate via sulfhydryl

like phosphatidylserine, have been used as models for lipid–protein interactions [65, 66]. Similar strategies can be utilized to conjugate antibodies or antibody fragments with cationic domains to liposomes.

The chemical conjugation strategies utilized to tether antibodies and antibody fragments to lipids, lipidic micelles and liposomes, can also be used to tether other protein-based ligands or therapeutic agents. For example, amine (-NH2) derivatized lipids can be used to conjugate to carboxyl groups of proteins or vice versa, utilizing carbodiimide (e.g., EDC) and succinimide (e.g., NHS) mediated amidation reactions [67–69]. The conjugated product can then be incorporated into liposomes by mixing with other structural lipids and cholesterol. An additional development here (also applicable to conjugation of antibodies, antibody fragments, and peptide ligands) is the incorporation of spacers between the ligand motifs and the liposomal surface, by using heterobifunctional PEG and other cross-linking agents. As described previously, PEGylation of liposomes and micelles (and other nanostructures) has been found to reduce macrophagic clearance and enhance the circulation period of the drug delivery systems. This property can be combined with the cell-selective or receptor-specific targeting property by using heterobifunctional PEG spacer molecules where one end of the PEG is conjugated onto the phospholipid (or liposome and micelle) molecule while the other end is conjugated onto the targeting moiety. For example, a heterobifunctional PEG derivative carrying a reactive succinimidyl carbonate (SC) group on one

end and a tert-butyloxycarbonyl (Boc) protected hydrazide group at the other end can be synthesized by an efficient four-step process, followed by reaction of the SC-end-group with the amino group of lipid DSPE forming a stable urethane linkage to get DSPE-PEG-hydrazide-Boc conjugate. Acidolytic removal of the Boc group can subsequently yield DSPE-PEG-hydrazide conjugate and the hydrazide termini can be further utilized to tether ligand functionalities [9]. Other important functionalities introduced via PEG tethers on lipids include amine, carboxyl, maleimide, and sulfhydryl [23, 56, 70, 71], all of which can be utilized to conjugate antibodies, their fragments and proteins.

Peptide conjugation to lipids and subsequent assembly liposomes and micelles (or conjugation to preformed vehicles) has become highly attractive for targeted drug delivery, since peptides have reduced immunogenicity compared to antibodies, yet can allow cell- or receptor-selectivity by virtue of reasonably high affinity and multivalent decoration-induced avidity. Regarding conjugation of peptides to lipids, a bioconjugation strategy that has become quite effective is the synthesis of peptides on a solid-phase resin, selectively deprotecting and reacting the N-termini of the peptide (while still tethered to the resin) to a carboxyl-derivatized PEGylated lipid (e.g., DSPE-PEG-COOH) to conjugate via amide bond formation, and finally cleaving the DSPE-PEG-peptide from the resin support by tri-fluoroacetic acid (TFA) (Scheme 6.6) [72, 73]. A similar strategy can be utilized by reacting a sulfhydryl terminus of resin tethered peptide

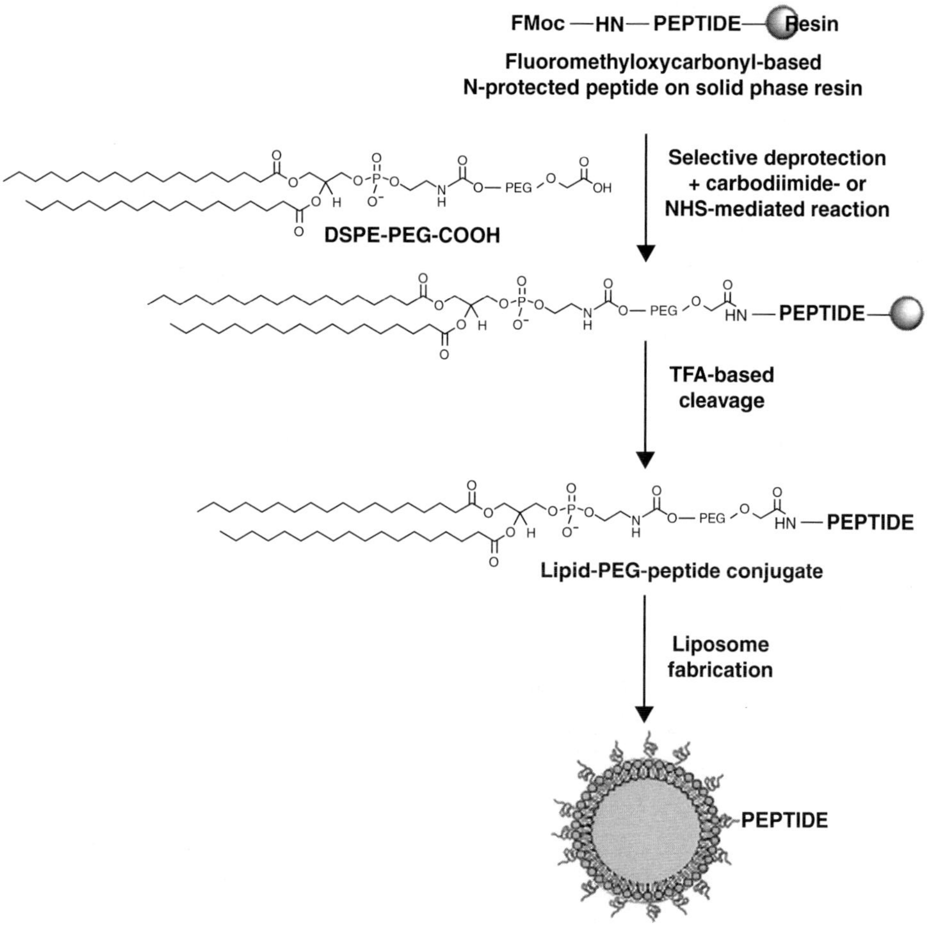

FMoc —HN— PEPTIDE—⬤Resin

**Fluoromethyloxycarbonyl-based
N-protected peptide on solid phase resin**

DSPE-PEG-COOH

**Selective deprotection
+ carbodiimide- or
NHS-mediated reaction**

**TFA-based
cleavage**

Lipid-PEG-peptide conjugate

**Liposome
fabrication**

PEPTIDE

SCHEME 6.6   liposome-peptide conjugation via amide

(e.g., peptides having a cysteine on the resin-free end) to lipid-bearing maleimide, bromo/iodo-acetyl, or 2-pyridylthio termini [74–76] as shown in Scheme 6.7. The hydrazide method and Schiff Base method that were described previously for conjugation of antibodies and proteins can also be used for conjugating peptides. All these strategies may allow for maintaining the directional specificity of the conjugated peptides toward achieving bioactivity and affinity to the corresponding target receptors. In addition, because of the small size of peptides compared to antibodies, multiple types of peptides can be conjugated onto the liposome or micelle surface without mutual interference of bioactivity, by first conjugating the various peptides to lipids using the above-described techniques and then coassembling the

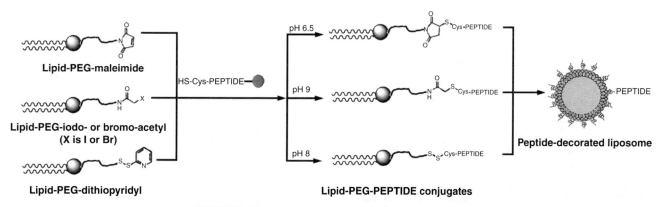

Lipid-PEG-maleimide

**Lipid-PEG-iodo- or bromo-acetyl
(X is I or Br)**

Lipid-PEG-dithiopyridyl

HS-Cys-PEPTIDE—⬤

pH 6.5

pH 9

pH 8

Lipid-PEG-PEPTIDE conjugates

Peptide-decorated liposome

PEPTIDE

SCHEME 6.7   liposome-peptide conjugate via sulfhydryl

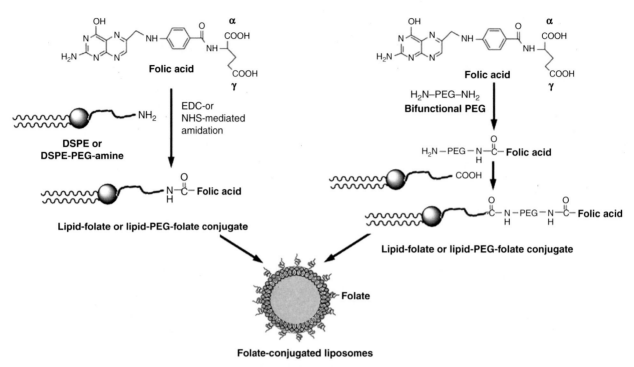

SCHEME 6.8 liposome-folate conjugation

various lipid–peptide conjugates into liposomes or micelles using a mixed lipid system [77–83].

Regarding conjugation of vitamins to lipids or liposomes for specific receptor targeting, the most important area of research has been the conjugation of folic acid (vitamin B9) for targeting folate receptors overexpressed on a variety of cancers [84–89]. Folic acid has an α-carboxyl and a γ-carboxyl group that can be utilized for conjugation to amine (-NH2) derivatized lipids or amine-decorated preformed liposomes and micelles (Scheme 6.8). It has been found that NHS can selectively activate the γ-carboxyl group for subsequent amide bond linkage [84]. An alternate strategy to conjugate folate to lipids or liposomes involves formation of folate-PEG-amine and subsequent reaction of the amine termini to carboxyl-derivatized lipids. In an additional strategy, folate-PEG-amines can also be reacted to cholesteryl chloroformates, and the resultant amide-linked cholesterol–folate conjugate can be incorporated in the liposomal bilayer. Lipid-soluble vitamins can also be hydrophobically incorporated within the liposome membrane bilayer.

The covalent and noncovalent approaches described in the previous sections for conjugation or incorporation of "stealth" property promoting polymers and cell-targeting ligands, have also been employed to conjugate chemical motifs on lipids and liposomes for biomedical imaging purposes. For example, chemical motifs like diethylenetriaminepentaacetic acid (DTPA) and tetraazacyclododecanetetraacetic acid (DOTA) that can form chelated complexes with the

T1-MRI contrast agent Gadolinium (Gd^{3+}) can be conjugated via a carboxyl (-COOH) terminus to the amine (-NH2) terminus of lipids like DSPE or DSPE-PEG-NH2 via carbodiimide-mediated chemistry [90–95]. These lipid–DTPA or lipid–DOTA conjugates can be assembled at control compositions to form liposomes that can be subsequently used to chelate multiple Gd-decorations on the liposome surface. In an extension of this strategy, polychelating polymers like DTPA-functionalized poly-L-lysine can be conjugated via amide linkage onto lipids like DSPE or stearylamide, thereby providing a way to amplify the amount of Gd that can be chelated onto the subsequent liposomes [19, 96]. Such a macromolecular assembly of MRI contrast agents is known to enhance the relaxivity (hence signal) of the imaging agent, compared to single molecule systems [97]. The same monochelating and polychelating agents can also be used for complexing with radioimaging agents like Indium (^{111}In) and Technetium (^{99}Tc) [96, 98].

Among other applications of the noncovalent strategies of surface bioconjugation, preformed liposomes made from cationic lipids like dioleoyltrimethylammonium propane (DOTAP), N-(2-hydroxyethyl)-N,N-dimethyl-2,3-bis(tetra decyloxy)-1-propanaminium (DMRIE), 2,3-dioleyloxy-N-(2-(sperminecarboxamido)ethyl)-N,N-dimethyl-1-propanim inium penta-hydrochloride (DOSPA), etc. have been used to complex with DNA for application as gene delivery vectors [99–102].

A novel bioconjugation technique that has become popular in recent years for ligand modifications of lipidic,

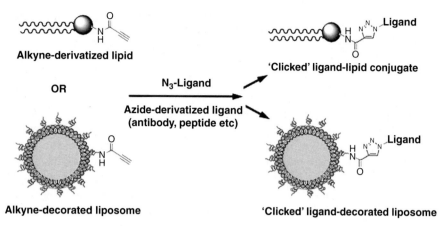

SCHEME 6.9   liposome-ligand click chemistry

polymeric, and inorganic vehicles is that involving azide–alkyne cycloaddition reaction, otherwise known as "click chemistry" [103–107]. The reaction is commonly catalyzed by copper, and usually involves lipid-bearing alkyne termini reacting with ligand motifs derivatized by azide termini to form a triazole linkage (Scheme 6.9). Another azide-based bioconjugation reaction that is recently generating interest among researchers for ligand modification of nanosystems is the Staudinger ligation mechanism where an azide terminus can react with a phosphine terminus to form a stable amide bond [108]. Although unique and promising in nature, these unconventional reactions may only have selective applications in ligand bioconjugation onto liposomes and other nanosystems, due to multiple steps involved in derivatizing the chemical moieties during scale-up.

While conjugation to lipids followed by self-assembly or conjugation onto surface-reactive functional groups on preformed liposomes or micelles are the most common strategies for presenting targeting motifs and bioactive agents on the vehicle surface, an alternative technique called "post insertion" has also been used [109–113]. In this technique, preformed unmodified (i.e., no surface bioconjugation) liposomes are incubated with ligand-conjugated lipid molecules such that over time, due to thermodynamic equilibrium between the liposome bilayer components and the ligand-conjugated lipid components (often present as small micellar aggregates), some molecules of the latter can get transferred and inserted within the former. This method is simple but has reduced control over ligand decoration density on the liposome surface.

6.3   MICROBUBBLES AND NANOBUBBLES

Another important nanomedicine area where lipids and lipidic bioconjugates have become important is that of microbubbles and nanobubbles used as contrast enhancement agents for ultrasound imaging [114–117]. For bioconjugation and targeting of these bubbles to disease-specific cells or sites for image contrast enhancement, two strategies have usually been used. In the first strategy, the lipidic and structural constituents of the bubble shell itself are engineered to facilitate their target attachment. For example, the incorporation of the negatively charged phospholipid phospatidylserine within the bubble shell has been shown to promote activation and adhesion of Complement system components to the shell and thereby induces bubble attachment to activated leukocytes [118–120]. In the second strategy, sterically stabilizing polymers (e.g., PEG) and target-specific molecular ligands like antibodies, antibody fragments, carbohydrates, peptides, and peptidomimetics are decorated on the bubble shell [121–125]. As before for liposomes and lipidic micelles, the ligands can be conjugated first to lipidic components followed by hydrophobic assembly around the bubble core, or the lipid-coated bubbles can be formed first followed by conjugation of ligands onto the surface. Also, the ligands can be presented directly on the bubble shell or can be presented tethered to PEG spacer moieties using covalent and noncovalent bioconjugation chemistry tools previously described for liposomes. Radioactive agent-derivatized lipids and MRI-agent-chelated lipids can be further incorporated along with ligand-modified lipids in the bubble shell, to allow for multimodal targeted molecular imaging [126, 127]. Also, using cationic and biotinylated lipid components in the bubble shell, DNA can be complexed for gene delivery applications [128]. Since air- or gas-filled bubbles have very limited stability in an *in vivo* environment and can collapse (cavitate) prematurely before reaching target site, some research efforts have recently been directed to conjugating cross-linkable polymers (e.g., polyvinylamine) onto the bubble surface to increase shell rigidity and hence bubble stability [129, 130]. These cross-linked shells can be further derivatized to chemically conjugate targeting motifs and inclusion compounds [131].

6.4 POLYMERSOMES

While the term "polymersome" was first coined in 1999, the prevalence of commercially available polymers with appropriate hydrophilic/total polymer mass ratios indicate that vesicles of this nature must have occurred previously, but their value in imaging and drug delivery was not recognized until the seminal discovery by the Hammer group in 1999. Other terms sometimes used to include polymersomes are dendrisomes and polymer vesicles. Since their introduction, publications generating and investigating new polymersomes have continued to rise with over 230 citations in PubMed by 2012 (Figure 6.1) including several reviews [5, 8, 132–137]. Comparatively this is a relatively new area of research (as opposed to over 1000 citations for "polymeric micelles"

and over 40,000 citations for liposomes in that same time period).

The main disadvantage of polymersomes in comparison to liposomes will always be questions about biocompatibility of the constituent polymers. However, the primary advantage of polymersomes over liposomes is that due to the synthetic nature of the constituent polymers they can be manufactured with a larger hydrophobic region than exists from using naturally occurring lipids. The hydrophobic region thickness of bilayers composed of naturally occurring lipids (e.g., C16 lipids) is about 3.5 nm. In one particular manifestation of polymersomes, this region thickness was calculated at 9.6 nm using synthetic block copolymers [138]. Tied to this is another advantage, namely ease of chemical conjugation or other functionalization of the polymers. Lipids

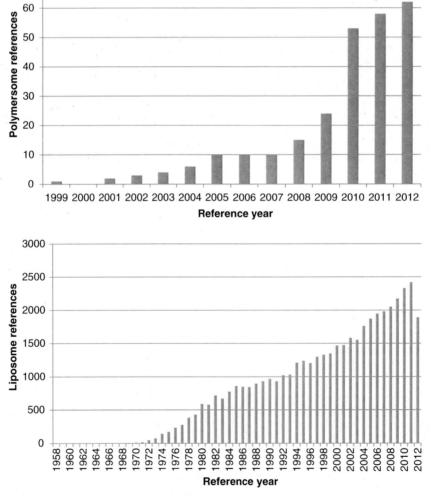

FIGURE 6.1 Steady increase in publications on liposomes and polymersomes. Polymersomes (top panel) were first proposed and generated in 1999, and now amounts to over 200 publications. Liposomes (bottom panel) were discovered in 1969 in electron miscopy studies and now amount to over 40,000 papers.

and liposomes are somewhat limited in chemistries available to add functionality such as stealth PEGylation, targeting functionalization, imaging capacity, etc. As discussed above, most functionalization of lipids occurs at the phosphate (i.e., hydrophilic end) with few opportunities for functionalization of the hydrophobic region (a strategy that helps maintain their fairly universal biocompatibility). In theory, polymersomes being generated from fully synthetic materials have the opportunity to be functionalized or otherwise modified anywhere within the molecule, however, in practice as of yet only a few functionalization strategies have been used.

While polymersomes and polymeric micelles both share this easy and facile chemical modification opportunity, the advantage polymersomes have over polymeric micelles is the combination of a defined hydrophobic region and defined hydrophilic region allowing easy opportunity for both kinds of payload. Hydrophilic payload is possible in polymeric micelles, but often requires complexation or covalent linkage of those hydrophilic payload molecules.

6.4.1 Conjugation to Polymer and/or Polymersome

Bioconjugation of polymersomes and polymers used in polymersomes has been a recent subject of an excellent review [139]. As with liposomes, by far the most common functionalization is through use of biotin–avidin/streptavidin/neutravidin for attachment of antibodies or antibody fragments [140–143]. Coupling chemistry is done by adding biotin, which has been either though:

1. N,N′-dicyclohexylcarbodiimide (DCC) esterification with 4-(dimethylamino)pyridine (DMAP) activation [144]
2. Preactivation of terminal hydroxyl with tresyl chloride [142], or 4-fluoro-3-nitrobenzoic acid [143]; followed by reaction with biocytin.

Although direct conjugation of antibodies has also occurred through amide conjugation [145], or cysteine thiol conjugation using maleimide [146], or vinyl sulfone [147]. Meng et al. also coupled antibodies by conjugation of Protein G and subsequent antibody association, and reported higher binding activity through this means than through direct antibody conjugation [145]. Egli et al. developed a novel hydrazone linkage to conjugate antibodies and other proteins; a bond with potential pH sensitivity. This reaction was done using succinimidyl ester-activated 4-formylbenzoate to attach to an amine at the polymer end, and succinimidyl ester-activated 6-hydrazinonicotinate at the protein end [141]. Christian et al. functionalized the hydroxyl end of PEG-containing polymers by use of a succinimidyl carbonate with DMAP activation. This resulted in a succinimidyl-functionalized polymer which was used

to conjugate peptides, in this case a TAT peptide to facilitate cell entry [148]. Conventional bifunctional cross-linkers are also frequently used such as SPDP (N-succinimidyl-3-(2-pyridyldithio) propionate) and SATA (N-succinimidyl-S-acetylthioacetate) [59].

As with liposomes, the most common exception to these basic functionalization chemistries is the use of the Huisgen 1, 3 polar cycloaddition between azides and alkynes in a click chemistry reaction. Work in this area has its own thorough review [149]. This has been done both with azide on the polymer side and alkyne on the ligand side [150]; and with alkyne on the polymer side and azide on the ligand side [151]. A few groups have begun to explore use of noncovalent metallochelate interactions (e.g., Histidine tag for the noncovalent association of oligonucleotides; and diethylene triamine penta acetic acid chelate for the chelation of the radiotracer (111)In)[152, 153]. Cyclodextrin-based host–guest complexes have also been explored as a means of generating noncovalent interactions through self-association [7, 154].

Di- and multifunctional conjugation has also been examined, such as Robbins et al. who used a dual ligand approach by adding antibodies through avidin as well as a carbohydrate ligand sialyl Lewis X [155]. Rajagopal et al. showed a trifunctional modification by conjugating three different fluorescent ligands: (a) an azide; (b) an amide conjugation with EDAC; and (c) a thiol conjugation with maleimide [156].

6.4.2 Intentional Incorporation of Functionality in the Synthesized Polymer

Another possibility to chemically tailor polymersomes is to incorporate functionality in the chemical synthesis of a vessel forming diblock, triblock, or multiarm copolymer. Perhaps the most prevalent functionality incorporated through polymer synthesis is the incorporation of PEG as the hydrophilic block, and in order to add stealth character to the final vesicle. Similar PEGylation strategies for liposomes have long been investigated to add stealth character to these vesicles and to thereby increase their circulation times and decrease aggregation. Some have argued that the "dyed-in-the-wool" approach of making polymersomes from PEG blocks could have a better result in reducing aggregation and increasing circulation when compared to a liposome which has been modified after the fact. There have been some observations to this effect [157]. However, a similar argument could be made that the high density of PEG chains when generating a bilayer of PEG block copolymers, reduces chain mobility and thereby decreases the stealth effect of the PEG [158]. Perhaps the best strategy to maintain a mixture of PEG blocks with different molecular weights, some shorter chained PEG blocks to be used solely for vesicle formation, with some chains with higher molecular weight PEG blocks to add stealth functionality [159].

Introduction of carbohydrate-containing blocks is also used for stealth purposes and has been generated using dextran [160, 161], hyaluronan [162, 163], maltoheptose [164], and others [160, 165].

Another functionality incorporated in the synthesized polymer is some manner of responsiveness. This could either be degradability, or response to input stimulus such as light, heat, pH change, etc. Nonspecific degradability function is generated by use of degradable polymers such as PCL [146, 166], PLA [145, 167, 168] and PLGA [169]; or through use of a disulfide linkage, which is expected to be reduced in the presence of cellular glutathione [135, 170]. Enzymatic degradability was specifically added by introducing peptide blocks with enzymatically degradable sequences such as lysozyme [171] or neutral proteases [166].

In closing, liposomes, lipids, polymersomes, and microbubbles are still a hot research topic in nanomedicine. The field represents a long history, cutting edge advances, and a promising future. Liposomes and microbubbles in many different formulations represent some of the few FDA approved micro and nanotherapeutics. Polymersomes are a more recent introduction but have also begun to see some of the functional diversity capable through bioconjugation, but have also seen new functionalization available due to their unique capacity of deriving fully from synthetic materials.

REFERENCES

1. Bangham AD, Horne RW, Glauert AM, Dingle JT, Lucy JA. *Nature* 1962;196:952–955. PMID: 13966357

2. Bangham AD, Horne RW. *J Mol Biol* 1964;8:660–668. PMID: 14187392

3. Discher BM, Won YY, Ege DS, Lee JC, Bates FS, Discher DE, Hammer DA. *Science* 1999;284:1143–1146. PMID: 10325219

4. Li F, Ketelaar T, Cohen Stuart MA, Sudhölter EJ, Leermakers FA, Marcelis AT. *Langmuir* 2008;24:76–82. PMID: 18052397

5. Onaca O, Enea R, Hughes DW, Meier W. *Macromol Biosci* 2009;9:129–139. PMID: 19107717

6. Zhou YF, Yan DY, Dong WY, Tian Y. *J Phys Chem B* 2007;111:1262–1270. PMID: ISI:000244039700004

7. Jin H, Liu Y, Zheng Y, Huang W, Zhou Y, Yan D. *Langmuir* 2012;28:2066–2072. PMID: 22129210

8. Discher DE, Ahmed F. *Annu Rev Biomed Eng* 2006;8:323–341. PMID: 16834559

9. Lasic DD, Martin F (eds). *Stealth Liposomes*, Boca Raton, FL: CRC Press; 1995.

10. Papahadjopoulos D, Allen TM, Gabizon A, Mayhew E, Matthay K, Huang SK, Lee KD, Woodle MC, Lasic DD, Redemann C. *Proc Natl Acad Sci USA* 1991;88:11460–11464. PMID: 1763060

11. Allen TM, Hansen C, Martin F, Redemann C, Yau-Young A. *Biochim Biophysica Acta* 1991;1066:29–36. PMID: 2065067

12. Moghimi SM, Szebeni J. *Prog Lipid Res* 2003;42:463–478. PMID: 14559067

13. Cullis PR, Chonn A, Semple SC. *Adv Drug Deliv Rev* 1998;32:3–17. PMID: ISI:000074610500002

14. Moghimi SM, Patel HM. *Adv Drug Deliv Rev* 1998;32:45–60. PMID: ISI:000074610500005

15. Papisov MI. *Adv Drug Deliv Rev* 1998;32:119–138. PMID: 10837639

16. Torchilin VP. *Handb Exp Pharmacol* 2010;197:3–53. PMID: 20217525

17. Torchilin VP, Trubetskoy VS. *Adv Drug Deliv Rev* 1995;16:141–155. PMID: ISI:A1995TD17100003

18. Woodle MC. *Adv Drug Deliv Rev* 1998;32:139–152. PMID: 10837640

19. Torchilin VP. *Nat Rev Drug Discov* 2005;4:145–160. PMID: 15688077

20. Zalipsky S. *Bioconjug Chem* 1993;4:296–299. PMID: 8218486

21. Veronese FM. *Biomaterials* 2001;22:405–417. PMID: ISI:000166722200001

22. Roberts MJ, Bentley MD, Harris JM. *Adv Drug Deliv Rev* 2002;54:459–476. PMID: 12052709

23. Wang R, Xiao R, Zeng Z, Xu L, Wang J. *Int J Nanomedicine* 2012;7:4185–4198. PMID: 22904628

24. Heath TD, Martin FJ. *Chem Phys Lipids* 1986;40:347–358. PMID: 3742677

25. Hashimoto Y, Endoh H, Sugawara M. Chemical methods for the modification of liposomes with proteins or antibodies.In: *Liposome Technology*, Vol. 3. Boca Raton, FL: CRC Press; 1993. pp 41–49.

26. Leserman LD, Machy P, Barbet J. Covalent coupling of monoclonal antibodies and protein A to liposomes: specific interaction with cells in vitro and in vivo. In: *Liposome Technology*, Vol. 3. Boca Raton, FL: CRC Press; 1993. pp 29–40.

27. Torchilin VP. Immobilization of specific proteins on a liposome surface: systems for drug targeting. In: *Liposome Technology*, Vol. 3. Boca Raton, FL: CRC Press; 1993. pp 75–90.

28. Murohara T, Margiotta J, Phillips LM, Paulson JC, DeFrees S, Zalipsky S, Guo LS, Lefer AM. *Cardiovasc Res* 1995;30:965–974. PMID: 8746213

29. Shimada K, Kamps JA, Regts J, Ikeda K, Shiozawa T, Hirota S, Scherphof GL. *Biochim Biophys Acta* 1997;1326:329–341. PMID: 9218563

30. Hsu MJ, Juliano RL. *Biochim Biophys Acta* 1982;720:411–419. PMID: 6896828

31. Vidal M, Sainte-Marie J, Philippot JR, Bienvenue A. *FEBS Lett* 1987;216:159–163. PMID: 3582664

32. Nishiya T, Sloan S. *Biochem Biophys Res Commun* 1996;224:242–245. PMID: 8694820

33. Zalipsky S, Mullah N, Harding JA, Gittelman J, Guo L, DeFrees SA. *Bioconjug Chem* 1997;8:111–118. PMID: 9095350

34. Lee RJ, Low PS. *J Biol Chem* 1994;269:3198–3204. PMID: 8106354

35. Harding JA, Engbers CM, Newman MS, Goldstein NI, Zalipsky S. *Biochim Biophys Acta* 1997;1327:181–192. PMID: 9271260

36. Phillips NC, Dahman J. *Immunol Lett* 1995;45:149–152. PMID: 7558165

37. Coulter A, Harris R. *J Immunol methods* 1983;59:199–203. PMID: 6341469

38. Rousseaux J, Rousseaux-Prevost R, Bazin H. *J Immunol methods* 1983;64:141–146. PMID: 6644029

39. Aragnol D, Leserman LD. *Proc Natl Acad Sci USA* 1986;83:2699–2703. PMID: 3458229

40. Winter G, Milstein C. *Nature* 1991;349:293–299. PMID: 1987490

41. Endoh H, Suzuki Y, Hashimoto Y. *J Immunol methods* 1981;44:79–85. PMID: 7252176

42. Dunnick JK, Mcdougall IR, Aragon S, Goris ML, Kriss JP. *J Nucl Med* 1975;16:483–487. PMID: ISI:A1975AD05200013

43. Manjappa AS, Chaudhari KR, Venkataraju MP, Dantuluri P, Nanda B, Sidda C, Sawant KK, Murthy RS. *J Controlled Release* 2011;150:2–22. PMID: ISI:000288641700002

44. Huang A, Huang L, Kennel SJ. *J Biol Chem* 1980;255:8015–8018. PMID: 7410345

45. Maruyama K, Takizawa T, Yuda T, Kennel SJ, Huang L, Iwatsuru M. *Biochim Biophys Acta* 1995;1234:74–80. PMID: 7880861

46. Koning GA, Morselt HW, Velinova MJ, Donga J, Gorter A, Allen TM, Zalipsky S, Kamps JA, Scherphof GL. *Biochim Biophys Acta* 1999;1420:153–167. PMID: 10446299

47. Hansen CB, Kao GY, Moase EH, Zalipsky S, Allen TM. *Biochim Biophys Acta* 1995;1239:133–144. PMID: ISI:A1995TF73100004

48. Chua MM, Fan ST, Karush F. *Biochim Biophys Acta* 1984;800:291–300. PMID: ISI:A1984TJ68500012

49. Sofou S, Sgouros G. *Expert Opin Drug Deliv* 2008;5:189–204. PMID: ISI:000259877800004

50. Torchilin V. *Expert Opin Drug Deliv* 2008;5:1003–1025. PMID: ISI:000259741600006

51. Sapra P, Tyagi P, Allen TM. *Curr drug deliv* 2005;2:369–381. PMID: 16305440

52. Sawant KK, Torchilin V. Design and synthesis of novel functional lipid-based bioconjugates for drug delivery and other applications. In: Mark SS (ed.), *Bioconjugation Protocols: Strategies and Methods*, Springer Science + Business Media, LLC; 2011. pp 357–378.

53. Domen PL, Nevens JR, Mallia AK, Hermanson GT, Klenk DC. *J Chromatogr* 1990;510:293–302. PMID: ISI:A1990DP51300032

54. Shahinian S, Silvius JR. *Biochim Biophys Acta* 1995;1239:157–167. PMID: ISI:A1995TF73100006

55. Kirpotin D, Park JW, Hong K, Zalipsky S, Li WL, Carter P, Benz CC, Papahadjopoulos D. *Biochemistry* 1997;36:66–75. PMID: ISI:A1997WB62300011

56. Martin F, Papahadjopoulos D. *J Biol Chem* 1982;37:286–288. PMID: ISI:A1982ND35600080

57. Barbet J, Machy P, Leserman LD. *J Supramol Struct Cell Biochem* 1981;16:243–258. PMID: ISI:A1981MR99900004

58. Jones MN, Hudson MJH. *Biochim Biophys Acta* 1993;1152:231–242. PMID: ISI:A1993MH05300005

59. Schwendener RA, Trüb T, Schott H, Langhals H, Barth RF, Groscurth P, Hengartner H. *Biochim Biophys Acta* 1990;1026:69–79. PMID: ISI:A1990DT10200010

60. Carlsson J, Drevin H, Axen R. *Biochem J* 1978;173:723–737. PMID: ISI:A1978FP83500008

61. Ansell SM, Tardi PG, Buchkowsky SS. *Bioconjug Chem* 1996;7:490–496. PMID: ISI:A1996VA23800013

62. Longman SA, Cullis PR, Bally MB. *Drug Deliv* 1995;2:156–165. PMID:

63. Longman SA, Cullis PR, Choi L, de Jong G, Bally MB. *Cancer Chemother Pharmacol* 1995;36:91–101. PMID: 7767956

64. Redelmeier TE, Guillet J-G, Ballyt MB. *Drug Deliv* 1995;2:98–109. PMID:

65. Kimelberg HK, Papahadjopoulos D. *J Biol Chemistry* 1971;246:1142–1148 (). PMID: ISI:A19711599400040

66. Hammes GG, Schullery SE. *Biochemistry* 1970;9:2555–2563. PMID: 5456729

67. Loughrey HC, Choi LS, Cullis PR, Bally MB. *J Immunol methods* 1990;132:25–35. PMID: ISI:A1990DW26300003

68. Loughrey HC, Wong KF, Choi LS, Cullis PR, Bally MB. *Biochim Biophys Acta* 1990;1028:73–81. PMID: ISI:A1990EC37100011

69. Takasaki J, Ansell SM. *Bioconjug Chem* 2006;17:438–450. PMID: ISI:000236226200025

70. Martin F. *Biophys J* 1981;33:A8–A8. PMID: ISI:A1981LA03900024

71. Huang A, Tsao YS, Kennel SJ, Huang L. *Biochim Biophys Acta* 1982;716:140–150. PMID: ISI:A1982NT08100004

72. Schuber F, Frisch B, Hassane FS. Coupling of peptides to surfaces of liposomes – applications to liposome-based synthetic vaccines. In: Gregoriadis G (ed.), *Liposome Technology, Volume II: Entrapment of Drugs and Other Materials into Liposomes*, Informa Healthcare: 2006, Vol. 2 pp 111–130.

73. Nobs L, Buchegger F, Gurny R, Allemann E. *J Pharm Sci* 2004;93:1980–1992. PMID: ISI:000223069700005

74. Yagi N, Yano Y, Hatanaka K, Yokoyama Y, Okuno H. *Bioorg Med Chem Lett* 2007;17:2590–2593. PMID: ISI:000246087900039

75. Riche EL, Erickson BW, Cho MJ. *J Drug Target* 2004;12:355–361. PMID: ISI:000224164300006

76. Wu HC, Chang DK. *Journal of oncology* 2010;2010:723–798. PMID: 20454584

77. Ravikumar M, Modery CL, Wong TL, Dzuricky M, Sen Gupta A. *Bioconjug Chem* 2012;23:1266–1275. PMID: ISI:000305358700020

78. Ravikumar M, Modery CL, Wong TL, Gupta AS. *Biomacromolecules* 2012;13:1495–1502. PMID: ISI:000303951600030

79. Modery CL, Ravikumar M, Wong TL, Dzuricky MJ, Durongkaveroj N, Sen Gupta A. *Biomaterials* 2011;32:9504–9514. PMID: ISI:000296684200035

80. Kibria G, Hatakeyama H, Ohga N, Hida K, Harashima H. *J Control Release* 2011;153:141–148. PMID: ISI:000293312900006

81. Kluza E, van der Schaft DW, Hautvast PA, Mulder WJ, Mayo KH, Griffioen AW, Strijkers GJ, Nicolay K. *Nano Lett* 2010;10:52–58. PMID: ISI:000273428700010

82. Saul JM, Annapragada AV, Bellamkonda RV. *J Control Release* 2006;114:277–287. PMID: ISI:000241177800001

83. Gunawan RC, Auguste DT. *Biomaterials* 2010;31:900–907. PMID: ISI:000273946100015

84. Duzgunes N (ed.). *Liposomes, Part A*, San Diego: Elsevier; 2003.

85. Low PS, Henne WA, Doorneweerd DD. *Acc Chem Res* 2008;41:120–129. PMID: ISI:000252419500014

86. Reddy JA, Dean D, Kennedy MD, Low PS. *J Pharm Sci* 1999;88:1112–1118. PMID: ISI:000083517500003

87. Turk MJ, Waters DJ, Low PS. *Cancer Lett* 2004;213:165–172. PMID: ISI:000224170900005

88. Sudimack J, Lee RJ. *Adv Drug Deliv Rev* 2000;41:147–162. PMID: ISI:000085933200003

89. Watanabe K, Kaneko M, Maitani Y. *Intl J Nanomedicine* 2012;7:3679–3688. PMID: ISI:000306409800001

90. Unger E, Needleman P, Cullis P, Tilcock C. *Invest* 1988;23:928–932. PMID: ISI:A1988R326600010

91. Grant CW, Karlik S, Florio E. *Magn Reson Med* 1989;11:236–243. PMID: 2779414

92. Kabalka GW, Davis MA, Moss TH, Buonocore E, Hubner K, Holmberg E, Maruyama K, Huang L. *Magn Reson Med* 1991;19:406–415. PMID: 1881329

93. Schwendener RA. *J Liposome Research* 1994;4:837–855. PMID:

94. Mulder WJ, Strijkers GJ, Griffioen AW, van Bloois L, Molema G, Storm G, Koning GA, Nicolay K. *Bioconjug Chem* 2004;15:799–806. PMID: 15264867

95. Bui T, Stevenson J, Hoekman J, Zhang S, Maravilla K, Ho RJ. *PLoS One* 2010;5: pii: e13082. PMID: ISI:000282312600017

96. Torchilin VP. *Adv Drug Deliv Rev* 1997;24:301–313. PMID: ISI:A1997WM65800022

97. Strijkers GJ, Mulder WJ, van Heeswijk RB, Frederik PM, Bomans P, Magusin PC, Nicolay K. *MAGMA* 2005;18:186–192. PMID: ISI:000232975400003

98. Torchilin VP. *Mol Med Today* 1996;2:242–249. PMID: ISI:A1996UQ80300006

99. Felgner PL, Rhodes G. *Nature* 1991;349:351–352. PMID: ISI:A1991EU50100060

100. Felgner PL, Gadek TR, Holm M, Roman R, Chan HW, Wenz M, Northrop JP, Ringold GM, Danielsen M. *Proc Natl Acad Sci USA* 1987;84:7413–7417. PMID: 2823261

101. Wasan EK, Fairchild A, Bally MB. *J Pharm Sci* 1998;87:9–14. PMID: 9452961

102. Safinya CR, Ewert K, Ahmad A, Evans HM, Raviv U, Needleman DJ, Lin AJ, Slack NL, George C, Samuel CE. *Philos Trans A Math Phys Eng Sci* 2006;364:2573–2596. PMID: 16973477

103. Huisgen R. *Angew Chem* 1963;2:633–645. PMID:

104. Frisch B, Hassane FS, Schuber F. *Methods Mol Biol* 2010;605:267–277. PMID: 20072887

105. Lallana E, Sousa-Herves A, Fernandez-Trillo F, Riguera R, Fernandez-Megia E. *Pharm Res* 2012;29:1–34. PMID: ISI:000298604200001

106. Hassane FS, Frisch B, Schuber F. *Bioconjug Chem* 2006;17:849–854. PMID: ISI:000237576000036

107. Lutz JF, Zarafshani Z. *Adv Drug Deliv Rev* 2008;60:958–970. PMID: ISI:000257135400003

108. van Berkel SS, van Eldijk MB, van Hest JC. *Angew Chem Int Ed Engl* 2011;50:8806–8827. PMID: 21887733

109. Moreira JN, Ishida T, Gaspar R, Allen TM. *Pharm Res* 2002;19:265–269. PMID: 11934232

110. Immordino ML, Dosio F, Cattel L. *Intl J Nanomedicine* 2006;1:297–315. PMID: 17717971

111. Allen TM, Sapra P, Moase E. *Cell Mol Biol Lett* 2002;7:217–219. PMID: 12097921

112. Pan X, Wu G, Yang W, Barth RF, Tjarks W, Lee RJ. *Bioconjug Chem* 2007;18:101–108. PMID: 17226962

113. Allen TM, Sapra P, Moase E, Moreira J, Iden D. *J Liposome Res* 2002;12:5–12. PMID: ISI:000177357000003

114. Calliada F, Campani R, Bottinelli O, Bozzini A, Sommaruga MG. *Eur J Radiol* 1998;27(Suppl 2):S157–S160. PMID: 9652516

115. Unger EC, Matsunaga TO, McCreery T, Schumann P, Sweitzer R, Quigley R. *Eur J Radiol* 2002;42:160–168. PMID: 11976013

116. Dijkmans PA, Juffermans LJ, Musters RJ, van Wamel A, ten Cate FJ, van Gilst W, Visser CA, de Jong N, Kamp O. *Eur J Echocardiogr* 2004;5:245–256. PMID: 15219539

117. Unger EC, Porter T, Culp W, Labell R, Matsunaga T, Zutshi R. *Adv Drug Deliv Rev* 2004;56:1291–1314. PMID: 15109770

118. Lindner JR, Dayton PA, Coggins MP, Ley K, Song J, Ferrara K, Kaul S. *Circulation* 2000;102:531–538. PMID: 10920065

119. Lindner JR, Song J, Xu F, Klibanov AL, Singbartl K, Ley K, Kaul S. *Circulation* 2000;102:2745–2750. PMID: 11094042

120. Christiansen JP, Leong-Poi H, Klibanov AL, Kaul S, Lindner JR. *Circulation* 2002;105:1764–1767. PMID: 11956115

121. Klibanov AL, Hughes MS, Marsh JN, Hall CS, Miller JG, Wible JH, Brandenburger GH. *Acta Radio Suppl* 1997;412:113–120. PMID: 9240089

122. Kaufmann BA, Sanders JM, Davis C, Xie A, Aldred P, Sarembock IJ, Lindner JR. *Circulation* 2007;116:276–284. PMID: 17592078

123. Klibanov AL. *Medical Biol Eng Comput* 2009;47:875–882. PMID: 19517153

124. Chen CC, Borden MA. *Langmuir* 2010;26:13183–13194. PMID: 20695557

125. Myrset AH, Fjerdingstad HB, Bendiksen R, Arbo BE, Bjerke RM, Johansen JH, Kulseth MA, Skurtveit R. *Ultrasound Med Biol* 2011;37:136–150. PMID: 21144962

126. Ferrara KW, Borden MA, Zhang H. *Acc Chem Res* 2009;42:881–892. PMID: 19552457

127. Cai W, Chen X. *J Nucl Med* 2008;49(Suppl 2):113S–128S. PMID: 18523069

128. Nomikou N, Tiwari P, Trehan T, Gulati K, McHale AP. *Acta Biomater* 2012;8:1273–1280. PMID: 21958669

129. Cavalieri F, El Hamassi A, Chiessi E, Paradossi G. *Langmuir* 2005;21:8758–8764. PMID: ISI:000231789800033

130. Cavalieri F, et al. *Macromolecular Symposia* 2006;234:94–101. PMID: ISI:000236986800014

131. Cavalieri F, El Hamassi A, Chiessi E, Paradossi G, Villa R, Zaffaroni N. *Biomacromolecules* 2006;7:604–611. PMID: ISI:000235538600030

132. Levine DH, Ghoroghchian PP, Freudenberg J, Zhang G, Therien MJ, Greene MI, Hammer DA, Murali R. *Methods* 2008;46:25–32. PMID: 18572025

133. Christian DA, Cai S, Bowen DM, Kim Y, Pajerowski JD, Discher DE. *Eur J Pharm Biopharm* 2009;71:463–474. PMID: 18977437

134. Meng F, Zhong Z, Feijen J. *Biomacromolecules* 2009;10:197–209. PMID: 19123775

135. Cheng R, Feng F, Meng F, Deng C, Feijen J, Zhong Z. *J Control Release* 2011;152:2–12. PMID: 21295087

136. Liao J, Wang C, Wang Y, Luo F, Qian Z. *Curr Pharm Des* 2012;18:3432–3441. PMID: 22632981

137. Gupta M, Agrawal GP, Vyas SP. *Curr Mol Med* 2013;13:179–204. PMID: 22834834

138. Ghoroghchian PP, Frail PR, Susumu K, Blessington D, Brannan AK, Bates FS, Chance B, Hammer DA, Therien MJ. *Proc Natl Acad Sci USA* 2005;102:2922–2927. PMID: 15708979

139. Egli S, Schlaad H, Bruns N, Meier W. *Polymers* 2011;3:252–280. PMID:

140. Hammer DA, Robbins GP, Haun JB, Lin JJ, Qi W, Smith LA, Ghoroghchian PP, Therien MJ, Bates FS. *Faraday Discuss* 2008;139:129–141. PMID: 19048993

141. Egli S, Nussbaumer MG, Balasubramanian V, Chami M, Bruns N, Palivan C, Meier W. *J Am Chem Soc* 2011;133:4476–4483. PMID: 21370858

142. Lin JJ, Silas JA, Bermudez H, Milam VT, Bates FS, Hammer DA. *Langmuir* 2004;20:5493–5500. PMID: 15986691

143. Lin JJ, Ghoroghchian PP, Zhang Y, Hammer DA. *Langmuir* 2006;22:3975–3979. PMID: 16618135

144. Broz P, Benito SM, Saw C, Burger P, Heider H, Pfisterer M, Marsch S, Meier W, Hunziker P. *J Control Release* 2005;102:475–488. PMID: 15653165

145. Meng FH, Engbers GHM, Feijen J. *J Control Release* 2005;101:187–198. PMID: ISI:000226086800017

146. Pang Z, Lu W, Gao H, Hu K, Chen J, Zhang C, Gao X, Jiang X, Zhu C. *J Control Release* 2008;128:120–127. PMID: 18436327

147. Petersen MA, Yin LG, Kokkoli E, Hillmyer MA. *Polym Chem* 2010;1:1281–1290. PMID: ISI:000282612700015

148. Christian NA, Milone MC, Ranka SS, Li G, Frail PR, Davis KP, Bates FS, Therien MJ, Ghoroghchian PP, June CH, Hammer DA. *Bioconjug Chem* 2007;18:31–40. PMID: 17226955

149. Opsteen JA, Brinkhuis RP, Teeuwen RL, Lowik DW, van Hest JC. *Chem Commun (Camb)* 2007;14:3136–3138. PMID: 17653366

150. van Dongen SF, Teeuwen RL, Nallani M, van Berkel SS, Cornelissen JJ, Nolte RJ, van Hest JC. *Bioconjug Chem* 2009;20:20–23. PMID: 19099498

151. van Dongen SFM, Nallani M, Schoffelen S, Cornelissen JJLM, Nolte RJM, van Hest JCM. *Macromolecular Rapid Communications* 2008;29:321–325. PMID: ISI:000253701900008

152. Brinkhuis RP, Stojanov K, Laverman P, Eilander J, Zuhorn IS, Rutjes FP, van Hest JC. *Bioconjug Chem* 2012;12. PMID: 22463082

153. Tanner P, Ezhevskaya M, Nehring R, Van Doorslaer S, Meier W, Palivan C. *J Phys Chem B* 2012;116:10113–10124. PMID: ISI:000307749100026

154. Felici M, Marza-Perez M, Hatzakis NS, Nolte RJ, Feiters MC. *Chemistry* 2008;14:9914–9920. PMID: 18810732

155. Robbins GP, Saunders RL, Haun JB, Rawson J, Therien MJ, Hammer DA. *Langmuir* 2010;26:14089–14096. PMID: 20704280

156. Rajagopal K, Christian DA, Harada T, Tian A, Discher DE. *Int J Poly Sci* doi:10.1155/2010/379286(2010). PMID:

157. Lee JC, Bermudez H, Discher BM, Sheehan MA, Won YY, Bates FS, Discher DE. *Biotechnol Bioeng* 2001;73:135–145. PMID: 11255161

158. Holland NB, Xu Z, Vacheethasanee K, Marchant RE. *Macromol* 2001;34:6424–6430. PMID: ISI: 000170662100044

159. Chang YW, Silas JA, Ugaz VM. *Langmuir* 2010;26:12132–12139. PMID: 20578755

160. Schatz C, Louguet S, Le Meins JF, Lecommandoux S. *Angew Chem Int Ed Engl* 2009;48:2572–2575. PMID: ISI:000264661600028

161. Upadhyay KK, Le Meins JF, Misra A, Voisin P, Bouchaud V, Ibarboure E, Schatz C, Lecommandoux S. *Biomacromolecules* 2009;10:2802–2808. PMID: 19655718

162. Upadhyay KK, Bhatt AN, Castro E, Mishra AK, Chuttani K, Dwarakanath BS, Schatz C, Le Meins JF, Misra A, Lecommandoux S. *Macromol Biosci* 2010;10:503–512. PMID: ISI:000278165200007

163. Upadhyay KK, Bhatt AN, Mishra AK, Dwarakanath BS, Jain S, Schatz C, Le Meins JF, Farooque A, Chandraiah G, Jain AK, Misra A, Lecommandoux S. *Biomaterials* 2010;31:2882–2892. PMID: ISI:000275777800020

164. Otsuka I, Fuchise K, Halila S, Fort S, Aissou K, Pignot-Paintrand I, Chen Y, Narumi A, Kakuchi T, Borsali R. *Langmuir* 2010;26:2325–2332. PMID: 20141199

165. Sun X, Zhang, N. *Mini Rev Med Chem* 2010;10:108–125. PMID: 20408796

166. Katz JS, Zhong S, Ricart BG, Pochan DJ, Hammer DA, Burdick JA. *J Am Chem Soc* 2010;132:3654–3655. PMID: 20184323

167. Ahmed F, Pakunlu RI, Srinivas G, Brannan A, Bates F, Klein ML, Minko T, Discher DE. *Mol Pharm* 2006;3:340–350. PMID: ISI:000203539400015

168. Meng FH, Hiemstra C, Engbers GHM, Feijen J. *Macromol* 2003;36:3004–3006. PMID: ISI:000182646600004

169. Yu Y, Pang Z, Lu W, Yin Q, Gao H, Jiang X. *Pharm Res* 2012;29:83–96. PMID: 21979908

170. Meng F, Hennink WE, Zhong Z. *Biomaterials* 2009;30:2180–2198. PMID: 19200596

171. Lee JS, Groothuis T, Cusan C, Mink D, Feijen J. *Biomaterials* 2011;32:9144–9153. PMID: 21872328

7

ORGANIC NANOPARTICLE BIOCONJUGATE: MICELLES, CROSS-LINKED MICELLES, AND NANOGELS

MARIA VAMVAKAKI

Institute of Electronic Structure and Laser, Foundation for Research and Technology, Hellas, Heraklion, Crete, Greece
Department of Materials Science and Technology, University of Crete, Heraklion, Crete, Greece

Major advances in modern materials science focus in the preparation of bioconjugates combining synthetic polymers with biological macromolecules such as peptides, proteins, polysaccharides, and oligonucleotides. These systems can direct structure formation of the material leading to complex, hierarchically organized structures that are inaccessible by the synthetic polymers alone, and are attractive candidates for use in nanobiotechnology, drug delivery, gene therapy, tissue engineering, and the development of biosensors. However, the use of such systems in biomedical applications requires that the bioconjugates are of uniform size and composition and water compatible. Besides, the precise control of the materials' properties is dictated by their design at a molecular level which allows control over their nanoscale organization and structure. Recent advances in polymer chemistry and a broad range of orthogonal methods, including controlled/"living" polymerization techniques as well as conjugation/activation chemistries, have been employed to prepare polymer–biomolecule hybrids of controlled structure and specific functionalities leading to the formation of well-defined self-assembled structures via the multilevel ordering of these materials, by long- and short-range interactions.

7.1 INTRODUCTION

One of the major challenges for modern materials scientists in the development of novel and more complex nanostructured materials and devices is to understand and use the mechanisms of "Mother Nature." Natural evolution has led to the selection of "smart" complex molecules, containing specific functional groups, which by both molecular self-assembly and additional physical processes create materials exhibiting ordered structures on a hierarchy of length scales from the molecular scale, up to the macroscopic level. This high order structuring provides the performance and functions needed in living matter. However, to date, synthetic materials scientists are still far from being able to imitate even the most basic biological processes. The major difference in the two approaches is that "Mother Nature" uses highly functional complex molecules (proteins, DNA, carbohydrates, etc.) and relatively simple and robust processing routes to create her objects, whereas traditional materials scientists apply sophisticated processing routes to achieve structure formation and transformation from simple elements. Recent progress in Materials Chemistry has focused on the development of a new class of hybrid materials, known as molecular "chimeras," that combine the functions of the biopolymers with the properties and economical advantages of artificial polymers [1, 2]. These molecules are linked together via a bioconjugation process to form a novel complex with superior properties. Two main types of hybrid bioconjugates have been developed: random bioconjugates, in which the polymer is bound on different sites of the biomolecule, and site-specific bioconjugates, where the synthetic polymer is covalently linked to specific groups of the biological molecules [3, 4].

Traditional bioconjugation chemistries employed carboxyl or amine functional groups introduced into the polymer and biomolecule and the use of cabrodiimide-activated

Chemistry of Bioconjugates: Synthesis, Characterization, and Biomedical Applications, First Edition. Edited by Ravin Narain.

coupling reactions to link them together, whereas lately clean, efficient, and bioorthogonal conjugation via click reactions have been developed for the modification of biomolecules [5, 6] and synthetic polymers [7, 8]. Due to their high efficiency, specificity, mild reaction conditions, and tolerance of functional groups [9, 10], the latter reactions have allowed to maximize efficacy and reproducibility and eliminate undesirable side reactions, and nonspecific binding. In general, a bioconjugation reaction should reach completion in short time, under mild aqueous conditions, should not be prone to competing reactions (i.e., hydrolysis), and should be highly chemoselective and orthogonal to other functionalization chemistries. Ideal conjugate chemistries proceed efficiently at 1:1 stoichiometries and at low concentrations, and have short reaction times at ambient or physiological temperature in aqueous media. To achieve this they make use of a catalyst or strained bonds in the reactive functional groups to drive the coupling reaction. On the other hand, highly selective reactions, known as bioorthogonal [5, 11, 12], are those that do not have significant reactivity toward functional groups found commonly in biomolecules (i.e., hydroxyl, carboxyl, amine, and thiol). The best candidates for such bioorthogonal reactions are the cycloaddition reactions which enable clean and efficient conjugation [13]. Classical synthetic organic chemistry cycloaddition and ligation reactions have been extensively employed in bioconjugation. These are often combined with standard bioconjugation approaches to provide an even greater degree of control in bioconjugate synthesis. Another approach uses synthetic biomolecules (oligonucleotides and oligopeptides) that incorporate functional groups that are not naturally occurring, thus providing high reaction specificity. In proteins the incorporation of unnatural amino acids possessing, for example, an azide or alkyne functionality on the side chain allows for such bioorthogonal conjugation reactions [14].

Given that much of biology is based upon hierarchy, with nanoscale structures playing key roles, synthetic nanostructured materials are being aggressively pursued. Structural characteristics and dimensions close to those found in biological materials are targeted which are then expected to translate into similarities in function. Another important feature that is vital for the fate of these materials when placed into a biological environment is surface chemistry. Polymeric materials conjugated with sugars [15, 16], peptides [17], oligonucleotides [18, 19], and other biological moieties have received particular attention to generate nanostructures that exhibit specific polyvalent interactions [20], and can function in complex matrices. Control over the nanostructure size (ten to a few hundred nanometers) and morphology allows to develop nanocontainers, delivery vehicles, and imaging agents of similar dimensions as biologically functional nanomaterials. However, the synthesis of such nanoparticulate bioconjugates requires ideal conjugation chemistries which allow control over the attachment point of the biomolecules

onto the nanoparticle, the orientation of the biomolecule, and the average number of molecules per nanoparticle. Control over these properties is of paramount importance in optimizing the function of nanoparticulate bioconjugates in applications such as diagnostic imaging, sensing, and drug delivery. Moreover, in certain applications (i.e., drug delivery) the linkage between the nanoparticle and the biomolecule can be labile under appropriate conditions, temperature pH, light, and others. In other cases, the conjugation of multiple functional groups is required to enhanced solubility, activity, targeting, etc. Therefore, the continuous development of new bioconjugate chemistries is important for the preparation of appropriate nanoparticulate bioconjugates for biological applications.

Biomolecules that can be covalently linked to synthetic polymers include proteins [21] and oligopeptides [22], lipids [23], drug molecules [24], sugars [25] and polysaccharides [26], oligonucleotides [19] and DNA [18]. Polymer–biomolecule bioconjugates can direct structure formation leading to complex, hierarchically organized nanostructures, that are inaccessible by the synthetic polymer or the biomolecule alone, and are attractive candidates for use in nanobiotechnology, such as drug delivery, gene therapy, tissue engineering, imaging, and biosensor development, but also in catalysis, water remediation, and other applications. The interest in these nanostructures is driven by the combination of nanoscale size, synthetic flexibility, and diverse physical and chemical properties including the response to external stimuli, the capture and release of hydrophilic or hydrophobic cargo, biodegradability, and cell targeting. For example, the targeted delivery of drug-encapsulated nanoparticles from controlled-release polymer systems to diseased cells may allow to maximize the therapeutic efficacy of the drug while reducing its side effects [27, 28].

This chapter gives a general overview on the design of nanoparticulate bioconjugates, and the orthogonal chemistries used for their materialization. Nanoparticulate materials of interest include micelles, nanogels, dendrimers, liposomes and polymersomes, and viral capsids [29–32] associated with one or more biologically relevant molecules at the interface to give a nanoparticulate bioconjugate. Micelles are spherical nanoassemblies comprising amphiphilic surfactant or block copolymer molecules with a hydrophobic interior core and a hydrated hydrophilic corona. Their advantage is the ability to tune their dimensionality across the nanoscale. Nanogels are colloidal nanoparticles, with size between 100 nm and 1 μm, composed of a cross-linked hydrophilic polymer network (hydrogel). Dendrimers are fractal-like, hyperbranched polymers that form spherical nanoparticles and display a high density of surface functional groups. Liposomes are unilamellar spherical nanoparticles comprising a bilayer of lipids that separate an internal aqueous phase from the bulk aqueous solvent. Polymersomes are bilayer structures analogous to liposomes, consisting of synthetic amphiphilic block copolymers. Finally, viral capsids are the protein shells of viruses which enclose the virus genetic material. The applications and effectiveness of the above systems are determined by the ability to control the size, shape, and biomolecule orientation and activity of the assembled nanostructures.

7.2 POLYMER–PROTEIN NANOPARTICULATE BIOCONJUGATES

Proteins exhibit unrivaled diversity in terms of their structures and functions, and high specificity which render them ideal candidates for use in numerous applications such as catalysts to accelerate chemical transformations, targeting agents to deliver a cargo to specific cells, recognition sites to bind specific analytes in complex mixtures and complex materials to generate and elaborate three-dimensional structures via self-assembly. The introduction of such capabilities into the context of new materials is highly desirable, and as a result proteins and their oligomeric analogs, oligopeptides, have been extensively employed as a library of building blocks for the synthesis of novel functional polymer–biomolecule conjugates which augment the native functions of proteins through the covalent attachment of linear or branched synthetic polymer chains [33–36].

The advantages of bioconjugates over unmodified proteins, include: (i) increased solubility and *in vivo* stability [37–39] conferred by the protective/stabilizing polymer layer by preventing the enzymatic degradation or unfolding of the protein/peptide segment [34, 40]; (ii) the ability to modulate protein activity [41] for applications in sensing and imaging; (iii) protein recovery [42]; (iv) use in catalysis for enzymatic processes in organic media owing to their improved solubility in nonaqueous solvents [43, 44]; (v) increased half-life profiles in the bloodstream, bioavailability and accumulation in a target tissue [45] for therapeutic applications [1, 33, 46]; (vi) the ability to control nanostructure formation in solution and on a surface [47, 48]. Bioconjugation and protein derivatization are fundamental for many biotechnological applications such as, for example, in medicine [33, 49–51], for the development of affinity chromatography media [52], immunoassays [53], biosensors [54], industrial biocatalysts [55], and in oil/water biphasic reactions [56].

Among the different strategies employed for the modification of proteins, the covalent attachment of poly(ethylene glycol) (PEG), known as PEGylation, has been most extensively employed for the development of advanced therapeutics. PEGylation was initially introduced in 1970 [57], and since then it has developed rapidly, yielding significant therapeutic benefits [58, 59]. PEGylated proteins are known to improve pharmacokinetics and biodistribution, prolong plasma half-life, increase protein solubility and reduce immunogenicity and proteolytic degradation [60].

Over the years, several different chain lengths of PEG have been examined with various linking chemistries have been employed [60–62]. In the initial studies PEG succinimidyl succinate and PEG succinimidyl carbonate were used to prepare randomly modified PEGylated proteins, by their conjugation on the lysine residues on the protein [63]. This was a relatively simple synthetic approach which however, presents certain disadvantages such as lack of selectivity in conjugation, diol contamination, unstable linkages, and loss of the protein bioactivity and receptor recognition [59]. Moreover, besides the choice of the target amino acid in the protein, the type of chemical bond and the polymer molecular weight can also influence the bioactivity of the final product. For example, when high molecular weight polymers are used, the polymeric environment can cause protein destabilization with partial to total unfolding, or it may prevent the interaction with the substrate or receptor. To overcome these problems, PEGs with midchain protein-reactive functional groups were developed. These bioconjugates were found to protect more effectively the protein surface by an "umbrella-like effect," thus enhancing the selectivity toward functionalities on the protein surface, which leads to higher bioactivity [64, 65], and extending the protein circulation half-life times [40]. The main disadvantage of this midfunctionalization of PEG is the complex synthetic procedures, involving multistep organic reactions of the hydroxyl groups of linear PEGs and branch agents such as lysine, which require multiple purification processes. Therefore, the size of the polymer [66, 67] and the chemistry and location for its grafting must be optimized in order to minimize the negative effects. This was confirmed by devising a chemical strategy to obtain mono-PEG polypeptide modification where both the site of polymer grafting (α-terminal or lysil-ε-primary amines) and the type of chemical bond (alkylation or acylation) were varied. This principle of tailored chemistry has been applied for the monomethoxypoly(ethylene glycol) (MePEG) derivatization of the somatostatin analog RC-160 (Vapreotide) an octapeptide with potential application as a therapeutic agent for the treatment of patients with growth hormone (GH) secreting adenomas or hormonal hypersecretion induced by various neuroendocrine tumors [68]. Selective BOC protection of the two available primary amines of the peptide, followed by reaction with two different MePEG reagents and removal of the protecting groups, was carried out. Four mono-PEG derivatives of RC-160 were synthesized in which the α- or ε-amino functions were modified either through an acyl or an alkyl bond. This approach allowed to study the effects of the type and location of polymer grafting on the biological activity of different mono-PEG derivatives. Moreover, the problem of the preservation of the biological activity of biomolecules after coupling with bulky polymers, in particularly when small peptides are employed, was addressed. By a combination of chemical, structural, and biological analysis by measuring their capability to inhibit GH release from rat anterior pituitary cells, only one of the four synthesized products was identified to have characteristics more suitable for further therapeutic development. More precisely, circular dichroism analyses demonstrated the importance of the integrity of α-amino function in the peptide, while *in vitro* biological tests proved that the preservation of lysine's charge is fundamental for the activity. Thus the most promising bioconjugate was the mono-ε-lysil-PEGylated form, obtained by reductive alkylation, where the amine's positive charge is preserved suggesting the importance of polymer chain length and linkage chemistry in the development of new polymer–peptide conjugates with improved pharmacological properties.

The most widely used synthetic approaches for the preparation of polymer–protein bioconjugates involve the reaction of preformed semitelechelic polymers with specific amino acid residues along the protein chain. A frequent chemical strategy involves the modification of lysil and/or α-terminal primary amines of the peptide/protein because their acylation leads to conjugates with desired properties and good residual biological activity [69]. Lysine side chains are usually targeted with amine-reactive polymer end-groups such as activated esters, isocyanates, or through reductive amination [60, 70]. However, a protein may contain several internal lysine residues in addition to the N-terminal amine. Therefore, this approach is often nonspecific, and the resultant protein–polymer conjugate is heterogeneous in the number and placement of the polymer chains.

Nowadays, it has been acknowledged that such heterogeneity in the structure reflects on the physical and biological properties of the conjugate and often results in decreased protein activity [71]. Hence, creating well-defined bioconjugates is important, and site-specific modification of the protein is a better approach to prepare such biomolecules. Techniques to prepare well-defined protein–polymer conjugates provide biomolecules with homogeneous structures and activities and various creative methods have been explored for their preparation. Besides, the site-specific modification of proteins is also important for their directed immobilization onto surfaces and ensures that the biorecognition sites are accessible [72]. Researchers have been using reversible protection group chemistry to preserve the critical amino acid residues during the chemical coupling reaction [73], or alternatively, the modification of sites that are known to be less important for the protein function are targeted [74, 75]. In general, two methods are used in tandem to accomplish the preparation of site-specific nanoparticulate bioconjugates: (i) synthesis of polymers with defined architectures and molecular weights and (ii) attachment of the polymers at specific locations on the protein. In this context, polymers containing protein ligands, such as biotin for interaction with streptavidin [76, 77], or Ni^{2+} for interaction with polyhistidine-tagged

recombinant proteins [78], have been demonstrated for site-specific polymer attachment. Other examples include oxime formation by reaction of ketone-modified tyrosine residues [79] or lysine residues [71, 80] with aminooxy end-functionalized polymers. Cysteine residues are also frequently targeted for site-specific conjugation by exploiting thiol chemistry. Proteins contain very few, if any, cysteines that do not participate in disulfide bonds. Therefore, by targeting free cysteines the number and placement of polymer chains on the protein can be precisely determined. In addition, if a protein lacks free thiols for conjugation, genetic engineering can incorporate cysteine residues in specific protein positions, for example away from the active site, such that a polymer can be attached without hindering protein activity [33, 81]. Reactive groups on the polymer which are often employed for thiol conjugation include iodoacetamide [82], vinyl sulfone [83–85], and maleimide [86, 87] which result in thioether bond formation via Michael addition. Reversible disulfide bonds can be also formed by modification of thiols with activated disulfides such as pyridyl [88], alkoxycarbonyl, or *o*-nitrophenyl [89].

The reaction of active thiol groups onto the protein with maleimide-functionalized PEG has been employed for the development of an artificial nanoparticulate O_2 carrier, comprising human serum albumin (HSA) [90]. HSA is found in our blood plasma at high concentrations and when complexed with tetrakis(*o*-amidophenyl)porphinatoiron(II) (FeXP) it can reversibly bind and release O_2, in a fashion similar to Hb, under physiological conditions. The administration of this synthetic O_2 carrier by rats has proved its safety and O_2-transporting efficacy [91]; however, the FeXP molecule can easily dissociate from HSA when introduced into the animals due to the presence of weak noncovalent interactions within the hydrophobic cavity of albumin. It was shown that the conjugation of maleimide-terminated PEG onto an artificial hemoprotein, composed of HSA including FeXP, by reacting 2-iminothiolane with the amino groups of Lys to create active thiol groups, which bind to α-maleimide-*ö*-methoxy PEG, resulted in the surface modification of HSA-FeXP and helped to prolong the circulation life of HSA-FeXP and thereby retain its O_2-transporting ability for a long period. This was attributed to the presence of flexible polymer chains on the HSA surface which retard the association reaction of O_2 to FeXP and stabilize the oxygenated complex rendering it a promising material for the development of synthetic O_2 carriers.

Alternatively, techniques [92, 93] for the incorporation of unnatural amino acids provide another powerful solution toward optimal nanoparticulate polymer–protein conjugates, as they allow the introduction of unique chemical functionalities which participate in reactions that disregard native protein groups. Microorganism protein engineering to include functionalities for polymer attachment, while preserving the protein function comprises a common approach for the development of such well-defined bioconjugates which demonstrate superior pharmacological properties *in vivo*. This method has been employed for the preparation of a novel recombinant arginine deiminase (ADI) from *Mycoplasma arthritidis* which allowed the design of an arginine-catabolizing enzyme bioconjugate bearing a stable succinimidyl carbonate linker for PEG derivatization [94]. Arginine is an important metabolic target in the normal function of several biological systems, and arginine deprivation has been investigated in animal models and human clinical trials for its effects on inhibition of tumor growth [95–99], angiogenesis [100], or nitric oxide synthesis [101]. The apparent arginine auxotrophy of melanoma and hepatocellular carcinoma cells may offer an effective method for selective growth inhibition by enzymatic depletion of plasma arginine [102, 103]. Besides, mammalian arginase has been shown to be effective in inhibiting growth of cancer cell lines *in vitro* as well as in experimental tumors in rodents; however, its therapeutic capacity is decreased due to its immunogenicity, limited stability and circulating life, and suboptimal biodistribution/bioavailability. Furthermore, all of these enzymes present drug delivery and immunogenicity challenges that may be addressed by bioconjugate chemistry. Multi-PEGylated derivatives of ADI have been prepared with 12 or 20 kDa MePEG polymers using linear succinimidyl carbonate linkers and were examined in terms of their enzymatic and biochemical properties *in vitro*, as well as pharmacokinetic and pharmacodynamic behavior in rats and mice. The bioconjugates were investigated via intravenous, intramuscular, or subcutaneous administration in rodents. In contrast to other studies [96, 104], no benefit was found when increasing total PEG mass, possibly due to the presence of the stable succinimidyl carbonate linker [105, 106]. However, a benefit in terms of both the catabolic activity and preparation of the experimental therapeutics was revealed when decreasing the PEG size and number of attached chains with 60% retention of the enzyme activity when 12 kDa PEG chains were attached on 33% of the primary amines.

Recently, other "click" reactions operating under mild conditions and with high coupling efficiencies have been employed for nanoparticulate bioconjugate formation.

Each of these methods is drawn from an important set of reactions that proceed in aqueous solution with excellent functional group tolerance [107]. Several of these strategies have been utilized in the biological context, including the condensation of ketones with hydrazides and alkoxyamines [108], the [3 + 2] cycloaddition of alkynes and azides to form triazoles [109–111], and the formation of amides through a modified Staudinger reduction of azides with triarylphosphines [112–115]. Huisgen 1,3 dipolar cycloaddition between azides and alkynes has been employed to prepare amphiphilic bovine serum albumin (BSA)-polystyrene

(PS) [116] and superoxide dismutase (SOD)-PEG [117] bioconjugates. The nonnaturally occurring groups were incorporated by reaction of a maleimide-functionalized alkyne with the free thiol of BSA and site-directed mutagenesis to install *p*-azidophenylalanine into SOD, respectively.

Another efficient protein modification reaction that targets a bioorthogonal functional group under mild reaction conditions is based on the oxidative coupling of anilines [118]. This group is particularly attractive for selective bioconjugation given the successful incorporation of 4-aminophenylalanine into proteins by a bacterium that can biosynthesize this amino acid and carry it through protein translation [119]. The development of a rapid, chemoselective, and high yield reaction with anilines under mild conditions involved the coupling of a phenylenediamine derivative with two alkyl groups with primary anilines. The final product resulted upon nucleophilic addition of water after subsequent oxidation and exhibited remarkable stability over the whole pH range. Moreover, the alkyl substituents block a second aniline addition thus preventing the formation of three-component products.

Finally, reaction between an aminooxy and a ketone/aldehyde is a "click" reaction which has the advantage that, no other reagents are required besides the two reacting molecules. Therefore, oxime bond formation is an attractive method for site-specific bioconjugate formation and various routes have been employed to incorporate these groups into proteins. For example, solid-phase protein synthesis has been utilized to incorporate both *O*-hydroxylamine- and ketone-functionalized lysines onto proteins using aminooxyacetic acid [80] and levulinic acid, respectively [120]. The resultant *O*-hydroxylamine or ketone groups were targeted for site-specific PEG attachment. Serine hydroxy side groups have also been oxidized to aldehydes using sodium periodate for site-specific polymer attachment [121], whereas pyridoxal-5-phosphate (PLP)-mediated N-terminal transamination was used to install the α-ketoamide moiety and form a site-specific PEG-functionalized protein via oxime bond formation with an aminooxy end-functional PEG [122–124]. However, up to date only PEG conjugates have been synthesized by oxime bond formation and other polymers have not been explored possibly due to the lack of efficient methods to prepare the required end-functionalized polymers.

Recently, various controlled polymerization techniques for synthesizing peptide–polymer hybrids in a well-defined manner have been utilized presenting an unprecedented flexibility which opens new avenues for a wide variety of nanoparticulate bioconjugates [125, 126]. Living radical polymerizations (LRPs) [127, 128] yield well-defined polymers with predetermined molecular weights and narrow polydispersity indices containing many different functionalities, desirable properties for well-defined polymer–protein conjugates. The high interest in modern bioconjugation chemistry using LRP can be exemplified by a number of innovative approaches reported by several groups [129–133]. Among them atom transfer radical polymerization (ATRP) [134–136] has proved to be the most powerful and facile method among all the living polymerization techniques for the synthesis of nanoparticulate bioconjugates.

ATRP has been used to prepare pyridyl disulfide semitelechelic poly(2-hydroxyethyl methacrylate) that without any postpolymerization modification conjugated to BSA via a reversible disulfide bond [137]. Maleimide end-functionalized poly(PEG methacrylate) has also been prepared by ATRP and conjugated to BSA and to glutathione through covalent C-S attachment of the polymer [138]. Polymerization from protein-reactive initiators circumvents postpolymerization reactions to install the desired moieties and guarantees that each polymer chain contains one reactive end group. These initiators enable facile synthesis of semitelechelic polymers for direct conjugation to proteins or after a simple deprotection step. Initiators for ATRP have been synthesized with protein-reactive groups resulting in α-functional polymers [129]. Pyridyl disulfide [137] and protected maleimide [138] ATRP initiators resulted in polymers for conjugation to thiol side chains, and biotinylated ATRP initiators [139–141] produced polymers that conjugated to streptavidin. The synthesis of Boc-protected aminooxy-functionalized ATRP initiators for chemoselective protein–polymer conjugate synthesis has been also reported. ATRP of poly(ethylene glycol) methacrylate (PEGMA), 2-hydroxyethyl methacrylate (HEMA), and N-isopropylacrylamide (NIPAAm) from these aminooxy-functionalized initiators have been carried out first, followed by subsequent deprotection of the α-functionalized polymer and attachment to N-levulinyl lysine-modified BSA via oxime bond formation. In another report the bromine chain ends of well-defined poly(oligo(ethylene glycol) acrylate) (POEGA), prepared by ATRP, were successfully transformed into various functional end groups (ö-hydroxy, ö-amino, and ö-Fmoc-amino acid) via a two-step pathway: (1) substitution of the bromine terminal atom by an azide function, (2) 1,3-dipolar cycloaddition of the terminal azide with functional alkynes. This two-step synthetic strategy was employed for the preparation of well-defined POEGA-b-GGRGDG bioconjugates in high yields via the "click" reaction of an azide functional POEGA with an alkyne functional oligopeptide GGRGDG [142].

However, each of these approaches to prepare protein–polymer conjugates attaches a preformed polymer to the protein. Most often an excess of polymer is used, which must be removed from the conjugate; this can be difficult to achieve if the polymer and protein are similar in size. In addition, it can be difficult to determine the number and location of the polymer chains in the final conjugate, particularly when multiple attachment sites are possible. An alternative way to prepare protein–polymer conjugates that circumvents these issues involves first modifying the protein with

initiation sites for polymerization and then polymerizing from the protein macroinitiator to form the bioconjugate *in situ*. Polymerization from initiator-conjugated proteins represents an advantageous alternative to the attachment of preformed polymer chains to proteins because it allows the facile purification and characterization of the synthesized bioconjugates and has been extensively exploited in bioconjugate synthesis. Polymers were initially grafted from proteins by randomly generating radicals on amino acid side chains [143–145], although in these examples, the number and sites of polymerization could not be controlled. Later studies involved the initiation of the polymerization from specific protein sites, and predetermined locations of polymer conjugation. In such an example, the protein streptavidin was modified with a biotinylated initiator and used for the formation of the conjugate *in situ* [146]. The initiation sites were defined by placement of the modified biotins, and conjugate formation was efficient. A drawback of this approach is that the macroinitiator is formed by the interaction of functionalized biotin, and thus the identity of the protein is limited to those that bind the ligand, namely streptavidin, avidin, and recombinant derivatives such as neutravidin. However, this methodology has the advantage that compared to traditional methods, bioconjugate purification is simplified since unreacted small molecules such as residual monomer are readily removed by simple dialysis or chromatography.

Recently, numerous studies have employed polymer/peptide initiators for the controlled growth of polymer chains *in situ* using "living" radical polymerization techniques. Biohybrid materials containing poly(meth)acrylates and PS have been prepared using amino acid ATRP initiators [147, 148]. The synthesis of different peptide–polymer bioconjugates by ATRP [149] and RAFT [150] of different monomers using oligopeptide-based initiators containing either 2-bromopropionate or dithioester moiety, respectively has been also reported. Another report described the synthesis of nanoparticulate peptide–polymer conjugates by a combination of self-assembling functional cyclic peptides and *in situ* surface-initiated ATRP [151]. An interesting approach involved the preparation of block bioconjugates of the protein transduction domain (PTD) of the HIV-1 TAT protein, by sequential condensation-based peptide growth and LRP from a solid support [125]. Standard solid-phase peptide synthesis for the preparation of the peptide chain was employed, from which nitroxide-mediated radical polymerization (NMRP) [152] was carried out, using initiating sites located on the chain termini of the peptides loaded on a solid support, to yield the synthetic segment of the bioconjugate (see Figure 7.1). This versatile synthetic strategy provides a route to create peptide-synthetic block copolymer bioconjugates with controlled stoichiometry and regioselectivity of each of the chain segments. Although the methodology is demonstrated using a certain protein and polymerization method, virtually any peptide sequence and LRP conditions can be employed.

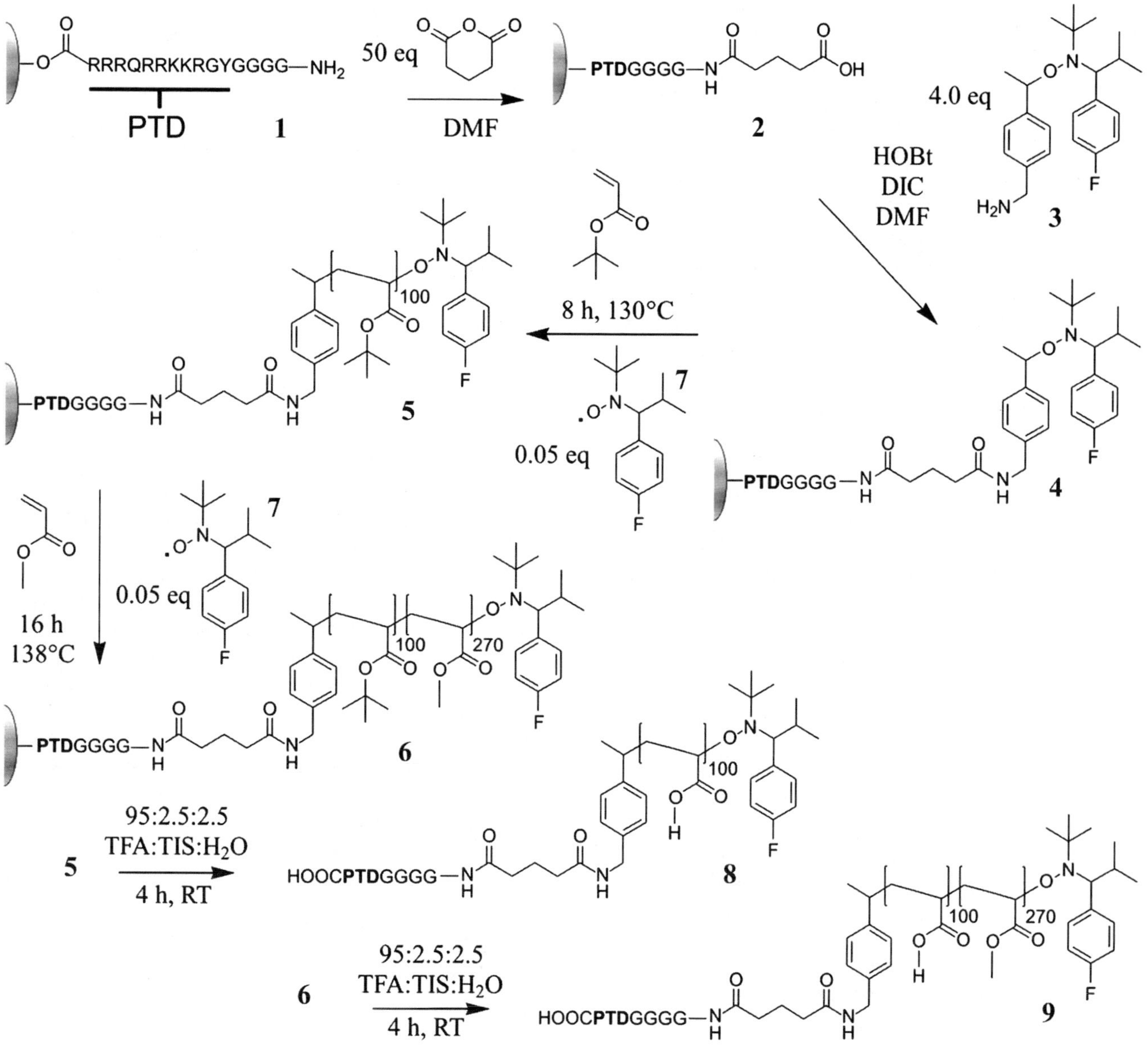

FIGURE 7.1 NMP polymerization from initiating sites on the chain termini of peptides loaded on a solid support to prepare peptidic–synthetic block copolymer bioconjugates. Reproduced with permission from Reference 125.

The materials obtained have defined surface localization and thus are expected to possess unique self-assembly and surface reorganization properties given the programmed secondary structure of the peptidic component.

An attractive class of synthetic nanoparticulate polymeric materials are the so-called nanogels. Nanogels are known to encapsulate proteins and thus improve protein activity and stability [153, 154]. Moreover, it has been found that proteins encapsulated in nanogels show superior temperature and organic solvent stability. Both of the above characteristics are very important and expand the catalytic and therapeutic potential of these systems. However, so far the synthesis

of protein–nanogel conjugates has been achieved by a two-step process. First the proteins are functionalized using N-hydroxysuccinimide (NHS) acrylate and subsequently the protein is copolymerized with acrylamides and cross-linkers using REDOX-initiated free radical polymerization [44, 155]. This approach leads to nonspecific protein functionalization and potential deactivation of the protein-active sites and denaturation [1, 40]. A controlled method which allows the synthesis of well-defined protein–nanogel conjugates by combining controlled radical polymerization with a genetically engineered protein containing a site-specific initiator was reported. A genetically engineered green fluorescent protein (GFP), containing a nonnatural amino acid bearing an ATRP initiator, was employed to prepare nanoparticulate protein–nanogel conjugates with a diameter of 240 nm by activator generated by electron transfer ATRP in an inverse miniemulsion [156]. These bioconjugates possess controlled size distributions, can be combined with a wide range of monomers, minimal loss of protein activity due to nanogel encapsulation, and control over protein loading per nanogel. Furthermore, genetic engineering allows precise control over the number of initiating sites and thus chains attached to the protein. Retention of the protein's tertiary structure was verified by measuring its fluorescent activity within the nanogels and opens new avenues for their use in controlled-release applications.

A unique property of the polymer–polypeptide conjugates is their rich morphological behavior, both in the bulk and in solution, which results in the formation of unpresented nanostructures. Bioconjugates combine the properties of both components in a single macromolecule and confer to the hybrid the self-assembling properties of the conventional block, in addition to a rich morphological behavior stemming from the conformational transition of the polypeptide segment from coil, to α-helix and to β-sheet with pH and/or temperature adjustment, due to the formation of intra- and/or intermolecular hydrogen bonds [157–159]. Recently, there have been several reports on the so-called bottom-up approaches for the preparation of such bioorganic nanostructured materials. Among these approaches, the self-organization/aggregation of peptide-polymer precursor molecules into complex macromolecular assemblies in various solvents have been of particular interest for use in biological applications [160–162]. Although most of these assemblies have been carried out in aqueous environment [151, 163] and in mixed solvent media [164, 165], the use of solely organic solvent as a medium has been also reported [166]. Their rich morphological behavior, derived from the α-helix and β-sheet conformations assumed by the peptide segment, promotes a hierarchical self-assembly of the material in the micron lengthscale and deviates strongly from the conventional aggregation behavior of synthetic polymers which is usually directed by parameters such as block compatibility, comonomer volume fraction, and solvent quality. Moreover, the peptide side chain enables the facile attachment of various moieties, that is, hydrophilic/hydrophobic segments, effectively altering the initial material properties. One such example has been reported by Kataoka et al. [167]. who was able to modulate the hydrophilic/hydrophobic balance in a PEO-*block*-PLL copolymer via the partial attachment of hydrocinnamoyl moieties on the ε-amino group of the PLL side chains. This resulted in a change in the micellization properties of the material, as well as in a dramatic change in the secondary structure of the peptide from random coil to the β-sheet, as a function of the degree of functionalization or the solution pH.

However, the self-assembly behavior of enzymes and proteins modified with synthetic polymer chains requires well-defined conjugates. Peptides used in this context include β-sheet fibrils, α-helical coiled coil peptides, and peptide-amphiphiles [168–171]. In all cases, the peptide or the polymer segment of the bioconjugate assembles through different supramolecular interactions and directs the size and shape of the nanostructures formed. Nanostructured materials with varieties of shapes including nanotubes [163, 172], nanoribbons [173], vesicles [164], nanospheres [174], and helices [175] have been prepared from different peptide–polymer conjugates by self-assembly in solution. Such nanostructured polymer–polypeptide carriers that undergo coil-to-helix transformations as the pH is lowered have been shown to induce disruption of the endosomes and are advantageous for drug delivery applications [176, 177]. Moreover, the self-organization of such polymer–protein bioconjugates into hydrogels, through the interaction of their peptide/protein component, has been proposed for use in drug delivery and tissue engineering applications [178, 179].

In another example, amphiphilic peptide–polymer bioconjugates containing poly(methyl methacrylate) chains attached to oligopeptide molecules have been prepared by ATRP using functional peptide initiators [22]. The self-assembly properties of the as-prepared peptide–polymer bioconjugates into hybrid micro/nanospheres in different polar organic solvents was studied and a mechanistic model was suggested for the aggregation of peptide–polymer conjugates into hybrid micro/nanospheres that correlates well with the experimental observation. A dye uptake into the micro/nanospheres was confirmed by fluorescence microscopy and time-correlated single-photon counting techniques rendering them attractive candidates for drug delivery applications.

Although, most of the polymer-polypeptide copolymers reported initially in the literature are based on one conventional hydrophobic block, polybutadiene (PBd) [180], PS [181], or polyisoprene (PI) [182], other attractive self-assembling systems include PEGylated proteins, and proteins specifically linked to other hydrophilic functional polymers [76, 183–188]. These materials have tremendous potential because they combine the responsive properties of the polypeptide with the ability to work in pure water which renders them advantageous for both biological and environmental applications.

A convenient approach for the synthesis of monodisperse large polypeptide copolymers of defined structure and compositions that operate in water involved the controlled denaturation of native proteins (serum albumin and lysozyme) followed by their *in situ* stabilization with PEO chains to yield bioconjugates of precise backbone lengths, net charges, and secondary structure [189]. Diverse functional groups available at precise positions along the protein backbone enabled further orthogonal modifications. The polypeptide copolymers revealed attractive physical and biological properties, that is, excellent solubility and stability in aqueous media and no significant cytotoxicity at relevant concentrations, and they could be degraded via proteolysis, which is very attractive for biomedical applications such as drug delivery, nanopatterning, and tissue engineering. Supramolecular micellar structures were formed in dilute solution because of the presence of alternating hydrophobic and hydrophilic "patches" along the polypeptide backbone which are attractive for use in drug encapsulation and delivery. This "semisynthetic chemistry" approach offers great opportunities for the preparation of diverse polypeptides with unique macromolecular architectures since the reaction scheme employed is generic and could, in principle, be applied for other precursor proteins, if the denaturing conditions are adjusted depending on the stability of the respective native protein.

Particular interest has been focused, on the use of oligo- and polypeptides as highly versatile and biodegradable building blocks for polymersome formation in water. The secondary structure formed in each case is primarily dictated by the primary sequence of amino acids. By selecting the appropriate amino acid residues a particular nanostructure can be obtained, whereas, stimulus-responsive behavior can be also introduced, such as a T/pH-dependent solubility [190]. One of the first reports on the formation of vesicular structures from peptide–polymer conjugates made use of an antibacterial hydrophobic helical peptide, Gramicidin A. Gramicidin A was coupled with PEG to give an amphiphilic block copolymer which formed readily a variety of aggregated nanostructures in aqueous solution [191]. Among the obtained structures a vesicular aggregate, named peptosome to denote a peptide-containing polymersome, was found [192]. This early example involved the use of a hydrophobic peptide which was buried under a hydrophilic PEG-shielding/stabilizing layer. However, in most recent cases the peptide component is the hydrophilic part, and thus forms the periphery of the vesicles, whereas the use of a hydrophobic polymer gives the core of the nanostructure. This was demonstrated by Dirks et al. [116] who coupled via the copper-catalyzed azide–alkyne Huisgen [3 + 2] cycloaddition reaction a hydrophilic tri-peptide, Gly–Gly–Arg–AMC, to hydrophobic PS. Another approach used active functionalities on the periphery of preformed polymerosomes to attach the hydrophilic peptide segments. For example, the attachment of several proteins via available amine residues onto NHS-activated carboxylic acid groups present on the periphery of polymersomes gave a protein-stabilizing layer [193]. The maleimide functionality has been also employed for vesicle bioconjugation. Maleimides readily react with thiols which can be made available in a controlled fashion in both peptides and proteins by the introduction of a cysteine residue. Using this approach [194] mouse–antirat monoclonal antibodies for brain delivery in

rats were coupled to the surface of polymersomes. Another study [195] adopted a strategy in which they mixed in up to 10% of an acetylene functional PS-PEG in polymersomes. These so-called anchors allowed for the immobilization of azido-functionalized enzymes on the vesicle surface. Very recently the vinyl sulfone moiety was introduced in polymersomes [196]. The fast and selective reaction of this functionality with thiols was employed to couple RGD peptides on the periphery of poly(methyl caprolactone)-*b*-PEG polymersomes.

Polymer–polypeptide polymersome-forming block copolymers consisting of a repetition of the same amino acid, known as homopolypeptides, are even more common nowadays. Such polypeptides are obtained in a controlled fashion by the polymerization of N-carboxyanhydrides (NCAs) using an amine initiator [197]. Thus the use of an amine-terminated conventional homopolymer as an initiator gives the functional block copolymer. In this way, PBd–poly(L-glutamate) [180, 198] and PBd–poly(L-lysine) [199] block copolymers have been prepared and extensively investigated in terms of their aggregation behavior to form polymersomes in water. Another apporach uses block copolymers in which both the hydrophobic and the hydrophilic part are composed of polypeptide blocks for the formation of polymersomes [200, 201]. This increasing interest in peptide-based polymersomes is driven by their usefulness for various applications. Peptosomes employing a block copolymer composed of poly(N-methylglycine) and poly(g-methyl-L-glutamate) (PMLG) and loaded with a near infrared dye, were shown to exhibit good *in vivo* circulation times and low recognition by the reticuloendothelial system (blood clearance mostly by liver and spleen) and were proposed for use in *in vivo* cancer imaging [202]. Another system based on oppositely charged block copolypeptides was reported by Kataoka to assemble in a semipermeable membrane and yielded PICsomes [203] that were proposed for use as biocatalytic nanoreactors.

Of particular interest upon polymersome formation is to retain protein function at the interface. Thiol–alkyne chemistry has been utilized to produce peptide-based A$_2$B star polymers through conjugation of acetylene-terminated poly(L-glutamic acid) with three different thiol functional lipophilic moieties (octadecane, cholesterol, and polyhedral oligomeric silsesquioxane) [204]. These molecules resemble a phospholipid and are shown to self-assemble in aqueous solution into pH-responsive vesicles. The proposed A$_2$B motif ensures that the hydrophobic (A) part is maximized, favoring bilayer formation while leaving the functional peptide (B) block at the vesicle interface. The aggregation number of the vesicles remains nearly constant as a function of pH, suggesting that the pH-responsiveness is a result of both the helix–coil transition as well as a change in chain packing at the vesicle interface. These tailored hydrophobic core properties allow to study membrane transport in various chemical/physical environments as well as to create drug delivery vehicles with various release profiles.

An alternative nanostructure which has attracted particular attention for drug delivery applications are the so-called shell cross-linked (SCK) micelles. Their surface functionalization with the oligomeric peptide sequence YGRKKRRQRRR, the protein transduction domain (PTD) from the human immunodeficiency virus TAT protein, has been reported [205]. A convergent synthetic strategy was employed, whereby the SCK nanoparticles and the PTD were prepared independently and then coupled together during immobilization of the PTD component on a solid support. The SCK nanoparticles were prepared by the micellization of amphiphilic block copolymers of poly(ϵ-caprolactone-b-acrylic acid), followed by amidation-based cross-linking of the carboxylic acid moieties in the micellar corona. The PTD sequence was constructed upon a solid support, from C-terminus to N-terminus, followed by extension with four glycine residues, leaving the amino chain end for subsequent coupling with remaining carboxylic acid functionalities present on the surface of the SCK. Upon cleavage from the solid support, deprotection of the peptide side chain functionalities as well as hydrolysis of the poly(ϵ-caprolactone) segments composing the SCK core domain was achieved, to yield PTD-based nanocages. Moreover, covalent binding of fluorescein-5-thiosemicarbazide onto the SCK precursor provided fluorescently tagged PTD-nanocages to allow for their detection by fluorescence microscopy. The fluorescent PTD-nanocages were found to interact with HeLa cells and were primarily located near the cell periphery although transduction into the cells to some extend was also observed.

Peptide–polymer bioconjugates that form networks via coiled coil multimerization provide several advantages over polymer matrices or biologically sourced ECM proteins, including defined composition and network topology and a stimulus-responsive behavior [206]. However, despite the increasing interest in the self-assembling behavior of these peptide-based block copolymers, the molecular determinants of these materials' immunogenicity, which are particularly important for their use in 3D cell culture, tissue engineering, and regenerative medicine [207, 208], have remained largely unexplored. Rudra et al. has recently addressed this issue by designing a set of molecules that self-assembled through coiled coil oligomerization and studying the immune responses against them in mice [209–212]. Amino acid substitutions were made at residues forming the hydrophobic core of the coiled coil, and these peptides were conjugated to PEG to give a triblock peptide–PEG–peptide [171]. Self-assembly of this triblock peptide–polymer-formed coiled coil multimers and supramolecular aggregates which afforded viscoelastic hydrogels. The extent to which the amino acid substitutions and supramolecular assembly influenced the material's immunogenicity was investigated. A native peptide

selected from mouse fibrinogen that did not form helical structures or self-assembled, and the engineered gKEI peptide which formed coiled coil bundles were used for comparison. The native and engineered peptides were not found to produce any detectable antibody response in mice, and none of the materials elicited detectable peptide-specific T-cell responses. However, despite the minimal changes in secondary structure found between the engineered peptide and the triblock peptide–PEG–peptide, specific antibody responses were induced in mice injected with the triblock peptide–PEG–peptide attributed to its multimerization. These results suggest that self-assembly and multimerization can influence the immunogenicity of peptide-based materials and thus strategies for modulating these processes are essential in tissue engineering applications.

In contrast to conventional linear polymers which only have one or two functional groups at the chain ends, and are randomly coiled and entangled with others, resulting in low reactivity of conjugation, dendritic polymers including dendrimers and hyperbranched polymers [213, 214] have a highly branched structure, providing a spherical shape and a high density of functional groups at the periphery. These structural advantages facilitate the interaction of the functional groups with other reagents, resulting in a high bioconjugation reactivity and thus dendritic polymers have been extensively used for enzyme immobilization [215] and conjugation [216]. The preparation of hyperbranched polymer–enzyme conjugates using lipase, an enzyme used extensively as a catalyst in chemical production both in aqueous and nonaqueous media, was reported [217, 218]. Conjugation of lipase with the hyperbranched aromatic polyamides was achieved using carbodiimide as a coupling reagent and a maximum of five to six lipase molecules were attached [219]. The conjugate exhibited a significantly enhanced stability at high temperature or in the presence of organic solvent, as compared to its native counterpart and a 20% increase in catalyst activity rendering it promising for industrial biocatalysis. An interesting study employed a boronated fourth-generation starburst dendrimer (SD) functionalized with thiol groups, using N-succinimidyl 3-(2-pyridyldithio)propionate, for cancer treatment. The dendrimer was conjugated with m-maleimidobenzoyl-N-hydroxysulfosuccinimide ester-derivatized EGF, a single-chain 53-residue polypeptide [220] via the reaction of the thiol groups of the derivatized BSD with the maleimide groups of the derivatized EGF to produce stable BSD–EGF bioconjugates containing 960 atoms of boron per molecule of EGF. The mild conditions used for bioconjugation preserved the tertiary structure of the polypeptide and thus its ability to bind to epidermal growth factor receptor on target cells. As a result the nanoparticulate bioconjugates were found to initially bind to the cell surface membrane and then endocytosed, which resulted in accumulation of boron in lysosomes. These favorable in $vitro$ properties of the bioconjugates

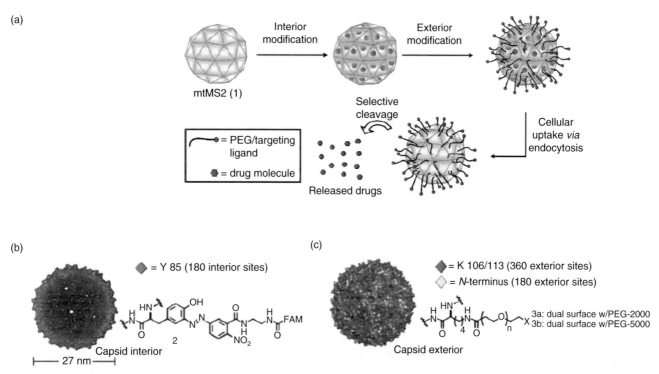

FIGURE 7.2   (a) General strategy for the modification of viral-like particles to serve as multifunctional drug delivery vessels. (b) Tyrosine 85 of the interior capsid surface undergoes rapid diazonium coupling with *p*-nitroaniline derivatives. (c) Accessible amino groups (lysines 106, 113, and the *N*-terminus) on the capsid exterior are readily modified with PEG-NHS esters. Reproduced with permission from Reference 222.

suggest their potential for use in boron neutron capture therapy via *in vivo* targeting of EGFR positive brain tumors. Finally, a peptidic-dendron conjugate has been also reported to form polymersome-like structures [221].

A particularly attractive platform toward polymer-protein self-assembled nanostructures is based on symmetrical icosahedral nanoscale virus-like particle (VLP) assemblies modified using two orthogonal modification strategies to decorate the exterior surface of the capsids with PEG chains, while installing a fluorescent dye, to serve as a drug mimic, inside the capsid (see Figure 7.2) [222, 223]. The polymer coating provides stabilization to the capsids by blocking the access of polyclonal antibodies to the capsid surface whereas, an additional chemical strategy for the attachment of small targeting groups to the outer end of the polymer chains, through an efficient oxime bond formation, was developed. These covalent modification strategies convert the viral capsids into modular drugs and imaging agent carrier systems.

In a similar work, a cell-free protein synthesis (CFPS) platform and a one-step, direct conjugation scheme was developed to produce VLPs that display multiple PEG chains on the VLP surface [224, 225]. Using a global methionine replacement approach, the bacteriophage MS2 and bacteriophage Qβ VLPs presented surface-exposed non-natural

methionine analogs (azidohomoalanine and homopropargylglycine) containing azide and alkyne side chains which were then directly conjugated with azide- and alkyne-containing PEG using Cu(I)-catalyzed click chemistry [226]. This platform facilitates conjugation of different ligands (drugs, nucleic acids, proteins, etc.) to the VLPs with control over the ratios and surface abundance of attached species resulting in the production of novel noninfectious and intrinsically stable VLP bioconjugates for use as drug delivery vehicles, diagnostics, and vaccines [227–229].

Besides the linking of inert polymer chains that provide stabilization and/or solubilization to the nanoparticulate bioconjugates, the use of stimuli-responsive, synthetic polymers has attracted particular attention lately for conjugation with biomolecules. These materials comprise an attractive class of the so-called "doubly smart" copolymers, since they combine the response of the biomolecule with that of the polymer, which can also undergo a reversible change in size, charge, and hydrophobicity in response to a variety of external stimuli such as temperature, pH, light, etc. [230–232]. Responsive polymer–protein bioconjugates provide many of the benefits of the conventionally employed polymers, while simultaneously allowing the responsive assembly or activity modulation of the attached biomolecule [131] and have been proposed for use in controlled capture/release applications, in protein separation, and in photoswitches [84, 85, 233]. Recent vast developments in the field of responsive bioconjugates, purposely designed to exhibit tunable switching properties triggered by an external stimulus, have employed stimuli of (bio-)chemical or physical nature such as temperature, pH, enzymes, ultrasound, and others [234, 235], whereas photo-sensitive hybrid materials that exhibit some sort of light-mediated response (i.e., shape/volume change, photochromism, sol–gel transition, photodegradation, etc.) have been also prepared. Light-responsive biohybrids carrying photo-sensitive moieties are highly desirable, because they allow to reversibly alter their properties by applying a remote stimulus with spatial and temporal control, such as light irradiation of specific wavelength and known energy. Goodman et al. [235] and more recently, Pieroni et al. [236] and Hamm et al. [237] have reported on the synthesis of chromophore-bearing (homo)polypeptides. It is noteworthy, that light was found to be able to induce a reversible change in the peptide conformation from α-helix to β-sheet. Moreover, it was reported that photochromism can result in helix reversal, random coil to α-helix transitions, modulation of the redox processes or change in the aggregation behavior of the system [238]. Such photoisomerizable chromophore (azobenzene or spiropyran)-bearing polypeptides were investigated as chiroptical switches [239], whereas the use of light-sensitive polypeptides as intelligent molecular materials has been proposed for various applications in biomedicine, micro- and nanoelectronics, ecology, and other advanced technological fields. Another advantage of these materials is that upon a change induced in the peptide conformation the size of the peptide segment changes significantly, which makes them intriguing for applications as nanosensors/actuators.

In this context, "smart" protein–polymer conjugates have been extensively developed by employing PEGMA and poly(N-isopropylacrylamide) (PNIPAm) as a temperature-responsive polymer which confers a phase transition temperature to the protein to which it is attached [131, 240–246]. These protein–polymer conjugates have been proposed for use as molecular sensors for diagnostic assays [72, 247], in enzyme recovery [248], for triggered blocking or release of substrates to protein-active centers [83, 184, 249], for the regulation of enzyme activity [87], and for chemical switches [84, 85] as a function of temperature. Placement of the polymer chain in close proximity to the protein-active center is critical for reversible activity control and "smart" switches [83].

The *in situ* synthesis of "smart" polymer conjugates by polymerization from defined initiation sites on the proteins is particularly attractive for the preparation of well-defined conjugates which form self-assembled nanostructures in solution and close to a solid surface. "Diblock" and "triblock" PEGMA–protein conjugates have been grown by aqueous-phase ATRP directly from a model protein, trypsin, using the "grafting from" approach, and characterized in terms of their temperature-dependent behavior in solution, and their structures at surfaces [250]. The phase transition temperature of the bioconjugates was dependent on the arrangement of their grafted polymer blocks and could be used to modulate the bioconjugate enzyme activity. Overall the hybrid bioconjugates were found to be more stable and substrate selective, but exhibit reduced proteolytic activity. Their use to target different receptors/substrates, through changes in polymer architecture and external stimulus, while switching the protein activity as required, is envisaged.

In another study, free cysteines of BSA lysozyme, were modified with initiators for ATRP either through a reversible disulfide linkage or irreversible bond by reaction with pyridyl disulfide- and maleimide-functionalized initiators, respectively [245]. Polymerization of NIPAm from the protein macroinitiators resulted in thermosensitive BSA-PNIPAm and lysozyme–PNIPAm bioconjugates in high yield. Lytic activities of the lysozyme conjugates determined by two standard assays did not exhibit any statistical differences in bioactivity compared to the unmodified enzyme suggesting that the linking of the polymer chains preserved protein activity.

"Smart" polymer-protein nanoparticulate bioconjugates are also attractive for use as nonviral transfection agents. A PNIPAm-poly(L-arginine) bioconjugate was prepared by a two-step reaction: first radical polymerization of NIPAm followed by EDC-activated coupling between the carboxyl-terminated PNIPAm and polyarginine [251]. The phase transition temperature of the bioconjugate was found to be slightly higher than that of the synthetic polymer and a temperature-dependent association and dissociation of the bioconjugate with DNA was found. The bioconjugate/DNA complexes formed uniform nanospheres with size about 50–120 nm which exhibited high transfection efficiency with COS-1 cells, equivalent to that of a commercial transfection agent. Moreover, the polypeptide cytotoxicity was found to decrease upon incorporation of the synthetic polymer chains which renders them attractive for gene transfer.

Finally, inspired by the function of metallothioneins found in nature that are able to bind toxic metal atoms at sub-ppb concentrations, protein–polymer conjugates have been developed as nanoscale materials for water remediation [21]. The synthesis of such metallothionein–polymer conjugates is hindered by the lack of an orthogonal conjugation reaction that will leave the cysteine, lysine, and carboxylate groups of the protein, required for metal binding, unaffected. To overcome this problem, proteins were expressed as part of an N- and C-terminal modification "cassette" which gave bioconjugates with high metal selectivity. This strategy involves the use of an intein domain that is fused to the C-terminus of a protein sequence allowing the chemical ligation with a ketone-labeled cysteine residue. The N-terminal ketone is next introduced using the biomimetic transamination reaction. The two ketone end groups serve as chemospecific attachment points for polymers that possess pendant alkoxyamine groups such as alkoxyamine-functionalized N-hydroxypropyl methacrylamide resulting in the formation of hydrogels that respond to the presence of toxic metal ions at ppb quantities.

Considering their uniquely broad range of functions and properties in the natural world, proteins clearly have much to offer in terms of creating new nanoparticulate materials. Accumulated experience in this area over the past decade has bolstered the confidence in the power of combining biomolecular structures with nonnatural components. We anticipate that, complex polymer–biomolecule conjugates bearing functional moieties that can function in environmentally benign aqueous media and which are rich in morphological and biological properties will open new possibilities toward the construction of elaborate nanostructures of similar or even rivaling complexity with those found in nature. Novel controlled polymerization methods combined with activation chemistries, are expected to lead to the development of well-defined bioconjugate materials that exhibit controlled self-assembly behavior leading to functional supramolecular structures. Future progress in the generation of such protein-based materials will require scalable synthetic techniques with improved yields and selectivities, inexpensive purification methods for bioconjugates, and theoretical and dynamical treatments for designing new materials through protein self-assembly.

7.3 POLYMER–ODN/DNA NANOPARTICULATE BIOCONJUGATES

Compared to their polymer-protein counterparts, reports on the synthesis of bioconjugates consisting of a DNA segment and a synthetic polymer moiety are rather rare. Since the DNA molecule exhibits a high specificity and chemical and biochemical stability, it is an attractive candidate for the development of novel functional polymers that can find applications in gene delivery, in targeted drug delivery, as reaction templates, etc. However, there are also several possible problems that require intensive research to be solved. For example, although the DNA molecule is water soluble, it is insoluble in most organic solvents and also, despite its higher stability as compared to proteins, and polysaccharides, it is not stable enough for the preparation of robust functional materials. The conjugation of DNA to artificial polymers appears to be a promising strategy to overcome the above drawbacks. To date, there are only few examples on DNA–polymer bioconjugates in the literature [252–254]. Most studies focus on the use of ODN or RNA aptamers for use in targeted drug delivery.

In this context, polymer-nucleic acid aptamer drug-encapsulated nano/microparticles have been prepared by a water-in-oil-in-water (W/O/W) double emulsion procedure using a PLA-PEG-COOH, block copolymer, and rhodamine-dextran as a model drug molecule [255]. The PLA was positioned in the interior of the particle, whereas the presence of the hydrophilic PEG polymer on the particle surface reduced the nonspecific interaction of the particles with other biomacromolecules, including cell surface antigens and increased the circulating half-life of the particles in plasma. The particles were conjugated to aptamers, by NHS ester activation chemistry of the exposed carboxylic acid groups on the particle surface, to covalently attach the particles to $5'$-NH_2-modified aptamers and obtain the nanoparticle–aptamer bioconjugates (see Figure 7.3a) [256].

These nanoparticulate polymer–aptamer bioconjugates were shown to increase the efficacy of chemotherapeutic treatment in prostate cancer epithelial cells (see Figure 7.3b) [255]. Moreover, docetaxel-encapsulated nanoparticles targeted prostate-specific membrane antigen and caused substantial reduction in both tumor size and treatment toxicity when compared to treatment with control nanoparticles lacking the aptamers, highlighting the importance of the targeting moieties on the nanoparticulate drug vectors [257].

In another interesting example, a bioconjugate between duplex oligodeoxynucleotides (dODNs) and a dendrimer (DEN) was designed and its feasibility as a novel delivery system for doxorubicin (DOX) in animal tumor models and against cancer cells *in vitro* was demonstrated [19]. The dendrimer acted as a core nanostructure for conjugation to obtain dODNs–DEN conjugates which formed stable complexes with ∼184 DOX molecules per conjugate due to the presence of drug-loading sites on the dODNs. The resulting DOX-loaded particles exhibited a sustained drug-release pattern both *in vitro* and *in vivo*. Pharmacokinetic studies showed that DOX-loaded dODNs–DEN conjugates were cleared from plasma at a much slower rate compared to free DOX. Furthermore, tumors retained a higher amount of DOX in mice treated with the bioconjugate compared to those treated with free DOX at the same dosage. No severe systemic toxicity or cardiotoxicity was found in mice treated with the bioconjugate indicating that the dODNs–DEN particles can be used as effective carriers to administer DOX *in vivo*. This generic platform is also promising for use in other drug delivery systems using a variety of oligonucleotide-based medicines as well as in targeted cancer therapy by introducing targeting ligands such as RNA or DNA aptamers.

An attractive concept for the synthesis of nanoparticulate bioconjugates was introduced using a CFPS platform and a one-step, direct conjugation scheme for the production of polyvalent virus-like particle (VLP) scaffolds displaying nucleic acids as well as other ligands including proteins, and synthetic polymers [224]. Viral bacteriophage MS2 and bacteriophage Qβ capsids with surface-exposed methionine derivatives containing azide and alkyne side chains were prepared, allowing the direct conjugation of azide- and

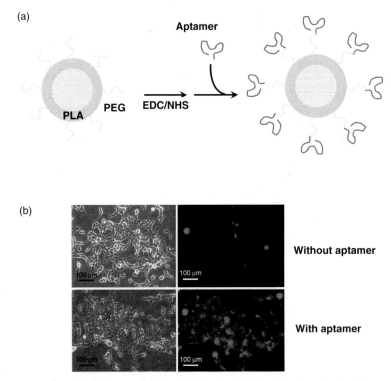

FIGURE 7.3 (a) Schematic representation of the surface modification of the PLA particles by covalent conjugation with RNA aptamers to obtain the nanoparticle–aptamer bioconjugates. (b) Binding of rhodamine-labeled dextran-encapsulated nanoparticles and particles conjugated to the PSMA aptamer, with LNCaP cells. Reproduced with permission from Reference 255. For a color version, see the color plate section.

alkyne-containing nucleic acids to the VLP surface using click chemistry. This direct attachment scheme facilitates the simultaneous conjugation of many other ligands to the VLPs, and can be employed for the production of novel nanoparticulate bioconjugates for use as drug delivery vehicles [223, 227], diagnostics [229], and vaccines [258, 259]. For example, a single-step assembly of VLPs displaying idiotypic antigens along with an immunostimulatory protein and stimulatory nucleic acids produced novel tumor idiotype vaccine candidates for treating B-cell lymphoma [224].

Besides ODN and DNA, synthetic small interfering RNAs (siRNAs) have been considered as a new class of nucleic acid therapeutics for treatment of various infectious and genetic diseases including cancer [260]. Recently, it was reported that direct conjugation of small drug molecules, aptamers, lipids, peptides, proteins, or polymers to siRNA could improve the *in vivo* pharmacokinetic behavior of the latter. To date siRNA has been chemically attached to a variety of bioactive organic molecules including lipids, polymers, antibodies, and peptides to enhance their pharmacokinetic behavior, cellular uptake, target specificity, and safety. Common conjugation strategies employ the 3′- or 5′-terminus of the RNA strand whereas, cleavable linkages including acid-labile and reducible disulfide bonds facilitate the release of

intact siRNA in the acidic endosome compartments and the reductive cytosolic space of the cells, respectively.

Various lipophiles, including cholesterol, bile acid, a series of lipids with different chain lengths, and tocopherol (vitamin E) have been conjugated to siRNA to produce lipophilic-siRNA bioconjugates [261, 262]. Cholesterol was covalently conjugated to the 3′-terminus of the sense strand of siRNA via a pyrrolidone linkage and the bioconjugates obtained formed particles in water which were used for systemic delivery with improved *in vivo* pharmacokinetic behavior [263]. Intravenous administration of the bioconjugates resulted in prolonged circulation times in tissues such as the liver, lungs, heart, fat tissues, and kidneys. On the other hand, bile acid–siRNA and lipid–siRNA bioconjugates were shown to interact with lipoproteins in the blood serum, (high and low density lipoprotein (HDL and LDL)), lipoprotein receptors, and transmembrane proteins, influencing tissue distribution and uptake [261]. Thus conjugates bound to LDL were mainly distributed to the liver, while the conjugates bound to HDL were taken up by various tissues, including the liver, kidneys, gut, and others. The affinity of the bioconjugates for certain lipoproteins, which affects their tissue distribution profiles, is determined by the degree of hydrophobicity and thus the length of the alkyl chain of the lipid. siRNA

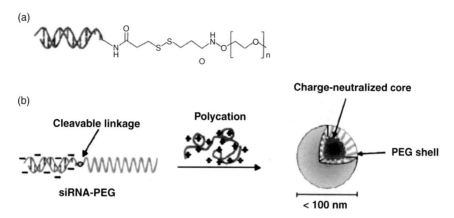

FIGURE 7.4 (a) PEG–siRNA conjugate with a reducible disulfide linkage. (b) Polyelectrolyte complex micelles of the bioconjugate with a cationic polymer. Reproduced with permission from Reference 18.

has been also attached to vitamin E via the 5′-terminus of the antisense strand and was used for systemic siRNA delivery to the liver [264]. After intracellular delivery, Dicer can cleave the bioconjugate [265] to give siRNA with the simultaneous release of vitamin E.

A promising direction in this area is the targeted intracellular delivery of antisense oligonucleotides [266, 267] and peptide nucleic acids (PNAs) [268, 269] and has been tackled using their conjugates to cell-penetrating peptides (CPPs). TAT(47-57), a CPP, was conjugated to the 3′-terminus of the antisense strand of an siRNA using a heterobifunctional cross-linker, sulfo-succinimidyl4-(p-maleimidophenyl)butyrate [270], which formed a noncleavable thioether linkage, and was employed for the receptor ligand-mediated delivery of siRNA. The bioconjugate induced the effective silencing of the target gene and exhibited a significant improvement in the intracellular delivery of siRNA compared to the parent RNA. In another example, the carboxylic acid group of D-(Cys-Ser-Lys-Cys), a mimetic peptide of IGF1, was activated and conjugated to the amine group of the 5′-sense strand of siRNA [271]. The resulting IGF1–siRNA conjugate inhibited about 60% of the target IRS1 gene expression without the use of a transfection reagent.

Lately, the use of cleavable linkages, such as disulfide bonds, between CPP and siRNA has attracted particular attention in minimizing the potential reduction of the RNAi activity by the presence of highly charged peptides. This approach has been employed to prepare bioconjugates that exhibit improved intracellular delivery of siRNA and the release of intact siRNA by cleavage of the disulfide linkages in the reductive cytoplasmic environment. Among the drawbacks of such CPP–siRNA bioconjugates are their similar degradation profiles to that of the parent siRNA and thus this conjugation strategy does not confer stability to the antisense oligonucleotides, whereas under certain circumstances

the peptide–siRNA conjugates may cause undesirable immune responses observed as the secretion of inflammatory cytokines [272]. To avoid this problem the conjugation of polymers to siRNA, via reversible/cleavable disulfide linkages, has been developed [273, 274]. It was shown that conjugation of PEG increased the circulation time of the bioconjugate and protected siRNA against degradation [274–276]. Moreover, multiple site conjugation of siRNA along the polymer backbone increased the therapeutic payload and its efficiency whereas, the conjugation of targeting folic acid moieties to the remaining free amines was also possible. Complete release of siRNA was achieved under reductive conditions with dithiothreitol which cleaved the disulfide bonds of the bioconjugates. Such PEG–siRNA conjugates with a reducible disulfide linkage (Figure 7.4a) have been used as nanoparticulate bioconjugates for cancer treatment [18, 275]. Core condensing agents, such as cationic polymers or peptides (PEI or KALA, a fusogenic peptide which destabilizes the endosomal membrane and facilitates the endosomal escape process [277]) were used to complex the conjugates and form micellar colloidal bioconjugate nanoparticles, the so-called polyelectrolyte complex (PEC) micelles (Figure 7.4b), which can serve as delivery vehicles for oligonucleic acid therapeutics [278–280]. The hydrophilic PEG chains in the micelle shell solubilize the hydrophobic polyelectrolyte core in the aqueous environment, prevent interparticle aggregation, and shield the surface charges of the core, rendering the micelles neutral nanoparticles with a higher blood compatibility, and also provides additional stability to the micelles when administered into the bloodstream, leading to prolonged circulation times. PEG–siRNA/PEI PEC micelles were used for local and systemic treatment of tumors in animals and exhibited enhanced transfection efficiencies [281]. The local and intravenous administration of PEC micelles comprising an siRNA-targeting vascular endothelial growth factor (VEGF) showed accumulation of the PEC micelles in the tumor region, after intravenous injection through a mouse tail vein, leading to significant retardation of tumor growth. This passive tumor targeting is attributed to the passive diffusion of the nanoparticles through the loosened endothelial vascular junctions in the vicinity of highly proliferative tumors, known as the enhanced permeation and retention (EPR) effect, allowing cellular entry to take place via adsorptive endocytosis [282].

Of great importance in this field is the conjugation of cell-specific ligands, including aptamers, antibodies, sugars, vitamins, and hormones, used to confer cell specificity to siRNA delivery systems leading to enhanced cellular uptake via a mechanism known as receptor-mediated endocytosis which improves the therapeutic efficacy at a much lower dose [283, 284]. In a simple case, a lactose moiety was coupled to the outer end of a PEG–siRNA conjugate and was used as a targeting ligand for the asialoglycoprotein receptors on hepatoma cells [276]. The lactosylated PEG–siRNA bioconjugate formed PEC micelles by interacting with poly(L-lysine) which exhibited growth inhibition in human hepatocarcinoma cells [285]. In another study, a luteinizing hormone-releasing hormone peptide analog was attached to the end of the PEG segment of a disulfide linked PEG–siRNA bioconjugate and was used for the targeted delivery of siRNA in the cytoplasm of ovarian cancer cells upon reductive cleavage of the disulfide bond.

Antibodies are very attractive targeting moieties, and antibody-mediated targeted drug delivery systems have gained much attention due to their superior stability during systemic circulation and high selectiveness toward a target protein on the cell surface. A brain-targeting system including two antibodies, one targeting a transferring receptor expressed in the blood–brain barrier and the second insulin receptors expressed in brain cancer, immobilized on the surface of liposomes was prepared [286]. *In vivo* studies on a rat with transplanted brain tumor showed that intravenous administration of the antibody–siRNA conjugate resulted in significant reduction of a reporter gene expression.

Finally, aptamers which target proteins onto the cell surface [287] have been also conjugated to siRNA and employed for cell-specific delivery. The aptamer used targeted a cellular membrane antigen expressed abundantly in prostate cancer cells and was bound to siRNA via a biotin–streptavidin-specific bond [288]. Biotinylated siRNA and aptamer were conjugated using streptavidin as a linkage molecule to prepare the bioconjugate which was bound onto the cells overexpressing the target receptor and was uptaken by the cells via receptor-mediated endocytosis [289]. Cleavage of the bioconjugate by endogenous Dicer produced siRNA which exhibited inhibition of tumor growth and tumor regression in a mouse tumor. Further work in this direction focuses on overcoming problems related to the nuclease susceptibility of the aptamers and the possibility of premature degradation of siRNA in the endolysosomal compartment which may

limit the application of aptamer–siRNA delivery systems in systemic treatments.

7.4 POLYMER–CARBOHYDRATE NANOPARTICULATE BIOCONJUGATES

A group of naturally occurring building blocks which have recently gained much attention are carbohydrates or polysaccharides [290–293]. Saccharides are ubiquitous elements in living systems and are conjugated to the surface of cells via the lipids that make up the cell membrane, the so-called glycolipids, or to proteins known as glycoproteins. Carbohydrate units play a very important role in many biological processes including cellular recognition and pathogen infections. In most cases these processes are mediated via interactions between saccharides and receptor proteins, called lectins which have a low affinity for saccharides, thus requiring multivalent interactions via the so-called cluster glycoside effect. Polymer–carbohydrate bioconjugates present attractive model systems in order to understand these multivalent interactions that take place in living systems and mimic their function. Moreover, carbohydrate-based drug and gene delivery carriers are becoming extremely popular for *in vitro* and *in vivo* applications. These carriers are found to be nontoxic and can play a significant role in targeted delivery. However, for the above highly specific functionalities, which will allow the generation of complex nanostructures from these bioconjugates by supramolecular assembly, are required. There are several examples of (co)polymers comprising carbohydrate functional units and either hydrophilic or hydrophobic moieties in the literature and their self-assembly in solution has been extensively studied [294–296].

In particular, micrometer or nanometer diameter particles labeled with carbohydrates are of particular interest because they mimic very closely the cell surface where lectins are located. In this context, multivalent carbohydrate liposomes [297–300] and polymer beads [301] have been prepared and used to elucidate fundamental properties of molecular biology. (Poly)saccharides have been also employed as end groups of block copolymer amphiphiles or as the hydrophilic part of block copolymers for the formation of highly functionalized and multivalent polymersomes. An example of a simple rod–coil amphiphile, tetra(p-phenylene)-*b*-PEG12-a-D-mannopyranoside was shown to form small polymersomes which interacted specifically with a bacterial strain of *Escherichia coli* (Figure 7.5) [302]. The carbohydrate specificity was proved by replacing the mannose head group for galactose to obtain amphiphiles that could still form polymersomes, but lacked the specific binding properties to the bacteria. Another example involved a b-cyclodextrin head group coupled to PS to give amphiphiles which also formed vesicles bearing hydrophobic fluorescent dyes in the hydrophobic interior of b-cyclodextrin

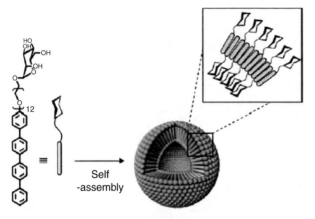

FIGURE 7.5 Self-assembly process of the rod–coil tetra(p-phenylene)-*b*-PEG12-*a*-D-mannopyranoside amphiphile to form small polymersomes in water. Reproduced with permission from Reference 301.

and the enzyme horseradish peroxidase (HRP) on their surface [303]. Besides small amphiphiles, polysaccharide block copolymers are also attractive candidates for polymersome formation [290]. Dextran-*b*-poly(benzyl-L-glutamate) [291], dextran-*b*-PS and hyaluronan-*b*-PBLG [26] have been employed to obtain polymeric vesicle for appropriate length of the synthetic block. These nanoparticulate bioconjugates were studied extensively in drug delivery applications [304, 305]. DOX-loaded hyaluronan-*b*-PBLG polymersomes exhibited effective *in vitro* and *in vivo* tumor targeting due to the hyaluronan shell as well as drug release and shrinkage of the tumors in mice upon delivery of the drug. An elegant example of a thermoresponsive polymer–saccharide bioconjugate comprised maltoheptaose coupled to PNIPAm via click chemistry [306]. The LCST of PNIPAm was conveniently used to tune the formation of polymersomes, with the nanoassemblies obtained above the transition temperature whereas, below the LCST they dissociated to their constituent chains.

An alternative attractive class of colloidal polymer particles are microgel particles which offer significant opportunities for targeted applications due to their tunable size, from nanometers to several micrometers, ability to incorporate biofunctional molecules such as drugs, proteins, oligonycleotides and others, and large surface area which is ideal for multivalent bioconjugation [307, 308]. Spherical micro- or nanoscale hydrogels coupled to saccharides are particularly attractive due to their facile preparation routes and upscaling potential. Bioconjugated cross-linked microgels, 0.5–5 μm in diameter, comprising a poly(*p*-phenyleneethynylene) core and monosaccharide (mannose, glucose, and galactose) surface functionalities have been synthesized via a Pd-catalyzed cross-coupling reaction [25]. Lectin binding onto the particles, assessed by confocal microscopy using a fluorophore-tagged Concanavalin A, was verified with some evidence

of agglutination, rendering them attractive for use as blood mimics, fluorescent probes, biocapture agents, or column packing material for affinity chromatography.

Numerous nanoparticulate polymer–carbohydrate bioconjugates based on synthetic polysaccharides, known as glycopolymers, carrying biocompatible sugar groups have been reported. Hyperbranched glycopolymers have been explored as potential nanocarriers and their interactions with blood cells and plasma components, to assess their blood biocompatibility, as well as their cytotoxicity was investigated [309]. These monomolecular nanoparticles exhibited good hemocompatibility and did not induce clot formation, red blood cell aggregation, or immune response. In addition, their very low cytotoxicity against primary and malignant cell lines suggested that these glycopolymer-functionalized carriers can serve as excellent candidates for various biomedical applications. An interesting example reports the preparation of core cross-linked (CCL) micelles possessing a thermoresponsive and degradable PNIPAm core cross-linked with bis(2-methacryloyloxyethyl) disulfide (DSDMA), a difunctional monomer, and a biocompatible and bioactive shell based on poly(2-aminoethylmethacrylamide) (PAEMA) which allows the functionalization with carbohydrate and biotin moieties [310]. These micelles exhibit a temperature-dependent swelling of the core based on the phase transition of PNIPAm and disintegrate into unimer chains following cleavage of the disulfide cross-linkers in the presence of dithiothreitol (DTT). As a result the release of active drugs from the core of these nanocarriers can be either triggered by temperature or the presence of thiols. Furthermore, the surface immobilized biotin and carbohydrate moieties of the micelles exhibit specific biomolecule recognition and are promising particles for controlled and targeted drug delivery applications. In a related study diblock copolymers bearing a galactose-containing block and a primary amine- or a galactose-containing block were prepared followed by the modification of the primary amine pendant groups of the copolymer with biotinyl-*N*-hydroxysuccinimide ester [311]. The polymers were immobilized onto gold nanoparticles to prepare multifunctional glyconanoparticles which exhibited specific biomolecular recognition toward avidin and *Ricinus communis* agglutinin lectin, based on their surface biotin and galactose moieties, respectively rendering them attractive for targeted biological processes.

Alternatively, the development of polymeric systems with tailored properties to be used as nonviral gene carriers is a very challenging field of research lately. There is a great need to synthesize new gene delivery vehicles that can deal with problems related to endosomal escape and nuclear entry. Nanogels are ideal candidates for both drug and gene delivery bacause they allow the facile encapsulation of biomolecules along with efficient formulation and storage. Biodegradable, temperature- and pH-sensitive carbohydrate-based cationic nanogels (Figure 7.6) have been shown to serve as effective

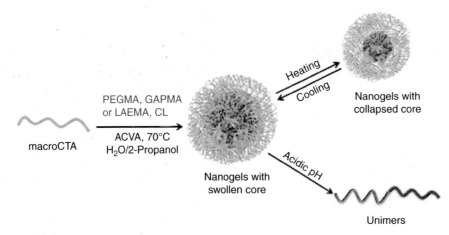

FIGURE 7.6 Schematic representation of the synthesis of biodegradable, responsive carbohydrate-based nanogels and their temperature-sensitive and pH-sensitive behavior. Reproduced with permission from Reference 313.

gene delivery systems to Hep G2 cells [312]. The temperature sensitivity of the nanogels allows their facile complexation of DNA, while their pH response is employed to degrade the nanogels and burst release the plasmid in the endosome. Moreover, these nanogels exhibit low toxicity and degradation in acidic environment which are desirable for systemic applications. Similar nanogels have been employed as an effective codelivery system for the release of both plasmid DNA and proteins [313]. First DNA–nanogel complexes are formed, by the interaction of the carbohydrate residues of the nanogels with DNA, and stabilized with linear cationic glycopolymers which improve cellular uptake and gene expression. Next, the DNA–nanogel complexes are loaded with protein and the controlled release of DNA and proteins upon degradation of the nanogels is studied.

7.5 POLYMER–BIOACTIVE MOLECULE NANOPARTICULATE BIOCONJUGATES

Bioactive molecules which are attractive for nanoparticulate bioconjugate development include biotin, lipids, and synthetic dye molecules. Binding of biotin to avidin, streptavidin, and NeutrAvidin gives nanoparticulate bioconjugates for application in targeted drug delivery, biological imaging, and emulsions. On the other hand, lipids constitute the molecular building blocks for liposomes which are extensively used in drug delivery applications. Finally, synthetic organic dyes present exciting properties for energy harvesting applications.

Bioactive molecule nanoparticulate conjugates based on polymer cross-linked micro- and nanoparticles with introduced biodegradability offer a number of benefits in biomedical applications and in particular as versatile platforms for controlled drug delivery scaffolds to target specific cells.

Responsive microgels that alter their chemical and physical properties in response to external stimuli such as enzymes and pH, and temperature changes are very attractive for use as selective drug delivery carriers, whereas the use of degradable cross-linkers including peptides [314, 315], anhydrides [316], and oligo(lactate) esters [317, 318] allow to introduce functionalized linkages to control biodegradability and thus active release. Such biodegradable nanogels based on PEGMA and comprising disulfide cross-links were prepared by inverse miniemulsion ATRP [319]. The release profile of encapsulated molecules as a function of microgel biodegradation in the presence of a biocompatible reducing glutathione tripeptide was assessed [320]. A fluorescent dye and the model anticancer drug DOX were loaded in the particles and their controlled release triggered by glutathione-induced intracellular biodegradation of the microgels into water-soluble polymers enabled their use for active release within cells. Biocompatibility using live/dead cytotoxicity assays on HeLa human cervix epithelial cells suggested that the nanogels were biocompatible and nontoxic to cells. The functionalization of the microgels with surface –OH groups enabled their facile bioconjugation with biotin, through a carbodiimide coupling reaction, whereas, upon binding of biotin to avidin, avidin–nanogel nanoparticulate bioconjugates were obtained and their bioavailability was investigated using an avidin–2-(4-hydroxyphenylazo)benzoic acid binding assay. The avidin bioconjugate formation was confirmed by fluorescence microscopy using FITC-labeled avidin for binding. Their potential use in targeted drug delivery is envisaged by conjugating the nanogels with cell receptor or ligands such as proteins and antibodies to facilitate their cell internalization through receptor-mediated endocytosis, followed by degradation by cell glutathione. The concept is especially compelling for targeting and suppressing the growth of oncologic cells by induced apoptosis. Moreover, this generic protocol can be used to obtain bioconjugated microgels that contain other bioactive molecules that could enhance normal cellular function, such as glucose or protein metabolism.

A similar biotin–streptavidin binding protocol was introduced to prepare a new class of ultrabright fluorescent nanoparticle bioconjugates for use in fluorescence-based biological detection. The strategy employs semiconducting polymer dots (Pdots) known for their high brightness, fast emission rate, excellent photostability, nonblinking, and nontoxic characteristics attributed to their large absorption cross sections, fast emission rates, and high fluorescence quantum yields [321–323]. However, their widespread use in *in vitro* and *in vivo* fluorescence studies is prevented by the difficulty to control their surface chemistry and bioconjugation to biological molecules. A smart approach was followed to address this challenge based on the entrapment of heterogeneous polymer chains into a single dot particle, driven by hydrophobic interactions [324]. A small amount of amphiphilic polymer-bearing functional groups is cocondensed with the semiconducting polymer to modify and functionalize the nanoparticle surface with carboxylic acid groups for subsequent covalent conjugation to amine-functionalized biomolecules, such as streptavidin and immunoglobulin G (IgG) and other antibodies using standard carbodiimide coupling chemistry (Figure 7.7). The Pdot bioconjugates can effectively and specifically label cellular surface receptors and subcellular structures, such as a cell surface marker in human breast cancer cells, without any detectable nonspecific binding. This functionalization and bioconjugation strategy can be easily applied for the facile covalent linkage of any hydrophobic, fluorescent, semiconducting polymer to biomolecules by specific antigen–antibody interactions opening new avenues for their application in modern biology and biomedicine.

However, besides their use in imaging applications, organic chromophore molecules are also attractive for developing synthetic light harvesting systems. This was tackled by positioning dye molecules into rigid periodic arrays obtained via self-assembly of the tobacco mosaic virus coat protein into well-defined disk and rod structures [21, 325]. The dye was coupled via thiol–maleimide click chemistry by introducing a highly reactive cysteine residue in position 123 of the protein. Multivalent bioconjugates were thus obtained generating effective light collection systems that present high synthetic efficiency, regularity of the chromophore attachment sites, and potential to alter their distances by changing the individual bioconjugation locations. The rod structures were found to achieve more efficient energy transfer than disks attributed to the orientation of the principal transition dipoles of the chromophores along the long axis in the former and the energy transfer through multiple redundant pathways, giving them a greater ability to circumvent assembly defects or photobleached sites [326]. The bioconjugation of suitable donor and acceptor dyes, enables the development of materials collecting a broad range of wavelengths within the solar spectrum with Forster resonance energy transfer (FRET) efficiencies between 33% and 55% [327]. However, the use of the collected energy requires additional modifications to attach photocatalytic and electron transfer groups (porphyrins, phthalocyanines), introduced in the inner pore or the external surface of the nanoassemblies in order to convert the collected light into useful chemical and electrical energy.

Bioconjugate amphiphilic molecules have also led to functionalized vesicles for targeted drug delivery application [328–330]. Such NH_2–PEG–lipid bioconjugates bearing a dithiobenzyl urethane linkage [331], cleavable by free thiols, between the PEG and the lipid were developed and used

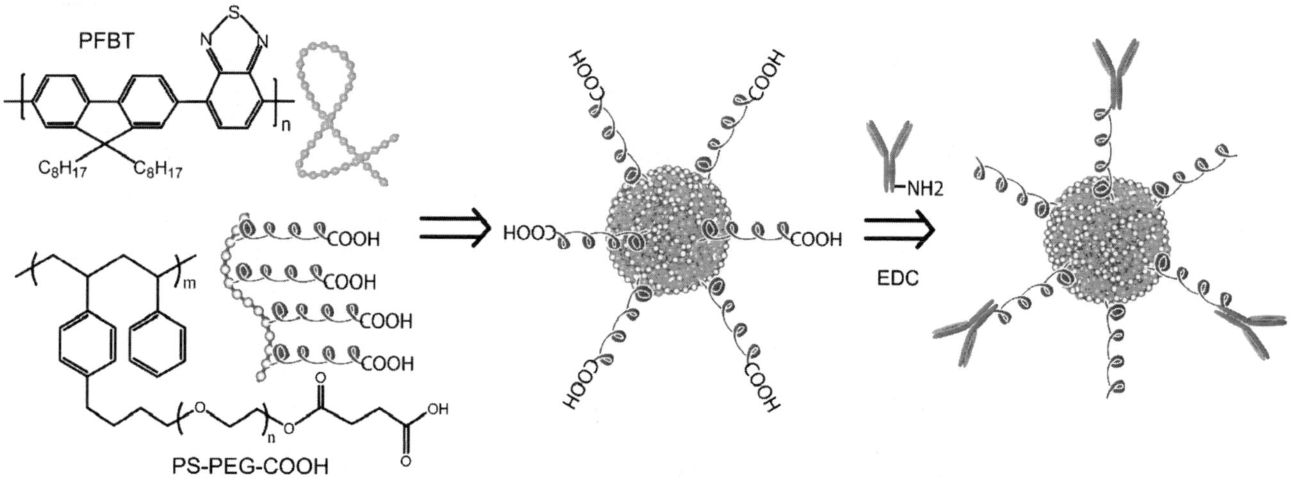

FIGURE 7.7 Surface functionalization of Pdots and subsequent bioconjugation via carbodiimide coupling reaction. Reproduced with permission from Reference 318.

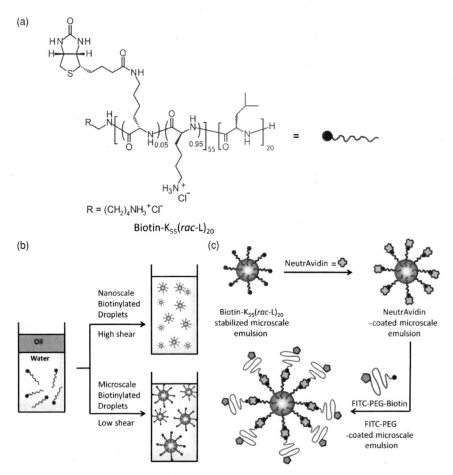

FIGURE 7.8 (a) Chemical structure of the biotin-PLL$_{55}$-*b*-poly(racemic-leucine)$_{20}$ bioconjugate. (b) Biotin-functionalized oil droplets using the block copolypeptide surfactant. (c) Binding of Nuetravidin and formation of PEG-coated droplets. Reproduced with permission from Reference 323. For a color version, see the color plate section.

for the generation of *in vivo* compatible agglomerated liposomes [23]. The amines at the distal ends of the PEG were employed to cross-link the liposomes into agglomerates by the addition of a suitable cross-linking agent reactive toward amines. This allowed to induce changes in the size distribution of the agglomerates, and as a result in the release rate of encapsulated compounds from the carriers upon the thiolytic cleavage of the urethane cross-linkages by cystein or glutathione. Besides the modulation of the drug-release rate from the agglomerates, the regeneration of the lipid was also possible by the addition of mild cleaving agents.

On the other hand, amphiphilic block copolypeptide surfactants can also serve as effective stabilizing molecules of oil-in-water and water-in-oil-in-water double emulsions [332]. An attractive feature of these polypeptide surfactants is their large hydrophilic segments which may bear many amine functional groups allowing their facile derivatization upon bioconjugation. This has been exploited for the development of copolypeptide surfactant-stabilized nano- and microscale emulsion droplets functionalized with biotin

at their outer surface, using standard carbodiimide coupling chemistry (Figure 7.8a) [333]. The biotin density on the droplet surface was varied by mixing biotinylated with nonbiotinylated surfactants, allowing to optimize the level of biotinylation, whereas the attachment of biotin to the hydrophilic domains of the block copolypeptides did not affect the emulsion formation or stability. Conjugated biotin to PLL$_{55}$-*b*-poly(racemic-leucine)$_{20}$ surfactants, enabled the functionalized oil droplets to bind complimentary functional molecules such as NeutrAvidin (Figure 7.8b) giving access to the preparation of droplets with different surface chemistries (e.g., anionic, neutral, or cationic) (Figure 7.8c) upon further binding of biotinylated molecules.

7.6 POLYMER–DRUG NANOPARTICULATE BIOCONJUGATES

Nanotechnology plays a very important role in developing effective drug delivery systems. Nanostructures and nanodevices tailored with specific functionalities have been extensively exploited for therapeutic treatment to transport the drug in the body from the site of administration to the therapeutic target and release it a controlled manner. Nanoparticulate carriers can improve drug properties by controlling release and distribution and enhancing drug absorption by cells and by protecting the drug from degradation, thus maximizing the local efficacy of drugs while concurrently minimizing systemic toxicity. Therefore, they offer many possibilities of improving the specificity of drug treatment. The rapid release (the so-called "burst release") of the encapsulated drug after administration, corresponding to the drug which is simply adsorbed at the surface of the nanocarrier, is another important limitation, since a significant fraction of the drug will be released before reaching the pharmacological target in the body, leading to lower activity and possible side effects.

The necessity to utilize specific components with the drugs, such as polymers, linkers, and cytotoxics, to improve drug-release profiles and targeting has brought a convergence of the fields of bioconjugate chemistry and drug delivery. There are several types of polymer–drug conjugates [334, 335] in the literature that have been exploited for this purpose. Moreover, the attachment of ligand biomolecules that can recognize specific cell receptors may further reduce the side effects of drugs. Polymeric nanoparticles with tunable chemical functionalities and dimensions have opened new avenues in controlled and targeted drug delivery. Nanoparticle's size, geometry, and surface properties can alter their uptake, accumulation, biological response, and toxicity [336–338]. The goal is to design bioactive molecule–polymer nanoparticulate bioconjugates able to deliver therapeutically significant doses of the drug to pathological tissues while avoiding solubility limitations. The key factors to establish pharmacokinetic profiles for multidrug cancer therapies are the degradation behavior of the nanoparticles and drug-release profile [339]. Such conjugates have also potential in the development of targeted drug delivery systems [340]. For example, the synthesis and drug-release behavior of bioconjugates based on functionalized polymeric nanoparticles linked with the monoclonal anti-Integrin aV CD51 (aI) antibody [341] were investigated [342]. The precursor spherical polymer colloids, poly(methylmethacrylate-co-acrylic acid) [P(MMA-co-AA)], poly(methylmethacrylate-co-dimethylpropargylamine) [P(MMA-co-DMPA)], poly(methylmethacrylate-co-allil mercaptane) [P(MMA-co-AM)], with different functionalities, acid, amine, or thiol, were synthesized by emulsion polymerization. Next the antibody, aI, which served as a targeted labeling agent [343] was attached with a simple and straightforward immobilization strategy based on adsorption. The biological efficacy of the nanoparticulate bioconjugates on cell migration was confirmed by the reduced migration potential of the

nanoparticle-treated human kidney cells suggesting that the bioconjugates may be used to interfere with specific cell functions in targeted delivery systems. PEGylated drugs are also attractive in designing nanoparticulate drug carriers. Antibody drugs are often linked to PEG [105]. For example, PEG–scFv compounds have been used versus anthrax exotoxin [344], whereas site-specific PEGylation in the hinge region of Abciximab, an FDA approved Fab effective in preventing coronary thrombosis, has been shown to hinder its binding to platelets resulting in inhibition of platelet aggregation and prolonged circulating times [345]. Anti-inflammatory therapeutics constitutes another category for application of PEGylated antibody drugs. Certolizumab (CDP870) is a TNF-a neutralizing PEG-Fab which has completed Phase III clinical trials in patients with rheumatoid arthritis and Crohn's disease [346, 347]. The conjugate exhibited an extended action in patients after intravenous administration. Another key cytokine in inflammatory responses, GM-CSF, was recently shown to be effectively neutralized by a PEGylated high affinity human scFv conjugate [348]. The use of PEG–scFv in oncology has been proposed using a PEGylated CEA-targeting Fab which exhibited improved pharmacokinetics of the small protein and tumor-specific binding [349]. The potential application of PEG–scFv as a platform technology for multivalent tumor targeting was reported for an anti-HER2 conjugate that demonstrated improved tumor accumulation [350]. Site-specific PEG conjugates [351, 352] and their use in immunoliposome and nanoparticle development have been extensively described [353–356].

An anticancer drug bioconjugate for the treatment of tumor and tumor vasculature was developed using sodium deoxycholate–heparin nanoparticles. Heparin is an anticoagulant agent known to inhibit the activity of growth factors which stimulate the smooth muscle cells around tumors [357–360]. Aminated sodium deoxycholate was prepared by modification of the C3-hydroxyl group with 4-nitrophenyl chloroformate (4-NPC) and ethylenediamine (EDA). Conjugation of heparin to sodium deoxycholate was achieved via amide bond formation between the carboxyl groups of heparin with the amine groups of sodium deoxycholate to bind several sodium deoxycholate moieties along the heparin chain [361]. The resulting bioconjugates formed nanoparticles exhibiting a significant decrease in endothelial tubular formation and were employed in *in vivo* tumor targeting and inhibition of angiogenesis. The self-assembled nanoparticles accumulate in tumors by passive-targeting mechanisms and presented superior antitumor activity and minimized side effects due to the enhanced permeability and retention (EPR) effect.

The development of drug bioconjugates via their covalent coupling to natural lipid derivatives has been proposed for the preparation of injectable prodrugs of different therapeutic compounds. Gemcitabine, used as a first-line treatment against solid tumors [362, 363], suffers low transport rates into the cells due to its high hydrophilicity. To overcome the problem amphiphilic prodrugs of gemcitabine conjugated to lipids were constructed which are expected to passively diffuse across cellular membranes. This was confirmed using gemcitabine conjugates with cardiolipin and phospholipidic derivatives which were shown to be effective transporters [364, 365]. Moreover, the linkage of gemcitabine to a derivative of squalene, a natural lipid precursor of the cholesterol's biosynthesis, by coupling the lipid to the amino group of the cytosine nucleus has been shown to result in bioconjugates that self-assemble in water into nanoparticles displaying a well-defined inner supramolecular organization on the nanoscale level with hexagonal molecular packing, resulting from the stacking of cylinders [366, 367]. The bioconjugates exhibited enhanced *in vitro* and *in vivo* anticancer efficacy [368–370]. Cells treated with radiolabeled squalenoyl gemcitabine were studied aiming to provide further insight into the *in vitro* subcellular localization and on the metabolization pathway of the nanoparticulate squalenoyl prodrug bioconjugates. This revealed that the nanoparticles diffused into cancer cells by an albumin-enhanced passive diffusion [371] and accumulated within cellular membranes due to strong interactions with phospholipids [372], especially those of the endoplasmic reticulum and delivered the prodrug directly in the cell cytoplasm [373]. Next, gemcitabine was either converted into its biologically active triphosphate metabolite or exported from the cells through membrane transporters. *In vitro* cytotoxicity assays revealed that the bioconjugate is more active than the prodrug on a transporter-deficient human-resistant leukemia model, attributed to the subcellular distribution of the drug and its metabolites. However, the nanoparticulate bioconjugates appeared to be considerably more cytotoxic compared to the parent drug and this was attributed to induce cell lysis. Other amphiphilic drugs that interact with cellular membranes [374] have been also shown to lead to toxic side effects [375–377]. The putative membrane-related cytotoxic activity of the nanoparticles on cancer cell lines and erythrocytes was shown. The bioconjugate-induced hemolysis in a time- and dose-dependent fashion, unlike the parent lipid or drug, which clearly proved that it permeabilized the cellular membranes. The mechanism of enhanced membrane permeability and disruption was verified by the insertion of the nanoparticles within model membranes and the resulting formation of nonlamellar structures upon their transfer to the phospholipid bilayers. Mechanistic studies revealed that the nanoassemblies did not enter cells by endocytosis, but rather exchanged bioconjugate molecules with plasmatic proteins, which resulted in a diffusive uptake of the prodrug modulated by the concentration and nature of extracellular proteins [371]. In a related study, squalenoylation was used for another anticancer drug, paclitaxel, to address in this case, the delivery of poorly soluble hydrophobic therapeutics. A variety of conjugates were synthesized by covalent coupling the 2'-hydroxyl group of paclitaxel as ester succinate or cleavable diglycolate ester to 1,1',2-tris-norsqualenoic acid [24]. To increase the hydrophilicity of the bioconjugates PEG chains of different lengths were placed between paclitaxel and the squalenoyl moiety and this resulted in a self-assembly of all prodrugs into stable nanosized aggregates with a negative net surface charge in water. The effect of the linkage bond on the release profile of the drug from these aggregates, in the absence and the presence of serum, was assessed. A high drug loading and the absence of burst release, thus avoiding the use of Cremophor, from the nanoassemblies was confirmed. Preliminary biological data showed the formation of microtubule bundles in HT-29 and KB-31 cells and high cytotoxicity on a lung tumor cell line dependent on the linker stability. The above studies verify the potential of squalenoylation in the design of nanoparticle vehicles based on hydrophilic and hydrophobic drugs although some issues related to the nanoparticulate cytotoxicity must be addressed before the wider use of this technology in nanomedicine.

Dendrimer–drug bioconjugates provide an alternative approach to influence drug transport and release profiles [378]. PAMAM dendrimer-bearing glutamic acid residues on their chain ends have been employed for the synthesis of nanomedicine based on DOX and PEG conjugation (Figure 7.9a) [379]. PEG chains were grafted onto the main amino acid chain, and DOX was covalently linked to the side chains of glutamic acid residues via amide or hydrazone bounds, resulting in a dendritic carrier in which the active drug was shielded by biocompatible PEG chains. *In vitro* evaluation of these dendritic conjugates indicated that their activity was high even in drug-resistant cell lines, whereas the lysosomal pH-labile hydrazone bond led to a considerably higher cytotoxicity than the amide bond, suggesting the endocytosis of the whole prodrug, followed by a progressive lysosomal drug release. Alternatively, an asymmetric biodegradable polyester dendrimer was grafted on one end with eight PEG chains and, on the other end, with DOX via hydrazone bonds (Figure 7.9b) [380]. Despite the asymmetric structure of the nanocarrier the PEG chains were considered to wrap around the whole structure and protect the drug-bearing part of the dendrimer, preventing conjugate aggregation and hepatic capture. This dendrimer–DOX conjugate exhibited excellent EPR-mediated tumor targeting and significantly improved efficiency against DOX-resistant colon tumor xenografts compared to free DOX.

Besides DOX, other anticancer agents have also been conjugated to dendrimers. Small PAMAM dendrimers conjugated to folic acid (FA) showed increased tumor accumulation in the cell's cytoplasm and higher *in vivo* anticancer efficacy against FA-R-overexpressing tumors [381]. In particular, the bioconjugates exhibited an excellent antitumor efficacy in mice-bearing human cervix carcinoma xenografts and induced no acute or chronic toxicity. Biotinylated

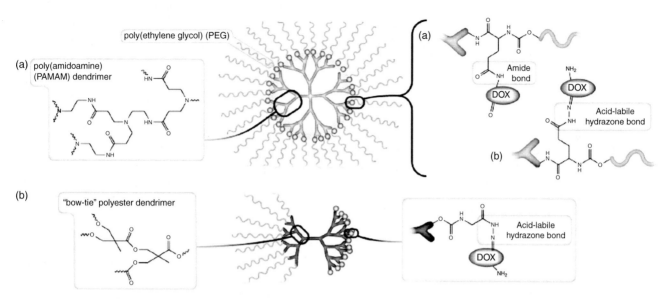

FIGURE 7.9   (a) Dendrimer–drug bioconjugate. DOX conjugated to a PAMAM–PEG dendrimer. (b) Asymmetric dendrimer conjugated to PEG and doxorubicin through hydrazone bonds. Reproduced with permission from Reference 372.

PAMAM dendrimers were found to exhibit an increased cellular uptake by the HeLa human cervix carcinoma cell line compared to the parent drug [382]. In other study, camptothecin (CPT) was bound onto an aspartic acid-functionalized symmetrical poly(lysine) dendrimer-bearing PEG chains [383]. The nanoparticulate bioconjugate exhibited a significantly higher circulation time and a greater tumor accumulation compared to free CPT, resulting in higher *in vivo* anticancer activity against colon carcinoma xenografts.

Block polymer systems able to self-assemble into polymeric micelles comprise another attractive system to serve as nanoparticulate drug carriers. A PEG–poly(aspartic acid) copolymer-bearing FA on the other end of the hydrophilic PEG chain, was conjugated to multiple adriamycin molecules at the hydrophobic PAA part via hydrazone linkers [384]. These FA-bearing micelles interact with the FA receptor (FA-R) overexpressed in cancer cells and lead to considerably enhanced cellular uptake, possibly by endocytosis. However, a drawback of these polymer micellar structures is their possible dissociation under certain conditions. To overcome this problem, a hybrid system based on a similar block copolymer, but bearing a biodegradable, globular, polyester dendritic core has been developed as a monomolecular nanocarrier to increase the construct stability and prevent interparticle aggregation [385].

Alternatively, liposome nanoparticles containing the drug molecules and surface cell-targeting moieties have been employed extensively as nanomedicine carriers [386]. *In vivo* assays confirmed the anticancer efficacy of liposomal prodrugs and the decrease of tumor growth. The mechanism of cell uptake was postulated as the passive diffusion of the prodrug through the cell membrane after leakage from the liposomal bilayers because of the large size of the construct which prevents its endocytosis [387]. PEGylated immunoliposomes were employed against metastatic colon cancer [388]. The drug-loaded immunoliposomes targeted colon cancer cell lines *in vitro* but bound to the plasma membrane and could not be internalized and/or metabolized within lysosomes due to their large size [389]. However, diffusion of the prodrug from the immunoliposomes into the plasma membrane was followed by endocytosis and activation in the lysosomes resulting in intracellular metabolization of the prodrug. An amphiphilic prodrug was also encapsulated into PEGylated liposomes sensitive to phospholipase A2 (PLA2), overexpressed in the interstitium of solid tumors, degradation [390]. This design allows the specific degradation of the carrier with subsequent release of the prodrug in close vicinity of the cancer cells. *In vitro* evaluation of the whole construct on a PLA2-secreting cell line confirmed such an anticancer activity. A drawback of this system is related to the diffusion of the conjugate from the liposomes to the cellular membranes which may lead to premature drug release with possible adverse effects on healthy tissue. Thus, an increased anchoring stability of the construct is required that would necessitate degradation of the carrier for the uptake of the prodrug. Most of the above approaches employ pH-sensitive hydrazone bonds for bioconjugation with linear polymers [391], Mabs [392] or polymeric micelles [384]. The advantage of such hydrazone-based prodrugs is the progressive intracellular drug release over several hours or days rather than a burst release effect [49].

An attractive approach was used to develop hollow nanoscale drug carriers with multiple attachment sites in both interior and exterior sites allowing the linkage of multiple targeting groups and offering the ability to deliver multiple copies of an imaging or drug cargo. The overall approach involves the use of self-assembling viral coat proteins to couple the targeting groups, which specifically bind to receptors on the surface of certain cell types, and the imaging and/or therapeutic agents [393]. These nano-objects were used in tissue-specific delivery strategies offering many promising avenues for the detection and treatment of disease. Specifically, the bacteriophage MS2 viral capsid, which possesses a protein shell consisting of 180 sequence-identical subunits that can be expressed recombinantly in *E. coli* hosts was isolated in fully assembled form. The 27 nm capsids have 32 pores with diameter of 1.8 nm which allow to access the attached molecules in the interior of the protein shell without requiring disassembly. Moreover, the presence of labile linkages allow to release the drug cargo upon endocytosis. Chemical modification of the protein shell to generate bioconjugates with this level of complexity, required the attachment of synthetic functional groups to specific sites on the interior and exterior protein surfaces and thus combinations of chemically orthogonal modification techniques used to instal different molecules in distinct locations on the same protein. Site-specific modification through reactions that target unique positions in the protein sequence and genetically introduced unnatural amino acids allowed to generate well-defined protein bioconjugates for use in materials applications. Moreover, these reactions were carried out in aqueous media at neutral pH, ambient temperatures, and low substrate concentrations. Two different chemical strategies have been developed for the attachment of synthetic cargo to the interior of the capsids. In the first approach the 180 copies of tyrosine on the internal surface of the capsids were reacted selectively with electron-deficient diazonium salts bearing aldehydes and other substituents used to install chromophores and imaging agents [223]. Second a reactive cysteine residue was placed in position 87 providing 180 sulfahydryl groups to react with maleimides for bioconjugate formation (Figure 7.10a) [394]. This allowed to attach 90 copies of a Gd complex or 120 copies of hydrophobic cryptophane cages in the interior of each capsid [395, 396]. Moreover, 120 copies of the chemotherapeutic agent, taxol, or an imaging agent were attached to the interior of the capsid using the cysteine modification strategy (Figure 7.10b) [397]. These well-defined nanoscale materials are also ideal in terms of the required monodispersity and biodegradability and were evaluated for the treatment of MDA-MB-231 cancer cells after endocytosis and linker cleavage and exhibited very promising results. Next, the installation of multiple copies of tissue-specific targeting groups on the external surface of the capsids was achieved by the introduction of p-aminophenylalanine at position 19 in each capsid monomer [119]. This provided 180 copies of the aniline group that were evenly distributed on the exterior surface of the assemblies and via the oxidative coupling reaction gave 135 copies of a breast cancer-targeting peptide in the exterior of the capsids [398]. Moreover, amino functional DNA aptamers that bind to tyrosine kinase 7, a receptor found in many cancer cell lines was coupled to the capsid surface via an N,N-diethylphenylene diamine group [394, 399]. The aptamer-modified capsids were found to bind to Jurkat cells at very low concentrations. Finally, a complex system for targeted therapy that displayed 20 copies of the DNA aptamer and 180 maleimide-functionalized porphyrins to serve as photodynamic therapy agents was developed (Figure 7.10c) [400]. Singlet oxygen production with minimal disassembly and the selective killing of a PTK7-positive cell line in the presence of other cells demonstrated the successful carrier operation.

These investigations provide a foundation for future progress in targeted drug conjugates. An understanding of the transport of the conjugates following internalization, and the controlled release of the drug within the desired intracellular compartment are critical. The greatest challenge in bioconjugate formation is linker design to provide stability to the drug in circulation, while efficiently releasing the drug in the chosen intracellular environment, possibly via specific enzymes or localized pH effects. The use of disulfide-linked drug conjugates may need to be re-visited in view of evidence that lysosomes exhibit an oxidizing environment [401, 402]. The location and number of site-specific attachments of drug molecules to the nanoparticle carrier should also be carefully designed and investigated. Other hurdles include unwanted exposure of the agents to normal tissues, insufficient intratumor uptake, and bioconjugate synthesis. Future developments in the field of nanoparticulate drug bioconjugates require also the design of stable prodrug constructs with appropriate targeting groups that will internalize into the cells and release the drug within the cellular environment. Recently, an interesting apporach toward multiple drug release for higher therapeutic efficacy employed a "self-immolating dendrimer" to trigger the release of several drug molecules upon cleavage of a single stimuli-responsive moiety which induces a chain scission reaction [403–405].

7.7 POLYMER-IMAGING AGENTS NANOPARTICULATE BIOCONJUGATES

Immunoassays have been widely used in medical diagnostics and environmental sensing [406, 407]. Assays often involve an immunocomplex formation resulting from antibody–antigen binding reactions. This process, however, does not generate signals that can be readily visualized. Advances in understanding such biological processes are based on applications of fluorescence microscopy, flow cytometry, and biosensors [408, 409], which make extensive use of

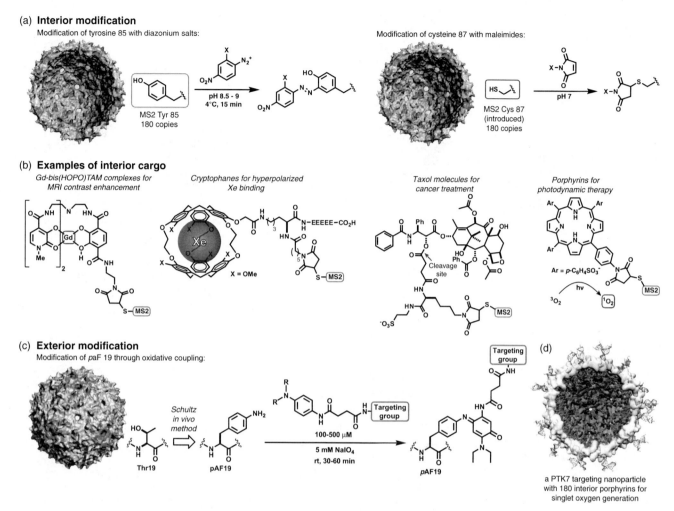

FIGURE 7.10 Synthesis of targeted delivery agents from bacteriophage MS2 capsids. (a) The interior surface can be modified by targeting a native tyrosine or an introduced cysteine residue. (b) Several types of cargo molecules (90–180 copies of each) have been attached to these sites. (c) Targeting groups can be installed on the external surface by modifying an artificial amino acid using a new oxidative coupling reaction. These strategies can be combined to yield targeted particles for therapeutic applications, such as the structure in (d), which combines protein tyrosine kinase7 (PTK7) binding aptamers with porphyrins for singlet oxygen generation. Reproduced with permission from Reference 21. For a color version, see the color plate section.

organic dye molecules as probes. Such signal generators or tag molecules (i.e., fluorescent dyes) are conjugated onto nanoparticles for real-time imaging. The sensitivity of this assay often directly correlates with the dye density or the number of dye molecules per nanoparticle. Higher dye density will result in better sensitivity. However, due to the hydrophobic nature of the fluorescent dyes, the attachment of multiple dye molecules onto a nanoparticle often causes severe precipitation problems in buffer. Thus, the sensitivity enhancement based on multiple dye attachment is greatly limited. Moreover, intrinsic limitations of the conventional dyes, such as low absorptivity and poor photostability, have posed great difficulties in further developments of high sensitivity imaging techniques and high throughout assays [410].

To overcome these problems, numerous approaches using nanoparticles, dendrimers, and self-assembling systems that can act simultaneously as carriers for therapeutic drugs and diagnostic molecules have been developed.

Polymersomes have exhibited very high potential in *in vivo* imaging applications [297]. The encapsulation of near infrared fluorescent dyes [202, 411, 412] and magnetic resonant imaging (MRI) contrast agents [413, 414] in the nanostructures provide nanocarriers with good *in vivo* circulation times, low recognition by the reticuloendothelial system (RES), and good cellular uptake [412]. However, the functionalization of the periphery of the particles with ligands or moieties that allow specific recognition is desirable. The use of the copper-catalyzed [2 + 3]

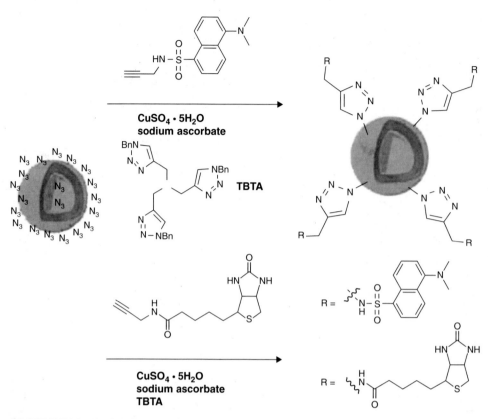

FIGURE 7.11   Surface functionalization of polymersomes assembled from PS-PAA-N₃ with biotin via click chemistry. The same methodology has been applied to immobilize green fluorescent protein (GFP) on the liposome surface. Reproduced with permission from Reference 297.

Huisgen cycloaddition reaction to immobilize fluorescent 5-(dimethylamino)naphthalene-1-sulfonyl groups, GFP, and biotin on the surface of azide end-functional poly(acrylic acid)-poly(styrene) vesicles has been reported (Figure 7.11). In another example, strong, metal ion complexation based on nickel–histidine interactions was employed to complex GFP and bone morphogenetic protein on the surface of PEG–PBd polymersomes functionalized with His-tags [415, 416].

Dendritic probes are particularly advantageous due to their monomolecular nature and surface multivalency. Polyamidoamine dendrimers have been exploited in immunodiagnostics and MRI. The ability to attach a large number of fluorophores through a dendrimer linker molecule thus forming a water-soluble fluorescein–dendrimer nanoparticulate bioconjugate allows a signal amplification strategy and an enhanced fluorescence signal. Fluorescent visualization of antibody–dendrimer–fluorescein conjugates, as well as their immunocomplexes with antigens allowed to probe the immunoreactions between the bioconjugates and antigens in real time [417]. An interesting approach employed dendritic polyglycerol sulfate to synthesize a polyanionic macromolecular conjugate with a near-infrared indocyanine green fluorescent dye which served as a multivalent polymeric probe for inflammation-specific molecular imaging [418]. The

polymer was readily synthesized by a polymerization and sulfation procedure and conjugated with either a near-infrared dye or a VIS dye to facilitate *in vitro* and *in vivo* detection in disease models. The high density sulfate groups generated from the polyol core targeted effectively inflammatory cells and exhibited localization in tissues in the inflammatory infiltrate in the synovial membrane. Fast and selective uptake resulted in *in vivo* accumulation of the construct and enabled the facile differentiation of diseased joints in an animal model of rheumatoid arthritis.

Besides small dye molecules, semiconducting polymers represent a new class of ultrabright fluorescent probes for biological imaging. These polymer dots, known as Pdots, have several important advantages for *in vitro* and *in vivo* fluorescence studies, such as their high brightness, fast emission rate, excellent photostability, nonblinking, and nontoxicity. A main challenge though is the ability to control their surface chemistry for attachment of targeting moieties, drugs, etc. This problem was addressed by condensing the semiconducting polymer with a small amount of another amphiphilic polymer via hydrophobic interactions to prepare polymer nanoparticles with appropriate surface functionalities for conjugation to biomolecules [324]. Single particle imaging, cellular imaging, and flow cytometry experiments indicate

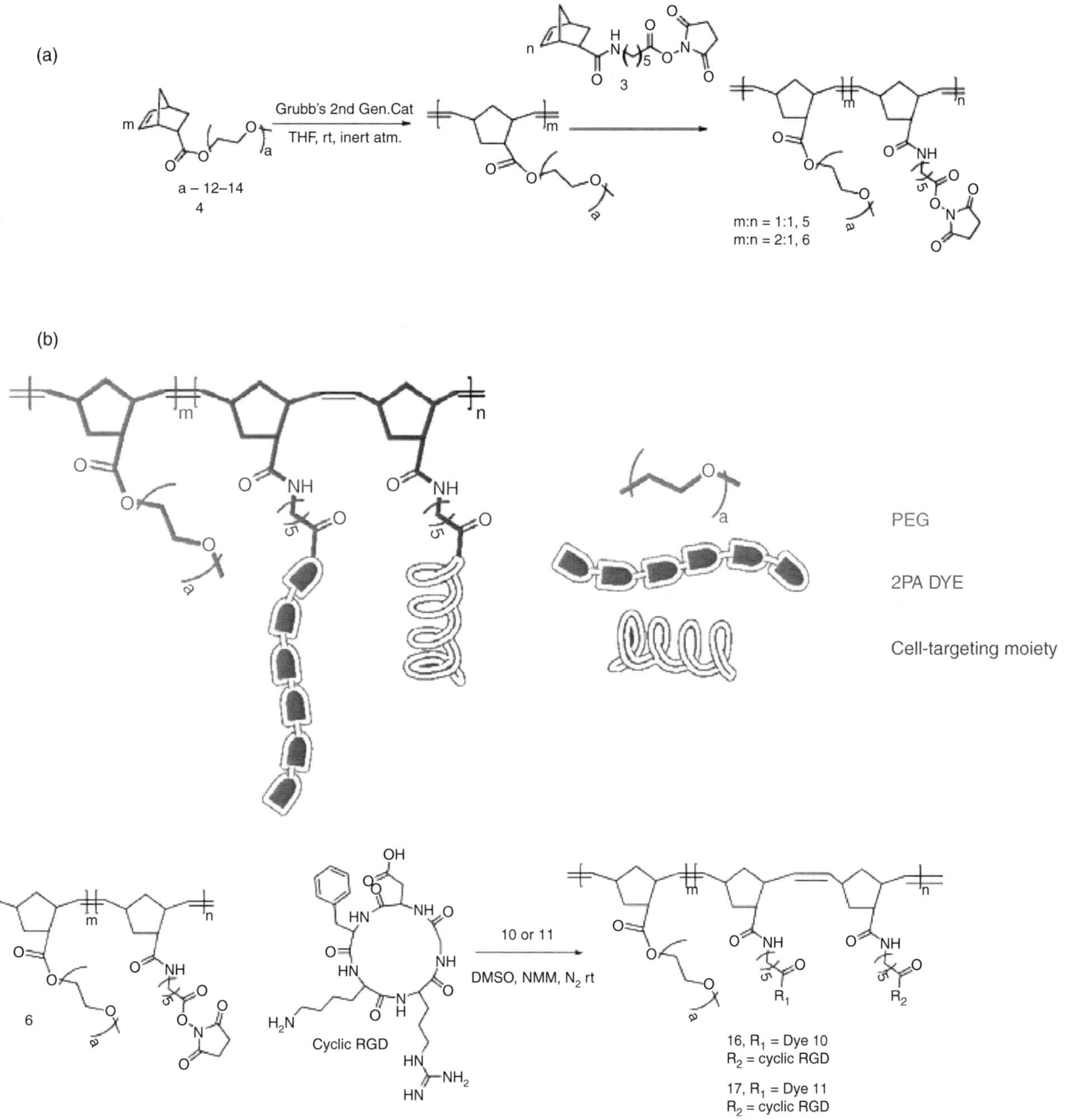

FIGURE 7.12 (a) Synthesis of norbornene-based block copolymers. (b) Structure of the multiscaffold copolymer. (c) Synthesis of polymer probes with targeting ability. Reproduced with permission from Reference 423.

a much higher fluorescence brightness of Pdots compared to those of organic fluorophore molecules and quantum dot probes rendering them very promising tools in fluorescence-based biological detection [410, 419].

Finally, two-photon fluorescence microscopy (2PFM) is a noninvasive and powerful technique used for diagnosis and treatment of diseases. However, fluorescent dyes used in 2PFM are hydrophobic and exhibit low solubility in water which hinders their use in biological imaging. A few 2PA

hydrophilic dyes [420–423] have been prepared, but their synthesis and purification are tedious. An alternative strategy employed dye-doped silica nanoparticles [424] or doping of amphiphilic copolymers with dyes [425, 426], to solubilize the hydropholic probes in the aqueous medium. Polymeric self-assembled nanostructures are ideal fluorescent probe delivery vehicles because they exhibit longer circulation times in the bloodstream and accumulate in the tissues via an EPR effect [427]. A drawback of these nanostructures is their inherent low stability in the blood-stream for long times. To circumvent this problem, the covalent attachment of the dye molecules onto the polymeric nanostructure to form a nanoparticulate bioconjugate has been proposed. Multifunctional nano-objects have been prepared using a norbornene-based block copolymer comprising PEG, a two-photon fluorescent dye and a targeting peptide (Figure 7.12a, 7.12b) [428]. PEG imparted solubility in water but also helped to prevent nonspecific adsorption thus reducing toxicity and immunogenicity [427, 429–431]. The succinimidyl ester groups in the second block allowed covalent conjugation with amine-terminated two-photon fluorenyl probes and an amine-terminated cyclic-RGD peptide to yield polymeric imaging probes with targeting ability (Figure 7.12c).

Due to the amphiphilic nature of the copolymer, micelles were formed in water with a size in the range of 100 nm. The biocompatibility, targeting ability, and 2PFM imaging of the new probes were assessed using the human epithelial U87MG cell line that overexpress $\alpha_v\beta_3$ integrin. The polymer bioconjugates were found to show minimal cytotoxicity and high targeting efficiency at the integrin-rich region of the U87MG cells both by one-photon and two-photon fluorescence microscopy. This high integrin selectivity and good bioimaging properties of the probes is very promising in targeted molecular imaging and in particular for angiogenesis imaging and early cancer detection.

Future developments in the field require to increase the sensitivity and photostability of the nanoparticulate imaging agents. The development of multifunctional nanoparticles that will act simultaneously as effective therapeutic carriers, for targeted and controlled drug delivery, as well as efficient diagnostic tools, will lead to great advances in the fields of nanomedicine and *in vivo* imaging.

REFERENCES

1. Veronese FM. *Biomaterials* 2001;22:405–417.

2. Jeong JH, Park TG. *Bioconjugate Chem* 2001;12:917–923.

3. Goodson RJ, Katre NV. *Nat Biotech* 1990;8:343–346.

4. Okano T. *Biorelated Functional Polymers:Controlled Release and Applications in Biomedical Engineering.* New York: Academic Press; 1998.

5. Sletten EM, Bertozzi CR. *Angew Chem Int Ed* 2009;48:6974–6998.

6. Li M, De P, Gondi SR, Sumerlin BS. *Macromol Rapid Commun* 2008;29:1172–1176.

7. Fournier D, Hoogenboom R, Schubert US. *Chem Soc Rev* 2007;36:1369–1380.

8. Iha RK, Wooley KL, Nyström AM, Burke DJ, Kade MJ, Hawker CJ. *Chem Rev* 2009;109:5620–5686.

9. Becer CR, Hoogenboom R, Schubert US. *Angew Chem Int Ed* 2009;48:4900–4908.

10. Theato P. *J Polym Sci, Part A: Polym Chem* 2008;46:6677–6687.

11. Best MD. *Biochemistry* 2009;48:6571–6584.

12. Prescher JA, Bertozzi CR. *Nat Chem Biol* 2005;1:13–21.

13. Devaraj NK, Weissleder R, Hilderbrand SA. *Bioconjugate Chem* 2008;19:2297–2299.

14. Bundy BC, Swartz JR. *Bioconjugate Chem* 2010;21:255–263.

15. Kiessling LL, Strong LE, Gestwicki JE. Principles for multivalent ligand design. In: *Annual Reports in Medicinal Chemistry*, Chapter 29, Vol. 35. Academic Press; 2000. pp 321–330.

16. Reuter JD, Myc A, Hayes MM, Gan Z, Roy R, Qin D, Yin R, Piehler LT, Esfand R, Tomalia DA, Baker JR. *Bioconjugate Chem* 1999;10:271–278.

17. Arap W, Pasqualini R, Ruoslahti E. *Science* 1998;279:377–380.

18. Jeong JH, Mok H, Oh Y-K, Park TG. *Bioconjugate Chem* 2008;20:5–14.

19. Lee I-H, Yu MK, Kim IH, Lee J-H, Park TG, Jon S. *J Controlled Release* 2011;155:88–95.

20. Mammen M, Choi S-K, Whitesides GM. *Angew Chem Int Ed* 1998;37:2754–2794.

21. Witus LS, Francis MB. *Acc Chem Res* 2011;44:774–783.

22. Paira TK, Banerjee S, Raula M, Kotal A, Si S, Mandal TK. *Macromolecules* 2010;43:4050–4061.

23. Karathanasis E, Ayyagari AL, Bhavane R, Bellamkonda RV, Annapragada AV. *J Controlled Release* 2005;103:159–175.

24. Dosio F, Reddy LH, Ferrero A, Stella B, Cattel L, Couvreur P. *Bioconjugate Chem* 2010;21:1349–1361.

25. Kelly TL, Lam MCW, Wolf MO. *Bioconjugate Chem* 2006;17:575–578.

26. Upadhyay KK, Meins JFL, Misra A, Voisin P, Bouchaud V, Ibarboure E, Schatz C, Lecommandoux S. *Biomacromolecules* 2009;10:2802–2808.

27. Langer R. *Science* 2001;293:58–59.

28. Langer R. *Nature* 1998;392:5–10.

29. Guo X, Szoka FC. *Acc Chem Res* 2003;36:335–341.

30. Torchilin VP. *Adv Drug Delivery Rev* 2006;58:1532–1555.

31. Discher DE, Ahmed F. *Annu Rev Biomed Eng* 2006;8:323–341.

32. Peer D, Karp JM, Hong S, Farokhzad OC, Margalit R, Langer R. *Nat Nano* 2007;2:751–760.

33. Duncan R. *Nat Rev Drug Discov* 2003;2:347–360.

34. Vandermeulen GWM, Klok H-A. *Macromol Biosci* 2004;4:383–398.

35. Le Droumaguet B, Nicolas J. *Polym Chem* 2010;1:563–598.

36. Li M, De P, Li H, Sumerlin BS. *Polym Chem* 2010;1:854–859.

37. de la Casa RM, Guisán JM, Sánchez-Montero JM, Sinisterra JV. *Enzyme Microb Technol* 2002;30:30–40.

38. Miyamoto D, Watanabe J, Ishihara K. *J Appl Polym Sci* 2005;95:615–622.

39. Murphy A, Ó Fágáin C. *Biotechnol Bioeng* 1998;58:366–373.

40. Harris JM, Chess RB. *Nat Rev Drug Discov* 2003;2:214–221.

41. Matsushima A, Kodera Y, Hiroto M, Nishimura H, Inada Y. *J Mol Catal B: Enzym* 1996;2:1–17.

42. Ding Z, Chen G, Hoffman AS. *J Biomed Mater Research* 1998;39:498–505.

43. Ito Y, Fujii H, Imanishi Y. *Biotechnol Progr* 1994;10:398–402.

44. Yan M, Ge J, Liu Z, Ouyang P. *J Am Chem Soc* 2006;128:11008–11009.

45. Chilkoti A, Dreher MR, Meyer DE. *Adv Drug Delivery Rev* 2002;54:1093–1111.

46. Hoffman AS, Stayton PS. *Macromol Symposia* 2004;207:139–152.

47. Hentschel J, Krause E, Börner HG. *J Am Chem Soc* 2006;128:7722–7723.

48. Pechar M, Kopečková P, Joss L, Kopeček J. *Macromol Biosci* 2002;2:199–206.

49. Bae Y, Fukushima S, Harada A, Kataoka K. *Angew Chem Int Ed* 2003;42:4640–4643.

50. Harris TJ, von Maltzahn G, Lord ME, Park J-H, Agrawal A, Min D-H, Sailor MJ, Bhatia SN. *Small* 2008;4:1307–1312.

51. Tirrell JG, Fournier MJ, Mason TL, Tirrel DA. *Chem Eng News Arch* 1994;72:40–51.

52. Chaiken IM. *Anal Biochem* 1979;97:1–10.

53. Self CH, Cook DB. *Curr Opin Biotechnol* 1996;7:60–65.

54. Yoo TJ. Fluorescent probes for antibody active sites. In: Chen RF, Edelhoch H (eds), *Biochemical Fluorescence: Concepts*, Vol. 2. 1976. pp 879–899.

55. Ito Y, Sugimura N, Kwon OH, Imanishi Y. *Nat Biotech* 1999;17:73–75.

56. Zhu G, Wang P. *J Am Chem Soc* 2004;126:11132–11133.

57. Abuchowski A, van Es T, Palczuk NC, Davis FF. *J Biol Chem* 1977;252:3578–3581.

58. Veronese F, Mero A. *BioDrugs* 2008;22:315–329.

59. Fishburn CS. *J Pharm Sci* 2008;97:4167–4183.

60. Roberts MJ, Bentley MD, Harris JM. *Adv Drug Delivery Rev* 2002;54:459–476.

61. Zalipsky S. *Bioconjugate Chem* 1995;6:150–165.

62. Greenwald RB. *J Controlled Release* 2001;74:159–171.

63. Kozlowski A, Milton Harris J. *J Controlled Release* 2001;72:217–224.

64. Ramon J, Saez V, Baez R, Aldana R, Hardy E. *Pharm Res* 2005;22:1375–1387.

65. Fee CJ. *Biotechnol Bioeng* 2007;98:725–731.

66. Veronese FM, Monfardini C, Caliceti P, Schiavon O, Scrawen MD, Beer D. *J Controlled Release* 1996;40:199–209.

67. Sherman MR, Williams LD, Saifer MGP, French JA, Kwak LW, Oppenheim JJ. Conjugation of high-molecular weight poly(ethylene glycol) to cytokines: granulocyte-macrophage colony-stimulating factors as model substrates. In: *Poly(ethylene glycol)*, Vol. 680. American Chemical Society; 1997. pp 155–169.

68. Morpurgo M, Monfardini C, Hofland LJ, Sergi M, Orsolini P, Dumont JM, Veronese FM. *Bioconjugate Chem* 2002;13:1238–1243.

69. Kinstler O, Brems D, Lauren S, Paige A, Hamburger J, Treuheit M. *Pharm Res* 1996;13:996–1002.

70. Tao L, Mantovani G, Lecolley F, Haddleton DM. *J Am Chem Soc* 2004;126:13220–13221.

71. Kochendoerfer GG, Chen S-Y, Mao F, Cressman S, Traviglia S, Shao H, Hunter CL, Low DW, Cagle EN, Carnevali M, Gueriguian V, Keogh PJ, Porter H, Stratton SM, Wiedeke MC, Wilken J, Tang J, Levy JJ, Miranda LP, Crnogorac MM, Kalbag S, Botti P, Schindler-Horvat J, Savatski L, Adamson JW, Kung A, Kent SBH, Bradburne JA. *Science* 2003;299:884–887.

72. Allard L, Cheynet V, Oriol G, Gervasi G, Imbert-Laurenceau E, Mandrand B, Delair T, Mallet F. *Bioconjugate Chem* 2004;15:458–466.

73. Caliceti P, Schiavon O, Sartore L, Monfardini C, Veronese FM. *J Bioact Compat Polym* 1993;8:41–50.

74. Ehrat M, Luisi PL. *Biopolymers* 1983;22:569–573.

75. Uchio T, Baudyš M, Liu F, Song SC, Kim SW. *Adv Drug Delivery Rev* 1999;35:289–306.

76. Hannink JM, Cornelissen JJLM, Farrera JA, Foubert P, De Schryver FC, Sommerdijk NAJM, Nolte RJM. *Angew Chem Int Ed* 2001;40:4732–4734.

77. Kulkarni S, Schilli C, Müller AHE, Hoffman AS, Stayton PS. *Bioconjugate Chem* 2004;15:747–753.

78. Griffith BR, Allen BL, Rapraeger AC, Kiessling LL. *J Am Chem Soc* 2004;126:1608–1609.

79. Schlick TL, Ding Z, Kovacs EW, Francis MB. *J Am Chem Soc* 2005;127:3718–3723.

80. Shao H, Crnogorac MM, Kong T, Chen S-Y, Williams JM, Tack JM, Gueriguian V, Cagle EN, Carnevali M, Tumelty D, Paliard X, Miranda LP, Bradburne JA, Kochendoerfer GG. *J Am Chem Soc* 2005;127:1350–1351.

81. Rosendahl MS, Doherty DH, Smith DJ, Carlson SJ, Chlipala EA, Cox GN. *Bioconjugate Chem* 2005;16:200–207.

82. Kogan TP. *Synth Commun* 1992;22:2417–2424.

83. Stayton PS, Shimoboji T, Long C, Chilkoti A, Ghen G, Harris JM, Hoffman AS. *Nature* 1995;378:472–474.

84. Ding Z, Fong RB, Long CJ, Stayton PS, Hoffman AS. *Nature* 2001;411:59–62.

85. Shimoboji T, Larenas E, Fowler T, Kulkarni S, Hoffman AS, Stayton PS. *Proc National Acad Sci* 2002;99:16592–16596.

86. Velonia K, Rowan AE, Nolte RJM. *J Am Chem Soc* 2002;124:4224–4225.

87. Pennadam SS, Lavigne MD, Dutta CF, Firman K, Mernagh D, Górecki DC, Alexander C. *J Am Chem Soc* 2004;126:13208–13209.

88. Li J-T, Carlsson J, Lin J-N, Caldwell KD. *Bioconjugate Chem* 1996;7:592–599.

89. Herman S, Loccufier J, Schacht E. *Macromol Chem Phys* 1994;195:203–209.

90. Huang Y, Komatsu T, Wang R-M, Nakagawa A, Tsuchida E. *Bioconjugate Chem* 2006;17:393–398.

91. Huang Y, Komatsu T, Yamamoto H, Horinouchi H, Kobayashi K, Tsuchida E. *J Biomed Mater Res Part A* 2004;71A:63–69.

92. Wang L, Schultz PG. *Angew Chem Int Ed* 2005;44:34–66.

93. Link AJ, Mock ML, Tirrell DA. *Curr Opin Biotechnol* 2003;14:603–609.

94. Wang M, Basu A, Palm T, Hua J, Youngster S, Hwang L, Liu H-C, Li X, Peng P, Zhang Y, Zhao H, Zhang Z, Longley C, Mehlig M, Borowski V, Sai P, Viswanathan M, Jang E, Petti G, Liu S, Yang K, Filpula D. *Bioconjugate Chem* 2006;17:1447–1459.

95. Miyazaki K, Takaku H, Umeda M, Fujita T, Huang W, Kimura T, Yamashita J, Horio T. *Cancer Res* 1990;50:4522–4527.

96. Ensor CM, Holtsberg FW, Bomalaski JS, Clark MA. *Cancer Res* 2002;62:5443–5450.

97. Beloussow K, Wang L, Wu J, Ann D, Shen W-C. *Cancer Lett* 2002;183:155–162.

98. Cheng PNM, Leung YC, Lo WH, Tsui SM, Lam KC. *Cancer Lett* 2005;224:67–80.

99. Izzo F, Marra P, Beneduce G, Castello G, Vallone P, De Rosa V, Cremona F, Ensor CM, Holtsberg FW, Bomalaski JS, Clark MA, Ng C, Curley SA. *J Clin Oncol* 2004;22:1815–1822.

100. Park IS, Kang SW, Shin YJ, Chae KY, Park MO, Kim MY, Wheatley DN, Min BH. *Br J Cancer* 2003;89:907–914.

101. Thomas JB, Holtsberg FW, Ensor CM, Bomalaski JS, Clark MA. *Biochem J* 2002;363:581–587.

102. Sugimura K, Ohno T, Kusuyama T, Azuma I. *Melanoma Res* 1992;2:191–196.

103. Wheatley DN, Kilfeather R, Stitt A, Campbell E. *Cancer Lett* 2005;227:141–152.

104. Holtsberg FW, Ensor CM, Steiner MR, Bomalaski JS, Clark MA. *J Controlled Release* 2002;80:259–271.

105. Zhao H, Yang K, Martinez A, Basu A, Chintala R, Liu H-C, Janjua A, Wang M, Filpula D. *Bioconjugate Chem* 2006;17:341–351.

106. Miron T, Wilchek M. *Bioconjugate Chem* 1993;4:568–569.

107. Kolb HC, Finn MG, Sharpless KB. *Angew Chem Int Ed* 2001;40:2004–2021.

108. Cornish VW, Hahn KM, Schultz PG. *J Am Chem Soc* 1996;118:8150–8151.

109. Wang Q, Chan TR, Hilgraf R, Fokin VV, Sharpless KB, Finn MG. *J Am Chem Soc* 2003;125:3192–3193.

110. Link AJ, Tirrell DA. *J Am Chem Soc* 2003;125:11164–11165.

111. Agard NJ, Prescher JA, Bertozzi CR. *J Am Chem Soc* 2004;126:15046–15047.

112. Saxon E, Bertozzi CR. *Science* 2000;287:2007–2010.

113. Kiick KL, Saxon E, Tirrell DA, Bertozzi CR. *Proc National Acad Sci* 2002;99:19–24.

114. Prescher JA, Dube DH, Bertozzi CR. *Nature* 2004;430:873–877.

115. Nilsson BL, Hondal RJ, Soellner MB, Raines RT. *J Am Chem Soc* 2003;125:5268–5269.

116. Dirks AJ, van Berkel SS, Hatzakis NS, Opsteen JA, van Delft FL, Cornelissen JJLM, Rowan AE, van Hest JCM, Rutjes FPJT, Nolte RJM. *Chem Commun* 2005;0:4172–4174.

117. Deiters A, Cropp TA, Summerer D, Mukherji M, Schultz PG. *Bioorg Med Chem Lett* 2004;14:5743–5745.

118. Hooker JM, Esser-Kahn AP, Francis MB. *J Am Chem Soc* 2006;128:15558–15559.

119. Mehl RA, Anderson JC, Santoro SW, Wang L, Martin AB, King DS, Horn DM, Schultz PG. *J Am Chem Soc* 2003;125:935–939.

120. Tumelty D, Carnevali M, Miranda LP. *J Am Chem Soc* 2003;125:14238–14239.

121. Garanger E, Boturyn D, Renaudet O, Defrancq E, Dumy P. *The J Org Chem* 2006;71:2402–2410.

122. Gilmore JM, Scheck RA, Esser-Kahn AP, Joshi NS, Francis MB. *Angew Chem Int Ed* 2006;45:5307–5311.

123. Jones DS, Cockerill KA, Gamino CA, Hammaker JR, Hayag MS, Iverson GM, Linnik MD, McNeeley PA, Tedder ME, Ton-Nu H-T, Victoria EJ. *Bioconjugate Chem* 2001;12:1012–1020.

124. Jones DS, Branks MJ, Campbell M-A, Cockerill KA, Hammaker JR, Kessler CA, Smith EM, Tao A, Ton-Nu H-T, Xu T. *Bioconjugate Chem* 2003;14:1067–1076.

125. Becker ML, Liu J, Wooley KL. *Chem Commun* 2003;180–181.

126. Hentschel J, Bleek K, Ernst O, Lutz J-F, Borner HG. *Macromolecules* 2008;41:1073–1075.

127. Wang J-S, Matyjaszewski K. *J Am Chem Soc* 1995;117:5614–5615.

128. Moad G, Rizzardo E, Thang SH. *Aust J Chem* 2005;58:379–410.

129. Heredia KL, Maynard HD. *Org Biomol Chem* 2007;5:45–53.

130. Nicolas J, Mantovani G, Haddleton DM. *Macromol Rapid Commun* 2007;28:1083–1111.

131. De P, Li M, Gondi SR, Sumerlin BS. *J Am Chem Soc* 2008;130:11288–11289.

132. Heredia KL, Grover GN, Tao L, Maynard HD. *Macromolecules* 2009;42:2360–2367.

133. Tao L, Kaddis CS, Ogorzalek Loo RR, Grover GN, Loo JA, Maynard HD. *Chem Commun* 2009;16:2148–2150.

134. Matyjaszewski K, Xia J. *Chem Rev* 2001;101:2921–2990.

135. Pintauer T, Matyjaszewski K. *Chem Soc Rev* 2008;37:1087–1097.

136. Kamigaito M, Ando T, Sawamoto M. *Chem Rev* 2001;101:3689–3746.

137. Bontempo D, Heredia KL, Fish BA, Maynard HD. *J Am Chem Soc* 2004;126:15372–15373.

138. Mantovani G, Lecolley F, Tao L, Haddleton DM, Clerx J, Cornelissen JJLM, Velonia K. *J Am Chem Soc* 2005;127:2966–2973.

139. Qi K, Ma Q, Remsen EE, Clark CG, Wooley KL. *J Am Chem Soc* 2004;126:6599–6607.

140. Bontempo D, Li RC, Ly T, Brubaker CE, Maynard HD. *Chem Commun* 2005, 4702–4704.

141. Vázquez-Dorbatt V, Maynard HD. *Biomacromolecules* 2006;7:2297–2302.

142. Lutz J-F, Börner HG, Weichenhan K. *Macromolecules* 2006;39:6376–6383.

143. Zhu J, Li P. *J Polym Sci, Part A: Polym Chem* 2003;41:3346–3353.

144. George A, Radhakrishnan G, Thomas Joseph K. *Polymer* 1985;26:2064–2068.

145. Imai Y, Iwakura Y. *J Appl Polym Sci* 1967;11:1529–1538.

146. Bontempo D, Maynard HD. *J Am Chem Soc* 2005;127:6508–6509.

147. Venkataraman S, Wooley KL. *Macromolecules* 2006;39:9661–9664.

148. Broyer RM, Quaker GM, Maynard HD. *J Am Chem Soc* 2007;130:1041–1047.

149. Rettig H, Krause E, Börner HG. *Macromol Rapid Commun* 2004;25:1251–1256.

150. ten Cate MGJ, Rettig H, Bernhardt K, Börner HG. *Macromolecules* 2005;38:10643–10649.

151. Couet J, Biesalski M. *Macromolecules* 2006;39:7258–7268.

152. Benoit D, Chaplinski V, Braslau R, Hawker CJ. *J Am Chem Soc* 1999;121:3904–3920.

153. Ge J, Lu D, Wang J, Liu Z. *Biomacromolecules* 2009;10:1612–1618.

154. Yan M, Liu Z, Lu D, Liu Z. *Biomacromolecules* 2007;8:560–565.

155. Ge J, Lu D, Wang J, Yan M, Lu Y, Liu Z. *J Phys Chem B* 2008;112:14319–14324.

156. Averick SE, Magenau AJD, Simakova A, Woodman BF, Seong A, Mehl RA, Matyjaszewski K. *Polym Chem* 2011;2:1476–1478.

157. Schlaad H. Solution properties of polypeptide-based copolymers. In: Klok H-A, Schlaad H (eds), *Peptide Hybrid Polymers*, Vol. 202. Berlin Heidelberg: Springer; 2006. pp 53–73.

158. Lim Y-b, Lee M. *J Mater Chem* 2008;18:723–727.

159. Khurana R, Uversky VN, Nielsen L, Fink AL. *J Biol Chem* 2001;276:22715–22721.

160. Hartgerink JD, Beniash E, Stupp SI. *Science* 2001;294:1684–1688.

161. Klok H-A. *Angew Chem Int Ed* 2002;41:1509–1513.

162. van Hest JCM, Tirrell DA. *Chem Commun* 2001;0:1897–1904.

163. Couet J, Samuel JDJS, Kopyshev A, Santer S, Biesalski M. *Angew Chem Int Ed* 2005;44:3297–3301.

164. Ayres L, Hans P, Adams J, Löwik DWPM, van Hest JCM. *J Polym Sci, Part A: Polym Chem* 2005;43:6355–6366.

165. Adams DJ, Atkins D, Cooper AI, Furzeland S, Trewin A, Young I. *Biomacromolecules* 2008;9:2997–3003.

166. Jenekhe SA, Chen XL. *Science* 1999;283:372–375.

167. Kataoka K, Ishihara A, Harada A, Miyazaki H. *Macromolecules* 1998;31:6071–6076.

168. Genové E, Shen C, Zhang S, Semino CE. *Biomaterials* 2005;26:3341–3351.

169. Jung JP, Jones JL, Cronier SA, Collier JH. *Biomaterials* 2008;29:2143–2151.

170. Silva GA, Czeisler C, Niece KL, Beniash E, Harrington DA, Kessler JA, Stupp SI. *Science* 2004;303:1352–1355.

171. Jing P, Rudra JS, Herr AB, Collier JH. *Biomacromolecules* 2008;9:2438–2446.

172. ten Cate MGJ, Severin N, Börner HG. *Macromolecules* 2006;39:7831–7838.

173. Collier JH, Messersmith PB. *Adv Mater* 2004;16:907–910.

174. Waku T, Matsusaki M, Kaneko T, Akashi M. *Macromolecules* 2007;40:6385–6392.

175. Hentschel J, Börner HG. *J Am Chem Soc* 2006;128:14142–14149.

176. Xu L, Anchordoquy T. *J Pharm Sci* 2011;100:38–52.

177. Christie RJ, Nishiyama N, Kataoka K. *Endocrinology* 2010;151:466–473.

178. Elemans JAAW, Rowan AE, Nolte RJM. *J Mater Chem* 2003;13:2661–2670.

179. Klok H-A. *J Polym Sci, Part A: Polym Chem* 2005;43:1–17.

180. Chécot F, Lecommandoux S, Gnanou Y, Klok H-A. *Angew Chem Int Ed* 2002;41:1339–1343.

181. Cornelissen JJLM, Fischer M, Sommerdijk NAJM, Nolte RJM. *Science* 1998;280:1427–1430.

182. Babin J, Rodriguez-Hernandez J, Lecommandoux S, Klok H-A, Achard M-F. *Faraday Discuss* 2005;128:179–192.

183. Hooftman G, Herman S, Schacht E. *J Bioact Compat Polym* 1996;11:135–159.

184. Ding Z, Long CJ, Hayashi Y, Bulmus EV, Hoffman AS, Stayton PS. *Bioconjugate Chem* 1999;10:395–400.

185. Harada A, Cammas S, Kataoka K. *Macromolecules* 1996;29:6183–6188.

186. Agut W, Brûlet A, Taton D, Lecommandoux S. *Langmuir* 2007;23:11526–11533.

187. Zhao C, Zhuang X, He C, Chen X, Jing X. *Macromol Rapid Commun* 2008;29:1810–1816.

188. Triftaridou AI, Chécot F, Iliopoulos I. *Macromol Chem Phys* 2010;211:768–777.

189. Wu Y, Pramanik G, Eisele K, Weil T. *Biomacromolecules* 2012;13:1890–1898.

190. Bellomo EG, Wyrsta MD, Pakstis L, Pochan DJ, Deming TJ. *Nat Mater* 2004;3:244–248.

191. Kimura S, Kim D-H, Sugiyama J, Imanishi Y. *Langmuir* 1999;15:4461–4463.

192. Fujita K, Kimura S, Imanishi Y. *Langmuir* 1999;15:4377–4379.

193. Meng F, Engbers GHM, Feijen J. *J Controlled Release* 2005;101:187–198.

194. Pang Z, Lu W, Gao H, Hu K, Chen J, Zhang C, Gao X, Jiang X, Zhu C. *J Controlled Release* 2008;128:120–127.

195. van Dongen SFM, Nallani M, Schoffelen S, Cornelissen JJLM, Nolte RJM, van Hest JCM. *Macromol Rapid Commun* 2008;29:321–325.

196. Petersen MA, Yin L, Kokkoli E, Hillmyer MA. *Polym Chem* 2010;1:1281–1290.

197. Deming TJ. *Nature* 1997;390:386–389.

198. Kukula H, Schlaad H, Antonietti M, Förster S. *J Am Chem Soc* 2002;124:1658–1663.

199. Gebhardt KE, Ahn S, Venkatachalam G, Savin DA. *J Colloid and Interface Sci* 2008;317:70–76.

200. Holowka EP, Pochan DJ, Deming TJ. *J Am Chem Soc* 2005;127:12423–12428.

201. Gaspard J, Silas JA, Shantz DF, Jan J-S. *Supramol Chem* 2009;22:178–185.

202. Tanisaka H, Kizaka-Kondoh S, Makino A, Tanaka S, Hiraoka M, Kimura S. *Bioconjugate Chem* 2007;19:109–117.

203. Kishimura A, Koide A, Osada K, Yamasaki Y, Kataoka K. *Angew Chem Int Ed* 2007;46:6085–6088.

204. Ray JG, Ly JT, Savin DA. *Polym Chem* 2011;2:1536–1541.

205. Liu J, Zhang Q, Remsen EE, Wooley KL. *Biomacromolecules* 2001;2:362–368.

206. Petka WA, Harden JL, McGrath KP, Wirtz D, Tirrell DA. *Science* 1998;281:389–392.

207. Mart RJ, Osborne RD, Stevens MM, Ulijn RV. *Soft Matter* 2006;2:822–835.

208. Semino CE. *J Dent Res* 2008;87:606–616.

209. Rudra JS, Tripathi PK, Hildeman DA, Jung JP, Collier JH. *Biomaterials* 2010;31:8475–8483.

210. Wang C, Stewart RJ, KopeCek J. *Nature* 1999;397:417–420.

211. Shen W, Zhang K, Kornfield JA, Tirrell DA. *Nat Mater* 2006;5:153–158.

212. Wang C, Kopeček J, Stewart RJ. *Biomacromolecules* 2001;2:912–920.

213. Kim YH, Webster OW. *J Am Chem Soc* 1990;112:4592–4593.

214. Gao C, Yan D. *Prog Polym Sci* 2004;29:183–275.

215. Yemul O, Imae T. *Biomacromolecules* 2005;6:2809–2814.

216. Chen G, Huynh D, Felgner PL, Guan Z. *J Am Chem Soc* 2006;128:4298–4302.

217. Schmid RD, Verger R. *Angew Chem Int Ed* 1998;37:1608–1633.

218. Carrea G, Riva S. *Angew Chem Int Ed* 2000;39:2226–2254.

219. Ge J, Yan M, Lu D, Zhang M, Liu Z. *Biochem Eng J* 2007;36:93–99.

220. Capala J, Barth RF, Bendayan M, Lauzon M, Adams DM, Soloway AH, Fenstermaker RA, Carlsson J. *Bioconjugate Chem* 1996;7:7–15.

221. Harada A, Nakanishi K, Ichimura S, Kojima C, Kono K. *J Polym Sci, Part A: Polym Chem* 2009;47:1217–1223.

222. Kovacs EW, Hooker JM, Romanini DW, Holder PG, Berry KE, Francis MB. *Bioconjugate Chem* 2007;18:1140–1147.

223. Hooker JM, Kovacs EW, Francis MB. *J Am Chem Soc* 2004;126:3718–3719.

224. Patel KG, Swartz JR. *Bioconjugate Chem* 2011;22:376–387.

225. Strable E, Prasuhn DE, Udit AK, Brown S, Link AJ, Ngo JT, Lander G, Quispe J, Potter CS, Carragher B, Tirrell DA, Finn MG. *Bioconjugate Chem* 2008;19:866–875.

226. Welsh JP, Patel KG, Manthiram K, Swartz JR. *Biochem Biophys Res Commun* 2009;389:563–568.

227. Wu M, Brown WL, Stockley PG. *Bioconjugate Chem* 1995;6:587–595.

228. Garcea RL, Gissmann L. *Curr Opin Biotechnol* 2004;15:513–517.

229. Liepold L, Anderson S, Willits D, Oltrogge L, Frank JA, Douglas T, Young M. *Magn Reson Med* 2007;58:871–879.

230. Lee AS, Bütün V, Vamvakaki M, Armes SP, Pople JA, Gast AP. *Macromolecules* 2002;35:8540–8551.

231. Vamvakaki M, Palioura D, Spyros A, Armes SP, Anastasiadis SH. *Macromolecules* 2006;39:5106–5112.

232. Achilleos DS, Hatton TA, Vamvakaki M. *J Am Chem Soc* 2012;134:5726–5729.

233. Fong RB, Ding Z, Long CJ, Hoffman AS, Stayton PS. *Bioconjugate Chem* 1999;10:720–725.

234. Gil ES, Hudson SM. *Prog Polym Sci* 2004;29:1173–1222.

235. Goodman M, Kossoy A. *J Am Chem Soc* 1966;88:5010–5015.

236. Pieroni O, Fissi A, Angelini N, Lenci F. *Acc Chem Res* 2000;34:9–17.

237. Ihalainen JA, Bredenbeck J, Pfister R, Helbing J, Chi L, van Stokkum IHM, Woolley GA, Hamm P. *Proc National Acad Sci* 2007;104:5383–5388.

238. Feringa BL, van Delden RA, Koumura N, Geertsema EM. *Chem Rev* 2000;100:1789–1816.

239. Ercole F, Davis TP, Evans RA. *Polym Chem* 2010;1:37–54.

240. Hoffman AS. *Clin Chem* 2000;46:1478–1486.

241. Magnusson JP, Bersani S, Salmaso S, Alexander C, Caliceti P. *Bioconjugate Chem* 2010;21:671–678.

242. Zarafshani Z, Obata T, Lutz J-Fo. *Biomacromolecules* 2010;11:2130–2135.

243. Ryan SM, Wang X, Mantovani G, Sayers CT, Haddleton DM, Brayden DJ. *J Controlled Release* 2009;135:51–59.

244. Lutz J-F. *Adv Mater* 2011;23:2237–2243.

245. Heredia KL, Bontempo D, Ly T, Byers JT, Halstenberg S, Maynard HD. *J Am Chem Soc* 2005;127:16955–16960.

246. Boyer C, Bulmus V, Liu J, Davis TP, Stenzel MH, Barner-Kowollik C. *J Am Chem Soc* 2007;129:7145–7154.

247. Ladavière C, Delair T, Domard A, Novelli-Rousseau A, Mandrand B, Mallet F. *Bioconjugate Chem* 1998;9:655–661.

248. Chen G, Hoffman AS. *Bioconjugate Chem* 1993;4:509–514.

249. Schering CA, Zhong B, Woo JCG, Silverman RB. *Bioconjugate Chem* 2004;15:673–676.

250. Yasayan G, Saeed AO, Fernandez-Trillo F, Allen S, Davies MC, Jangher A, Paul A, Thurecht KJ, King SM, Schweins R, Griffiths PC, Magnusson JP, Alexander C. *Polym Chem* 2011;2:1567–1578.

251. Cheng N, Liu W, Cao Z, Ji W, Liang D, Guo G, Zhang J. *Biomaterials* 2006;27:4984–4992.

252. Li Z, Zhang Y, Fullhart P, Mirkin CA. *Nano Lett* 2004;4:1055–1058.

253. Alemdaroglu FE, Ding K, Berger R, Herrmann A. *Angew Chem Int Ed* 2006;45:4206–4210.

254. Ergen E, Weber M, Jacob J, Herrmann A, Müllen K. *Chem Eur J* 2006;12:3707–3713.

255. Farokhzad OC, Khademhosseini A, Jon S, Hermmann A, Cheng J, Chin C, Kiselyuk A, Teply B, Eng G, Langer R. *Anal Chem* 2005;77:5453–5459.

256. Farokhzad OC, Jon S, Khademhosseini A, Tran T-NT, LaVan DA, Langer R. *Cancer Res* 2004;64:7668–7672.

257. Farokhzad OC, Cheng J, Teply BA, Sherifi I, Jon S, Kantoff PW, Richie JP, Langer R. *Proc National Acad Sci* 2006;103:6315–6320.

258. Jennings GT, Bachmann MF. *Biol Chem* 2008;389:521–536.

259. Reddy ST, Rehor A, Schmoekel HG, Hubbell JA, Swartz MA. *J Controlled Release* 2006;112:26–34.

260. Elbashir SM, Harborth J, Lendeckel W, Yalcin A, Weber K, Tuschl T. *Nature* 2001;411:494–498.

261. Wolfrum C, Shi S, Jayaprakash KN, Jayaraman M, Wang G, Pandey RK, Rajeev KG, Nakayama T, Charrise K, Ndungo EM, Zimmermann T, Koteliansky V, Manoharan M, Stoffel M. *Nat Biotech* 2007;25:1149–1157.

262. Lorenz C, Hadwiger P, John M, Vornlocher H-P, Unverzagt C. *Bioorg Med Chem Lett* 2004;14:4975–4977.

263. Soutschek J, Akinc A, Bramlage B, Charisse K, Constien R, Donoghue M, Elbashir S, Geick A, Hadwiger P, Harborth J, John M, Kesavan V, Lavine G, Pandey RK, Racie T, Rajeev KG, Rohl I, Toudjarska I, Wang G, Wuschko S, Bumcrot D, Koteliansky V, Limmer S, Manoharan M, Vornlocher H-P. *Nature* 2004;432:173–178.

264. Nishina K, Unno T, Uno Y, Kubodera T, Kanouchi T, Mizusawa H, Yokota T. *Mol Ther* 2008;16:734–740.

265. Kim D-H, Behlke MA, Rose SD, Chang M-S, Choi S, Rossi JJ. *Nat Biotech* 2005;23:222–226.

266. Turner JJ, Arzumanov AA, Gait MJ. *Nucleic Acids Res* 2005;33:27–42.

267. Astriab-Fisher A, Sergueev D, Fisher M, Shaw B, Juliano RL. *Pharm Res* 2002;19:744–754.

268. Turner JJ, Ivanova GD, Verbeure B, Williams D, Arzumanov AA, Abes S, Lebleu B, Gait MJ. *Nucleic Acids Res* 2005;33:6837–6849.

269. Tripathi S, Chaubey B, Ganguly S, Harris D, Casale RA, Pandey VN. *Nucleic Acids Res* 2005;33:4345–4356.

270. Chiu Y-L, Ali A, Chu C-y, Cao H, Rana TM. *Chem Biol* 2004;11:1165–1175.

271. Cesarone G, Edupuganti OP, Chen C-P, Wickstrom E. *Bioconjugate Chem* 2007;18:1831–1840.

272. Moschos SA, Jones SW, Perry MM, Williams AE, Erjefalt JS, Turner JJ, Barnes PJ, Sproat BS, Gait MJ, Lindsay MA. *Bioconjugate Chem* 2007;18:1450–1459.

273. Valade D, Boyer C, Davis TP, Bulmus V. *Aust J Chem* 2009;62:1344–1350.

274. Heredia KL, Nguyen TH, Chang C-W, Bulmus V, Davis TP, Maynard HD. *Chem Commun* 2008;0:3245–3247.

275. Kim SH, Jeong JH, Lee SH, Kim SW, Park TG. *J Controlled Release* 2006;116:123–129.

276. Oishi M, Nagasaki Y, Itaka K, Nishiyama N, Kataoka K. *J Am Chem Soc* 2005;127:1624–1625.

277. Lee SH, Kim SH, Park TG. *Biochem Biophys Res Commun* 2007;357:511–516.

278. Jeong JH, Kim SH, Kim SW, Park TG. *Bioconjugate Chem* 2005;16:1034–1037.

279. Jeong JH, Kim SW, Park TG. *J Controlled Release* 2003;93:183–191.

280. Oishi M, Nagatsugi F, Sasaki S, Nagasaki Y, Kataoka K. *ChemBioChem* 2005;6:718–725.

281. Kim SH, Jeong JH, Lee SH, Kim SW, Park TG. *J Controlled Release* 2008;129:107–116.

282. Maeda H, Wu J, Sawa T, Matsumura Y, Hori K. *J Controlled Release* 2000;65:271–284.

283. Ikeda Y, Taira K. *Pharm Res* 2006;23:1631–1640.

284. Wu G, Yang W, Barth RF, Kawabata S, Swindall M, Bandyopadhyaya AK, Tjarks W, Khorsandi B, Blue TE, Ferketich AK, Yang M, Christoforidis GA, Sferra TJ, Binns PJ, Riley KJ, Ciesielski MJ, Fenstermaker RA. *Clin Cancer Res* 2007;13:1260–1268.

285. Oishi M, Nagasaki Y, Nishiyama N, Itaka K, Takagi M, Shimamoto A, Furuichi Y, Kataoka K. *ChemMedChem* 2007;2:1290–1297.

286. Zhang Y, Zhang Y-f, Bryant J, Charles A, Boado RJ, Pardridge WM. *Clin Cancer Res* 2004;10:3667–3677.

287. Hicke BJ, Stephens AW. *J Clin Invest* 2000;106:923–928.

288. Chu TC, Twu KY, Ellington AD, Levy M. *Nucleic Acids Res* 2006;34:e73.

289. McNamara JO, Andrechek ER, Wang Y, Viles KD, Rempel RE, Gilboa E, Sullenger BA, Giangrande PH. *Nat Biotech* 2006;24:1005–1015.

290. Schatz C, Lecommandoux S. *Macromol Rapid Commun* 2010;31:1664–1684.

291. Schatz C, Louguet S, Le Meins J-F, Lecommandoux S. *Angew Chem Int Ed* 2009;48:2572–2575.

292. Lundquist JJ, Toone EJ. *Chem Rev* 2002;102:555–578.

293. Seeberger PH, Werz DB. *Nature* 2007;446:1046–1051.

294. Spaltenstein A, Whitesides GM. *J Am Chem Soc* 1991;113:686–687.

295. Li Z-C, Liang Y-Z, Li F-M. *Chem Commun* 1999;0:1557–1558.

296. Narain R, Armes SP. *Chem Commun* 2002;0:2776–2777.

297. Brinkhuis RP, Rutjes FPJT, van Hest JCM. *Polym Chem* 2011;2:1449–1462.

298. Du J, O'Reilly RK. *Soft Matter* 2009;5:3544–3561.

299. Blanazs A, Armes SP, Ryan AJ. *Macromol Rapid Commun* 2009;30:267–277.

300. Discher DE, Eisenberg A. *Science* 2002;297:967–973.

301. Du J, Chen Y. *Macromolecules* 2004;37:5710–5716.

302. Kim B-S, Yang W-Y, Ryu J-H, Yoo Y-S, Lee M. *Chem Commun* 2005;0:2035–2037.

303. Felici M, Marzá-Pérez M, Hatzakis NS, Nolte RJM, Feiters MC. *Chem Eur J* 2008;14:9914–9920.

304. Upadhyay KK, Bhatt AN, Mishra AK, Dwarakanath BS, Jain S, Schatz C, Le Meins J-F, Farooque A, Chandraiah G, Jain AK, Misra A, Lecommandoux S. *Biomaterials* 2010;31:2882–2892.

305. Upadhyay KK, Bhatt AN, Castro E, Mishra AK, Chuttani K, Dwarakanath BS, Schatz C, Le Meins J-F, Misra A, Lecommandoux S. *Macromol Biosci* 2010;10:503–512.

306. Otsuka I, Fuchise K, Halila S, Fort Sb, Aissou K, Pignot-Paintrand I, Chen Y, Narumi A, Kakuchi T, Borsali R. *Langmuir* 2009;26:2325–2332.

307. Jung T, Kamm W, Breitenbach A, Kaiserling E, Xiao JX, Kissel T. *Eur J Pharm Biopharm* 2000;50:147–160.

308. Zhang H, Mardyani S, Chan WCW, Kumacheva E. *Biomacromolecules* 2006;7:1568–1572.

309. Ahmed M, Lai BFL, Kizhakkedathu JN, Narain R. *Bioconjugate Chem* 2012;23:1050–1058.

310. Jiang X, Liu S, Narain R. *Langmuir* 2009;25:13344–13350.

311. Deng Z, Li S, Jiang X, Narain R. *Macromolecules* 2009;42:6393–6405.

312. Sunasee R, Wattanaarsakit P, Ahmed M, Lollmahomed FB, Narain R. *Bioconjugate Chem* 2012;23:1925–1933.

313. Ahmed M, Narain R. *Mol Pharm* 2012;9:3160–3170.

314. Plunkett KN, Berkowski KL, Moore JS. *Biomacromolecules* 2005;6:632–637.

315. Kim S, Healy KE. *Biomacromolecules* 2003;4:1214–1223.

316. Muggli DS, Burkoth AK, Keyser SA, Lee HR, Anseth KS. *Macromolecules* 1998;31:4120–4125.

317. Martens PJ, Bryant SJ, Anseth KS. *Biomacromolecules* 2003;4:283–292.

318. Eichenbaum KD, Thomas AA, Eichenbaum GM, Gibney BR, Needham D, Kiser PF. *Macromolecules* 2005;38:10757–10762.

319. Oh JK, Tang C, Gao H, Tsarevsky NV, Matyjaszewski K. *J Am Chem Soc* 2006;128:5578–5584.

320. Oh JK, Siegwart DJ, Lee H-i, Sherwood G, Peteanu L, Hollinger JO, Kataoka K, Matyjaszewski K. *J Am Chem Soc* 2007;129:5939–5945.

321. Wu C, Bull B, Szymanski C, Christensen K, McNeill J. *ACS Nano* 2008;2:2415–2423.

322. Pu K-Y, Li K, Shi J, Liu B. *Chem Mater* 2009;21:3816–3822.

323. Rahim NAA, McDaniel W, Bardon K, Srinivasan S, Vickerman V, So PTC, Moon JH. *Adv Mater* 2009;21:3492–3496.

324. Wu C, Schneider T, Zeigler M, Yu J, Schiro PG, Burnham DR, McNeill JD, Chiu DT. *J Am Chem Soc* 2010;132:15410–15417.

325. Klug A. *Philos Trans R Soc London, Ser B: Biol Sci* 1999;354:531–535.

326. Miller RA, Stephanopoulos N, McFarland JM, Rosko AS, Geissler PL, Francis MB. *J Am Chem Soc* 2010;132:6068–6074.

327. Miller RA, Presley AD, Francis MB. *J Am Chem Soc* 2007;129:3104–3109.

328. Rivnay B, Bayer EA, Wilchek M. [11] Use of avidin — biotin technology for liposome targeting. In: Ralph Green KJW

(ed.), *Methods in Enzymology*, Vol. 149. Academic Press; 1987. pp 119–123.

329. Mart R, Liem K, Webb S. *Pharm Res* 2009;26:1701–1710.

330. Hammer DA, Robbins GP, Haun JB, Lin JJ, Qi W, Smith LA, Peter Ghoroghchian P, Therien MJ, Bates FS. *Faraday Discuss* 2008;139:129–141.

331. Zalipsky S, Qazen M, Walker JA, Mullah N, Quinn YP, Huang SK. *Bioconjugate Chem* 1999;10:703–707.

332. Hanson JA, Chang CB, Graves SM, Li Z, Mason TG, Deming TJ. *Nature* 2008;455:85–88.

333. Hanson JA, Deming TJ. *Polym Chem* 2011;2:1473–1475.

334. Vicent MJ, Greco F, Nicholson RI, Paul A, Griffiths PC, Duncan R. *Angew Chem Int Ed* 2005;44:4061–4066.

335. Khandare J, Minko T. *Prog Polym Sci* 2006;31:359–397.

336. Albanese A, Sykes EA, Chan WCW. *ACS Nano* 2010;4:2490–2493.

337. Zhang S, Li J, Lykotrafitis G, Bao G, Suresh S. *Adv Mater* 2009;21:419–424.

338. Toy R, Hayden E, Shoup C, Baskaran H, Karathanasis E. *Nanotechnology* 2011;22:115101.

339. Sheikh Hassan A, Sapin A, Lamprecht A, Emond E, El Ghazouani F, Maincent P. *Eur J Pharm Biopharm* 2009;73:337–344.

340. Quiles S, Raisch KP, Sanford LL, Bonner JA, Safavy A. *J Med Chem* 2009;53:586–594.

341. Cai W, Chen X. *J Nucl Med* 2008;49:113S–128S.

342. Laganà A, Venditti I, Fratoddi I, Capriotti AL, Caruso G, Battocchio C, Polzonetti G, Acconcia F, Marino M, Russo MV. *J Colloid and Interface Sci* 2011;361:465–471.

343. Eck W, Craig G, Sigdel A, Ritter G, Old LJ, Tang L, Brennan MF, Allen PJ, Mason MD. *ACS Nano* 2008;2:2263–2272.

344. Mabry R, Rani M, Geiger R, Hubbard GB, Carrion R, Brasky K, Patterson JL, Georgiou G, Iverson BL. *Infection and Immunity* 2005;73:8362–8368.

345. Knight DM, Jordan RE, Kruszynski M, Tam SH, Giles-Komar J, Treacy G, Heavner GA. *Platelets* 2004;15:409–418.

346. Choy EHS, Hazleman B, Smith M, Moss K, Lisi L, Scott DGI, Patel J, Sopwith M, Isenberg DA. *Rheumatology* 2002;41:1133–1137.

347. Schreiber S, Rutgeerts P, Fedorak RN, Khaliq–Kareemi M, Kamm MA, Boivin M, Bernstein CN, Staun M, Thomsen OØ, Innes A. *Gastroenterology* 2005;129:807–818.

348. Krinner E-M, Hepp J, Hoffmann P, Bruckmaier S, Petersen L, Petsch S, Parr L, Schuster I, Mangold S, Lorenczewski G, Lutterbüse P, Buziol S, Hochheim I, Volkland J, Mølhøj M, Sriskandarajah M, Strasser M, Itin C, Wolf A, Basu A, Yang K, Filpula D, Sørensen P, Kufer P, Baeuerle P, Raum T. *Protein Eng Des Selection* 2006;19:461–470.

349. Delgado C, Pedley RB, Herraez A, Boden R, Boden JA, Keep PA, Chester KA, Fisher D, Begent RHJ, Francis GE. *Br J Cancer* 1996;73:175–182.

350. Kubetzko S, Balic E, Waibel R, Zangemeister-Wittke U, Plückthun A. *J Biol Chem* 2006;281:35186–35201.

351. Yang K, Basu A, Wang M, Chintala R, Hsieh MC, Liu S, Hua J, Zhang Z, Zhou J, Li M, Phyu H, Petti G, Mendez M, Janjua H, Peng P, Longley C, Borowski V, Mehlig M, Filpula D. *Protein Eng* 2003;16:761–770.

352. Xiong C-Y, Natarajan A, Shi X-B, Denardo GL, Denardo SJ. *Protein Eng Des Selection* 2006;19:359–367.

353. Hayes ME, Drummond DC, Hong K, Zheng WW, Khorosheva VA, Cohen JA, Noble CO IV, Park JW, Marks JD, Benz CC, Kirpotin DB. *Mol Pharm* 2006;3:726–736.

354. Kirpotin DB, Drummond DC, Shao Y, Shalaby MR, Hong K, Nielsen UB, Marks JD, Benz CC, Park JW. *Cancer Res* 2006;66:6732–6740.

355. Mamot C, Drummond DC, Noble CO, Kallab V, Guo Z, Hong K, Kirpotin DB, Park JW. *Cancer Res* 2005;65:11631–11638.

356. Wartlick H, Michaelis K, Balthasar S, Strebhardt K, Kreuter J, Langer K. *J Drug Targeting* 2004;12:461–471.

357. Hejna M, Raderer M, Zielinski CC. *J National Cancer Inst* 1999;91:22–36.

358. Smorenburg SM, Van Noorden CJF. *Pharmacol Rev* 2001;53:93–106.

359. Jayson GC, Gallagher JT. *Br J Cancer* 1997;75:9–16.

360. Soker S, Goldstaub D, Svahn CM, Vlodavsky I, Levi BZ, Neufeld G. *Biochem Biophys Res Commun* 1994;203:1339–1347.

361. Cho KJ, Moon HT, Park G-e, Jeon OC, Byun Y, Lee Y-k. *Bioconjugate Chem* 2008;19:1346–1351.

362. Hertel LW, Boder GB, Kroin JS, Rinzel SM, Poore GA, Todd GC, Grindey GB. *Cancer Res* 1990;50:4417–4422.

363. Hui Y, Reitz J. *Am J Health-Syst Pharm* 1997;54:162–170.

364. Chen P, Chien P-Y, Khan AR, Sheikh S, Ali SM, Ahmad MU, Ahmad I. *Anti-Cancer Drugs* 2006;17:53–61.

365. Alexander R, Greene B, Torti S, Kucera G. *Cancer Chemother Pharmacol* 2005;56:15–21.

366. Couvreur P, Reddy LH, Mangenot S, Poupaert JH, Desmaële D, Lepêtre-Mouelhi S, Pili B, Bourgaux C, Amenitsch H, Ollivon M. *Small* 2008;4:247–253.

367. Bekkara-Aounallah F, Gref R, Othman M, Reddy LH, Pili B, Allain V, Bourgaux C, Hillaireau H, Lepêtre-Mouelhi S, Desmaële D, Nicolas J, Chafi N, Couvreur P. *Adv Funct Mater* 2008;18:3715–3725.

368. Couvreur P, Stella B, Reddy LH, Hillaireau H, Dubernet C, Desmaële D, Lepêtre-Mouelhi S, Rocco F, Dereuddre-Bosquet N, Clayette P, Rosilio V, Marsaud V, Renoir J-M, Cattel L. *Nano Lett* 2006;6:2544–2548.

369. Reddy LH, Dubernet C, Mouelhi SL, Marque PE, Desmaele D, Couvreur P. *J Controlled Release* 2007;124:20–27.

370. Reddy LH, Ferreira H, Dubernet C, Mouelhi SL, Desmaele D, Rousseau B, Couvreur P. *Anti-Cancer Drugs* 2008;19:999–1006. doi: 1010.1097/CAD.1000b1013e3283126585.

371. Bildstein L, Marsaud V, Chacun H, Lepetre-Mouelhi S, Desmaele D, Couvreur P, Dubernet C. *Soft Matter* 2010;6:5570–5580.

372. Pili B, Bourgaux C, Amenitsch H, Keller G, Lepêtre-Mouelhi S, Desmaële D, Couvreur P, Ollivon M. *Biochim Biophys Acta (BBA) - Biomembranes* 2010;1798:1522–1532.

373. Bildstein L, Dubernet C, Marsaud V, Chacun H, Nicolas V, Gueutin C, Sarasin A, Bénech H, Lepêtre-Mouelhi S, Desmaële D, Couvreur P. *J Controlled Release* 2010;147:163–170.

374. Schreier S, Malheiros SVP, de Paula E. *Biochim Biophys Acta (BBA) - Biomembranes* 2000;1508:210–234.

375. Ahyayauch H, Gallego M, Casis O, Bennouna M. *J Physiol Biochem* 2006;62:199–205.

376. Jaromin A, Żarnowski R, Kozubek A. *Cell Mol Biol Lett* 2006;11:438–448.

377. Stasiuk M, Kozubek A. *FEBS Lett* 2008;582:3607–3613.

378. Bildstein L, Dubernet C, Couvreur P. *Adv Drug Delivery Rev* 2011;63:3–23.

379. Kono K, Kojima C, Hayashi N, Nishisaka E, Kiura K, Watarai S, Harada A. *Biomaterials* 2008;29:1664–1675.

380. Lee CC, Gillies ER, Fox ME, Guillaudeu SJ, Fréchet JMJ, Dy EE, Szoka FC. *Proc National Acad Sci* 2006;103:16649–16654.

381. Kukowska-Latallo JF, Candido KA, Cao Z, Nigavekar SS, Majoros IJ, Thomas TP, Balogh LP, Khan MK, Baker JR. *Cancer Res* 2005;65:5317–5324.

382. Yang W, Cheng Y, Xu T, Wang X, Wen L-p. *Eur J Med Chem* 2009;44:862–868.

383. Fox ME, Guillaudeu S, Fréchet JMJ, Jerger K, Macaraeg N, Szoka FC. *Mol Pharm* 2009;6:1562–1572.

384. Bae Y, Jang W-D, Nishiyama N, Fukushima S, Kataoka K. *Mol BioSyst* 2005;1:242–250.

385. Prabaharan M, Grailer JJ, Pilla S, Steeber DA, Gong S. *Biomaterials* 2009;30:5757–5766.

386. Crosasso P, Brusa P, Dosio F, Arpicco S, Pacchioni D, Schuber F, Cattel L. *J Pharm Sci* 1997;86:832–839.

387. Hillaireau H, Couvreur P. *Cell Mol Life Sci* 2009;66:2873–2896.

388. Koning GA, Gorter A, Scherphof GL, Kamps JAAM. *Br J Cancer* 1999;80:1718–1725.

389. Koning GA, Kamps JAAM, Scherphof GL. *Cancer Detection and Prevention* 2002;26:299–307.

390. Kaasgaard T, Andresen TL, Jensen SS, Holte RO, Jensen LT, Jørgensen K. *Chem Phys Lipids* 2009;157:94–103.

391. Choi W-M, Kopečková P, Minko T, Kopeček J. *J Bioact Compat Polym* 1999;14:447–456.

392. Trail P, Willner D, Lasch S, Henderson A, Hofstead S, Casazza A, Firestone R, Hellstrom I, Hellstrom K. *Science* 1993;261:212–215.

393. Koudelka KJ, Manchester M. *Curr Opin Chem Biol* 2010;14:810–817.

394. Tong GJ, Hsiao SC, Carrico ZM, Francis MB. *J Am Chem Soc* 2009;131:11174–11178.

395. Hooker JM, Datta A, Botta M, Raymond KN, Francis MB. *Nano Lett* 2007;7:2207–2210.

396. Meldrum T, Seim KL, Bajaj VS, Palaniappan KK, Wu W, Francis MB, Wemmer DE, Pines A. *J Am Chem Soc* 2010;132:5936–5937.

397. Wu W, Hsiao SC, Carrico ZM, Francis MB. *Angew Chem Int Ed* 2009;48:9493–9497.

398. Carrico ZM, Romanini DW, Mehl RA, Francis MB. *Chem Commun* 2008;0:1205–1207.

399. Shangguan D, Li Y, Tang Z, Cao ZC, Chen HW, Mallikaratchy P, Sefah K, Yang CJ, Tan W. *Proc National Acad Sci* 2006;103:11838–11843.

400. Stephanopoulos N, Tong GJ, Hsiao SC, Francis MB. *ACS Nano* 2010;4:6014–6020.

401. Austin CD, Wen X, Gazzard L, Nelson C, Scheller RH, Scales SJ. *Proc National Acad Sci USA* 2005;102:17987–17992.

402. Wu AM, Senter PD. *Nat Biotech* 2005;23:1137–1146.

403. Meijer EW, van Genderen MHP. *Nature* 2003;426:128–129.

404. Avital-Shmilovici M, Shabat D. *Bioorg Med Chem Lett* 2009;19:3959–3962.

405. Shabat D. *J Polym Sci, Part A: Polym Chem* 2006;44:1569–1578.

406. Harris B. *Trends Biotechnol* 1999;17:290–296.

407. Van Emon JM, L, Gerlach C. *J Microbiol Methods* 1998;32:121–131.

408. Pepperkok R, Ellenberg J. *Nat Rev Mol Cell Biol* 2006;7:690–696.

409. Giepmans BNG, Adams SR, Ellisman MH, Tsien RY. *Science* 2006;312:217–224.

410. Fernandez-Suarez M, Ting AY. *Nat Rev Mol Cell Biol* 2008;9:929–943.

411. Christian DA, Garbuzenko OB, Minko T, Discher DE. *Macromol Rapid Commun* 2010;31:135–141.

412. Christian NA, Milone MC, Ranka SS, Li G, Frail PR, Davis KP, Bates FS, Therien MJ, Ghoroghchian PP, June CH, Hammer DA. *Bioconjugate Chem* 2006;18:31–40.

413. Cheng Z, Tsourkas A. *Langmuir* 2008;24:8169–8173.

414. Cheng Z, Thorek DLJ, Tsourkas A. *Adv Funct Mater* 2009;19:3753–3759.

415. Nehring R, Palivan CG, Casse O, Tanner P, Tüxen J, Meier W. *Langmuir* 2008;25:1122–1130.

416. Nehring R, Palivan CG, Moreno-Flores S, Mantion A, Tanner P, Toca-Herrera JL, Thunemann A, Meier W. *Soft Matter* 2010;6:2815–2824.

417. Ong KK, Jenkins AL, Cheng R, Tomalia DA, Durst HD, Jensen JL, Emanuel PA, Swim CR, Yin R. *Anal Chim Acta* 2001;444:143–148.

418. Licha K, Welker P, Weinhart M, Wegner N, Kern S, Reichert S, Gemeinhardt I, Weissbach C, Ebert B, Haag R, Schirner M. *Bioconjugate Chem* 2011;22:2453–2460.

419. Resch-Genger U, Grabolle M, Cavaliere-Jaricot S, Nitschke R, Nann T. *Nat Meth* 2008;5:763–775.

420. Morales AR, Luchita G, Yanez CO, Bondar MV, Przhonska OV, Belfield KD. *Org Biomol Chem* 2010;8:2600–2608.

421. Wang X, Nguyen DM, Yanez CO, Rodriguez L, Ahn H-Y, Bondar MV, Belfield KD. *J Am Chem Soc* 2010;132:12237–12239.

422. Briñas RP, Troxler T, Hochstrasser RM, Vinogradov SA. *J Am Chem Soc* 2005;127:11851–11862.

423. Chung S-J, Zheng S, Odani T, Beverina L, Fu J, Padilha LA, Biesso A, Hales JM, Zhan X, Schmidt K, Ye A, Zojer E, Barlow S, Hagan DJ, Van Stryland EW, Yi Y, Shuai Z, Pagani GA, Brédas J-L, Perry JW, Marder SR. *J Am Chem Soc* 2006;128:14444–14445.

424. Kim S, Ohulchanskyy TY, Pudavar HE, Pandey RK, Prasad PN. *J Am Chem Soc* 2007;129:2669–2675.

425. Tian Y, Chen CY, Cheng YJ, Young AC, Tucker NM, Jen AKY. *Adv Funct Mater* 2007;17:1691–1697.

426. Chen C-Y, Tian Y, Cheng Y-J, Young AC, Ka J-W, Jen AKY. *J Am Chem Soc* 2007;129:7220–7221.

427. van Hest JCM. *Polym Rev* 2007;47:63–92.

428. Biswas S, Wang X, Morales AR, Ahn H-Y, Belfield KD. *Biomacromolecules* 2010;12:441–449.

429. Kim J-H, Park K, Nam HY, Lee S, Kim K, Kwon IC. *Prog Polym Sci* 2007;32:1031–1053.

430. Lutz J-F, Börner HG. *Prog Polym Sci* 2008;33:1–39.

431. Tian Y, Wu W-C, Chen C-Y, Strovas T, Li Y, Jin Y, Su F, Meldrum DR, Jen AKY. *J Mater Chem* 2010;20:1728–1736.

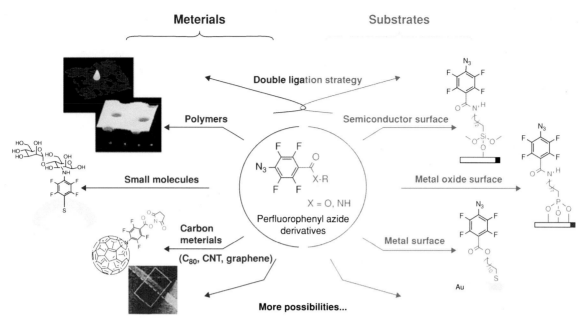

FIGURE 1.16 Applications of perfluorophenyl azide derivatives. Reprinted with permission from Reference 253, Copyright 2010, American Chemical Society.

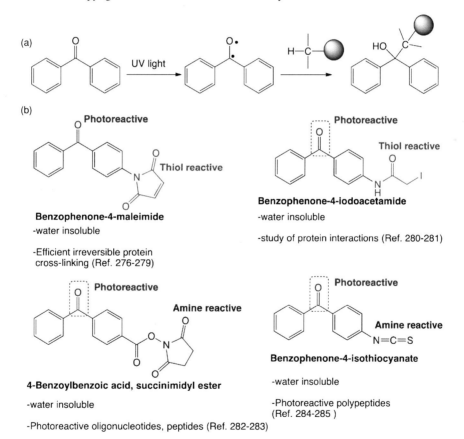

FIGURE 1.17 Benzophenone-based photoreactive cross-linking reagents reactive toward amines and thiols.

Chemistry of Bioconjugates: Synthesis, Characterization, and Biomedical Applications, First Edition. Edited by Ravin Narain.
© 2014 John Wiley & Sons, Inc. Published 2014 by John Wiley & Sons, Inc.

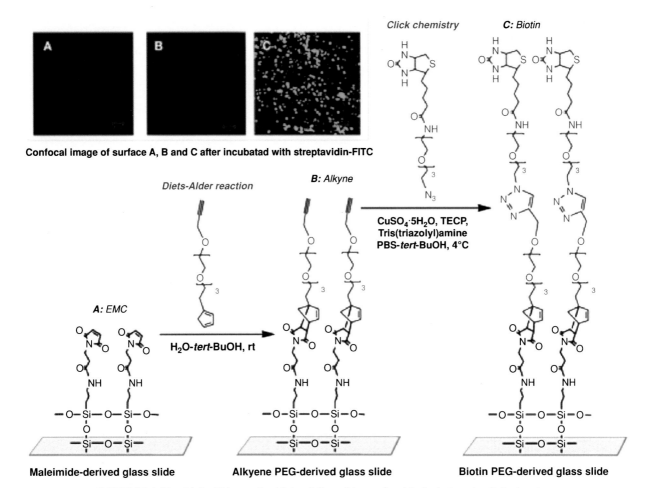

Confocal image of surface A, B and C after incubatad with streptavidin-FITC

Diets-Alder reaction

B: Alkyne

Click chemistry

C: Biotin

A: EMC

H₂O-*tert*-BuOH, rt

CuSO₄·5H₂O, TECP,
Tris(triazolyl)amine
PBS-*tert*-BuOH, 4°C

Maleimide-derived glass slide

Alkyene PEG-derived glass slide

Biotin PEG-derived glass slide

SCHEME 1.47 Diels–Alder cycloaddition followed by surface biotinylation via click chemistry. Reprinted with permission from Reference 360, Copyright 2006, American Chemical Society.

$k = 0.25$
$M^{-1}s^{-1}$

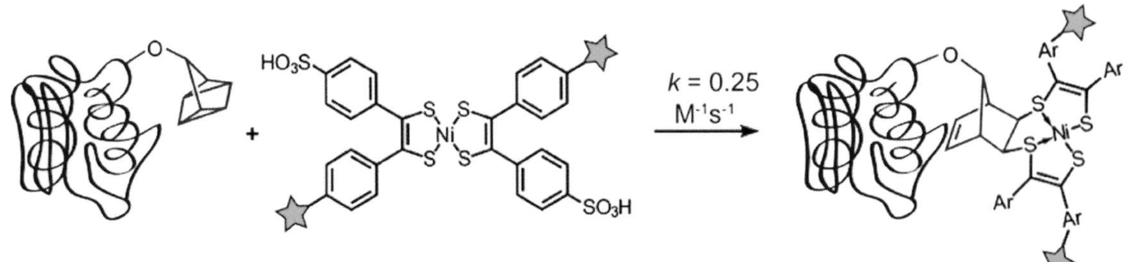

SCHEME 1.51 Quadricyclane ligation for labeling of proteins. Reprinted with permission from Reference 419, Copyright 2011, American Chemical Society.

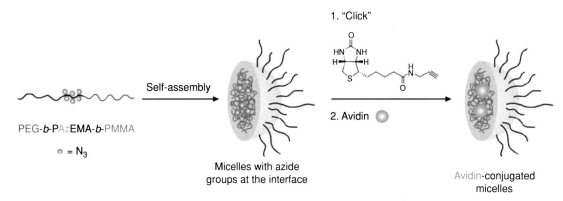

SCHEME 1.68 Bioconjugation of biotin to the interfaces of polymeric micelles and subsequent binding with avidin. Reprinted from Reference 485 with permission from the Royal Society of Chemistry

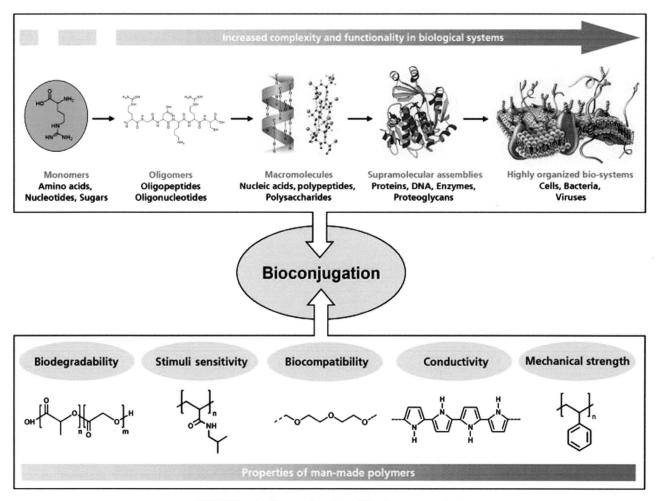

FIGURE 3.1 Schematics of possible bioconjugates [1].

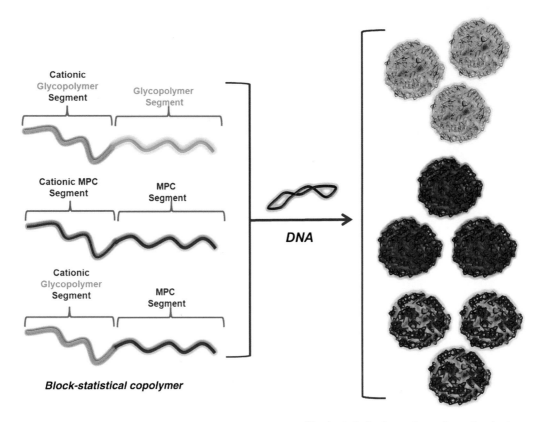

FIGURE 3.3 Formulation of polyplexes using *"block-statistical"* cationic glycopolymers [42].

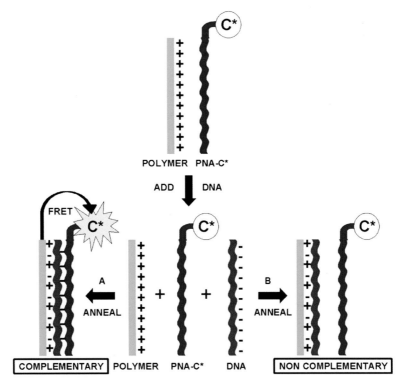

FIGURE 3.4 A representation of CP-PNA reporter conjugates to detect ssDNA [49].

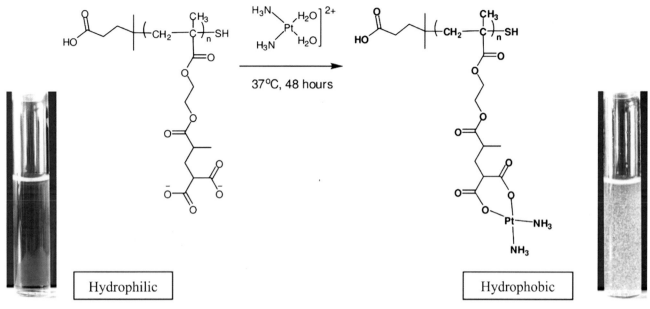

FIGURE 3.5   Synthesis of Pt–polymer conjugates, as shown by change in water solubility [64].

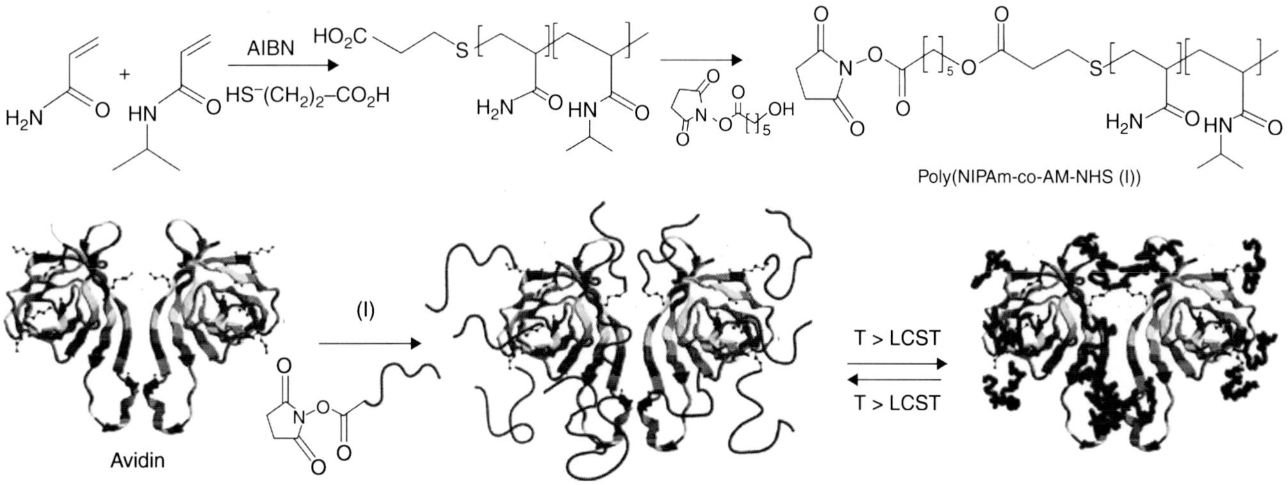

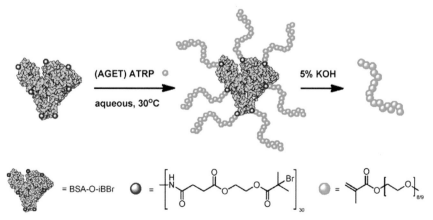

Poly(NIPAm-co-AM-NHS (I))

Avidin

SCHEME 3.9   Synthesis of thermoresponsive copolymers and their attachment with avidin [7].

= BSA-O-iBBr

SCHEME 3.10   Synthesis of PPH-BSA hybrids prepared by ATRP [29].

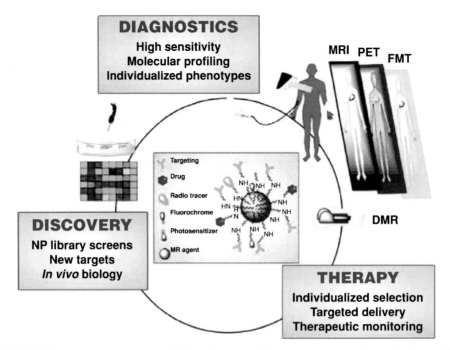

FIGURE 4.9 A superparamagnetic ION with a cross-linked dextran coating. From Reference 66.

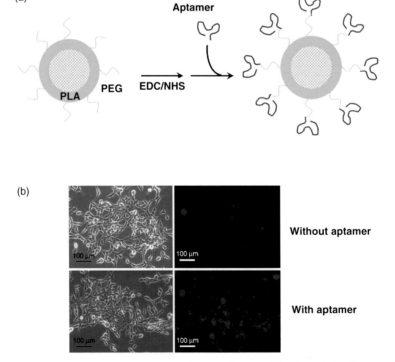

FIGURE 7.3 (a) Schematic representation of the surface modification of the PLA particles by covalent conjugation with RNA aptamers to obtain the nanoparticle–aptamer bioconjugates. (b) Binding of rhodamine-labeled dextran-encapsulated nanoparticles and particles conjugated to the PSMA aptamer, with LNCaP cells. Reproduced with permission from Reference 255.

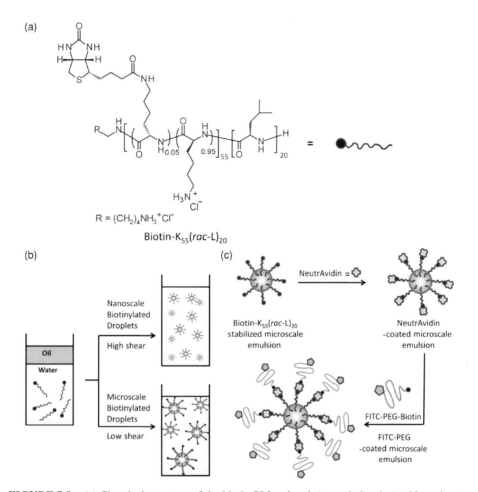

FIGURE 7.8 (a) Chemical structure of the biotin-PLL$_{55}$-*b*-poly(racemic-leucine)$_{20}$ bioconjugate. (b) Biotin-functionalized oil droplets using the block copolypeptide surfactant. (c) Binding of Nuetravidin and formation of PEG-coated droplets. Reproduced with permission from Reference 323.

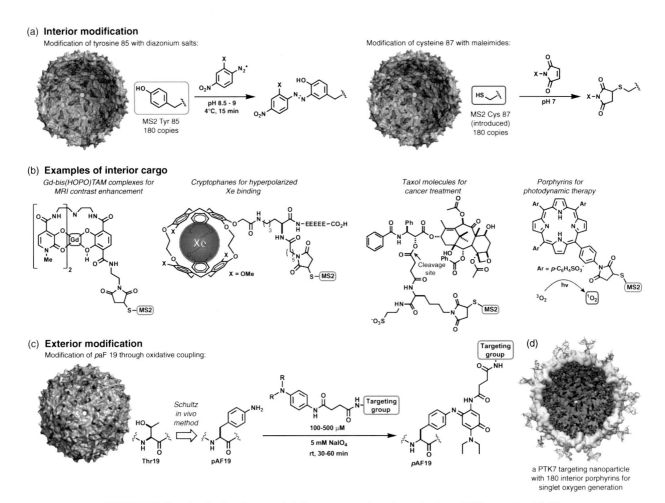

(a) Interior modification

Modification of tyrosine 85 with diazonium salts:

MS2 Tyr 85
180 copies

pH 8.5 - 9
4°C, 15 min

Modification of cysteine 87 with maleimides:

MS2 Cys 87
(introduced)
180 copies

pH 7

(b) Examples of interior cargo

Gd-bis(HOPO)TAM complexes for MRI contrast enhancement

Cryptophanes for hyperpolarized Xe binding

X = OMe

Taxol molecules for cancer treatment

Cleavage site

Porphyrins for photodynamic therapy

Ar = p-C₆H₄SO₃⁻

$^3O_2$ $^1O_2$

hv

(c) Exterior modification

Modification of paF 19 through oxidative coupling:

Schultz in vivo method

Thr19   pAF19

100-500 μM

5 mM NaIO₄
rt, 30-60 min

pAF19

Targeting group

(d)

a PTK7 targeting nanoparticle with 180 interior porphyrins for singlet oxygen generation

FIGURE 7.10 Synthesis of targeted delivery agents from bacteriophage MS2 capsids. (a) The interior surface can be modified by targeting a native tyrosine or an introduced cysteine residue. (b) Several types of cargo molecules (90–180 copies of each) have been attached to these sites. (c) Targeting groups can be installed on the external surface by modifying an artificial amino acid using a new oxidative coupling reaction. These strategies can be combined to yield targeted particles for therapeutic applications, such as the structure in (d), which combines protein tyrosine kinase7 (PTK7) binding aptamers with porphyrins for singlet oxygen generation. Reproduced with permission from Reference 21.

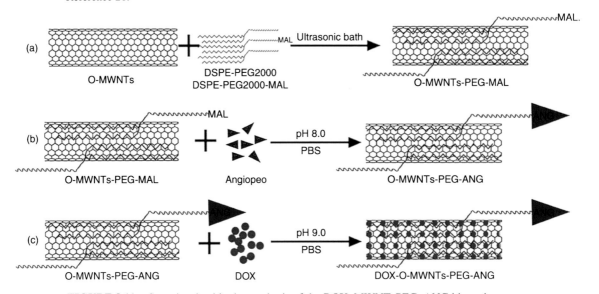

(a) O-MWNTs + DSPE-PEG2000 DSPE-PEG2000-MAL —MAL **Ultrasonic bath** → O-MWNTs-PEG-MAL —MAL

(b) O-MWNTs-PEG-MAL + Angiopeo →(pH 8.0 PBS) O-MWNTs-PEG-ANG

(c) O-MWNTs-PEG-ANG + DOX →(pH 9.0 PBS) DOX-O-MWNTs-PEG-ANG

FIGURE 8.11 Steps involved in the synthesis of the DOX–MWNT–PEG–ANG bioconjugates as potential dual-targeting drug delivery vehicles Reproduced with permission from Reference 55.

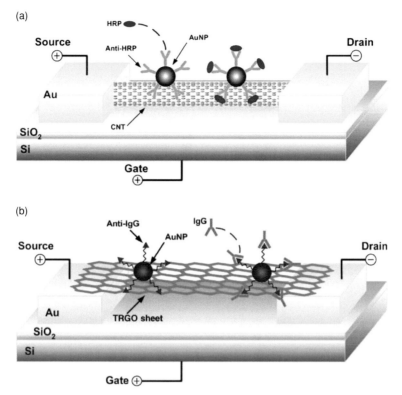

FIGURE 9.3 Detection of HRP by anti-HRP-conjugated AuNPs with (a) carbon nanotubes or (b) GO sheet as field-effect transistor. Reprinted with permission from References 50 and 51, Copyright 2010, Elsevier and John Wiley & Sons, Inc., respectively.

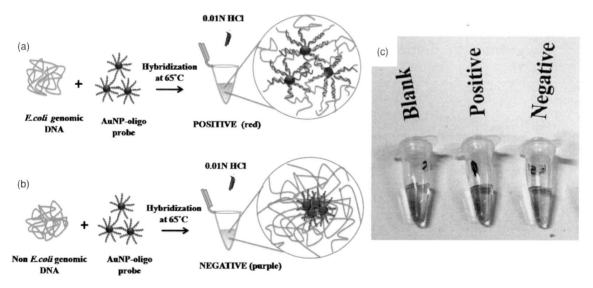

FIGURE 9.5 A schematic representation of the working of oligonucleotide–AuNP conjugate-based *E. coli* detection. (a) In the presence of *E. coli* genomic DNA, AuNPs are stabilized and prevented from aggregation upon acid addition. The red color of the solution indicates presence of target DNA (positive). (b) In the presence of non-*E. coli* genomic DNA, AuNPs tend to aggregate and the color is purple (negative). (c) Visual observation of blank, positive, and negative samples. Adapted from Reference 75.

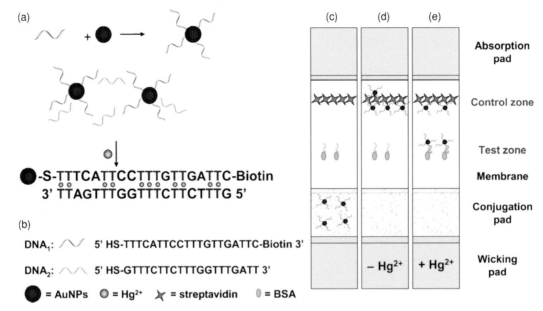

FIGURE 9.6 Design of the test strip format. (a) A schematic illustration of sensors and theranostic agents for Hg^{2+}. (b) Descriptions of DNA$_1$ sequence, DNA$_2$ sequence, AuNPs, Hg^{2+}, streptavidin, and BSA. (c) Blank test strip loaded with DNA-functionalized AuNPs (on the conjugation pad), DNA$_2$–BSA (test zone) and streptavidin (control zone). (d) Negative test: in the absence of Hg^{2+}, the DNA-functionalized AuNPs were captured at the control zone through streptavidin–biotin interaction, producing a single red line. (e) Positive test: in the presence of Hg^{2+}, the DNA-functionalized AuNPs were captured at the test zone by T-Hg^{2+}-T coordination to complementary DNA in addition to the control zone, resulting in two red lines. Reprinted with permission from Reference 76, Copyright 2012, Elsevier.

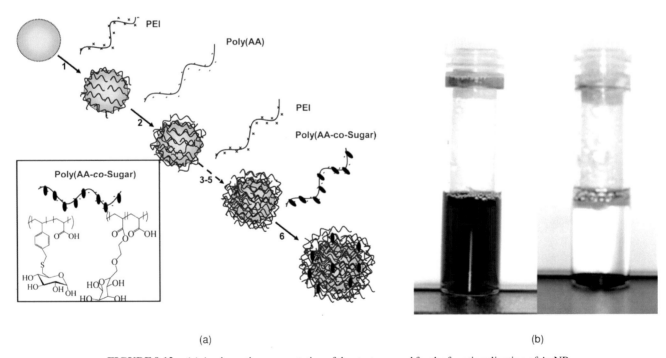

FIGURE 9.12 (a) A schematic representation of the strategy used for the functionalization of AuNP using layer-by-layer methodology; (b) AuNPs/poly(AA-*co*-glucose) conjugates in the presence of Con A before (right) and after (left) the addition of free glucose. Reprinted with permission from Reference 100, Copyright 2010, American Chemical Society.

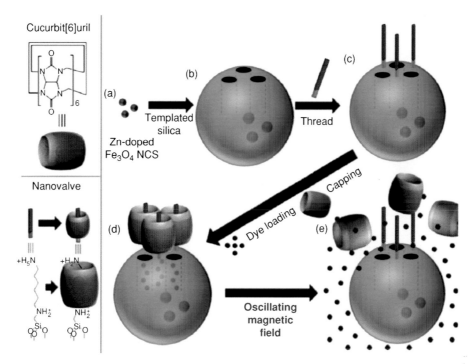

FIGURE 10.11 Synthesis of NPs capable of magnetically triggered drug release. (a) Zn-doped Fe$_3$O$_4$ nanocrystals were synthesized and (b) positioned into the core of the mesoporous silica NPs. (c) The positively charged thread was installed to the NP surface. (d) Drug (or dye) was loaded into the pores of the NPs and capped with the cucurbit[6]uril cap. Drug release was realized when the whole complex was placed in an oscillating magnetic field (e). Reproduced from Reference 152 with permission from American Chemical Society.

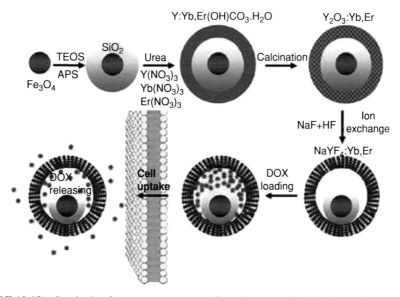

FIGURE 10.18 Synthesis of a mesoporous magnetic and upconversion luminescent nanorattle for DOX delivery. Reproduced with permission from Reference 192. (Copyright American Chemical Society.)

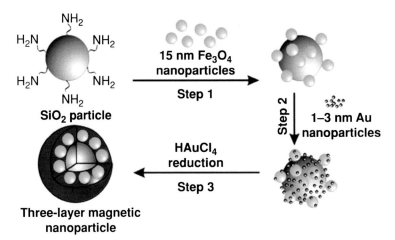

FIGURE 10.22 Formation of gold-coated three-layered magnetic NPs for DNA immobilization. Reproduced with permission from Reference 232. (Copyright American Chemical Society).

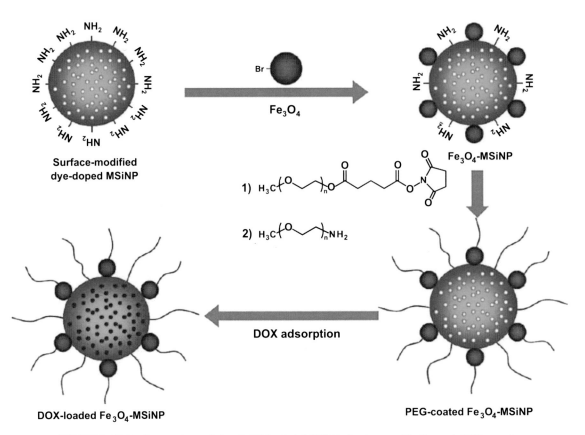

FIGURE 12.2 Preparation of the DOX-loaded MSiNPs having an Fe_3O_4 and PEG on the surface [35].

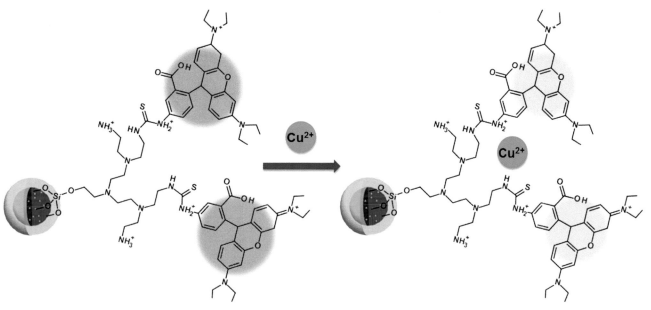

FIGURE 12.5 The fluorescent SiNPs for ultrasensitive detection of Cu^{2+} ion [66].

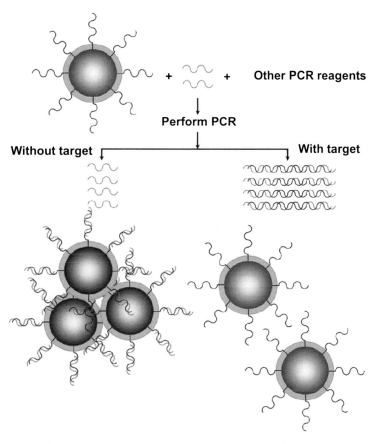

FIGURE 12.7 The aggregation of nanoparticles in PCR system for the detection of the presence/absence of a target oligonucleotide [73].

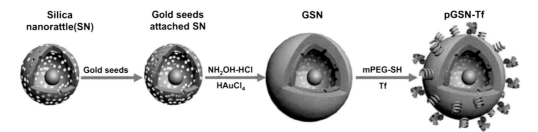

FIGURE 12.8 The Au-coated silica nanorattles (Au-SiNRs) having the PEG and transferrin [95].

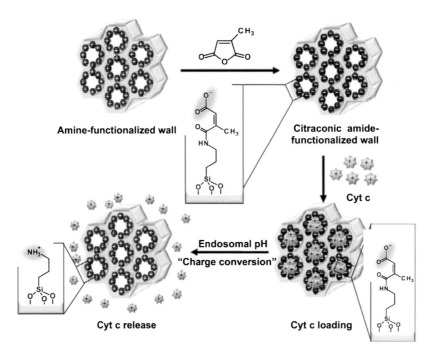

FIGURE 12.9 The charge-convertible mesopore surface [99].

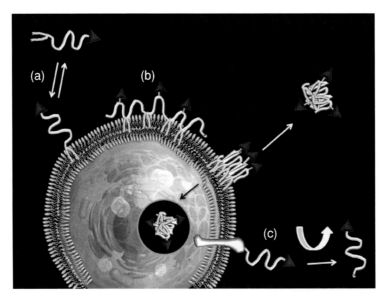

FIGURE 14.2 Schematic illustration of surface dynamics of the conjugated polymers on the cell membrane.

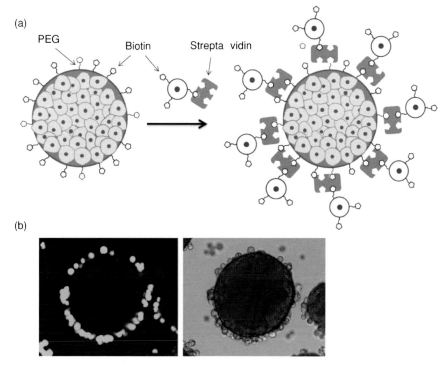

FIGURE 14.5 Schematic illustration of the immobilization of SA-immobilized HEK293 cells on the surface of biotin-PEG-modified islets (a). Confocal laser scanning (left) and differential interference (right) microscope images of surface-modified islets with HEK293 cells.

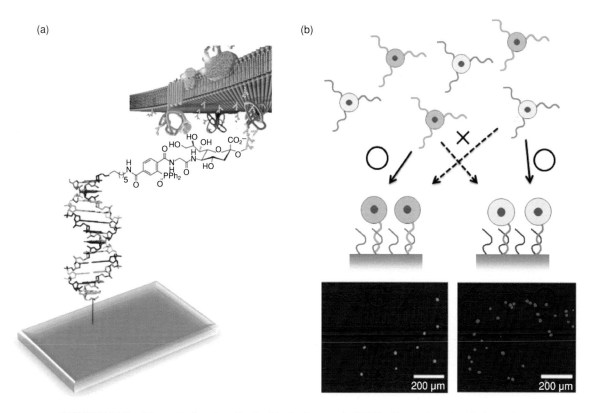

FIGURE 14.7 Live cells functionalized with single-stranded DNA oligonucleotides bind to substrates that bear complementary DNA strands in a sequence-specific manner (a). Only cells bearing ssDNA strands that were complementary to the immobilized DNA were observed to bind to the surface, whereas otherwise identical cells bearing mismatched sequences were washed away (b).

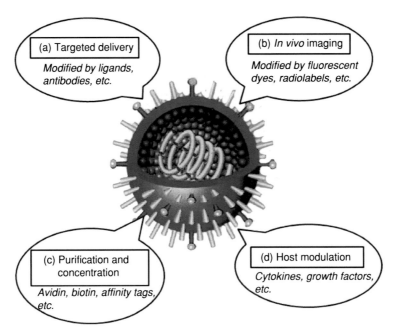

FIGURE 14.8   Overview of applications for surface modifications of viral vectors in gene therapy.

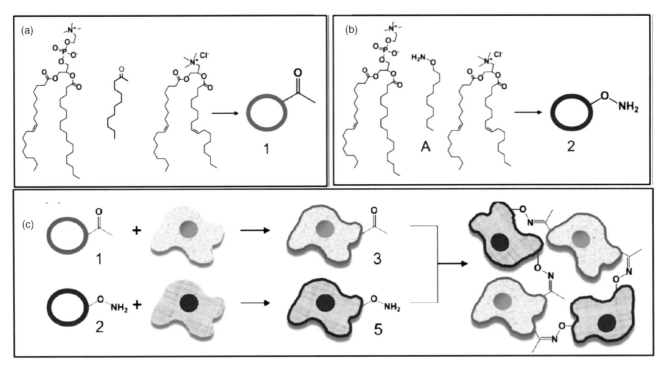

FIGURE 15.19   (a) Dodecanone molecules were incorporated into neutral, eggpalmitoyl-oleoyl phosphatidylcholine (POPC) and cationic, 1,2-dioleoyl-3-trimethylammonium-propane (DOTAP) at a ratio of 5:93:2 to form ketone-presenting liposomes (1). (b) O-Dodecyloxyamine (a) molecules were incorporated into POPC and DOTAP at a ratio of 5:93:2 to form oxyamine-presenting liposomes (2). (c) Two fibroblast populations were cultured separately with ketone-containing (1) or oxyamine-containing (2) liposomes. Because of the presence of a positively charged liposome, fusion occurred, producing ketone-tethered (3) and oxyamine-tethered (5) cells. Upon mixing these cell populations, clustering and tissue-like formation, based on chemoselective oxime conjugation, occurred. Reproduced with permission from Reference 115, Copyright (2011), American Chemical Society.

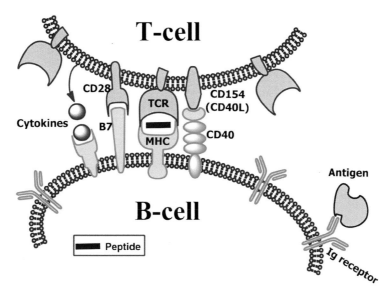

FIGURE 16.1 Immune responses involved in antigen presentation (peptide) by class II MHC proteins.

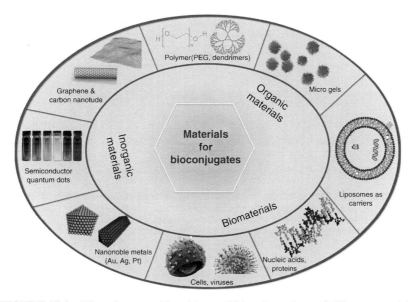

FIGURE 17.4 Bioconjugates achieved by modifying the surfaces of different materials.

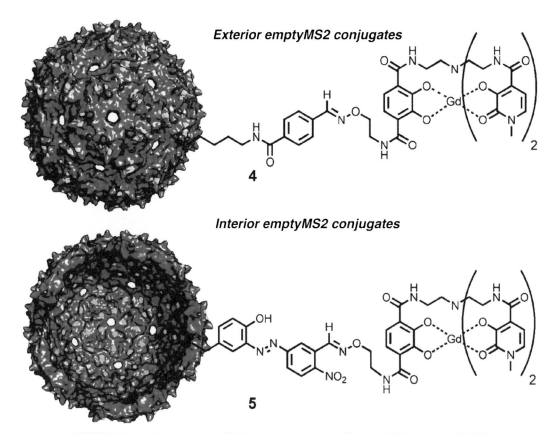

FIGURE 18.2 Gd complex modified externally or internally with MS2 virus capsid [10].

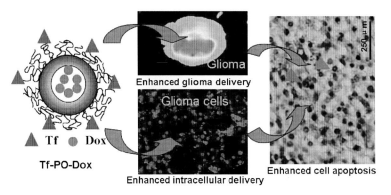

FIGURE 19.7 Uptake of transferring-conjugated, DOX-loaded polymersomes in rat glioma cells and their potential as anticancer agent [31].

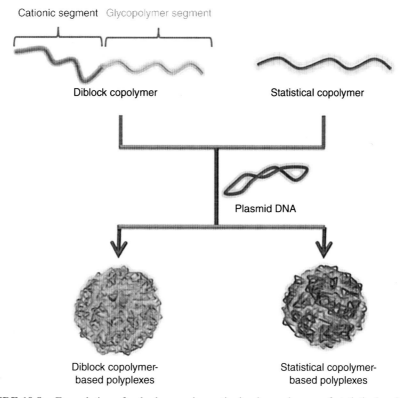

FIGURE 19.8 Formulation of polyplexes using cationic glycopolymers of statistical and diblock architectures and β-galactosidase plasmid DNA [38].

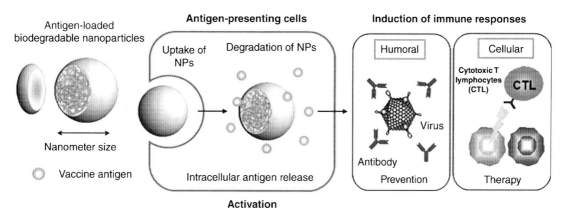

FIGURE 19.10 Formulation of nanoparticle-based vaccines and induction of immune response [51].

8

CARBON NANOTUBES AND FULLERENE C$_{60}$ BIOCONJUGATES

KESHWAREE BABOORAM AND RAVIN NARAIN
Department of Chemical and Materials Engineering, Alberta Glycomics Centre, University of Alberta, Edmonton, AB, Canada

8.1 INTRODUCTION

The discovery of the fullerenes is certainly one of the most intriguing given not only how it has helped in the expansion of the number of carbon allotropes once limited only to graphite, diamond, and amorphous carbon, but also the intense scientific interest that these materials have sparked for their potential applications in areas such as electronics, materials science, and nanotechnology. Composed exclusively of carbon, more specifically of graphene sheets, the fullerene family of materials comprises different structures ranging from spheres to ellipsoids and tubes. Of all the fullerenes, the buckminsterfullerene C$_{60}$ (spherical), also called buckyballs, and carbon nanotubes (CNTs) (cylindrical) have received the most attention and have been extensively explored for their inherent physical and chemical properties thanks to their unique morphology and structure [1]. Considerable progress made over the past years in the functionalization of these fullerenes has addressed the main problem that was in the way of these fascinating materials, that is their lack of solubility, hence opening major venues for their further exploration and potential application. As a consequence, CNTs and buckminsterfullerenes have been rigorously studied for biological and biomedical applications over the past few years. Another major driving factor for these studies has been the fact that these materials, being of nano size range (from 1 nm to a few hundred nanometers), are comparable to various biological molecules such as enzymes, antibodies, and DNA plasmids [2].

This chapter, therefore, focuses on the progress made to date on the study of fullerene (buckminsterfullerene C$_{60}$ and CNTs) bioconjugates for biological and biomedical applications. The types of bioconjugates that are discussed include those made from proteins, nucleic acids, and carbohydrates, and their potential uses as bioactive (macro)molecules, drug carriers, biosensors, and imaging agents are explored. A brief description of each of these fascinating nanomaterials is also first given to help the reader understand the structures and properties that form the basis for the substantial interest in their application in the understanding of biological systems, as well as the detection and treatment of diseases.

8.2 CARBON NANOTUBES—THE CYLINDERS

CNTs are graphene sheets that have been seamlessly rolled up into cylinders that exhibit extraordinary physical, mechanical, and chemical properties. The different angles and curvatures in which the graphene sheet is rolled up to give a single-walled carbon nanotube (SWNT), govern the exact structure (which can either be armchair, zigzag, or chiral) and hence, the final properties (metallic or semiconducting) of the CNT. Figure 8.1 shows the three different structures of the SWNTs [3]. The other type of CNTs refers to the multiwalled carbon nanotubes (MWNTs) which are either formed by one large graphene sheet rolled in around itself, or of several sheets of graphene arranged in concentric cylinders (as shown in Figure 8.2) [4].

Chemistry of Bioconjugates: Synthesis, Characterization, and Biomedical Applications, First Edition. Edited by Ravin Narain.
© 2014 John Wiley & Sons, Inc. Published 2014 by John Wiley & Sons, Inc.

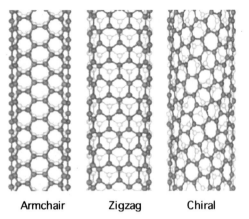

Armchair   Zigzag   Chiral

FIGURE 8.1 The different types of single-walled carbon nanotubes (SWNTs) resulting from the different ways that the graphene sheet closes up [3].

While CNTs exhibit unique properties which make them attractive for numerous applications in nanotechnology, nanoelectronics, and various areas of materials science [5–8], their actual usage has for a long time been hindered by their insolubility and potential toxicity. The surface functionalization of CNTs to improve their dispersion and solubility in aqueous media, and hence their biocompatibility, has led to a revolution in the countless potential applications of these materials, including in the biomedical field. The size and shape of the CNTs are the key parameters that have made these materials attractive for biomedical research. For instance, an appropriately functionalized 1D SWNT is flexible and can therefore, bend to display an array of binding sites to a cell. This causes the multivalence effect which is

key in the cell recognition process, by improving the binding affinity of the functionalized CNTs. Moreover, the shape and structure of the SWNT is such that all the carbon atoms at the surface are effectively exposed, offering an ultrahigh surface area for an array of molecules/biomolecules to bind to [2, 9]. This invaluable property is magnified in the MWNTs given the presence of multiple layers of graphene, and therefore, their larger sizes. MWNTs are thus more attractive than SWNTs for biological applications, namely as carriers of large biomolecules like DNA [10, 11].

The reason for the insolubility of as-prepared CNTs in aqueous environments is their extremely hydrophobic surface. Therefore, as mentioned earlier, the chemical modifications or functionalization of the CNTs' surfaces is a crucial step in making the latter soluble, biocompatible, and potentially less toxic or nontoxic. Two types of approaches have been exploited in the surface functionalization of CNTs, namely the covalent and the noncovalent methods. While the latter approach is based on the hydrophobic interactions between the CNT's surface and that of an amphiphilic molecule to produce water-soluble surfactant-wrapped CNTs, covalent functionalization uses chemical reactions to attach certain types of molecules on the CNT surface in order to solubilize them. The various strategies used for the covalent functionalization of CNTs, namely oxidation followed by conjugation with hydrophilic polymers or other functional moieties, the photoinduced addition of azide compounds, the Bingel reaction, and the 1,3-dipolar cycloaddition reaction, are all summarized in Scheme 8.1 [2]. On the other hand, Scheme 8.2 provides an overview of the different techniques employed in the noncovalent functionalization of CNTs, which involves three main approaches as described by Yang et al. [12]: (a) the noncovalent attachment of a biomolecule directly onto the surface of the nanotube [13]; and the hybrid approach, where (b) a small molecule is first noncovalently anchored on the CNT followed by (c) its chemical reaction with biomolecules [14].

Preservation (or not) of the intrinsic physical properties of the CNTs completely depends on the distinctiveness of these two functionalization strategies. Because covalent functionalization leads to the chemical modification and hence, the disruption of the nanotube structure, it has been observed that many of the intrinsic physical as well as electronic properties are eradicated at the end of the process. On the other hand, the noncovalent approach leaves the nanotube structure unharmed, which results in the preservation of these interesting physical and electronic properties. As such, the type of functionalization technique that is adopted dictates the end use of the CNTs [12, 15]. Covalently functionalized CNTs lose their valuable properties like Raman scattering and photoluminescence, which therefore limits their optical applications. CNTs that undergo noncovalent functionalization find many more applications in the biomedical areas, for example, in detection, imaging, and sensing.

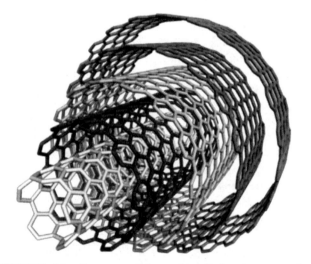

FIGURE 8.2 Structure of the multi-walled carbon nanotube (MWNT): sheets of graphite arranged in concentric cylinders. Reproduced with permission from Reference 4.

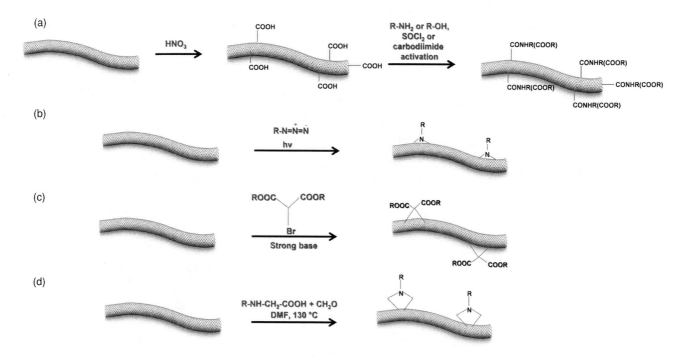

SCHEME 8.1   The different strategies for the covalent functionalization of CNTs: (a) CNTs are oxidized and then conjugated with hydrophilic polymers (e.g., PEG) or other functional moieties, (b) photoinduced [1, 2] addition of azide compounds with CNTs, (c) Bingel reaction on CNTs, (d) 1,3-dipolar cylcoaddition on CNTs. For biological applications, "R" in the figure is normally a hydrophilic domain which renders CNTs water soluble. Further conjugation of bioactive molecules can be applied based on such functionalization [2].

8.3 THE BUCKMINSTERFULLERENE, C$_{60}$—THE BUCKYBALL

Discovered in 1985, the buckminsterfullerene has a molecular formula of C$_{60}$, and is also commonly called buckyball because of the structural arrangement of a total of 60 carbon atoms in such a way that makes it look very much like a soccer ball [16]. Figure 8.3 shows the cage-like fused ring structure of the buckyball, which contains twenty hexagons and twelve pentagons made by interconnected carbon atoms [17].

The interesting photophysical properties of C$_{60}$ and its ability to generate reactive oxygen species when exposed to visible light stem from the unique chemical structure presented by this molecule. These features, along with the physicochemical properties of C$_{60}$, have made it a strong candidate for photodynamic therapy in biological systems, and in the development of tumor theranostics, respectively [18, 19]. For such applications however, it is of prime importance to produce C$_{60}$ (or derivatives thereof) that are soluble in the aqueous phase. Interestingly, buckminsterfullerene is the only soluble form of carbon, but only sparingly and only in aromatic organic solvents such as toluene. However, just like the CNTs, the C$_{60}$ molecules are particularly insoluble

in polar solvents and hence form aggregates in aqueous solutions. To overcome the problem of the lack of solubility in the aqueous phase, which limits the use of C$_{60}$ in medicinal chemistry and for biological applications, considerable amount of effort has been invested into the functionalization of the latter molecule in order to adjust its properties. As in the case of the nanotubes, functionalization of the buckminsterfullerene can also be achieved through chemical or supramolecular approaches, and in recent years, functionalized C$_{60}$ has found (or has been targeted for) numerous applications in drug delivery, radiation protection, gene therapy, photodynamic therapy, and as MRI contrast agents [20–23]. Functionalization, in this particular case, can be divided into two classes: exohedral, in which substituents are placed outside the buckyball and endohedral, in which they are inside the cage. The empty cavity of the C$_{60}$ fullerene, for instance, is a perfect place to include a metal atom in order to produce a metallofullerene, found to have novel electronic properties perfectly fit for magnetic resonance imaging (MRI) [24]. Although various C$_{60}$ solubilizing techniques, such as entrapment in cyclodextrins and calixarenes and cosolvation with polyvinylpyrrolidone (PVP) in chloroform, have been successfully put forward in the past years, the most versatile method of solubilizing and producing C$_{60}$ derivatives with

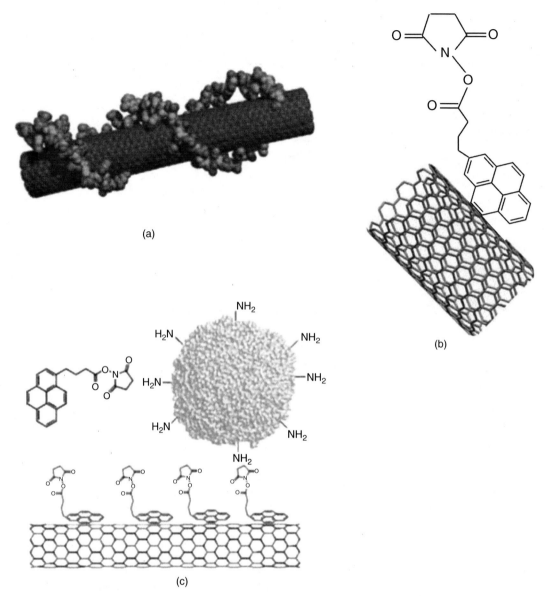

SCHEME 8.2   The different ways to noncovalently functionalize CNTs: (a) A SWNT coated by a single-stranded DNA via π–π stacking. Reproduced with permission from Reference 13. (b) Pyrene molecules anchored on a SWNT by noncovalent interactions, (c) proteins are anchored on the SWNT surface via pyrene π–π stacked on a nanotube surface. Reproduced with permission from Reference 14.

interesting and useful physical and chemical properties, especially for applications in the biological and biomedical fields, remains the chemical modification of the buckyball [25–28]. For example, Hirsch et al. have reported a dendrimeric C_{60} derivative (the structure of which is shown in Figure 8.4) carrying 18 carboxylic groups, hence giving it a very high solubility of 34 mg/mL in water at around physiological pH [29].

The process of functionalization has immensely facilitated the attachment of biomolecules such as proteins, carbohydrates, and nucleic acids to the surfaces of CNTs and buckminsterfullerene, C_{60}. As we can see, the functionalization of these carbon entities has set the milestone for the production of what is commonly referred to as the CNT- or C_{60}-bioconjugates, which has opened up an entirely new path toward advanced development in chemical biology with a focus on targeting cell behavior. For specific biological functions, the way these entirely carbon-based materials are distributed *in vivo* depends heavily on the functionalization of the latter, as well as on their size. If properly functionalized, the nanotubes have been shown to successfully excrete from the body through the biliary pathway [30, 31]. As for the C_{60},

FIGURE 8.3   The structure of the buckminster fullerene, C_{60} [17].

although the presence of functional groups on the sphere has shown to reduce cytotoxicity of the water-soluble derivatives in some cases, a lack of understanding of the mechanisms behind their observed activities points to an urgent need for studies into the biocompatibility and safe applications of functionalized C_{60} in the biomedical arena [19, 32, 33].

In the coming sections, the latest achievement in the production of CNT- and C_{60}-bioconjugates will be discussed. Particular emphasis will be on the types of bioconjugates (SWNT, MWNT, and buckminsterfullerene C_{60}-based), the methods used in their synthesis, and the specific applications that they have been targeted to.

8.4   CARBON NANOTUBES—BIOCONJUGATES THROUGH COVALENT FUNCTIONALIZATION

The production of bioconjugates of CNTs with the fundamental components of living organisms, namely carbohydrates, proteins, and nucleic acids, has allowed the assessment of the biocompatibility and the various prospective biological and biomedical applications of the CNTs. The covalent conjugation of proteins onto the surface of CNTs is often a vital step toward their application in the biological field, especially in biosensing. Experiments, in which proteins were immobilized on CNTs (both on the interior and the exterior) through covalent functionalization, were carried out as early as 1995 by Sadler et al. [34]. The latter used acid treatment to prepare open-ended CNTs with acidic sites (consisting mainly of carboxylic groups) present both on the inner and outer layers of the CNTs. The small proteins, Zn_2Cd_5-metallothionein, cytochrome c_3, and β-lactamase I, were then grafted to the CNTs' surfaces through covalent bonding to the acidic groups, and were found to significantly retain their catalytic activity even after their attachment to the nanotubes. Sun and coworkers used diimide-activated amidation to functionalize CNTs with bovine serum albumin (BSA) proteins, under ambient conditions [35]. Highly water-soluble CNT–BSA conjugates were thus prepared, in which about 90% of the BSA proteins were reported to remain bioactive. Once again, the CNTs were treated with an oxidizing acid in order to introduce carboxylic acid groups on their surface. 1-ethyl-3-(3-dimethylaminopropyl) carbodiimide (EDAC) was then used to activate these carboxylic groups (through the formation of an active O-acylisourea intermediate) toward nucleophilic attack from the amino groups found on the BSA protein. This resulted in the complexation of the protein molecules to the CNTs via stable amide bonds under mild reaction conditions that preserved the protein's bioactivity. Figure 8.5 shows the general strategy employed in the functionalization of CNTs with proteins via diimide-activated amidation.

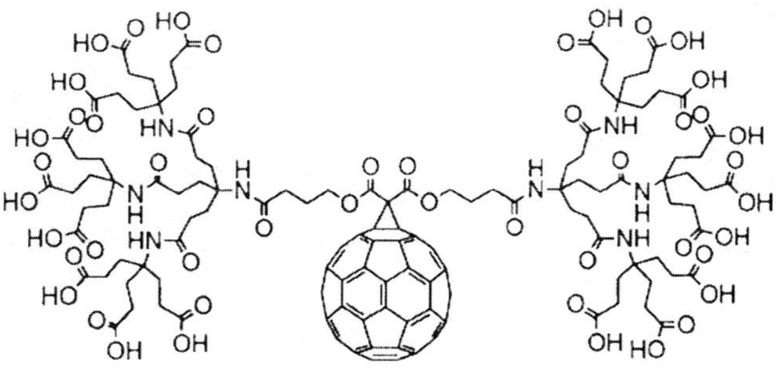

In their attempt to conjugate the antibiotic amphotericin B (used in the treatment of fungal infections) to MWNTs for drug delivery purposes, Zeinali et al. used the diimide-activated amidation method, but in a slightly different approach [36]. After the introduction of carboxylic acid groups on the MWNT surfaces, the latter was reacted with ethylene diamine in order to form an amide bond, leaving one amine group free. The carboxylic acid group on the amphotericin B was, on the other hand, activated with the EDAC molecule and then reacted with the functionalized MWNT to form an amide linkage between the CNT and the bioactive molecule.

Using the *in situ* ring-opening polymerization of ε-caprolactone, the biodegradable poly(ε-caprolactone) (PCL) has been covalently attached onto MWNT surfaces by Zeng et al. [37]. MWNTs bearing carboxylic acid groups were first reacted with an excess of thionyl chloride, followed by an excess of ethylene glycol in order to introduce hydroxyl end groups on the nanotubes' surfaces. *In situ* ring-opening polymerization of ε-caprolactone was then carried out in the presence of a tin catalyst, stannous octoate, in order to produce MWNT-poly(ε-caprolactone) conjugates. Figure 8.6 illustrates the procedure used for grafting PCL onto the MWNTs.

Biodegradation experiments carried out on the MWNT–PCL conjugates showed that the PCL moieties retained their

FIGURE 8.4   Structure of a highly soluble dendrimeric C_{60} derivative. Reproduced with permission from Reference 29.

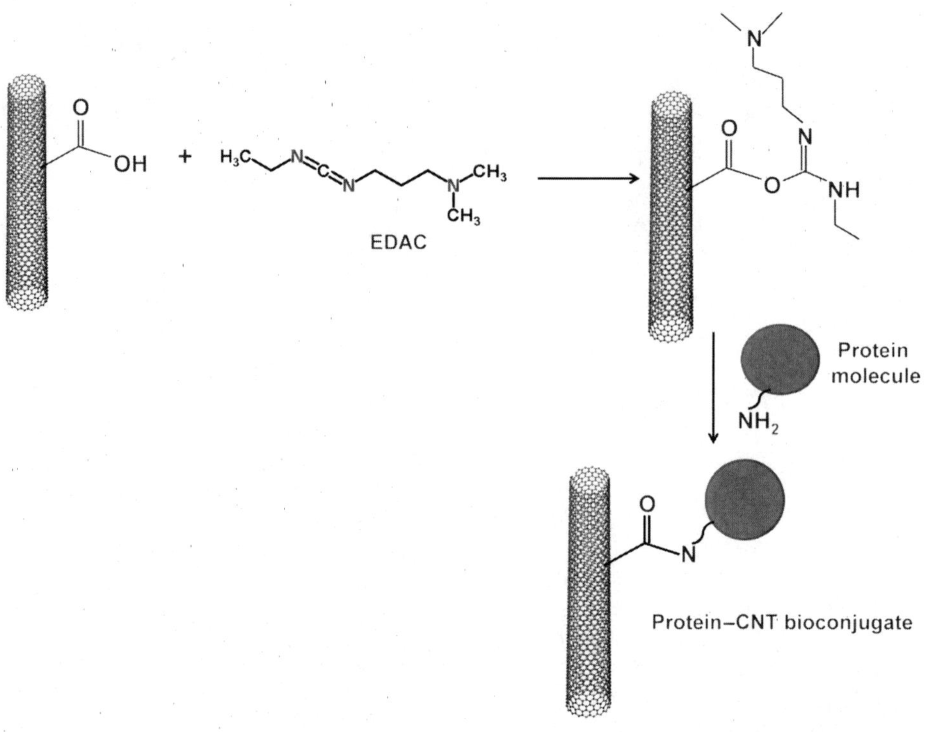

FIGURE 8.5   Functionalization of carbon nanotubes (CNTs) through diimide-activated amidation.

biodegradability by showing complete enzymatic degradation, within 4 days, in a buffer solution containing pseudomonas lipase. The latter was also found to be biologically unaltered by the MWNTs. The results therefore point to the potential applications of these MWNT–PCL conjugates in bionanomaterials and biomedicine, as well as in artificial organs and bones.

Asuri and coworkers have successfully prepared SWNT–enzyme bioconjugates through the covalent attachment of three different enzymes (horseradish peroxidase, subtilisin Carlsberg, and chicken egg white lysozyme, respectively) to acid-treated SWNTs [38]. N-hydroxysuccinimide (NHS) was first coupled to the carboxylic acid groups on the SWNTs using the diimide-activated amidation process. The NHS functionalized nanotubes were then treated with each of the aforementioned enzyme solutions in order to covalently

attach them to the SWNT surface. These SWNT–enzyme bioconjugates were not only soluble in aqueous solution, but also showed high active loading of the enzymes on the nanotubes. The selected enzymes were found to retain a considerable fraction of both their native activity and their structure even after attachment to the SWNTs, and they also displayed improved stability under denaturing environments such as high temperatures. The results of this study are indicative of the applications of these CNT–protein bioconjugates in sensing devices and in self-assembly.

In another study, MWNTs were covalently coupled with folic acid, an essential and bioavailable water-soluble B complex vitamin, to produce a biocompatible CNT bioconjugate that can potentially be used in the improvement of cancer treatments [39]. The authors used acid treatment to functionalize the MWNTs with phenolic and carboxylic groups

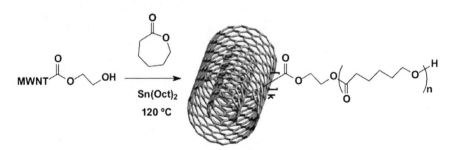

FIGURE 8.6   Procedure for the grafting of PCL onto MWNTs prefunctionalized with hydroxyl groups. Reproduced with permission from Reference 37.

and then further functionalized the nanotubes with aspartic acid (an amino acid) in order to increase the number of carboxylic acid groups on the nanotubes' surface, and enhance the latter's dispersibility in aqueous solutions [40]. Folic acid molecules were then attached to the functionalized MWNTs through its amino group, using the coupling agent, 2-(1H-benzotriazol-1-yl)-1,1,3,3-tetramethyluronium fluorophosphates (HBTU). The MWNT–folic acid conjugates were shown to have greater versatility owing to the bioreversibility of the amide bond in these bioconjugates, as a result of which they were anticipated as target drug carriers.

Dai et al. have demonstrated how the large surface area that exists on covalently as well as noncovalently prefunctionalized SWNTs can be effectively used to perform supramolecular chemistry [41]. In their covalent approach, these researchers have prepared water-soluble SWNTs by the attachment of polyethylene glycol (PEG) chains to the carboxylic acid groups of the acid-treated nanotubes. Interestingly, the remaining space on these covalently functionalized SWNTs were then used for the π stacking of various aromatic molecules, which included doxorubicin (DOX), a cancer chemotherapy drug and fluorescein, a fluorescence molecule. The drug loading efficiency was shown to be extremely high, and the binding and the release of these molecules to and from the SWNT surface were found to be dependent on the diameter of the nanotubes, and could be controlled by changes in pH. Moreover, the presence of other functional groups on the SWNT–DOX conjugates opens up the way for further attachment of other molecules such as peptides or antibodies, hence making the latter useful for targeted drug delivery. The results of this work offer exciting insight into the development of CNT vehicles for various applications in drug delivery, and in chemical and biological imaging and sensing. Figure 8.7 illustrates the two types of species attached to an SWNT: the PEG chain that is covalently bonded through COOH groups to impart solubility to the nanotubes, and the DOX molecule that are noncovalently attached through π-stacking and to be released in targeted cancer cells.

Peptide nucleic acid (PNA) which is an artificial analog of DNA has been used for the first time by Singh and coworkers to link together SWNTs [42]. The nanotubes were made to undergo mild acid treatment in order to acid functionalize only their ends. The SWNTs were then cross-linked to one another via the PNA molecules to produce highly functionalized SWNT–PNA–SWNT conjugates with molecular recognition. The steps followed in the synthesis of these SWNT–PNA complexes are shown in Figure 8.8. The carbodiimide coupling chemistry was again employed to link PNA to the CNTs. The as-prepared CNT bioconjugates are targeted for the development of future nanoelectronic devices with higher order assembly.

Rusling and coworkers have reported a novel amplification strategy for their initially developed SWNT

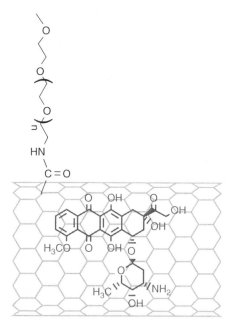

FIGURE 8.7 A PEG–SWNT–DOX conjugate prepared through both covalent and noncovalent functionalization of SWNT. Reproduced with permission from Reference 41.

immunosensors [43–45]. In their work published in 2006 they have, for the first time, applied these SWNT immunosensors to the detection of a protein cancer biomarker in real and complex biomedical samples such as human serum and tissue lysates. The strands of nanotube found in a vertical forest assembly of SWNTs were acid functionalized and then bonded to horseradish peroxidase (HRP) labels and secondary antibodies (Ab2) through the carbodiimide coupling chemistry using 1-(3-(dimethylamino)-propyl)-3-ethylcarbodiimide hydrochloride (EDC) and N-hydroxysulfosuccinimide (NHSS). This assembly of SWNT bioconjugates were found to offer accurate and highly amplified sensitivity in the detection of the prostate-specific antigen (PSA) in human serum. It is strongly believed that future medical devices based on this strategy will be very promising in the reliable point-of-care diagnostics of cancer as well as other diseases.

Narain et al. have carried out the surface functionalization of SWNTs with cationic glycopolymers, and have studied the resulting CNT bioconjugates as *in vitro* transfer agents [46]. Biocompatible and low-toxicity SWNT–glycopolymer conjugates were synthesized by the covalent bonding of the amino groups of the poly-3-aminopropylmethacrylamide (PAPMA) glycopolymer to the carboxyl groups present on the nanotubes' surface. 1-(3-(dimethylaminopropyl)-3-ethylcarbodiimide hydrochloride (EDC) and NHS were used as coupling agents in the process. Two different cationic block polymers, P(APMA$_{31}$-b-LAEMA$_{32}$) and P(APMA$_{38}$-b-GAPMA$_{20}$), synthesized from 3-aminopropyl

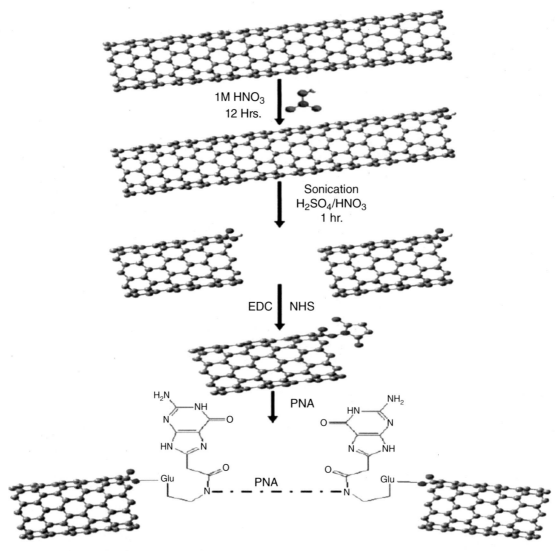

FIGURE 8.8   Synthesis of the SWNT–PNA–SWNT conjugates. Reproduced with permission from Reference 42.

methacrylamide (APMA), 3-gluconamidopropyl methacrylamide (GAPMA), and 2-lactobionamidoethyl methacrylamide (LAEMA), were grafted onto the SWNT surface following the described procedure. The resulting SWNT–glycopolymer conjugates were further complexed with DNA to yield SWNT–polymer–DNA bioconjugates which were studied as gene delivery vehicles. Interestingly, these novel SWNT bioconjugates showed good biocompatibility and high transfection efficiencies that make them good candidates to potentially replace other commercially available gene delivery vehicles. Figure 8.9 shows the steps that were employed in the preparation of the SWNT–polymer–DNA bioconjugates.

Gao and coworkers have reported the synthesis of MWNT–papain bioconjugates from MWNTs initially functionalized with carboxyl and amine groups, respectively [47]. The authors' interest in studying papain-based bioconjugates

lies in the fact that papain is a well-characterized enzyme that has various industrial applications in areas such as food, pharmaceutical, and cosmetics among others [48, 49]. As depicted in Figure 8.10, the papain molecules were attached to the MWNTs in two different ways—first through the carboxylic acid groups initially introduced on the MWNT surface by acid treatment, and second via amine groups that were introduced on the MWNTs by the procedure described in the onset of Figure 8.10 (The carboxylic acid groups were first acyl-chlorinated with thionyl chloride, followed by reaction with hexamethylenediamine to produce MWNT–NH_2 functionalities). In order to improve the stability of the MWNT–papain bioconjugates, the latter were further coated with silica by treatment with acid-hydrolyzed trimethoxylsilane (TMOS). Since the papain molecules immobilized on the MWNT–NH_2 showed a relatively higher activity compared to those immobilized on the acid-functionalized nanotubes,

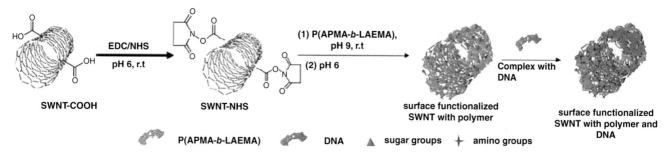

FIGURE 8.9 The functionalization of carboxyl-bearing SWNTs with glycopolymers, and DNA. Reproduced with permission from Reference 46.

the former were chosen to prepare the silica-coated bioconjugates. Coating with silica resulted in bioconjugates that showed enhanced stability and a high retention of papain activity under denaturing conditions such as elevated temperatures and extreme pH, making them attractive materials for applications in biocatalysis and sensing.

8.5 CARBON NANOTUBES—BIOCONJUGATES THROUGH NONCOVALENT FUNCTIONALIZATION

As mentioned earlier, the noncovalent approach for the functionalization of CNTs as a means to prepare CNT bioconjugates is often preferred over the covalent approach, especially when the latter are targeted for applications in

detection, imaging, and sensing. While noncovalent functionalization largely maintains the structural and optical properties of the nanotubes, the stability and biocompatibility of noncovalently functionalized CNTs are often compromised [50]. Ideally, the functionalization of CNTs, especially for the preparation of bioconjugates, should make the nanotubes highly water soluble and biocompatible, cause minimal damage to their structure, and have functional groups readily available for the attachment of biomolecules. Intensive research in this direction, in particular by the research groups of Liu and Dai, has led to remarkable progress and success [13, 50–52].

Dai, Liu, and coworkers designed a novel scheme for the noncovalent functionalization of SWNTs, which were consequently bonded to biomolecules to yield bioconjugates with interesting properties [51]. Their methodology

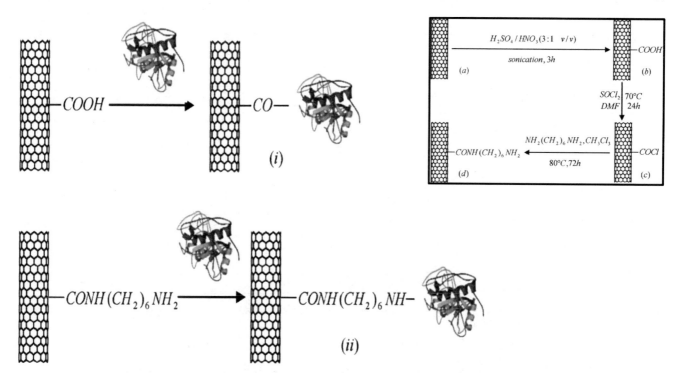

FIGURE 8.10 The different strategies used in the preparation of MWNT–papain bioconjugates. Reproduced with permission from Reference 47.

was based on the *nonwrapping* approach that was first introduced by Chen et al., who used short and rigid polymers, poly(aryleneethynylene)s (PPE) that adsorbed onto CNT surfaces through π-stacking to produce highly water-soluble nanotubes [53]. In their work, Dai's and Liu's group first stabilized and solubilized the SWNTs in the aqueous phase through the noncovalent adsorption of phospholipids (PLs) containing poly(ethylene glycol) (PEG) chains, and terminal amine group (PL–PEG–NH$_2$) or maleimide group (PL–PEG–maleimide) on their surface. The PL alkyl chains attach themselves strongly to the SWNT's surface through van der Waals and hydrophobic interactions, and the PEG chains extend out into the solution, making the SWNTs soluble in water. Once the PL–PEG chains were immobilized on the nanotubes, their terminal functionalities were used to couple them to a range of biological molecules. The resulting SWNT–PL–PEG-based bioconjugates were then used in transportation, release, and nuclear translocation of DNA oligonucleotides in mammalian cells with promising results. Such molecular CNT-based transporters have great potentials in gene and protein therapy. In another of their work, Kam and Dai have successfully demonstrated the spontaneous adsorption of various proteins such as streptavidin (SA), protein A (SpA), BSA, and cytochrome *c* (cyt*c*) on the sidewalls of acid-oxidized SWNTs [52]. The proteins bound nonspecifically and noncovalently to the nanotubes and were easily transported into various mammalian cells, where they could perform biological tasks. This work opened up a window onto a new class of CNT-based molecular transporters potentially useful for the *in vitro* and *in vivo* delivery of proteins. In yet another report, Dai et al. have carried

out the noncovalent functionalization of SWNTs to prepare highly specific electronic biosensors capable of selectively detecting and responding to specific proteins in solution [54]. Polyethylene oxide chains were first immobilized on the nanotubes' surface in order to prevent the nonspecific binding of proteins. The latter were then selectively recognized by and bound to specific functional groups on the polyethylene oxide chain ends. This allowed for highly sensitive and specific detection of important biomolecules, such as antibodies associated with human autoimmune diseases, without the need for biomarkers.

A dual-targeting drug delivery system that is based on PEGylated MWNTs has very recently been developed by Jiang et al. for the treatment of brain glioma [55]. The noncovalent functionalization of oxidized MWNTs was performed by dispersing the latter in a solution of 1,2-distearoyl-snglycero-3-phosphoethanolamine-N-[methoxy (polyethylene glycol)-2000] (DSPEPEG2000) and 1,2-distearoyl-sn-glycero-3-phosphoethanolamine-N-[maleimide (polyethylene glycol)-2000] (DSPE–PEG2000–MAL). The resulting MWNT–PEG–MAL complex was then reacted with angiopep-2, a ligand of lipoprotein receptor-related protein (LRP). The MWNT–PEG–MAL–Angiopep bioconjugate was first labeled with fluorescein isothiocyanate (FITC) and then loaded with DOX, an antitumor drug. This is a very interesting piece of work in which the nanotubes were employed not only as a drug delivery vehicle but also as the carrier of a targeting ligand. The synthesis scheme for the DOX–MWNT–PEG–ANG bioconjugate is shown in Figure 8.11. The dual-targeting drug delivery system was studied *in vitro* and *in vivo*, and its biological safety tests

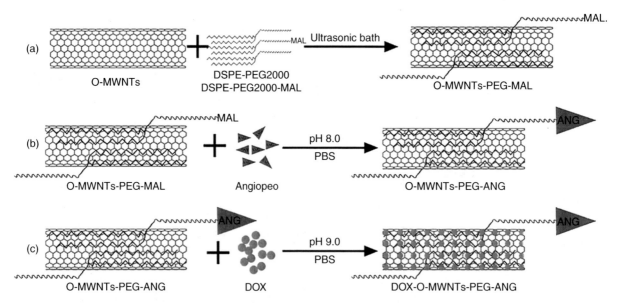

FIGURE 8.11 Steps involved in the synthesis of the DOX–MWNT–PEG–ANG bioconjugates as potential dual-targeting drug delivery vehicles Reproduced with permission from Reference 55. (For a color version, see the color plate section.)

prove that it is biocompatible and has low toxicity, making it promising in the treatment of brain tumor.

Based on the noncovalent functionalization approach developed by Dai and coworkers, Dekker et al. have successfully prepared enzyme-coated SWNTs as single-molecule biosensors [14, 56]. A linker, which on one side noncovalently attaches itself to the surface of the nanotube through its pyrene group and on the other end covalently bonds to the enzyme (glucose oxidase), was used to immobilize the latter on the SWNTs. The glucose oxidase–SWNT conjugates were found to function as pH sensors as well as enzyme-activity sensors a change in conductance when glucose was added.

Bertozzi and coworkers have reported the use of glycodendrimers that function as homogeneous bioactive coatings for CNTs in the production of biocompatible SWNTs [57]. The choice of dendrimers for this purpose follows the intensive amount of work carried out on the biomedical applications of these molecules as well as their role in improving the solubility of CNTs [58–61]. A new class of bifunctional glycodendrimers based on 2,2-bis(hydroxymethyl)propionic acid, which is a biocompatible building block, has been developed. These glycodendrimers, as shown in Figure 8.12, consist of carbohydrate moieties on the peripherals and at the same time, carry a pyrene tail which can noncovalently

attach itself to the SWNTs through π–π interactions. The as-formed glycodendrimer-functionalized SWNTs were completely soluble in aqueous media, showed great stability for up to several months, and also proved to be biocompatible through the positive results obtained from the cytotoxicity tests. The carbohydrate units on the glycodendrimer-coated SWNTs were then bound to different FITC-conjugated lectins in order to evaluate the specific binding of the SWNT–glycodendrimer complexes to cell-surface receptors. The promising results point to the potential development of versatile glycodendrimers that can be applied to future biosensors.

A highly efficient SWNT-based cancer drug delivery system has been described by Dai et al. [31]. The researchers have performed the noncovalent functionalization of SWNTs with PLs that contain branched PEG chains. Paclitexal (PTX), a drug that is widely used in cancer treatment, was then conjugated to the PEG chain ends through an amide bond and using succinic anhydride as the linker. The resulting linkage being a cleavable ester bond, this allowed for the easy release of the PTX molecules *in vivo*. Figure 8.13 shows the SWNT–PTX conjugate, highlighting the effective loading of the drug onto the CNTs through the use of branched PEG chains on the PLs. The SWNT–PTX bioconjugates were water soluble, and showed a much higher efficiency in destroying tumor growth compared to clinical

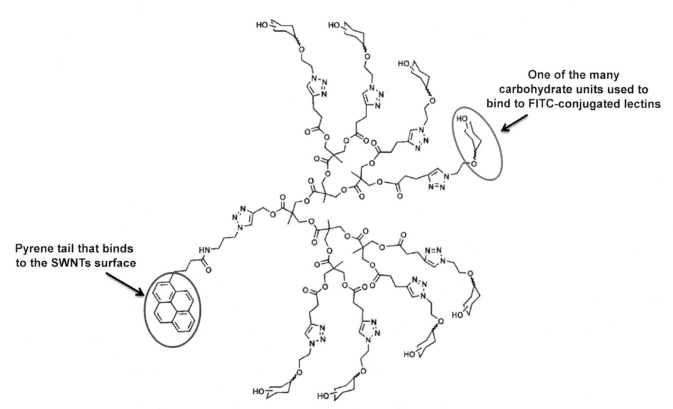

FIGURE 8.12   Glycodendrimer used to coat SWNTs through noncovalent interaction of the pyrene tail to the nanotubes' surface. Reproduced with permission from Reference 57.

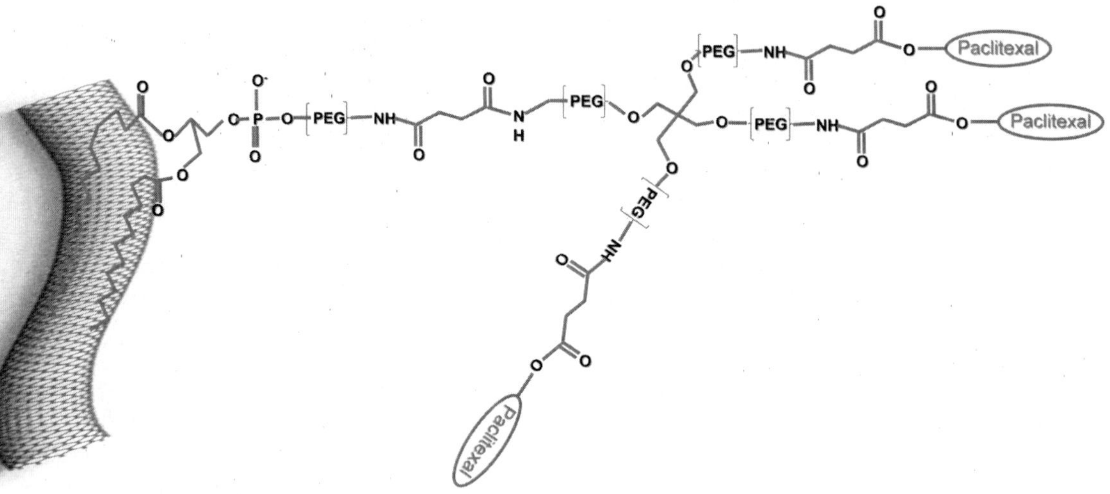

FIGURE 8.13 Branched PEG chains that are noncovalently adsorbed on SWNTs through phospholipids (PLs), allowing for a higher loading of the drug, PTX [31].

Taxol in a murine 4T1 breast cancer model [62]. In this work, the authors emphasize on the importance of the functionalization chemistry to tailor the efficacy of the drug delivery, and hence, the treatment efficiency.

8.6 BUCKMINSTERFULLERENE—BIOCONJUGATES THROUGH FUNCTIONALIZATION OR CHEMICAL MODIFICATION

The very interesting physical and chemical properties of the buckminsterfullerene C_{60} has inspired the design and the synthesis of a large number of C_{60}-based compounds through functionalization of the buckyball, targeting various promising applications such as drug delivery, gene therapy, antiviral activity, enzyme inhibition, DNA photocleavage, anti-HIV activity, antibacterial activity, and as antioxidants [21, 22, 63, 64]. Once the hurdle of the lack of solubility in polar media was overcome by various methodologies, fullerene C_{60} could be investigated for biological and biomedical applications with greater certainty. After all, it is an undeniable fact that the potential biological benefits of these molecules can be realized only through novel alterations to their existing composition. The research works of Samai and Geckeler, and Filippone and coworkers were based on the covalent attachment of cyclodextrins to buckminsterfullerene C_{60} to produce water-soluble derivatives of the latter [25, 65]. Other functionalization techniques that have been used to increase the hydrophilicity of C_{60} include the introduction of functional groups such as hydroxyl ($-OH$), carboxyl ($-COOH$), and amine ($-NH_2$), which also facilitate the linking of other ligands, such as important biomolecules, onto the surface of the sphere [22, 29]. Polyhydroxylated C_{60}, known as fullerenols, have been found to be very good antioxidants owing to their ability to sift for and trap free radicals [66]. On the other hand, amino acids that contain the C_{60} fullerene have shown to significantly regulate the activity of enzymes and hence, help in obtaining important information on the relationship between the structure and function of proteins and enzymes [67, 68]. Given the fast increase in the number of novel functionalized fullerenes, Conyers et al. have put together a list of functionalized fullerenes along with their respective potential biomedical applications, in their review on the roles of these materials in biomedical applications [23].

The highly water-soluble C_{60} derivative (Figure 8.4) prepared by Hirsch and Brettreich by the attachment of a dendrimer carrying eighteen carboxylic groups, to the surface of C_{60}, has been found by Daroczi and coworkers to protect zebrafish embryos against the hazardous effects of ionizing radiations and acts as an oxygen radical scavenger [69]. It is believed that the antioxidant properties of C_{60} are at the center of the radioprotective ability of this dendritic fullerene. The amphiphilic fullerene-1 (known as AF-1), which is derived from this highly water-soluble dendritic fullerene described earlier, has been reported as a novel functionalized C_{60} molecule that can interestingly self-assemble into either a hydrophilic vesicle with a hollow interior, or a spherical nanostructure possessing a hydrophobic interior [70, 71]. These works have demonstrated that a temperature increase during the self-assembly process, allows AF-1 to form a vesicular structure that is capable of trapping molecules of hydrophobic nature [71]. Such a discovery points toward the potential application of C_{60}-based dendrimers to the development of nanocarriers for the effective delivery of drugs to specific sites in the body.

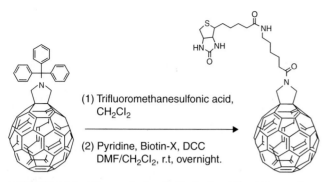

FIGURE 8.14 The synthesis of a biotinylated C_{60} fullerene from *N*-(triphenylmethyl)-3,4-fulleropyrrolidine. Reproduced with permission from Reference 72.

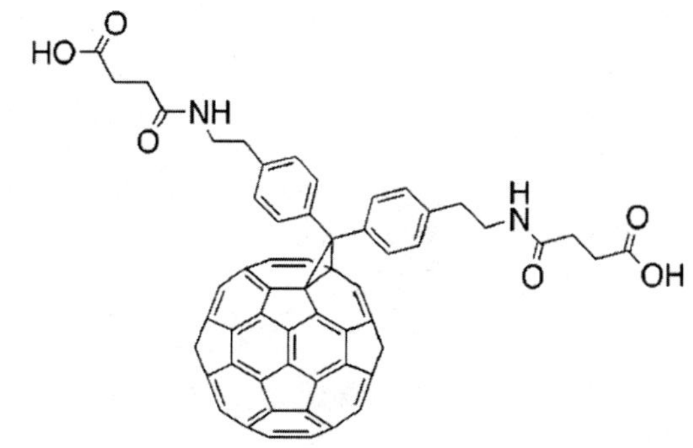

FIGURE 8.15 Structure of the first C_{60} derivative employed in the inhibition of the HIV enzyme. Reproduced with permission from Reference 76.

Bachas et al. have performed the attachment of biotin-conjugated proteins to fullerene C_{60} using SA as a molecular adaptor [72]. In their approach, a prefunctionalized C_{60} fullerene, *N*-(Triphenylmethyl)-3,4-fulleropyrrolidine, was coupled to biotin using a procedure described in Figure 8.14. Given the high binding affinity between biotin and SA, the resulting biotinylated C_{60} was then coupled to SA. With three vacant binding sites still available on the latter, it then beautifully served as a molecular adaptor by allowing the attachment of a biotinylated enzyme to the functionalized fullerene. This method has opened up a new window onto the possibility for the creation of SA-carrying biotinylated building blocks, to which a range of biotinylated biomolecules can be easily attached.

Gozin and coworkers have used the eee-isomer of tris-malonic acid-C_{60} to successfully prepare and characterize a complex with the human serum albumin (HSA), making it the first stable protein–fullerene complex ever observed with a native protein [73]. In addition to an improved water solubility and biocompatibility, this HAS–C_{60} conjugate also showed the potential of attracting and leading further research into the possible effects of these carbon-containing materials on the health of humans, as well as toward the development of novel therapeutic agents and forming the basis of future bioelectronics devices.

One of the very important applications of functionalized C_{60} fullerene has been in the development for an effective treatment against HIV. Wudl, Friedman, Kenyon, and coworkers were the first to design and synthesize a water-soluble, diamido diacid diphenyl C_{60} fulleroid derivative, shown in Figure 8.15, to specifically inhibit the HIV enzyme [74–76]. *In vitro* studies performed using this C_{60} derivative showed that inhibition of acutely and chronically affected peripheral blood mononuclear cells (PBMCs) could be achieved at half maximal effective concentration (EC_{50}) of 7 μM.

With a view to developing compounds with enhanced anti-HIV activity, Prato et al. later reported the design and synthesis of novel C_{60} derivatives as potential HIV aspartic protease inhibitors [76]. The novel compounds were finely designed to contain the two ammonium groups (required in HIV inhibition) tactically positioned on the functionalized fullerene so they would be available to interact with the HIV aspartates (through electrostatic and/or hydrogen bonding), but without attaching them directly onto the C_{60} spherical framework. Scheme 8.3 shows the synthetic route employed to obtain this water-soluble C_{60} derivative.

Recently, Krishna et al. have demonstrated the use of the water-soluble, biocompatible and biodegradable, and antioxidant polyhydroxy C_{60} fullerenes (PHFs) for noninvasive cancer imaging and therapy [77]. PHF-containing nanoparticles were found to provide excellent photoacoustic contrast, and after injection and exposure to near-infrared laser, were able to reduce a tumor's cross-sectional area by up to 72% within only 2 hours of treatment. The photoacoustic and photothermal properties of the PHF have established a way for the exploitation of noninvasive image-guided treatment of

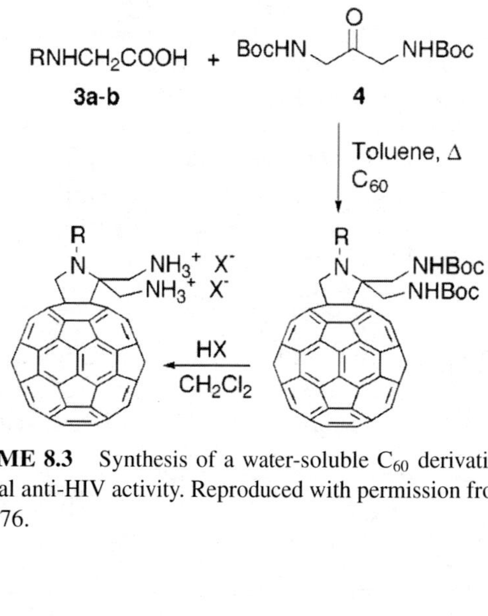

SCHEME 8.3 Synthesis of a water-soluble C_{60} derivative with potential anti-HIV activity. Reproduced with permission from Reference 76.

cancer that entails the least possible side effects. The incorporation of other elements into the C_{60} cage of PHF is also possible and it adds to the imaging capacity of these molecules. For example, gadolinium atoms have been successfully incorporated inside the C_{60} cage to produce fullerene derivatives with an MRI contrast which is 20 times higher than that of commercial contrast agents [78].

8.7 CONCLUSIONS AND FUTURE TRENDS

The element carbon forms the basis of life, and there is no question that those molecules made of carbon atoms only, whether in the tubular or the spherical form, have further magnified the importance of carbon in human lives. By overcoming one of the biggest hurdles (the lack of solubility) which came in the way of CNTs and the buckminsterfullerene C_{60} toward their biological and biomedical applications, scientists from all around the world have brought these fascinating materials a long way through functionalization and countless thoughtful products of engineering. Like it has so far been observed in the literature, the potential applications and the toxicity profiles are very dependent on the methods of functionalization employed in the preparation of these bioconjugates, and under other circumstances, vice versa. In the case of CNTs, although chemical modification and functionalization are well known to affect some of their intrinsic properties, the effective impart of biocompatibility into these tubular carbon molecules through these pathways is just too important to forgo. Functionalization of fullerenes in general with biological and bioactive molecules such as proteins, carbohydrates, and DNA, for example, has had tremendous impact on the design of much needed drug-delivery systems, biosensors, and other biomedical devices. While CNTs and C_{60} have been and are being widely explored for biological and biomedical uses such as drug delivery or the treatment of diseases like cancer and AIDS, concerns about potential risks related to their toxicity have still not been fully addressed. The future trend in this field of research should, therefore, be focused mainly on studies that can create consistent toxicity profiles of functionalized CNTs and C_{60} in order to allow for their exploitation in clinical settings. For example, one important aspect of CNTs that needs to be looked at is their length, which tends to affect their biological performance like *in vitro* cellular uptake and *in vivo* pharmacokinetics [2]. As such, the preparation of CNT samples with narrow length distributions might be a good starting point. To reflect broadly on all the fullerenes, a better understanding of the mechanistic pathways involved in the formation of their bioconjugates in order to obtain a better control of the whole bioconjugation process, and more advanced studies into their physicochemical traits must become the main focus of scientists aiming for the applications of fullerene bioconjugates in biomedicine.

REFERENCES

1. Whitesides GM. *Nat Biotech* 2003;21:1161–1165.
2. Liu Z, Tabakman S, Welsher K, Dai HJ. *Nano Res* 2009;2:85–120.
3. Weiman RB. *The Industrial Physicist* 2004;10:24–29.
4. White AA, Best SM, Kinloch IA. *Int J Appl Ceram Technol* 2007;4:1–13.
5. Ago H, Petritsch K, Shaffer MSP, Windle AH, Friend RH. *Adv Mater* 1999;11:1281–1285.
6. Javey A, Guo J, Wang Q, Lundstrom M, Dai HJ. *Nature* 2003;424:654–657.
7. Cao Q, Rogers JA. *Nano Res* 2008;1:259–272.
8. Fan SS, Chapline MG, Franklin NR, Tombler TW, Cassell AM, Dai HJ. *Science* 1999;283:512–514.
9. Liu Z, Sun X, Nakayama N, Dai HJ. *ACS Nano* 2007;1:50–56.
10. Liu Y, Wu DC, Zhang WD, Jiang X., He CB, Chung TS, Goh SH, Leong KW. *Angew Chem Int Ed* 2005;44:4782–4785.
11. Singh R, Pantarotto D, McCarthy D, Chaloin O, Hoebeke J, Partidos CD, Briand JP, Prato M., Bianco A., Kostarelos K. *J Am Chem Soc* 2005;127:4388–4396.
12. Yang W, Thordarson P, Gooding JJ, Ringer SP, Braet F. *Nanotechnology* 2007;18:1–12.
13. Kam NWS, O'Connell M, Wisdom JA, Dai HJ. *Proc Natl Acad Sci USA* 2005;102:11600–11605.
14. Chen RJ, Zhang YG, Wang DW, Dai HJ. *J Am Chem Soc* 2001;123:3838–3839.
15. Karousis N, Tagmatarchis N. *Chem Rev* 2010;110:5366–5397.
16. Kroto HW, Heath JR, O'Brien SC, Curl RF, Smalley RE. *Nature* 1985;318:162–163.
17. Hedberg J. The Structure of the Buckminster Fullerene, C_{60}. Available at http://www.jameshedberg.com/scienceGraphics.php?sort=all&id=c60-buckyball-atoms-red (accessed April 15, 2013).
18. Arbogast JW, Darmanyan AP, Foote CS, Diederich FN, Whetten RL, Rubin Y, Alvarez MM, Anz SJ. *J Phys Chem* 1991;95:11–12.
19. Chen Z, Ma L, Liu Y, Chen C. Applications of functionalized fullerenes in tumor theranostics. *Theranostics* 2012;2:238–250.
20. Kadish KM, Ruoff RS. *Fullerenes: Chemistry, Physics and Technology*. New York: John Wiley & Sons, Inc.; 2000. pp 431–436.
21. Da Ros T, Prato M. *Chem Commun* 1999;8:663–669.
22. Bosi S, Da Ros T, Spalluto G, Prato M. *Eur J Med Chem* 2003;38:913–923.
23. Partha R, Conyers JL. *Int J Nanomedicine* 2009;4:261–275.
24. Kato H, Kanazawa Y, Okumura M, Taninaka A, Yokawa T, Shinohara H. *J Am Chem Soc* 2003;125:4391–4397.
25. Samai S, Geckeler KEJ. *Chem Soc Chem Commun* 2000;13:1101–1102.
26. Ikeda A, Suzuki Y, Yoshimura M, Shinkai S. *Tetrahedron* 1998;54:2497–2508.

27. Yamakoshi YN, Yagami T, Fukuhara K, Sueyoshi S, Miyata N. *J Chem Soc Chem Commun* 1994;4:517–518.

28. Hirsch A. *The Chemistry of the Fullerenes.* Stuttgart, Germany: Thieme; 1994.

29. Brettreich M, Hirsch A. *Tetrahedron Lett* 1998;39:2731–2734.

30. Meng L, Zhang X, Lu Q, Fei Z, Dyson PJ. *Biomaterials* 2012;33:1689–1698.

31. Liu Z, Chen K, Davis C, Sherlock S, Cao Q, Chen X, Dai H. *Cancer Res* 2008;68:6652–6660.

32. Bullard-Dillard R, Creek KE, Scrivens WA, Tour JM. *Bioorg Chem* 1996;24:376–385.

33. Yamawaki H, Iwai N. *Am J Physiol Cell Physiol* 2006; 290:C1495–C1502.

34. Tsang SC, Davis JJ, Green MLH, Hill HA, Leung YC, Sadler PJ. *J Chem Soc Chem Commun* 1995;17:1803–1804.

35. Huang W, Taylor S, Fu K, Lin Y, Zhang D, Hanks TW, Rao AM, Sun YP. *Nano Lett* 2002;2:311–314.

36. Vossoughi M, Gojginia S, Kazemi A, Alemzadeha I, Zeinalic M. *Engineering Letters* 2009;17:293–296.

37. Zeng H, Gao C, Yan D. *Adv Funct Mater* 2006;16:812–818.

38. Asuri P, Bale SS, Pangule RC, Shah DA, Kane RS, Dordick JS. *Langmuir* 2007;23:12318–12321.

39. Ngoy JM, Iyuke SE, Neuse WE, Yah CS. *J Applied Sci* 2011;11:2700–2711.

40. Tasis D, Tagmatarchis N, Bianco A, Prato M. *Chem Rev* 2006;106:1105–1136.

41. Liu Z, Sun X, Nakayama-Ratchford N, Dai H. *ACS Nano* 2007;1:50–56.

42. Singh KV, Pandey RR, Wang X, Lake R, Ozkan CS, Wang K, Ozkan M. *Carbon* 2006;44:1730–1739.

43. Yu X, Munge B, Patel V, Jensen G, Bhirde A, Gong JD, Kim SN, Gillespie J, Gutkind S, Papadimitrakopoulos F, Rusling JF. *J Am Chem Soc* 2006;128:11199–11205.

44. Yu X, Chattopadhyay D, Galeska I, Papadimitrakopoulos F, Rusling JF. *Electrochem Comm* 2003;5:408–411.

45. Yu X, Kim S, Papadimitrakopoulos F, Rusling JF. *Mol Biosys* 2005;1:70–78.

46. Ahmed M, Xiaoze J, Deng Z, Narain R. *Bioconjugate Chem* 2009;20:2017–2022.

47. Wang Q, Zhou L, Jiang Y, Gao J. *Enzyme Microb Tech* 2011;49:11–16.

48. Vernet T, Tessier D, Chatellier J, Plouffe C, Lee TS, Thomas DY, Storer AC, Ménard R. *J Biol Chem* 1995;270:16645–16652.

49. Li FY, Xing YJ, Ding X. *Enzyme Microb Technol* 2007; 40:1692–1697.

50. Liu Z, Tabakman SM, Chen Z, Dai H. *Nat Protoc* 2009;4:1372–1382.

51. Kam NWS, Liu Z, Dai H. *J Am Chem Soc* 2005;127:12492–12493.

52. Kam NWS, Dai H. *J Am Chem Soc* 2005;127:6021–6026.

53. Chen J, Liu HY, Weimer WA, Halls MD, Waldeck DH, Walker GC. *J Am Chem Soc* 2002;124:9034–9035.

54. Chen RJ, Bangsaruntip S, Drouvalakis KA, Kam NWS, Shim M, Li Y, Kim W, Utz PJ, Dai H. *Proc Natl Acad Sci USA* 2003;100:4984–4989.

55. Ren J, Shen S, Wang D, Xi Z, Guo L, Pang Z, Qian Y, Sun X, Jiang X. *Biomaterials* 2012;33:3324–3333.

56. Besteman K, Lee J-O, Wiertz FGM, Heering HA, Dekker C. *Nano Lett* 2003;3:727–730.

57. Wu P, Chen X, Hu N, Tam UC, Blixt O, Zettl A, Bertozzi CR. *Angew Chem Int Ed* 2008;47:5022–5025.

58. Lee CC, MacKay JA, Frechet JMJ, Szoka FC. *Nat Biotechnol* 2005;23:1517–1526.

59. Kim Y, Zimmerman SC. *Curr Opin Chem Biol* 1998;2:733–742.

60. Svenson S, Tomalia DA. *Adv Drug Delivery Rev* 2005; 57:2106–2129.

61. Sun YP, Huang WJ, Lin Y, Fu KF, Kitaygorodskiy A, Riddle LA, Yu YJ, Carroll DL. *Chem Mater* 2001;13:2864–2869.

62. Kulamarva A, Raja P, Bhathena J, Chen H, Talapatra S, Ajayan P, Nalamasu O, Prakash S. *Nanotechnology* 2009;20(2):025612.

63. Jensen AW, Wilson SR, Schuster DI. *Bioorg Med Chem* 1996;4:767–779.

64. Tagmatarchis N, Shinohara H. *Mini Rev Med Chem* 2001; 1:339–348.

65. Filippone S, Heimann F, Rassat A. *Chem Commun* 2002; 14:1508–1509.

66. Tsai MC, Chen YH, Chiang LY. *J Pharm Pharmacol* 1997; 49:438–445.

67. Bianco A, Da Ros T, Prato M, Toniolo C. *J Pept Sci* 2001; 7:208–219.

68. Yang J, Wang K, Driver J, Yang J, Barron AR. *Org Biomol Chem* 2007;5:260–266.

69. Daroczi B, Kari G, McAleer MF, Wolf JC, Rodeck, U, Dicker AP. *Clin Cancer Res* 2006;12:7086–7091.

70. Partha R, Lackey M, Hirsch A, Casscells SW, Conyers JL. *J Nanobiotechnology* 2007;5:6.

71. Partha R, Mitchell LR, Joshi PP, Conyers JL. *ACS Nano* 2008;2:1950–1958.

72. Capaccio M, Gavalas VG, Meier MS, Anthony JE, Bachas LG. *Bioconjugate Chem* 2005;16:241–244.

73. Belgorodsky B, Fadeev L, Ittah V, Benyamini H, Zelner S, Huppert D, Kotlyar AB, Gozin M. *Bioconjugate Chem* 2005;16:1058–1062.

74. Friedman SH, Decamp DL, Sijbesma RP, Srdanov G, Wudl F, Kenyon GL. *J Am Chem Soc* 1993;115:6506–6509.

75. Sijbesma R, Srdanov G, Wudl F, Castoro JA, Wilkins C, Friedman SH, Decamp DL, Kenyon GL. *J Am Chem Soc* 1993;115:6510–6512.

76. Marcorin GL, Da Ros T, Castellano S, Stefancich G, Bonin I, Miertus S, Prato M. *Org Lett* 2000;2:3955–3958.

77. Krishna V, Singh A, Sharma P, Iwakuma N, Wang Q, Zhang Q, Knapik J, Jiang H, Grobmyer SR, Koopman B, Moudgil B. *Small* 2010;6:2236–2241.

78. Bolskar RD. *Nanomedicine* 2008;3:201–213.

SECTION IV

INORGANIC NANOMATERIALS BIOCONJUGATES (METALS, METAL OXIDES—QUANTUM DOTS, IRON-OXIDE)

9

GOLD NANOMATERIALS BIOCONJUGATES

QIAN YANG

Department of Civil and Environmental Engineering, University of Maryland, College Park, MD, USA

Gold nanoparticle (AuNP) is one of the most investigated nanomaterials because of the promising and unique optical, chemical, and electrical properties it possesses [1, 2]. The biocompatibility (low short-term toxicity), well-established chemistry, and ease of characterization find AuNPs' various applications [3]. In the meantime, the AuNP surface supplies one of the most stable and facile platforms for (bio)molecular conjugation. AuNP-based bioconjugates, associating with proteins, DNAs, carbohydrates, etc., have been developed and intensively studied. Drugs, bioactive molecules, and labeling agents are also conjugated with AuNPs and these conjugates are applied, or have high potential to be applied in drug delivery, sensing, diagnostic imaging, and therapeutic treatment.

Although the use of AuNPs (as gold colloids) can be traced back to the fifth or fourth century BC, it did not grab modern scientific attention until Michael Faraday's well-known work in 1857 [2, 4]. He developed a method to produce colloidal gold in a two-phase system by reducing an aqueous solution of chloroaurate ($AuCl_4^-$) using phosphorus in CS_2. In the twentieth century that followed and in the last decade, various methods were established for the preparation of gold colloids. The most popular one is the chemical reduction of the precursor, tetrachloroauric acid ($HAuCl_4$), by various reducing agents such as citrate [5], hydrazine [6], sodium/potassium borohydride [7, 8], and dimethyl formamide [9]. When a reducing agent is added to the solution containing the metal salt, the metal ions are reduced and metallic solid particles are nucleated. Turkevitch pioneered the citrate reduction method in 1951 and this route resulted in AuNPs with diameter of 10–20 nm [5]. This method was later improved by Frens in 1970s and realized the possibility of obtaining AuNPs with a prechosen size [10, 11]. In 1994, Brust and Schiffrin introduced a two-phase method for AuNP formation which had strong impact in the whole field and inspired many subsequent researches [12]. Most importantly, this method, for the first time, allowed the facile synthesis of thermally stable and air-stable AuNPs of reduced size dispersity and controlled size for the first time [13].

Generally speaking, AuNP bioconjugates are formed by associating (bio)molecules covalently or noncovalently to AuNPs (see Figure 9.1) [14]. For the noncovalent binding, there are two major approaches: electrostatic adsorption and specific affinity. Electrostatic adsorption is the easiest way and no chemical reaction is needed [15]. It needs certain conditions to occur since the AuNP and the conjugating molecule should be oppositely charged. Normally, the pH and ionic strength in the medium are the most important parameters to achieve efficient adsorption, that is, conjugation. On the other hand, this approach is completely a random process, and the AuNP and the targeting molecule can bind each other in any orientation and arrangement. Interactions involved in specific affinity conjugation can be either typical biologically inter-related pairs such as biotin–streptavidin and sugar–lectin, or interactions between synthetic ligands and biomolecules such as nickel-mediated interaction with His-tagged proteins. These interactions are highly selective and are, naturally or by clustering to high ligand density, strong enough to maintain stable binding under mild conditions. One of the components in these pairs is immobilized to AuNPs which can already be considered a conjugate and bind to targets with the other member of the pair on the surface. Biotin–streptavidin is the most widely adopted linkage due to the high strength and ease of decorating functional group to biotin which is subsequently used to tether biotin to AuNPs [16]. The multivalent streptavidin then serves as a bridge to connect to

Chemistry of Bioconjugates: Synthesis, Characterization, and Biomedical Applications, First Edition. Edited by Ravin Narain.
© 2014 John Wiley & Sons, Inc. Published 2014 by John Wiley & Sons, Inc.

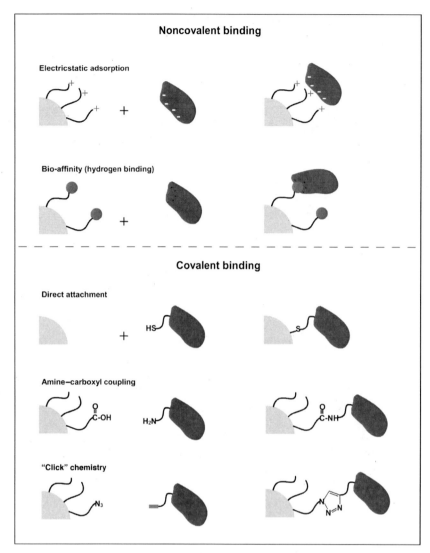

FIGURE 9.1   Different approaches to form AuNP bioconjugates.

the target molecule labeled with biotin. Recently, binding of nickel-nitrilotriacetic acid (Ni-NTA) groups to His-tagged proteins became an important method with very high binding strength (dissociation constant K_d of approximately 10^{-13}M at pH 8 [17]). Advantages of this strategy include the ease of introducing His tag into protein sequence by genetic engineering, controlled stoichiometry, as well as possibility of site-specific labeling. One of the inherent advantages of non-covalent conjugation is that the reversibility of the linkage can be achieved in a relatively easier way and this renders specific applications like sensing and drug delivery.

Covalent attachment leads to much more stable conjugation. The irreversibility provided by covalent binding ensures robust performances for applications in complex chemical and biological circumstances. Covalent conjugation to AuNPs can be achieved by direct attaching or through surface ligands (linkers) and, actually, both methods rely on the Au–thio chemistry. Direct attachment is preferably

applied to obtain protein–AuNP conjugate from those proteins/peptides containing cysteine residues which possess thiol group. Though it is a straightforward and convenient way to form protein–AuNP conjugates, a potential risk is breaking up of disulfide bonds in protein structure due to the comparable strength of Au–S and S–S bonds which can further lead to protein denaturation [14, 18, 19]. The surface ligands strategy involves self-assembly of the ligand with thiol group at one end to form Au–S bond and the functional group at the other which undergoes chemical reaction forming covalent bond with the target molecule. Various chemical reactions can be used to produce linkage between AuNP surface ligands and the conjugating objects. Among them, the amine–carboxyl coupling, often facilitated by succinimide and carbodiimide, is the most popular and widely used one. More recently, a "click" reaction, azide–alkyne Huisgen cycloaddition, was introduced by Sharpless et al. [20] and was then intensively applied to conjugate various biomolecules with AuNPs. It is highly selective and does not interfere with most other organic groups existing in biological system. Moreover, the reaction is quick and high yielding. Unlike amine–carboxyl coupling, "click" chemistry is not pH sensitive and works well in the pH range of 4–11 which helps satisfy different pH values needed to maintain biological and other fragile structures [20]. Comprehensive discussion on conjugation chemistry can be found in the review by Russ Algar et al. [21].

In the following sections, we will introduce AuNP-based bioconjugates, including peptides/proteins, oligodeoxynucleotides (ODN)/DNA, carbohydrates, drugs, imaging agents, and bioactive (macro)molecule-conjugated AuNPs.

9.1 PEPTIDES/PROTEINS

As one of the most important biomacromolecules and the second most abundant substance in the body, proteins are essential parts of organisms and play an important role in virtually every biological process. Proteins are polypeptide from amino acids and the sequence of amino acids in a certain protein is defined by the sequence of corresponding gene. They function as catalysts, binding sites, structural building blocks, and are involved in cell signaling. Combining the advantages in nanotechnology and the important functions of proteins, conjugation of peptides/proteins to AuNPs have found many applications in different fields such as biocatalysis [22, 23], biosensing [24, 25], biomedical imaging [26, 27], delivery system [28, 29], and fundamental understanding in protein folding [30, 31]. Enzymes and antibodies are of special interest for protein–AuNP conjugation because of their particular importance and functions. The conjugates of enzymes and antibodies with AuNPs are particularly useful in biosensing and catalytic tuning. In this section, we will focus on enzyme- and antibody-conjugated AuNPs and discuss their catalytic and biosensing applications.

Enzymes and catalytic peptides are preferably immobilized to NPs because of overwhelming advantages associated with the nanoscale size compared with conventional macroscopic supports: extremely high amount of enzyme loadings, low mass transfer resistance, and no external agitation is needed since NPs undergo Brownian dispersion efficiently [23, 32]. On the other hand, compared with free enzymes or those immobilized on conventional supports, enzymes conjugated with AuNP showed enhanced enzymatic activity and stability [33–35] and change in substrate specificity [36, 37].

Zaramella et al. obtained a catalytically active peptide–AuNP conjugate by electrostatically assembling small peptide sequences on the surface of cationic monolayers on AuNP [38]. The peptide molecule was designed to contain aspartic acid residues for conjugating to the cationic AuNP, fluorescent tryptophan residue for confirming the conjugation, and histidine residues for catalyzing the transesterification reaction of the p-nitrophenyl ester of N-carboxybenzylphenylalanine. Conjugation of this peptide to AuNP resulted in two orders of magnitude rate acceleration of transesterification reaction, attributed to the facilitated contact of peptide with substrate and the local chemical environment that enhances the catalytic activity.

For many applications, especially for enzyme–AuNP conjugate-related applications, protein–AuNP conjugates are preferably linked to a particular position on protein surface. Proteins have a complex three-dimensional structure, which is always important to preserve their function, and foldings in protein structure are easily denatured by various interactions with a solid surface, for example AuNPs. Moreover, randomly conjugating proteins to AuNP makes it more likely that AuNP is located at any possible binding site. For example, when the conjugation chemistry is done with amine groups on a protein surface, there could be many possible binding sites and a random orientation of proteins on NP surface can be expected. For catalytic or sensing applications of protein–AuNP conjugates, it is crucial that NP binding does not occur at/near the active site that may either denature the protein structure or even block the active site. Moreover, random conjugation of AuNP with protein results in an uncertain activity due to the nonrepeatability of the protein orientation on the AuNP surface, that is, although every single batch of protein–AuNP conjugates may have the same amount of protein loading, the amount of protein binding at the ideal position may differ from batch to batch. Furthermore, it is impossible to obtain reliable quantitative information, for example enzyme activity, from such a system. Therefore, site-specific conjugating is desirable and some techniques have been developed.

A well-known and simple example of site-specific conjugating is using cysteine at position 102 in cytochrome c [39–42]. This cysteine is near the C-terminus and on the surface of the protein, and the thiol is facile to conjugate to AuNP. Aubin-Tam et al. showed that this protocol is ideal for maintaining a folded protein structure [41]. Although it is convenient and straightforward, this method can hardly be extended to other proteins as there are always more than one cysteine available in protein structure for conjugation.

Abad and coworkers developed a method to conjugate enzymes to AuNP in a site-specific and oriented way that preserves catalytic functionality and avoids nonspecific adsorption [43]. Genetic engineering of horseradish peroxidase was taken out to introduce six-histidine tag at the N-terminus of the enzyme and this tag was used to conjugate enzyme to AuNP with cobalt(II)-terminated ligand on the surface (see Figure 9.2a). With similar principle, lipase was site-specifically immobilized to AuNP [44]. Lipase was first genetically engineered to express only one solvent-accessible surface lysine residue. This lysine was then modified by carbodiimide chemistry resulting in an acetylene group. After that, the acetylene group was used to couple to AuNPs

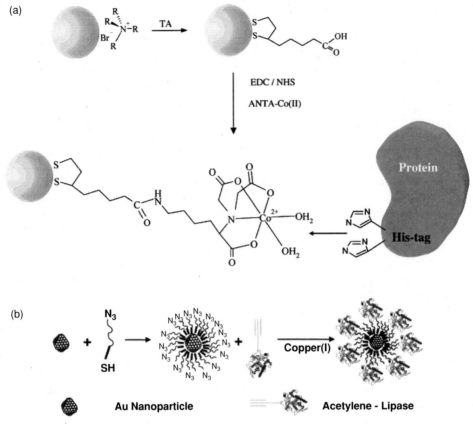

FIGURE 9.2   Site-specific conjugation of proteins to AuNPs by (a) Co^{2+}/NTA–histidine recognition and (b) click chemistry. Reprinted with permission from References 43 and 44, Copyright 2005, 2006, American Chemical Society.

functionalized with azide ligands via click chemistry as shown in Figure 9.2b.

Aubin et al. demonstrated a method to site-specifically conjugate ribonuclease S (RNase S) with AuNP [45]. RNase S is a two-part form of RNase A, an enzyme cleaves RNA. RNase S contains a short piece of 20 residues, the S-peptide, and the rest of the enzyme, S-protein. When S-protein and S-peptide are incubated together, they spontaneously associate to form an active enzyme [46]. However, when exist alone, S-peptide or S-protein exhibits no enzymatic activity. Therefore, AuNP can be incubated first with a peptide, which is extended from S-peptide with cysteine residue, to form an AuNP–peptide conjugation. Then, the NP–S–peptide conjugate can site-specifically associate with the S-protein to form an NP–RNase S complex and exert catalytic function. This strategy does not require genetic engineering to introduce unique sites for conjugation; it can only be applied to certain systems.

Besides conjugation site, another important parameter on the activity of enzyme–AuNP conjugates is the surface charges on the NP surface. The work mentioned above from Aubin-Tam et al. presented the influence of surface charge on protein structure in which AuNPs with positive, negative, and neutral ligands were conjugated with cytochrome c [41]. Their results demonstrated that AuNPs with neutral polyethylene ligands have the best ability to maintain the folded structure of conjugated cytochrome c (35% α-helix). On the contrary, AuNP with charged surface denatured cytochrome c upon conjugation due to the electrostatic interactions with the Ω-loops and α-helix near the linking site. You et al. tuned catalytic behavior of R-chymotrypsin (ChT) toward substrates with different electric charges [36]. ChT–AuNP conjugates showed a threefold increase in substrate specificity for the cationic substrate but a decrease by 95% for the anionic substrate when compared with those of free ChT.

Antibody can be a clue to the development of reliable and accurate biosensors for various applications in a broad range such as pathogen detection and monitoring of diseases [47, 48]. AuNPs conjugated to antibodies combine the high affinity and specificity of antibodies to antigen and the novel photo and electrochemical properties of AuNP. Efficient light scattering and surface plasmon resonance properties of AuNP make it possible to detect NPs at concentrations as low as 10^{-16}M, and have been exploited in an effort to overcome the limitations of conventional colorimetric assays [49].

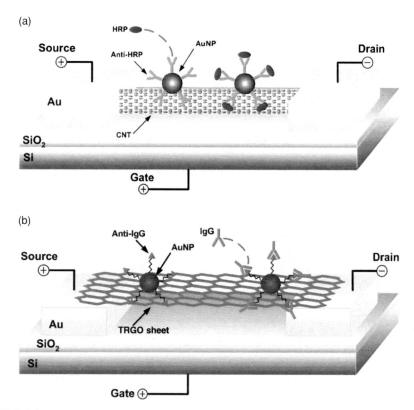

FIGURE 9.3   Detection of HRP by anti-HRP-conjugated AuNPs with (a) carbon nanotubes or (b) GO sheet as field-effect transistor. Reprinted with permission from References 50 and 51, Copyright 2010, Elsevier and John Wiley & Sons, Inc., respectively. For a color version, see the color plate section.

Therefore, antibody–AuNP conjugates are widely used as sensitive biosensors in the field of light and electron microscopy, for visualizing disease-related protein, DNA, virus, and bacteria in biological samples.

Chen's group reported that antibody–AuNP conjugates enhanced field-effect transistor (FET) for protein detection [50]. Antihorseradish peroxidase (anti-HRP) was conjugated to AuNP and then decorated to carbon nanotubes (CNTs) (see Figure 9.3a). Binding HRP to AuNP led to the amplitude change in the drain current which was sensitively detected by FET measurements. The specificity of this sensor was confirmed by the negligible response to mismatched proteins such as IgG. In their following work, the system was further developed using graphene oxide (GO) sheet as FET (Figure 9.3b) [51]. Anti-IgG–AuNP conjugate was used to functionalize GO sheet for detection of IgG. Due to the larger carrier mobility and specific surface area of thin GO sheet compared with those of CNT, the new system showed much higher detecting sensitivity by improving the lower detection limit to the order of ng/mL. These results give a strong suggestion that these sensors can be used for detecting various pathogens by decorating FET with particular AuNP–antibody conjugates, and these new sensors are promisingly capable of *in vitro* diagnostics.

Maier et al. designed an antibody–AuNP-based optical immunochip biosensor for detecting allergen proteins ovalbumin (OVA) and ovomucoid (OVO) [52]. Anti-OVA and anti-OVO were conjugated with AuNP and used as signal transducers in both direct immunoassay and sandwich assay (see Figure 9.4). Compared with aforementioned electric methods, this is a simple and rapid colorimetric solid-phase immunoassay on a planar chip substrate with similar lower detection limitation (1 ng/mL) and it does not require any instrumentation for readout.

Ambrosi and coworkers synthesized a novel double-codified nanolabel based on an AuNP modified with anti-human IgG HRP-conjugated antibody that simultaneously allows both enhanced spectrophotometric (HRP-based) and electrochemical detection (AuNP-based) of antigen human IgG [53]. The detection limits for this double-codified AuNP-based assay were 52 and 260 pg of human IgG/mL for the spectrophotometric (HRP-based) and electrochemical (AuNP-based) detections, respectively that are much lower than those typically achieved by ELISA tests.

Early and accurate detection of cancer often requires time-consuming techniques and expensive instrumentation. Antibody–AuNP could be useful in developing simple and inexpensive methods for detecting cancer cells and targeted

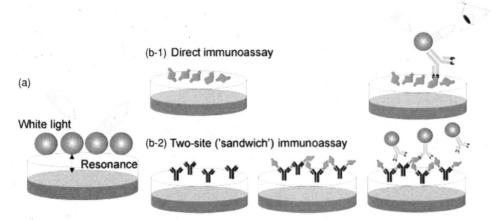

FIGURE 9.4 (a) AuNPs are deposited above a highly reflective mirror surface. Interlayer distance-dependent light that is reflected by the mirror is in phase with the incident field. (b) Schematic representation of the two immunoassay formats. (b-1) Direct immunoassay. The antigen is immobilized onto the surface of the optically transparent distance layer of the immunochips and screened by antibody–AuNP conjugates. (b-2) Two-site (sandwich) assay. The chips are precoated with the first antibody. The antigen is then captured by the antibody and detected with a second antibody–AuNP conjugate. Reprinted with permission from Reference 52, Copyright 2008, American Chemical Society.

cancer treatment. El-Sayed et al. prepared anti-epidermal growth factor receptor (anti-EGFR) antibody-conjugated AuNP and used it for the diagnosis of oral epithelial living cancer cells by surface plasmon resonance (SPR) scattering [49]. They found that the antibody-conjugated AuNPs bind homogeneously and specifically to the surface of the cancer cells with an absorption maximum at 545 nm and the binding to noncancerous cells seems to be nonspecific with an absorption maximum mostly around 552 nm. Therefore, both SPR scattering images and SPR absorption spectroscopy from anti-EGFR antibody-conjugated AuNPs were found to distinguish between cancerous and noncancerous cells. Stuchinskaya and coworkers targeted breast cancer cells with antibody–AuNP conjugates for photodynamic cancer therapy. Anti-HER2 monoclonal antibodies were conjugated to AuNPs through amine–carboxyl coupling [54]. This AuNP conjugate can selectively target breast cancer cells with overexpressed HER2 epidermal growth factor on the cell surface and enhance the efficacy of photodynamic therapy cell death.

Antibody-AuNP conjugates also provide an opportunity for detecting virus in high sensitivity. Driskell and coworkers demonstrated a simple, rapid, and quantitative method for detecting influenza A virus in a single-step homogeneous format using antibody–AuNP conjugate probe and dynamic light scattering (DLS) for readout [55]. The mechanism of this system is quite simple: influenza-specific antibodies are conjugated to AuNPs, and the aggregation of the AuNP probes upon addition of the target virus is evaluated by DLS. Interestingly, the amount of influenza virus can be quantified as a result of the increase in mean hydrodynamic radius of the AuNP aggregates without separation of unbound AuNP probes. The detection limit of this assay was found to be 8.6×10^1 50% tissue culture infectious dose ($TCID_{50}$)/mL which is significantly improved over the lower detection limits of commercial influenza test kits ranging from 2.5×10^3 to 1.0×10^4 $TCID_{50}$/mL.

9.2 ODN/DNA

DNA is a molecule containing and transmitting genetic information. Basically, DNA is a long polymer with sugar–phosphate backbones and four different bases, Adenine (A), Cytosine (C), Guanine (G), and Thymine (T), can be attached to each sugar unit. As genetic material, two antiparallel single strands form double-stranded DNA through hydrogen bond and π-stack following specific base-pairing recognition rules, that is, A with T and C with G. The use of DNAs as building blocks and growth templates was introduced about two decades ago [56–61] and the incorporation of DNA as a scaffold to construct new materials from AuNPs became possible later [62–65]. Profiting from the routine technique of solid-phase synthesis of DNA oligomers, conjugation between DNA and AuNP became feasible. Normally, thiol group is decorated to the 5′- or the 3′-end of the DNA chain and then combined with AuNP surface as described in the pioneering work from Mirkin and coworkers [63]. An issue that should be taken into consideration in DNA–AuNP conjugation is nonspecific binding. Mirkin et al. tested adsorption of the four deoxynucleotides onto AuNP surface and found that the T deoxynucleotide has the lowest nonspecific binding [66]. This is because T has only one amine group, as the amine on the deoxynucleotide surface dominates the nonspecific adsorption. Therefore, it is easy to deduce that DNA chains with high T content can suppress nonspecific interaction. Similarly, Brown et al. investigated the influence of DNA sequence on the conjugation to AuNP [67]. They concluded that DNA sequence with A and C resulted in lower maximum coverages than those with T and G due to the nonspecific adsorption. Technically, there are two means to reduce the nonspecific binding of DNA to AuNPs. One is to increase the density of DNA linked to the surface by Au–S bond, that is, cover more surface with desired covalent binding. This can be achieved by using excess of DNA at high salt concentration as salt can shield charges on DNA and lower the repulsion between chains [68]. Another way to reduce nonspecific binding of DNA to AuNP was developed by Herne and Tarlov by treating the gold surface with mercaptohexanol, after DNA conjugation and XPS measurement confirmed the removal of most of the nonspecifically adsorbed DNA from the surface [69].

These early works on DNA–AuNP conjugates explored a new area of material science and a technology based on colorimetric DNA–AuNP assays [70]. DNA–AuNP conjugate-based sensor systems are normally used in homogeneous solution with a simple process, and the aggregation is accompanied with visible color change which can be sensitively detected even with the naked eye. The aggregation of DNA–AuNP conjugates can be triggered by either adding a linker DNA which is complementary to the one conjugated on AuNP surface (the so-called three-strand system), or directly by two DNA strands on AuNPs that are complementary to each other (the so-called two-strand system). Upon hybridization, the red AuNP suspensions form purple aggregates because of reduction in the interparticle distance. Heating the aggregated DNA–AuNP conjugate suspensions leads to controlled disassembly of the aggregates which brings a change in color from purple (aggregated) to red (dispersed) [71] and, interestingly, this process was proven to be reversible. The hybridization-induced aggregation of DNA–AuNP conjugates is sensitive to the density of DNA on the AuNP surface. A research by Song and coworkers showed that the aggregation only occurs when DNA surface density on AuNPs is greater than 34 pmol cm^{-2}, and when the DNA surface density is less than 26 pmol cm^{-2}, the DNA–AuNP conjugates self-aggregate at a certain ionic strength [72]. Another factor that affects DNA density on AuNP surface is the size of AuNP, that is, the curvature of the Au surface [73]. Mirkin et al. pointed out that by increasing the curvature of the AuNP surface, considerably more oligonucleotides can be packed into a given area [74]. They tested AuNPs with diameters ranging from 10 to 200 nm and found that the molecular area occupied by each 25-nucleotide DNA strand decreased to one-third with the NP size, from a maximum of 15 nm^2 (surface density $6.8 \times 10^{12}/cm^2$) to only

5 nm^2 (surface density $2.0 \times 10^{13}/cm^2$). The aggregation of DNA–AuNP by hybridization has been widely applied to detect various species of bacteria and viruses, and DNAs, proteins, metal ions, and other organic compounds.

Padmavathy et al. reported the use of single-stranded oligonucleotide-conjugated AuNPs as visual detection probes for rapid and specific detection of *Escherichia coli* [75]. The oligonucleotide was carefully designed to be specific to *E. coli* and does not share any sequence homology with non-*E. coli* family members. The mechanism is shown in Figure 9.5a and 9.5b. In the presence of complementary *E. coli* genomic DNA, the AuNP–oligonucleotide conjugates are stabilized and prevented from aggregation caused by adding acid which makes the solution red in color. Without *E. coli* genomic DNA or in the presence of non-*E. coli* genomic DNA, the AuNP-oligonucleotide conjugates aggregate together and result in a purple color. This method is highly convenient as the color change can be observed by the naked eye (see Figure 9.5c) and it is ultrasensitive with a detecting limit of 11.4 ng.

Guo et al. developed a DNA–AuNP conjugate-based test strip for Hg^{2+} detection and this colorimetric mercury sensor exhibited high sensitivity and selectivity in response to mercury [76]. As shown in Figure 9.6, the mercury detection is based on the formation of thymine-Hg^{2+}-thymine (T-Hg^{2+}-T) complex and streptavidin–biotin interaction. DNA1 was functionalized with thiol group at one end and biotin at the other and assembled to AuNP. In the presence of Hg2 +, the DNA–AuNP conjugates were captured by DNA2-tailored BSA via intertwisting two strands of DNAs, leaving a red line at the test zone. The detection sensitivity for Hg^{2+} of this strip can be as high as 3 nM, which is significantly lower than the 10 nM maximum contaminant limit defined by the US Environmental Protection Agency (EPA) for drinking water.

On the other hand, the ability of single-stranded DNA molecules to assemble onto specific regions on the complementary strand in a highly precise way makes them a useful tool to assemble one-, two- and three-dimensional DNA–AuNP conjugates with exquisite control. One-dimensional DNA–AuNP conjugates can be achieved either by assembly of single-stranded DNA-decorated AuNP to the complementary strand or by electrostatic/covalent binding of AuNP to DNA backbone [77, 78]. The pioneering research on one-dimensional DNA–AuNP was done by Alivisatos and coworkers in 1996 [62]. They conjugated AuNPs with single-stranded DNAs and then simply organized them to the complementary DNA template in head-to-head and head-to-tail configurations (see Figure 9.7a). This work proved the potential of DNA to assemble AuNP into well-defined and homogeneous aggregates and enlightened AuNP-based two- and three-dimensional structures. More recently, Simon et al. developed a method to construct one-dimensional chain-like DNA–AuNP conjugate by "click-chemistry" [79]. As shown

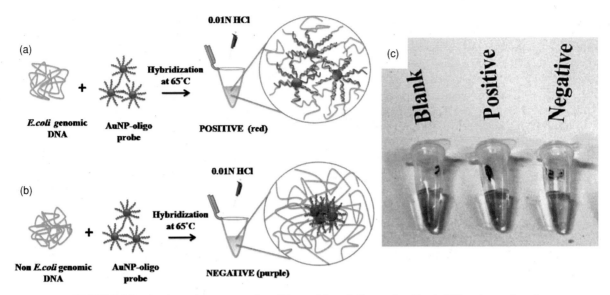

FIGURE 9.5 A schematic representation of the working of oligonucleotide–AuNP conjugate-based *E. coli* detection. (a) In the presence of *E. coli* genomic DNA, AuNPs are stabilized and prevented from aggregation upon acid addition. The red color of the solution indicates presence of target DNA (positive). (b) In the presence of non-*E. coli* genomic DNA, AuNPs tend to aggregate and the color is purple (negative). (c) Visual observation of blank, positive, and negative samples. Adapted from Reference 75. For a color version, see the color plate section.

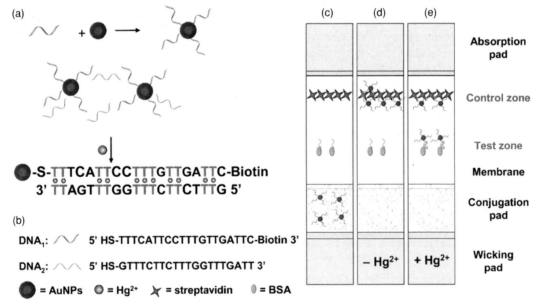

FIGURE 9.6 Design of the test strip format. (a) A schematic illustration of sensors and theranostic agents for Hg^{2+}. (b) Descriptions of DNA_1 sequence, DNA_2 sequence, AuNPs, Hg^{2+}, streptavidin, and BSA. (c) Blank test strip loaded with DNA-functionalized AuNPs (on the conjugation pad), DNA_2–BSA (test zone) and streptavidin (control zone). (d) Negative test: in the absence of Hg^{2+}, the DNA-functionalized AuNPs were captured at the control zone through streptavidin–biotin interaction, producing a single red line. (e) Positive test: in the presence of Hg^{2+}, the DNA-functionalized AuNPs were captured at the test zone by T-Hg^{2+}-T coordination to complementary DNA in addition to the control zone, resulting in two red lines. Reprinted with permission from Reference 76, Copyright 2012, Elsevier. For a color version, see the color plate section.

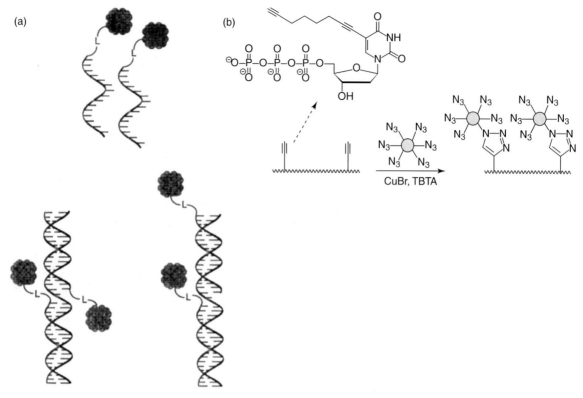

FIGURE 9.7 (a) Two nanoparticles modified with unique single strands of DNA are organized into head-to-head and head-to-tail dimers using a complementary DNA template. Reprinted with permission from Reference 62, Copyright 1996, *Nature,* Macmillan Publishers Ltd. (b) A schematic illustration of incorporating alkyne to DNA and following attachment of azide-modified AuNP. (b) Adapted from Reference 79.

in Figure 9.7b, AuNP was decorated with azide groups and was bound to duplex DNA with alkyne-modified thymine bases in the presence of Cu(I). The DNA–AuNP conjugates created by this approach showed dense AuNP coverage on DNA and uniform interparticle distance.

Two-dimensional assembly of DNA–AuNP conjugates can be obtained by hybridization of the NP to a preformed DNA scaffold, into a two-dimensional array or supermolecular patterns, or to surface-tethered complementary oligonucleotides. Seeman and coworkers reported the first example of a DNA-templated two-dimensional AuNP array [80]. A two-dimensional DNA scaffolding was constructed from a set of 21 synthetic oligonucleotides that were designed to assemble into four different double-crossover tiles and one of them contained an open hybridization site complementary to the DNA sequence conjugated to AuNP. Then the DNA-conjugated AuNPs were hybridized to the DNA scaffolding and organized into lines. Later on, the same group obtained a DNA–AuNP array with two different size AuNPs in alternating lines by creating an additional hybridization site on another tile (Figure 9.8) [81]. Similarly, Yan et al. constructed a two-dimensional AuNP array by hybridizing

DNA–AuNP conjugates to DNA grid from a two-tile system containing four-branch junctions [82]. Although DNA–AuNP conjugates are promising in organizing NPs according to a deliberately designed pattern, the intrinsic flexibility

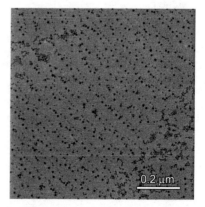

FIGURE 9.8 A TEM image of the two-particle array. The pattern of alternating parallel rows of small and large gold particles is clearly visible. Reprinted with permission from Reference 81, Copyright 2005, American Chemical Society.

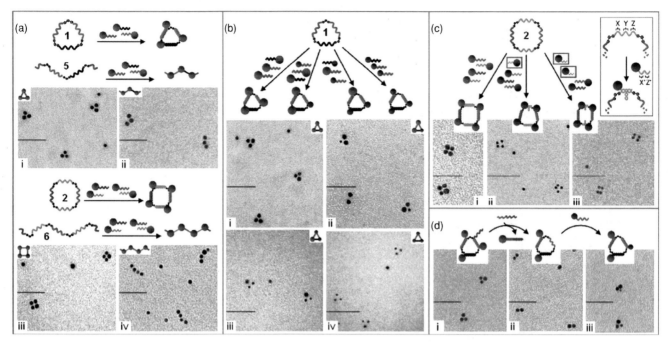

FIGURE 9.9   (a) **1** and **2** organize Au particles into triangles and squares; **5** and **6** result in open linear assemblies of three and four particles. (b) **1** generates triangles of (i) three large (15 nm), (ii) two large/one small (5 nm), (iii) one large/two small, and (iv) three small particles. (c) **2** assembles four Au particles into (i) squares (15 nm particles), (ii) trapezoids, and (iii) rectangles (5 nm). Inset: use of a loop shortens the template's arm. (d) Write/erase function with **1** by (i) writing three Au particles (15 nm) into triangles, (ii) removal of a specific particle using an eraser strand, and (iii) rewriting with a 5 nm particle. Bar is 50 nm. Reprinted with permission from Reference 84, Copyright 2007, American Chemical Society.

of DNA strands diminished their maneuverability. Sleiman group solved this problem by introducing a rigid organic vertex and realized a discrete AuNP assembly [83]. They conjugated AuNPs with a set of hybrid DNA molecules with rigid organic vertices and double-stranded DNA arms, and arranged these AuNP conjugates into well-defined hexamers. The group further developed this novel approach using cyclic and single-stranded DNA templates to guide the assembly of DNA–AuNP conjugates (see Figure 9.9) [84]. Modularity of this approach was indicated by the ability to assemble AuNPs with different sizes in precise locations on the same template into all possible combinations. They demonstrated the ability of their system in forming different two-dimensional structures such as triangle, trapezoid, rectangle, and square. Moreover, this system also showed high addressability. One AuNP can be selectively removed from a triangular assembly (three-AuNP system) without affecting the connectivity between the remaining AuNPs, and after that a different-sized AuNP can be "written" to the exact position of the one that was removed.

The second method refers to the formation of an NP layer on the solid substrate surface and is useful in constructing biosensors. The main concern of this method is the relatively low NP density assembled on the surface and, subsequently,

a metal enhancement step has to be incorporated in sensing applications. Simon and coworkers proposed to solve this problem by creating a connection between AuNPs to force them align densely [85]. They prepared DNA–AuNP conjugates containing two independently addressable oligonucleotide sequences. One of the sequences was used to attach the AuNP to the solid support, while the other sequence was used to establish cross-links between adjacent AuNPs on the surface. Though the feasibility of this method was proven, the attempt to construct a continuous and flawless AuNP layer failed. Koplin et al. developed a method to give DNA–AuNP-covered surface with high enough density to be electrically conducting [86]. First, dendrimer was attached to silica surface and functionalized with DNA oligomers. Then AuNPs conjugated with complementary DNA strand were hybridized to the surface, yielding a uniform and dense (≥ 850 particles μm^{-2}) NP-covered surface.

In 1997, Mirkin et al. reported the first example of a three-dimensional assembly of DNA–AuNP conjugates and they extended it to a binary (two-component) system comprising two different-sized AuNPs later [71, 87]. AuNPs were modified with thiol-terminated oligonucleotides and the addition of the complementary linker resulted in formation of three-dimensional NP aggregates upon hybridization. The color change associated with hybridization, attributed to surface plasmon resonance of the Au, visualized the aggregation and led to a highly sensitive detection of the complementary DNA with lower limit down to 10 fmol. Moreover, the three-dimensional DNA–AuNP network showed higher dehybridization temperature than free DNA with the same sequence and chain length.

9.3 CARBOHYDRATES

Carbohydrates are ubiquitous in all living entities and have been found on the surface of nearly every cell in the form of polysaccharides, glycoproteins, glycolipids, or/and other glycoconjugates [88, 89]. Based on the carbohydrate–protein interactions, they serve as sites for the docking of other cells, molecules, and pathogens in a more or less specific recognition process [90]. It has been proven that carbohydrates play a key role in a variety of biological processes and the carbohydrate–protein recognition is always the first step in numerous phenomena based on cell–cell interactions, such as blood coagulation, immune response, viral infection, inflammatory reaction, and cellular signal transfer [88, 91, 92]. Despite the importance of the carbohydrate–protein interactions and their high specificity, the affinity between these proteins and simple (monomeric) sugar moieties is low ($K_a = 10^3$–$10^4 M^{-1}$) [93]. Though the mechanism has not been fully understood, it is clear that the presentation of sugar groups on an appropriate scaffold creates a multivalent display that can efficiently mimic the mode of affinity enhancement in nature, resulting in higher affinities than expected from the addition of the individual interactions; this is called glycocluster effect [91, 94].

Despite the importance of carbohydrates in biological system, as one of the three most important biomolecules, design and preparation of sugar-conjugated nanomaterials are much

slower than that of nucleic acid and protein. Conjugating carbohydrates to AuNPs, the so-called gold glyconanoparticles, combines the recognition property of the sugar and the advantages of NPs. Gold glyconanoparticles are generally formed in two ways: covalently conjugating sugar ligands to AuNP surface or encapsulating AuNP by polysaccharides. Carbohydrates were assembled onto AuNP surface as polyvalent ligands and the carbohydrate–AuNP conjugates were used as tools for both fundamental study of understanding the glycobiology and more practical applications in diagnosis and therapeutic treatment.

As the early motivation of developing carbohydrate-conjugated AuNP, glyconanoparticles have been proven to be a powerful tool in studying carbohydrate–carbohydrate and carbohydrate–protein interactions. As shown in Figure 9.10, Penadés et al. [95–97] synthesized a series of thiol-derivatized neoglycoconjugates of lactose (1–3), maltose (4), glucose (5), and Lewis X (6). These glycoconjugates were then directly tethered to AuNP and were used to study carbohydrate interactions and to interfere in cell–cell adhesion processes. Thermodynamic evidence was found in the study of Ca^{2+}-mediated carbohydrate–carbohydrate interactions by gold glyconanoparticles functionalized with the disaccharides (lactose and maltose) and trisaccharide Lewis X (Le^X-Au) [97]. Isothermal titration calorimetry results revealed that the Ca^{2+}-mediated aggregation of Le^X-Au is a slow process and takes place with a decrease in enthalpy of 160 ± 30 kcal mol^{-1}, while the heat evolved in the case of lactose and maltose glyconanoparticles was very low and thermal equilibrium was quickly achieved. Moreover, the aggregation of Le^X-Au is selectively mediated by Ca^{2+} and the heat emission was observed to be one-fifth of that in the presence of Mg^{2+} and Na^+ cations.

Similarly, Reynolds and coworkers studied ion-mediated carbohydrate–carbohydrate interactions by gold glyconanoparticles [98]. They synthesized a series of lactose

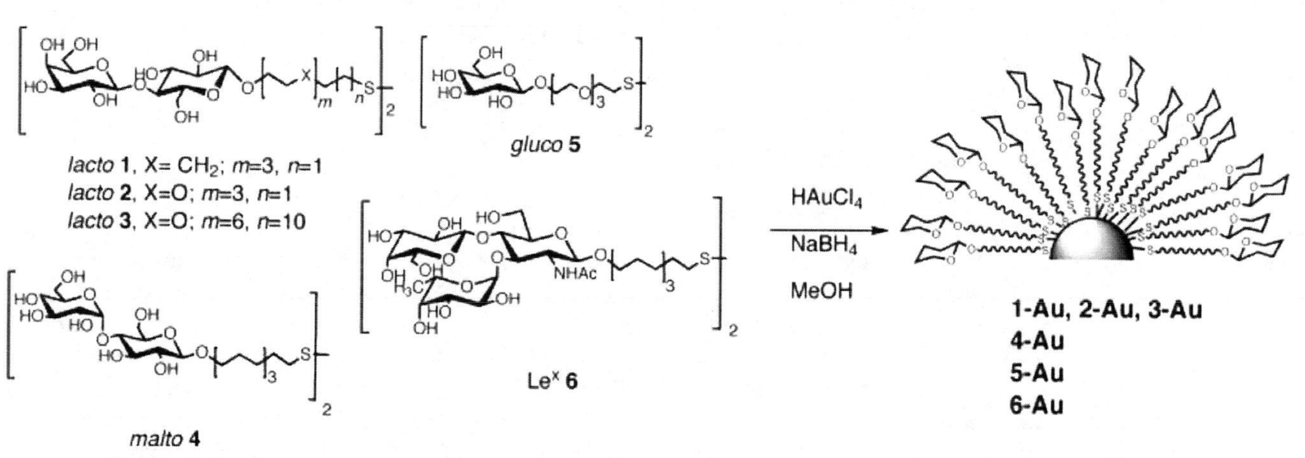

lacto **1**, X= CH$_2$; *m*=3, *n*=1
lacto **2**, X=O; *m*=3, *n*=1
lacto **3**, X=O; *m*=6, *n*=10

gluco **5**

malto **4**

Le^X **6**

HAuCl$_4$
NaBH$_4$
MeOH

1-Au, 2-Au, 3-Au
4-Au
5-Au
6-Au

FIGURE 9.10 Lactose (1–3), maltose (4), glucose (5), and Lewis X (6)-decorated AuNPs. Reprinted from Reference 96, Copyright 2003, with permission from John Wiley & Sons, Inc.

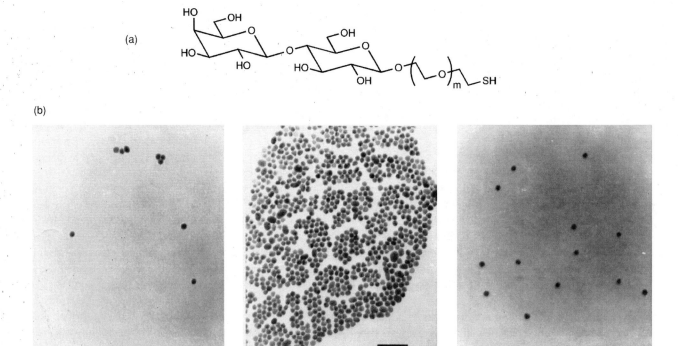

FIGURE 9.11 (a) Structures of lactose derivatives; (b) TEM images of the lactose derivative **14** conjugated AuNPs (left), following addition of 10 mM Ca^{2+} ions (middle), and following subsequent addition of 10 mM EDTA (right), the scale bar represents 50 nm in each instance. Reprinted with permission from Reference 98, Copyright 2006, American Chemical Society.

derivatives varied by the length of the thiolated ethylene glycol anchor chain as shown in Figure 9.11a and directly immobilized these molecules to AuNP surface. The aggregation of lactose-derivatized AuNPs caused by calcium-mediated carbohydrate–carbohydrate interaction was measured via surface plasmon resonance and the aggregation was found to be quantitatively dependent on calcium ion concentration and the ethylene glycol chain length. Calcium binds in an octa-coordinate fashion surrounded by two lactose molecules and four water molecules and causes the aggregation of lactose-derivatized AuNPs. Similar to the inhibition effect of free sugars in carbohydrate–protein interaction, this aggregation was disassociated by adding ethylenediaminetetraacetic acid (EDTA), a calcium ion chelating agent, and NPs were redispersed (see Figure 9.11b). Interestingly, they found that the chain length of the ethylene glycol-derived spacer has an obvious effect on calcium–lactose complex formation. The shorter the ethylene glycol spacer chain length is, the faster is the rate of aggregation forming with the increase of calcium concentration. On the other hand, the derivative with the longest ethylene glycol chain length (**14** in Figure 9.11a) provided a chance to measure the calcium ion concentration in the broadest range (10–35 mM Ca^{2+}).

Gold glyconanoparticles were also intensively used to study carbohydrate–protein interaction since the first work by Kataoka group [99]. A colorimetric assay was generally used to monitor the protein-induced inter-NP interactions. It is based on the color change of AuNP dispersion associated with aggregation state of AuNPs. Boyer et al. first reported the decoration of AuNP with glycopolymer *via* layer-by-layer assembly method [100]. They synthesized citrate-stabilized AuNPs with 20 nm diameter, and positively charged polyethylenimine (PEI) and negatively charged polyacrylic acid (PAA) were used to form multilayers on NP surface. The citrate imparted during AuNP synthesis resulted in negative charges on NP surface that, on the one hand, exert electrostatic repulsion preventing their aggregation/precipitation and, on the other, supply the first "docking" sites to assemble the cationic polymer (PEI). At last, copolymers with PAA and glucose or galactose units [poly(AA-*co*-glucose) and poly(AA-*co*-galactose); see Figure 9.12a] were adsorbed to AuNP. With poly(AA-*co*-glucose) on the surface, AuNPs were precipitated by adding Con A to the dispersion in which Con A, presenting four glucose/mannose binding sites, served as a cross-linker. Moreover, the recognition between glucose and Con A could be inhibited by adding free glucose and the nanoparticles were redispersed (Figure 9.12b). In the meantime, Con A showed no effect on dispersion of poly(AA-*co*-galactose)-functionalized AuNP since galactose is not the specific sugar of Con A. Lin et al. presented quantitative evaluation of the interaction between gold glyconanoparticles and lectin for the first time [101].

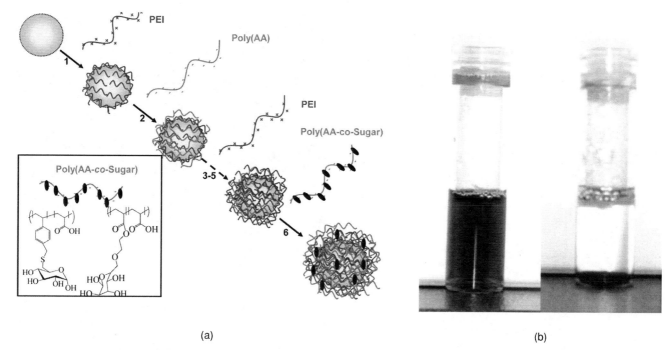

(a)

(b)

FIGURE 9.12 (a) A schematic representation of the strategy used for the functionalization of AuNP using layer-by-layer methodology; (b) AuNPs/poly(AA-*co*-glucose) conjugates in the presence of Con A before (right) and after (left) the addition of free glucose. Reprinted with permission from Reference 100, Copyright 2010, American Chemical Society. For a color version, see the color plate section.

AuNPs were decorated by mannose, glucose, and galactose and the inhibition effect of these gold glyconanoparticles competing sugar monolayer on gold surface was determined by SPR. More recently, Bogdan et al. [102] made a quantitative analysis of binding constant between glycodendrimer (mannose)-coated AuNP and Con A by surface energy transfer process. A binding constant of $(5.6 \pm 0.1) \times 10^6 M^{-1}$ was found for Con A and glycodendrimer-modified AuNP interaction which is 100 times higher than the binding constant of the interaction between the protein and glycodendrimer alone. Sugar-decorated AuNP was also used as absorber to selectively isolate protein through carbohydrate–protein interaction. Suda group developed a method for one-step purification of lectin from crude plant extract by AuNP with surface-presented glucose [103]. Compared with conventional methods, AuNP offers several advantages, including small amount of source materials (<1 g), no further purification or concentration step as well as high product purity.

For diagnostic applications, NPs have been proven to be very promising [104, 105]. AuNPs with multivalent glycoligands can effectively target receptors displayed on various pathogen surfaces with high sensitivity. As mentioned before, the infection of pathogens is directed by carbohydrate-involved interaction. The adhesion, which is the first step of infection, of pathogens to target cells is accomplished by carbohydrate–protein and/or carbohydrate–carbohydrate

interactions [106]. Glycoligands (glycoproteins, glycopeptides, etc.) on virus or bacteria envelop bind to lectins or lectin-like proteins on host cell surface and, reversely, sugar-binding domains on pathogen surface are also recognized by glycoligands on host cell to facilitate attachment. For example, densely packed hemagglutinin on influenza virus plays a critical role in infecting bronchial epithelial cell by the interaction with N-acetylneuraminic acid on the host cell [107, 108]. Huang et al. [109] introduced a fast and relatively convenient method to detect *E. coli* with mannose-decorated gold glyconanoparticles (Figure 9.13a). Brightly fluorescent cell clusters were found by incubating mannose-conjugated AuNP with *E. coli* (see Figure 9.13b). This *E. coli* cluster was formed by the multivalent interaction between the mannosylated AuNP and mannose receptors (FimH lectin [110]) on the bacterial pili. Within the cell density range of 1.00×10^6 to 5.00×10^7 cells/mL, the fluorescence signal intensity showed a linear ($R^2 = 0.96$) relationship with the *E. coli* suspension concentration and the lower detecting limit was found to be 7.20×10^5 cells/mL. In a more recent work from the same group, the limit of detection was further enhanced by two orders of magnitude and they found that the growth of *E. coli* was selectively inhibited by AuNP-induced agglutination [111].

The fact, as mentioned above, that the entry of different pathogens depends on the adhesion to host cell through

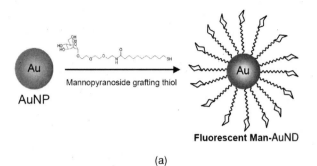

Fluorescent Man-AuND

(a)

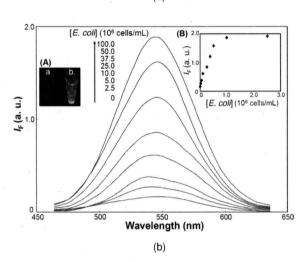

(b)

FIGURE 9.13 (a) Preparation of mannose-decorated AuNPs. (b) Fluorescence spectra of mannose-conjugated AuNPs (25 nM) used as probes for the detection of *E. coli* [(2.50×10^6)–(1.00×10^8)] cells/mL]. (Inset A): Visualization of Man-Au NDs (25 nM) in the (a) absence and (b) presence of *E. coli* (2.50×10^8 cells/mL) upon excitation (365 nm) under a handheld UV lamp. (Inset B): Plot of fluorescence intensity (545 nm) versus *E. coli* concentration. Reprinted with permission from Reference 109, Copyright 2009, American Chemical Society.

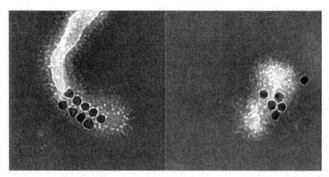

FIGURE 9.14 TEM images of influenza A virions after 60 min incubation with 14 nm sialic acid-conjugated AuNPs. Reprinted from Reference 114, Copyright 2010, with permission from John Wiley & Sons, Inc.

carbohydrate-mediated interactions brings us an opportunity to interfere in this process using man-made artificial glycoconjugates with multivalent glycoligands [112]. Conventional antibacteria treatments, like antibiotics, always built up bacteria resistance and cause critical problems to human society. Antiadhesion agents do not kill pathogens; therefore, the risk of urging mutation is much less and it is believed that the spread of resistance to antiadhesion agent is low [113]. On the other hand, antiadhesion therapy is also valid for cancer treatment, especially in preventing metastasis. The key step of cancer metastasis is adhesion of cancer cell to vascular endothelium in which carbohydrate–protein and carbohydrate–carbohydrate interactions are involved. Inhibition of metastasis by antiadhesion sugar antigens provides an effective way to suppress tumor growth. Glycoligand-conjugated AuNPs offer a powerful tool for antiadhesion therapy by presenting high density glycoligands on the

surface. Sialic acid-terminated glycerol dendron-conjugated AuNP was used to inhibit the infection of influenza virus by competitively binding to the viral fusion protein hemagglutinin on the host cell surface [114]. TEM images (see Figure 9.14) clearly show the multiple attachment of AuNPs to virions, and after incubation, with sialic acid-modified AuNP, virus reduced the infection by 40%. Penadés group dedicated themselves to developing antiadhesion gold glyconanoparticles against HIV infection [115–119]. They reported the design and synthesis of multivalent gold NPs conjugated with high mannose undecasaccharide $Man_9(GlcNAc)_2$ (manno-GNPs; see Figure 9.15 for the oligosaccharide structure) which is present in the HIV-1 envelope glycoprotein gp120 [118, 120]. These gold glyconanoparticles were employed to inhibit the binding between GP120 and dendritic cell-specific intercellular adhesion molecule 3-grabbing nonintegrin (DC-SIGN) which is essential for HIV infection, and mediates HIV transfer to T-lymphocytes. Later on, the same group applied manno-GNPs as a mimic of carbohydrate epitope on 2G12, the HIV-1 monoclonal antibodies, and studied their binding by SPR [116]. Their results showed that manno-GNPs bound to 2G12 with high affinity and by pre-incubating with manno-GNPs, the neutralization of 2G12 was reduced. This work demonstrates that manno-GNP could be used as a promising tool for developing HIV vaccine.

Rojo et al. tested lactose-bearing gold glyconanoparticles, which were supposed to inhibit the binding of melanoma cells to endothelium, as inhibitor to suppress metastasis of tumor cell [121]. A well-designed *ex/in vivo* experiment sequence was carried out for the evaluation of the antimetastasis potential of the glyconanoparticles. Melanoma cells were pre-incubated with lactose gold glyconanoparticles for a short time and then injected into mice. Both lungs of the animals were evaluated after 3 weeks under the microscope for analysis of tumor foci and they found up to 70% of tumor inhibition compared with the group inoculated with melanoma cells without pretreatment.

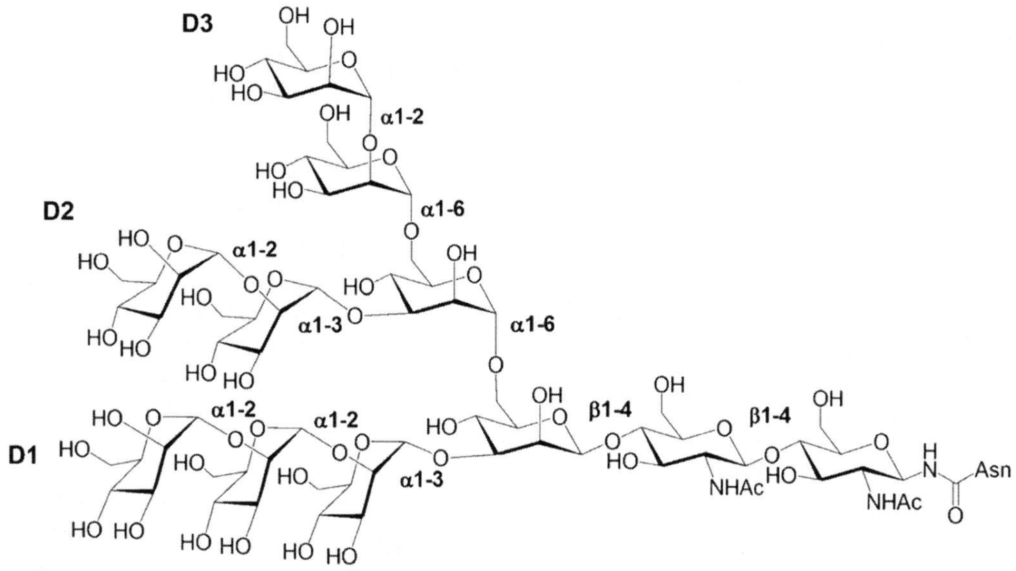

FIGURE 9.15 Structure of Man$_9$(GlcNAc)$_2$. Reprinted from Reference 116, Copyright 2011, with permission from Elsevier.

9.4 DRUGS

Advances in nanoscience and NP engineering provide new opportunities for therapeutic applications, especially for drug delivery. AuNPs, conjugated with drugs, play an important role in nanoscale drug delivery system [122, 123]. The endocytosis is facilitated at nanometer scale and the size of AuNP ensured sufficient uptake by cells [124]. On the other hand, the huge surface area, comparing to same mass of bulk materials, possessed by AuNPs gives the possibility of extremely high drug loading. At the same time, functions expressed on AuNP will be amplified which then benefits the drug efficacy (e.g., reduction in needed dose) and the selectivity in targeted delivery. On the other hand, the versatile surface of AuNP is easy to be functionalized and the size of AuNP is highly tunable. Covalent and noncovalent attachments are the two major means to conjugate drugs to AuNPs [120, 125–130]. The covalent attachment offers stable connection between drug molecule and AuNP and guarantees successful delivery through complicated environments (salts, pH, etc.) in body. However, in covalent approach, drugs need to be modified to endow functional groups for conjugation with AuNP which may cause reduced therapeutic effects. On the other hand, "triggers" are needed to release covalently bound drugs from AuNP once the conjugates reach the target. Trigger can be either *in situ* parameters, such as chemicals and enzymes existing in target tissue/cell and specific pH value on site, or external signal-like photo irradiation. Agasti et al. reported a 2 nm AuNP conjugated with a mixed photocleavable and zwitterionic thiol ligands [131]. An anticancer drug, fluorouracil (5-FU), was attached to the photocleavable ligand through a terminally anchored orthonitrobenzyl group

(Au_PCFU; see Figure 9.16a). As shown in Figure 9.16b, no significant cell death was observed in cells treated with only light or Au_PCFU, indicating no obvious toxicity from light or the drug-conjugated AuNP itself. After being exposed to 365 nm UV light for 0, 1, 6, and 15 minutes, the cell viability decreased gradually and the cell viability correlated with the dose of liberated drug very well. In contrast, at the expense of stability, noncovalent conjugation of drugs to AuNP avoids modification of drugs and no trigger is needed for release. The drug delivery approach must be chosen depending on the respective therapeutics and targets [120].

Nevertheless, there are two major issues related to AuNP-based drug delivery; one is nonspecific adsorption and the other is the specificity of targeting. Depending on the treatment type and intake route, NPs are exposed to various biological fluids, such as blood, gastric fluid and bile, during delivery through human body. Anyway, NPs are surrounded by numerous biomacromolecules in the body and the high surface energy of bare AuNP makes it a prime target of nonspecific adsorption. For example, in blood stream, there are more than 3000 kinds of proteins and the high ionic strength in physiological environment makes it an even more complicated condition. Therefore, stabilization of AuNP by surface modification becomes particularly important. It is worth noting that conventional stabilization of NPs achieved by electrostatic repulsion (e.g., presenting charges on AuNP surface) is not sufficient because of the aforementioned high ionic strength which may screen surface charges on NPs. Physical separation by steric repulsion from polymer ligands then receive intense interest. PEG is the most used ligand for this purpose and it has been proven to be highly efficient. Besides PEG, other polymers such as glycopolymer,

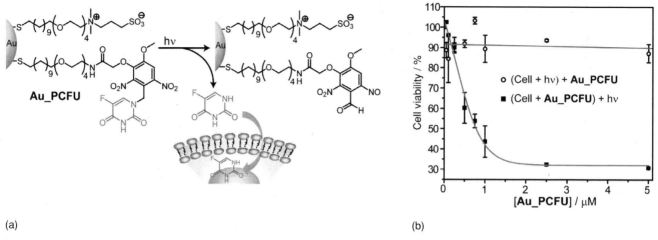

(a)   (b)

FIGURE 9.16   (a) Photochemical reaction of Au_PCFU and delivery of payload to cell. (b) Cell viability in the presence of only light or Au_PCFU and exposing to UV with Au_PCFU. Reprinted with permission from Reference 131, Copyright 2009, American Chemical Society.

polyvinylpyrrolidone, and zwitterionic polymers also showed good stabilizing effect. Grafting these polymers also helps decrease immunogenic response of the human body and avoid clearance by reticuloendothelial system. This can prolong the existence of NPs in circulation and increase the exposure time of NPs to the target tissue/organ. However, polymer ligands also block the particle surface and decrease the surface accessibility to other species such as drugs.

Another concern in AuNP-based drug delivery is specificity which is a common issue that also stands for other drug delivery systems. Increasing specificity helps deliver drugs to intended target, for example, organs and tissues, and reduce the side effect by minimizing undesired drug existing in other places. This is normally achieved by targeting receptors overexpressed on certain cells by ligands on AuNP. Among them, folic acid/folate receptor (FA/FR) pair is widely used to target different types of cancer [7, 132, 133]. It is known that FRs are overexpressed in various malignant cancer cells in kidney, lung, breast, endometrial, colorectal, brain, etc. Dixit et al. prepared a PEG chain with thioctic acid and folic acid coupled on opposite ends and conjugated it with AuNP [133]. A TEM study showed significant uptake of these FA-conjugated AuNPs by KB cell which is human cancer (carcinoma nasopharynx) cell with overexpressed FR. Simultaneously, no uptake occurred for WI-38 cell (human lung fibroblast) on which FR is poorly expressed indicating the successful selective targeting against cancer cell by these FA-conjugated AuNPs. Similarly, Li et al. [134] conjugated FA with AuNP to target KB cell but they, in parallel, attached another ligand, glucose, resulting in a so-called dual-ligand AuNP. This dual-ligand AuNP obviously enhanced the contrast between FR overexpressing cancer cells and FR low expressing cells and the uptake of AuNPs was increased

by 3.9 and 12.7 times compared with mono-ligand AuNP-FA and AuNP-glucose, respectively. Human transferrin/receptor pair is also a promising candidate for AuNP-based cancer treatment and has received significant attention [135–138]. A sophisticated AuNP drug delivery system should normally be composed of conjugates with drug, stabilizing ligand, and targeting ligand. Heo et al designed a theranostic drug delivery system by direct attachment of thiol-derivatized PEG, biotin, beta-cyclodextrin (β-CD), and Paclitaxel (PTX) to AuNP surface [139]. PEG served as an antifouling shell to reduce nonspecific adsorption of proteins and other biomacromolecules to the AuNP conjugate surface. Biotin, a vitamin required by all living cells, especially by rapidly dividing cancer cells for their rapid proliferation and overexpression of biotin-specific receptors, is common on the surface of cancer cells [140]. Therefore, biotin is considered as a good ligand to target cancer cell for drug delivery. PTX is one of the most effective chemotherapeutic drugs for treating various cancers [141, 142]. β-CD was also incorporated to AuNP as a drug-holding pocket to form inclusion complexation (IC) interactions with PTX which improves PTX's water solubility. This AuNP conjugate showed higher affinity to cancer cells such as HeLa, A549, and MG63 and PTX can be effectively released from the conjugate. Most interestingly, the conjugates exhibited significant anticancer effects against HeLa cancer cells with almost no cytotoxicity to normal NIH3T3 cells.

Though it is clear that by attaching targeting ligand, one can achieve selective binding of AuNP to a specific cell, the actual situation is not so positive. The main challenge lies on the penetration of AuNPs into target tissue and distributing evenly over the whole region. Studies also showed size-dependent distribution of NPs *in vivo* in different organs [143–146]. For example, De Jong et al. demonstrated that small AuNPs (10 nm) were present in various organs/body components including blood, liver, spleen, kidney, testis, thymus, heart, lung, and brain, whereas the larger NPs were only found in blood, liver, and spleen [144]. This makes *in vivo* delivery of AuNP even more unclear as to whether the actual effect is from targeting or preferential accumulation by particle size [136]. Though methods to improve AuNP penetration are very limited, some work showed improved delivery of NPs into certain tissues. Guerrero et al. conjugated AuNP with amphipathic peptide and increased the delivery of NPs to brain which could be promising in treating Alzheimer's disease [147].

9.5 IMAGING AGENTS

Recently, AuNPs have received much attention for biological detection with a wide range of analytes, from proteins to DNAs and bacteria to viruses. In particular, AuNP-based cancer detecting and analysis has been a hot topic in this field in recent years. The intrinsic light scattering and SPR properties of AuNP make the NP itself a very useful imaging probe for sensing applications such as MRI, CT, and photoacoustic tomography (PAT). For example, AuNP has been applied to develop new CT contrast agents to replace conventional iodine. This is because, higher molecular weight of AuNPs presents an enhanced absorption coefficient (gold: $5.16 \text{ cm}^2\text{g}^{-1}$; iodine: $1.94 \text{ cm}^2\text{g}^{-1}$ at 100 keV), which results in 2.7 times higher contrast than typical iodine agents [148, 149]. For this purpose, AuNPs are always coated with antibiofouling agent like PEG to extend their lifetime in bloodstream [150, 151]. Kim et al. found that PEG-coated AuNPs have 5.7 times higher X-ray absorption coefficient than the current iodine-based CT contrast agent and a much

longer blood circulation time (>4 h) [152]. Zhang et al. demonstrated that AuNPs (20 and 50 nm) have high photoacoustic contrast and can be clearly visualized in mice *in vivo* using PAT [153]. Accumulation of AuNPs in tumors was effectively imaged with PAT and this method is promising in noninvasive early diagnosis and imaging of cancers.

As mentioned in earlier sections, the strong adsorption of AuNP at visible light region makes it easy to be observed by the naked eye and facilitates applications in biosensing. Both light scattering and SPR are based on the property or the interaction between Au cores. Conjugating AuNP with imaging agents provides a chance to build signal transduction systems like surface-enhanced Raman scattering (SERS). SERS is a technique based on the enhancement of Raman scattering on the target molecule by the AuNP metal core, in which the scattered light is detected from the Raman active dye molecules conjugated to the surface of AuNPs. Wang et al. used SERS for detecting circulating tumor cells (CTCs) [154]. AuNP was first conjugated with Raman reporter and then with epidermal growth factor peptide as a targeting ligand. This AuNP imaging probe was successfully employed to identify CTCs in the peripheral blood of 19 patients with squamous cell carcinoma of the head and neck in a range of 1 to 720 CTCs per milliliter of whole blood. Guerrero et al. [155] conjugated two peptides, CLPFFD and CK, covalently with AuNP (12 ± 0.5 nm) and then labeled the CK peptide with *N*-succinimidyl-4-[^{18}F]-fluorobenzoate ([^{18}F]-SFB) which serve as radioisotope probe allowed to perform AuNP biodistribution studies with positron emission tomography (PET) imaging. Both the peptides were designed with a cysteine unit providing thiol group for direct conjugation with AuNP. The amine group, on the other end of CK, reacted with the probe and gave a stable conjugate as shown in Figure 9.17. PET scans of rat after being injected with this

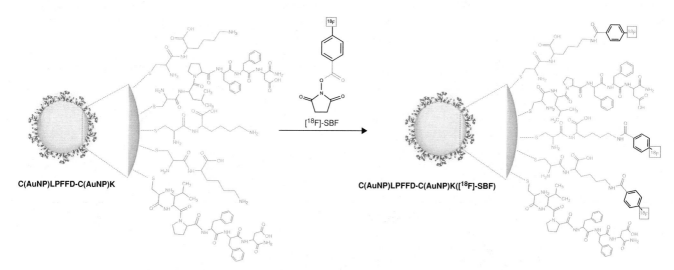

FIGURE 9.17   Conjugation of [18F]-SFB probe with AuNP by the reaction between the exposed amino group of CK peptide and the carbonyl group of the [18F]-SFB probe. Reprinted with permission from Reference 155, Copyright 2012, American Chemical Society.

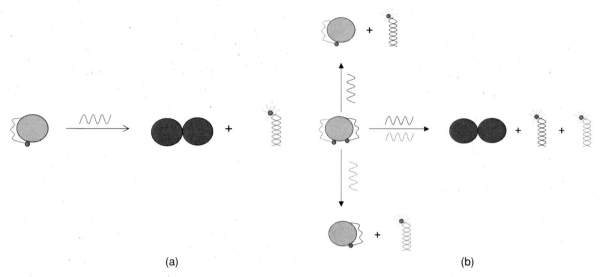

FIGURE 9.18 A schematic representation of the single (a) and multiple (b) DNA hybridization processes. Reprinted from Reference 157, Copyright 2007, with permission from IOP Publishing Ltd.

AuNP conjugates demonstrated the trapping of the NPs in liver, kidney, spleen, intestine, and bladder.

On the other hand, by tuning the distance from the fluorophore to AuNP, the fluorescence can be either enhanced or quenched. Based on this phenomenon, metal-enhanced fluorescence (MEF) and metal-induced fluorescence quenching are employed as sensitive imaging methods for biomedical diagnosis. Bessem et al. developed a rapid detection assay for malaria diagnosis from infected blood culture [156]. This assay is based on the fluorescence quenching of cyanine 3B-labeled *Plasmodium falciparum* heat shock protein 70 (*Pf*Hsp70) upon binding to AuNPs conjugated with anti-Hsp70 antibody. When competing with free antigens, the fluorescence-labeled *Pf*Hsp70 is released to the solution, leading to an increase in fluorescence intensity. The assay showed a lower detection limit of 2.4 $\mu g/cm^2$ and a linear antigen response in a concentration range of 8.2 to 23.8 $\mu g/cm^2$.

Ray et al. developed a special ultrasensitive AuNP-based DNA sensor as the concept shown in Figure 9.18 [157]. A fluorescence dye-labeled probe-DNA was reversibly adsorbed to the AuNP and the fluorescence dye was quenched. In the presence of the complementary DNA strand, the DNA probe on AuNP was hybridized with it and released from AuNP. Therefore, the fluorescence was restored and detected by spectrometer. Most interestingly, this system can detect multiple DNA hybridization. As shown in Figure 9.18b, after hybridization of both DNAs on AuNPs, different fluorescence can be seen. At the same time, AuNPs undergo aggregation, and a color change can be observed. On the other hand, with one DNA hybridized and only one fluorescence

released, either of the two color fluorescence will be detected. Moreover, no color change was observed in this case because there was a DNA still adsorbed on AuNP and the NPs were not able to undergo aggregation.

Nitin and coworkers reported using AuNP as multimodality reporter in both fluorescence and reflectance modes which can extend the range of sensitivity, penetration depth, and resolution [158]. AuNP was conjugated with both fluorescence dye and targeting molecule and then the binding of this AuNP to SiHA cell surface was imaged by both reflectance and fluorescence confocal microscopy (see Figure 9.19). Similar distribution of reflectance and fluorescence signal on the surface of the cells was observed.

Similarly, Geng et al. reported AuNP conjugates for dual-modal targeted cellular imaging [159]. Poly[9,9-bis(6′-N,N,N-trimethylammonium)hexyl) fluorenyldivinylene-alt-4,7-(2,1,3-benzothiadiazole) dibromide] (PFVBT) was conjugated to AuNPs and showed far-red fluorescence. The fluorescence of PFVBT was well maintained in the conjugates due to the eccentric location of AuNPs in the system. The far-red fluorescence from PFVBT and the scattering property of AuNPs give the conjugates a dual-modal probe for targeted cellular imaging.

9.6 BIOACTIVE (MACRO)MOLECULES

Bioactive (macro)molecule is a rather broad term, and there is not a clear definition of it. Actually, those species, proteins, (poly)saccharides, etc., mentioned in earlier sections can be cataloged in bioactive (macro)molecule. Regarding their

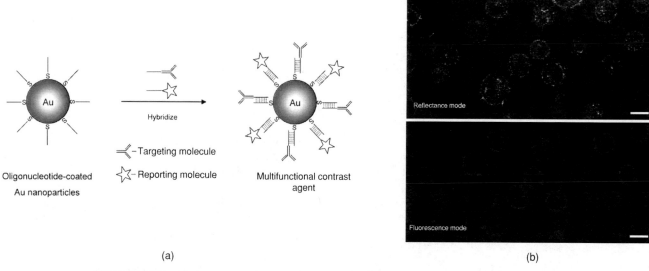

(a)

(b)

FIGURE 9.19   (a) A conceptual approach of using oligonucleotide-coated gold nanoparticles to develop multifunctional contrast agents for targeting, reporting, and delivery. (b) Proof of concept study to illustrate development of bimodal reporting molecular contrast agents. Fluorescence and reflectance confocal images of SiHa cells labeled with bimodal contrast agent targeting the EGF receptor. The scale bar represents 15 μm. Reprinted with permission from Reference 158, Copyright 2007, American Chemical Society.

conjugates with AuNPs, it is difficult to make a summary in terms of applications. However, the purpose of conjugating such molecules with AuNP is either to interfere in the interaction between AuNP and biological systems such as cell and tissue or to endow AuNP with a certain biofunction such as bacteria-inhibiting ability. Therefore, in this section, we will give several examples of bioactive (macro)molecules other than those compounds aforeintroduced, conjugated AuNPs.

Stiti et al. developed a carbonic anhydrase inhibitor (CAI)-conjugated AuNP which showed excellent human carbonic anhydrase (CA) IX inhibitory properties [160]. Human CA IX is an extracellular transmembrane isoform and is overexpressed in many tumor tissues. It is correlated with a bad response to classical chemo- and radiotherapies. As shown in Figure 9.20, CAI-conjugated AuNP (GNP-1) and a control sample (GNP-2) were prepared and the inhibiting effect

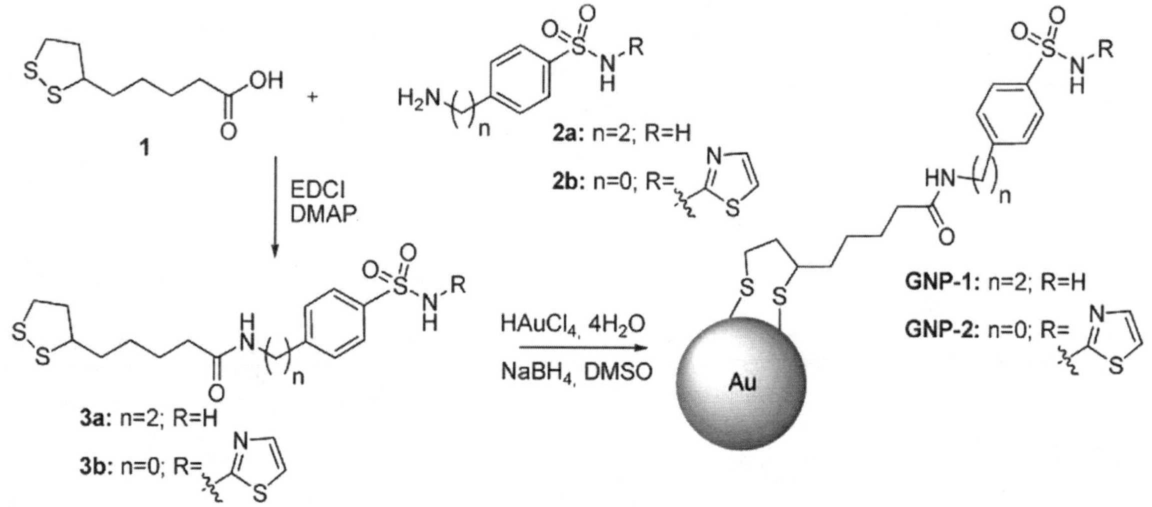

FIGURE 9.20   Synthesis and conjugation of CAI to AuNP. Reprinted with permission from Reference 160, Copyright 2008, American Chemical Society.

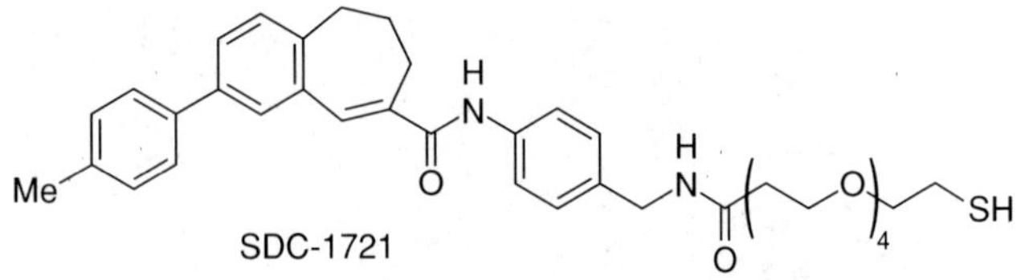

FIGURE 9.21 Structure of SDC-1721. Reprinted with permission from Reference 161, Copyright 2008, American Chemical Society.

TABLE 9.1   Percent Growth Inhibition of *E. coli* by Mixed Ligand-coated AuNPs

Conjugate	20	19	30	20a	20b	20c	20d
Thiols	5,6,8	5,6,7	1,5,6	6	5,8	5,6	6,8
Inhibition (%)	99.9	99.5	99.9	40	20	90	99.9
Conc. (μM)	0.5	0.5	5.0	0.5	0.5	0.5	0.5

Source: Adapted from Reference 162.

was tested. GNP-1 exhibited significant inhibition on CA IX and the specificity was confirmed over CA I and CA II. This AuNP conjugate is promising for both imaging and treatment purposes of tumors overexpressing CA IX. Bowman et al. prepared SDC-1721-conjugated (see Figure 9.21 for structure) AuNPs that effectively inhibit HIV-1 fusion to human T cells [161]. AuNP-SDC-1721 conjugates inhibited viral entry and at 48 hours, viral production was found insignificant. SDC-1721 is a low affinity and biologically inactive molecule toward HIV infection and, however, conjugating it to AuNP in a multivalent manner makes it therapeutically viable. This work has proven that AuNP can be used as a scaffold to exploit and select biologically active molecules for therapeutic applications.

Bresee and coworkers described a simple one-pot method to assemble a library of mixed ligand-conjugated AuNPs [162]. The inhibition activity toward *E. coli* growth was determined. As can be seen from Figure 9.22 and Table 9.1, some of the combinations of ligands on AuNP surface showed 99.9% growth inhibition at 0.5 μM. Moreover, the inhibition activity appeared to depend on the specific combination of ligands displayed on the AuNP surface. Chamundeeswari et al. conjugated ampicillin to AuNP and the antimicrobial activity was evaluated [163]. The optimum concentration of inhibition for *E. coli* was 27.4 μg/mL and for *Staphylococcus aureus* and *Klebsiella mobilis,* 20.6 μg/mL. Also, the AuNP conjugates showed a twofold increase in activity when compared with that of free ampicillin.

Yang and Murphy made lipid-conjugated AuNPs which showed remarkable stability against aggregation in buffers and at different pHs [164]. This certainly benefits applications of AuNPs in biological environments for cellular imaging, targeting, or therapeutics. Hao et al. studied uptake of phospholipid-coated AuNPs by cells and compared with PEG-conjugated AuNPs [165]. The uptake speed, as well as the uptake amount of phospholipid–AuNP conjugates, was faster than that of PEG-AuNP and the maximum uptake was found at 8 and 16 hours after incubation respectively. This can be attributed to the structural similarity of the lipid layer on AuNPs and on the cell membrane. Interestingly, aggregation of phospholipid–AuNP was observed by the fusion of the lipid bilayer structure surrounding AuNPs (see Figure 9.23).

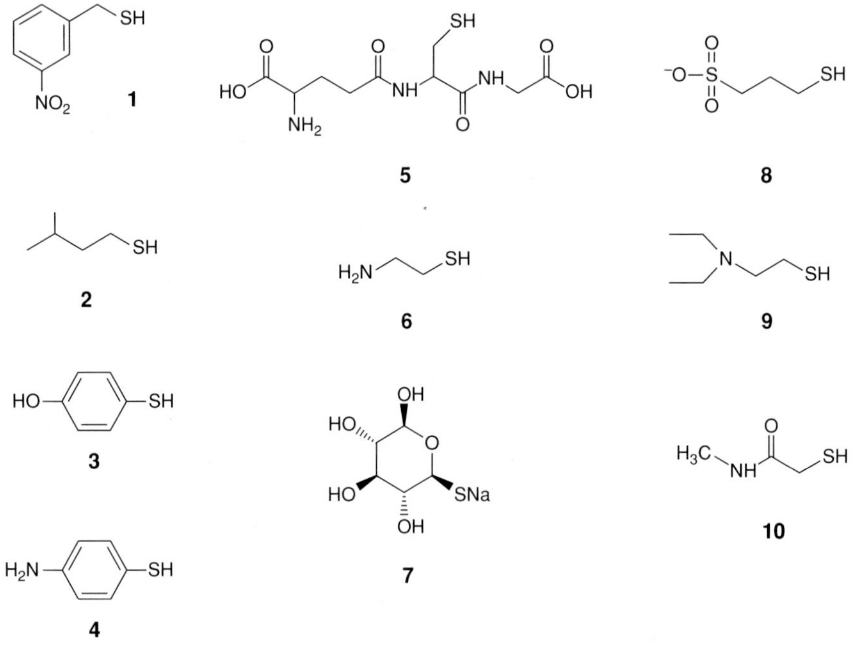

FIGURE 9.22 Thiol ligands used to construct the AuNP conjugate library. Adapted from Reference 162.

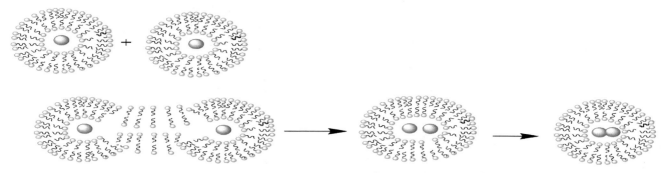

FIGURE 9.23   Aggregation of phospholipid-coated AuNPs. Adapted from Reference 165.

9.7 CONCLUSION

The unique physical properties, precise control over size, ease of functionalization, and well-established surface chemistry for conjugating make AuNPs a versatile platform for conjugation of various bio(macro)molecules. Both covalent and noncovalent methods have been developed to conjugate peptides/proteins, DNA, carbohydrate ligands, drugs, imaging agents, and other bioactive molecules to AuNPs. Among them the Au–thio chemistry is the most investigated and applied one because of the high strength of the Au–S bond and availability of diverse functional groups. These bioconjugates possess the potential for clinical and biomedical applications in both diagnostic and therapeutic aspects. On the other hand, AuNP bioconjugates also supply us important tools for detecting certain molecules in extremely low concentration and complicated chemical environment. Some issues and challenges still remain in this field. One is site-specific labeling; although some approaches have been reported to achieve site-specific conjugation, they are restricted to several specific target molecules (mainly proteins) and a universal protocol for other species is lacking. Another is multifunctional conjugation which benefits practical application, for example, increase target specificity or optimize distribution *in vivo*. Especially, a method for precise control of different functions on AuNP surface (density, distribution, and quantitative amount) is not available. Therefore, with a view to realizing further and broader applications of AuNP bioconjugates, efforts are still needed to solve existing problems.

REFERENCES

1. Jans H, Huo Q. *Chem Soc Rev* 2012;41:2849–2866.

2. Daniel MC, Astruc D. *Chem Rev* 2004;104:293–346.

3. Hakkinen H. *Nat Chem* 2012;4:443–455.

4. Dreaden EC, Mackey MA, Huang XH, Kang B, El-Sayed MA. *Chem Soc Rev* 2011;40:3391–3404.

5. Turkevich J, Stevenson PC, Hillier J. *Discuss Faraday Soc* 1951;11:55–75.

6. Chen DH, Chen CJ. *J Mater Chem* 2002;12:1557–1562.

7. Patra CR, Bhattacharya R, Mukherjee P. *J Mater Chem* 2010;20:547–554.

8. Liu YL, Male KB, Bouvrette P, Luong JHT. *Chem Mater* 2003;15:4172–4180.

9. Pastoriza-Santos I, Liz-Marzan LM. *Langmuir* 2002;18:2888–2894.

10. Frens G. *Kolloid-Zeitschrift and Zeitschrift Fur Polymere* 1972;250:736–741.

11. Frens G. *Nat Phys Sci* 1973;241:20–22.

12. Brust M, Walker M, Bethell D, Schiffrin DJ, Whyman R. *J Chem Soc Chem Commun* 1994;801–802.

13. Brust M, Fink J, Bethell D, Schiffrin DJ, Kiely C. *J Chem Soc Chem Commun* 1995;1655–1656.

14. Aubin-Tam ME, Hamad-Schifferli K. *Biomed Mater* 2008;3:034001.

15. Geoghegan WD, Ackerman GA. *J Histochem Cytochem* 1977;25:1187–1200.

16. Green NM. *Methods in Enzymology* 1990;184:51–67.

17. Schmitt J, Hess H, Stunnenberg HG. *Mol Biol Rep* 1993;18:223–230.

18. Ulman A. *Chem Rev* 1996;96:1533–1554.

19. Whitesides GM, Laibinis PE. *Langmuir* 1990;6:87–96.

20. Kolb HC, Finn MG, Sharpless KB. *Angew Chem Int Ed* 2001;40:2004–2021.

21. Algar WR, Prasuhn DE, Stewart MH, Jennings TL, Blanco-Canosa JB, Dawson PE, Medintz IL. *Bioconjug Chem* 2011;22:825–858.

22. Katz E, Willner I. *Angew Chem Int Ed* 2004;43:6042–6108.

23. Wang P. *Curr Opin Biotechnol* 2006;17:574–579.

24. Cao XD, Ye YK, Liu SQ. *Anal Biochem* 2011;417:1–16.

25. Saha K, Agasti SS, Kim C, Li X, Rotello VM. *Chem Rev* 2012;112:2739–2779.

26. Nune SK, Gunda P, Thallapally PK, Lin YY, Forrest ML, Berkland CJ. *Expert Opin Drug Deliv* 2009;6:1175–1194.

27. Hainfeld JF, Powell RD. *J Histochem Cytochem* 2000;48:471–480.

28. Ghosh P, Han G, De M, Kim CK, Rotello VM. *Adv Drug Deliv Rev* 2008;60:1307–1315.

29. Murakami T, Tsuchida K. *Mini RevMed Chem* 2008;8:175–183.

30. Bhattacharya J, Jasrapuria S, Sarkar T, GhoshMoulick R, Dasgupta AK. *Nanomedicine* 2007;3:14–19.

31. Wang FA, Wang JL, Liu XQ, Dong SJ. *Talanta* 2008;77:628–634.

32. Jia HF, Zhu GY, Wang P. *Biotechnol Bioeng* 2003;84:406–414.

33. Ardao I, Comenge J, Benaiges MD, Alvaro G, Puntes VF. *Langmuir* 2012;28:6461–6467.

34. Gole A, Dash C, Ramakrishnan V, Sainkar SR, Mandale AB, Rao M, Sastry M. *Langmuir* 2001;17:1674–1679.

35. Wu CS, Wu CT., Yang YS, Ko FH.*Chem Commun* 2008; 42:5327–5329.

36. You CC, Agasti SS, De M, Knapp MJ, Rotello VM. *J Am Chem Soc* 2006;128:14612–14618.

37. Simard JM, Szymanski B, Erdogan B, Rotello VM. *J Biomed Nanotech* 2005;1:341–344.

38. Zaramella D, Scrimin P, Prins LJ. *J Am Chem Soc* 2012;134:8396–8399.

39. Chah S, Kumar CV, Hammond MR, Zare RN. *Anal Chem* 2004;76:2112–2117.

40. Heering HA, Wiertz FGM, Dekker C, de Vries S. *Journal of the American Chemical Society* 2004;126:11103–11112.

41. Aubin-Tam ME, Hamad-Schifferli K. *Langmuir* 2005;21:12080–12084.

42. Aubin-Tam ME, Hwang W, Hamad-Schifferli K. *Proc Natl Acad Sci USA* 2009;106:4095–4100.

43. Abad JM, Mertens SFL, Pita M, Fernandez VM, Schiffrin DJ. *J Am Chem Soc* 2005;127:5689–5694.

44. Brennan JL, Hatzakis NS, Tshikhudo TR, Dirvianskyte N, Razumas V, Patkar S, Vind J, Svendsen A, Nolte RJM, Rowan AE, Brust M. *Bioconjug Chem* 2006;17:1373–1375.

45. Aubin ME, Morales DG, Hamad-Schifferli K. *Nano Lett* 2005;5:519–522.

46. Labhardt AM. *Proc Natl Acad Sci USA* 1984;81:7674–7678.

47. Lee W, Oh BK, Lee WH, Choi JW. *Colloids Surf B Biointerfaces* 2005;40:143–148.

48. Karyakin AA, Presnova GV, Rubtsova MY, Egorov AM. *Anal Chem* 2000;72:3805–3811.

49. El-Sayed IH, Huang XH, El-Sayed MA. *Nano Lett* 2005;5:829–834.

50. Mao S, Lu GH, Yu KH, Chen JH. *Carbon* 2010;48:479–486.

51. Mao S, Lu G, Yu K, Bo Z, Chen J. *Adv Mate* 2010;22:3521–3526.

52. Maier I, Morgan MRA, Lindner W, Pittner F. *Anal Chem* 2008;80:2694–2703.

53. Ambrosi A, Castaneda MT, Killard AJ, Smyth MR, Alegret S, Merkoci A. *Anal Chem* 2007;79:5232–5240.

54. Stuchinskaya T, Moreno M, Cook MJ, Edwards DR, Russell DA. *Photochem Photobiol Sci* 2011;10:822–831.

55. Driskell JD, Jones CA, Tompkins SM, Tripp RA. *Analyst* 2011;136:3083–3090.

56. Seeman NC. *Nanotechnology* 1991;2:149–159.

57. Wang YL, Mueller JE, Kemper B, Seeman NC. *Biochemistry* 1991;30:5667–5674.

58. Chen JH, Seeman NC. *Electrophoresis* 1991;12:607–611.

59. Zhang YW, Seeman NC. *J Am Chem Soc* 1994;116:1661–1669.

60. Braun E, Eichen Y, Sivan U, Ben-Yoseph G. *Nature* 1998;391:775–778.

61. Keren K, Krueger M, Gilad R, Ben-Yoseph G, Sivan U, Braun E. *Science* 2002;297:72–75.

62. Alivisatos AP, Johnsson KP, Peng XG, Wilson TE, Loweth CJ, Bruchez MP, Schultz PG. *Nature* 1996;382:609–611.

63. Mirkin CA, Letsinger RL, Mucic RC, Storhoff JJ. *Nature* 1996;382:607–609.

64. Niemeyer CM. *Angew Chem Int Ed* 2001;40:4128–4158.

65. Mirkin CA. *Inorg Chem* 2000;39:2258–2272.

66. Storhofff JJ, Elghanian R, Mirkin CA, Letsinger RL. *Langmuir* 2002;18:6666–6670.

67. Brown KA, Park S, Hamad-Schifferli K. *J Phys Chem C* 2008;112:7517–7521.

68. Demers LM, Mirkin CA, Mucic RC, Reynolds RA, Letsinger RL, Elghanian R, Viswanadham G. *Anal Chem* 2000;72:5535–5541.

69. Herne TM, Tarlov MJ. *J Am Chem Soc* 1997;119:8916–8920.

70. Rosi NL, Mirkin CA. *Chem Rev* 2005;105:1547–1562.

71. Elghanian R, Storhoff JJ, Mucic RC, Letsinger RL, Mirkin CA. *Science* 1997;277:1078–1081.

72. Song J, Li Z, Cheng Y, Liu C. *Chem Commun* 2010;46:5548–5550.

73. Cederquist KB, Keating CD. *ACS Nano* 2009;3:256–260.

74. Hill HD, Millstone JE, Banholzer MJ, Mirkin CA. *ACS Nano* 2009;3:418–424.

75. Bakthavathsalam P, Rajendran VK, Mohammed JAB. *J Nanobiotechnology* 2012;10:8.

76. Guo Z, Duan J, Yang F, Li M, Hao T, Wang S, Wei D. *Talanta* 2012;93:49–54.

77. Storhoff JJ, Elghanian R, Mucic RC, Mirkin CA, Letsinger RL. *J Am Chem Soc* 1998;120:1959–1964.

78. Deng ZX, Tian Y, Lee SH, Ribbe AE, Mao CD. *Angew Chem Int Ed* 2005;44:3582–3585.

79. Fischler M, Sologubenko A, Mayer J, Clever G, Burley G, Gierlich J, Carell T, Simon U. *Chem Commun* 2008;2:169–171.

80. Le JD, Pinto Y, Seeman NC, Musier-Forsyth K, Taton TA, Kiehl RA. *Nano Lett* 2004;4:2343–2347.

81. Pinto YY, Le JD, Seeman NC, Musier-Forsyth K, Taton TA, Kiehl RA. *Nano Lett* 2005;5:2399–2402.

82. Zhang JP, Liu Y, Ke YG, Yan H. *Nano Lett* 2006;6:248–251.

83. Aldaye FA, Sleiman HF. *Angew Chem Int Ed* 2006;45:2204–2209.

84. Aldaye FA, Sleiman HF. *J Am Chem Soc* 2007;129:4130–4131.

85. Niemeyer CM, Ceyhan B, Noyong M, Simon U. *Biochem Biophys Res Commun* 2003;311:995–999.

86. Koplin E, Niemeyer CM, Simon U. *J Mater Chem* 2006;16:1338–1344.

87. Mucic RC, Storhoff JJ, Mirkin CA, Letsinger RL. *J Am Chem Soc* 1998;120:12674–12675.

88. Gruner SAW, Locardi E, Lohof E, Kessler H. *Chem Rev* 2002;102:491–514.

89. Wang Q, Dordick JS, Linhardt RJ. *Chem Mater* 2002;14:3232–3244.

90. Lindhorst TK. Artificial multivalent ligands to understand and manipulate carbohydrate-protein interaction. *Top Curr Chem* 2002;218:201–235.

91. Lee YC, Lee RT. *Acc Chem Res* 1995;28:321–327.

92. Dwek RA. *Chem Rev* 1996;96:683–720.

93. Nagahori N, Nishimura SI. *Biomacromolecules* 2001;2:22–24.

94. Lundquist JJ, Toone EJ. *Chem Rev* 2002;102:555–578.

95. de la Fuente JM, Penades S. *Tetrahedron-Asymmetry* 2002;13:1879–1888.

96. Barrientos AG, de la Fuente JM, Rojas TC, Fernandez A, Penades S. *Chemistry* 2003;9:1909–1921.

97. de la Fuente JM, Eaton P, Barrientos AG, Menendez M, Penades S. *J Am Chem Soc* 2005;127:6192–6197.

98. Reynolds AJ, Haines AH, Russell DA. *Langmuir* 2006;22:1156–1163.

99. Otsuka H, Akiyama Y, Nagasaki Y, Kataoka K. *J Am Chem Soc* 2001;123:8226–8230.

100. Boyer C, Bousquet A, Rondolo J, Whittaker MR, Stenzel MH, Davis TP. *Macromolecules* 2010;43:3775–3784.

101. Lin CC, Yeh YC, Yang CY, Chen GF, Chen YC, Wu YC, Chen CC. *Chem Commun* 2003;23:2920–2921.

102. Bogdan N, Roy R, Morin M. *RSC Adv* 2012;2:985–991.

103. Nakamura-Tsuruta S, Kishimoto Y, Nishimura T, Suda Y. *J Biochem* 2008;143:833–839.

104. Perez-Lopez B, Merkoci A. *Anal Bioanal Chem* 2011;399:1577–1590.

105. Liu SQ, Tang ZY. *J Mater Chem* 2010;20:24–35.

106. Marradi M, Garcia I, Penades S. *Prog Mol Biol Transl Sci* 2011;104:141–173.

107. Lees WJ, Spaltenstein A, Kingery-Wood JE, Whitesides GM. *J Med Chem* 1994;37:3419–3433.

108. Mammen M, Dahmann G, Whitesides GM. *J Med Chem* 1995;38:4179–4190.

109. Huang CC, Chen CT, Shiang YC, Lin ZH, Chang HT. *Anal Chem* 2009;81:875–882.

110. Abraham SN, Sun DX, Dale JB, Beachey EH. *Nature* 1988;336:682–684.

111. Tseng YT, Chang HT, Chen CT, Chen CH, Huang CC. *Biosens Bioelectron* 2011;27:95–100.

112. Kitov PI, Sadowska JM, Mulvey G, Armstrong GD, Ling H, Pannu NS, Read RJ, Bundle DR. *Nature* 2000;403:669–672.

113. Ofek I, Hasy DL, Sharon N. *FEMS Immunol Med Microbiol* 2003;38:181–191.

114. Papp I, Sieben C, Ludwig K, Roskamp M, Bottcher C, Schlecht S, Herrmann A, Haag R. *Small* 2010;6:2900–2906.

115. Arnaiz B, Martinez-Avila O, Falcon-Perez JM, Penades S. *Bioconjug Chem* 2012;23:814–825.

116. Marradi M, Di Gianvincenzo P, Enriquez-Navas PM, Martinez-Avila OM, Chiodo F, Yuste E, Angulo J, Penades S. *J Mol Biol* 2011;410:798–810.

117. Martinez-Avila O, Hijazi K, Marradi M, Clavel C, Campion C, Kelly C, Penades S. *Chemistry* 2009;15:9874–9888.

118. Martinez-Avila O, Bedoya LM, Marradi M, Clavel C, Alcami J, Penades S. *Chembiochem* 2009;10:1806–1809.

119. Di Gianvincenzo P, Marradi M, Martinez-Avila OM, Bedoya LM, Alcami J, Penades S. *Bioorg Med Chem Lett* 2010;20:2718–2721.

120. Cheng Y, Meyers JD, Broome AM, Kenney ME, Basilion JP, Burda C. *J Am Chem Soc* 2011;133:2583–2591.

121. Rojo J, Diaz V, de la Fuente JM, Segura I, Barrientos AG, Riese HH, Bernade A, Penades S. *Chembiochem* 2004;5:291–297.

122. Brown SD, Nativo P, Smith JA, Stirling D, Edwards PR, Venugopal B, Flint DJ, Plumb JA, Graham D, Wheate NJ. *J Am Chem Soc* 2010;132:4678–4684.

123. Chanda N, Kattumuri V, Shukla R, Zambre A, Katti K, Upendran A, Kulkarni RR, Kan P, Fent GM, Casteel SW, Smith CJ, Boote E, Robertson JD, Cutler C, Lever JR, Katti KV, Kannan R. *Proc Natl Acad Sci USA* 2010;107:8760–8765.

124. Khlebtsov N, Dykman L. *Chem Soc Rev* 2011;40:1647–1671.

125. Cheng Y, Samia AC, Meyers JD, Panagopoulos I, Fei BW, Burda C. *J Am Chem Soc* 2008;130:10643–10647.

126. Kim CK, Ghosh P, Pagliuca C, Zhu ZJ, Menichetti S, Rotello VM. *J Am Chem Soc* 2009;131:1360–1361.

127. Ock KS, Ganbold EO, Park J, Cho K, Joo SW, Lee SY. *Analyst* 2012;137:2852–2859.

128. Vivero-Escoto JL, Slowing II, Wu CW, Lin VSY. *J Am Chem Soc* 2009;131:3462–3463.

129. Luo YL, Shiao YS, Huang YF. *ACS Nano* 2011;5:7796–7804.

130. Adeli M, Sarabi RS, Farsi RY, Mahmoudi M, Kalantari M. *J Mater Chem* 2011;21:18686–18695.

131. Agasti SS, Chompoosor A, You CC, Ghosh P, Kim CK, Rotello VM. *J Am Chem Soc* 2009;131:5728–5729.

132. Patra CR, Verma R, Kumar S, Greipp PR, Mukhopadhyay D, Mukherjee P. *J Biomed Nanotechnol* 2008;4:499–507.

133. Dixit V, Van den Bossche J, Sherman DM, Thompson DH, Andres RP. *Bioconjug Chem* 2006;17:603–609.

134. Li X, Zhou HY, Yang L, Du GQ, Pai-Panandiker AS, Huang XF, Yan B. *Biomaterials* 2011;32:2540–2545.

135. Wang J, Tian SM, Petros RA, Napier ME, DeSimone JM. *J Am Chem Soc* 2010;132:11306–11313.

136. Choi CHJ, Alabi CA, Webster P, Davis ME. *Proc Natl Acad Sci US A* 2010;107:1235–1240.

137. Yang PH, Sun XS, Chiu JF, Sun HZ, He QY. *Bioconjug Chem* 2005;16:494–496.

138. Li JL, Wang L, Liu XY, Zhang ZP, Guo HC, Liu WM, Tang SH. *Cancer Lett* 2009;274:319–326.

139. Heo DN, Yang DH, Moon HJ, Lee JB, Bae MS, Lee SC, Lee WJ, Sun IC, Kwon IK. *Biomaterials* 2012;33:856–866.

140. Minko T, Paranjpe PV, Qiu B, Lalloo A, Won R, Stein S, Sinko PJ. *Cancer Chemother Pharmacol* 2002;50:143–150.

141. Rowinsky EK, Donehower RC. *N Engl J Med* 1995;332:1004–1014.

142. Yvon AMC, Wadsworth P, Jordan MA. *Mol Biol Cell* 1999;10:947–959.

143. Fujita K, Sakamoto S, Ono Y, Wakao M, Suda Y, Kitahara K, Suganuma T. *J Biol Chem* 2011;286:5143–5150.

144. De Jong WH, Hagens WI, Krystek P, Burger MC, Sips AJ, Geertsma RE. *Biomaterials* 2008;29:1912–1919.

145. Tawaratsumida K, Furuyashiki M, Katsumoto M, Fujimoto Y, Fukase K, Suda Y, Hashimoto M. *Biol Chem* 2009;284:9147–9152.

146. Wijelath ES, Rahman S, Namekata M, Murray J, Nishimura T, Mostafavi-Pour Z, Patel Y, Suda Y, Humphries MJ, Sobel M. *Circ Res* 2006;99:853–860.

147. Guerrero S, Araya E, Fiedler JL, Arias JI, Adura C, Albericio F, Giralt E, Arias JL, Fernandez MS, Kogan MJ. *Nanomedicine* 2010;5:897–913.

148. Kattumuri V, Katti K, Bhaskaran S, Boote EJ, Casteel SW, Fent GM, Robertson DJ, Chandrasekhar M, Kannan R, Katti KV. *Small* 2007;3:333–341.

149. Hainfeld JF, Slatkin DN, Focella TM, Smilowitz HM. *Br J Radiol* 2006;79:248–253.

150. Papahadjopoulos D, Allen TM, Gabizon A, Mayhew E, Matthay K, Huang SK, Lee KD, Woodle MC, Lasic DD, Redemann C, Martin FJ. *Proc Natl Acad Sci USA* 1991;88:11460–11464.

151. Lee H, Lee E, Kim DK, Jang NK, Jeong YY, Jon S. *J Am Chem Soc* 2006;128:7383–7389.

152. Kim D, Park S, Lee JH, Jeong YY, Jon S. *J Am Chem Soc* 2007;129:7661–7665.

153. Zhang Q, Iwakuma N, Sharma P, Moudgil BM, Wu C, McNeill J, Jiang H, Grobmyer SR. *Nanotechnology* 2009;20.

154. Wang X, Qian XM, Beitler JJ, Chen ZG, Khuri FR, Lewis MM, Shin HJC, Nie SM, Shin DM. *Cancer Res* 2011;71:1526–1532.

155. Guerrero S, Herance JR, Rojas S, Mena JF, Gispert JD, Acosta GA, Albericio F, Kogan MJ. *Bioconjug Chem* 2012;23:399–408.

156. Guirgis BSS, Cunha CSE, Gomes I, Cavadas M, Silva I, Doria G, Blatch GL, Baptista PV, Pereira E, Azzazy HME, Mota MM, Prudencio M, Franco R. *Anal Bioanal Chem* 2012;402:1019–1027.

157. Ray PC, Darbha GK, Ray A, Hardy W, Walker J. *Nanotechnology* 2007;18:375504.

158. Nitin N, Javier DJ, Richards-Kortum R. *Bioconjug Chem* 2007;18:2090–2096.

159. Geng JL, Li K, Pu KY, Ding D, Liu B. *Small* 2012;8:2421–2429.

160. Stiti M, Cecchi A, Rami M, Abdaoui M, Barragan-Montero V, Scozzafava A, Guari Y, Winum J-Y., Supuran CT. *J Am Chem Soc* 2008;130:16130–16131.

161. Bowman MC, Ballard TE, Ackerson CJ, Feldheim DL, Margolis DM, Melander C. *J Am Chem Soc* 2008;130:6896–6897.

162. Bresee J, Maier KE, Melander C, Feldheim DL. *Chem Commun* 2010;46:7516–7518.

163. Chamundeeswari M, Sobhana SSL, Jacob JP, Kumar MG, Devi MP, Sastry TP, Mandal AB. *Biotechnol Appl Biochem* 2010;55:29–35.

164. Yang JA, Murphy CJ. *Langmuir* 2012;28:5404–5416.

165. Hao YZ, Yang XY, Song S, Huang M, He C, Cui MY, Chen J. *Nanotechnology* 2012;23:045103.

10

METHODS FOR MAGNETIC NANOPARTICLE SYNTHESIS AND FUNCTIONALIZATION

MOHAMMAD H. EL-DAKDOUKI[1,2,†], KHEIREDDINE EL-BOUBBOU[1,3,†], JINGGUANG XIA[1], HERBERT KAVUNJA[1], AND XUEFEI HUANG[1]

[1]*Department of Chemistry, Michigan State University, East Lansing, Michigan, USA*
[2]*Department of Chemistry, Beirut Arab University, Beirut, Lebanon*
[3]*College of Science and Health Professions, King Saud bin AbdulAziz University for Health Sciences, National Guard Health Affairs, Riyadh, Saudi Arabia*

With their nanometer dimensions and unique magnetic properties, magnetic nanoparticles (NPs) have been used in a variety of biological applications including bioseparation, biosensing, noninvasive imaging, targeted drug delivery as well as disease treatment based on hyperthermia [1–8]. The recognition of the many potential applications of magnetic NPs has generated large interests in their fabrication and functionalization. In this chapter, we will review the methods developed to produce functionalized magnetic NPs. We start the discussion by describing the major synthetic methods to form select magnetic cores of the NPs. Following the core synthesis, we will focus on the strategies used to conjugate various ligands (carbohydrates, drugs, nucleic acids, and antibodies) to the surface of magnetic NPs. As NP synthesis and functionalization is a rapidly developing research area, we apologize for not being able to include all the examples from the literature due to the length limit of this chapter. Furthermore, we will not describe the biological applications in detail. Interested readers can refer to other chapters in this book.

10.1 SYNTHESIS OF MAGNETIC NP CORES

Several classes of magnetic NPs have been synthesized [9, 10]. These include iron oxide (Fe_xO_y) [11–15], various types of ferrites ($MO.Fe_2O_3$ where M = Ni, Co, Mg, Zn, Mn) [2, 16, 17], metallic colloidal nanocrystals (Fe, Ni, Co) [18], alloys (FePt) [19] as well as gadolinium oxide-based NPs [20–28]. During the last decade, many publications describing efficient synthetic routes to prepare shape/size-controlled, highly stable, monodispersed, and biocompatible magnetic NPs have appeared. In this section, we will present representative discussion of major synthetic pathways utilized in preparation of the two commonly used magnetic nanocomposites, that is, iron oxide NPs (IONPs) and gadolinium oxide NPs.

Particular attention should be directed to the synthetic methods of magnetic NPs as they can significantly affect the size, shape, structure, dimensions, properties, and hence the applications of nanocomposites [9, 29]. It is important to select a synthetic approach that will lead to monodispersed

---

[†]These authors contributed equally to this work.

Chemistry of Bioconjugates: Synthesis, Characterization, and Biomedical Applications, First Edition. Edited by Ravin Narain.
© 2014 John Wiley & Sons, Inc. Published 2014 by John Wiley & Sons, Inc.

nanocomposites. Furthermore, strategies must be developed to stabilize the "*naked*" magnetic NPs against aggregation and degradation. These strategies are comprised of coating with organic species (i.e., surfactants or polymers) or an inorganic layer (i.e., silica), which can be further functionalized with biological and synthetic ligands to modulate their binding and recognition properties.

10.1.1 Synthesis of Iron Oxide NPs

Iron (Fe), the fourth most common element in the Earth's crust, exists in a wide range of forms and oxidation states (-2 to $+8$, the most common being $+2$ and $+3$). Due to its four (i.e., Fe or Fe^{2+}) or five (i.e., Fe^{3+}) unpaired electrons in the 3d orbital, iron has a strong magnetic moment. When crystals are formed from iron, they can thus be in ferromagnetic, antiferromagnetic, or ferrimagnetic states. Superparamagnetism is exhibited by small ferromagnetic or ferrimagnetic NPs (for the explanation of the different magnetization phenomena, readers are directed to several excellent reviews) [9, 10, 30]. The most common of all are iron oxides, which exist in sixteen different forms. IONPs, mainly magnetite (Fe_3O_4) or its oxidized form maghemite (γ-Fe_2O_3), are tiny iron oxide particles with diameters ranging between 1 and 200 nm. Magnetite is the most magnetic of all the naturally occurring minerals on Earth. Magnetite particles having single domains (sizes $\sim$ 30–70 nm) or pseudo-single domain magnetization states ($\sim$70 nm–20 μm) are the dominant carriers of remnant magnetization. When the size of the NPs is below a critical value ($\sim$20 nm), the individual NPs behave like a giant paramagnetic atom with a single magnetic domain, exhibiting superparamagnetic behavior. Superparamagnetic IONPs respond rapidly to an applied magnetic field, but exhibit negligible residual magnetism when the field is removed [31], making them especially attractive in biological applications.

IONPs can be easily synthesized from inexpensive iron salts and can serve as highly efficient negative contrast agents for magnetic resonance imaging (MRI) applications (enhancing the T_2 or/and T_2^* relaxation times of water molecules in their vicinity) [6, 32–34]. Clinical applications of nontargeted IONPs for liver and lymph node imaging made such particles particularly appealing for further research [30, 35, 36].

Most commonly, IONPs are prepared by coprecipitation of iron salt solution in basic media and stabilized using biocompatible surfactant [11, 37, 38]. Alternatively, high temperature thermal decomposition of organometallic precursors in the presence of surfactants can be employed to produce IONPs with marked improvements in size/shape control, size distribution, and crystallinity [17, 39, 40]. In addition, microemulsion and hydrothermal approaches are utilized to a lesser extent, and will be briefly summarized. All other techniques were excellently reviewed elsewhere [7, 9].

10.1.1.1 Coprecipitation Method The coprecipitation procedure is probably the simplest and fastest methodology to obtain IONPs. Magnetite (Fe_3O_4) or its oxidized form maghemite (γ-Fe_2O_3) are prepared by aging appropriate stoichiometric ratios of ferrous (Fe^{2+}) and ferric (Fe^{3+}) salts in an aqueous basic media (typically NH_4OH or $NaOH$) (Figure 10.1a). This protocol was first explored by Massart in 1981 and is, therefore, known as the "Massart method" [41]. A typical precipitation procedure to produce magnetite consists of mixing a solution of $FeCl_3$ and $FeSO_4$ (2:1 molar ratio) with mechanic agitation of about 2000 rpm under an inert atmosphere. For magnetite preparation, the inert atmosphere is important to ensure the formation of high quality of the nanocrystal and protect against its oxidation to maghemite or hematite (α-Fe_2O_3). The resulting solution is heated to 70–80°C with elevated stirring speed ($\sim$7500 rpm) while a solution of NH_4OH (10% by volume) is added dropwise. A dark precipitate of Fe_3O_4 magnetite NPs will be formed. Although a large amount of NPs can be synthesized using this process, the control of particle size distribution in aqueous media is limited due to particle agglomerations, since only kinetic factors are controlling the growth of the crystal. The yield, size, shape, surface properties, and polydispersity of the final material can be affected by many factors including the base, pH, temperature, nature, and concentration of the counterions (chlorides, sulfates, nitrates, etc.), and Fe^{3+}/Fe^{2+} ratio [42].

To reduce the agglomeration, compounds containing chelating organic anions (i.e., carboxylate ions such as acrylic or oleic acid) or surface-complexing polymers (e.g., dextran, polyethylene glycol (PEG), polyvinylpyrrolidone, polyglutamic acid, or 3-aminopropyltriethoxysilane (APTES)) can be added during or after the formation of NPs. In addition to imparting stability, these ligands can help in controlling the size distribution of the NPs [13]. Characteristic TEM images of IONPs prepared through the coprecipitation method are shown in Figure 10.1d, e. Depending on the coating agent used, there can be various degrees of particle aggregation.

Dextran polymer and silica are among the two common coatings for IONPs. Dextran-coated iron oxide-based magnetic materials (commercialized as Feridex, Resovist®, and Combidex) [30, 35, 36] have been approved for usage in humans for detection of liver cancer as well as cancer metastasis to lymph nodes. The overall size of the dextran-coated IONPs (core + shell) can be controlled by the amounts of dextran utilized in the synthesis [43], with higher amounts of dextran leading to larger particles. The overall size of the particles can affect the biodistribution of the particles with larger ones favoring liver uptake and smaller particles used for lymph node imaging [35, 44]. Another advantage of using the dextran coating is that dextran can be derivatized to introduce other types of functional groups such as amines and carboxylic acids, which can become useful for further

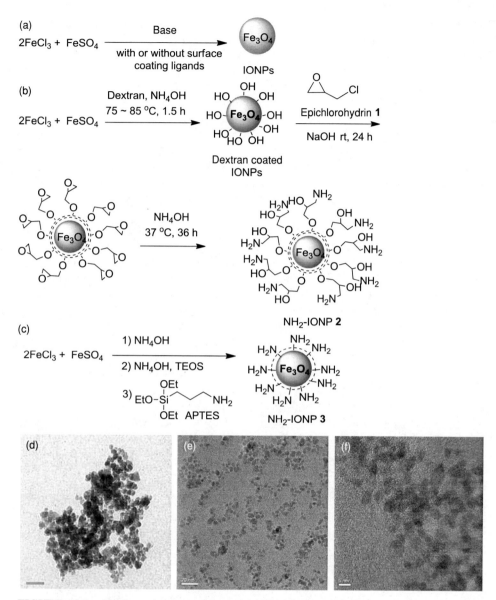

FIGURE 10.1   (a) General scheme for synthesis of IONPs using the coprecipitation method, (b) synthesis of dextran-coated IONPs and subsequent amine functionalization, (c) synthesis of silica-coated IONPs and subsequent amine functionalization with APTES, TEM images of (d) bare IONPs (the scale bar is 20 nm), (e) IONPs formed with dextran coating (the scale bar is 20 nm), and (f) IONPs coated with APTES (the scale bar is 5 nm).

functionalization of the NPs. An example of this is shown in Figure 10.1b, where dextran coating is cross-linked with epichlorohydrin **1** under basic condition. This is followed by ammonia treatment to introduce amines onto the external surface. The aminated NP **2** is a versatile platform for ligand functionalization in further biological applications.

Another common method to functionalize magnetic NPs is to coat them with silica. The procedure is simple and typically occurs by adding tetraethyl orthosilicate (TEOS) under basic condition following formation of the iron oxide core (Figure 10.1c). The hydroxyl groups on the surface of the NPs

attack the TEOS to immobilize the orthosilicate, which can undergo condensation/polymerization reaction with TEOS catalyzed by the hydroxide anion to form silica coating (commonly referred to as Stöber process) on the NP. An advantage of this approach is the well established and versatile method that can be adopted for further modification of the surface using various functionalized siloxanes. For example, upon formation of the silica coating, APTES could be added to the reaction mixture, which can react with the silica coating on the NPs and introduce amino moieties on the external surface (Figure 10.1c) [45–48].

10.1.1.2 Thermal Decomposition Method With the coprecipitation method, it is difficult to precisely control the size of the NPs. To improve the quality of the NPs, it was found that the classical LaMer model [49] for the formation of monodispersed colloidal dispersions, that is, burst of nucleation, diffusion-controlled growth, and Ostwald ripening (redeposition of small crystals onto larger ones) could be extended to NP synthesis.

There are two stages in NP growth: (1) nucleation where the monomers reach critical supersaturation and form seed for crystal growth; and (2) growth of the produced nuclei, which is controlled by the rate of diffusion of the solutes to the surface of the crystal [50, 51]. To produce fine monodispersed NPs, ideally the two stages should be separated with fast nucleation rate and short nucleation period to limit the NP growth during the nucleation stage. Furthermore, nucleation should be avoided in the growth period, which is difficult to control in an aqueous media as in the coprecipitation method. Mechanistic insights and theoretical explanations on how to form monodispersed and uniform nanocrystals, including metals and their oxides, have been reviewed [14, 52].

Several researchers report that the use of elevated reaction temperatures can significantly enhance the uniformity of crystal formation. Two high temperature protocols have been developed. The first one is the hot-injection method by rapid introduction of reagents into the surfactant containing hot solution ($\geq 300°C$) to start the reaction. Upon injection, the extremely fast reaction leads to a sudden increase of the monomer concentration to the supersaturation level and, hence, the rapid formation of nanocrystal nuclei [53, 54]. The second strategy is the "heating-up" method, where a solution of precursors, surfactants, and solvent is mixed at a lower temperature to form the monomer by thermal decomposition. The reaction mixture is then heated to high temperatures (up to 300°C) at which the formation of nanocrystals takes place.

The "heating-up" procedure is simple, which has been successfully utilized to synthesize monodispersed nanocrystals of a wide range of nanomaterials [14]. Among the many examples, thermal decomposition of organometallic compounds (e.g., iron–oleate complex) in high boiling-point nonpolar organic solvents has proven to be an attractive route for the production of nanocrystals with high yield, uniformity, good crystallinity, and reproducibility. Hyeon and coworkers synthesized superparamagnetic maghemite γ-Fe_2O_3 nanocrystals by forming iron–oleic acid metal complex via heating iron pentacarbonyl ($Fe(CO)_5$) and oleic acid in dioctyl ether at 100°C, which was followed by heating to 300°C to form nanocrystals [40]. Instead of using the toxic and expensive $Fe(CO)_5$, they later reported the elegant ultra-large-scale synthesis of monodispersed nanocrystals via the slow heating of metal–oleate complex and oleic acid (Figure 10.2a) [55]. Furthermore, well-defined single-crystalline γ-Fe_2O_3 nanocrystals with different shapes ranging from diamonds, triangles to spheres were synthesized using the thermolysis of $Fe(CO)_5$ in the presence of various capping ligands [56]. Various metal NPs were also synthesized using the heating-up thermal decomposition method. Sun and his colleagues prepared highly uniform monodispersed FePt NPs by heating a reaction mixture containing platinum acetylacetonate, $Fe(CO)_5$, 1,2-hexadecanediol, oleic acid, and oleylamine [19]. Based on this method, the Sun group reported the synthesis of highly monodispersed controlled-sized magnetite Fe_3O_4 NPs (4 nm) from a high temperature (200–300°C) 1,2-hexadecanediol solution of iron(III) acetylacetonate ($Fe(acac)_3$) in the presence of oleic acid and oleylamine (Figure 10.2b) [15]. Larger-sized NPs (8, 12, and 16 nm) can

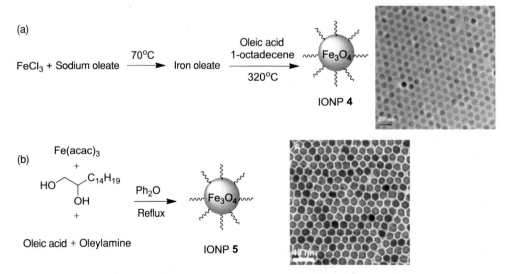

FIGURE 10.2   (a) IONPs prepared via thermal decomposition of iron–oleate complex and a TEM image of formed IONPs (the scale bar is 20 nm) [55], (b) IONPs synthesized by thermal decomposition of $Fe(acac)_3$ and a TEM image of formed IONPs (the scale bar is 48 nm). Reproduced with permission from Reference 15. (Copyright American Chemical Society and Nature Publishing Group.)

be formed via seed-mediated growth of the smaller nanocrystals. By controlling the quantity of NP seeds, different sizes of NPs can be synthesized. Furthermore, various sizes of monodispersed metal ferrites (MFe_2O_4, where M = Co, Fe, Mn, etc.) were prepared by the seed-mediated growth process using $Fe(acac)_3$ and $M(acac)_2$ as reactants [17]. In another example, Jana et al. reported the synthesis of size- and shape-controlled magnetic metal (Fe, Cr, Mn, Co, and Ni) oxide nanocrystals based on high temperature pyrolysis of metal fatty acid salts as the precursors and alkylamines when the activation reagents in noncoordinating solvents (octadecene) heated at 300°C under an inert atmosphere [57].

The high temperature thermolysis protocol has been very successful. However, a limitation of this approach is that the nanocrystals obtained are coated with hydrophobic ligands on the external surface and hence only dispersible in nonpolar organic solvents. For biological applications, it is highly desirable that these high quality IONPs can be brought into aqueous media. A "ligand-exchange" strategy, where the hydrophobic chains on the particle surface are replaced by molecules containing polar groups, has been employed [58–60].

A variety of ligands have been used to exchange oleic acid from the surface of IONPs, which include hydrophilic alkoxysilanes, poly(acrylic acid), 2,3-dimercaptosuccinic acid (DMSA), and polysaccharide such as hyaluronic acid. The polyelectrolyte ligand forms a stable coating to the nanocrystal particle surface through multiple anchoring groups. As an example, Jun et al. prepared well-controlled magnetic Fe_3O_4 nanocrystals (ranging from 4 to 12 nm) through thermal decomposition of $Fe(acac)_3$ in a hot organic solvent, and exchanged the hydrophobic capping ligands with DMSA to render them water soluble [61]. DMSA can chelate with the NP surface, with the remaining free thiol groups allowing the attachment of targeting ligands such as Herceptin as discussed in Section 10.5.

In an effort to better enhance the MR signal caused by conventional IONPs, a series of metal-doped magnetism-engineered IONPs (MEIONPs) of spinel MFe_2O_4 (where M = Mn, Fe, Co, or Ni) were fabricated through thermal decomposition of divalent metal salt (M^{2+}) and iron tris-2,4-pentadionate in the presence of oleic acid and oleylamine, which were exchanged by DMSA [16]. The Mn-doped MEIONPs, possessing strong magnetic properties, not only enhanced the sensitivity for cancer cell detection and *in vivo* imaging of small tumors, but also were successfully used for simultaneous molecular imaging and targeted small interfering RNA (siRNA) delivery *in vitro* [62].

As the method of thermal decomposition followed by ligand exchange entails an additional synthetic step to access the water-soluble magnetic NPs, the development of one-pot procedures where the two steps are combined has been investigated. Li and coworkers reported the one-pot thermal decomposition of $Fe(acac)_3$ or inexpensive hydrated ferric salts using the polar 2-pyrrolidone as a coordinating solvent to produce water-soluble magnetite nanocrystals [63, 64]. Ge et al. synthesized water-soluble magnetite nanocrystals by rapidly injecting a preheated NaOH/diethylene glycol (DEG) solution into a mixture of poly(acrylic acid) (PAA)/$FeCl_3$/DEG at 220°C [54]. Nevertheless, the quality of nanocrystals created by these methods was not as high as those prepared via the organic-based thermal decomposition method and are, thus, used less frequently by researchers.

In summary, the thermal decomposition technique allows for high control over size, shape, and size distribution. Representative TEM images of IONPs prepared via the thermal decomposition route are shown in Figure 10.2.

10.1.1.3 Hydrothermal Method

Hydrothermal method is another technique used to synthesize metal oxide nanomaterials. This heterogeneous reaction is typically performed in aqueous solvents in reactors or autoclaves under high pressure and temperature (pressure ≥ 2000 psi and temperature ≥ 150°C) with or without reducing agents or additives. Hydrothermal processing can produce materials with good crystallinity. Nanocrystalline magnetite [65], maghemite [66], hematite [67], and hollow spheres of various metal oxides [68, 69] were all successfully prepared using this method. Magnetite Fe_3O_4 octahedral particles (~500 nm) were fabricated from iron powders through a simple one-step alkali-assisted hydrothermal process by autoclaving reduced iron powders in a NaOH solution at 180°C for 24 hours [70]. Recently, Ge et al. reported a hydrothermal synthesis of Fe_3O_4 NPs with controllable diameters (15–30 nm), narrow size distribution, and tunable magnetic properties by oxidation of $FeCl_2 \cdot 4H_2O$ in aqueous NH_4OH solution under an elevated temperature (134°C) and pressure (2 bar) [71]. Furthermore, Co-doped zinc-ferrite ($Co_xZn_{1-x}Fe_2O_4$) NPs (7–14 nm) were prepared by a PEG-assisted hydrothermal method [72]. Although hydrothermal method produces well-crystallized iron oxide particles with high saturation magnetization, slow reaction kinetics with no control over monodispersity (nucleation and growth) remains a major disadvantage of this approach.

10.1.1.4 Microemulsion Method

The water-in-oil (w/o) microemulsion and/or reverse micelle method has been used to synthesize NPs of various kinds, including magnetic nanomaterials [13, 73, 74]. Microemulsion is a thermodynamically stable single-phase system that consists of three components: two of them are immiscible (water, oil) and the third with an amphiphilic behavior (surfactant, i.e., Triton X-100). Using this method, single-crystalline spinel ferrites, such as $MnFe_2O_4$, were successfully prepared using sodium dodecylbenzenesulfonate via the water-in-toluene reverse micelle technique [73]. In another example, Santra et al. prepared silica-coated IONPs where they varied the surfactant, the oil, and the base used [75]. Upon mixing two separate microemulsions (one containing iron salts + surfactant and the other containing the base + surfactant), the water droplets collided and coalesced, allowing the mixing of the reactants to produce the NPs. The NPs produced were single crystalline with an average particle size of 4–15 nm, but poor monodispersity. Gupta and coworkers utilized the inverse microemulsion process to synthesize spherical PEGylated superparamagnetic IONPs with narrow size (40–50 nm) inside the aqueous cores of aerosol-OT/n-hexane reverse micelles [12]. Since the size of the inner core of the reverse micelles is in the nanometer range, the uncoated magnetic NPs prepared inside these nanoreactors were found to be approximately 15 nm. In general, control over shape and monodispersity using this method is not high, which limits their use for biomedical applications. Moreover, a large amount of solvents is needed with low yields of nanomaterial.

10.1.2 Synthesis of Gadolinium Oxide NPs

Gadolinium ions (Gd^{3+}) have one of the highest magnetic moments in the periodic table because of its seven unpaired electrons in the 4f shells that yield S-state electron magnetic moments. Gd^{3+} ions can induce the longitudinal relaxation of protons of the bound water molecules, which are useful to enhance the contrast for T_1-based MRI [76]. While free Gd^{3+} ions are toxic, their toxicities could be significantly minimized by chelating the ion with ligands. This led to the development of several commercially available and clinically used Gd^{3+} chelates as T_1-based magnetic resonance (MR) contrast agents such as Magnevist® where a Gd^{3+} ion is bound to diethylenetriaminepentacetate (DTPA) [77]. However, transmetallation by ions such as Zn^{2+} and Ca^{2+} in biological systems can release the complexed Gd^{3+} ions leading to systemic toxicity [78]. Furthermore, without additional targeting ligands, Gd-chelates are usually poorly taken up and retained by cells, rendering it unsuitable for cellular imaging. To address these drawbacks, gadolinium (III) oxide (Gd_2O_3) NPs are currently undergoing intensive research as alternative contrast agents [20–28]. In this section, we will describe the multiple methods that have been devised for the synthesis of Gd_2O_3 NPs, and discuss the various parameters that influence the longitudinal and transverse relaxivities of the particles.

10.1.2.1 The Polyol Method
The polyol method is the most commonly used route for the synthesis of various sized Gd_2O_3 NPs [79–82]. This method generates highly reproducible and uniform NPs with sizes ranging between 1 and 5 nm. In general, this method calls for reacting Gd^{3+} salts such as $GdCl_3$ and $Gd(NO_3)_3$ with sodium hydroxide in DEG to generate Gd (III) hydroxide. The latter is then dehydrated at elevated temperatures (180–200°C) leading to the formation of a transparent colloid of fine and narrow size-distributed Gd_2O_3 NPs. In this technique, DEG is used as both solvent and surfactant that prevents the agglomeration of NPs. Substoichiometric amounts of sodium hydroxide are usually used to avoid the formation of larger undesired

aggregates that precipitate out of the solution. Faucher et al. were able to access multiple NP sizes (3, 60, 75, and 105 nm) by varying the heating rate (1–5°C/min) [83]. Bridot et al. reported the synthesis of three sets of Gd_2O_3 NPs (3.3, 5.2, and 8.9 nm) by using the smaller NPs as nucleation seeds for the formation of the larger NPs [84]. As a modification to the polyol method, Kattel et al. indicated that Gd_2O_3 NPs, as well as other lanthanide oxide NPs, could be generated without the need to heat to elevated temperatures [85]. They found that the addition of hydrogen peroxide to a solution of Gd^{3+} precursor and sodium hydroxide in triethylene glycol at 80°C efficiently generated Gd_2O_3 NPs.

10.1.2.2 The Combustion Method
The combustion method presents a simple and easy approach for the preparation of Gd_2O_3 NPs [80, 86]. This solid-state reaction involves boiling an aqueous mixture of $Gd(NO_3)_3$ and the amino acid glycine to dryness. When the temperature increased further to 250°C, the solid mixture ignited leading to the production of Gd_2O_3 NPs as white solid in about 10 seconds. The chemical reaction involved is depicted in the following chemical reaction:

$$6Gd(NO_3)_3 + 10NH_4^+CH_3CO_2^- + 18O_2 \rightarrow 3Gd_2O_3 + 20CO_2 + 5N_2 + 25H_2O + 18NO_2$$

TEM of the NPs synthesized by the combustion method indicated aggregates of nanocrystals. Although the NPs were crystalline, traces of starting materials, that is, $Gd(NO_3)_3$ and glycine, were detected on their surface.

10.1.2.3 Decomposition of Gadolinium Acetate
Yudasaka and coworkers have been interested in the fabrication of single-walled carbon nanohorns (SWNHs) for potential applications in drug delivery and imaging [87–91]. To assist in the intracellular localization of SWNHs, they proposed embedding Gd_2O_3 inside the nanohorns [87]. Aggregates of SWNHs (SWNHag) were prepared by CO_2 laser ablation of graphite. Oxidization of SWNHag with oxygen gas at 500°C opened holes that entrapped molecules such as gadolinium acetate by simple mixing. Heating the Gd acetate-loaded SWNHag at 1200°C decomposed gadolinium acetate into Gd_2O_3, sealed the openings, and generated Gd_2O_3@SWNHag. Excess Gd_2O_3 adsorbed on the surface of the NPs was removed by sonication in 35–37% aqueous HCl solution.

10.1.2.4 Other Gadolinium-based Hybrids
The Kennedy and Chen groups published several reports for the development of lanthanide oxide NPs for optical imaging [92–97]. In one report, Gd_2O_3 NPs were explored as excellent platforms for doping europium (Eu^{3+}) ions [92]. Eu:Gd_2O_3 NPs were synthesized by flame spray pyrolysis, where an ethanol solution containing $Eu(NO_3)_3$

and $Gd(NO_3)_3$ was sprayed into a hydrogen diffusion flame with a temperature of 2100°C. The formation of the NPs occurred in the flame and the NPs were collected by a cold finger thermophoretically. The synthesized NPs were sonicated in methanol where larger aggregates (>200 nm) precipitated, while small NPs (<200 nm) remained in the supernatant. While the optical properties of Eu:Gd_2O_3 NPs were good, the NPs were polydispersed with a size between 5 and 200 nm.

Hifumi et al. synthesized dextran-coated gadolinium phosphate ($GdPO_4$) nanorods as T_1-based MRI contrast agents following the hydrothermal synthesis method [98, 99]. An aqueous solution of $Gd(NO_3)_3$, ammonium hydrogen phosphate, dextran, and sodium hydroxide was placed in a sealed tube and heated to 200°C. The collected nanorods were monodispersed and highly crystalline which exhibited a diameter of 20–30 nm along the major axis and 6–15 nm along the minor axis as determined by TEM. More importantly, the paramagnetic NPs showed high r_1 and r_2 values of 13.9 and 15.0 s^{-1} (mM Gd^{3+})$^{-1}$ ($r_2/r_1 = 1.1$), highlighting its superiority as MR-positive contrast agents compared to Magnevist ($r_1 = 3.8$ s^{-1} (mM Gd^{3+})$^{-1}$).

Recently, Dumont et al. synthesized gadolinium phosphate NPs as T_2-MR-negative contrast agents and DNA delivery vehicles [100]. The authors relied on a water-in-oil microemulsion approach by mixing two precursor solutions. The first was prepared by adding an aqueous solution of $Gd(NO_3)_3$ to an organic solution of the surfactant IGEPAL CO-520 in cyclohexane, while the other was a mixture of aqueous NaH_2PO_4 solution in cyclohexane with IGEPAL CO-520. The nanorods were 50 nm in length and 10 nm in width. On contrary to the $GdPO_4$ nanorods synthesized by Hifumi et al. [98], the nanorods prepared by Dumont et al. proved to be good negative MR contrast agents with $r_2/r_1 = 60.9$ when evaluated at 14.1 T.

10.1.2.5 Factors Affecting the Relaxometric Properties of Gd_2O_3 NPs
One of the major applications of Gd-based NPs is to function as MR contrast agents. Factors that affect the relaxometric and physiochemical properties of gadolinium oxide NPs have been discussed in several reports [79, 83, 101, 102], and a brief analysis will be provided in this section. In general, a good T_1-MR contrast agent should have as high r_1 values as possible, while keeping r_2/r_1 ratio close to one. The fact that Gd NPs exhibit higher r_1 values than the Gd-chelates was explained by the ability of multiple Gd^{3+} ions on the surface of the NP to cooperatively accelerate the longitudinal relaxation of water protons. This effect is not observed for Gd-chelates, which are usually dispersed in solution. In addition, an ideal Gd-based NP should have large surface area-to-volume ratio. Ultrasmall NPs (<5 nm) with larger surface area-to-volume ratios ensure that more Gd ions are present on the surface to interact with protons in immediate proximity that is essential for achieving high r_1. Therefore, Gd_2O_3 aggregates with Gd^{3+} buried in the NPs

and not accessible for interaction with protons display poor T_1-positive contrast. It is important to emphasize that r_1 is not the only key for high signal intensities as an optimal r_2/r_1 (ideally should be 1) is also required. This becomes critical at high Gd concentrations. When measurements are conducted at higher magnetic fields, elevated transverse relaxivities are encountered causing an increase in r_2/r_1 ratio [103]. This issue should be considered when assessing positive contrast enhancement at the cellular level where the efficient cellular uptake of Gd NPs can lead to an increase in the local concentration of Gd^{3+} ions.

10.2 IMMOBILIZATION OF CARBOHYDRATES ONTO MAGNETIC NPs

Carbohydrates are involved in many important biological recognition events [104–107]. Based on this knowledge, carbohydrate-coated NPs have been explored for a wide range of interesting biological applications [1]. In the following section, we will discuss the methods for conjugating carbohydrates onto magnetic NPs, which will be divided according to noncovalent and covalent approaches taken.

10.2.1 Noncovalent Conjugation of Carbohydrates

The noncovalent approach relies on electrostatic interactions, chelation, hydrogen bonding, or hydrophobic interactions to immobilize the glyco-conjugates onto NP surface. Carbohydrates can be directly or indirectly incorporated onto the surface of the NPs by noncovalent interactions.

10.2.1.1 Direct Adsorption of Carbohydrates onto NPs
One example of direct adsorption of carbohydrates onto NP surfaces was reported by Horak et al., where magnetic iron oxide magnetic NPs were functionalized with D-mannose [108]. IONPs were prepared through the coprecipitation method in the presence of an aqueous solution of concentrated D-mannose (*in situ* coating method). Due to its multiple hydroxyl groups, mannose could chelate with iron oxide thus coating the NP surface. Alternatively, the mannose coating could be introduced postsynthetically by incubating the already formed IONPs with a mannose solution [108]. The *in situ* coating method yielded smaller NPs (2 nm) compared to the postsynthetic route (6 nm) which was presumably due to the presence of mannose capping the NPs during NP growth period. Besides mannose, magnetic NPs coated with various monosaccharides and polysaccharides including glucuronic acid, lactobionic acid, ficoll (a sucrose polymer), heparin, dextran, or chitosan (CH) have been synthesized through direct absorption of the carbohydrate ligands using either *in situ* coating [30, 35, 36, 109, 110] or postsynthetic modification [110–115]. Particularly noteworthy is the

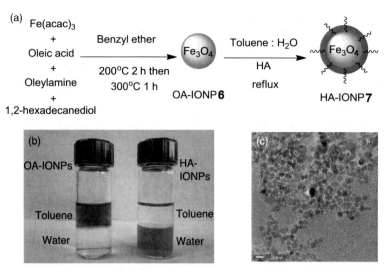

FIGURE 10.3   (a) Synthesis of HA-IONPs via ligand exchange of OA-IONPs with hyaluronan, (b) pictures of OA-IONPs and HA-IONPs in a toluene–water two-phase system demonstrating the changes in solubilities in organic solvent toluene and water, (c) TEM of HA-IONPs (the scale bar is 10 nm). Reprinted with permission from Reference 116. (Copyright Royal Society of Chemistry.)

dextran-coated superparamagnetic IONPs discussed in Section 10.1.1.1, which have been widely utilized for biological studies including clinical applications [30, 35, 36].

Besides the IONPs prepared through the coprecipitation method, methods have been developed to attach carbohydrates onto the hydrophobic NPs synthesized by thermal decomposition through the ligand-exchange method [116]. We explored a biphasic reaction system by heating a solution of the oleic acid-coated IONPs (OA-IONPs) in an organic solvent with an aqueous solution of hyaluronan (HA) (Figure 10.3). With the multiple hydroxyl and carboxylic acid moieties, HA can displace the oleic acid off the NP surface, thus rendering the particles (HA-IONPs) water soluble as shown in Figure 10.3b. The organic solvent utilized, ratio of HA to NP, and heating temperature are important parameters for the preparation of colloidal stable HA-coated NPs [116].

10.2.1.2   *Indirect Carbohydrate Linkage to NPs Through Noncovalent Interactions*   Although the direct absorption approach is operationally simple, biological recognition of the immobilized carbohydrates, especially monosaccharides, may be impaired since the carbohydrates are directly involved in NP chelation. The indirect immobilization approach can address this problem by derivatizing carbohydrates with functional groups, which have high affinities with the NPs. This can also help control the ligand orientation on NP surface.

Water-soluble rhamnose-coated IONPs were prepared by Lartigue and coworkers [117]. The rhamnoside was anchored onto the OA-IONP surface via a phosphonate-containing triethyleneglycol linker that facilitated the chemisorption of the glycoside through phosphonate binding with iron oxide.

The acetate protective groups were removed by base treatment rendering the NP water soluble. A similar methodology was applied for the synthesis of mannose and ribose-functionalized magnetic NPs for hyperthermia applications [118].

Inspired by the adhesive properties of mussel proteins, the Park group took a biomimetic approach by immobilizing HA derivative on magnetic NPs via a dopamine anchor [114]. The key feature of dopamine is the ortho-dihydroxylphenyl (catechol) functional group that forms strong bonds with inorganic/organic surfaces. To attach HA, the OA-IONPs were transferred into the aqueous phase using a detergent cetyltrimethylammonium bromide (CTAB), which provided cationic surface charges to the NPs. The HA–dopamine conjugates were then incubated with the CTAB-coated particles to introduce the HA onto the NPs. Although no direct proof was provided regarding catechol chelation with NPs, the positive charges of the CTAB NPs were found to be important to yield well-dispersed HA-coated NPs [114].

The Penades group has pioneered the conjugation of carbohydrates to magnetic NPs through sulfhydryl groups [119]. They synthesized a variety of glycosides including lactoside **8** containing disulfide at the reducing ends (Figure 10.4) [120]. The immobilization of carbohydrates was performed through the *in situ* coating approach, where gold-doped iron oxide magnetic NPs were formed in the presence of the thioglycoside under reducing condition (Figure 10.4). The structure of the glyco-conjugates as well as the isolation procedures adopted influenced the shape and size of the glyco-NPs obtained. The success of carbohydrate conjugation to NPs was confirmed by nuclear magnetic resonance (NMR) spectroscopy. Elemental analysis and inductive coupled plasma

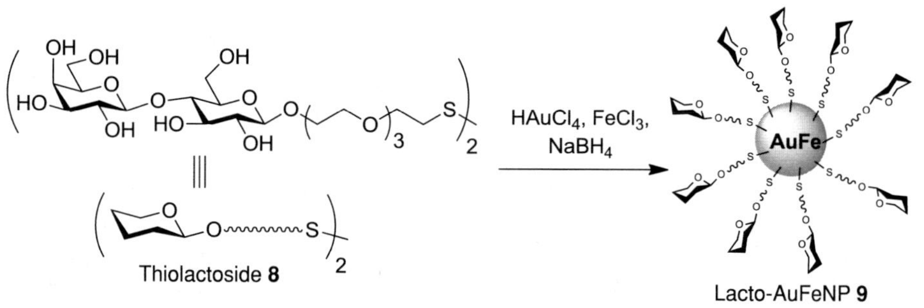

FIGURE 10.4 Disulfide-terminated glycoside can be immobilized onto gold-coated magnetic IONPs using the *in situ* coating protocol.

atomic emission spectrometry were used to estimate the number of carbohydrate molecules on the surface.

To avoid the impact of carbohydrate ligand on the core diameter of the NPs formed, the ligand-exchange method was also investigated. Bimetallic superparamagnetic NPs XFe_2O_4 (X = Fe, Mn, and Co) were synthesized using the thermal decomposition procedure, which were coated with gold to form XFe_2O_4@Au (X = Fe, Mn, and Co) NPs [121]. Thiolated lactose and *N*-acetyl glucosamine with various linker lengths successfully exchanged the hydrophobic ligands on the NPs forming water-soluble magnetic glyco-NPs. These glyco-NPs were found to have high stability, low cytotoxicity, and low immunogenicity with superior magnetic relaxivities [122].

10.2.2 Covalent Conjugation of Carbohydrates

Conjugation of carbohydrate to magnetic NPs through covalent bond formation is a common approach, because it leads to strong bonds between the carbohydrate and the magnetic NPs. Covalent functionalization of NPs typically starts from aminated magnetic NPs. Amines can be introduced onto the magnetic NPs through two main approaches. As shown in Figure 10.1, these include: (1) functionalization of IONPs with amine-containing siloxanes through the Stöber process (Figure 10.1c) [123–127]; (2) derivatization of dextran-coated IONPs by treatment with epichlorohydrin followed by ammonia (Figure 10.1b) [128–130]. In addition, aminated magnetic NPs can be prepared by coating with free amine-containing proteins such as apoferritin [131]. To covalently conjugate carbohydrates with aminated magnetic NPs, a variety of strategies have been developed, which include alkyne–azide cycloaddition reaction, amide and amine bond formation, thiol–ene reaction and photochemistry.

10.2.2.1 *[2 + 3] Alkyne–Azide Cycloaddition Reaction*
The Huisgen [2 + 3] alkyne–azide cycloaddition reaction has been proven as an effective method for conjugation of carbohydrates to magnetic NPs [123, 124, 132–134]. Lin and coworkers prepared azide- and alkyne-modified NPs from

aminated silica-coated iron oxide magnetic NPs through the bifunctional linker **10** (Figure 10.5a) [132]. Mannoside containing the azide (**13**) or alkyne (**14**) was immobilized onto the NPs through Huisgen [2 + 3] alkyne–azide cycloaddition reaction catalyzed by Cu(I) (Figure 10.5b). Higher conjugation efficiency was observed when the azide-functionalized NP **11** was coupled with mannosyl alkyne **13**. This was attributed to the higher reaction rates observed in systems containing clustered azides rather than clustered alkynes [135].

A more simplified protocol using the Huisgen cycloaddition reaction was reported by Prosperi and coworkers [133]. The key feature in their method was the one-pot conversion of the IONPs to the glyco-NPs (Figure 10.6). The free amines on the aminated NPs were converted to azides catalyzed by Cu(II) sulfate. Upon completion of this diazo-transfer reaction, sodium ascorbate was added to reduce Cu(II) to Cu(I), which promoted the subsequent Huisgen cycloaddition with the alkynated mannosides and lactosides [133]. Elemental analysis, inductive coupled plasma–atomic emission spectrometry, infrared spectroscopy, and high resolution magic-angle-spinning NMR experiments confirmed the successful conjugation of carbohydrates on NP surface using this one-pot method.

To increase the amount of azide on the NP surface, Xiong et al. introduced branched PEG brushes onto the NPs through atom transfer radical polymerization reaction [134]. The terminal hydroxyl groups of the PEG chains were then converted to azido groups, which were subsequently coupled with alkyne-modified maltosides via the Huisgen cycloaddition reaction. The amount of maltose on the surface of the NPs obtained through this route was 88.56 μmol/m^2. In comparison, only 5.58 μmol/m^2 of maltose could be immobilized on the NPs without the formation of PEG brushes.

10.2.2.2 *Reductive Amination* Besides Huisgen cycloaddition reactions, a wide range of other reactions have been utilized to immobilize carbohydrates. The most direct approach to attach carbohydrate onto aminated magnetic NP is through reactions with the hemiacetal reducing end of

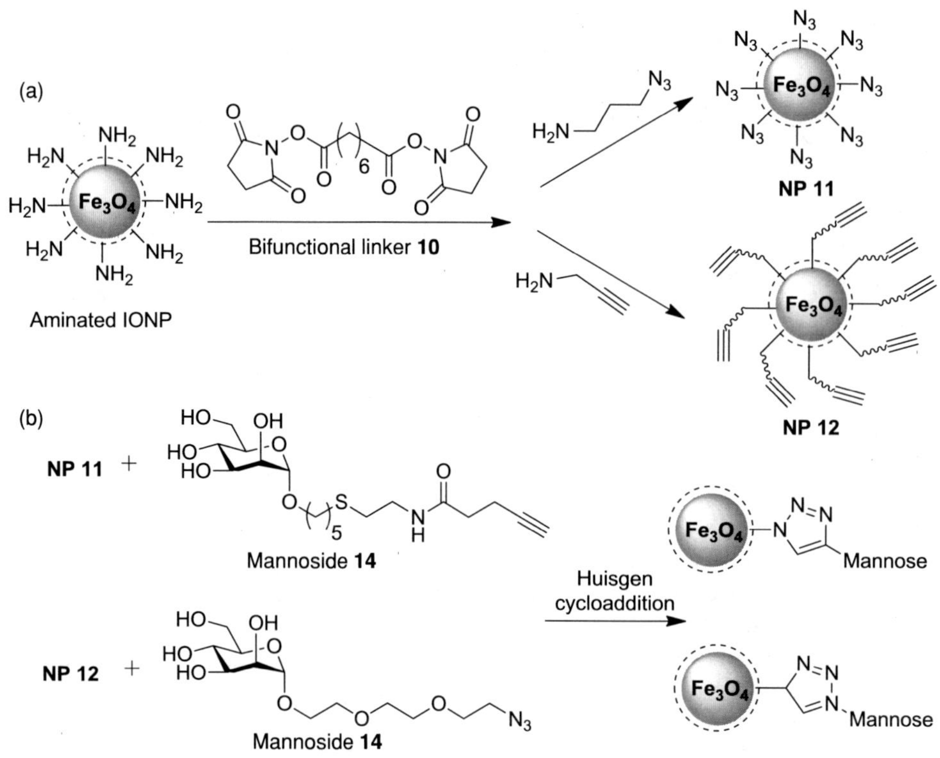

FIGURE 10.5   (a) Synthesis of azide- and alkyne-functionalized magnetic NPs and (b) their conjugation to azide- and alkyne-modified mannosides.

the sugar. Chen and coworkers demonstrated that maltose **15** could be grafted onto NPs by imine formation with amines on the NPs, followed by reduction with $NaCNBH_3$ (Figure 10.7) [136]. The advantage of this approach is that no synthetic modification of the carbohydrate is required. However, one drawback is that the glycopyranosyl ring is opened resulting from the reductive amination process. Therefore, biological recognition of carbohydrates can be adversely affected especially for monosaccharides.

10.2.2.3   Amide Formation

In order to avoid the opening of the glycopyranosyl ring, a very common strategy is to derivatize carbohydrates with a carboxylic acid or to use naturally existing carboxylic acid-bearing carbohydrates to conjugate with aminated NPs through amide bonds. Several coupling agents such as N-(3-dimethylaminopropyl)-N′-ethyl-carbodiimide (EDC)/N-hydroxysuccinimide (NHS)

[137, 138], benzotriazole-1-yl-oxy-tris-(dimethylamino)-phosphonium hexafluorophosphate (BOP) [123, 124], and 2-chloro-4,6-dimethoxy-1,3,5-triazine (CDMT) [128, 129, 139] have been successfully applied to facilitate the amide bond formation. Examples of carbohydrates immobilized include carboxymethylated CH [137], carboxymethylated cyclodextrin [138–140], HA [128, 129], and functionalized monosaccharides [123–125]. Alternatively, NPs can be functionalized with carboxylic acid on the surface, which could couple with glycosyl amines promoted by reagents such as EDC [141, 142].

10.2.2.4   Nucleophilic Substitution Reactions

Amines on magnetic NPs can function as nucleophiles in nucleophilic substitution reactions to link with carbohydrates. Badruddoza et al. demonstrated that the primary alcohols

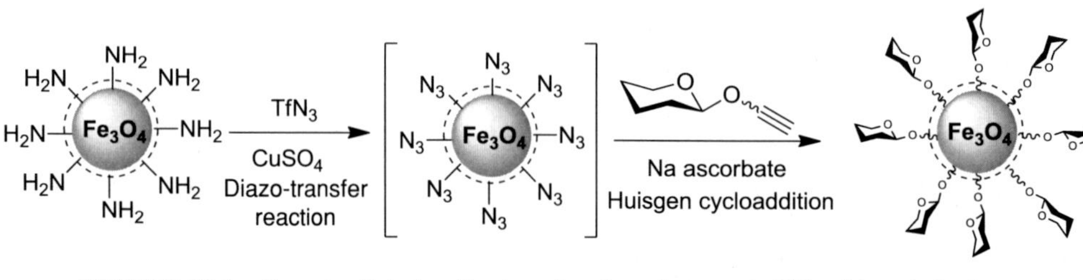

FIGURE 10.6   One-step functionalization of aminated magnetic NPs with carbohydrates.

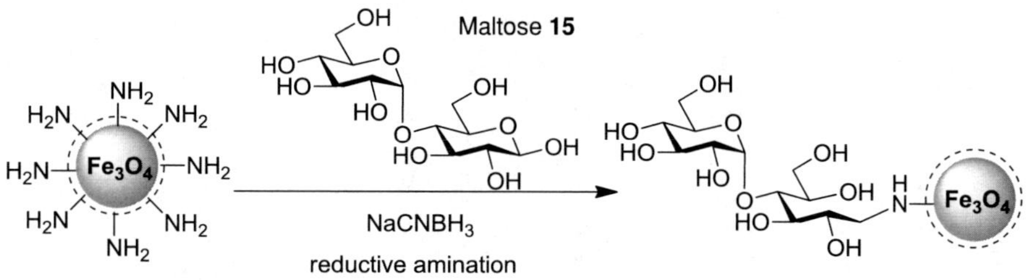

FIGURE 10.7   Immobilization of carbohydrates via reductive amination.

of β-cyclodextrin could be converted to a good leaving group tosylate using toluene sulfonyl chloride [143]. Subsequent nucleophilic attack by amine-functionalized silica-coated magnetic NPs displaced the tosylates thus covalently bonded with β-cyclodextrin.

Davis and coworkers used an innovative amidine linker to immobilize carbohydrates onto magnetic NPs [130]. They synthesized an *N*-acetyl glucosamine derivative bearing an S-cyanomethyl (SCM) moiety at the reducing end (glucosamine **16**), which functioned as both an anomeric protecting group and a masked linker hence reducing the synthetic steps in carbohydrate derivatization (Figure 10.8). The SCM was activated with sodium methoxide to 2-imido-2-methoxy-ethyl group (compound **17**) that subsequently reacted with aminated magnetic NPs to form the amidine linker immobilizing the *N*-acetyl glucosamine onto NPs. The *N*-acetyl glucosamine can be enzymatically glycosylated on the NPs to form sialyl LewisX (sLe^x) tetrasaccharide, an important carbohydrate involved in leukocyte recruitment during inflammatory processes. Alternatively, the SCM *N*-acetyl glucosamine **16** was converted to SCM sLe^x tetrasaccharide in solution via enzyme reactions, which was subsequently attached onto magnetic NPs via the amidine linker.

10.2.2.5 Cross-linking of Aminated Magnetic NPs with Carbohydrates Using Bifunctional Linkers   Aminated NPs can be coupled with carbohydrates through bifunctional

linkers. Xing and coworker treated aminated magnetic NPs with a dicarboxylic acid-terminated poly(ethylene oxide) (HOOC-PEG-COOH) promoted by the coupling agent dicyclohexylcarbodiimide (DCC) [127]. This was followed by EDC/NHS promoted amidation with aminated maltoside leading to maltose-functionalized magnetic NPs. A variety of other linkers have been utilized including glutaraldehyde for Schiff base formation [144, 145], hexamethylene diisocyanate [140], or disuccinimidyl carbonate [126] for carbonate/carbamate formation to cross-link the aminated NPs with carbohydrates.

10.2.2.6 Nucleophilic Addition with Vinyl Sulfone Linkers   Instead of introducing amines through chemical reactions, amine-containing proteins can be used to coat magnetic NPs. For example, apoferritin could be induced to assemble on the surface of NPs at neutral pH [131]. To functionalize the NPs with carbohydrates, *N*-acetyl-glucosamine (e.g., **18**) and mannose were derivatized with a vinyl sulfone at the reducing ends. 1,4-Conjugate addition of amine on the apoferritin-coated NPs to the vinyl sulfones led to the formation of glyco-NPs.

10.2.2.7 Photochemical Reaction   Yan and coworkers developed a novel strategy to immobilize carbohydrates onto magnetic NPs using photochemistry. Phosphate-terminated

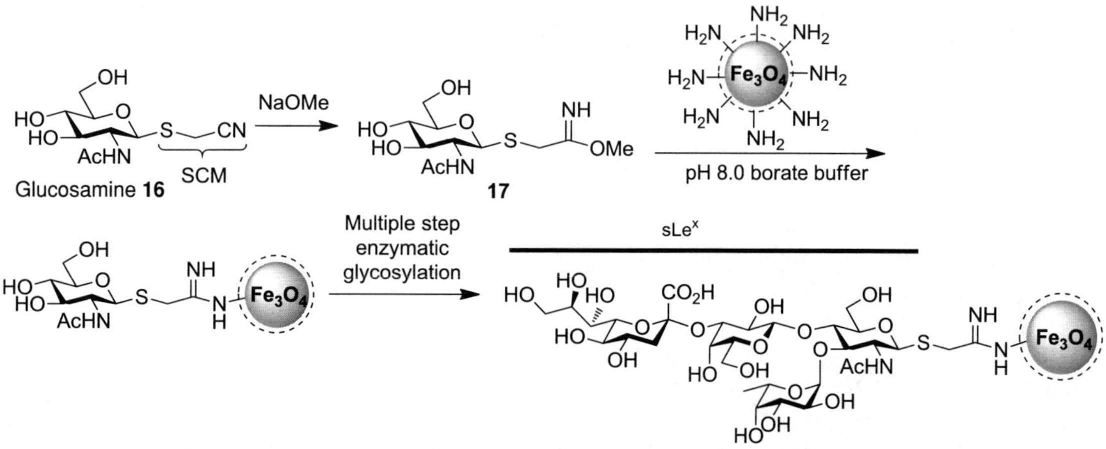

FIGURE 10.8   Synthesis of sLe^x-functionalized glyco-NPs via the amidine linkage.

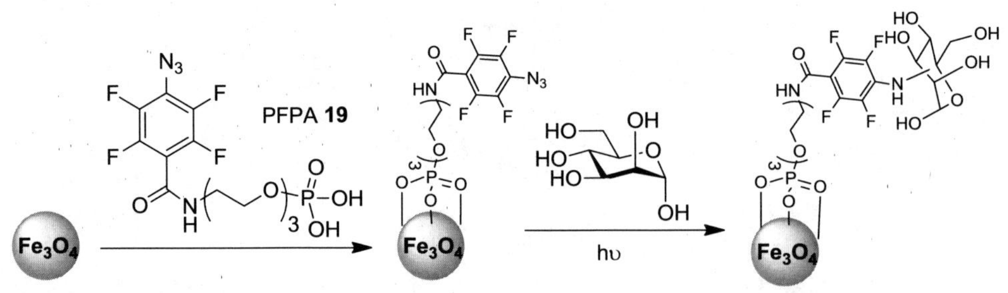

FIGURE 10.9   Immobilization of carbohydrates onto PFPA-functionalized magnetic NPs through photochemical reactions.

perfluorophenylazide **19** (PFPA) was synthesized, which bound to the surface of IONPs through phosphate chelation. Upon photoirradiation, the azido group was converted to reactive nitrene species that could react with underivatized carbohydrates through insertion into the C-H bonds of carbohydrates (Figure 10.9) [146, 147]. The major advantage of this approach is that naturally occurring carbohydrates can be directly utilized without synthetic modifications. Although disruption of carbohydrate structures can be a potential concern, lectin-binding studies showed that monosaccharide immobilized through this method retained their biological recognition specificities.

10.2.2.8   Thiol–ene Reaction for Immobilization of Carbohydrates   The Muller group investigated the utility of thiol–ene reaction for synthesis of magnetic glyco-NPs. Magnetic IONPs were prepared by the thermal decomposition method, which were then coated with methacrylate siloxane **20** (Figure 10.10) [148]. A fluorescent glyco-copolymer **21** containing 6-*O*-methacryloylgalactopyranose was produced by atom transfer radical polymerization (ATRP) followed by

deprotection to generate a glycopolymer with a chain end-free thiol. Thiol–ene reaction between the thiol in polymer **21** and the methacrylate on the NPs led to the formation of the glyco-NP.

10.3   FUNCTIONALIZATION OF MAGNETIC NPs FOR DELIVERY OF THERAPEUTIC AGENTS

Magnetic NPs have been shown to be promising platforms for delivery of therapeutic agents. In this section, we will discuss the various methods developed to immobilize therapeutic agents onto magnetic NPs. This will include (1) magnetic NPs as hyperthermia-inducing agents; (2) doxorubicin; (3) curcumin; (4) paclitaxel; (5) bleomycin; (6) cisplatin; (7) tamoxifen; and (8) silver ion.

10.3.1   Induction of Hyperthermia by Oscillating Magnetic Field

Hyperthermia refers to the method that generates heat for therapy, which can be employed in combination with other

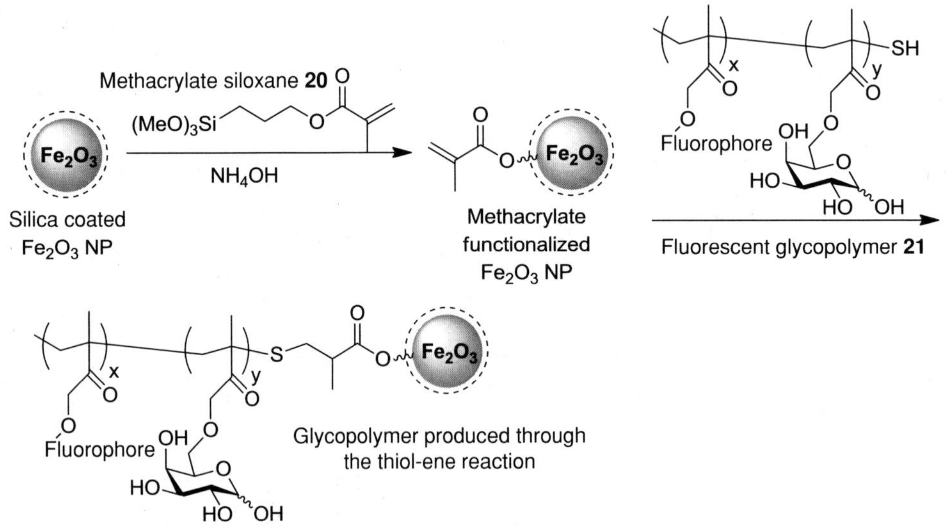

FIGURE 10.10   Synthesis of fluorescent magnetic glycopolymer-grafted NPs; in the process, galactose moieties were introduced on to the NP surface.

cancer treatments such as chemotherapy and/or radiotherapy. In addition to thermal ablation of tumors upon the generation of high temperatures of heat (T > 46°C), moderate hyperthermia (46°C > T > 41°C) has been exploited in cancer therapy to cause irreversible intracellular transformations such as protein denaturation and DNA cross-linking, and to induce changes such as improved perfusion and oxygenation in the resistant tumor-cell microenvironment to sensitize it against conventional tumor therapies [149]. When placed in a rapidly oscillating magnetic field, magnetic NPs have the capability to convert the magnetic energy into heat due to their magnetic properties, leading to cell death. In this sense, magnetic NPs can be considered therapeutics by themselves. For more discussion on magnetic NP-induced hyperthermia, interested readers can refer to the reviews on this subject [150].

Besides killing cells through the increase of temperatures, hyperthermia has been used as an innovative method to trigger the release of drugs incorporated in "smart," thermoresponsive materials. Hoare et al. recently reported the development of magnetically triggered biocompatible nanocomposite membranes to achieve zero-order release of drugs with membrane-to-membrane and cycle-to-cycle reproducibility [151]. These membranes were formulated by mixing an ethylcellulose solution with a temperature-sensitive nanogel composed of the poly(ethylene glycol)-coated superparamagnetic IONPs and a copolymer of poly(N-isopropylacrylamide-co-acrylamide) (PNIPAm). At low temperatures, the PNIPAm exists in a fully swollen form and thus can trap drugs inside the polymer coating. However, when the temperature is raised beyond a critical point, the PNIPAm becomes hydrophobic and collapses upon itself. When placed in an oscillating magnetic field, the MNPs generate heat that is absorbed by the thermoresponsive nanogel causing its collapse and the release of drugs. The phase transition temperature was controlled by copolymerizing various ratios of isopropylacrylamide and acrylamide, thus allowing the formation of nanogels with transition temperatures between 32°C and 46°C.

A study published by Zink and coworkers represents a good example of magnetically triggered irreversible release of drugs [152]. They synthesized zinc-doped Fe_3O_4 nanocrystals (ZnNCs) via the thermal decomposition method, which were incorporated into mesoporous silica NPs (MSNs) (Figure 10.11). This was followed by installing a nanovalve onto the surface of the NP by anchoring a positively charged thread, N-(6-N-aminohexyl)aminomethyl-triethoxysilane through the Stöber process. The anticancer

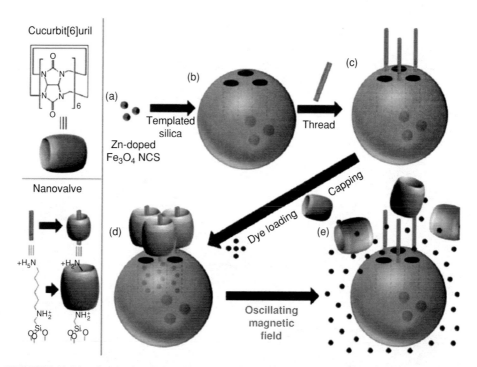

FIGURE 10.11 Synthesis of NPs capable of magnetically triggered drug release. (a) Zn-doped Fe_3O_4 nanocrystals were synthesized and (b) positioned into the core of the mesoporous silica NPs. (c) The positively charged thread was installed to the NP surface. (d) Drug (or dye) was loaded into the pores of the NPs and capped with the cucurbit[6]uril cap. Drug release was realized when the whole complex was placed in an oscillating magnetic field (e). Reproduced from Reference 152 with permission from American Chemical Society. For a color version, see the color plate section.

drug DOX was allowed to diffuse into the mesoporous structures and the pores of the MSNs were capped using cucurbit[6]uril, which electrostatically bound with the thread within its interior cavity. When an external magnetic field was applied, ZnNCs converted the magnetic energy into heat causing a local increase in the temperature, and triggering the opening of the cap and the release of the drug cargo. The drug-loaded mechanized NPs were tested for its ability to release the drug cargo. An oscillating magnetic field was applied to MBA-MB-231 breast cancer cells that have been transfected with NPs. The ZnNCs containing constructs, but not ZnNCs-free ones, demonstrated the release of the drug causing higher toxicities.

10.3.2 Photothermal Therapy

Besides using alternating magnetic field, another approach to generate heat is to coat magnetic NPs with materials that can absorb light in the near infrared (NIR) region [153–155]. It is attractive to use NIR radiation for *in vivo* applications, where attenuation of light by blood and soft tissue at NIR wavelength is greatly reduced leading to deeper tissue penetration by the light.

In this regard, gold-based NPs have attracted much attention due to its chemical inertness and biocompatibility [156–159]. Wang et al. engineered targeted gold nanorod/Fe_3O_4 "Nano-Pearl-Necklaces" that permitted the imaging of cancer cells by MRI aided by the Fe_3O_4 and its photothermal destruction when laser irradiation was applied [160]. Synthesis of this type of multifunctional hybrid started by preparing uniform, carboxylic acid-functionalized 15 nm IONPs via the thermal decomposition method [161]. An amphiphilic polymer, synthesized from poly(isobutylene-alt-maleic anhydride) and dodecylamine, was anchored onto the NPs through hydrophobic interactions, which introduced carboxylic acids onto the surface and improved the aqueous solubility. The carboxylic acid-functionalized magnetic NPs were then coupled with amine-functionalized Au nanorods mediated by EDC/NHS to produce $Au_{rod}-(Fe_3O_4)_n$ NPs. Herceptin, a humanized IgG monoclonal antibody (mAb) that binds to the extracellular domain of the HER-2 receptor, was conjugated to the hybrid nanoconstruct using EDC/NHS chemistry. These targeted nanoconjugates were selectively taken up by HER-2 receptor overexpressing breast cancer cells as demonstrated by MRI and fluorescence correlation spectroscopy, and caused much cell death when a laser was applied.

Along similar lines, Kim et al. manufactured multifunctional magnetic gold nanoshells (Mag-GNS) as potential candidates for cancer MRI and photothermal therapy [162]. The OA-IONPs were ligand-exchanged with 2-bromo-2-methylpropionic acid (BMPA), where the carboxylate group coordinated with iron oxide core leaving the bromo-group exposed on the surface. Mixing the BMPA-capped magnetic NPs with amine coated silica NP resulted in the covalent immobilization of magnetic NP onto SNP through nucleophilic substitution reactions by displacing the bromide with amines. The negatively charged 1–3 nm Au seeds were next electrostatically attached to the positively charged Fe_3O_4-SNP. Finally, a continuous Au shell was formed around the silica sphere particles by a one-step reduction of "$HAuCl_4$ growth solution" with formaldehyde assisted by the Au seeds acting as nucleation centers.

The Dai group developed an FeCo/graphitic carbon shell nanocrystals (Figure 10.12) [163, 164]. These nanocrystals enabled T_1- and T_2-weighted imaging by MR assisted by the FeCo alloy core, and photothermally enhanced drug release due to the presence of the graphitic shell that can generate heat under NIR irradiation [164]. To synthesize the FeCo core, fumed silica was impregnated with iron (III) nitrate ($Fe(NO_3)_3.9H_2O$) and cobalt (II) nitrate ($Co(NO_3)_2.6H_2O$) in

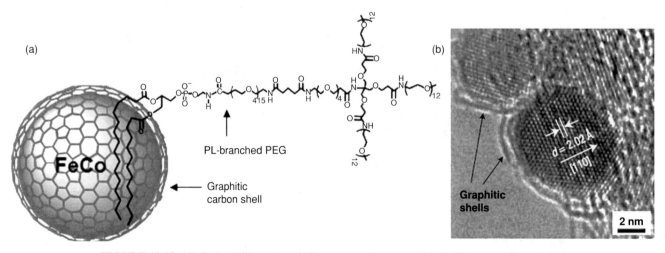

FIGURE 10.12   FeCo/graphitic carbon shell nanocrystals prepared by the Dai group. Reproduced with permission from Reference 163. (Copyright John Wiley & Sons, Inc.)

methanol. To form the single graphitic shell on the FeCo core, methane chemical vapor deposition was implemented where the impregnated silica powder was heated at 800°C under a flow of methane [165, 166]. The removal of silica by HF etching afforded the desired FeCo/graphitic carbon nanocrystals. Sonicating the nanocrystals with a phospholipid branched-PEG carboxylate (PL-branched PEG) greatly enhanced colloidal stability of the nanocrystals in aqueous media. The authors rationalized that PL-branched PEG adsorbed to the nanocrystals via van der Waals and hydrophobic interactions.

10.3.3 Photosensitizers on Magnetic NPs

Photodymanic therapy (PDT) has emerged as a powerful tool in cancer therapy that can complement existing therapies [167–169]. PDT relies on light-sensitive molecules called photosensitizers that are excited to higher energy states when irradiated with light. The excited photosensitizer will transfer energy to the surrounding molecular oxygen causing the formation of reactive oxygen species that can damage nearby living cells [170, 171]. Huang et al. conjugated a photosensitizer, that is, chlorin e6 (Ce6), on the surface of an IONP to assemble a dual imaging and magnetically targeted therapeutic NP [172]. The amine-functionalized iron oxide core, prepared using the coprecipitation method, was coupled with Ce6 mediated by EDC/NHS resulting in the formation of Ce6-NPs. Excess Ce6 was used to ensure only one carboxylic group of Ce6 participated in the reaction leaving the others exposed to enhance the water solubility of the final nanoconstruct. Conjugating Ce6 to magnetic NP did not affect the absorption and fluorescence emission spectra of the molecule. When evaluated *in vivo*, the magnetically targeted and PDT treated mice showed significant tumor regression compared to control groups.

10.3.4 Doxorubicin Immobilization

The anticancer drug DOX (trade name Adriamycin) belongs to the anthracyclin family and has long been used in the treatment of several cancers such as ovarian cancer, breast cancer, leukemia, and Hodgkin's lymphoma [173, 174]. It exerts its anticancer effect by intercalating and wedging between DNA bases thus inhibiting topoisomerase II activity. To eliminate some of the acute side effects of the drug, especially cardiotoxicity, a PEGylated liposomal formulation has been developed and commercialized as Doxil. Unfortunately, the development of hand–foot syndrome by patients receiving the formulation and the moderate therapeutic outcomes limited the use of Doxil. Thus, tremendous efforts are disposed toward the selective delivery of DOX to tumors, which ultimately can reduce off-site adverse effects, improve the therapeutic window, and overcome multidrug resistance [175, 176]. One of these approaches involves loading DOX into magnetic NPs resulting in a theranostic platform that

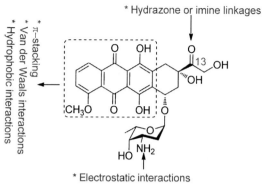

FIGURE 10.13 Potential modification sites of DOX.

offers both chemotherapy and the possibility of monitoring the progress of the treatment via noninvasive imaging techniques such as MRI [177–179].

DOX has been conjugated to magnetic NPs through the following means (Figure 10.13): (i) a hydrozone linkage formed between the C-13 carbonyl group of DOX and hydrazide functionality on the nanocarrier. A hydrazide moiety is preferred over a primary amine (imine linkage) due to the higher nucleophilicity of the hydrazide; (ii) formation of amide bond between the primary amine in DOX and the carrier; (iii) electrostatic attraction between the positively charged hydrochloride salt of DOX and negatively charged NP; (iv) π-stacking, van der Waals and hydrophobic interactions between the naphthacenequinone nucleus of DOX and a hydrophobic coating on the NP; (v) physical entrapment of DOX in the core of a porous carrier. The examples provided below present illustrations on each of the conjugation approaches and were chosen based on the conjugation method.

10.3.4.1 Hydrazone Linkage The formation of a pH-sensitive hydrazone bond between the C-13 carbonyl of DOX and a hydrazide moiety on the NP constitutes a popular route to covalently immobilize the drug [180–183]. Recently, we reported the development of a multifunctional IONP to deliver DOX to CD44-expressing cancer cells [129]. The carboxylic acid-containing HA-coated superparamagnetic IONP (HA-IONP) was synthesized by amidation of NH$_2$-IONP [43], which was followed by functionalization with adipic dihydrazide (ADH) via EDC-mediated coupling (Figure 10.14). Fivefold excess of ADH was used to prevent cross-linking of particles. The formation of the hydrazone bond between DOX and the hydrazide moiety on the NP occurred at an optimal pH 6.0. The successful immobilization of DOX was confirmed by Fourier Transform infrared spectroscopy, high resolution magic-angle-spin NMR, and

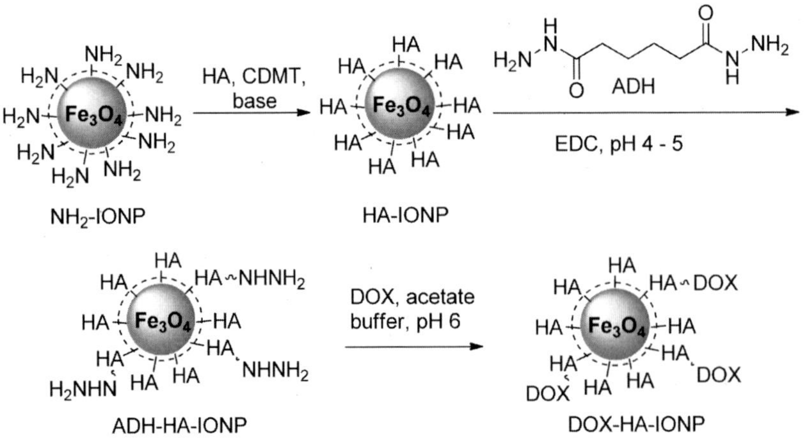

FIGURE 10.14 Immobilization of DOX onto IONPs through the hydrazone linkage via ADH linker.

UV–Vis spectrophotometry. *In vitro* drug release studies indicated that at an acidic pH of 4.5, but not at pH 7.4 or 9.0, DOX was released from the NPs due to hydrolysis of the hydrazone linkage. Cell viability studies showed that toxicities of DOX were greatly enhanced against both DOX-sensitive and DOX-resistant cancer cells through NP delivery.

10.3.4.2 Linking Through the Amino Group of DOX

Fang et al. exploited the ligand-exchange approach to anchor a biodegradable and pH-sensitive DOX-loaded poly(β-amino ester) (PBAE) copolymer onto the surface of IONPs [184]. The copolymer was synthesized by cross-linking DEG diacrylate (DEGDA) with the primary amines in PEG amine, dopamine (DA), and DOX through Michael additions (Figure 10.15). The PBAE polymer contains DA for NP binding, PEG for enhancing aqueous solubility, and DOX as the anticancer drug. PBAE successfully exchanged oleic acid on the surface of IONPs presumably through DA anchors leading to monodispersed and highly stable nanoconjugates with a hydrodynamic diameter of approximately 80 nm. The efficiency of DOX loading was quite high (~90%) resulting in 679 µg DOX/mg-Fe. At pH 7.4, 6.5, and 5.5, 30%, 40%, and 55% of DOX were released respectively from NPs after 72 hours through degradation of the polymer backbone.

The study by Liu et al. represents an example where epirubicin (EPI), a DOX analog with an equatorial 4′-hydroxyl group in the daunosamine sugar, was immobilized on the surface of an IONP through an amide bond between the primary amine of EPI and carboxylic acid functionality on the NP [185]. In their study, the magnetic properties of IONPs and the focused ultrasound were synergistically combined to actively deliver the therapeutic NP across the blood–brain barrier to brain tumor. The poly[aniline-*co*-sodium *N*-(1-one-butyric acid)aniline] polymer (SPAnNa) was synthesized using supercritical carbon dioxide [186]. The addition of HCl solution caused the neutralization of SPAnNa, which aggregated on the surface of bare IONPs. EDC and *N*-hydroxysulfosuccinimide (NHSS)-mediated amide formation between the polymer-coated NP and EPI resulted in the immobilization of the cytotoxic drug on the NP. Using this protocol, 452 µg of EPI was immobilized per 1 mg of iron. The combination of focused ultrasound, which can temporarily and locally open the blood–brain barrier to allow the crossing of macromolecules, with magnetic targeting, which enables accumulation of drug-loaded NPs at the desired site, resulted in increased deposition of the therapeutic NP in the brain.

10.3.4.3 Electrostatic Interactions

Yu et al. hypothesized that the positively charged DOX hydrochloride can be attached electrostatically to the surface of magnetic NPs coated with a negatively charged polymer

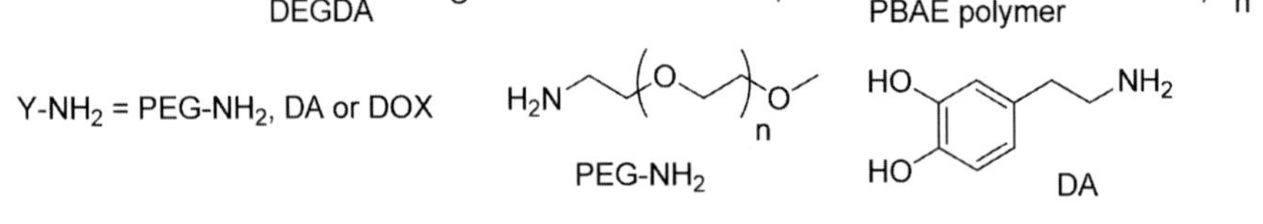

FIGURE 10.15 Synthesis of the DOX-containing PBAE polymer for NP immobilization.

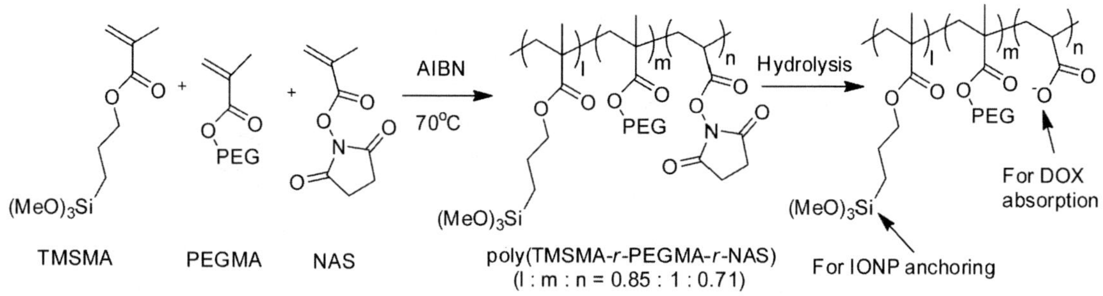

FIGURE 10.16   Synthesis of negatively charged polymer for DOX absorption on IONPs.

[187]. They engineered thermally cross-linked super-paramagnetic IONPs (TCL-IONPs) coated with an anti-biofouling polymer poly(TMSMA-r-PEGMA-r-NAS). The poly(TMSMA-r-PEGMA-r-NAS) was formed by copolymerizing 3-(trimethoxysilyl)propyl methacrylate (TMSMA), poly(ethyleneglycol) methyl ether methacrylate (PEGMA) with N-acryloxysuccinimide (NAS) at 70°C in the presence of 2,2′-azobisisobutyronitrile (AIBN) as the polymerization initiator (Figure 10.16) [188]. The N-hydroxysuccinimide esters in the polymer were hydrolyzed and the polymer was immobilized onto IONPs prepared via the coprecipitation method. Thermal cross-linking between entangled polymer chains was achieved by heating the polymer-coated NPs at 80°C leading to TCL-IONPs containing surface carboxylic acids. The simple mixing of DOX and TCL-IONP resulted in the electrostatic adsorption of the drug onto NP surface. The successful incorporation of DOX was assessed by monitoring the fluorescence quenching of DOX upon adsorption to NP due to the fluorescence resonance energy transfer. Maximal fluorescence quenching was achieved when 4 μg of DOX were mixed with 200 μg of TCL-IONP.

10.3.4.4 Hydrophobic Interaction DOX, in its free amine form, is hydrophobic, which can be entrapped within a lipophilic coating on magnetic NPs through hydrophobic

or $\pi-\pi$ interactions [189, 190]. Lim et al. developed an intelligent magnetic nanoplatform with the amphiphilic fluorescent surfactant pyrenyl polyethyleneglycol (pyrenyl PEG) anchored on oleic acid-coated MnFe$_2$O$_4$ NPs [191]. It was envisioned that at physiological conditions (pH 7.4), $\pi-\pi$ stacking between the DOX core and the pyrenyl moiety should favor the efficient encapsulation of the drug. When the nanocarrier was taken up by cells through endocytosis, the acidic pH encountered inside endosomes and late lysosomes would induce the protonation of DOX and reduce the $\pi-\pi$ interactions, thus triggering the release of the drug. The amphiphilic polymer pyrenyl PEG **22** was synthesized by reacting NHS-activated pyrenebutyric acid **23** with the hetero-functional PEG **24** (COOH-PEG-NH$_2$) (Figure 10.17). Sonication of a mixture of DOX, pyrenyl PEG, and MnFe$_2$O$_4$ nanocrystals in chloroform/hexane (organic phase), and a buffer solution (pH 9.8) followed by evaporation of organic solvents at room temperature afford the desired pH-sensitive drug-releasing magnetic NPs. To aid in tumor targeting, Herceptin was covalently introduced through EDC/NHS-mediated coupling with the carboxyl groups of the polymer. The feasibility of the proposed approach was validated by testing DOX release at different pH conditions. More than 80% of loaded DOX was released after 24 hours at pH 5.5, although 40% of the drug was also released within the same time frame at pH 7.4.

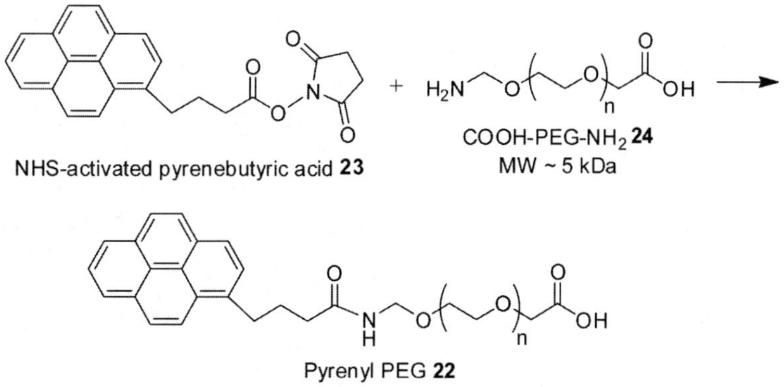

NHS-activated pyrenebutyric acid **23**

COOH-PEG-NH$_2$ **24**
MW ~ 5 kDa

Pyrenyl PEG **22**

FIGURE 10.17   Synthesis of an amphiphilic-PEG polymer for the noncovalent immobilization of DOX via hydrophobic interactions.

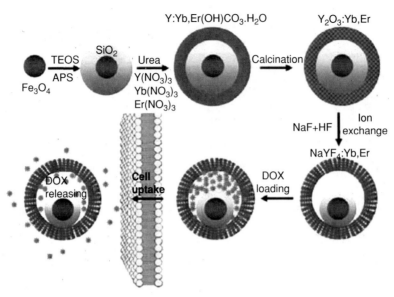

FIGURE 10.18 Synthesis of a mesoporous magnetic and upconversion luminescent nanorattle for DOX delivery. Reproduced with permission from Reference 192. (Copyright American Chemical Society.) For a color version, see the color plate section.

10.3.4.5 Entrapment of DOX in the Core of a Porous Carrier

The diffusion of molecular DOX into the channels/cores of mesoporous NPs presents another approach for DOX loading and delivery [192–194]. In this regard, Zhang et al. constructed a novel mesoporous magnetic and upconversion (UC) luminescent nanorattle for targeted imaging and drug delivery [192]. Compared to traditional luminescent materials, UC platforms exhibit better tissue-penetration depth, improved photo and chemical stability, and higher signal-to-noise ratios.

IONPs, synthesized by a hydrothermal method, were coated with silica using a reverse-microemulsion technique with TEOS as the silica precursor (Figure 10.18). The coating of $Fe_3O_4@SiO_2$ NPs with a layer of amorphous Y/Yb and $Er(OH)CO_3$ followed by calcination at $550°C$ resulted in the generation of the cubic phase Y_2O_3:Yb, Er coating. The formation of $NaYF_4$, one of the most efficient NIR-to-vis UC luminescent material, was achieved through an ion-exchange process in the presence of NaF and HF. The thickness of the inner SiO_2 shell was found to be dependent on HF concentration, reaction temperature, and time. The TEM images and N_2 adsorption–desorption isotherms of the as-synthesized $Fe_3O_4@SiO_2@\alpha$-$NaYF_4$/Yb, Er magnetic UC-fluoride nanorattle (MUC-F-NR) revealed the hollow core–shell structure with a diameter of 115 nm and the average pore size of 4.8 nm. MUC-F-NR displayed excellent superparamagnetic and UC-luminescent properties. Simple soaking of the UC NPs with DOX resulted in efficient drug loading, which was found to be highly dependent on the SiO_2 shell thickness, that is, more drug was loaded when the volume of the inner hollow space was larger. When the NPs

were administered *in vivo*, placing an external magnet at the tumor site led to enhanced local accumulation of the nanorattle and better therapeutic outcomes.

10.3.5 Delivery of Curcumin by Magnetic NPs

Besides DOX, magnetic NPs have been exploited to improve the bioavailability, selectivity, and water solubility of a wide variety of drugs, and to overcome multidrug resistance [195–203]. Curcumin (Cur) is a natural phenol that has demonstrated anticancer activity by blocking the transcriptional factor, nuclear factor κB (NFκB), which is known to regulate cell proliferation, apoptosis, and resistance in cells [204, 205]. However, like many other hydrophobic anticancer drugs, the poor aqueous solubility and bioavailability of the molecule limited its therapeutic outcome, in addition to the development of MDR by reactivating NFκB [206]. To overcome these shortcomings, Manju et al. proposed loading Cur onto IONPs via a layer-by-layer assembly (LbL) [207]. They covalently linked Cur to oppositely charged polymers, that is, HA as the negatively charged [208] and polyvinylpyrrolidone (PVP) as the positively charged polymer (Figure 10.19a, b) [209]. IONPs were coated with *N*-3-trimethoxysilylpropylethylenediamine (TMSPEDA) to plant positive charges on the surface (TMSPEDA@MNPs) (Figure 10.19c). The LbL assembly started by electrostatically anchoring the negatively charged HA-Cur onto the positively charged surface. The success of the reaction was monitored by measuring the zeta potential (ζ-potential) of the NPs. While TMSPEDA@MNPs exhibited ζ of $+25.7$,

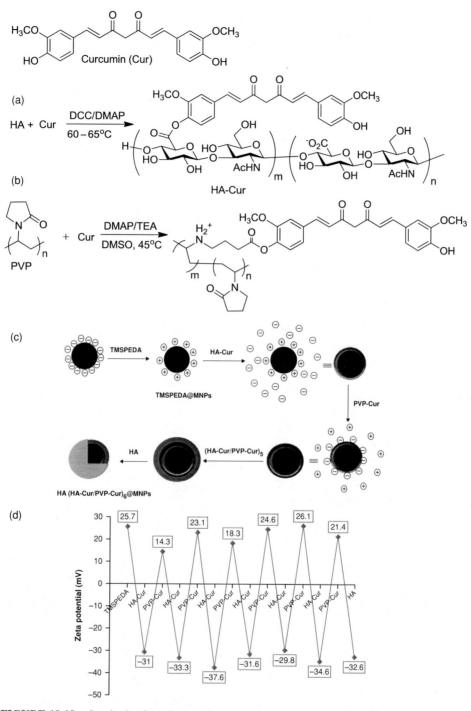

FIGURE 10.19   Synthesis of (a) the negatively charged HA-Cur and (b) positively charged PVP-Cur, (c) coating of the magnetic NPs by the LBL approach, (d) ζ-potential changes during the LBL process. Reproduced with permission from Reference 207. (Copyright American Chemical Society).

the addition of the first HA-Cur layer resulted in a negative ζ (-31 mV). Subsequently, when the positively charged PVP-Cur was added, the NPs displayed a positive ζ ($+14$ mV). During the sequential addition of six layers of each of HA-Cur and PVP-Cur, ζ alternated from positive to negative (Figure 10.19d). The final HA(HA-Cur/PVP-Cur)$_6$@MNPs exhibited a hydrodynamic diameter of 132 nm and polydispersity index (PDI) of 0.145. The NPs were biocompatible and displayed enhanced cytotoxicity compared to free Cur against cancer cells.

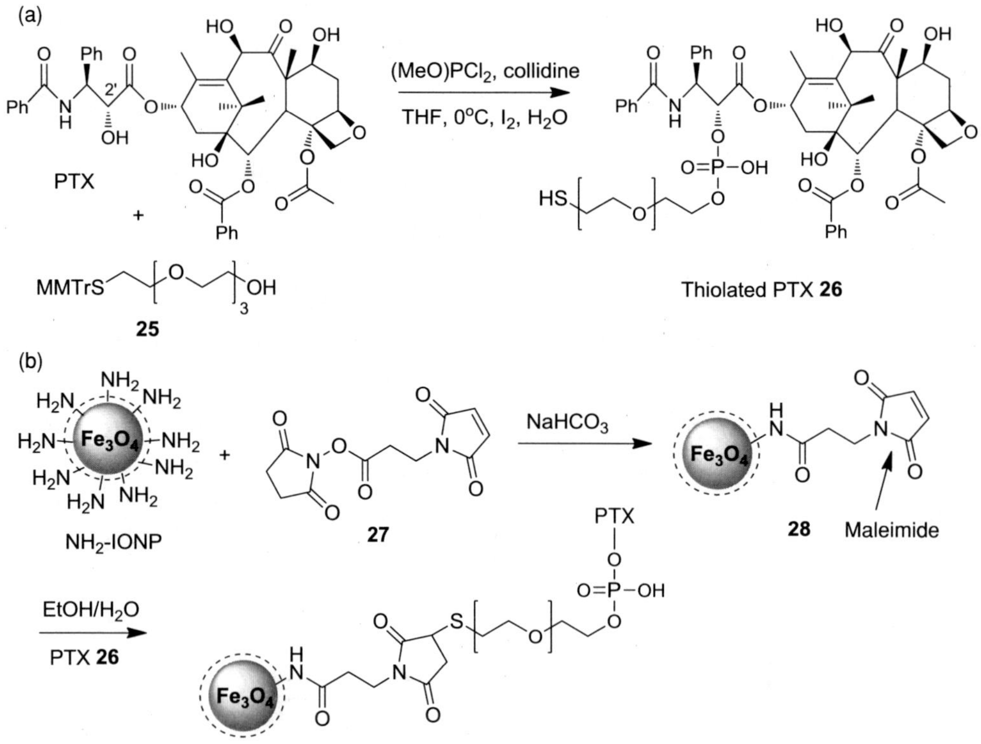

FIGURE 10.20   (a) Derivatization of PTX with a thiol linker, (b) immobilization of thiolated PTX onto maleimide-functionalized magnetic NPs.

10.3.6   Delivery of Paclitaxel, Bleomycin, Cisplatin, and Tamoxifen by Magnetic NPs

The mitotic inhibitor paclitaxel (PTX) can be considered as the pioneer drug for the class of agents that exert cytotoxicity by stabilizing microtubules thus arresting cell cycle at the G2/M phase and inducing apoptosis [210]. It is used for the treatment of various types of cancer including ovarian, breast, and lung cancer. To improve its low aqueous solubility, PTX is usually administered intravenously in a 50% Cremophor EL® and 50% dehydrated ethanol formulation. Unfortunately, the presence of such a high level of Cremophor EL can trigger acute hypersensitivity reactions, and it likely contributes to the less than ideal, nonlinear pharmacokinetic behavior of the drug observed in both mice and humans. To improve the water solubility of PTX and attain selectivity in targeting to cancer cells, PTX, or its analog docetaxel, was conjugated to magnetic NPs [211–214]. For example, Hwu et al. reported the covalent conjugation of PTX at the 2′-hydroxyl position to an IONP through a phosphodiester linkage [215]. It was envisioned that the phosphodiester moiety would favor interactions with cancer cells, and the release of the drug due to the prominent dephosphorylation in cancer cells in comparison to normal cells.

To introduce phosphate into PTX, a one-pot reaction was performed by conjugating a protected tetraethyleneglycol monothiol **25** to the 2′-hydroxyl of PTX mediated by methyl dichlorophosphite, followed by the oxidation of the phosphite and deprotection (Figure 10.20). Amine-coated IONP was functionalized with maleimide through a hetero-bifunctional linker **27** (Figure 10.20b). The thiolated PTX **26** was coupled to NP through the selective reaction between the thiol in PTX and the maleimide moiety on NP. Analysis of TGA data indicated that 83 PTX molecules were attached on the NP surface. The synthesized PTX-NP conjugate exhibited a 780- and 96-fold increase in hydrophilicity compared to free PTX and the PEGylated PTX adduct respectively. As the 2′-hydroxyl group is crucial for PTX biological activity, it was essential to evaluate PTX release profile with phosphodiesterase. HPLC analysis revealed that 91% of conjugated PTX were released after 10 days, compared to minimal release (<1%) in the absence of the enzyme.

In an attempt to reduce the side effect associated with the use of bleomycin (BLM) as an anticancer agent, especially pulmonary toxicity, Georgelin et al. fabricated a silica core–shell maghemite NP (γ-Fe$_2$O$_3$@SiO$_2$-PEG-NH$_2$) for the passive targeting of BLM to cancer cells [216]. BLM A$_5$ is a glycopeptide anticancer compound that imposes cytotoxicity by triggering the formation of reactive oxygen species that will induce single-stranded and double-stranded DNA breaks [217]. BLM was grafted on the γ-Fe$_2$O$_3$@SiO$_2$-PEG-NH$_2$ NP through a glutaraldehyde linker followed by sodium cyanoborohydride reduction, with a grafting efficiency of 17%. Gel electrophoresis was used to evaluate the ability of

BLM A_5 to cleave supercoiled DNA. About 2 μM of free BLM A_5 and 10 μM of conjugated BLM A_5 were needed respectively to cleave 90% of the plasmid, indicating that the drug on NP retained its biological activity. The decrease in antitumor activity was explained by the possible limited accessibility of the immobilized drug to DNA, which can be potentially improved by using a longer linker.

The anticancer benefits of cisplatin were first demonstrated by Barnett Rosenberg and coworkers [218]. This square planar molecule is a DNA cross-linker that proved to be effective against various types of sacromas, carcinomas, and lymphomas. The clinical use of cisplatin is restricted by its adverse side effects, mainly nephrotoxicity and neurotoxicity, and the rapid development of resistance [219]. Sonoda et al. reported a simple method for the conjugation of cisplatin onto IONPs, which can be used to combine the hyperthermia effect from the magnetic NPs with the chemotherapeutic effect of cisplatin [220]. Resovist, a commercially available IONP clinically approved for MRI of liver, was utilized as its modified dextran coating contains carboxylic acid groups necessary for the conjugation of cisplatin. Mixing of aqueous solutions of cisplatin and Resovist allowed the displacement of the chlorine ligand of cisplatin by carboxyl groups on the NPs. Evaluation of cisplatin release from the NPs showed that 30% of the loaded cisplatin was released after 24 hours. Such a slow release profile can be beneficial for enhancing the antitumor activity of cisplatin. After administration of the NP–cisplatin conjugate in rats, applying a magnet to the animals resulted in augmented accumulation of the NPs in the area underneath the magnet, which also led to an increase in the local temperature when the animals were placed in an induction-heating device.

The anticancer drug TAX is widely used for the treatment of estrogen-receptor (ER) positive breast cancer cells [221]. TAX itself has weak affinity to ER. It is metabolized in the liver to hydrotamoxifen with high binding affinity to ERs, which inhibits the binding of estrogen that is required for the growth of breast cancer cells. The resulting TAX–ER complex binds with DNA thus blocking transcription and causing cell cycle arrest at the G_0 and G_1 phases [222]. A major side effect of the hydrophobic TAX is an increased risk for the development of endometrial cancer due to the accumulation of the drug in patient's body [223]. One approach to overcome this side effect is to formulate TAX in a nanoplatform that improves the drug's bioavailability and increases its local concentration in tumoral sites. Hu et al. incorporated TAX into a composite NP composed of magnetite and the biodegradable poly(lactic acid) (PLLA) [224]. A solvent/extraction technique was used in an oil/water emulsion to assemble the designed nanocomposites. A mixture of oleic acid-stabilized Fe_3O_4 NPs, PLLA, and TAX in dichloromethane (organic phase) was added into an aqueous solution containing 1% polyvinylalcohol as an emulsifier. Sonication of the heterogeneous mixture formed the oil/water

emulsions, which led to the desired NPs after evaporation of organic solvent. Drug-release studies indicated that 57% of the entrapped drug was released after 24 hours, and 88% after 6 days. The uptake of the NPs was time and concentration dependent, and the TAX-loaded NPs were much more cytotoxic than free TAX.

10.3.7 Silver-coated Magnetic NPs for Antibiotic Applications

With the development of bacterial resistance against currently used antibiotics, the search for new effective antibacterial agents continues to attract much attention worldwide. Silver ions (Ag^+) are known antibacterial and antimicrobial agents. It is believed that silver ions can electrostatically attach to the bacterial cell wall, penetrate into the cell, and interact with thiol-containing proteins that are essential for the bacterial growth causing the organism's death. However, the tendency of the silver ions to interact with common anions such as chloride and form insoluble precipitate limited its application. Silver (Ag^0) NPs have been investigated as a reservoir of the slow release of Ag^+ ions resulting from the oxidation of Ag^0 in the presence of cellular oxidants. One of the limitations of silver NPs is that NPs have inadequate penetration through the biofilms formed by bacteria, thus lowering the therapeutic efficacy.

To overcome this problem, Mahmoudi et al. envisioned that magnetic field can be used to facilitate the penetration of silver-coated magnetic NPs into the biofilm [225]. To test this, they prepared OA-IONPs and installed carboxylated dextran onto the NP surface via ligand exchange. Ethanediyl bis(isonicotinate), which has high binding affinity with silver ions, was absorbed onto the carboxylated IONPs via electrostatic interactions. Silver ions were then bound onto the NPs followed by sodium borohydride reduction to generate a thin, 2–3 nm silver shell on the NPs. When evaluated on two common strains of bacteria, the silver-coated magnetic NPs killed significantly more bacteria than free silver ions, the silver-free control NPs, as well as Kanamycin (a commonly used antibiotic). The antibacterial efficacy was further enhanced when an external magnetic field was applied.

10.4   FUNCTIONALIZATION OF MAGNETIC NPS FOR NUCLEIC ACID DELIVERY

Nucleic acids can be delivered to cells to correct a genetic defect, trigger the expression of a specific protein, or silence the expression of target genes. Due to their negatively charged nature, nucleic acids cannot enter the cells by themselves and vectors are needed to achieve this goal. Vectors can be broadly classified as viral and nonviral. Although viral delivery of genetic material is highly efficient, the toxicity, immunogenicity, low loading capacities, and associated high

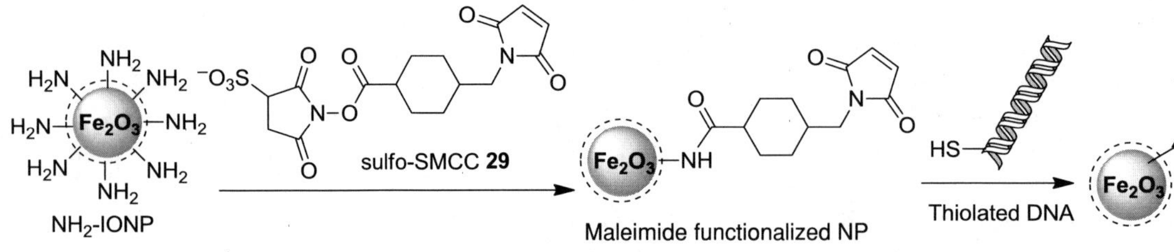

FIGURE 10.21   Chemoselective ligation of thiolated DNA with maleimide-functionalized NP.

cost have prompted scientists to explore nonviral delivery vectors such as NPs [226].

Magnetic NPs are investigated as generally nontoxic and easily functionalizable carrier systems for the safe and efficient delivery of genetic materials [227, 228]. An advantage of magnetic NPs compared to other classes of nanoplatforms is that the accumulation of nucleic acid-loaded magnetic NPs can be significantly enhanced in the target tissue when an external magnetic field is applied. This magnetofection process can lead to higher cellular uptake and better gene expression or silencing [229]. Alternatively, a ligand that will recognize a receptor expressed on target cells can be immobilized on the NPs to enhance selectivity in delivery. In the following section, we will discuss various approaches that have been exploited to load genetic materials onto magnetic NPs.

10.4.1   Conjugation of Thiolated DNA

A popular approach to immobilize DNA onto magnetic NPs is by derivatizing DNA with a free thiol. The sulfhydryl moiety is much more nucleophilic than many common nucleophiles such as amines and hydroxyl groups, thus can chemoselectively ligate with suitably functionalized magnetic NPs. The first example of this takes advantage of the reaction between thiol and maleimide. Ruiz-Hernandez et al. reacted aminated maghemite (γ-Fe_2O_3) NPs with a cross-linker sulfosuccinimidyl 4-N-maleimidomethyl cyclohexane-1-carboxylate **29** (sulfo-SMCC) (Figure 10.21) [230]. This introduced maleimides onto the magnetic NPs, which selectively reacted with a thiolated DNA. The DNA-bearing magnetic NPs functioned as a cap to mesoporous silica NPs in an innovative design of a "smart" construct capable of releasing drugs in response to magnet-induced hyperthermia.

Besides the sulfhydryl and maleimide chemistry, thiolated genetic material can be introduced into NPs modified with pyridyldisulfide functionality. Lee et al. engineered manganese-doped magnetism-engineered iron oxide (MnMEIO) NPs for the delivery of siRNA sequences to silence the expression of specific target genes [62]. The MnMEIO exhibited higher magnetic moments compared to conventional magnetic IONPs [16]. Cationized bovine serum albumin (BSA, isoelectric point 6.1), prepared by the reaction of BSA with 2,2'-(ethylenedioxy)bis(ethylamine), was

ligand-exchanged onto MnMEIO leading to free amines on particle surface [231]. A pyridyldisulfide functionality was introduced onto the NP with N-succinimidyl-3-(2-pyridyldithio)propionate (SPDP) through amide bond formation. A thiolated anti-GFP siRNA (siGFP) for the suppression of green fluorescent protein (GFP) gene and a thiolated PEG modified with cyclic Arg-Gly-Asp (c-RGD) for the active targeting of $\alpha_v\beta_3$ integrin receptor overexpressed on cancer cells were introduced via the formation of disulfide linkages to provide the targeted MnMEIO–siGFP/PEG–RGD NPs with a hydrodynamic size of 75 nm. The drop in ζ-potential values from -5 mV for MnMEIO to -30 mV for the siRNA-coated NP and the results from agarose gel electrophoresis suggested the successful conjugation of siRNA.

Stoeva et al. took advantage of the strong interaction between a free thiol and gold to load thiolated DNA onto a three-layered core–shell composite magnetic NP [232]. Oleic acid-coated Fe_3O_4 NPs were prepared and transferred into aqueous medium using the amphiphilic surfactant tetramethylammonium 11-aminoundecanoate. The negatively charged NPs with the ζ-potential of -35 mV electrostatically interacted with amine-functionalized SiO_2 NPs bearing a ζ-potential $+50$ mV (Step 1 in Figure 10.22). The SiO_2-Fe_3O_4 NPs with a ζ-potential of $+17$ mV were able to attract negatively charged 1–3 nm Au NPs (Step 2), which acted as nucleation sites for the formation of a gold layer on

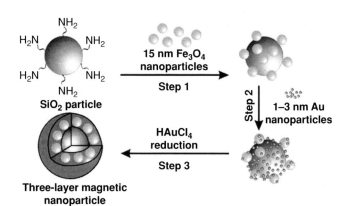

FIGURE 10.22   Formation of gold-coated three-layered magnetic NPs for DNA immobilization. Reproduced with permission from Reference 232. (Copyright American Chemical Society. For a color version, see the color plate section.

the nanocomposites (Step 3). 3′- and 5′ thiol-modified DNAs were then immobilized onto the three-layered NPs through the binding of sulfhydryl group with gold.

10.4.2 Carbodiimide-mediated Coupling

In addition to thiols, amines can be introduced on the terminal of DNA or RNA for conjugation with magnetic NPs. The posttranscriptional regulator microRNA (miRNA) is believed to regulate every cellular process, and is involved in various vital functions such as cell development, differentiation, and apoptosis [233]. Inhibition of miRNA functions through the development of anti-miRNA sequences (miR) offers an attractive route to combating cancer. Kim et al. reported the fabrication of fluorescent magnetic NPs loaded with miR to selectively inhibit miRNA-221 in cancer cells [234]. To target tumor, the small oligonucleotide aptamer AS1411, which binds to nucleolin protein overexpressed on the surface of cancer cells, was utilized. Both miRNA-221 and AS1411 were synthesized with a free amino group at the 5′ end. Carboxylic acid-functionalized $CoFe_2O_4$ NPs were coupled with miRNA-221 and AS1411 using EDC. The success of the conjugation was validated by reduction of free RNA concentration in solution following coupling and changes in migratory abilities of NPs in gel electrophoresis.

10.4.3 Electrostatic Interactions

Another approach to condense genetic material onto magnetic NPs is to introduce a cationic coating on NPs that can interact electrostatically with the anionic backbone of nucleic acids. Two types of cationic polymers, that is, polyethylene imine (PEI) and CH, have been intensively studied as coating reagents. These polymers can be immobilized onto NPs through covalent linkages, electrostatic attractions, or hydrophobic interactions, which can facilitate not only the delivery of genetic materials into the cells, but also the escape of NPs from endosomes by acting as a proton sponge that disrupts the acidity in the endosomes.

In an attempt to devise a carrier for the delivery of genetic material with enhanced uptake by cancer cells, Zhang and coworkers constructed a nanovector composed of an iron oxide core coated with two cationic polymers, that is, PEG-linked chitosan (CP) and PEI [235, 236]. This was accomplished by coprecipitation of Fe^{2+} and Fe^{3+} salts under basic conditions in the presence of CP to generate CP-stabilized IONPs. The free amines on the NPs were then derivatized with 2-iminothiolane (Traut's reagent) to introduce thiols onto the NP (Figure 10.23a). The sulfhydryl moieties of thiolated IONPs reacted with iodoacetate-modified PEI to afford NP-CP-PEI. Enhanced green fluorescent protein (EGFP) encoding DNA was condensed onto the NP-CP-PEI via electrostatic attraction (Figure 10.23b). As a comparison, PEI was also directly coated on NP during NP synthesis

(NP-PEI). Although NP-PEI was more efficient in transfection, NP-CP-PEI was much more biocompatible.

To enhance the cellular uptake of the DNA–NP complex, a glioma-targeting polypeptide chlorotoxin (CTX) was thiolated by Traut's reagent (Figure 10.23c). The DNA–NPs were treated with a heterobifunctional PEG linker **30** (NHS–PEG–maleimide) to introduce the maleimide moiety onto the NPs through amide bonds (Figure 10.23b). The thiolated CTX reacted with maleimide-functionalized DNA–NPs generating DNA–NP–CTX complex (Figure 10.23c) for gene therapy to glioma, a deadly form of brain tumor.

Wang et al. reported the development of polymeric micelles encapsulating magnetic IONPs, which were coated with cationic polymers for DNA delivery [237]. A block amphiphilic copolymer methoxy polyethylene glycol (mPEG)-poly(lactide) was prepared through a stannous 2-ethyl-hexanoate-catalyzed ring-opening polymerization reaction between mPEG and dilactide [238]. The solvent evaporation method was used to cluster IONPs in the core of a micelle that self-assembled from the amphiphilic copolymer. The clustering of the IONPs would increase the T2 relaxivity leading to enhanced MRI sensitivity. The resulting magnetic micelle (mag-micelle) displayed a negative ζ-potential (-16.7 mV), thus enabling the binding of CH through electrostatic attractions and facilitating the formation of CH-mag-micelle (ζ-potential $+6.58$ mV). The binding of PEI onto the CH-mag-micelle further increased the ζ-potential to $+17.89$ mV, rendering it a good binder of DNA.

Park et al. developed PEGylated magnetite nanocrystal clusters (PMNCs) cross-linked with branched PEI as a magnetically responsive platform for siRNA delivery [239]. Hydrocaffeic acid (HCA), an analog of dopamine, was utilized to anchor PEI on the magnetic NPs. HCA was conjugated to PEI through amide bond formation with a degree of substitution of 13% corresponding to circa 64 HCA molecules per PEI chain (Figure 10.24). Sonication of an aqueous phase containing PEI-HCA conjugate and an organic phase containing the hydrophobic OA-IONPs generated oil-in-water emulsions. The multiple copies of HCA per PEI rendered PEI-HCA an excellent cross-linker of several IONPs. PEI-HCA/NP ratio of 5:1 led to the formation of spherical nanoclusters, which upon introduction of PEGs through reactions with amines in the PEI backbone produced the PMNCs with hydrodynamic size of 152 nm, ζ-potential of $+30$ mV and good colloidal stability. With the high positive charges on the surface, the PMNCs absorbed siGFP through electrostatic interactions. When tested for siGFP delivery, PMNCs showed superior silencing of GFP expression in the presence of an external magnet.

Majewski et al. introduced cationic functionalities onto the surface of magnetic NPs by grafting tertiary amine-containing thermo- and pH-responsive polymers such as poly(2-(dimethylamino)ethyl methacrylate) (PDMAEMA) [240]. The synthesis of the magnetic core–shell NPs started

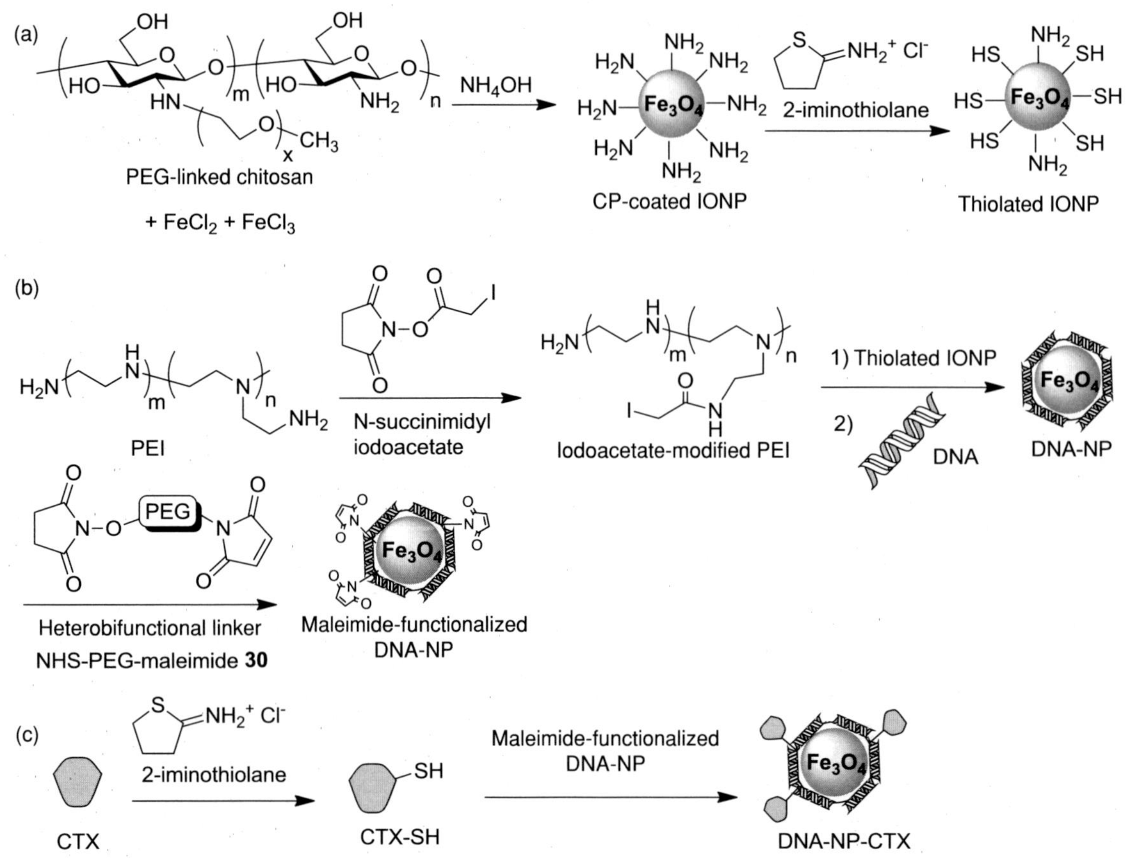

FIGURE 10.23 (a) Synthesis of thiolated IONP, (b) immobilization of DNA onto the NPs and subsequent functionalization with maleimide, (c) functionalization of the DNA–NP with targeting agent CTX.

by immobilizing the dopamine-modified ATRP initiator 2-bromoisobutyryl dopamide **30** (BIBDA) onto the surface of maghemite NPs prepared by the thermal decomposition of $Fe(CO)_5$ (Figure 10.25). DMAEMA **31** was polymerized on the NPs under ATRP conditions thus introducing multiple tertiary amines onto the NPs. The molecular weight of the grafted polymer (Mn = 93,000 g/mol; PDI = 1.22) was determined by hydrolyzing the iron core under acidic condition and analyzing the resulting solution by size exclusion chromatography. The condensation of DNA-encoding EGFP onto the NPs via electrostatic interactions led to better transfection efficiency compared to PEI vector.

In their quest to find a versatile tunable pH-responsive polymeric coating for the controlled loading and release of cationic and anionic payloads [241], Bigall et al. reported a facile method for the preparation of magnetic nanobeads with different isoelectric points by introducing various amino side chains to the polymer coating [242]. They modified poly(maleic anhydride-alt-1-octadecene) (polymer I) with 2-(2-pyridyl)ethylamine to introduce a pyridine moiety (polymer II), 2,2(ethylene dioxy)-bis-(ethylamine) to insert a primary amine group (polymer III), and *N,N*-dimethylethylenediamine to add a ternary amine group (polymer IV) (Figure 10.26). The coprecipitation of polymers II,

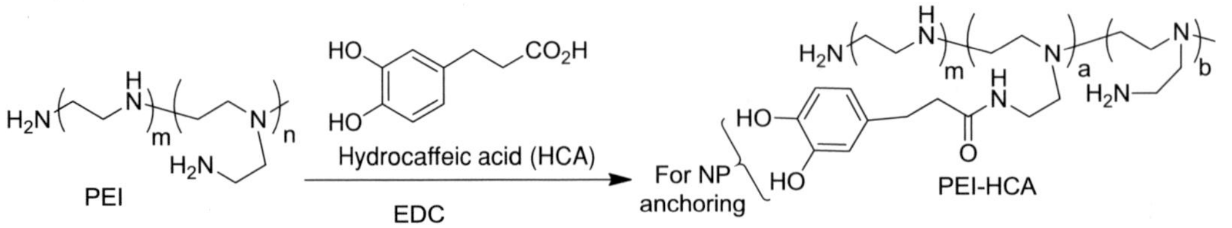

FIGURE 10.24 Synthesis of hydrocaffeic acid-functionalized PEI for IONP coating and siRNA immobilization.

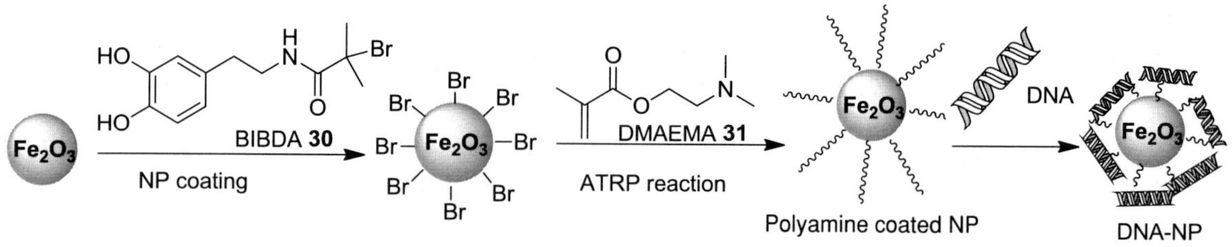

FIGURE 10.25 Grafting of a tertiary amine-containing polymer onto the magnetic NP facilitated the condensation of DNA on the NP.

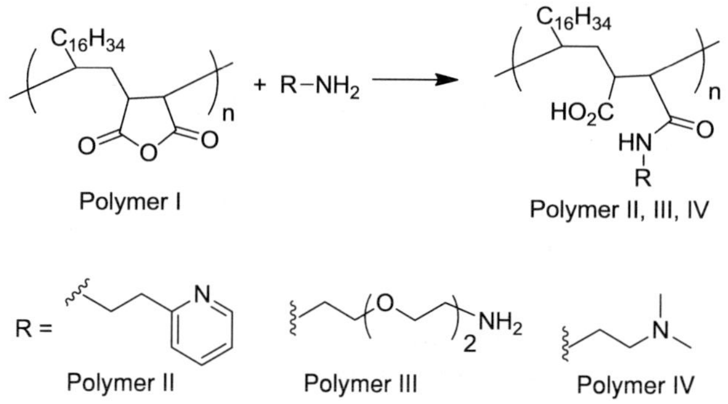

FIGURE 10.26 Polymers prepared with various isoelectric points.

III, or IV with manganese IONPs generated magnetic nanobeads with isoelectric points of 4.8, 6.5, and 8, respectively. For example, the nanobeads from polymer IV inverts its charge from negative to positive at pH less than 8. These nanobeads were suspended in pH 6.5 media, which were incubated with anti-GFP-siRNA to study the loading and release of the cargo in basic environment (pH 9.0). The loading of siRNA onto the positively charged nanobeads was monitored by photoluminescence spectroscopy with saturation occurring at 50 nM. When the pH was increased to 9.0, the nanobeads became negatively charged causing the release of the bound siRNA due to electrostatic repulsion. Such a system can be potentially used to deliver anionic therapeutics to slightly basic environments such as the ileum section of the intestine.

Biswas et al. described a new approach for the generation of positively charged magnetic NPs containing hydrophobic chains to enhance its transfection potential [243] via magnetofection or magnet-assisted gene delivery [244, 245]. *N,N*-Bis-(2-aminooxyethyl)-*N,N*-dimethylammonium iodide **32**

(BADI) was immobilized on the negatively charged bare Fe_3O_4 NPs through hydrogen bonding and electrostatic interactions. With the free aminooxy moieties on the surface, myristaldehyde could be ligated to NPs through chemoselective formation of oxime. On an average, there was one myristaldehyde per BADI moiety, and the resulting magnetic-lipid NPs (MLP) had a ζ-potential of $+40$ mV. Although the transfection efficiency using MLP was very promising, it was highly dependent on the presence of an external magnetic field.

To improve the transfection efficiency, BADI **32** was pretreated with excess myristaldehyde, thus introducing two myristaldehydes onto each BADI (Figure 10.27). The bisoxime derivative of BADI coated the Fe_3O_4 NPs leading to dMLP, which gave superb transfection performance. Not only was dMLP more efficient in DNA transfection assays than MLP, but also the transfection was less dependent on the presence of an external magnet. In addition, dMLP displayed three orders of magnitude of higher transfection efficiency than the widely used transfection reagent Lipofectamine 2000. These results clearly demonstrated that not only the chemical composition of an NP but also the mode of formulation, is important for its potential application.

Another approach for PEI immobilization is through hydrophobic interactions between a lipophilic chain covalently attached to PEI and hydrophobic coating on magnetic NPs. For example, Liu et al. alkylated PEI with iododecane and immobilized the amphiphilic polymer on OA-IONPs through insertion of the hydrophobic chains into the oleic acid coating [246].

Instead of immobilizing genetic materials directly, Chorny et al. reported recently the delivery of adenovirus-encoding GFP gene by magnetically guided gene/cell delivery of magnetic NPs [247]. BSA-coated hydrophobic magnetic NPs were prepared through the coprecipitation method followed

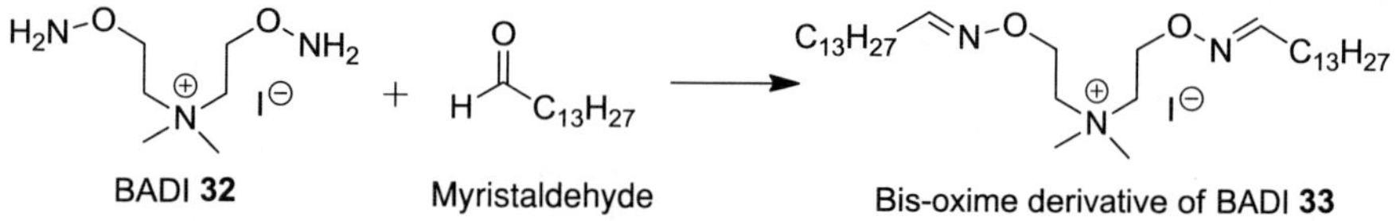

FIGURE 10.27 Oxime formation of BADI **32** with myristaldehyde for coating of negatively charged magnetic NPs.

by surface ligand modification. A poly(allylamine)-derived trifunctional surface-active polymer (PAA-Chol-PDT-SO$_3^-$) was synthesized containing a thiol-reactive pyridyldithio functionality for covalent adaptor protein attachment, a cholesterol moiety to aid in anchoring of the polymer onto NPs through hydrophobic interactions, and a negatively charged sulfonate group for enhanced particle stabilization. NPs were surface modified with PAA-Chol-PDT-SO$_3^-$ using the emulsification-solvent evaporation method. A thiolated human recombinant D1 domain of Coxsackie-adenovirus receptor (CAR) with affinity for adenovirus was introduced to the NP formulation via disulfide bond formation. The CAR-modified NP subsequently bound with GFP-encoding adenovirus for gene delivery.

10.5 CONJUGATION OF MAGNETIC NPs WITH ANTIBODIES

10.5.1 Direct Covalent Connection

With the large diversity of antigen structures and potential high specificity, antibodies present attractive candidates as targeting moieties for biological recognition. Antibodies are consisted of two heavy chains linked with two light chains through disulfide bonds, which are typically shaped like letter Y. Each antibody molecule contains two variable regions at the branches of the Y, which are termed fragment antigen binding (Fab fragment) for antigen binding. The stem of the Y is called fragment constant (Fc fragment), which is constant within one class of antibodies.

As presented earlier in this chapter in Herceptin conjugation, one of the easiest and most straightforward strategies to immobilize antibodies onto magnetic NPs is through the covalent attachment by amide bonds. Carboxylic acid-functionalized magnetic NPs are typically activated with EDC/NHS to form activated NHS esters, which react with

the amine groups on the antibodies to form amide bonds [248–250]. This reaction is mild and can be carried out in aqueous buffers. Besides amide formation, bifunctional linker such as glutaraldehyde can be used to cross-link amine-functionalized magnetic NPs with amines on antibody surface through Schiff base formation [251]. Alternatively, antibodies can be directly coated onto aldehyde-functionalized NPs [252]. As antibodies contain multiple surface amine groups, this direct functionalization method does not require chemical modification of the antibodies. However, one drawback is that amine groups are also present in the Fab fragment. As there is little selectivity toward location of the amines, the random nature of this strategy will inevitably link some NPs through amines in the Fab fragment. This will likely block the antigen recognition site and impede biological recognition.

10.5.2 Antibody Immobilization Through Biotin–Avidin Interactions

Avidin is a glycoprotein found in the egg white and tissues of birds, reptiles, and amphibians, which contains four identical subunits with a combined molecular weight around 67 kDa. Streptavidin is a tetrameric protein isolated from *Streptomyces avidinii* with a mass of 60 kDa. While avidin and streptavidin are loosely related, they both have very high affinities for biotin ($K_d \approx 10^{-15}$M). Combined with the use of specific cross-linkers, magnetic NPs functionalized with Abs can be prepared by taking advantage of biotin–avidin interactions [253, 254]. For example, starting from OA-IONPs, the Zhang group installed iodoacetamide moieties on the surface (NP-SIA). [254] The amine groups on streptavidin were derivatized with *N*-succinimidyl-*S*-acetylthioacetate **34** through amide formation, which upon mild hydroxylamine treatment produced thiolated streptavidin (Figure 10.28). The free thiol groups on the thiolated streptavidin reacted with iodoacetamide on the NP-SIA through nucleophilic displacement thus coating the NPs with streptavidin. A

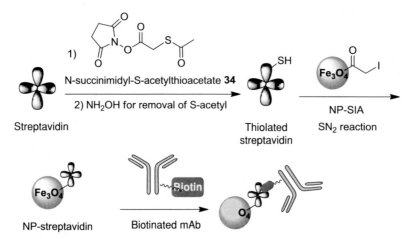

FIGURE 10.28 NP functionalization with streptavidin can be used to immobilize biotinated mAbs.

glypican-binding mAb was biotinylated by the NHS ester of biotin, which bound to the magnetic NPs through streptavidin/biotin interactions.

The biotin–avidin/streptavidin system is attractive as it is widely used in protein purification, immunoassays, and bioconjugation. The avidin/streptavidin-coated NPs can serve as a useful platform for immobilization of many types of biotinylated biomacromolecules. However, one disadvantage is that as the location of biotin functionalization is random, the orientation of antibody immobilized on the NPs cannot be controlled.

10.5.3 Thiol Functionalization of Antibody

Instead of thiolated streptavidin, antibodies can be modified with thiols. One common approach is to treat antibodies or targeting peptides with Traut's reagent, which converts a free amine group on the antibody to a sulfhydryl group [255–257]. Another method is to selectively reduce the disulfide bonds linking the heavy chains of the antibody, producing half-antibody with thiol groups on the constant region of the heavy chains [258]. Magnetic NPs can be functionalized to contain the thiol-reactive maleimide moieties, which can chemoselectively ligate with the free thiols on the antibody [255–257]. Alternatively, antibody (e.g., Herceptin) was derivatized with SPDP, which linked with free thiol-containing magnetic NPs [61].

Although we discuss antibody conjugation in this section, these methods can be applied to immobilization of targeting peptides and proteins in a similar manner. Selected examples of peptide/protein conjugation include direct amide coupling [259, 260] and thiolated peptide/protein with thiol-reactive NPs [235, 261–267].

These aforementioned methods in general, lack the control of location of functionalization on the antibody and the orientation of antibodies on the NPs. To address this problem, several conjugation methods have been developed to reduce the undesirable functionalization of Fab fragments.

10.5.4 Electrostatic Interactions with Antibodies

Instead of directly linking antibody with NPs through amide bonds, de la Fuente and coworkers developed a two-step procedure by first absorbing the antibody onto NPs through electrostatic interactions followed by the formation of covalent amide bond [268]. The rational is that as the charge density on the antibody is highest in the plane involving all two heavy chains and two light chains, the absorption of antibody on charged NPs will be most likely through this plane. Thus, due to its proximity to the NP surface, the amide bond will most likely be formed with residues present in this plane rather than those in the Fab fragment. This hypothesis was tested by partially converting carboxylic acid moieties on magnetic NPs to activated esters through EDC/NHS. The degree of NHS ester formation on MNPs must be optimized to balance between a sufficient number of unmodified carboxylic groups to assure the ionic adsorption of the antibody and enough activated COOH groups for covalent attachment of the antibody once it has been ionically adsorbed. The antibody was incubated with the NPs under slightly acidic pH (pH ∼ 6) to enhance the amount of positive charges, leading to amide bond formation between the NHS ester and amino groups in the vicinity of the antibody area involved during the ionic absorption. The antibodies immobilized through this two-step protocol showed improved capabilities in biological recognition compared to those produced through the one-step direct attachment procedure.

10.5.5 Conversion of the Carbohydrates on the Heavy Chain of Fc Region for NP Immobilization

Many mAbs contain an N-linked glycosylation site in the Fc fragments. The vicinal diols present in carbohydrates can be selectively oxidized to dialdehydes by sodium periodate without affecting other functional groups in the antibodies. A heterobifunctional linker terminated with a hydrazide moiety can be used to conjugate the oxidized antibody with magnetic NPs [269]. As an example, Fe_2O_3 magnetic NPs were prepared by the coprecipitation method followed by nitric acid oxidation [270]. The surface of the NPs was then coated with a layer of gold. An anti-epidermal growth factor receptor mAb was oxidized with sodium periodate, which was incubated with a hydrazide-containing thiol linker **35** (Figure 10.29). The hydrazide can chemoselectively react with the aldehydes on the oxidized antibody through hydrazone formation, thus introducing free thiols onto the antibody. The thiol-bearing antibody can be anchored on the gold-coated magnetic NPs by the strong interactions of gold with the

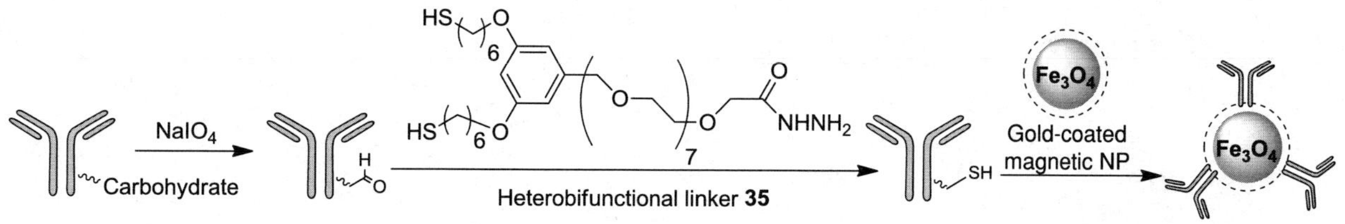

FIGURE 10.29   Oxidation of glycans on antibodies introduced aldehyde functionalities, which can be used to immobilize the antibody onto gold-coated magnetic NPs.

sulfhydryl groups. Through this methodology, antibody is immobilized through the Fc domain. This is advantageous as it orients the antibody with the Fab fragment pointing outward from the NP surface, which can enhance the functional availability of the antibody for targeting.

10.5.6 Using Protein A or Protein G

Protein A or protein G is commonly used for antibody purification, which has high affinity with Fc fragments of antibodies (Protein A can bind with human IgG1, IgG2 antibodies as well as mouse IgG2, while protein G can bind all major Ig classes except IgM). Protein A/G can be attached onto magnetic NPs for antibody immobilization. NPs with carboxyl groups could react with protein A/G mediated by EDC/NHS [122, 271]. For aminated magnetic NPs, heterobifunctional cross-linkers such as SPDP could be utilized to couple with sulfhydryl-functionalized protein G or A. In addition, polyhistidine tag has been used for the conjugation of protein G with magnetic NPs. Chung and coworkers constructed an engineered protein G with a polyhistidine tag in the N-terminal [272]. Nitrilotriacetic acid-nickel (NTA-Ni) functionalized magnetic NPs were synthesized, which bound with protein G through the binding of NTA-Ni with the polyhistidine tag. Antibodies were then successfully immobilized onto the protein G-coated NPs. Alternatively, NTA-Ni-functionalized magnetic NPs can be used to directly link with polyhistidine-tagged antibodies [273].

The diameter of NPs can be an important parameter for biological applications. Protein A contains five homologous domains with molecular weight of 56 kDa, which can significantly increase the size of NPs upon conjugation. Instead of using the whole protein, Prosperi and coworkers engineered the B domain of protein A for antibody immobilization [274]. Iron oxide magnetic NPs were prepared through the thermal decomposition method, which were then modified to become water soluble and bear the thiol-reactive pyridyl disulfide moiety. The engineered domain B of protein A contains a C-terminal polyhistidine tag and retains strong affinities for human IgGs. The polyhistidine tag reacted with the magnetite nanocrystals thus immobilizing protein A onto the magnetic NPs, which successfully captured antibodies such as Herceptin. This strategy displays several advantages: (1) all immobilized antibodies are presented in the same orientation on the NPs with the intact Fab fragments for antigen binding; and (2) the use of a single B domain of protein A may reduce the immunogenicity of the whole hybrid NP.

10.6 CONCLUSION

It is evident that magnetic NP is a powerful platform to deliver biologically active compounds, which range from small organic entities to biomacromolecules. The specific method adapted for NP functionalization will depend on the structure of the ligands, biological applications intended as well as synthetic expertise available. Innovative chemistry has been applied to produce magnetic NPs with well-controlled chemical, physical, and biological properties. To realize the full potential of this interesting class of nanomaterials, a multi-disciplinary approach needs to be continuously pursued to bring together biologists and synthetic chemists and to bridge the gap between basic science and translational applications.

REFERENCES

1. El-Boubbou K, Huang X. *Curr Med Chem* 2011;18:2060–2078.
2. Jun YW, Seo JW, Cheon J. *Acc Chem Res* 2008;41:179–189.
3. Jun YW, Lee JH, Cheon J. *Angew Chem Int Ed* 2008;47:5122–5135.
4. Na HB, Song IC, Hyeon T. *Adv Mater* 2009;21:2133–2148.
5. Shukoor MI, Natalio F, Tahir MN, Ksenofontov V, Therese HA, Theato P, Schroeder HC, Mueller WEG, Tremel W. *Chem Commun* 2007;4677–4679.
6. Gao J, Gu H, Xu B. *Acc Chem Res* 2009;42:1097–1107.
7. Laurent S, Forge D, Port M, Roch A, Robic C, Vander Elst L, Muller RN. *Chem Rev* 2008;108:2064–2110.
8. Kim J, Piao Y, Hyeon T. *Chem Soc Rev* 2009;38:372–390.
9. Lu AH, Salabas EL, Schüeth F. *Angew Chem Int Ed* 2007;46:1222–1244.
10. Jeong Y, Teng X, Wang Y, Yang H, Xia Y. *Adv Mater* 2007;19:33–60.
11. Cheng FY, Su CH, Yang YS, Yeh CS, Tsai CY, Wu CL, Wu MT, Shieh DB. *Biomaterials* 2005;26:729–738.
12. Gupta AK, Wells S. *IEEE Trans Nanobiosci* 2004;3:66–73.
13. Gupta AK, Gupta M. *Biomaterials* 2005;26:3995–4021.
14. Park J, Joo J, Kwon SG, Jang Y, Hyeon T. *Angew Chem Int Ed* 2007;46:4630–4660.
15. Sun S, Zeng H. *J Am Chem Soc* 2002;124:8204–8205.
16. Lee J-H, Huh Y-M, Jun Y-W, Seo J-W, Jang J-T, Song H-T, Kim S, Cho E-J, Yoon H-G, Suh J-S, Cheon J. *Nat Med* 2007;13:95–99.
17. Sun S, Zeng H, Robinson DB, Raoux S, Rice PM, Wang SX, Li G. *J Am Chem Soc* 2004;126:273–279.
18. Puntes VF, Krishan KM, Alivisatos AP. *Science* 2001;291:2115–2117.
19. Sun S, Murray CB, Weller D, Folks L, Moser A. *Science* 2000;287:1989–1992.
20. Faucher L, Tremblay M, Lagueux J, Gossuin Y, Fortin M-A. *ACS Appl Mater Interfaces* 2012;4:4506–4515.
21. Kryza D, Taleb J, Janier M, Marmuse L, Miladi I, Bonazza P, Louis C, Perriat P, Roux S, Tillement O, Billotey C. *Bioconjug Chem* 2011;22:1145–1152.

22. Faure A-C, Dufort S, Josserand V, Perriat P, Coll J-L, Roux S, Tillement O. *Small* 2009;5:2565–2575.

23. Samuel J, Raccurt O, Mancini C, Dujardin C, Amans D, Ledoux G, Poncelet O, Tillement O. *J Nanopart Res* 2011;13:2417–2428.

24. Bridot J-L, Dayde D, Riviere C, Mandon C, Billotey C, Lerondel S, Sabattier R, Cartron G, Le Pape A, Blondiaux G, Janier M, Perriat P, Roux S, Tillement O. *J Mater Chem* 2009;19:2328–2335.

25. Mowat P, Mignot A, Rima W, Lux F, Tillement O, Roulin C, Dutreix M, Bechet D, Huger S, Humbert L, Barberi-Heyob M, Aloy MT, Armandy E, Rodriguez-Lafrasse C, Le Duc G, Roux S, Perriat P. *J Nanosci Nanotechnol* 2011;11:7833–7839.

26. Woo BK, Joly AG, Chen W. *J Lumin* 2010;131:49–53.

27. Ahren M, Selegard L, Klasson A, Soeuderlind F, Abrikossova N, Skoglund C, Bengtsson T, Engstroem M, Kaell P-O, Uvdal K. *Langmuir* 2010;26:5753–5762.

28. Faucher L, Guay-Begin A-A, Lagueux J, Cote M-F, Petitclerc E, Fortin M-A. *Contrast Media Mol Imaging* 2011;6:209–218.

29. Choi HS, Liu W, Misra P, Tanaka E, Zimmer JP, Ipe BI, Bawendi MG, Frangioni JV. *Nat Biotechnol* 2007;25:1165–1170.

30. Wang YX, Hussain SM, Krestin GP. *Eur Radiol* 2001;11:2319–2331.

31. Tartaj P. *Eur J Inorg Chem* 2009;333–343.

32. Xie J, Lee S, Chen X. *Adv Drug Deliver Rev* 2010;62:1064–1079.

33. Thorek DLJ, Chen AK, Czupryna J, Tsourkas A. *Ann Biomed Eng* 2006;34:23–38.

34. Bjørnerud A, Johansson L. *NMR Biomed* 2004;17:465–477.

35. Reimer P, Balzer T. *Eur Radiol* 2003;13:1266–1276.

36. Weissleder R, Elizondo G, Wittenberg J, Rabito CA, Bengele HH, Josephson L. *Radiology* 1990;175:489–493.

37. Ma M, Zhang Y, Yu W, Shen H-y, Zhang H-q, Gu N. *Colloids Surf A* 2003;212:219–226.

38. Kang YS, Risbud S, Rabolt JF, Stroeve P. *Chem Mater* 1996;8:2209–2211.

39. Park J, Lee E, Hwang N-M, Kang M, Kim SC, Hwang Y, Park J-G, Noh H-J, Kim J-Y, Park J-H, Hyeon T. *Angew Chem Int Ed* 2005;44:2872–2877.

40. Hyeon T, Lee SS, Park J, Chung Y, Na HB. *J Am Chem Soc* 2001;123:12798–12801.

41. Massart R. *IEEE Trans Magn* 1981;17:1247–1248.

42. Itoh H, Sugimoto T. *J Colloid Interface Sci* 2003;265:283–295.

43. Palmacci S, Josephson L, Groman EV. Synthesis of Polysaccharide-covered Superparamagnetic Oxide Colloids for Magnetic Resonance Contrast Agents or Other Applications, (Advanced Magnetics, Inc., USA). 1995, U.S. Patent 9505669.

44. Barentsz JO, Tekkis PP. Use of USPIOs for clinical lymph node imaging. In: Bulte JWM, Modo, MMJ (eds),

Nanoparticles in Biomedical Imaging: Emerging Technologies and Applications. New York: Springer; 2008. pp 25–40.

45. Smith EA, Chen W. *Langmuir* 2008;24:12405–12409.

46. Koh I, Wang X, Varughese B, Isaacs L, Ehrman SH, English DS. *J Phys Chem B* 2006;110:1553–1558.

47. Bruce IJ, Sen T. *Langmuir* 2005;21:7029–7035.

48. Yamaura M, Camilo RL, Sampaio LC, Macedo MA, Nakamura M, Toma HE. *J Magn Magn Mater* 2004;279:210–217.

49. LaMer VK, Dinegar RH. *J Am Chem Soc* 1950;72:4847–4854.

50. Murray CB, Kagan CR, Bawendi MG. *Annu Rev Mater Sci* 2000;30:545–610.

51. Frey NA, Peng S, Cheng K, Sun S. *Chem Soc Rev* 2009;38:2532–2542.

52. Kwon SG, Hyeon T. *Small* 2011;7:2685–2702.

53. Herman DAJ, Ferguson P, Cheong S, Hermans IF, Ruck BJ, Allan KM, Prabakar S, Spencer JL, Lendrum CD, Tilley RD. *Chem Commun* 2011;47:9221–9223.

54. Ge J, Hu Y, Biasini M, Dong C, Guo J, Beyermann WP, Yin Y. *Chem Eur J* 2007;13:7153–7161.

55. Park J, An K, Hwang Y, Park J-G, Noh H-J, Kim J-Y, Park J-H, Hwang N-M, Hyeon T. *Nat Mater* 2004;3:891–895.

56. Cheon J, Kang N-J, Lee S-M, Lee J-H, Yoon J-H, Oh SJ. *J Am Chem Soc* 2004;126:1950–1951.

57. Jana NR, Chen Y, Peng X. *Chem Mater* 2004;16:3931–3935.

58. Dong A, Ye X, Chen J, Kang Y, Gordon T, Kikkawa JM, Murray CB. *J Am Chem Soc* 2011;133:998–1006.

59. Yu WW, Chang E, Sayes CM, Drezek R, Colvin VL. *Nanotechnology* 2006;17:4483–4487.

60. Gittins DI, Caruso F. *Angew Chem Int Ed* 2001;40:3001–3004.

61. Jun YW, Huh YM, Choi JS, Lee JH, Song HT, Kim S, Yoon S, Kim KS, Shin JS, Suh JS, Cheon J. *J Am Chem Soc* 2005;127:5732–5733.

62. Lee J-H, Lee K, Moon SH, Lee Y, Park TG, Cheon J. *Angew Chem Int Ed* 2009;48:4174–4179.

63. Li Z, Chen H, Bao H, Gao M. *Chem Mater* 2004;16:1391–1393.

64. Li Z, Sun Q, Gao M. *Angew Chem Int Ed* 2005;44:123–126.

65. Xuan S, Hao L, Jiang W, Gong X, Hu Y, Chen Z. *J Magn Magn Mater* 2007;308:210–213.

66. Zhu H, Yang D, Zhu L, Yang H, Jin D, Yao K. *J Mater Sci* 2007;42:9205–9209.

67. Giri S, Samanta S, Maji S, Ganguli S, Bhaumik A. *J Magn Magn Mater* 2005;285:296–302.

68. Wang H, Geng W, Wang Y. *Res Chem Intermed* 2011;37:389–395.

69. Titirici MM, Antonietti M, Thomas A. *Chem Mater* 2006;18:3808–3812.

70. Mao B, Kang Z, Wang E, Lian S, Gao L, Tian C, Wang C. *Mater Res Bull* 2006;41:2226–2231.

71. Ge S, Shi X, Sun K, Li C, Uher C, Baker JR, Holl MMB, Orr BG. *J Phys Chem C* 2009;113:13593–13599.

72. Gözüak F, Köseoğlu Y, Baykal A, Kavas H. *J Magn Magn Mater* 2009;321:2170–2177.

73. Liu C, Zou B, Rondinone AJ, Zhang ZJ. *J Phys Chem B* 2000;104:1141–1145.

74. Lopez Perez JA, Lopez Quintela MA, Mira J, Rivas J, Charles SW. *J Phys Chem B* 1997;101:8045–8047.

75. Santra S, Tapec R, Theodoropoulou N, Dobson J, Hebard A, Tan W. *Langmuir* 2001;17:2900–2906.

76. Caravan P, Ellison JJ, McMurry TJ, Lauffer RB. *Chem Rev* 1999;99:2293–2352.

77. Runge VM, Ai T, Hao D, Hu X. *Invest Radiol* 2011;46:807–816.

78. Cacheris WP, Quay SC, Rocklage SM. *Magn Reson Imaging* 1990;8:467–481.

79. Park JY, Baek MJ, Choi ES, Woo S, Kim JH, Kim TJ, Jung JC, Chae KS, Chang Y, Lee GH. *ACS Nano* 2009;3:3663–3669.

80. Soderlind F, Pedersen H, Petoral Rodrigo M, Kall P-O, Uvdal K. *J Colloid Interface Sci* 2005;288:140–148.

81. Feldmann C. *Adv Funct Mater* 2003;13:101–107.

82. Fortin M-A, Petoral RM Jr, Soederlind F, Klasson A, Engstroem M, Veres T, Kaell P-O, Uvdal K. *Nanotechnology* 2007;18:395501.

83. Faucher L, Gossuin Y, Hocq A, Fortin M-A. *Nanotechnology* 2011;22:295103.

84. Bridot J-L, Faure A-C, Laurent S, Riviere C, Billotey C, Hiba B, Janier M, Josserand V, Coll J-L, Vander Elst L, Muller R, Roux S, Perriat P, Tillement O. *J Am Chem Soc* 2007;129:5076–5084.

85. Kattel K, Park JY, Xu W, Kim HG, Lee EJ, Bony BA, Heo WC, Lee JJ, Jin S, Baeck JS, Chang Y, Kim TJ, Bae JE, Chae KS, Lee GH. *ACS Appl Mater Interfaces* 2011;3:3325–3334.

86. Zhang W-W, Zhang W-P, Xie P-B, Yin M, Chen H-T, Jing L, Zhang Y-S, Lou L-R, Xia S-D. *J Colloid Interface Sci* 2003;262:588–593.

87. Miyawaki J, Matsumura S, Yuge R, Murakami T, Sato S, Tomida A, Tsuruo T, Ichihashi T, Fujinami T, Irie H, Tsuchida K, Iijima S, Shiba K, Yudasaka M. *ACS Nano* 2009;3:1399–1406.

88. Iijima S, Yudasaka M, Yamada R, Bandow S, Suenaga K, Kokai F, Takahashi K. *Chem Phys Lett* 1999;309:165–170.

89. Ajima K, Murakami T, Mizoguchi Y, Tsuchida K, Ichihashi T, Iijima S, Yudasaka M. *ACS Nano* 2008;2:2057–2064.

90. Hashimoto A, Yorimitsu H, Ajima K, Suenaga K, Isobe H, Miyawaki J, Yudasaka M, Iijima S, Nakamura E. *Proc Natl Acad Sci USA* 2004;101:8527–8530.

91. Ajima K, Yudasaka M, Murakami T, Maigne A, Shiba K, Iijima S. *Mol Pharm* 2005;2:475–480.

92. Dosev D, Nichkova M, Liu M, Guo B, Liu G-y, Hammock BD, Kennedy IM. *J Biomed Opt* 2005;10:064006.

93. Nichkova M, Dosev D, Perron R, Gee SJ, Hammock BD, Kennedy IM. *Anal Bioanal Chem* 2006;384:631–637.

94. Nichkova M, Dosev D, Gee SJ, Hammock BD, Kennedy IM. *Anal Chem* 2005;77:6864–6873.

95. Ju Q, Tu D, Liu Y, Li R, Zhu H, Chen J, Chen Z, Huang M, Chen X. *J Am Chem Soc* 2012;134:1323–1330.

96. Ju Q, Liu Y, Tu D, Zhu H, Li R, Chen X. *Chem Eur J* 2011;17:8549–8554.

97. Yang LW, Zhang YY, Li JJ, Li Y, Zhong JX, Chu PK. *Nanoscale* 2010;2:2805–2810.

98. Hifumi H, Yamaoka S, Tanimoto A, Citterio D, Suzuki K. *J Am Chem Soc* 2006;128:15090–15091.

99. Hifumi H, Yamaoka S, Tanimoto A, Akatsu T, Shindo Y, Honda A, Citterio D, Oka K, Kuribayashi S, Suzuki K. *J Mater Chem* 2009;19:6393–6399.

100. Dumont MF, Baligand C, Li Y, Knowles ES, Meisel MW, Walter GA, Talham DR. *Bioconjugate Chem* 2012;23:951–957.

101. Engstroem M, Klasson A, Pedersen H, Vahlberg C, Kaell P-O, Uvdal K. *Magn Reson Mater* 2006;19:180–186.

102. McDonald Michael A, Watkin Kenneth L. *Acad Radiol* 2006;13:421–427.

103. Helm L. *Future Med Chem* 2010;2:385–396.

104. Dove A. *Nature Biotech* 2001;19:913–917.

105. Varki A. *Glycobiology* 1993;3:97–130.

106. Dwek RA. *Chem Rev* 1996;96:683–720.

107. Koeller KM, Wong C-H. *Nature Biotech* 2000;18:835–841.

108. Horak D, Babic M, Jendelova P, Herynek V, Trchova M, Pientka Z, Pollert E, Hajek M, Sykova E. *Bioconjugate Chem* 2007;18:635–644.

109. Kekkonen V, Lafreniere N, Ebara M, Saito A, Sawa Y, Narain R. *J Magn Magn Mater* 2009;321:1393–1396.

110. Villanueva A, Canete M, Roca AG, Calero M, Veintemillas-Verdaguer S, Serna CJ, Morales MdP, Miranda R. *Nanotechnology* 2009;20:115103.

111. Sun J, Su Y, Rao S, Yang Y. *J Chromatogr, B* 2011;879:2194–2200.

112. Zhang X, Niu H, Pan Y, Shi Y, Cai Y. *Anal Chem* 2010;82:2363–2371.

113. Sui Y, Cui Y, Nie Y, Xia G-M, Sun G-X, Han J-T. *Colloids Surf, B* 2012;93:24–28.

114. Lee Y, Lee H, Kim YB, Kim J, Hyeon T, Park H, Messersmith PB, Park TG. *Adv Mater* 2008;20:4154–4157.

115. Babiuch K, Wyrwa R, Wagner K, Seemann T, Hoeppener S, Becer CR, Linke R, Gottschaldt M, Weisser J, Schnabelrauch M, Schubert US. *Biomacromolecules* 2011;12:681–691.

116. El-Dakdouki MH, El-Boubbou K, Zhu DC, Huang X. *RSC Adv* 2011;1:1449–1452.

117. Lartigue L, Oumzil K, Guari Y, Larionova J, Guerin C, Montero J-L, Barragan-Montero V, Sangregorio C, Caneschi A, Innocenti C, Kalaivani T, Arosio P, Lascialfari A. *Org Lett* 2009;11:2992–2995.

118. Lartigue L, Innocenti C, Kalaivani T, Awwad A, Sanchez Duque MdM, Guari Y, Larionova J, Guerin C, Montero J-LG, Barragan-Montero V, Arosio P, Lascialfari A, Gatteschi D, Sangregorio C. *J Am Chem Soc* 2011;133:10459–10472.

119. Barrientos AG, de la Fuente JM, Rojas TC, Fernandez A, Penades S. *Chem Eur J* 2003;9:1909–1921.

120. De la Fuente JM, Alcantara D, Eaton P, Crespo P, Rojas TC, Fernandez A, Hernando A, Penades S. *J Phys Chem B* 2006;110:13021–13028.

121. Gallo J, Garcia I, Padro D, Arnaiz B, Penades S. *J Mater Chem* 2010;20:10010–10020.

122. Garcia I, Gallo J, Genicio N, Padro D, Penades S. *Bioconjugate Chem* 2011;22:264–273.

123. El-Boubbou K, Gruden C, Huang X. *J Am Chem Soc* 2007;129:13392–13393.

124. El-Boubbou K, Zhu DC, Vasileiou C, Borhan B, Prosperi D, Li W, Huang X. *J Am Chem Soc* 2010;132:4490–4499.

125. Liu H-Z, Tang J-J, Ma X-X, Guo L, Xie J-W, Wang Y-X. *Anal Sci* 2011;27:19–24.

126. Earhart C, Jana NR, Erathodiyil N, Ying JY. *Langmuir* 2008;24:6215–6219.

127. Zhou L, Wu J, Zhang H, Kang Y, Guo J, Zhang C, Yuan J, Xing X. *J Mater Chem* 2012;22:6813–6818.

128. Kamat M, El-Boubbou K, Zhu DC, Lansdell T, Lu X, Li W, Huang X. *Bioconjugate Chem* 2010;21:2128–2135.

129. El-Dakdouki MH, Zhu DC, El-Boubbou K, Kamat M, Chen J, Li W, Huang X. *Biomacromolecules* 2012;13:1144–1151.

130. van Kasteren SI, Campbell SJ, Serres S, Anthony DC, Sibson NR, Davis BG. *Proc Natl Acad Sci USA* 2009;106:18–23.

131. Valero E, Tambalo S, Marzola P, Ortega-Munoz M, Lopez-Jaramillo FJ, Santoyo-Gonzalez F, de Dios Lopez J, Delgado JJ, Calvino JJ, Cuesta R, Dominguez-Vera JM, Galvez N. *J Am Chem Soc* 2011;133:4889–4895.

132. Lin P-C, Ueng S-H, Yu S-C, Jan M-D, Adak AK, Yu C-C, Lin C-C. *Org Lett* 2007;9:2131–2134.

133. Polito L, Monti D, Caneva E, Delnevo E, Russo G, Prosperi D. *Chem Commun* 2008;621–623.

134. Xiong Z, Zhao L, Wang F, Zhu J, Qin H, Wu Ra, Zhang W, Zou H. *Chem Commun* 2012;48:8138–8140.

135. Rodionov VO, Fokin VV, Finn MG. *Angew Chem Int Ed* 2005;44:2210–2215.

136. Banerjee SS, Chen D-H. *Chem Mater* 2007;19:3667–3672.

137. Liang Y-Y, Zhang L-M. *Biomacromolecules* 2007;8:1480–1486.

138. Badruddoza AZM, Tay ASH, Tan PY, Hidajat K, Uddin MS. *J Hazard Mater* 2011;185:1177–1186.

139. Li H, El-Dakdouki MH, Zhu DC, Abela GS, Huang X. *Chem Commun* 2012;48:3385–3387.

140. Fuhrer R, Herrmann IK, Athanassiou EK, Grass RN, Stark WJ. *Langmuir* 2011;27:1924–1929.

141. Moros M, Hernaez B, Garet E, Dias JT, Saez B, Grazu V, Gonzalez-Fernandez A, Alonso C, de la Fuente JM. *ACS Nano* 2012;6:1565–1577.

142. Moros M, Pelaz B, Lopez-Larrubia P, Garcia-Martin ML, Grazu V, de la Fuente JM. *Nanoscale* 2010;2:1746–1755.

143. Badruddoza AZM, Hidajat K, Uddin MS. *J Colloid Interface Sci* 2010;346:337–346.

144. Liu X, Hu Q, Fang Z, Zhang X, Zhang B. *Langmuir* 2009;25:3–8.

145. Samra ZQ, Athar MA. *Biotechnol Bioprocess Eng* 2009;14:651–661.

146. Wang X, Liu L-H, Ramstrom O, Yan M. *Exp Biol Med* 2009;234:1128–1139.

147. Liu L-H, Dietsch H, Schurtenberger P, Yan M. *Bioconjugate Chem* 2009;20:1349–1355.

148. Pfaff A, Schallon A, Ruhland TM, Majewski AP, Schmalz H, Freitag R, Muller AHE. *Biomacromolecules* 2011;12:3805–3811.

149. Goldstein LS, Dewhirst MW, Repacholi M, Kheifets L. *Int J Hyperthermia* 2003;19:373–384.

150. Wust P, Hildebrandt B, Sreenivasa G, Rau B, Gellermann J, Riess H, Felix R, Schlag PM. *Lancet Oncol* 2002;3:487–497.

151. Hoare T, Timko BP, Santamaria J, Goya GF, Irusta S, Lau S, Stefanescu CF, Lin D, Langer R, Kohane DS. *Nano Lett* 2011;11:1395–1400.

152. Thomas CR, Ferris DP, Lee J-H, Choi E, Cho MH, Kim ES, Stoddart JF, Shin J-S, Cheon J, Zink JI. *J Am Chem Soc* 2010;132:10623–10625.

153. Park H, Yang J, Lee J, Haam S, Choi I-H, Yoo K-H. *ACS Nano* 2009;3:2919–2926.

154. Park J-H, von Maltzahn G, Ong LL, Centrone A, Hatton TA, Ruoslahti E, Bhatia SN, Sailor MJ. *Adv Mater* 2010;22:880–885.

155. Ren J, Shen S, Pang Z, Lu X, Deng C, Jiang X. *Chem Commun* 2011;47:11692–11694.

156. Choi Won I, Sahu A, Kim Young H, Tae G. *Ann Biomed Eng* 2012;40:534–546.

157. Choi J, Yang J, Jang E, Suh J-S, Huh Y-M, Lee K, Haam S. *Anti-Cancer Agents Med Chem* 2011;11:953–964.

158. Huang X, El-Sayed IH, Qian W, El-Sayed MA. *J Am Chem Soc* 2006;128:2115–2120.

159. Yuan H, Fales AM, Vo-Dinh T. *J Am Chem Soc* 2012;134:11358–11361.

160. Wang C, Chen J, Talavage T, Irudayaraj J. *Angew Chem Int Ed* 2009;48:2759–2763.

161. Yu WW, Falkner JC, Yavuz CT, Colvin VL. *Chem Commun* 2004;2306–2307.

162. Kim J, Park S, Lee JE, Jin SM, Lee JH, Lee IS, Yang I, Kim J-S, Kim SK, Cho M-H, Hyeon T. *Angew Chem Int Ed* 2006;45:7754–7758.

163. Lee JH, Sherlock SP, Terashima M, Kosuge H, Suzuki Y, Goodwin A, Robinson J, Seo WS, Liu Z, Luong R, McConnell MV, Nishimura DG, Dai H. *Magn Res Med* 2009;62:1497–1509.

164. Sherlock SP, Tabakman SM, Xie L, Dai H. *ACS Nano* 2011;5:1505–1512.

165. Seo WS, Lee JH, Sun X, Suzuki Y, Mann D, Liu Z, Terashima M, Yang PC, McConnell MV, Nishimura DG, Dai H. *Nat Mater* 2006;5:971–976.

166. Lee JH, Sherlock Sarah P, Terashima M, Kosuge H, Suzuki Y, Goodwin A, Robinson J, Seo Won S, Liu Z, Luong R, McConnell Michael V, Nishimura Dwight G, Dai H. *Magn Reson Med* 2009;62:1497–1509.

167. Celli JP, Spring BQ, Rizvi I, Evans CL, Samkoe KS, Verma S, Pogue BW, Hasan T. *Chem Rev* 2010;110:2795–2838.

168. Bugaj AM. Targeted photodynamic therapy - a promising strategy of tumor treatment. *Photochem Photobiol Sci* 2011;10:1097–1109.

169. Paszko E, Ehrhardt C, Senge MO, Kelleher DP, Reynolds JV. *Photodiagn Photodyn Ther* 2011;8:14–29.

170. Donnelly RF, McCarron PA, Morrow DIJ, Sibani SA, Woolfson AD. *Expert Opin Drug Delivery* 2008;5:757–766.

171. Sibani SA, McCarron PA, Woolfson AD, Donnelly RF. *Expert Opin Drug Delivery* 2008;5:1241–1254.

172. Huang P, Li Z, Lin J, Yang D, Gao G, Xu C, Bao L, Zhang C, Wang K, Song H, Hu H, Cui D. *Biomaterials* 2011;32:3447–3458.

173. Jiang W, Lionberger R, Yu LX. *Bioanalysis* 2011;3:333–344.

174. Cui J, Li C, Guo W, Li Y, Wang C, Zhang L, Zhang L, Hao Y, Wang Y. *J Controlled Release* 2007;118:204–215.

175. Gullaiya S, Swamy AHMV. *Int J Pharm Technol* 2010;2:474–496.

176. Chatterjee K, Zhang J, Honbo N, Karliner JS. *Cardiology* 2010;115:155–162.

177. Kim J, Kim HS, Lee N, Kim T, Kim H, Yu T, Song IC, Moon WK, Hyeon T. *Angew Chem Int Ed* 2008;47:8438–8441.

178. Jain TK, Morales MA, Sahoo SK, Leslie-Pelecky DL, Labhasetwar V. *Mol Pharm* 2005;2:194–205.

179. Yang J, Lee C-H, Ko H-J, Suh J-S, Yoon H-G, Lee K, Huh Y-M, Haam S. *Angew Chem Int Ed* 2007;46:8836–8839.

180. Chen F-H, Zhang L-M, Chen Q-T, Zhang Y, Zhang Z-J. *Chem Commun* 2010;46:8633–8635.

181. Yang X, Grailer JJ, Rowland IJ, Javadi A, Hurley SA, Matson VZ, Steeber DA, Gong S. *ACS Nano* 2010;4:6805–6817.

182. Chang Y-L, Meng X-L, Zhao Y-L, Li K, Zhao B, Zhu M, Li Y-P, Chen X-S, Wang J-Y. *J Colloid Interface Sci* 2011;363:403–409.

183. Yang X, Hong H, Grailer JJ, Rowland IJ, Javadi A, Hurley SA, Xiao Y, Yang Y, Zhang Y, Nickles RJ, Cai W, Steeber DA, Gong S. *Biomaterials* 2011;32:4151–4160.

184. Fang C, Kievit Forrest M, Veiseh O, Stephen Zachary R, Wang T, Lee D, Ellenbogen Richard G, Zhang M. *J Control Release* 2012;162:233–241.

185. Liu H-L, Hua M-Y, Yang H-W, Huang C-Y, Chu P-C, Wu J-S, Tseng IC, Wang J-J, Yen T-C, Chen P-Y, Wei K-C. *Proc Natl Acad Sci USA* 2010;107:15205–15210.

186. Chen H-C, Hua M-Y, Liu Y-C, Yang H-W, Tsai R-Y. *J Mater Chem* 2012;22:13252–13259.

187. Yu MK, Jeong YY, Park J, Park S, Kim JW, Min JJ, Kim K, Jon S. *Angew Chem Int Ed* 2008;47:5362–5365.

188. Lee H, Yu MK, Park S, Moon S, Min JJ, Jeong YY, Kang H-W, Jon S. *J Am Chem Soc* 2007;129:12739–12745.

189. Yang X, Grailer JJ, Rowland IJ, Javadi A, Hurley SA, Steeber DA, Gong S. *Biomaterials* 2010;31:9065–9073.

190. Rapoport N, Gao Z, Kennedy A. *J Natl Cancer Inst* 2007;99:1095–1106.

191. Lim E-K, Huh Y-M, Yang J, Lee K, Suh J-S, Haam S. *Adv Mater* 2011;23:2436–2442.

192. Zhang F, Braun GB, Pallaoro A, Zhang Y, Shi Y, Cui D, Moskovits M, Zhao D, Stucky GD. *Nano Lett* 2012;12:61–67.

193. Lee JE, Lee N, Kim H, Kim J, Choi SH, Kim JH, Kim T, Song IC, Park SP, Moon WK, Hyeon T. *J Am Chem Soc* 2010;132:552–557.

194. Kong SD, Zhang W, Lee JH, Brammer K, Lal R, Karin M, Jin S. *Nano Lett* 2010;10:5088–5092.

195. Mikhaylov G, Mikac U, Magaeva AA, Itin VI, Naiden EP, Psakhye I, Babes L, Reinheckel T, Peters C, Zeiser R, Bogyo M, Turk V, Psakhye SG, Turk B, Vasiljeva O. *Nat Nanotechnol* 2011;6:594–602.

196. Corem-Salkmon E, Ram Z, Daniels D, Perlstein B, Last D, Salomon S, Tamar G, Shneor R, Guez D, Margel S, Mardor Y. *Int J Nanomed* 2011;6:1595–1602.

197. Mejias R, Perez-Yaguee S, Gutierrez L, Cabrera LI, Spada R, Acedo P, Serna CJ, Lazaro FJ, Villanueva A, Morales MdP, Barber DF. *Biomaterials* 2011;32:2938–2952.

198. Wang C, Zhang H, Chen Y, Shi F, Chen B. *Int J Nanomed* 2012;7:781–787.

199. Subbiahdoss G, Sharifi S, Grijpma DW, Laurent S, van der Mei HC, Mahmoudi M, Busscher HJ. *Acta Biomater* 2012;8:2047–2055.

200. Liong M, Lu J, Kovochich M, Xia T, Ruehm SG, Nel AE, Tamanoi F, Zink JI. *ACS Nano* 2008;2:889–896.

201. Cinti C, Taranta M, Naldi I, Grimaldi S. *PLoS One* 2011;6:e17132.

202. Hua M-Y, Liu H-L, Yang H-W, Chen P-Y, Tsai R-Y, Huang C-Y, Tseng IC, Lyu L-A, Ma C-C, Tang H-J, Yen T-C, Wei K-C. *Biomaterials* 2011;32:516–527.

203. Vermisoglou EC, Pilatos G, Romanos GE, Devlin E, Kanellopoulos NK, Karanikolos GN. *Nanotechnology* 2011;22:355602.

204. Aggarwal BB, Kumar A, Bharti AC. *Anticancer Res* 2003;23:363–398.

205. Yallapu MM, Jaggi M, Chauhan SC. *Drug Discovery Today* 2011;17:71–80.

206. Padhye S, Chavan D, Pandey S, Deshpande J, Swamy KV, Sarkar FH. *Mini-Rev Med Chem* 2010;10:372–387.

207. Manju S, Sreenivasan K. *Langmuir* 2011;27:14489–14496.

208. Manju S, Sreenivasan K. *J Colloid Interface Sci* 2011;359:318–325.

209. Manju S, Sreenivasan K. *J Pharm Sci* 2011;100:504–511.

210. El-Dakdouki MH, Erhardt PW. *Pure Appl Chem* 2012;84:1479–1542.

211. Luo B, Xu S, Luo A, Wang W-R, Wang S-L, Guo J, Lin Y, Zhao D-Y, Wang C-C. *ACS Nano* 2011;5:1428–1435.

212. Ernsting MJ, Foltz WD, Undzys E, Tagami T, Li S-D. *Biomaterials* 2012;33:3931–3941.

213. Liu Q, Zhang J, Sun W, Xie QR, Xia W, Gu H. *Int J Nanomed* 2012;7:999–1013.

214. Chorny M, Fishbein I, Yellen BB, Alferiev IS, Bakay M, Ganta S, Adamo R, Amiji M, Friedman G, Levy RJ. *Proc Natl Acad Sci USA* 2010;107:8346–8351.

215. Hwu JR, Lin YS, Josephrajan T, Hsu M-H, Cheng F-Y, Yeh C-S, Su W-C, Shieh D-B. *J Am Chem Soc* 2009;131:66–68.

216. Georgelin T, Bombard S, Siaugue J-M, Cabuil V. *Angew Chem Int Ed* 2010;49:8897–8901.

217. Chen J, Stubbe J. *Nat Rev Cancer* 2005;5:102–112.

218. Rosenberg B, VanCamp L, Trosko JE, Mansour VH. *Nature* 1969;222:385–386.

219. Stordal B, Pavlakis N, Davey R. *Cancer Treat Rev* 2007;33:688–703.

220. Sonoda A, Nitta N, Nitta-Seko A, Ohta S, Takamatsu S, Ikehata Y, Nagano I, Jo J-i, Tabata Y, Takahashi M, Matsui O, Murata K. *Int J Nanomed* 2010;5:499–504.

221. Osborne CK. *N Engl J Med* 1998;339:1609–1618.

222. Shiau AK, Barstad D, Loria PM, Cheng L, Kushner PJ, Agard DA, Greene GL. *Cell* 1998;95:927–937.

223. de Lima GR, Facina G, Shida JY, Chein MBC, Tanaka P, Dardes RC, Jordan VC, Gebrim LH. *Eur J Cancer* 2003;39:891–898.

224. Hu FX, Neoh KG, Kang ET. *Biomaterials* 2006;27:5725–5733.

225. Mahmoudi M, Serpooshan V. *ACS Nano* 2012;6:2656–2664.

226. Luo D, Saltzman WM. *Nat Biotechnol* 2000;18:33–37.

227. McBain SC, Yiu HHP, Dobson J. *Int J Nanomed* 2008;3:169–180.

228. Sokolova V, Epple M. *Angew Chem Int Ed* 2008;47:1382–1395.

229. Plank C, Zelphati O, Mykhaylyk O. *Adv Drug Delivery Rev* 2011;63:1300–1331.

230. Ruiz-Hernandez E, Baeza A, Vallet-Regi M. *ACS Nano* 2011;5:1259–1266.

231. Choi J-s, Park JC, Nah H, Woo S, Oh J, Kim KM, Cheon GJ, Chang Y, Yoo J, Cheon J. *Angew Chem Int Ed* 2008;47:6259–6262.

232. Stoeva SI, Huo F, Lee J-S, Mirkin CA. *J Am Chem Soc* 2005;127:15362–15363.

233. Iorio MV, Croce CM. *J Clin Oncol* 2009;27:5848–5856.

234. Kim JK, Choi K-J, Lee M, Jo M-h, Kim S. *Biomaterials* 2012;33:207–217.

235. Kievit FM, Veiseh O, Fang C, Bhattarai N, Lee D, Ellenbogen RG, Zhang M. *ACS Nano* 2010;4:4587–4594.

236. Kievit FM, Veiseh O, Bhattarai N, Fang C, Gunn JW, Lee D, Ellenbogen RG, Olson JM, Zhang M. *Adv Funct Mater* 2009;19:2244–2251.

237. Wang C, Ravi S, Martinez GV, Chinnasamy V, Raulji P, Howell M, Davis Y, Seehra MS, Mohapatra S. *J Controlled Release* 2012;163:82–92.

238. Lucke A, Tessmar J, Schnell E, Schmeer G, Gopferich A. *Biomaterials* 2000;21:2361–2370.

239. Park JW, Bae KH, Kim C, Park TG. *Biomacromolecules* 2011;12:457–465.

240. Majewski AP, Schallon A, Jerome V, Freitag R, Mueller AHE, Schmalz H. *Biomacromolecules* 2012;13:857–866.

241. Di Corato R, Bigall NC, Ragusa A, Dorfs D, Genovese A, Marotta R, Manna L, Pellegrino T. *ACS Nano* 2011;5:1109–1121.

242. Bigall NC, Curcio A, Leal MP, Falqui A, Palumberi D, Di Corato R, Albanesi E, Cingolani R, Pellegrino T. *Adv Mater* 2011;23:5645–5650.

243. Hecker JG, Berger GO, Scarfo KA, Zou S, Nantz MH. *ChemMedChem* 2008;3:1356–1361.

244. Scherer F, Anton M, Schillinger U, Henke J, Bergemann C, Kruger A, Gansbacher B, Plank C. *Gene Ther* 2002;9:102–109.

245. Biswas S, Gordon LE, Clark GJ, Nantz MH. *Biomaterials* 2011;32:2683–2688.

246. Liu G, Xie J, Zhang F, Wang Z, Luo K, Zhu L, Quan Q, Niu G, Lee S, Ai H, Chen X. *Small* 2011;7:2742–2749.

247. Chorny M, Alferiev IS, Fishbein I, Tengood JE, Folchman-Wagner Z, Forbes SP, Levy RJ. *Pharm Res* 2012;29:1232–1241.

248. Cruz LJ, Tacken PJ, Bonetto F, Buschow SI, Croes HJ, Wijers M, de Vries IJ, Figdor CG. *Mol Pharm* 2011;8:520–531.

249. Hadjipanayis CG, Machaidze R, Kaluzova M, Wang L, Schuette AJ, Chen H, Wu X, Mao H. *Cancer Res* 2010;70:6303–6312.

250. Cheng K, Peng S, Xu C, Sun S. *J Am Chem Soc* 2009;131:10637–10644.

251. Yang H-M, Park CW, Woo M-A, Kim MI, Jo YM, Park HG, Kim J-D. *Biomacromolecules* 2010;11:2866–2872.

252. Toma A, Otsuji E, Kuriu Y, Okamoto K, Ichikawa D, Hagiwara A, Ito H, Nishimura T, Yamagishi H. *Br J Cancer* 2005;93:131–136.

253. Artemov D, Mori N, Okollie B, Bhujwalla ZM. *Magn Res Med* 2003;49:403–408.

254. Park JO, Stephen Z, Sun C, Veiseh O, Kievit FM, Fang C, Leung M, Mok H, Zhang M. *Mol Imaging* 2011;10:16.

255. Steinhauser I, Spänkuch B, Strebhardt K, Langer K. *Biomaterials* 2006;27:4975–4983.

256. Kievit FM, Stephen ZR, Veiseh O, Arami H, Wang T, Lai VP, Park JO, Ellenbogen RG, Disis ML, Zhang M. *ACS Nano* 2012;6:2591–2601.

257. Koyama T, Shimura M, Minemoto Y, Nohara S, Shibata S, Iida Y, Iwashita S, Hasegawa M, Kurabayashi T, Hamada H, Kono K, Honda E, Aoki I, Ishizaka Y. *J Controlled Release* 2012;159:413–418.

258. Hu C-MJ, Kaushal S, Cao HST, Aryal S, Sartor M, Esener S, Bouvet M, Zhang L. *Mol Pharm* 2010;7:914–920.

259. Zhang C, Jugold M, Woenne EC, Lammers T, Morgenstern B, Mueller MM, Zentgraf H, Bock M, Eisenhut M, Semmler W, Kiessling F. *Cancer Res* 2007;67:1555–1562.

260. Qiao R, Jia Q, Hüwel S, Xia R, Liu T, Gao F, Galla H-J, Gao M. *ACS Nano* 2012;6:3304–3310.

261. Reddy GR, Bhojani MS, McConville P, Moody J, Moffat BA, Hall DE, Kim G, Koo Y-EL, Woolliscroft MJ, Sugai JV,

Johnson TD, Philbert MA, Kopelman R, Rehemtulla A, Ross BD. *Clinical Cancer Res* 2006;12:6677–6686.

262. Zhang Y, Yang M, Park J-H, Singelyn J, Ma H, Sailor MJ, Ruoslahti E, Ozkan M, Ozkan C. *Small* 2009;5:1990–1996.

263. Chen K, Xie J, Xu H, Behera D, Michalski MH, Biswal S, Wang A, Chen X. *Biomaterials* 2009;30:6912–6919.

264. Veiseh O, Kievit FM, Fang C, Mu N, Jana S, Leung MC, Mok H, Ellenbogen RG, Park JO, Zhang M. *Biomaterials* 2010;31:8032–8042.

265. Sun C, Veiseh O, Gunn J, Fang C, Hansen S, Lee D, Sze R, Ellenbogen RG, Olson J, Zhang M. *Small* 2008;4:372–379.

266. Veiseh O, Sun C, Fang C, Bhattarai N, Gunn J, Kievit F, Du K, Pullar B, Lee D, Ellenbogen RG, Olson J, Zhang M. *Cancer Res* 2009;69:6200–6207.

267. Högemann-Savellano D, Bosy E, Blondet C, Sato F, Abe T, Josephson L, Weissleder R, Gaudet J, Sgroi D, Petersy PJ, Basilion JP. *Neoplasia* 2003;5:495–506.

268. Puertas S, Batalla P, Moros Ma, Polo E, del Pino P, Guisán JM, Grazú V, de la Fuente JsM. *ACS Nano* 2011;5:4521–4528.

269. Kumar S, Aaron J, Sokolov K. *Nat Protocols* 2008;3:314–320.

270. Larson TA, Bankson J, Aaron J, Sokolov K. *Nanotechnology* 2007;18:325101.

271. Kaittanis C, Santra S, Perez JM. *J Am Chem Soc* 2009;131:12780–12791.

272. Lim YT, Cho MY, Lee JM, Chung SJ, Chung BH. *Biomaterials* 2009;30:1197–1204.

273. Yang L, Mao H, Wang YA, Cao Z, Peng X, Wang X, Duan H, Ni C, Yuan Q, Adams G, Smith MQ, Wood WC, Gao X, Nie S. *Small* 2009;5:235–243.

274. Mazzucchelli S, Colombo M, De Palma C, Salvadè A, Verderio P, Coghi MD, Clementi E, Tortora P, Corsi F, Prosperi D. *ACS Nano* 2010;4:5693–5702.

11

QUANTUM DOTS BIOCONJUGATES

Xiaoze Jiang[1], Meifang Zhu[1], and Ravin Narain[2]

[1]*State Key Laboratory for Modification of Chemical Fibers and Polymer Materials, College of Material Science and Engineering, Donghua University, Shanghai, PR China*

[2]*Department of Chemical and Materials Engineering, Alberta Glycomics Centre, University of Alberta, Edmonton, AB, Canada*

11.1 INTRODUCTION

Semiconductor nanocrystals (quantum dots, QDs), especially luminescent QDs, have received tremendous attention in the past decades [1–10]. The motivation to use these QDs has been quite diverse due to their unique properties such as their remarkable photochemical stability, size-dependent broad absorption, high extinction coefficients, size-tunable narrow emission, and high fluorescence quantum yield. Such features render the QDs as very attractive candidates for use, in particular, in biological assays, biolabeling, and imaging, but the inherent problem arising from their hydrophobic property and toxicity needs to be resolved by carrying out surface functionalization to enhance their solubility under physiological conditions and reduce their toxicity. Therefore, libraries of methods have been explored to address the water solubility and stability of QDs, while maintaining their excellent features. This chapter mainly discusses the preparation process of water-soluble QDs via various methods and briefly mentions their bioconjugation.

11.2 SYNTHESIS OF WATER-SOLUBLE QDs

To realize the real applications in biology, the QDs have been designed to be water stable and soluble over a broad range of biological conditions, while maintaining their fluorescence quantum yield and the sizes the solubilized QDs have obviously remained unchanged. Several strategies have been reported in the preparation of water-soluble QDs, and the solubilization could be summarized to three main approaches: ligand exchange, ligand encapsulation, and direct encapsulation [3, 5, 9]. Through these three methods, water-soluble QDs have been achieved and reported to be highly stable with acceptable quantum yield and have been further modified to extend their applications in cellular uptake and imaging.

11.2.1 Ligand Exchange

This method has been applied to prepare water-soluble QDs mainly by two steps, the highly luminescent nanocrystals were first prepared by the presence of hydrophobic organic surfactants such as trioctylphosphine/trioctylphosphine oxide (TOP/TOPO) via organometallic synthetic route, and the surfactants were then exchanged by water-soluble ligands bearing stronger binding ability to QDs to solubilize and stabilize the hydrophobic inorganic core in physiological conditions. After the replacement of ligands, the obtained QDs possess the required water solubility and long-term stability with high quantum yield.

11.2.1.1 Preparation of TOP/TOPO QDs In the preliminary works, the high quality nearly monodisperse CdE (E = S, Se, Te) nanometer size crystallites, in particular CdSe, have been prepared via a relatively simple synthetic route [11, 12]. Briefly, the organometallic reagents in TOP solution were rapidly injected into the hot coordinating solvent TOPO to produce a temporally discrete homogeneous

Chemistry of Bioconjugates: Synthesis, Characterization, and Biomedical Applications, First Edition. Edited by Ravin Narain.
© 2014 John Wiley & Sons, Inc. Published 2014 by John Wiley & Sons, Inc.

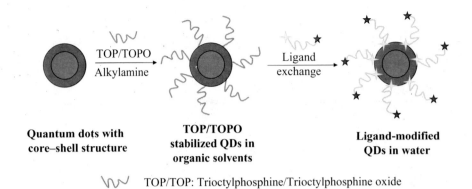

Quantum dots with **TOP/TOPO** Ligand-modified
core–shell structure stabilized QDs in QDs in water
organic solvents

TOP/TOP: Trioctylphosphine/Trioctylphosphine oxide

★ Functional ligands

SCHEME 11.1 The preparation of water-soluble QDs by two steps via ligand exchange method.

nucleation. The size and shape of crystallites were adjusted by the slow growth and annealing process in the coordinating solvent, and nearly monodisperse nanocrystallites could be obtained in powder state via size-selective precipitation and can be redispersed in various organic solvents.

The prepared nanocrystals by the TOPO/TOP route and size separated after synthesis meet most requirements for high quantity QDs except low luminescence quantum yield (around 5–15%) [13]. The low quantum efficiency was addressed by growing hetero-epitaxially an inorganic shell of the wide-band gap semiconductor such as ZnS around the crystals [13, 14]. The quantum fluorescent yield of core–shell QDs was improved up to ~50%. However, the whole process required two steps and was limited by the production scale and irreproducibility of the growth dynamics, and the shape of the obtained nanocrystals conditioned by an uncertain composition of the coordinating solvent.

To improve the above conventional organometallic TOPO/TOP synthesis, recent developments of the organometallic synthetic routes to II–VI semiconductor nanocrystals have focused on the three-component mixtures as the capping agents by introducing an additional coordinating component such as alkylamine or hexylphosphonic acid to the TOPO/TOP solution, thus stable and reproducible luminescence quantum efficiencies (QEs) of 60% could be prepared easily in one step [15–19].

11.2.1.2 Preparation of Water-soluble QDs via Ligand Exchange
The TOP/TOPO-stabilized QDs with high quantum yield could be dispersed in organic solvents such as pyridine or hexane and stored for long time, but they were insoluble in water and are unsuitable for biological systems. Therefore, the surface surfactants have to be replaced with water-soluble ligands to prevent the particle aggregation of QDs by the steric hindrance effect and confer their water dispersion ability. The choice of ligands was crucial for the preparation of water-stable QDs and the used ligands were reported ranging from water-soluble small organic molecules to macromolecules such as polymers,

proteins, peptides, or DNA [2, 5]. Two basic requirements need to be considered in the reported studies for an ideal ligand, which should contain both functional binding sites and hydrophilic groups or moieties. The strong binding ability with the core of QDs was essential to replace the original TOP/TOPO and alkylamine surfactants, and hence the hydrophilic shell on the surface of the QDs would provide the water solubility. After replacement of the surfactants, the obtained QDs could be dispersed well into aqueous media and were used effectively as biomarkers as shown in Scheme 11.1.

Therefore, various ligands with different structures, binding and hydrophilic groups have been investigated as suitable capping agents for the preparation of water-soluble QDs [2, 5, 9].

11.2.1.3 Functional Binding Groups
To realize the effective replacement on the surface of QDs, the functional binding groups were generally located on the terminal of ligand and have the strong binding ability, mainly containing thiol groups [20–30], histidine (His) [31–37], or phosphine groups [38, 39], to interact with the surface of the inner QD core. Three categories of binding groups are listed and shown in Scheme 11.2.

(a) Thiol ligands
Thiol groups could effectively interact with the surface of QDs as strongly as covalent bonds, therefore this chemistry was used for the ligand replacement with water-soluble thiol-derivatized alkyl acid, sugar, and biotin molecules as shown in Scheme 11.3.

In the preliminary works, the ligands containing one thiol group and alkyl acid were chosen as the thiol and carboxyl groups could provide the attachment to the surface of the QDs [20–25]. These kinds of ligands not only provide the water solubility and stability of QDs, but also offer the ability to further functionalize the surface to achieve the bioconjugation via the well-known EDC coupling reaction. The

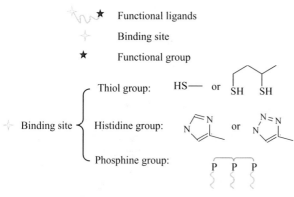

★ Functional ligands
     Binding site
★ Functional group

Binding site
- Thiol group: HS— or   SH   SH
- Histidine group: [structure] or [structure]
- Phosphine group: P   P   P

★ Functional group: Carboxyl, hydroxyl, or biotin

SCHEME 11.2   Representative ligands with different binding sites and functional groups.

representative work was reported by Nie et al. [20]. They first prepared the ZnS-capped QDs with a CdSe core size of 4.2 nm and then functionalized the surface with the glacial mercaptoacetic acid in chloroform. The resulting QDs were then extracted from chloroform to PBS buffer solution and purified to remove excess mercaptoacetic acid by centrifuge. The water-soluble QDs were obtained with the carboxyl groups on the surface. It should be noted that this kind of QD has two characteristics: first, the free carboxyl groups could be covalent or electrostatic conjugate to various functional groups or biomolecules such as proteins, peptides, or nucleic acid. Second, the alkyl acid of surface layer not only controls the water solubility

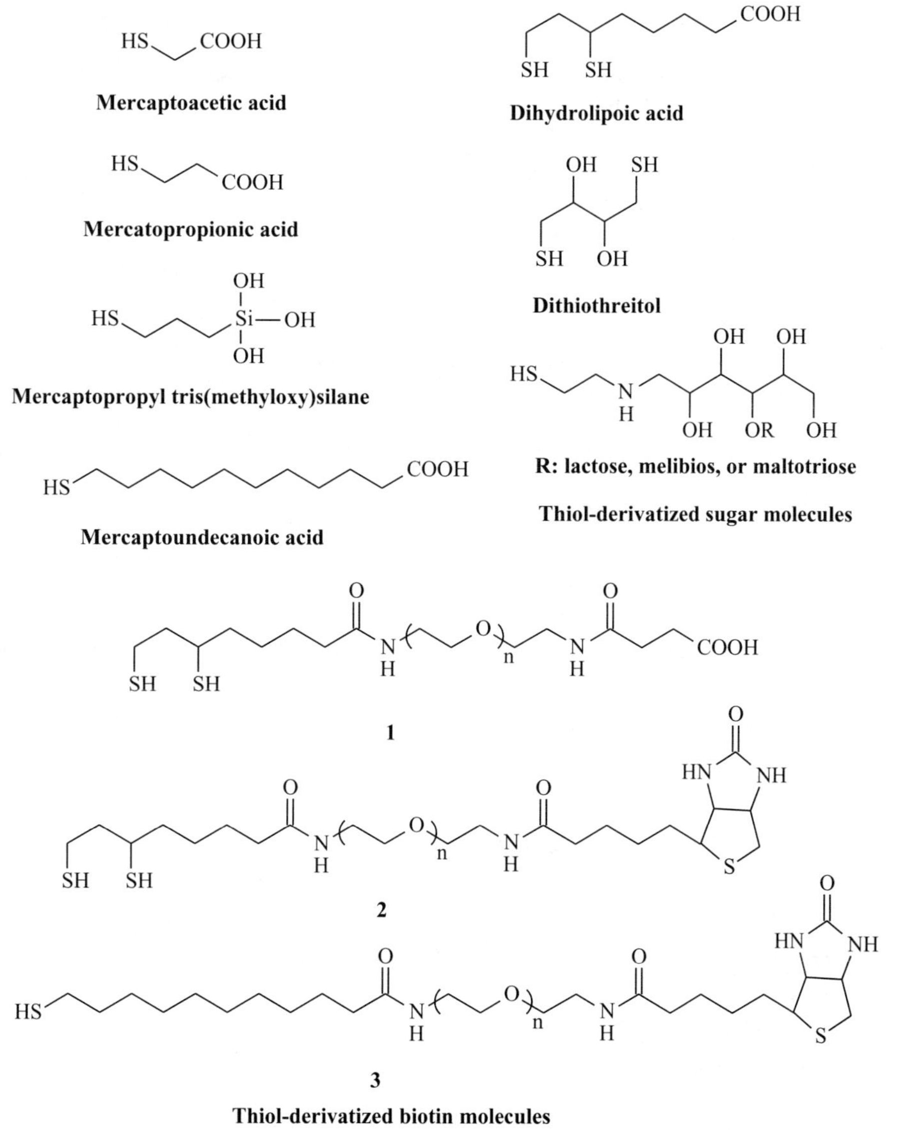

HS␣␣␣COOH

Mercaptoacetic acid

HS␣␣␣COOH

Mercatopropionic acid

HS — Si(OH)(OH) — OH

Mercaptopropyl tris(methyloxy)silane

HS␣␣␣COOH

Mercaptoundecanoic acid

COOH
SH   SH

Dihydrolipoic acid

OH   SH
SH   OH

Dithiothreitol

HS␣␣N(H)␣␣OH OH / OH OR OH

R: lactose, melibios, or maltotriose

Thiol-derivatized sugar molecules

1

2

3

Thiol-derivatized biotin molecules

SCHEME 11.3   Representative ligands from thiol-derivatized alkyl acid, sugar, and biotin molecules.

and stability by adjusting the solution pH, but also reduces passive protein adsorption of QDs.

Following the same principle, Letsinger et al. reported the preparation of propionic acid-stabilized QDs [21]. The 3-mercaptopropionic acid was used to replace the surface TOP/TOPO surfactants on ZnS-capped QDs with CdSe core in N,N-dimethyl formamide (DMF) solvent, and the solubility of the resulting QDs with propionic acid could be significantly improved by the presence of 4-(dimethylamino) pyridine (DMAP) for the deprotonation of surface carboxyl groups, thus the final QDs have the well-dispersed ability in aqueous media and could be stored for up to 1 week at room temperature. Based on the above results, more thiol–alkyl acids could be used as new ligands, but the stabilized QDs were only soluble under basic conditions, and their further functionalization was definitely limited with the long carbon chain of ligands.

With the recent developments, QDs prepared from the above thiol–alkyl acid ligands have limited water stability as the linkage of mono thiol to the Zn on the surface of QDs is dynamic. Further chemical reaction to those ligands was found to cause the slow degradation of the QDs with diffusion of cadmium atoms into the aqueous solution. There are two ways to solve this issue. The first one is by adding a new layer such as the siloxane shell [24, 25] on the surface of QDs and the second approach is to find novel ligands with strong binding ability such as polymer chains [26–30] which will also provide the long-term stability, solubility, and can prevent the release of cadmium atoms from the prepared water-soluble QDs.

To add an additional layer on the QDs, two methods have been exploited and the basic principle used is similar. The first one is to use the mercaptopropyl tris(methyloxy)silane (MPS) as the new ligand instead of 3-mercaptopropionic acid as reported by Weiss and Alivisatos [25]. Butanol/TOPO ZnS-CdSe shell–core QDs were first prepared, and then mixed the precursor QDs with MPS in methanol to replace the TOPO molecules after basification. The primary siloxane layer was formed by the hydrolyzation of methoxysilane groups of MPS into silanol groups, and the latter was then further converted to siloxane bond by heating, the polymerized siloxane layer, namely cross-linked siloxane layer, could effectively "frozen" the core of QDs and prevent the release of heavy metal ions. This layer could also be further modified with other water-soluble functional groups on the surface of QDs. The final silanized QDs have strong stability in water and PBS buffers at physiological conditions and could be stored for months, and retain their optical properties and showed enhanced photochemical stability over organic fluorophores.

The strategy of designing the siloxane layer on the surface of QDs helped in stabilizing the shell–core structure of QDs, but the potential loss of monothiol-terminated ligands is still present. Another way to solve this problem is by using bidentate or multidentate ligands bearing two or more thiol groups [26–30]. Mattoussi group has reported the use of dihydrolipoic acid (DHLA) as surface ligands to stabilize and solubilize the QDs, and the obtained DHLA-capped QDs have enhanced solubility and stability in basic conditions and could be stored for long time due to the bidentate chelate effect [26]. However, the QDs obtained by this strategy were not stable enough and were easily aggregated to form large particles when the solution pH changed to acidic conditions.

To solve the pH limitation problem, water-soluble QDs with hydroxyl groups on the surface were prepared by Pathak et al. [30] instead of the carboxyl groups. The dithiothreitol (DTT) was used as the new ligand to replace the TOP/TOPO to generate the water-soluble QDs with the hydroxyl groups on the surface. This surface modification leads to the aqueous solubility of QDs, but also reduces the nonspecific binding of protein adsorption and the terminal hydroxyl groups could be further activated by reacting with 1,1'-carbonyl diimidazole (CDI) to form the imidazole-carbamate groups and were then attached with other functional compounds.

This pH limitation problem could also be easily solved by water-soluble chains such as the poly(ethylene glycol) (PEG) chains. Mattoussi's group reported the synthesis of PEG-terminated DHLA (DHLA-PEG) by reacting various chain length PEGs with thioctic acid and followed by the reduction of the 1,2-dithiolane to create a bidentate thiol motif [27–29] The DHLA-PEG showed enhanced affinity with the surface of QDs and successfully replaced the TOP/TOPO surfactants as discussed before to render the QDs soluble in aqueous environments. The prepared QDs exhibited the greatest resistance to the environmental changes and were chemically stable over long periods of time up to several months due to the presence of polymer layer.

The key feature of this method is the usage of water-soluble polymers bearing bidentate thiol groups, which not only provides the strong affinity with the surface of QDs by the multichelate effect to avoid the potential loss of thiol group, but also renders them soluble in water environment. This strategy was very useful in the development of a new series of organic ligands that are capable of changing the QD property from hydrophobic to hydrophilic and provided a new approach to create the water-soluble and biocompatible QDs.

The stability and solubility of QDs were addressed by the introduction of water-soluble PEG chains, but these kinds of modified QDs were not very useful for QD bioconjugation due to the lack of accessible functionalities. To add more functional groups on the terminal of the DHLA-PEG chains, more reaction steps were carried out in a stepwise approach. The diazide-terminated PEG was prepared by reacting the PEG chain with methanesulfonyl chloride and then sodium azide, the azide of diazide-terminated PEG could be converted to amine group in biphasic acidic solution, and followed by EDC coupling with thioctic acid to form new PEG derivatives bearing the azide and 1,2-dithiolane moieties. The azide group could be also further converted to amine groups to form the new functional site for further functionalization of PEG chains. Through this strategy, various functional groups such as biotin, amino, hydroxyl, and carboxyl groups were attached on the terminal of 1,2-dithiol-terminated PEG to extend the QD bioconjugation [28].

(b) Histidine ligands

It is well known that the His residue can coordinate to metal zinc ions through metal affinity coordination. Therefore, ZnS-capped CdSe QDs offered surface Zn ions to interact with and bind to water-soluble compounds bearing His moieties [31–37]. When the water-soluble compounds bearing His residues were chosen as His-expressing proteins or peptides, the surface coordination simultaneously renders the aqueous dispersion and biofunctionalization of QDs.

Himmel's group reported this approach of surface coordination of QDs by using the genetically engineered multiple poly-His-tagged cohesin/dockerin protein polymer [31, 32]. The protein–polymer conjugates were prepared to contain 6X His tags by the self-assembly of protein monomer bearing the His-terminated dockerin and clostridium thermocellum cohesin. TOP/TOPO-passivated QDs were first prepared, and the capping agents were then exchanged by MPA and titrated with Tris base to confer the QD's solubility in an aqueous environment. Then, the excess protein polymer and MPA-Tris-modified QDs were incubated together under Bis-Tris buffer solution; the resulting QDs were purified by passing through high pressure size exclusion chromatography (HPSEC).

Similar works were reported by both Mattousis' and Wright's group to investigate the ligand exchange of DHLA-passivated luminescent ZnS-CdSe shell–core QDs to different proteins containing the His tails through the coordination of engineered C-terminal oligo-His sequences via metal affinity interaction [33–37].

(c) Phosphine ligands

TOP and TOPO surfactants were used as the major components to prepare alkyl phosphine-passivated QDs, but this kind of ligand has two main limitations: The modified QDs are only soluble in organic solvents and ligands, are monodentate which lacks the strong binding ability with the QDs and hence the modified QDs are not very stable after dissolution. To address those limitations, Kim et al. [38, 39] have reported a new kind of ligand composed of oligomeric phosphine to stabilize the QDs. These phosphine-based ligands were precisely designed by increasing more affinity sites and additional functional groups, thus providing better solubility and stability. With this aim, oligomeric phosphines were prepared by two steps: the oligomerization of monodentate trishydroxypropylphosphine (THPP) with the addition of crosslinker diisocyanatohexane, and then further reaction of the residual hydroxyl groups with other compounds bearing the isocyanate group or other functional groups.

The oligomeric phosphines bearing octyl alkyl chain and pentanoic acid were synthesized to replace the original surfactants on QDs in order to prepare different phosphine-capped QDs. Compared to the TOP-stabilized QDs, the passivated QDs with octyl chain-based oligomeric phosphines possess more efficiency and could keep the quantum yield for a longer period of time. QDs replaced by oligomeric phosphines bearing pentanoic acid was identified to have higher quantum yield up to 20% higher than with mercaptoundecanoic acid (MUA)-modified QDs, and the binding ability was also strengthened as compared to QDs modified with monomeric phosphine bearing one carboxylic functional group, although all QDs were prepared via the same ligand exchange process. To further confirm the solubility and chemical flexibility of pentanoic acid attached on the surface of QDs, the prepared QDs were dissolved in various buffer solutions and in high salt concentration solutions, and they were also cross-linked by 2,6-diaminopimelic acid to increase their water stability and to prevent the quantum yield loss in serum at 37°C.

Those oligomeric phosphines proved to be suitable ligands for the preparation of water-soluble and stable QDs with acceptable quantum yields through the ligand exchange process, and these types of QDs were also flexible for further conjugation without altering their optical properties.

From those discussions above, an ideal ligand for the preparation of water-soluble and stable QDs via ligand exchange method must have three requirements: strong binding sites, soluble moieties, and the availability of reactive functional groups. The strong binding site to the surface of hydrophobic QDs is essential to provide the stability of the QDs and to prevent the potential loss of ligands by further functionalization. The soluble moieties are composed of long polymer chains or cross-linked siloxane shell, which could render the QDs stable for long periods of time and soluble not only at physiological conditions, but also over the broad pH range from weakly acidic to strongly basic conditions.

Those two former aspects only produce water-soluble and stable QDs and were utilized in several biological fields, but the availability of various reactive functional groups is also essential to broaden the applications of these materials, Therefore, a range of (macro)molecules such as biotin, avidin, protein, peptide, and DNA have been successfully conjugated on the surface of QDs.

11.2.2 Ligand Encapsulation

The dispersion and further functionalization of QDs could be effectively solved by the ligand exchange method as discussed before, but this method generally affects the quantum fluorescent yield of original QDs. Therefore, another strategy was exploited to realize the preparation of water-soluble and stable QDs without affecting their optical properties. Hydrophobic van der Waals interactions were used in this strategy to form a third layer to stabilize the TOP/TOPO-modified QDs without any surface modifications, [40–44] which were influenced by the surface encapsulation of other inorganic nanoparticles such as iron, gold, or silver nanoparticles to render their water solubility (Scheme 11.4) [45–48].

Benoit et al. [40] reported the encapsulation of QDs in phospholipid micelles by a simple and facile method. The TOP/TOPO-passivated QDs were prepared as discussed before and the single QDs could be encapsulated into the hydrophobic core of micelles prepared from the mixture of n-poly(ethylene glycol) phosphatidylethanolamine (PEG-PE) and phosphatidylcholine (PC). The encapsulation of QDs critically relied on the properties of PEG-PE block copolymer, only when all the three components, PEG chains, two alkyl chains, and the linker PE, were present, then the QDs were effectively stabilized. QD particles could be encapsulated into the core of micelles for the copolymer bearing only one alkyl chain; however, the encapsulated QDs were not stable and aggregated under high salt concentration. Furthermore, QDs could not be encapsulated in the micelles when the PC is used alone. Thus, the QD micelles could be obtained by mixing the PC and PEG-PE with various PEG chain lengths, and these types of QDs could be used under biological conditions even *in vivo*. Further functionalization of QDs was achieved by replacing up to 50% of the PEG-PE phospholipids with amino-functionalized PEG-PE.

Following the same principle, Fan et al. [41] successfully achieved the individual incorporation of TOP and oleylamine (OA)-passivated QDs into the hydrophobic interiors of surfactant/lipid micelles composed of a mixture of 1,2-dioctanoyl-sn-glycero-3-phosphocholine (C8-lipid) and hexadecylamine (HDA), or 1,2-distearoyl-sn-glycero-3-phosphoethanolamine-*N*-amino poly(ethylene glycol) (aPEG) and dipalmitoyl phosphatidylchloline (DPPC). Water-soluble QDs were therefore obtained via a simple and rapid synthesis. After incorporated into the core of surfactant/lipid-formed micelles, the optical properties of QDs were maintained and their stability was confirmed.

The advantage of the above methods to provide water-soluble QD micelles is its capability to utilize a wide variety of surfactant/lipid mixtures with different functional groups on their terminal. It should be also noted that the synthesis is fast and simple, and the encapsulation process involved no chemical substitution, the obtained QDs after encapsulation

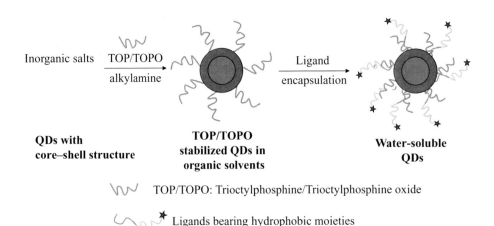

SCHEME 11.4   The preparation of water-soluble QDs by two steps via ligand encapsulation method.

preserve their optical properties such as quantum yield and their stability in water is enhanced.

To further improve the stability and solubility of QDs, the capped agents were then chosen from amphiphilic polymers instead of surfactant/lipid mixtures as polymers have higher water solubilities and also provide higher stability of the QDs in water. Wu's group [43] used octylamine-modified polyacrylic acid, as an extra water-soluble shell on the TOPO/TOP-passivated QDs. The hydrophobic moieties of polymer were found to interact with the hydrophobic ligands on the QDs. To further enhance their dispersion and stability, the functional carboxyl groups on the surface were further cross-linked to fix the structure of polymer-modified QDs.

Parak et al. also reported the same strategy to transfer the hydrophobically capped QDs from organic to aqueous solution by choosing another amphiphilic polymer, poly(maleic anhydride-*alt*-1-tetradecene) [44]. The hydrophobic tails of this amphiphilic polymer could intercalate with the surfactants of QDs to form the third polymer-based shell via hydrophobic interactions, and this polymer coating on the outer surface of QD renders the water solubility of the QDs and effectively prevents their aggregation. The hydrophilic groups of the polymer not only confers the dispersion ability but the polymer could be further cross-linked which make the QDs stable even under harsh conditions such as the high salt concentration.

11.2.3   Direct Encapsulation

The dispersion and functionalization of QDs in physiological conditions were generally carried out by the ligand exchange or ligand encapsulation methods as discussed above. The water-soluble QDs were prepared via both methods at least by two steps: the attachment of surfactants and then the replacement or encapsulation of second ligand on surface of QDs. To simplify the process, water-soluble ligands from small molecules to polymers were used to prepare water-soluble QDs as shown in Scheme 11.5 [49, 50].

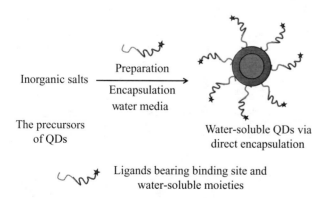

SCHEME 11.5   The preparation of water-soluble QDs by one step via direct encapsulation in water media.

The QDs were prepared from water-soluble ligands with small molecular weight and have the limited stability. To enhance their stability, polymeric ligands were studied, and Liu's group [50] investigated a new bidentate 1,2,3-triazole-based ligand to directly stabilize and solubilize CdS QD particles based on similar chemical structures of 1,2,3-triazole-based ligand with His without the usage of TOP/TOPO surfactants in the whole process. The new bidentate ligands [triazole based ligand (TA)] were synthesized from alkyne PEG and 4-{4-[bis(2-azidoethyl)-amino]-phenylazo}-benzoic acid via click chemistry, which possesses excellent water solubility, biocompatible PEG chain, and two 1,2,3-triazole groups. The water-soluble ligand-capped CdS QDs were then directly produced by mixing the bidentate TA ligand with cadmium chloride and sulfurated hydrogen gas at room temperature in deionized water. Due to the characteristics of TA ligand, the obtained QDs were confirmed to be water-soluble, biocompatible, and multifunctional.

The big differences of this strategy from other reports are: first, the new ligand was designed to have bidentate to increase its interaction with the surface of QDs and synthesized from click chemistry which opened a new way to induce libraries of biocompatible, soluble, and functional polymer into this system via mild conditions. Second, the water-soluble CdS QDs were directly obtained just by mixing the ligand and the metal salt under mild conditions, and the prepared QDs were stable in aqueous environments under different ionic strengths and over a broad pH range. The QDs have photoswitching ability of the fluorescence due to the azobenzene group introduced in the terminal of TA ligand

11.3   BIOCONJUGATION OF WATER-SOLUBLE QDs

Biomolecules such as biotin, carbohydrate, protein, DNA, or peptide were bioconjugated onto the surface of QDs either via covalent bond or electrostatic interaction. The synthesized details were listed as follows.

11.3.1   Biotinylated QDs

To introduce biotin on the surface of QDs, two strategies were used to prepare the biotinylated QDs for the study of site-specific interactions between biotin and avidin [28–55]. One is to synthesize biotinylated alkylthiol compounds, which contains thiol and biotin groups on each terminal of the compounds as shown in Scheme 11.3 [28, 51]. Reiss et al. [51] reported the synthesis of compound 3 via three steps. The biotin was first activated by reacting with the N-hydroxysuccinimide (NHS) to form biotin NHS ester, the NHS group was then replaced by 4,7,10-trioxa-1,13-tridecanediamine to form the amine group on the terminal. The amine group could also be used to covalently attach to the

NHS-activated 11-mercaptoundecanoic acid to form the final compound 3 bearing biotin and thiol moieties. This biotinylated alkylthiol was used as the new ligand to stabilize and solubilize QDs via the ligand exchange method to prepare biotinylated QDs, and the obtained QDs have high stability and solubility, comparable optical properties, and detectable availability of biotin groups. Similarly, Mattoussi's group reported another biotin-terminated TA-PEG, compound 2 as shown in Scheme 11.1, which contains two thiol groups, biotin groups, and water-soluble PEG linker [28, 52]. This compound 2 not only possesses the biotin moiety, but also has stronger binding sites and long PEG chains to bind onto the surface of QDs and to provide high stability, respectively, as discussed before. Those two methods were successfully utilized to introduce biotin on the surface of QDs and the availability of biotin could also be detected, but the biotin conjugation was realized via the ligand exchange method.

To exploit the biotinylated process via other way, the biotin was linked onto the pendant chains of polymer, and then attached onto the surface of QDs via combination of polymer and QDs [55, 56]. Narain et al. [55] prepared the biotinylated glycopolymer via reversible addition–fragmentation chain transfer (RAFT) polymerization of the three monomers bearing biotin, sugar, and amine groups as pendant groups, and the biotin was then attached onto the surface of QDs via the EDC coupling of biotinylated glycopolymer and carboxyl-modified QDs, and the obtained QDs have well colloidal stability and photostability. The surface modification did not significantly affect the optical properties of the original QDs, and the availability of biotin could be also quantified by the 4-hydroxyazobenzene 2-carboxylic acid (HABA)/avidin binding assay. More importantly, the carbohydrate groups linked on the polymer side chains were also attached onto the surface of QDs, and the cytotoxicity of QDs was improved by the introduction of both biotin and carbohydrate moieties.

11.3.2 Carbohydrate-functionalized QDs

Protein–carbohydrate interactions are important reactions for the study of cellular recognition process, and QDs were confirmed as a useful tool to investigate their interaction after the conjugation of carbohydrate moieties onto the surface of QDs. There are several methods to encapsulate QDs using carbohydrate residues [56–62]. Rosenzweig et al. [58] reported the preparation of glyconanospheres containing CdSe-ZnS QDs. The QDs with CdSe core and ZnS shell were first prepared and stabilized via TOPO/TOP surfactants in organic solvents. The surface ligands were then exchanged by mercaptosuccinic acid to obtain the negatively charged with high water solubility. Negatively charged carboxymethyl dextran (CM-dextran) was prepared by mixing dextran with chloroacetic acid in alkaline solution, and the combination of negatively charged QDs and CM-dextran with positively charged polylysine resulted in the formation of luminescent glyconanospheres. These polysaccharide-modified QDs were further cross-linked for higher stability of the nanospheres, and the final QDs revealed to be highly water soluble, stable, and possess high affinity to specific lectins.

The hydrophobic interaction was also used to prepare the glyco-functional QDs instead of the electrostatic interaction to prevent their dissociation under the high salt concentration [59]. The calyx[4]resorcarene macrocycle containing four long alkyl chains and eight saccharide moieties on the opposite sides [60] were chosen as the glyco-based ligand to be intercalated into the TOPO/TOPO-formed shell of QDs to convert the QDs to be water soluble though the hydrophobic interaction between the cone-shaped amphiphile with TOP/TOPO surfactants. Although the novel QDs conjugated sugar ball could be obtained by the surface coverage of TOPO/TOP-stabilized QDs with saccharide-based amphiphiles to study its cellular uptake, the complicated preparation of glyco-based ligand limits their potential applications.

To enhance the interactions between the QDs and carbohydrate moieties via covalent bond, thiol-appended sugar molecules shown in Scheme 11.1 were prepared by mixing the sugars with 2-aminoethanethiol to form the neoglycoconjugates and was then used to covalently couple the QDs to fabricate the water-soluble and glyco-functional QDs [61]. Different glyco-functional QDs could be achieved by using different sugars bearing lactose, melibiose, and maltotriose and then utilized in the study of lectin detection via agglutination or deagglutination. This preparation process of sugar derivatives avoids the multiple steps of protection and deprotection, and the neoglycoconjugation method could be applied to higher oligosaccharides as well.

The above methods focused on the encapsulation of QDs by the small molecules bearing carbohydrate moieties; however, the binding of QDs with the studied compounds is weak and the glyco-functional QDs would be disassociated via the dilution of solution or the addition of high salt concentration. To improve their stability, glycopolymers were exploited to prepare novel glyco-modified QDs. Sun et al. prepared polyacrylamide-bearing carbohydrate and amine residues as pendant chains using the biotinylated radical initiator via free radical polymeration [56]. Due to the presence of biotin groups, the glycopolymer could be attached onto the surface of streptavidin-derivatized QDs to form the nanocrystal–glycopolymer conjugates. The obtained conjugates have the carbohydrate groups on the surface like antennae and were used as a useful tool to study carbohydratelectin-specific interactions.

The preparation of biotin-terminated glycopolymer avoids the complicated protection and deprotection process and this polymer could be attached to the surface of avidin-modified nanocrystals in principle, but the glycopolymer have the broad molecular distribution and multiple synthesis steps. To

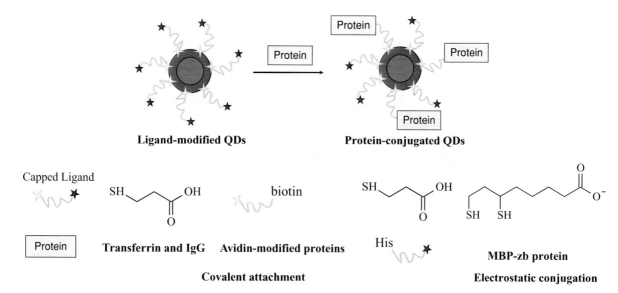

SCHEME 11.6   The preparation of protein-modified QDs using different ligands and proteins via covalent attachment or electrostatic conjugation.

address this problem, Narain's group [55] reported the preparation of glycopolymer-bearing biotin and carbohydrate as pendant groups via the RAFT process as discussed above.

11.3.3   Protein-functional QDs

To introduce protein molecules on the surface of QDs, several approaches have been studied such as designing QDs with positive charges [26], primary amine groups [20], His groups [31–37], or other reactive functional groups on the surface [28]. QDs with proteins on the surface were therefore achieved via covalent interaction, ligand exchange, or electrostatic interactions (Scheme 11.6).

Among the studied protein-modified QDs, the simplest way was to react reactive amine on the protein such as transferrin onto the mercapto-solubilized QDs using the EDC coupling reaction. The protein-conjugated QDs were demonstrated to be biocompatible *in vitro* and *in vivo* as reported by Nie et al. [20]. Another simple way reported by Mattoussi et al. [26] was achieved by electrostatic interactions using proteins having positively charged lysine residues and carboxyl-capped QDs under basic conditions.

With the research development of zinc–His coordination, proteins bearing His moieties were utilized instead of thiol end-terminated alkyl acid to stabilize the QDs via the ligand exchange [31–37]. The obtained protein-modified QDs have excellent solubility under aqueous conditions, enhanced stability, and comparatively similar quantum yields as the original QDs. Protein-modified QDs may also be prepared by exploiting reactive functional groups but also by using QDs with specific ligands such as biotin to interact with specific protein such as avidin or streptavidin. Indeed, Mattoussi et al. reported on the preparation of biotin-functionalized QDs,

and the specific interaction of the QDs with streptavidin-b-phycoerythrin (b-PE) was achieved to form the protein- modified QDs [28]. Therefore, different protein-modified QDs were prepared and utilized in various biological applications [63–73].

11.3.4   DNA, Peptide, or Other Biomolecules Functionalized QDs

With the similar principles for the preparation of biotin- or protein-functional QDs, the other biomolecules such as DNA [21, 30, 74–77], peptide [78–80], antibody [81–83], avidin [84, 85], enzyme [86], BSA [87, 88] etc. were precisely designed to contain the sulfhydryl group charges, or other functional groups and then were mainly used as effective ligands to encapsulate the QDs via ligand exchange or encapsulation method to prepare the water-soluble and stable QDs and utilize them as the effective biomarker to investigate their potentially biological applications (Scheme 11.7).

Here the DNA molecules were mentioned and discussed as the representative biomolecules to conjugate QDs. As discussed before, the QDs were designed to be water soluble by the modification of surface ligands by the introduction of alkylthiol acid, then the alkylthiol-functionalized DNA was generally added into the above QD aqueous solution and then the thiol groups on the terminal of DNA react with QDs by the thiol–metal interaction to form the DNA-functionalized QDs. The important step for their preparation is to render the QDs water soluble, otherwise, the DNA is very hard to be conjugated onto the surface of QDs due to the well-known solubility problem of QDs in water media.

The DNA-conjugated QDs could be also obtained by reacting thiol-modified DNA with PEG/PE-modified QDs

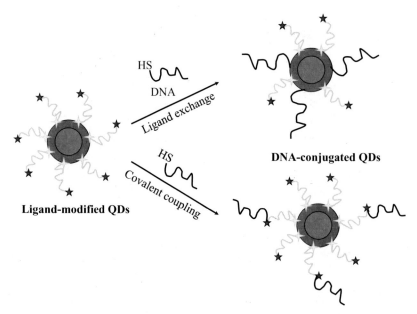

SCHEME 11.7   The preparation of DNA-modified QDs via ligand exchange or covalent coupling by using thiol-modified DNA.

bearing the amine groups on their surface due to covalent coupling of thiol and amine groups using a heterobifunctional coupling agent, sulfosuccinimidyl 4-(*N*-maleimidomethyl) cyclohexane-1-carboxylate (Sulfo-SMCC) [40]. Such prepared DNA-conjugated QDs remained strongly fluorescent and water stable, and the successful preparation of DNA-modified QDs has significantly improved the material research in biology.

11.4   CONCLUSION

QDs have been identified as a useful biomarker in biology, but the inherent hydrophobic property of QDs hinders their biomedical applications, so tremendous works have focused on the water dispersion, photostability, and bioconjugation of QDs by libraries of ligands bearing different functional groups. Among those reported works, the ligands are generally composed of biomolecules such as DNA, biotin, carbohydrate, peptide, protein, BSA, etc. or water-soluble long polymer chains such as PEG or glycopolymer, and functional groups such as amine or sulfhydryl groups. Due to the presence of functional groups, the ligand can be utilized to encapsulate the QDs by the surface interaction or reaction between the ligand and surfactants of QDs, which possess acceptable quantum yield (QY) and were usually stabilized by TOPO/TOP ligands.

According to the encapsulation process, the attachment of ligand onto the surface of QDs was categorized to three strategies: ligand exchange, ligand encapsulation, and direct encapsulation. The most common way is the ligand exchange, which usually modifies the QDs with

mercaptoalkyl acid and then the thiol end-terminated ligands are attached to the QDs to render their solubility by the strong binding between the thiol and metal elements. Most biomolecules such as DNA, peptide, or protein are conjugated onto the surface of QDs via this method, especially for protein-bearing His moieties, the His–metal affinity are extended to prepare the protein-modified QDs. However, this simple way usually affects the quantum fluorescent yield as compared with that of original QDs after ligand exchange.

Recent studies thus tend to the surface modification of QDs by different functional ligands or biomolecules via various organic chemistry or electrostatic interactions. The surface of QDs was first activated by functional groups such as the CDI, NHS, carboxyl groups, or hydrophobic moieties, and the biomolecules bearing amine groups, positive charge, or hydrophobic moieties were then easily attached onto the QDs via the amine replacement or specific interaction, charge interaction, or hydrophobic interaction to realize the introduction of biomolecules onto the surface of QDs. Therefore, the functional polymers bearing new functional groups such as biotinylated glycopolymers or functional proteins would be become the new ligands to stabilize QDs and extend their biological applications in the preparation of novel water-soluble functional QDs in future.

REFERENCES

1. Alivisators AP. *Science* 1996;271:933–937.
2. Huo Q. *Colloids Surf B Biointerfaces* 2007;59:1–10.
3. Galian RE, Guardia MDL. *Trend Anal Chem* 2009;28:279–291.

4. Vinayaka AC, Thakur MS. *Anal Bioanal Chem* 2010;397:1445–1455.

5. Smith AM, Duan HW, Mohs AM, Nie SM. *Adv Drug Deliv Rev* 2008;60:1226–1240.

6. Algar WR, Tavares AJ, Krull UJ. *Anal Chim Acta* 2010;673:1–25.

7. Medintz IL, Mattoussi H. *Phys Chem Chem Phys* 2009;11:17–45.

8. Algar WR, Prasuhn DE, Stewart MH, Jennings TL, Blanco-Canosa JB, Dawson PE, Medintz IL. *Bioconjug Chem* 2011;22:825–858.

9. Jamieson T, Bakhshi R, Petrova D, Pocock R, Imani M, Seifalian AM. *Biomaterials* 2007;28:4717–4732.

10. Michalet X, Pinaud FF, Bentolila LA, Tsay JM, Doose S, Li JJ, Sundaresan G, Wu AM, Gambhir SS, Weiss S. *Science* 2005;28:538–544.

11. Bailey R, Nie S. *J Am Chem Soc* 2003;125:7100–7106.

12. Murray CB, Norris DJ, Bawendi MG. *J Am Che Soc* 1993;115:8706–8715.

13. Hines MA, Guyot-Sionnest P. *J Phys Chem* 1996;100:468–471.

14. Dabbousi BO, Rodriguez-Viejo J, Mikulec FV, Heine JR, Mattoussi H, Ober R, Jensen KF, Bawendi MG. *J Phys Chem B* 1997;101:9463–9475.

15. Peng ZA, Peng X. *J Am Chem Soc* 2001;123:183–184.

16. Talapin DV, Rogach AL, Kornowski A, Haase M, Weller H. *Nano Lett* 2001;1:207–211.

17. Hines MA, Guyot-Sionnest P. *J Phys Chem B* 1998;102:3655–3657.

18. Manna L, Scher EC, Alivisator AP. *J Am Chem Soc* 2000;122:12700–12706.

19. Peng X, Manna L, Yang W, Wickham J, Scher E, Kadavanich A, Alivisator AP. *Nature* 2000;404:59–61.

20. Chan WC, Nie S. *Science* 1998;281:2016–2018.

21. Mitchell GP, Mirkin CA, Letsinger RL. *J Am Chem Soc* 1999;121:8122–8123.

22. Kloepfer JA, Mielke RE, Wong MS, Nealson KH, Stucky G, Nadeau JL. *Appl Environ Microbiol* 2003;69:4205–4213.

23. Wuister SF, Swart I, Driel FV, Hickey SG, Donega CDM. *Nano Lett* 2003;3:503–507.

24. Bruchez M Jr, Moronne M, Gin P, Weiss S, Alivisators AP. *Science* 1998;281:2013–2016.

25. Gerion D, Pinaud F, Williams SC, Parak WJ, Zanchet D, Weiss S, Alivisators AP. *J Phys Chem B* 2001;105:8861–8871.

26. Mattoussi H, Mauro JM, Goldman ER, Anderson GP, Sundar VC, Mikulec FV, Bawendi MG. *J Am Chem Soc* 2000;122:12142–12150.

27. Uyeda HT, Medintz IL, Jaiswal JK, Simon SM, Mattoussi H. *J Am Chem Soc* 2005;127:3870–3878.

28. Susumu K, Uyeda HT, Medintz IL, Pons T, Delehanty JB, Mattoussi H. *J Am Chem Soc* 2007;129:13987–13996.

29. Pons T, Uyeda HT, Medintz IL, Mattoussi H. *J Phys Chem B* 2006;110:20308–20316.

30. Pathak S, Choi S-K, Arnheim N, Thompson ME. *J Am Chem Soc* 2001;123:4103–4104.

31. Ding S-Y, Rumbles G, Jones M, Tucker MP, Nedeljkovic J, Simon MN, Wall JS, Himmel ME. *Macromol Mater Eng* 2004;289:622–628.

32. Ding S-Y, Jones M, Tucker MP, Nedeljkovic JM, Wall J, Simon MN, Rumbles G, Himmel ME. *Nano Lett* 2003;3:1581–1585.

33. Goldman ER, Medintz IL, Hayhurst A, Anderson GP, Mauro JM, Iverson B, Georgiou G, Mattoussi H. *Anal Chim Acta* 2005;534:63–67.

34. Medintz IL, Clapp AR, Mattoussi H, Goldman ER, Fisher B, Mauro JM. *Nat Mater* 2003;2:630–638.

35. Delehanty JB, Medintz IL, Pons T, Brunel FM, Dawson PE, Mattoussi H. *Bioconjug Chem* 2006;17:920–927.

36. Slocik JM, Moore JT, Wright DW. *Nano Lett* 2002;3:169–173.

37. Medintz IL, Clapp AR, Brunel FM, Tiefenbrunn T, Uyeda HT, Chang EL, Deschamps JR, Dawson PE, Mattoussi H. *Nat Mater* 2006;5:581–589.

38. Kim S, Bawendi MG. *J Am Chem Soc* 2003;125:14652–14653.

39. Kim S, Lim YT, Soltesz EG, Grand AMD, Lee J, Nakayama A, Parker JA, Mihaljevic T, Laurence RG, Dor DM, Cohn LH, Bawendi MG, Frangioni JV. *Nat Biotechnol* 2004;22:93–97.

40. Dubertret B, Skourides P, Norris DJ, Noireaux V, Brivanlou AH, Libchaber A. *Science* 2002;298:1759–1762.

41. Fan HY, Yang K, Boye DM, Sigmon T, Malloy KJ, Xu HF, López GP, Brinker CJ. *Science* 2004;304:567–571.

42. Fan HY, Leve EW, Scullin C, Gabaldon J, Tallant D, Bunge S, Boyle T, Wilson MC, Brinker CJ. *Nano Lett* 2005;5:645–648.

43. Wu XY, Liu HJ, Liu JQ, Haley KN, Treadway JA, Larson JP, Ge NF, Peale F, Bruchez MP. *Nat Biotechnol* 2003;21:41–46.

44. Pellegrino T, Manna L, Kudera S, Liedl T, Koktysh D, Rogach AL, Keller S, Rädler J, Natile G, Parak WJ. *Nano Lett* 2004;4:703–707.

45. Shen LF, Laibinis PE, Hatton TA. *Langmuir* 1999;15:447–453.

46. Lala N, Lalbegi SP, Adyanthaya SD, Sastry M. *Langmuir* 2001;17:3766–3768.

47. Swami A, Kumar A, Sastry M. *Langmuir* 2003;19:1168–1172.

48. Wang Y, Wong JF, Teng XW, Lin XZ, Yang Y. *Nano Lett* 2003;3:1555–1559.

49. Gaponik N, Talapin DV, Rogach AL, Hoppe K, Shevchenko EV, Kornowski A, Eychmüller A, Weller H. *J Phys Chem* 2002;106:7177–7185.

50. Shen R, Shen XQ, Zhang ZM, Li YS, Liu SY, Liu HW. *J Am Chem Soc* 2010;132:8627–8634.

51. Charvet N, Reiss P, Roget A, Dupuis A, Grünwald D, Carayon S, Chandezon F, Livache T. *J Mater Chem* 2004;14:2638–2642.

52. Susumu K, Uyeda HT, Medintz IL, Mattoussi H. *J Biomed Biotech* 2007;2007:1–7.

53. Lingerfelt B, Mattoussi H, Goldman ER, Mauro JM, Anderson GP. *Anal Chem* 2003;75:4043–4049.

54. Howarth M, Takao K, Hayashi Y, Ting AY. *Proc Natl Acad Sci* 2005;24:7583–7588.

55. Jiang XZ, Ahmed M, Deng ZH, Narain R. *Bioconjug Chem* 2009;20:994–1001.

56. Sun XL, Grande D, Baskaran S, Hanson SR, Chaikof EL. *Biomacromolecules* 2002;3:1065–1070.

57. Sun XL, Cui WX, Haller C, Chaikof EL. *Chembiochem* 2004;5:1593–1596.

58. Chen YF, Ji TH, Rosenzweig Z. *Nano Lett* 2003;4:581–584.

59. Osaki F, Kanamori T, Sando S, Sera T, Aoyama Y. *J Am Chem Soc* 2004;126:6520–6521.

60. Nakai T, Kanamori T, Sando S, Aoyama Y. *J Am Chem Soc* 2003;125:8465–8475.

61. Babu P, Sinha S, Surolia A. *Bioconjug Chem* 2007;18:146–151.

62. Tamura J, Fukuda M, Tanaka J, Kawa M. *J Carbohydr Chem* 2002;21:445–449.

63. Ishii D, Kinbara K, Ishida Y, Ishii N, Okochi M, Yohda M, Aida T. *Nature* 2003;423:628–632.

64. Clapp AR, Medintz IL, Mauro JM, Fisher BR, Bawendi MG, Mattoussi BR. *J Am Chem Soc* 2004;126:301–310.

65. Medintz IL, Konnert JH, Clapp AR, Stanish I, Twigg ME, Mattoussi H, Mauro JM, Deschamps JR. *Proc Natl Acad Sci* 2004;101:9612–9617.

66. Geho D, Lahar N, Gurnani P, Huebschman M, Herrmann P, Espina V, Shi A, Wulfkuhle J, Garner H, Petricoin E, Liotta LA, Rosenblatt KP. *Bioconjug Chem* 2005;16:559–566.

67. Irrgang J, Ksienczyk J, Lapiene V, Niemeyer CM. *Chemphyschem* 2009;10:1483–1491.

68. Medintz IL, Clapp AR, Mattoussi H, Goldman ER, Fisher B, Mauro JM. *Nat Mater* 2003;2:630–638.

69. Pons T, Medintz IL, Wang X, English DS, Mattoussi H. *J Am Chem Soc* 2006;128:15324–15331.

70. Algar WR, Wenger D, Huston AL, Blanco-Canosa JB, Stewart MH, Armstrong A, Dawson PE, Hildebrandt N, Medintz IL. *J Am Chem Soc* 2012;134:1876–1891.

71. Ji XJ, Zheng JY, Xu JM, Rastogi VK, Cheng TC, DeFrank JJ, Leblanc RM. *J Phys Chem B* 2005;109:3793–3799.

72. Zhang P. *J Fluorescence* 2006;16:349–353.

73. Xie HZ, Li YF, Kagawa HK, Trent JD, Mudalige K, Cotlet M, Swanson BI. *Small* 2009;5:1036–1042.

74. Patolsky F, Gill R, Weizmann Y, Mokari T, Banin U, Willner I. *J Am Chem Soc* 2003;125:13918–13919.

75. Gill R, Willner I, Shweky I, Banin U. *J Phy Chem B* 2005;109:23715–23719.

76. Zhou D. *Biochem Soc Trans* 2012;40:635–639.

77. Peng H, Zhang LJ, Kjällman THM., Soeller C, Sejdic JT. *J Am Chem Soc* 2007;129:3048–3049.

78. Akerman ME, Chan WCW, Laakkonen P, Bhatia SN, Ruoslahti E. *Proc Natl Acad Sci* 2002;99:12617–12621.

79. Delehanty JB, Bradburne CE, Boeneman K, Susumu K, Farrell D, Mei BC, Blanco-Canosa JB, Dawson G, Dawson PE, Mattoussi H, Medintz IL. *Integr Biol* 2010;2:265–277.

80. Pinaud F, King D, Moore HP, Weiss S. *J Am Chem Soc* 2004;126, 6115–6123.

81. Goldman ER, Anderson GP, Tran PT, Mattoussi H, Charles PT, Maruo JM. *Anal Chem* 2002;74:841–847.

82. Hua XF, Liu TC, Cao YC, Liu B, Wang HQ, Wang JH, Huang ZL, Zhao YD. *Anal Bioanal Chem* 2006;386:1665–1671.

83. Kanwal S, Traore Z, Zhao CF, Su XG. *J Luminescence* 2010;13:1901–1906.

84. Pattani VP, Li CF, Desai TA, Vu TQ. *Biomed Microdevices* 2008;10:367–374.

85. Mansson A, Sundberg M, Balaz M, Bunk R, Nicholls IA, Omling P, Tagerud S, Montelius L. *Biochem Biophys Res Commun* 2004;314:529–534.

86. Ipe BI, Shukla A, Lu HC, Zou B, Rehage H, Niemeyer CM. *Chemphyschem* 2006;7:1112–1118.

87. Chouhan RS, Vinayaka AC, Thakur MS. *Anal Bioanal Chem* 2010;397:1467–1475.

88. Trapiella-Alfonso L, Costa-Fernandez JM, Pereiro R, Sanz-Medel A. *Biosens Bioelectron* 2011;26:4753–4759.

12

SILICA NANOPARTICLE BIOCONJUGATES

YOHEI KOTSUCHIBASHI[1], MITSUHIRO EBARA[2], AND RAVIN NARAIN[1]

[1]*Department of Chemical and Materials Engineering, Alberta Glycomics Centre, University of Alberta, Edmonton, AB, Canada*
[2]*Biomaterials Unit, International Center for Materials Nanoarchitectonics (WPI-MANA), National Institute for Materials Science (NIMS), 1-1 Namiki, Tsukuba, Ibaraki, Japan*

12.1 INTRODUCTION

Silica nanoparticles (SiNPs) have been actively researched as biomaterials due to the ease of controlling the size and shape, easy surface modification, and their low toxicity. The two main techniques used for the preparation of SiNPs are reverse microemulsion and Stöber methods [1, 2]. The reverse microemulsion can control the diameter in the nanoorder range simply by changing the concentration of water, oil, and surfactant. On the other hand, the Stöber method is used to prepare SiNPs having nano or submicron size using tetraethyl orthosilicate (TEOS) as the main chemicals in water, ethanol, and ammonia. The desired size is obtained by controlling the concentrations and weight ratios. The biocompatibility and low toxicity of the SiNPs are of great importance for *in vitro* and *in vivo* applications. First, the pure SiNPs with different sizes are researched for investigation of the relationship between size and toxicity. In general, SiNPs of small sizes less than 100 nm showed more toxicity than larger particles, possibly due to their high surface area and higher uptake in cells. Napierska et al. investigated the toxicity of the size-different amorphous SiNPs (14–335 nm) on endothelium cell (EA.hy 926 cell line) by MTT and lactate dehydrogenase (LDH) assays [3]. The larger SiNPs of sizes 104 and 335 nm showed low toxicity ($TC_{50} > 1000$ μg/cm^2 on MTT and LDH assay) as compared to nanoparticles of sizes ranging from 14 to 16 nm ($TC_{50} < 50$ μg/cm^2). These particles having nano-order can passively extravasate and accumulate in tumor parenchyma is known as EPR effect [4]. However, larger SiNPs are transported along the vasculature and accumulate in

various organs through different mechanisms. Decuzzi et al. prepared size- and shape-different SiNPs such as spherical (700–3000 nm), hemispherical, discoidal, and cylindrical, and their accumulations in different organs after intravenous injection at different concentrations were studied (Figures 12.1b–12.1d) [5]. Interestingly, the observed particles in the organs are affected by their size and shape. These relationships of size/shape and toxicity are also reported for metal particles, which suggest that the surface area and property are of great importance to obtain the desired SiNPs for biomedical uses. SiNPs tend to aggregate each other due to their large surface area if not functionalized. Besides, once aggregated, it is difficult to redisperse the SiNPs into solvent. To solve this problem, Lin et al. prepared hydrothermally treated mesoporous SiNPs (MSiNPs) with dual-organosilane modification (hydrophilic silane for increasing water dispersity and hydrophobic silane for reducing silica hydrolysis) [6]. The MSiNPs can be stored as a dried powder, and can easily be redispersed into biological fluids. The redispersion is also improved by coating with hydrophilic materials. The mesoporous structure of SiNPs is one of the most studied structures due to their large surface area and small pore size [7–9]. Figure 12.1 shows the SiNPs with unique structure such as sphere, mesoporous, dual-mesoporous [10], hemispherical, discoidal, cylindrical, hollow [11], ellipsoidal [12], raspberry-like [13], chrysanthemum-like [14], Janus [15], silica-coated liposomes [16], and so on. It is interesting to note that the various shapes of the SiNPs with similar volume show the different interactions with biological matter both *in vitro* and *in vivo*. Zhao and coworkers investigated the interaction between human red blood cell (RBC)

Chemistry of Bioconjugates: Synthesis, Characterization, and Biomedical Applications, First Edition. Edited by Ravin Narain.
© 2014 John Wiley & Sons, Inc. Published 2014 by John Wiley & Sons, Inc.

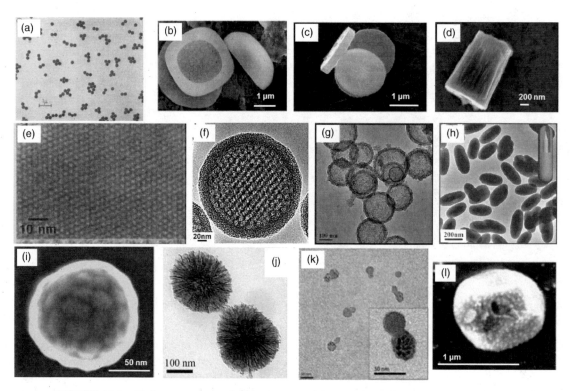

FIGURE 12.1   Various structures of silica nanomaterials: (a) sphere [2], (b) quasi-hemispherical [5], (c) discoidal [5], (d) cylindrical [5], (e) mesoporous [8], (f) dual-mesoporous [10], (g) hollow [11], (h) ellipsoidal [12], (i) raspberry-like [13], (j) chrysanthemum-like [14], (k) Janus [15], (l) silica-coated liposomes [16].

membranes and MSiNPs having different size and surface properties. The affinity of RBC to poly(ethylene glycol) (PEG)-coated MSiNPs is decreased due to the PEG layer which prevents nonspecific binding. On the other hand, charged MSiNPs show high affinity with RBC via electrostatic interaction [17]. To obtain the high stability and low toxicity, the surface of SiNPs is modified to introduce new functional groups such as NH_2, COOH, and SH group. By the conventional Stöber method, these functional groups are easily introduced on the silica surface by the reaction of the silanol groups by different molecules or macromolecules. In general, PEG-coated nanoparticles are more resistant to removal from the body by the reticuloendothelial system (RES) as compared to non-PEGylated nanomaterials that are readily accumulated into the liver and the spleen [18]. Cauda et al. investigated the stability of PEGylated MSiNPs (size = 47 nm, pore size = 4 nm) having different PEG chains on the surface in simulated body fluid (SBF) [19]. The PEGylated MSiNPs having longer/denser PEG chains kept their original structure after 1 month in the SBF as compared to the unfunctionalized MSiNPs that showed a degraded structure and needle-like crystals.

Modified SiNPs are also often used as the reaction site for living radical polymerization (LRP) such as nitroxyl-mediated polymerization (NMP) [20], atom transfer radical polymerization (ATRP) [21, 22], and reversible addition–fragmentation chain transfer (RAFT) polymerization [23, 24]. These surface-initiated LRPs of SiNPs have gained a lot of interest in the last few decades as they can be exploited to generate SiNPs with dense well-defined polymer chains on the surface. Moreover, SiNPs with stimuli-responsive polymers can also be synthesized which are shown to respond with temperature, pH, light, and molecules, etc. Recently, click chemistry [25] have gained significant attention for the modification of SiNPs with polymers. For example, Chen and coworkers prepared temperature-responsive poly(N-isopropylacrylamide) (PNIPAAm)-coated SiNPs by click chemistry by reacting azide-modified SiNPs with alkyne-terminated PNIPAAm prepared via RAFT polymerization [26]. The diameter of the PNIPAAm-coated SiNPs changed from 130 to 67 nm reversibly between 24° and 40°C due to their temperature-responsive properties. However, the azide–alkyne type of click chemistry usually needs a toxic metal catalyst. Therefore, the metal-free click chemistry such as the thiol–ene reaction has often been considered for fabrication of bioconjugate materials. SiNPs having stimuli-responsive PNIPAAm (M_n = 6300 g/mol, M_w/M_n = 1.15) and the pH-responsive poly(2-(diethylamino)ethyl methacrylate) (PDEAEMA) (5200 g/mol, M_w/M_n = 1.22) synthesized via the thiol–ene reaction [27]. The polymer-coated SiNPs

(127 ± 11 nm) showed both temperature and pH-responsive behaviors and the diameter changes were depended on the polymer contents. The combinations of LRPs and click chemistry are useful to create unique materials. In this chapter, we will describe the current bioconjugation strategies to create new materials with SiNPs.

12.2 DRUGS

SiNPs are extensively used as drug carriers due to their ease of synthesis and unique surface properties. Besides, SiNPs can be designed to have all the other essential requirements of an ideal drug delivery system (DDS) such as low toxicity, high loading/encapsulation of desired drug, zero premature release, recognition of the target cells or tissue, and controlled release of the drug [28]. Since the discovery of mesoporous material in 1990s [7–9], the MSiNPs have particularly attracted attention not only in DDS but also in other fields such as adsorption, separation, catalysis, sensors, photonics, and nanodevices for their specific properties such as tunable pore diameter (2–30 nm), narrow pore-size distribution, and high surface area (>700 m^2/g) [29, 30]. By using template methods for instance surfactant assembly, colloidal crystal, emulsions, and latex, it is possible to obtain well-defined mesoporous structure. The properties of the MSiNPs are first compared with SiNPs because the large surface area is sometimes affected by the nonspecific interactions with small molecules, proteins, and cell surfaces, which may lead to their toxicity. Lee and coworkers investigated the toxicity including inflammation and apoptosis between the SiNPs and MSiNPs having a similar spherical structure [31]. The diameters of both particles are 100 nm, and there is large difference in their surface areas on SiNPs and MSiNPs that are 40 m^2/g and 1150 m^2/g, respectively. The MSiNPs do not affect the viability of macrophage cells as studied by MTT assay test at a concentration of 100 μg/mL. However, the SiNPs showed high toxicity under same condition. Moreover, the MSiNPs have significantly lower inflammation and apoptotic cell death than SiNPs, and it was concluded that the low activation of mitogen-activated protein kinases, nuclear factor-κB, and caspase 3 is responsible for the higher biocompatibility of the MSiNPs. The toxicities of micro-order mesoporous silica (1 ∼ 160 μm) are observed on human colon carcinoma (Caco-2) cell by Heikkilä and coworkers, which suggested that these silica materials show high toxicity depending on the concentrations (>2 mg/mL) and incubation time [32]. Liu and coworkers prepared mesoporous hollow SiNPs (MHSiNPs) whose diameters are 110 nm, and investigated the single and repeated dose toxicity by intravenous injection in mice [33]. The lethal dose 50 (LD_{50}) of the single dose is over 1000 mg/kg, and the continuous intravenous administration (20, 40, and 80 mg/kg) for 14 days caused no death, which concluded the high biocompatibility

of the MHSiNPs. The MHSiNPs were particularly accumulated into the mononuclear phagocytic cell in liver and spleen, and the alanine aminotransferase (ALT) and aspartate aminotransferase (AST) are increased at 80 mg/kg of continuous injection of the MHSiNPs while no increase was observed at 20 and 40 mg/kg. After 4 weeks, the accumulated MHSiNPs in liver and spleen were decreased by 7% and 41% respectively, and were excreted from the body. Moreover, Hudson et al. reported that the injected MSiNPs (∼150 and ∼800 nm) into mice through intraperitoneal and intravenous caused death or euthanasia [34]. Depending on the conditions, MSiNPs can exhibit the high toxicity. Therefore, the size and shape of the MSiNPs or conjugation with biocompatibility materials are critical. Multifunctionality is of great importance for these drug carriers to achieve the treatment and diagnosis simultaneously. Lee and coworkers created a dye-doped MSiNPs immobilized with magnetite nanocrystals on the surface. The MSiNPs can encapsulate drugs into the core, and the magnetite crystal and dyes are used for magnetic resonance (MR) and fluorescent imaging, respectively. First, they prepared the dye-doped MSiNPs using cetyltrimethylammonium bromide (CTAB) as template surfactant, and modify the surface by Fe_3O_4 nanocrystal and PEG chain using their reactive amino group on the MSiNPs after removal of the surfactant (Figure 12.2). The diameter of MSiNPs and Fe_3O_4 nanocrystal were 70 nm and 8.5 nm respectively as shown by transmission electron microscopy (TEM) image. Doxorubicin (DOX), an anticancer drug, was loaded into the pore and the drug-release property; toxicity *in vitro* and *in vivo* was investigated. The DOX-loaded fluorescent MSiNPs having the Fe_3O_4 and PEG chain showed high toxicity *in vitro*, and the accumulation by enhanced permeation and retention (EPR) effect in tumor part were observed by MR and fluorescent imaging when the material was injected intravenously in mice [35]. To control the drug release, MSiNPs were coated by stimuli-responsive polymers that can be responsive to chemical and/or physical properties via the external environment such as temperature, pH, light, magnetic field, and so on. For example, Yavuz et al. prepared gold nanocages with hollow interiors and porous walls. The gold nanocages were modified by the PNIPAAm copolymers having an lower critical solution temperature (LCST) around 39°C, and achieved the controlled drug release from the hollow part through irradiation with the near-infrared laser [36]. In this way, the precise-triggered drug release from MSiNPs *in vivo* has been achieved. Singh and coworkers prepared the two types of the polymer-coated MSiNPs that are PNIPAAm-PEG (temperature-responsive) and PEG-peptide possessing matrix metalloproteinase (MMP) substrate polypeptides (protease responsive) [37]. To synthesize these MSiNPs, first an acrylamide was adsorbed to the MSiNPs electrostatically and then, the acryl groups were utilized to synthesize a covalently cross-linked PEG shell. The PEG-coated MSiNPs were modified by the stimuli-responsive

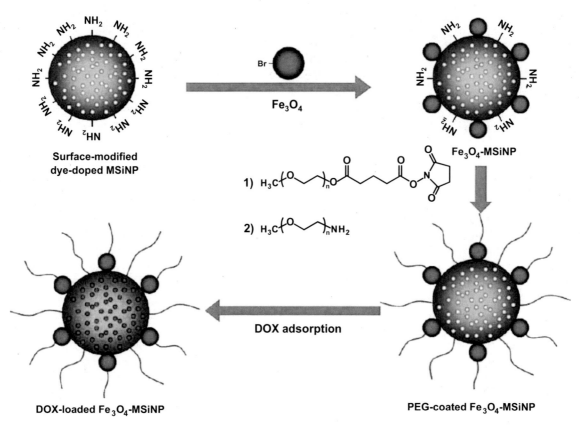

FIGURE 12.2 Preparation of the DOX-loaded MSiNPs having an Fe_3O_4 and PEG on the surface [35]. For a color version, see the color plate section.

copolymers using the simple radical polymerization in mild reaction conditions. At 37°C, the PNIPAAm-PEG-coated MSiNPs were observed to release about 40% of DOX for the first 2 hours, which was greater than at room temperature. By comparison, the uncoated MSiNPs released the same amount of DOX at both temperatures. Moreover, it was reported that the high levels of chemotoxicity of DOX on the PEG-peptide MSiNPs occurred when the drug was loaded into the polymer-shell part (~15–20% cell viability) as compared to the core loaded. Finally, to check the protease-triggered release of DOX, the polymer-coated MSiNPs were studied *in vivo* on the mouse injected with a human sarcoma cell line (HT-1080) which is known to have elevated levels of MMPs. The polymer-coated MSiNPs showed a higher chemotoxicity due to the DOX release than noncoated MSiNPs, which were analyzed to estimate the apoptosis markers. In general, the super-molecular materials are known to have a sensitive stimuli-responsive behavior because of their weak interactions. The MSiNPs having a 1-methyl-1*H*-benzimidazol (MBI) on the surface were prepared by Meng and coworkers [38]. The MBI can interact with β-cyclodextrin (β-CD), which operated as the pH-responsive nanovalves for the controlled release of the incorporated drugs (Figure 12.3). The MBI was selected for their desired pKa of 5.6 that

was utilized for the triggered drug release at the endosomal acidification conditions in cells. The diameter and pore size were approximately 100 nm and 2 nm respectively, and the DOX-release amounts were increased by increasing the solution pH (about 40% of DOX release at pH 5). They also investigated the cell toxicity due to the triggered release of drug in human-differentiated myeloid (THP-1) and squamous carcinoma (KB-31) cell line in which the MSiNPs were efficiently taken into acidifying endosomal compartments, and release the DOX. Moreover, they focused on the zinc-doped iron oxide nanocrystals (Zn-Fe_3O_4) that showed 3-fold hyperthermic effect and 10-fold MRI contrast relative to undoped iron crystals, and prepare Zn-Fe_3O_4-doped MSiNPs having super-molecular valves on the surface that were composed of *N*-(6-*N*-aminohexyl)aminomethyltriethoxysilane and cucurbit[6]uril [39]. The incorporated Zn-Fe_3O_4 enabled the remote-controlled drug release from the MSiNPs by the magnetic field, which achieves effective cancer-killing properties on breast cancer cell (MDA-MB-231). To coat these exothermic materials that are iron and gold nanomaterial by SiNPs can topically heat only their surface environment, which suggested adding nondamage to normal cell including protein and other molecular. Under this concept, Yagi, Techawanitchai, and coworkers prepared a magnetite/silica

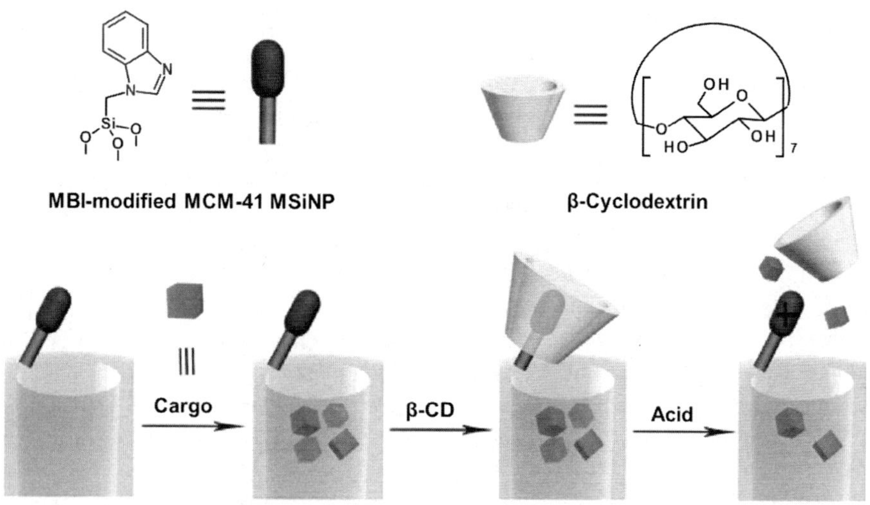

FIGURE 12.3   Preparation of the pH-responsive MSiNPs with nanovalve [38].

particle with temperature-responsive copolymers by simple radical reaction as fillers for column chromatography to separate some steroid drugs [40, 41]. The elution time of hydrophobic steroid drugs were successfully controlled by "on–off" switching of magnetic field, because the polarity of the grafted surface on the magnetite/silica particles could be altered from hydrophilic to hydrophobic. This system enables the more accurate, prompt, and simple separation of various biomolecules because it does not require heating and cooling of the entire mobile phase. Guo et al. reported a unique MSiNPs having chiral structure inside to apply the controlled chiral drug release [42]. The chiral MSiNPs with various pore sizes and structures were prepared using the surfactants and a chiral cobalt complex (CCC) as cotemplate. The strong peak approximately 1080 cm^{-1} that assigns the Si–O asymmetric vibration was observed in all chiral MSiNPs by circular dichroism spectroscopy. On the other hand, there was no peak on achiral MSiNPs prepared by the same protocol expect in the absence of CCC-template. The called CSBA-15 of chiral MSiNPs synthesized using TEOS, Pluronic P123, CTAB, and CCC showed the differential kinetic release profiles of the chiral drug of metoprolol. The S-enantiomers of the metoprolol showed a stronger adhesive capability than R-enantiomers, one of the drugs released from the CSBA-15, which was attributed to the chiral interactions between the chiral drug and local chiral bonding site of the chiral MSiNP matrix. Recently, silica nanomaterials having the unique structure such as MSiNPs with cubic pore structure and ellipsoidal SiO_2/Fe_3O_4 nanomaterial have been developed for the new controlled drug-release system [43, 44]. Wei and coworkers proposed a three-dimensional ordered MSiNPs (3D MSiNPs) [45]. These 3D MSiNPs were prepared using poly(ethylene oxide)-*b*-poly(methyl methacrylate) (PEO-*b*-PMMA) with different polymer composites as template materials. First, the

block copolymers and TEOS were dissolved in THF/2M-HCl solution. After that, the THF was evaporated at 25°C to assemble the micelle composite structures that were composed of block copolymer micelle surrounded by TEOS trough hydrogen bond (PEO_{125}-*b*-$PMMA_{174}$/THF/2M-HCl/TEOS mass ratio = 1:175:50:7.5). The micelle composite formed 3D structure by evaporation, and the 3D MSiNPs were obtained after removal of the polymer templates. The properties of the 3D MSiNPs that were prepared using PEO_{125}-*b*-$PMMA_{174}$ (hydrothermal treatment at 100°C and then calcination at 550°C) as templates were face-centered cubic (fcc) mesostructure, large pore size of up to 37.0 nm, large window size (8.7 nm), high surface area (508 m^2/g), and large pore volume (1.46 cm^3/g). By modification of gold nanoparticles, the 3D MSiNPs were utilized as catalyst for efficient reduction of 4-nitrophenol. These unique mesoporous structures of SiNPs were expected to carve a new stage for the drug carriers to combine with several materials.

12.3   IMAGING AGENTS

The development of imaging materials is a great challenge for biotechnology and medical sciences. These materials have to show high stability and brightness to make useful for any applications [46]. In this section, the conjugation of SiNPs with several imaging agents will be discussed. In the clinical fields, the diagnosis systems can be of different types such as optical imaging, magnetic resonance imaging (MRI), computed tomography (CT), ultrasound (US), positron emission tomography (PET), or single photon emission computed tomography (SPEC), which have their own unique advantages but also have intrinsic limitations [47, 48]. Organic and metal-organic fluorescent dye molecules are often used for optical imaging applications. Photobleaching and quenching are observed in biological environment possibly due to their aggregations and interaction with solvent, molecules, oxygen, ion, and protein [49]. Quantum dots (QDs) are highly stable fluorescent materials consisting of II–VI or III–V elements or other semiconductors having large absorption cross-sections, large Stokes shifts, and narrow emission bands. However, the high toxicity of these QDs is a major concern for biological applications [50, 51]. Metal nanoparticles are usually coated with hydrophilic materials to prevent their aggregation [52–54]. SiNPs are ideal materials to protect these imaging materials from aggregation, improve their photostability, and biocompatibility. These imaging materials were easily incorporated into the SiNP core by reverse microemulsion and Stöber methods. Lee et al. prepared three kinds of fluorescent (red, green, and blue) SiNPs with highly monodisperse gold nanoparticles (1–2 nm) by water-in-oil (W/O) microemulsion and intensive ultrasound irradiation [55]. The gold nanoparticles were modified by gold-binding polypeptide (GBP) from avian influenza virus surface antigen (AIa) to detect their specific antibody and a magnet was studied. Fluorescent hollow/rattle-type mesoporous $Au@SiO_2$ nanocapsule were prepared by Wang and coworkers [56]. The optical property was controlled by changing the size of Au nanorods. The Au nanorods completely disappeared after 150 minutes of etching, and became a hollow mesoporous material. The DOX-loaded mesoporous $Au@SiO_2$ nanocapsule showed the continuous drug release for 50 hours in PBS at 37°C, and the toxicity was increased depending on the concentration at HepG2 cell in spite of the fact that the empty nanocapsule kept the low toxicity between 0.6 and 5.1 nM. The size, shape, and surface property of SiNPs are known to dictate the accumulation in different organs. Drug delivery to the brain through blood flow is especially difficult due to the special intracerebral defense systems, for instance blood–brain barrier (BBB), and it is one of the great challenges to establish the efficient delivery system [57]. In general, PEGylation of nanoparticles have been shown to increase the circulation lifetime in blood. For example, the PEGylation of liposomes lead to a significant increase in blood circulating time (up to 90 hours) and reduced clearance by the RES system (200-fold decrease) [58]. Ku et al. prepared PEGylated polyamidoamine dendrimers (PAMAM, G2) conjugated fluorescein-doped magnetic SiNPs (PEGylated PF-MSiNPs, diameter = 80–90 nm, ζ-potential = + 1.49 mV) that can penetrate the BBB in rat brain [59]. As the result, the PEGylated PF-MSiNPs were found to penetrate the BBB through transcytosis of vascular endothelial cells, subsequently diffused into the cerebral parenchyma and distributed in the neurons. However, the non-PEGylated PF-MSiNPs were unable to cross the BBB. Bardi et al. developed an amino-functionalized CdSe/ZnS QD-doped SiNPs (25 and 50 nm) for imaging and gene carrier in MIH-3T3 and human neuroblastoma (SH-SY5Y) cell lines [60]. The

QD-doped SiNPs were found to be a good carrier of DNA with low toxicity. (There were no cell death in the culture for 24 hours at different concentrations from 0.1 to 10 μg/mL.) Our group prepared fluorescein isothiocyanate (FITC)-doped SiNPs having pH-responsive polymers by surface-initiated ATRP. The SiNPs showed high fluorescence intensity. Ow and coworkers investigated the brightness of free tetramethyl-rhodamine isothiocyanate (TRITC, 1.0 nm), TRITC core with (15 nm) or without (2.2 nm) SiNPs, and the silica-coated TRITC core showed high brightness and stability in solutions [61]. It is interesting to note that the TRITC core without silica layer have low fluorescence intensity as compared to free TRITC due to quenching. These results suggest that the dyes are protected from external environment by coated silica layers. However, using only one fluorescent dye into SiNPs can limit their bioapplication. Therefore, the SiNPs having more than two fluorescent materials have been the focus recently. Wu et al. prepared MSiNPs with two different fluorescent dyes that are rhodamine-lactam and FITC into the core [62]. They investigated the interaction of the fluorescent SiNPs in cells with or without bafilomycin A1 (BFA). The BFA is known as an ATP-H1 pump inhibitor and is able to alkalinize the lysosomal pH. The fluorescent properties were found to change by the solution pH and the fluorescent SiNPs showed different fluorescence in presence of the BFA as shown by confocal microscopic images. Those results suggest their potential use in studies of lysosome-involved cell biology or evaluation of lysosome-targeted cancer therapy. Moreover, Burns and coworkers designed a core/shell fluorescent SiNPs as the pH sensor [63]. The 70 nm SiNPs have TRITC-rich core as reference, and FITC-rich shell as sensor layer by covalent bond. The shell's silica matrices also worked as a filter for diffusion of the analyte molecules and a protector for the core part from interaction with proteins and organic quenches. Rat basophilic leukemia mast cells (RBL-2H3) were used to investigate the pH value of various intracellular compartments, and they found that the various intracellular locations showed different pH values ranging from pH 6.5 to 5.0.

Fluorinated compounds are also a versatile probe for monitoring the biological molecules and events such as gene repression, protein existence, enzymatic activity, environmental alteration, and other biological reactions with an ^{19}F MRI or an ^{19}F NMR spectroscopy. ^{19}F NMR probes based on SiNPs prepared by Tanaka and coworkers [64]. The SiNPs were modified with water-soluble perfluorinated dendrimers via the disulfide linkers for quantitative measurements of glutathione reductase activity (Figure 12.4). He et al. synthesized oxygen-deficient luminescent MSiNPs [65]. The MSiNPs were prepared by bottom–up self-assembly of triethoxysilane (TES) and P123 micelles, and then they were baked at over 500°C for dehydrogenation between O_3Si-H groups. In the report, a high calcination temperature (600°C) was used for the formation of the highly luminescent and mesoporous

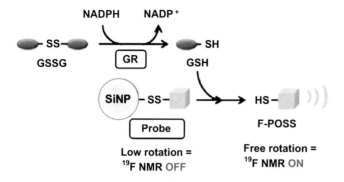

FIGURE 12.4   The ^{19}F NMR probes based on SiNPs [64].

structure. DOX was loaded into the MSiNPs (128 mg/g) due to their large specific surface area (356 m^2/g) and pore volume (0.56 m^3/g). The DOX-loaded MSiNPs were largely uptaken by human breast cancer cell of MCF-7. Moreover, these SiNPs were also used as a sensitive detector. Copper is one of the toxic heavy metals that is used for ATRP and click chemistry to create a unique material, so it is important to check the residual copper before considering the bioapplications. According to U.S. environmental protection agency (EPA), the maximum level (1.3 ppm, ~20 μM) of Cu^{2+} in drinking water is permitted. Zong et al. prepared fluorescent SiNPs for ultrasensitive detection of Cu^{2+} ion [66]. The SiNPs have an FITC-doped core and polyethyleneimine (PEI) on the surface. The PEI was partially reacted with rhodamine B isothiocyanate (RBITC) (Figure 12.5). These SiNPs could achieve the ultrasensitive detection of Cu^{2+}

(detection limit, 10 nM), which were nearly 2×10^3 times lower than the maximum level for the U.S. EPA. The high sensitivity was attributed to the chelation between Cu^{2+} and PEI, and a signal amplification effect, which can show a good reversibility by adding the EDTA. SiNP-based rapid and ultrasensitive sensor was also prepared for *Bacillus anthracis* [67].

12.4   NUCLEIC ACID (DNA/RNA)

Since Mirkin and Alivisatos proposed the first two reports of self-assembly system on DNA-controlled nanoparticles in 1996, DNA has been widely used for sensor, imaging, and gate through combining with several kinds of nanoparticles including Au, Ag, iron, silica, and QD [68–70]. The aggregation behaviors of these conjugated materials are easily controlled for the modified DNA [71]. Wu et al. achieved the construct of 3D networks from size-different SiNPs having a DNA shell [72]. These SiNPs (120 and 290 nm in diameters) were silanized with 3-mercaptopropyltrimethoxysilane (MPTMS) to obtain the SH group on the surface. The SiNPs with SH group were dispersed into NaHCO$_3$/NaCO$_3$ buffer, and the selected oligonucleotide having disulfide was added for a thiol–disulfide exchange reaction (the DNA density in the SiNPs, $1.194 \times 10^{-7} \sim 7.686 \times 10^{-8}$ mol/m^2). Two methods for building the 3D network were investigated: (1) using a three-strand system where two SiNPs having noncomplementary sequences, and an extra DNA can interact with both grafted DNA on the SiNPs, (2) two SiNPs have complementary sequences of DNA (Figure 12.6).

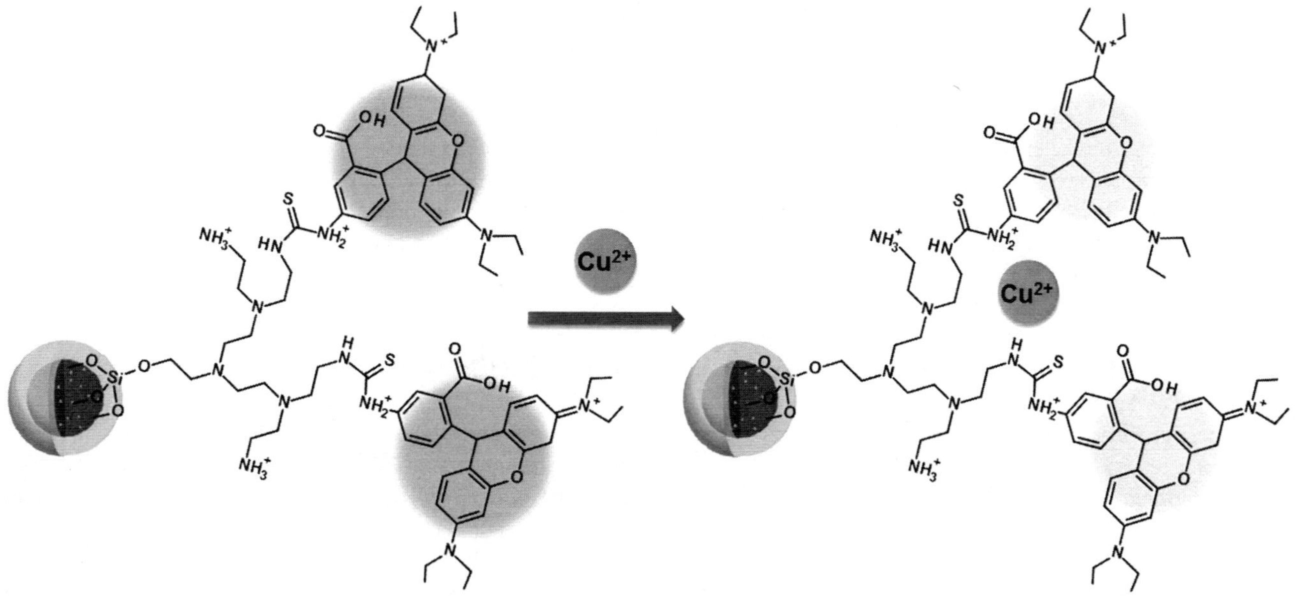

FIGURE 12.5   The fluorescent SiNPs for ultrasensitive detection of Cu^{2+} ion [66]. For a color version, see the color plate section.

Method A

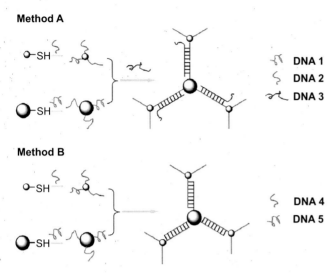

Method B

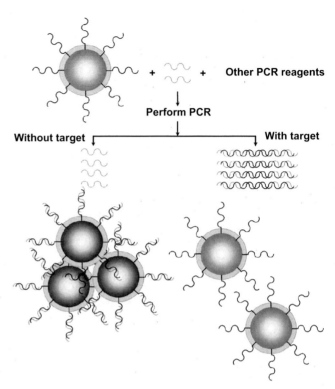

FIGURE 12.6   The three-dimensional (3D) networks from size-different SiNPs having a DNA shell [72].

FIGURE 12.7   The aggregation of nanoparticles in PCR system for the detection of the presence/absence of a target oligonucleotide [73]. For a color version, see the color plate section.

In both cases, raspberry-like 3D structures of DNA-SiNPs were observed. Wong and coworkers prepared silica-coated oligonucleotide-Au nanoparticles by chemisorption of thiol-modified oligonucleotides [73]. The silica shells helped in the thermal stability in the absence of silica coating, 66% of oligonucleotide desorption from Au surface were observed after 2 hours of incubation at 94°C. However, with the silica shell, the desorption of oligonucleotides was reduced to 15%. This result suggested that the thermal stability of the oligonucleotide-Au was improved by the silica coating. These gold nanoparticles changed their dispersion color from red to purple upon aggregation due to a red shift in the surface plasmon resonance (SPR) absorption band. The color change of the nanoparticles was useful in polymerase chain reaction (PCR) system for the detection of presence/absence of a target oligonucleotide (Figure 12.7). The precise interaction of DNA is expected to trigger the controlled release of a drug. MSiNPs (100 nm) were coated with amino groups that can interact with oligonucleotide via the electrostatic interaction. Loaded fluorescent dye was released from the pore triggered by a highly effective displacement reaction in the presence of a target complementary strand [74]. Chen et al. prepared the MSiNPs (diameter = 100 nm, surface area = 1006 m^2/g, and pore size = 3.0 nm) with nucleic acids that are connected on the MSiNPs by azide–alkyne click chemistry reaction [75]. These nucleic acid–MSiNPs have a dual stimuli-responsive system, that is, the incorporated drug can be released from the pore by increasing solution temperature or presence of deoxyribonuclease I (DNase I) for DNA degradation. In fact, the encapsulated rhodamine B is released by almost 100% at 50°C after 4 hours (at 20°C, the release amount reach to under 10% for 2 hours), and the release behavior depended on the DNase I concentrations of which the released dye

reached about 81% and 55% after 24 hours on the introduction of 20 UmL^{-1} and 10 UmL^{-1}, respectively. Endonucleases can hydrolyze the internal phosphodiester bonds in DNA or RNA, which are ubiquitous in most organisms. Therefore, the uptake of DNA-MSiNPs into cells is followed by the degradation of the DNA by endonucleases and subsequent release of the loaded drugs. Under this system, they achieved the cellular uptake of camptothecin (CPT) anticancer drug into HepG2 cells, and the high cell toxicities were dependent on the concentration of the CTP-loaded DNA-MSiNPs. Moreover, Chen and coworkers proposed a pH-driven DNA nanoswitch system based on the MSiNP (diameter = 350 nm, surface area = 1070 m^2/g, and pore size = 2.5 nm) that were prepared by modified base-catalyzed sol–gel process using CTAB as a template. The MSiNPs were combined with a single-strand DNA (ssDNA) through a covalent bond. To generate the DNA nanoswitch, another ssDNA having four stretches of the cytosine-rich domain and the thiol group at the chain end was designed, which can efficiently change their structure between a folded quadruplex i-motif and an extended random conformation by changing the solution pH. The thiol groups were used to interact with gold nanoparticles (3.6 nm) that were located at the entrance of pore site. Therefore, these DNA-MSiNPs can release the incorporated drug in acid environment such as tumors and inflammatory tissues via the structural transformation. They investigated the release profile as a function of pH using rhodamine B. At pH 8.0, no drug release was observed. When the solution pH was adjusted to 5.0, the intensities of rhodamine B were significantly increased. Moreover, the on–off switching for the drug release can be repeated by adjusting the solution pH between 8.0 and 5.0 [76]. These nucleic acids can be used not only for controlled drug release but also for gene therapy. It is a great challenge to protect the DNA before entry to the nucleus. To achieve the aim, several materials such as polymers, micelles, vehicles, and gel were designed for the effective gene delivery [77–79]. SiNPs were also focused as the gene carrier due to their biocompatibility, and ease of the design of the structures including surface modification [80]. By a swelling agent incorporation method, Kim and coworkers synthesized MSiNPs with large pore size (23 nm) that was used to incorporate plasmid DNA (pDNA) [81]. The aminated MSiNPs with large pore have enough space and interaction force for the pDNA, so a high loading capacity of the pDNA was achieved without supplementary polymers such as PEI. The MSiNPs with large pore show the high loaded pDNA amount as compared to normal MSiNPs (pore size = 2 nm). Interestingly, the ζ-potential of the MSiNPs with pDNA in the pore sizes of 23 and 2 nm were 4.35 and -34.5 mV (MSiNPs were 17.5 (pore 23 nm) and 12.3 (pore 2 nm) mV), which suggested that the pDNA were loaded into the large pore site in contrast to the surface adsorption on the MSiNPs having small pore. Keeping the positive charge after loading the pDNA was essential for the effective cellular uptake. Yang and coworkers reported on the adsorption and protection of pDNA on MSiNPs (diameter <100 nm, pore size = 2.6 nm) that have various composition ratios of TEOS/APTEOS [82]. To control the surface positive charge, the TEOS/APTEOS contents were selected as 15:1, 10:1, 7.5:1, 5:1, 4:1, 3:1, and 2:1, respectively. The ζ-potentials were increased from 26.9 to 46.1 mV by increase of the cationic APTEOS content. Moreover, the adsorbed amounts of pDNA reached to over 75 µg/mg at 3:1 and 2:1 of TEOS/APTEOS ratios. These strong interactions with SiNPs/MSiNPs can protect the pDNA from degradation. Bhakta et al. prepared a gadolinium oxide-doped SiNPs (50 nm) for imaging and gene delivery [83]. The APTEOS were coated on the gadolinium oxide-doped SiNPs for electrostatic interaction with pDNA. As for the pDNA, a marker plasmid of pSVβgal was immobilized onto the surface, and the transfection efficiencies in COS-7 and 293T cells were evaluated *in vitro*. Moreover, the high transfection efficiencies were recorded in COS-7 and 293T cells. The solution property is also important to control the DNA desorption behavior. High salt concentration (NaCl) and low pH value were found to cause a large desorption of DNA from MSiNPs, as reported by Fujiwara and coworkers [80]. Li et al. investigated the DNA adsorption into MSiNPs in aqueous solution with differential salt concentrations, salt types, and pH

values. Besides the main driving forces such as the shielded electrostatic force, the dehydration effect, and the intermolecular hydrogen bonds, they also showed that high salt concentration and decreasing pH value promoted the DNA adsorption on the MSiNPs. Moreover, the MSiNPs can adsorb twice the amount of DNA as compared to SiNPs in 2M-guanidine-HCl at pH 5.2 [84]. They also tried to pack a small interfering RNA (siRNA) into magnetic-MSiNPs [85]. The siRNA participate in the RNA interference (RNAi), which is a powerful tool to inhibit the specific gene function for disease treatment. The siRNA was successfully loaded into mesopore of the magnetic-MSiNPs through a strongly dehydrated condition. After siRNA loading, the surfaces were covered with PEI (magnetic-MSiNPs-siRNA@PEI), which is 213 nm in diameter and 17.7 ± 1.7 mV in ζ-potential. The results showed that the magnetic-MSiNPs-EGFP-siRNA@PEI can effectively knockdown the EGFP expression. They also achieved the knockdown on the endogenous B-cell lymphoma 2 (*Bcl-2*) under the same system. Meng and coworkers proposed double DDS of DOX and P-glycoprotein (Pgp) siRNA using MSiNPs for drug-resistant cancer cell line (KB-V1 cells) [86]. The P-glycoprotein is known as one of the mechanisms for multiple drug resistance (MDR) in cancer cells [87]. These MSiNPs were coated with PEIs (1.8, 10, and 25 kD) after modification of the phosphonate groups on the surface for the efficient drug loading and the sizes were 294, 261, and 252 nm in diameters respectively in DMEM cell culture medium. The fluorescence intensity of DOX in the KB-V1 cells was clearly increased by the siRNA-PEI-MSiNP carrier and was approximately 3 times more than the PEI-MSiNPs without siRNA. The combinations of siRNA and drug were expected to expand a new DDS. The nucleic acid–SiNP systems for the gene therapy *in vivo* have been also reported. Suwalski and coworkers prepared a conjugate nanomaterial consisting of MSiNPs and pDNA (including plate-derived growth factor (PDGF)) for the accelerated Achilles tendon healing [88]. The carboxyl modified MSiNPs (10 mg/mL in water) were added in 150 mM NaCl solution containing the pDNA, and the pDNA-loaded MSiNPs were obtained after stirring for 30 minutes. To check the transfection efficiency, first, they used a pDNA-encoding luciferase gene (5 µg)–MSiNPs (6.25 µg) that were directly injected in rat Achilles tendons. The luciferase activity showed no significant difference between pDNA-MSiNPs and free pDNA after 1 day. However, there were clear differences after 10 and 15 days, which were detected as 10- and 100-fold less than the measured activity after 3 days respectively in the pDNA-MSiNPs. Comparatively, the luciferase activity of free DNA was decreased in 114-fold (10 days) and 1500-fold (15 days) lower than that at 3 days. Moreover, the conjugated MSiNPs with plasmid-encoding PDGF gene (pPDGF) successfully created the structural higher organization for the injured Achilles tendons with aligned fibers as compared to free pPDGF *in vivo* testing.

12.5   PEPTIDES/PROTEINS

Peptide and protein are composed of the 20 kinds of amino acids. The short polypeptides are used not only as cellular recognition but also as drug carrier due to the ease of their chemical properties. Yokoyama et al. prepared a block copolymer consisting of PEG and poly(aspartic acid (Asp)), and these Asp blocks were reacted with DOX to form a micelle via their hydrophobic interactions with free DOX [89]. These block copolymers can be assembled with the anionic DNA via the electrostatic interaction. The introduction of specific proteins into cells is a powerful tool as genetic materials for their expression of the specific protein. Mao and coworkers focused on an HIV-1 Tat peptide [90] that has a highly cationic cluster composed of six arginine and two lysine residues in the middle of the sequence. These Tat peptides are known as the membrane permeability peptide, which have been achieved in delivering various cargoes such as metal particles, protein, peptide, and nucleic acids [91, 92]. FITC-SiNPs-Tat peptide (200 nm) was prepared and their cellular penetration and ability of nucleus targeting were studied. The Tat peptide (H-Try-Gly-Arg-Lys-Lys-Arg-Arg-Gln-Arg-Arg-Arg-OH) was reacted with aldehyde enriched FITC-SiNPs at different modification amount of 2 and 3.5 µg/mg SiNPs. When the FITC-SiNPs-Tat peptide were incubated with HepG2 cells in DMEM/10% FBS at 37°C, the fluorescence intensity due to the cellular uptake was clearly increased depending on the Tat peptide amount. The fluorescence intensity of the FITC-SiNPs-Tat peptide having 2 and 3.5 µg showed 3–4.5 times higher than FITC-SiNPs-NH$_2$ after 8 hours, respectively. Moreover, the effective localizations in the cell nucleus of the FITC-SiNPs-Tat peptide were observed after *in vitro* culture for 24 hours by confocal laser scanning microscopy (CLSM) images [93]. Various peptides/proteins were utilized in view of increasing the stability, biocompatibility, cell recognition ability, and cellular uptake of the materials. A collagen-coated MSiNPs having a lactobionic acid (LA) was introduced by Luo and coworkers for targeted drug delivery [94]. The collagen was employed as a cap of mesoporous site for the encapsulation of the drug, and connected with the surface of MSiNPs (diameter = 130 nm, pore size = 3.8 nm) through the disulfide bond. The

DDS is designed to have the following characteristics: (1). cell recognition for the galactose group in LA; (2). decomposition of the disulfide bond of the collagen in the acidic environment of the endosomes; (3). release of the loaded drug. The model drug release (FITC) in MSiNPs-collagen-LA was restrained at 6.5% for 2 hours due to the capped collagen. In presence of the dithiothreitol (DTT), 80% of the model drug was released which suggested that the system exhibited a good response to the reducing environment. Furthermore, the targeting drug delivery was achieved in HepG2 cells. Around 60% of FITC was released from MSiNPs-collagen-LA in endocytosed particles after 24 hours. These fluorescence intensities in the HepG2 cells were 2.2 times higher than that of endothelial cells after incubation for 4 hours due to their targeting properties. Liu et al. prepared Au-coated silica nanorattles (Au-SiNRs) having the PEG and transferrin (Tf) on the surface for cancer therapy via a near-infrared laser light (NIR) [95]. The Tf is one of the major targeting ligands of which many types of cancer cells have abundant expression of the receptors [96]. With those materials longer irradiation time and repeated irradiation of the NIR were not needed. They investigated the targeting ability; cancer therapy via double-effective treatment of drug and hyperthermia, and clearance rate via feces and urine. The silica nanorattles (diameter ∼ 101 nm) having the mesoporous surface were coated by approximately 30 nm of gold shells having methoxy-poly(ethylene glycol)-thiol (mPEG-SH, 5 kD) on the surface for the longer circulation. Moreover, the Tf was linked to the Au-SiNRs via the carbodiimide chemistry between the carboxylic end of the Au-SiNRs and amino groups of the Tf (Figure 12.8). The hydrodynamic diameter was changed from 158 to 185 nm after modification of the mPEG-SH and Tf on the Au-SiNRs. The Au-SiNRs-PEG-Tf saline solution (200 µL, 1 mg/mL) was intravenously injected into MCF-7 tumor-bearing BALB/c nude mouse before 6 hours of the irradiation of 808 nm NIR laser. The temperature at the tumor site was clearly increased from 30.5 to 45.7°C only 3 minutes of the laser irradiation while no temperature variation was observed in the control site. A combination of cancer drug (Docetaxel, Doc) and NIR for the cancer therapy *in vivo* was also tested. About 60% of Doc was released from the carrier for 1 week, and no

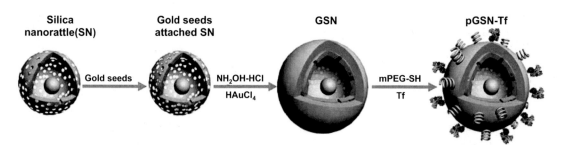

FIGURE 12.8   The Au-coated silica nanorattles (Au-SiNRs) having the PEG and transferrin [95]. For a color version, see the color plate section.

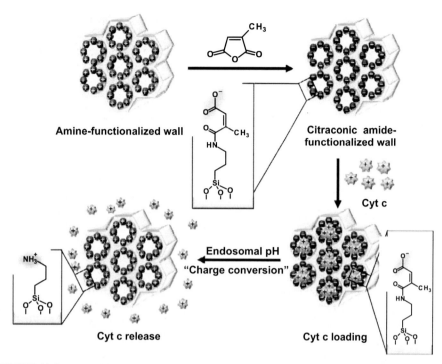

FIGURE 12.9 The charge-convertible mesopore surface [99]. For a color version, see the color plate section.

cumulative release of the Doc was observed by irradiation of NIR laser. The tumors treated with Au-SiNRs-PEG-Tf (Doc and NIR) were completely regressed by 17 days. However, the tumors were relapsed under the same treatment condition but in the absence of the Tf on the surface. Karlsson and Carlsson proposed that it is possible to estimate the adsorption direction using the fluorescent labeling by spectroscopic method, and it was shown that the protein adsorption orientations with SiNPs were strongly affected by the pH value [97]. Shrivastava and coworkers reported the protein orientation on the silica surface measured by proteolysis-mass spectrometry [98]. As the protein, the cytochrome c (Cyt c), RNase A, and lysozyme were selected for adsorption on the 4 and 15 nm of SiNPs. A mixed sample consisting of free proteins and protein-SiNPs were prepared and reacted with heavy acetic anhydride (AcH) and light acetic anhydride (AcL) respectively through the lysine residues. The lysine located near the silica surface was blocked due to their reaction with the AcH. A large decrease of the acetylation and presumably stronger protein–SiNP interactions were observed with SiNPs of 15 nm as compared to the 4 nm particles. Park and coworkers modified the pore walls of MSiNPs using cationic citraconic amide (Cit) having carboxylate end groups for trapping the anionic protein Cyt c as a model protein drug (Figure 12.9) [99]. The pore size of the MSiNPs-Cit was 4.77 nm, which was enough encapsulation for the Cyt c (M_w = 12.4 kDa, size = 2.6 × 3.2 × 3.0 nm^3) [100]. The calculated Cyt c loading presurface area (g/m^2) shows a highest value at MSiNPs-Cit (4.04 × 10^{-4}) as

compared to MSiNPs (9.69 × 10^{-5}) and MSiNPs-succinic anhydride (Suc) (2.82 × 10^{-4}) as negative control. Moreover, the MSiNPs-Cit can release the loading protein at endosomal pH due to the hydrolyzable Cit group that exhibited a charge conversion to expose the amino group, and their electrostatic repulsion promoted the effective protein release. The ζ-potentials of MSiNPs-Cit changed from −10 (pH 7.4) to 16 mV (pH ~5), which was in agreement with Cyt c release property of the MSiNPs-Cit. At pH 7.4, the released Cyt c was reached at 10% after 10 hours. On the other hand, the release amount was increased to over 30% at pH 5 due to their cationic pore walls. Moreover, the Cyt c (labeled with Alexa Fluor 488) release was observed in the cytoplasm via the acidic endosomal environment using CLSM image. Bale and coworkers also reported protein delivery into the MCF-7 and rat neural stem cell (NSCs) using the conjugated materials that were composed of hydrophobic SiNPs (diameter = 15 nm) and the proteins. Importantly, these proteins were delivered into the cytosol without entrapment in the endosomes, and the uptake route using the fluorescent dye-doped SiNPs were determined [101]. Sano and coworkers introduced a continuous release of the anticancer protein from the SiNPs prepared simply [102]. An artificial protein having the abilities of penetration of cell membranes and trigger of apoptosis was selected [103]. The protein-loaded SiNPs were synthesized in PBS solution consisting of the protein and prehydrolyzed tetramethoxysilane (TMOS) stirred for 20 minutes at room temperature, and the protein content of the composite materials were determined to be approximately 30 wt%. The encapsulated proteins were protected from the denaturation and degradation via the physical environment. The release of the protein from the composite was observed to increase gradually with incubate time due to the slowly hydrolyzed silica site in aqueous media. Moreover, the protein-loaded SiNPs exhibited high cell toxicity in MCF-7 where 40% cells were killed after 12 hours via the continuous release of the protein, while they showed low toxicity by 6 hours.

12.6 CARBOHYDRATES

Carbohydrates are one of the most abundant, easily accessible, and cheap biomolecules in nature, which are involved in many important cellular recognition processes including cell growth regulation, differentiation, adhesion, cancer cell metastasis, cellular trafficking, inflammation by bacteria and viruses, and immune response [104–106]. The bio-specific carbohydrate–protein interaction are usually weak but can be increased dramatically to create the topical high concentration of these carbohydrates on a material, commonly known as the "glyco-cluster effect" [107]. Under the concept, carbohydrates are combined with several materials for instance polymers, micelles, hyperbranches, dendrimers, and gels as carbohydrate-based materials for biomedical applications [104]. The carbohydrate–inorganic composite materials have been also attention for their unique properties such as the carbohydrate-protein interaction, biocompatibility, and high stability. FITC-doped SiNPs with glucose and galactose-derived residues on the surface were prepared in our group to investigate the interaction phenomena with cells in the presence or absence of lectins that have specific interaction with carbohydrates, and it was shown that the cell uptake of the SiNPs having galactose was prevented by RCA_{120} due to their aggregations [108]. Galactose-displaying core–shell SiNPs were synthesized by Pfaff and

coworkers for intranuclear optical imaging [109]. The core part were shown to have magnetic properties due to the incorporated γ-Fe_2O_3 and the surfaces have vinyl groups that were used for thiol–ene click chemistry with the P(6-O-methacryloylgalactopyranose (MAGal))-b-P(MAGal-co-4-(pyrenyl)butyl methacrylate (PyMA)) ($M_n = 14{,}800$ g/mol, $M_w/M_n = 1.18$) having a thiol group at the polymer chain end. Therefore, polymer composites have both fluorescent and magnetic properties (Figure 12.10). The galactose-displaying SiNPs (6–40 µg/mL, <10% nonvital cells) were observed by microscopy images in the cytoplasm and nucleus of lung cancer cells possibly due to the enhanced interactions between the galactose and galectin receptors on the cell surface. Bacteria are also known to express the various sugar-specific binding sites on their surface. For imaging and detection of bacteria, Wang et al. prepared a dye-doped SiNPs with various carbohydrates by photo-coupling reaction that is a general coupling chemistry based on perfluorophenyl azides (PFPAs) [110]. The FITC-doped SiNPs modified by PFPAs (~100 nm in diameter) and the carbohydrates were irradiated with a medium-pressure Hg lamp for 10 minutes. The D-mannose-labeled FITC-doped SiNPs (D-mannose-SiNPs) were treated with Concanavalin A (Con A) which is a lectin exhibiting specific affinity to α-D-mannopyranoside, α-D-glucopyranoside, and their derivatives. The decreased fluorescence intensity and aggregations were observed when the D-mannose-SiNPs were incubated with Con A for 1 hour. However, the fluorescence intensity of the incubated D-galactose-SiNPs with Con A was only slightly decreased. Moreover, the interaction between D-mannose-SiNPs and *Escherichia coli* bacteria strain ORN 178 (with mannose-specific binding domain, that is, the FimH lectin, on type 1 pili) and strain ORN 208 (without the FimH lectin) were investigated. The treated bacteria strain ORN 178 with the D-mannose-SiNPs showed clear fluorescence indicating the specific affinity. On the other hand, no fluorescence was observed with bacteria strain ORN 208. Carbohydrates were

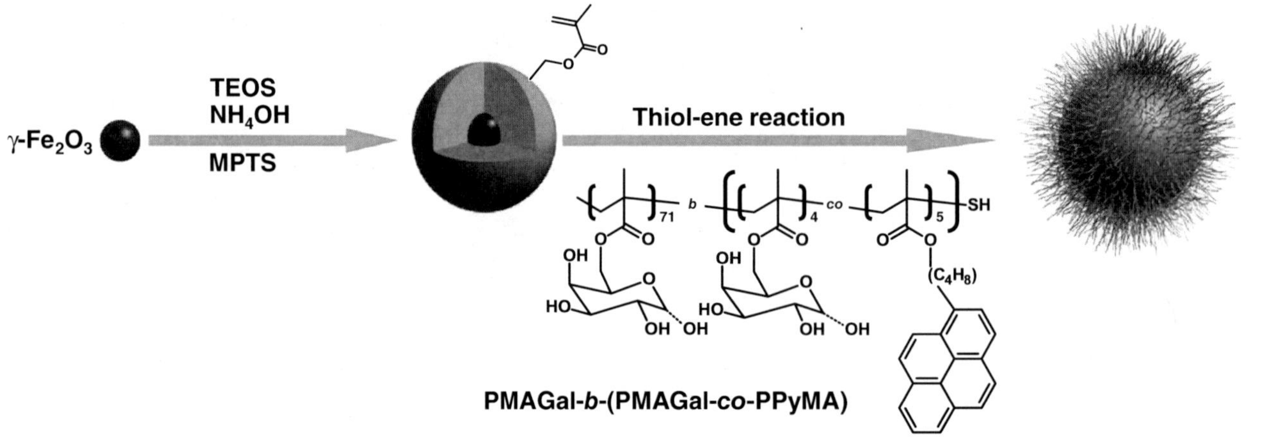

FIGURE 12.10   Synthesis of fluorescent, magnetic, glycopolymer-grafted nanoparticles [109].

also utilized for the triggered controlled release of materials. Glucose is the most studied sugar for the triggered release of insulin from several carrier materials, but the insulin release is usually decreased through the repeated cycles. To solve this problem, Zhao et al. focused on the cyclic adenosine monophosphate (cAMP), which activates Ca^{2+} channels of pancreas β-cells and hence stimulates insulin secretion. The cAMP was encapsulated in MSiNPs (diameter = 120 nm, pore size = 2.3 nm) to increase their membrane permeability and the surfaces of the MSiNPs were covered via the gluconic acid-modified insulin (G-Ins). Therefore, these MSiNPs offered the delivery of both insulin and cAMP for continuous secretion of insulin by the presence of glucose. In fact, the efficient release on both drugs was achieved in the presence of 50 mM of glucose and fructose at pH 7.4, and the release properties were affected from the concentration of the carbohydrate and solution pH. Moreover, the encapsulated cAMP in the rat pancreatic islet tumor cells (RIN-5F) was enhanced as compared to free solution of cAMP due to their poor membrane permeability [111]. Guo et al. synthesized well-defined lactose-containing polymers by RAFT polymerization, which were reacted with SiNPs at various graft densities depending on their molecular weights [112]. Leirose and coworkers synthesized silica–maltose composite nanoparticles of 250–750 nm in diameter by sol–gel chemistry approach [113]. The diameters of the nanoparticles were increased by increasing the maltose contents, and these maltose residues were located in the interstitial space between elementary silica particles. Natural polysaccharides were also applied in the modification of SiNPs. Earhart et al. prepared silica-coated metal nanoparticles (Ag, Fe_3O_4, and ZnS-CdSe) of diameter 2–10 nm with a dextran shell [114]. All dextran-coated nanoparticles were found to aggregate in the presence of Con A due to their strong interaction with the glucose residues of the dextran. A film material consisting of triacetylcellulose (TAC) which are cellulose derivative containing approximately 60 mol% of acetylated group have been widely used for polarizing plates and photographic film via their high transparency, smooth surface, and optical isotropy. Kim and coworkers prepared the TAC polymer–SiNP nanocomposite film. The TAC-modified SiNPs (diameter = ~33 nm) were expected to show strong interaction with TAC polymer and maintaining its optical properties due to their reduced light scattering. The nanocomposite films having several TAC-modified SiNPs (5, 10, 20, and 40 wt%) were successfully prepared, and the transparencies were kept over 80% on all samples between 400 and 800 nm of light wavelength. TGA measurement showed that the T_g and T_c were increased depending on the TAC-modified SiNP content, which suggested that these SiNPs have affected the crystallization behavior of the nanocomposite films [115]. Bernardos et al. used three kinds of hydrolyzed starch derivatives on the MSiNPs for the enzyme-responsive drug release [116]. These hydrolyzed starch derivatives that were Glucidex

47 (5% glucose, 50% maltose, 45% oligosaccharides and polysaccharides), Glucidex 39 (3% glucose, 37% maltose, 60% oligosaccharides and polysaccharides) and Glucidex 29 (10% glucose, 9% maltose, 81% oligosaccharides and polysaccharides) can be decomposed in the presence of pancreatin including amylases (Figure 12.11). The pure MSiNPs have diameter, pore size, and surface area of 100–200 nm, 2.29 nm, and 975 m^2/g, respectively. The dye (ruthenium complex) release properties of these Glucidex 47, 39, and 29 were 63%, 48%, and 31% at pH 7.5 after 5 hours in the presence of the pancreatin. The release of the dye was less than 2% in the absence of the enzyme. The Glucidex 47 was incubated in tumoral human cervix adenocarcinoma (HeLa) and nontumoral pig kidney (LLC-PK1) cell lines at both temperatures 4° and 37°C. The high ruthenium complex release was observed at 37°C as compared to 4°C due to the active uptake and decomposition of the starch derivatives. Moreover, the tumoral HeLa cells showed the highest ruthenium complex release as compared to nontumoral LLC-PK1 cells. Encapsulation of DOX into the MSiNPs carrier allowed the effective cell death of the HeLa cells.

12.7   BIOACTIVE (MACRO)MOLECULES

The conjugation of bioactive (macro)molecules in synthetic materials is extensively studied as such strategies offer several advantages. Rosenholm and coworkers focused on a folate receptor (FR) that is overexpressed on various kinds of cancer cell lines [117]. They prepared an amino-functionalized MSiNPs (diameter = 400 nm) using cetyltrimethylammonium chloride (CTACl) as template material, and PEI was grown onto the MSiNPs by hyperbranching surface polymerization. Moreover, the methotrexate (MTX), anticancer drug having a similar structure with FA, was reacted with NH_2 groups of the PEI or amino-functionalized MSiNPs [118]. Therefore, the MTX was expected to exhibit dual roles due to their high affinity with the FR, and their cytotoxicity. The MSiNPs-PEI-MTX was incubated in human embryonic kidney cell (HEK 293 cells; low expression level of FR) and FR-expressing HeLa cells, and compared their targeting and toxicity properties with MSiNPs-PEI-FA as a control. The MSiNPs-PEI-MTX was highly uptaken in HeLa cells as compared to HEK 293 cells due to the enhanced interaction of the FR with MTX. Moreover, the apoptosis level of HeLa cells treated by MSiNPs-PEI-MTX was about five times higher than that in HEK 293 cells, which reached 33% after 72 hours of incubation. Importantly, the MSiNPs-PEI-FA showed a very low cytotoxicity under similar incubation condition and the apoptosis level was less than 10% in HeLa cells. These bioactive materials were applied not only as a therapy but also as a way of killing bacteria. N-halamine is known as an antibacterial material with unique properties such as water solubility,

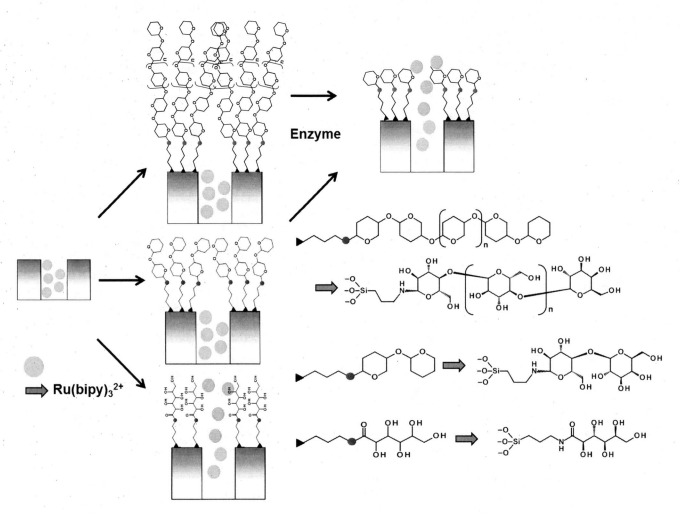

FIGURE 12.11 The carbohydrate gates on the MSiNPs for the enzyme-responsive drug release [116].

high stability, low toxicity, and low cost. The positive halogen from the *N*-halamine was shown to participate in the ionic reaction and interaction with bacteria, which lead to the damage of the metabolic processes in micro-organisms [119]. However, it is difficult to separate the *N*-halamine molecule efficiently from their mixture solution due to their low molecular weight. To solve this problem, Dong and coworkers prepared magnetite/*N*-halamine silica nanocomposites that showed high toxicity to bacteria [120]. First, iron oxide nanoparticles of 6–14 nm were coated with SiNPs, and the poly(3-allyl-5,5-dimethylhydantoin(ADMH)-*co*-methyl methacrylate(MMA)) was used to modify the magnetic-SiNP surface. The ADMH segments were transformed to *N*-halamine by the treatment of 10% commercial aqueous sodium hypochlorite (NaClO) solution buffered for 1 hour at pH 7 (Figure 12.12). The hydrodynamic diameters of the magnetic-SiNPs were changed from 100–160 nm to 120–190 nm via the modification of the copolymers having the *N*-halamine groups, and it was possible to collect the magnetic-

SiNPs easily using the magnet. The sprinkled *N*-halamine magnetic-SiNPs on the culture plate successfully prevented the growth of two types of bacteria that were *S. aureus* and *P. aeruginosa*. The antibacterial activities in *S. aureus* were dependent on the *N*-halamine content which showed the faster antibacterial action with the increase of *N*-halamine (0.96, 1.17, and 1.70%). Moreover, the antibacteria material having 0.96% of *N*-halamine completely killed the *P. aeruginosa* within 5 minutes. Yang et al. prepared the MSiNPs coated by lipid vesicles having a hypocrellin B (HB: photosensitizer) into pore for photodynamic therapy (PDT). The HB can generate reactive oxygen species via the irradiation of light. The composite materials that were composed of lipid-MSiNPs-HB were effectively uptaken into MCF-7 cells through an endocytic mechanism, and showed a high toxicity via the irradiation of 480 nm light [121]. Febvay and coworkers proposed an endosome disruption system using the light irradiation. The MSiNPs having photoactive compound in the pore were used, and were modified with biocompatible

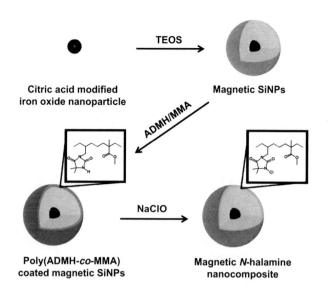

FIGURE 12.12   The magnetite/*N*-halamine silica nanocomposite [120].

polymers and antibody for high stability and cell recognition [122].

12.8   CONCLUSION AND FUTURE TRENDS

The bioconjugation of silica nanomaterials have been extensively studied as these materials offer a range of unique advantages such as high dispersibility, low toxicity, size control and shape, and easy surface modification. The unique chemical and physical properties of silica nanomaterials have allowed the conjugation of a range of biomolecules either on the surface or inside the mesoporous structure. Moreover, it was also possible to trigger the release of the loaded materials by simple engineering of the surface or the pore. The SiNP bioconjugates have been designed for several applications as medical imaging, cellular targeting/recognition, drug delivery, and gene therapy. Silica nanomaterials with multiple functionalities (for instance, having both the abilities of diagnosis and therapy) have been developed. Moreover, the bioconjugation approach and synthesis of the silica nanomaterials have been improved to allow large-scale production and greater impact of these materials.

REFERENCES

1. Arriagada FJ, Osseo-Asare K. *J Colloid Interface Sci* 1999;211:210–220.

2. Stöber W, Fink A, Bohn E. *J Colloid Interface Sci* 1968;26:62–69.

3. Napierska D, Thomassen LCJ, Rabolli V, Lison D, Gonzalez L, Kirsch-Volders M, Martens JA, Hoet PH. *Small* 2009;7:846–853.

4. Matsumura Y, Maeda H. *Cancer Res* 1986;46:6387–6392.

5. Decuzzi P, Godin B, Tanaka T, Lee S-Y, Chiappini C, Liu X, Ferrari, M. *J Control. Release* 2010;141:320–327.

6. Lin Y-S, Abadeer N, Hurley KR, Haynes CL. *J Am Chem Soc* 2011;133:20444–20457.

7. Kresge CT, Leonowicz ME, Roth WJ, Vartuli JC, Beck JS. *Nature* 1992;359:710–712.

8. Beck JS, Vartuli JC, Roth WJ, Leonowicz ME, Kresge CT, Schmitt KD, Chu CTW, Olson DH, Sheppard EW, McCullen SB, Higgins JB, Schlenker JL. *J Am Chem Soc* 1992;114:10834–10843.

9. Yanagisawa T, Shimizu T, Kuroda K, Kato C. *Bull Chem Soc JPN* 1990;63:988–992.

10. Niu D, Ma Z, Li Y, Shi, J. *J Am Chem Soc* 2010;132:15144–15147.

11. Chen Y, Chen H, Guo L, He Q, Chen F, Zhou J, Feng J, Shi, J. *ACS NANO* 2010;4:529–539.

12. Chen Y, Chen H, Zeng D, Tian Y, Chen F, Feng J, Shi, J. *ACS NANO* 2010;4:6001–6013.

13. Ishii H, Sato K, Nagao D, Konno, M. *Colloids Surf B* 2012;92:372–376.

14. Zhang H, Li Z, Xu P, Wu R, Jiao Z. *Chem Commun* 2010;46:6783–6785.

15. Teo BM, Suh SK, Hatton TA, Ashokkumar M, Grieser F. *Langmuir* 2011;27:30–33.

16. Mohanraj VJ, Barnes TJ, Prestidge CA. *Int J Pharm* 2010;392:285–293.

17. Zhao Y, Sun X, Zhang G, Trewyn BG, Slowing II, Lin VS-Y. *ACS NANO* 2011;5:1366–1375.

18. Owens III DE, Peppas NA. *Int J Pharm* 2006;307:93–102.

19. Cauda V, Argyo C, Bein T. *J Mater Chem* 2010;20:8693–8699.

20. Georges MK, Veregin RPN, Kazmaier PM, Hamer GK. *Macromolecules* 1993;26:2987–2988.

21. Kato M, Kamigaito M, Sawamoto M, Higashimura T. *Macromolecules* 1995;28:1721–1723.

22. Wang JS, Matyjaszewski K. *Macromolecules* 1995;28:7572–7573.

23. Chiefari J, Chong YK, Ercole F, Krstina J, Jeffery JTPT, Mayadunne RTA, Meijs GF, Moad CL, Moad G, Rizzardo E, Thang SH. *Macromolecules* 1998;31:5559–5562.

24. Chong YK, Le TPT, Moad G, Rizzardo E, Thang SH. *Macromolecules* 1999;32:2071–2074.

25. Kolb HC, Finn MG, Sharpless KB. *Angew Chem Int Ed* 2001;40:2004–2021.

26. Chen J, Liu M, Chen C, Gong H, Gao C. *ACS Appl Mater Interfaces* 2011;3:3215–3223.

27. Kotsuchibashi Y, Ebara M, Aoyagi T, Narain R. *Polymer Chemistry* 2012;3:2545–2550.

28. Slowing II, Vivero-Escoto JL, Wu C-W, Lin VS-Y. *Adv Drug Delivery Rev* 2008;60:1278–1288.

29. Wan Y, Zhao D. *Chem Rev* 2007;107:2821–2860.

30. Trewyn BG, Giri S, Slowing II, Lin VS-Y. *Chem Commun* 2007;31:3236–3245.

31. Lee S, Yun H-S, Kim S-H. *Biomaterials* 2011;32:9434–9443.

32. Heikkilä T, Santos HA, Kumar N, Murzin DY, Salonen J, Laaksonen T, Peltonen L, Hirvonen J, Lehto V-P. *Eur J Pharm Biopharm* 2010;74:483–494.

33. Liu T, Li L, Teng X, Huang X, Liu H, Chen D, Ren J, He J, Tang F. *Biomaterials* 2011;32:1657–1668.

34. Hudson SP, Padera RF, Langer R, Kohane DS. *Biomaterials* 2008;29:4045–4055.

35. Lee JE, Lee N, Kim H, Kim J, Choi SH, Kim JH, Kim T, Song IC, Park SP, Moon WK, Hyeon T. *J Am Chem Soc* 2010;132:552–557.

36. Yavuz MS, Cheng Y, Chen J, Cobley CM, Zhang Q, Rycenga M, Xie J, Kim C, Song KH, Schwartz AG, Wang LV, Xia Y. *Nature Mater* 2009;8:935–939.

37. Singh N, Karambelkar A, Gu L, Lin K, Miller JS, Chen CS, Sailor MJ, Bhatia SN. *J Am Chem Soc* 2011;133:19582–19585.

38. Meng H, Xue M, Xia T, Zhao Y-L, Tamanoi F, Stoddart JF, Zink JI, Nel AE. *J Am Chem Soc* 2010;132:12690–12697.

39. Thomas CR, Ferris DP, Lee J-H, Choi E, Cho MH, Kim ES, Stoddart JF, Shin J-S, Cheon J, Jeffrey I,Zink JI. *J Am Chem Soc* 2010;132:10623–10625.

40. Yagi H, Yamamoto K, Aoyagi T. *J Chromatogr B* 2008; 876:97–102.

41. Techawanitchai P, Yamamoto K, Ebara M, Aoyagi T. *Sci Technol Adv Mater* 2011;12:044609.

42. Guo Z, Du Y, Liu X, Ng S-C, Yuan Chen Y, Yang Y. *Nanotechnology* 2010;21:165103.

43. Suteewong T, Sai H, Cohen R, Wang S, Bradbury M, Baird B, Gruner SM, Wiesner U. *J Am Chem Soc* 2011;133:172–175.

44. Chen Y, Chen H, Zhang S, Chen F, Zhang L, Zhang J, Zhu M, Wu H, Guo L, Feng J, Shi, J. *Adv Funct Mater* 2011;21:270–278.

45. Wei J, Wang H, Deng Y, Sun Z, Shi L, Tu B, Luqman M, Zhao D. *J Am Chem Soc* 2011;133:20369–20377.

46. Burns A, Ow H, Wiesner H. *Chem Soc Rev* 2006;35:1028–1042.

47. Willmann JK, van Bruggen N, Dinkelborg LM, Gambhir, SS. *Nat Rev Drug Discovery* 2008;7:591–607.

48. Lee D-E, Koo H, Sun I-C, Ryu JH, Kim K, Kwon IC. *Chem Soc Rev* 2012;41:2656–2672.

49. Wang F, Tan WB, Zhang Y, Fan X, Wang M. *Nanotechnology* 2006;17:R1-R13.

50. Medintz IL, Uyeda HT, Goldman ER, Mattoussi H. *Nat Matter* 2005;4:435–446.

51. Hoshino A, Fujioka K, Oku T, Suga M, Sasaki YF, Ohta T, Yasuhara M, Suzuki K, Yamamoto K. *Nano Lett* 2004;4:2163–2169.

52. Jain PK, El-Sayed IH, El-Sayed MA. *Nanotoday* 2007;2: 18–29.

53. Pinho SLC, Pereira GA, Voisin P, Kassem J, Bouchaud V, Etienne L, Peters JA, Carlos L, Mornet S, Geraldes CFGC, Rocha J, Delville M-H. *ACS NANO* 2010;4:5339–5349.

54. Giovanetti LJ, Ramallo-López JM, Foxe M, Jones LC, Koebel MM, Somorjai GA, Craievich AF, Salmeron MB, Requejo FG. *Small* 2012;8:468–473.

55. Lee KG, Wi R, Park TJ, Yoon SH, Lee J, Lee SJ, Kim DH. *Chem Commun* 2010;46:6374–6376.

56. Wang T-T, Chai F, Wang C-G, Li L, Liu H-Y, Zhang L-Y, Su Z-M, Liao Y. *J Colloid Interface Sci* 2011;358:109–115.

57. Chavanpatil MD, Khdair A, Panyam J. *J Nanosci Nanotechnol* 2006;6:2651–2663.

58. Allen TM. *Trends Pharmacol Sci* 1994;15:215–220.

59. Ku S, Yan F, Wang Y, Sun Y, Yang N, Ye L. *Biochem Biophys Res Commun* 2010;394:871–876.

60. Bardi G, Malvindi MA, Gherardini L, Costa M, Pompa PP, Cingolani R, Pizzorusso T. *Biomaterials* 2010;31:6555–6566.

61. Ow H, Larson DR, Srivastava M, Baird BA, Webb WW, Wiesner U. *Nano Lett* 2005;5:113–117.

62. Wu S, Li Z, Han J, Han S. *Chem Commun* 2011;47:11276–11278.

63. Burns A, Sengupta P, Zedayko T, Baird B, Wiesner U. *Small* 2006;2:723–726.

64. Tanaka K, Kitamura N, Chujo Y. *Bioorg Med Chem* 2012;20:96–100.

65. He Q, Shi J, Cui X, Wei C, Zhang L, Wu W, Bu W, Chen H, Wu H. *Chem Commun* 2011;47:7947–7949.

66. Zong C, Ai K, Zhang G, Li H, Lu L. *Anal Chem* 2011;83:3126–3132.

67. Ai K, Zhang B, Lu L. *Angew Chem Int Ed* 2009;48:304–308.

68. (a) Winfree E, Liu F, Wenzler LA, Seeman NC. *Nature* 1998;394:539–544. (b) Shih WM, Quispe JD, Joyce GF. *Nature* 2004;427:618–621. (c) Rothemund PWK. *Nature* 2006;440:297–302.

69. Mirkin CA, Letsinger RL, Mucic RC, Storhoff JJ. *Nature* 1996;382:607–609.

70. Alivisatos AP, Johnssonm KP, Peng X, Wilson TE, Loweth CJ, Bruchez MP, Schultz PG. *Nature* 1996;382:609–611.

71. Sato K, Hosokawa K, Maeda M. *J Am Chem Soc* 2003;125:8102–8103.

72. Wu J, Silvent J, Coradin T, Aimé C. *Langmuir* 2012;28:2156–2165.

73. Wong JKF, Yip SP, Lee TMH. *Small* 2012;8:214–219.

74. Climent E, Martínez-Máñez R, Sancenón F, Marcos MD, Soto J, Maquieira A, Amorós P. *Angew Chem Int Ed* 2010;49:7281–7283.

75. Chen C, Geng J, Pu F, Yang X, Ren J, Qu X. *Angew Chem Int Ed* 2011;50:882–886.

76. Chen L, Di J, Cao C, Zhao Y, Ma Y, Luo J, Wen Y, Song W, Song Y, Jiang L. *Chem Commun* 2011;47:2850–2852.

77. Kakizawa Y, Harada A, Kataoka K. *Biomacromolecules* 2001;2:491–497.

78. Toita S, Morimoto N, Akiyoshi K. *Biomacromolecules* 2010;11:397–401.

79. Ahmed M, Bhuchar N, Ishihara K, Narain R. *Bioconjug Chem* 2011;22:1228–1238.

80. Fujiwara M, Yamamoto F, Okamoto K, Shiokawa K, Nomura R. *Anal Chem* 2005;77:8138–8145.

81. Kim M-H, Na H-K, Kim Y-K, Ryoo S-R, Cho HS, Lee KE, Jeon H, Ryoo R, Min D-H. *ACS NANO* 2011;5:3568–3576.

82. Yang H, Zheng K, Zhang Z, Shi W, Jing S, Wanga L, Zheng W, Zhao D, Xu J, Zhang P. *J Colloid Interface Sci* 2012;369:317–322.

83. Bhakta G, Sharma RK, Gupta N, Cool S, Nurcombe V, Maitra A. *Nanomed Nanotech Biol Med* 2011;7:472–479.

84. Li X, Zhang J, Gu H. *Langmuir* 2012;28:2827–2834.

85. Li X, Xie QR, Zhang J, Xia W, Gu H. *Biomaterials* 2011;32:9546–9556.

86. Meng H, Liong M, Xia T, Li Z, Ji Z, Zink JI, Nel AE. *ACS NANO* 2010;4:4539–3550.

87. Gottesman MM. *Annu Rev Med* 2002;53:615–627.

88. Suwalski A, Dabboue H, Delalande A, Bensamoun SF, Canon F, Midoux P, Saillant G, Klatzmann D, Salvetat J-P, Pichon C. *Biomaterials* 2010;31:5237–5345.

89. Yokoyama M, Okano T, Sakurai Y, Fukushima S, Okamoto K, Kataoka K. *J Drug Target* 1999;7:171–186.

90. Loret EP, Vives E, Ho PS, Rochat H, Rietschoten JV, Johnson WC Jr. *Biochemistry* 1991;30:6013–6023.

91. Tkachenko AG, Xie H, Coleman D, Glomm W, Ryan J, Anderson MF, Franzen S, Feldheim DL. *J Am Chem Soc* 2003;125:4700–4701.

92. Sethuraman VA, Bae YH. *J Control Release* 2007;118:216–224.

93. Mao Z, Wan L, Hu L, Ma L, Gao C. *Colloids Surf B* 2010;75:432–440.

94. Luo Z, Cai K, Hu Y, Zhao L, Liu P, Duan L, Yang W. *Angew Chem Int Ed* 2011;50:640–643.

95. Liu H, Liu T, Wu X, Li L, Tan L, Chen D, Tang F. *Adv Mater* 2012;24:755–761.

96. Gatter KC, Brown G, Trowbridge IS, Woolston RE, Mason DY. *J Clin Pathol* 1983;36:539–545.

97. Karlsson M, Carlsson C. *Biophysic J* 2005;88:3536–3544.

98. Shrivastava S, Nuffer JH, Siegel RW, Dordick JS. *Nano Lett* 2012;12:1583–1587.

99. Park HS, Kim CW, Lee HJ, Choi JH, Lee SG, Yun Y-P, Kwon IC, Lee SJ, Jeong SJ, Lee SC. *Nanotechnology* 2010;21:225101.

100. Vinu A, Murugesan V, Tangermann O, Hartmann M. *Chem Mater* 2004;16:3056–3065.

101. Bale SS, Kwon SJ, Shah DA, Banerjee A, Dordick JS, Kane RS. *ACS NANO* 2010;4:1493–1500.

102. Sano K, Minamisawa T, Shiba K. *Langmuir* 2010;26:2231–2234.

103. Saito H, Honma T, Minamisawa T, Yamazaki K, Noda T, Yamori T, Shiba K. *Chem Biol* 2004;11:765–773.

104. Narain R (ed.). *Engineered carbohydrate-based materials for biomedical applications.* John Wiley & Sons, Inc.; 2011.

105. (a) Ahmed M, Narain R. *Biomaterials* 2011;32:5279–5290. (b) Ahmed M, Narain R. *Biomaterials* 2012;32:3990–4001.

106. (a) Maruyama A, Ishihara T, Kim J-S, Kim SW, Akaike T. *Bioconjug Chem* 1997;8:735–742. (b) Hasegawa U, Shinichiro M, Nomura SM, Kaul SC, Hirano T, Akiyoshi K. *Biochem Biophys Res Commun* 2005;331:917–921.

107. Lee YC, Lee RT. *Acc Chem Res* 1995;28:321–326.

108. Kotsuchibashi Y, Zhang Y, Ahmed M, Ebara M, Aoyagi T, Narain R. *J Biomed Mater Res A* 2013;101:2090–2096.

109. Pfaff A, Schallon A, Ruhland TM, Majewski AP, Schmalz H, Freitag R, Müller, AHE. *Biomacromolecules* 2011;12:3805–3811.

110. Wang X, Ramström O, Yan M. *Chem Commun* 2011;47:4261–4263.

111. Zhao Y, Trewyn BG, Slowing II, Lin VS-Y. *J Am Chem Soc* 2009;131:8398–8400.

112. Guo T-Y, Liu P, Zhu J-W, Song M-D, Zhang B-H. *Biomacromolecules* 2006;7:1196–1202.

113. Leirose GDS, Cardoso MB. *J Pharmaceut Nanotech* 2011; 100:2826–2834.

114. Earhart C, Jana NR, Erathodiyil N, Ying JY. *Langmuir* 2008;24:6215–6219.

115. Kim Y-J, Ha S-W, Jeon S-M, Yoo DW, Chun S-H, Sohn B-H, Lee J-K. *Langmuir* 2010;26:7555–7560.

116. Bernardos A, Mondragón L, Aznar E, Marcos MD, Martínez-Máñez R, Sancenón F, Soto J, Barat JM, Pérez-Payá E, Guillem C, Amorós P. *ACS NANO* 2010;4:6353–6368.

117. Elnakat H, Ratnan M. *Adv Drug Delivery Rev* 2004;56:1067–1084.

118. Rosenholm JM, Peuhu E, Bate-Eya LT, Eriksson JE, Sahlgren C, Lindén M. *Small* 2010;6:1234–1241.

119. Liu S, Sun G. *Ind Eng Chem Res* 2006;45:6477–6482.

120. Dong A, Lan S, Huang J, Wanga T, Zhao T, Wanga W, Xiao L, Zheng X, Liu F, Gao G, Chen Y. *J Colloid Interface Sci* 2011;364:333–340.

121. Yang Y, Song W, Wang A, Zhu P, Fei J, Li J. *Phys Chem Chem Phys* 2010;12:4418–4422.

122. Febvay S, Marini DM, Belcher AM, Clapham DE. *Nano Lett* 2010;10:2211–2219.

13

POLYHEDRAL OLIGOMERIC SILSESQUIOXANES (POSS) BIOCONJUGATES

WEIAN ZHANG, ZHENGHE ZHANG, AND LIZHI HONG

Shanghai Key Laboratory of Functional Materials Chemistry, East China University of Science and Technology, Shanghai, PR China

13.1 INTRODUCTION

In the last two decades, polyhedral oligomeric silsesquioxane (POSS) has attracted considerable attention, although silsesquioxane chemistry has been well studied in the last century [1–10]. POSS molecules have a cage-shaped three-dimensional (3D) structure with the formula $(RSiO_{1.5})_n$, ($n \geq$ 6) [11–17]. Among them, octasilsesquioxanes $(R_8Si_8O_{12})$ have been mostly investigated; they consist of a rigid, cubic inorganic silica core with a 0.53 nm side length and eight corner organic groups (Figure 13.1). Thus, POSS is different with some conventional inorganic nanoparticles, and it is a class of unique inorganic components with a definite nanostructure and chemical composites. POSS molecules can be modified into a variety of POSS derivatives based on their corner groups. These corner groups can be reactive (–OH, –COOH, –NH$_2$, –SH, acrylate, etc.) or unreactive unit (alkyl, benzyl, etc.), and provide the POSS molecules with the higher reactivity and solubility. In the past several years, POSS molecules have been incorporated into polymer matrices easily using chemical coupling, copolymerization, cross-linking, or physical blending to produce novel POSS-containing hybrid polymers with promising properties such as improved mechanical and thermal properties, oxidation resistance, and reduced flammability [1–10, 18–34].

POSS molecule has been considered to be the promising material in biological fields, according to the unique structure and properties of POSS derivatives. The novel biomedical devices, tissue engineering materials, and dental materials based on POSS have been well developed, which effectively improve mechanical properties and leads to a long service life of these materials based on POSS [35–37]. Additionally, from the chemical structures of POSS molecules, they are composed of Si-O-Si and Si-C bonds, which are inert and nontoxic, and they can be readily used to construct POSS-containing bioconjugates using the functionalization of corner groups. For example, POSS molecules have been developed into functional dendrimers with POSS as the core, and these dendrimers have been used as drug delivery, biological imagines, biosensors, etc. [9, 38–41]. In this chapter, we give some examples to describe the construction and properties of POSS-containing bioconjugates based on peptides, carbohydrates, and DNA and highlight their potential applications as biomedical devices, tissue engineering materials, dental materials, drug delivery, and biological images.

13.2 PREPARATION OF POSS-CONTAINING BIOCONJUGATES

13.2.1 POSS-Containing Peptides/Proteins

Peptide/protein hybrid materials incorporated with POSS molecules have been developed by several groups, and their secondary structures, self-assembly, and application were also studied [42, 43]. In the preparation of POSS-containing hybrid polymers, the POSS molecules can be introduced into polymer matrices mainly by physical blending and chemical reactions, which is also used to prepare the POSS-containing peptide/protein conjugates. Recently, with the great progress in polymer synthesis, especially for living polymerization technique and click chemistry, well-defined

Chemistry of Bioconjugates: Synthesis, Characterization, and Biomedical Applications, First Edition. Edited by Ravin Narain.

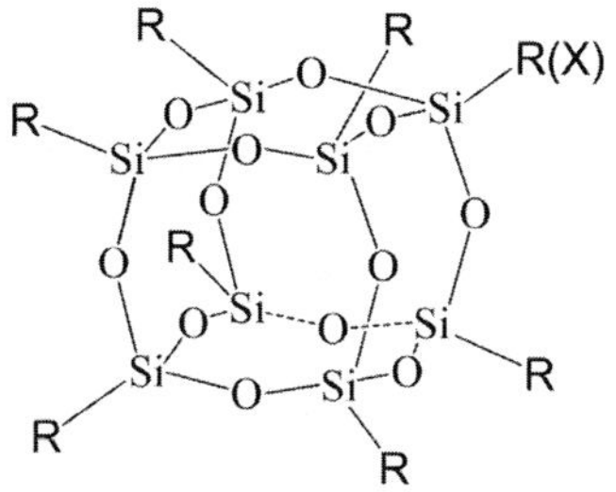

FIGURE 13.1 The structure of the typical POSS molecule, where R(X) is the reactive or unreactive group.

polymers can be constructed in various kinds of architectures such as telechelic, block, star, branch, and dendritic polymers [44–46]. POSS-containing hybrid polymers with the above-mentioned architectures also have been prepared using advanced polymer synthesis techniques. For example, we can prepare hemitelechelic POSS-containing hybrid polymers using living polymerization such as ring-opening polymerization (ROP), atom transfer radical polymerization (ATRP), and reversible addition–fragmentation chain transfer (RAFT) polymerization using mono-functionalized POSS molecules [47, 48]. Similarly, the multi-functionalized POSS molecules also can be modified into initiators for living polymerization to produce star-shaped POSS-containing hybrid polymers [49, 50]. POSS-containing peptide hybrids with different architectures have been constructed using living polymerization. For example, Savin and coworkers

prepared the hemitelechelic POSS-containing hybrids via ROP of the *N*-carboxyanhydride of lysine(Z) directly using aminopropylheptaisobutyl-POSS (POSS-NH$_2$) as the initiator, and they further used these POSS-containing poly(Z-lysine) hybrids to construct the thermoreversible physical gel in tetrahydrofuran [51]. More recently, the discovery of "click chemistry" has made a deep and wide impact on the fields of chemistry, biology, and material sciences, since it provides a simple method to conjugate the variety of components together [52]. "Click chemistry" has also been proven to be a fast and efficient approach to prepare POSS-containing peptide with different architectures as it can be performed in high yields under mild reaction conditions and it has a good tolerance of functional groups. POSS molecules were often modified into "click chemistry agents" such as azido-functional POSS and mercapto-functional POSS, and then were attached to peptide chains as the end-group or pendant group. Kuo and his coworkers did a lot of work in this field. They prepared azido-mono-functionalized POSS (POSS-N$_3$) using trisilanolheptaisobutyl polyhedral oligomeric silsesquioxane (POSS-(OH)$_3$) corner-capped with trichloro[4-(chloromethyl)phenyl]silane to afford 4-(chloromethyl)phenyl POSS (POSS-Cl), followed by the transfer of the chlorine group of Cl-POSS into the azido group with sodium azide in DMF (Figure 13.2). The POSS-N$_3$ was further used to prepare hemitelechelic POSS-containing peptide hybrids with preprepared alkyne-mono-functionalized poly(γ-benzyl-L-glutamate) (alkyne-PBLG) [47]. The

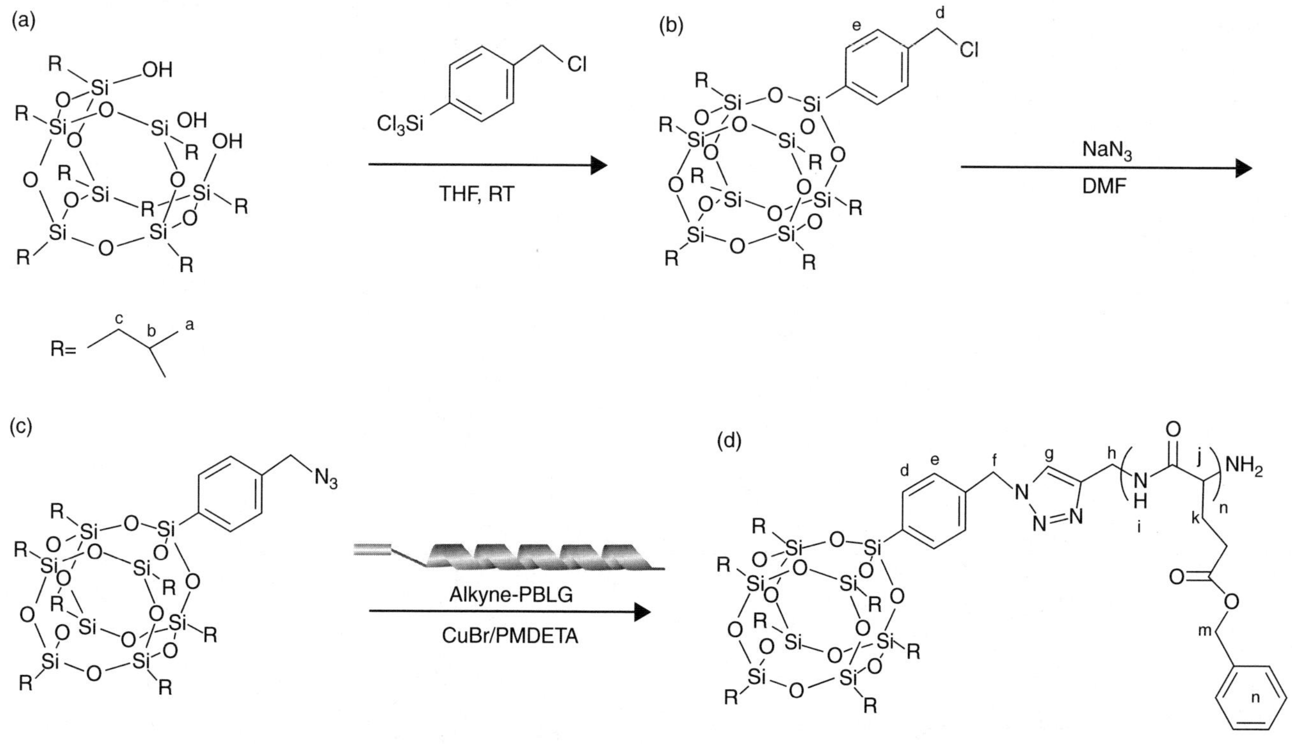

FIGURE 13.2 Synthesis of POSS-containing poly(γ-benzyl-L-glutamate) (POSS-PBLG).

incorporation of the POSS unit at the chain end of the PBLG moiety enhanced the α-helical conformation of PBLG in the solid state by the intramolecular hydrogen bonding between the POSS and PBLG units, which is determined by Fourier transform infrared spectroscopy, solid-state nuclear magnetic resonance, and wide-angle X-ray diffraction analysis. The POSS-N$_3$ was also introduced as the pendant group of polypeptide, poly(γ-propargyl-L-glutamate) (PPLG), to construct POSS-containing PPLG homopolymer and block copolymers which were prepared by the ROP of γ-propargyl-L-glutamate N-carboxyanhydride (PLG-NCA) respectively using butylamine and amine-terminated polystyrene as the initiators [53, 54]. The incorporation of the POSS as the pendant groups of the PPLG moiety also enhanced the α-helical conformation in the solid state, and POSS-containing PS-b-(PPLG-g-POSS) diblock copolymers can form a hexagonal cylinder packing nanostructure. They also prepared star-shaped POSS-containing polypeptide hybrids from octa-azido-functionalized POSS with alkyne-poly(γ-benzyl-L-glutamate) (alkyne-PBLG) via a click reaction [55]. Except for the above click reaction catalyzed by copper, photoinduced free-radical thiol–ene click reaction also utilized to prepare POSS-containing peptide. The star-shaped POSS-containing peptide hybrids can be prepared using octavinyl octasilsesquioxane with commercial peptide such as tripeptide glutathione, Glu-Cys-Gly (GSH), and tetrapeptide Arg-Gly-Asp-Cys via thiol–ene click reaction [56]. In addition, Savin and his coworkers synthesized POSS-containing poly(L-glutamic acid) using mercapto-mono-functionalized POSS (POSS-SH) with alkyne-terminated PBLG via the thiol–alkyne chemistry, and POSS-containing PBLG can self-assemble into pH-responsive vesicles in aqueous solution [51, 57]. Fabritz recently developed a new bioorthogonal POSS scaffold bearing eight aminooxy coupling sites allowing for the conjugation of diverse peptides via

oxime ligation, and they found that the coupling efficacy depends on the ligand in view of steric hindrance and electrostatic repulsion [58]. For the first time POSS-containing conjugation of cystine-knot miniproteins bearing a backbone of about 30 amino acids was successfully accomplished without loss of bioactivity.

13.2.2 POSS-Containing Carbohydrates

As the bioconjugates, carbohydrates are referred to as sugars or saccharides. POSS-containing carbohydrate hybrids have been recently developed using octa-functionalized POSS. Feher and coworkers first prepared POSS-containing carbohydrate-functionalized silsesquioxane which demonstrated the selective and reversible complexation to carbohydrate-binding proteins [59]. The glycodendrons were attached to the octa-amine-functionalized POSS via standard amide bond formation with carbohydrate-derived lactones. Although the strategy was useful for producing POSS-containing glycoclusters, its yield is not high (20–53%). Lee and coworkers recently developed an efficient synthetic route for a variety of POSS-containing glycoclusters from unprotected mannosides and lactosides by photocatalyzed thiol–vinyl click chemistry (Figure 13.3) [39]. The thiol-radical addition reaction was carried out between thiol-terminated glycoside residues and octavinyl POSS in H$_2$O/THF (1:1) under irradiation with UV light (254 nm) in the presence of a catalytic amount of AIBN. After 24–48 hours, the crude product was purified by Sephadex G-15 gel filtration (H$_2$O as eluent) to afford pure POSS-containing glycoclusters with a high yield (66–73%). This strategy should be easily extended to prepare other kinds of POSS-containing glycoclusters, including those of more complicated oligosaccharides to POSS cores with possible variations in spacers. The click reaction of copper-catalyzed 1,3-dipolar azide–alkyne

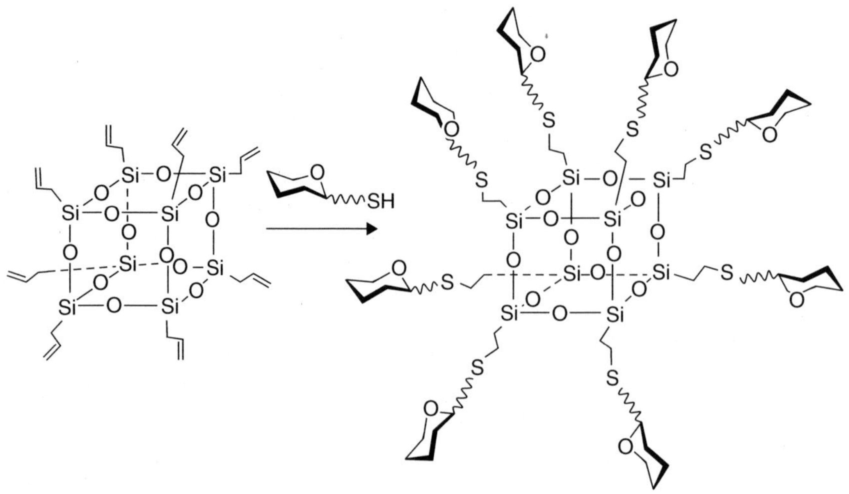

FIGURE 13.3   Preparation of POSS-containing carbohydrate cluster compounds using thiol-radical addition reaction.

cycloaddition was also used to prepare POSS-containing carbohydrates. Octakis(3-azidopropyl)octasilsesquioxane has been proved to be an excellent nanobuilding block for the efficient synthesis of new functional cubic POSS through copper-catalyzed click reaction [60, 61]. Trastoy and coworkers also used the same strategy to synthesize POSS-containing glycoclusters with different structures under the optimized click reaction conditions (cat. $CuSO_4 \cdot H_2O$, sodium ascorbate in CH_2Cl_2/H_2O 1:1 at room temperature) using the appropriate alkyne-functionalized carbohydrate derivatives with linkers of different chain lengths between the anomeric carbon and the terminal alkyne group, and they further studied the specific binding interactions of these glyco-POSS compounds with a model plant lectin using an array of complementary biophysical techniques [62].

13.2.3 POSS-Containing DNA Complexes

POSS molecule was also attempted to construct POSS-containing DNA complex, where the complexation is formed by the electrostatic interaction between negatively charged DNA and positively charged cationic POSS molecules. Zhu and coworkers constructed the POSS-containing DNA complexes using a cationic lipid of the POSS imidazolium salt with double-stranded DNA, and further studied their self-assembly behaviors [63]. The complexes can form an inverted hexagonal phase above the melting point of POSS crystals, due to the negative spontaneous curvature of the POSS imidazolium cationic lipid. They also studied the effect of POSS crystallization process on the self-assembled morphologies, since the competition existed between the crystallization of POSS molecules and the negative spontaneous curvature of cationic POSS imidazolium lipids [64]. A lamellar phase was obtained when the crystallization was relatively slow (e.g., isothermal crystallization at 130°C), where the crystallization of POSS molecules predominated the self-assembly process. When the crystallization happened very fast (e.g., quenched to 0°C), lipid negative curvature predominated the self-assembly process, and the inverted hexagonal phase was retained with POSS lamellar crystals growing in the interstitials of DNA cylinders. Interestingly, double-stranded DNA adopted the B-form helical conformation in the inverted hexagonal phase, whereas the helical conformation was largely destroyed in the lamellar phase.

13.3 APPLICATIONS OF POSS-CONTAINING BIOCONJUGATES

13.3.1 Medical Devices

POSS molecules with a unique nanocage structure were incorporated into the polymeric matrices to produce novel POSS-containing hybrid polymers with promising properties such as improved mechanical, thermal, and superior surface properties, which allow POSS-containing hybrid polymers to be suitable for biological applications [10, 65–67]. Seifalian and coworkers have contributed to a lot of pioneering work on the application of POSS-containing hybrid polymers in biological fields, and they predicted that POSS-containing hybrid polymers could be the next generation biological materials [68, 69]. Polyurethanes (PUs) are segmented polymers consisting of alternating sequences of soft segments and hard segments that constitute a unique microphase-separated structure, which has been wildly used in a variety of medical devices [21, 70]. The hard segments of the traditional PUs were of urethanes or urea, but recently, organic silicon was introduced into PUs to enhance both hydrolytic and oxidative stabilities [71, 72]. Seifalian and coworkers prepared POSS-containing PUs, and they found that these POSS nanocores shield the soft segment(s) of the PU, responsible for its compliance and elasticity from all forms of degradation, oxidation, and hydrolysis [73]. Thus, these nanocomposites provide an optimal method by which these polymers may be strengthened while maintaining their elasticity, making them ideal for use as medical devices such as cardiovascular grafts, heart valves, or artificial scaffolds in cartilage tissue engineering [25, 74]. For example, they evaluated the calcification property, mechanical characteristics, and surface integrity of POSS–PU in an *in vitro* accelerated physiological pulsatile pressure system model. The results revealed that POSS–PU has a significant resistance to calcification, and its mechanical properties and surface all remained unchanged in comparison with the control PU. In addition, significantly less platelet adhered to the POSS–PU. These findings suggested that POSS–PU is a potential application for the fabrication of artificial heart valves [75]. It is well known that silicone was used in breast implants, since it was thought to be a relatively inert biomaterial with minimal complications and inflammation rates. Nevertheless, in the long-term practice, silicone had some obvious shortcomings such as delaying wound healing and pseudoinflammation [76, 77]. POSS-containing hybrid polymers *in vivo* study did not demonstrate significant inflammation and material degradation, which could be an alternative option in breast implant products instead of conventional silicone [78].

13.3.2 Tissue Engineering Scaffolds

Tissue engineering which is combining of cells, medicine, engineering, molecular biology, and materials science and engineering to improve or replace biological functions, has attracted great interest as a rapidly emerging research field. POSS derivative is much better for tissue engineering materials, since it has a 3D unique structure and biological properties. POSS-containing hybrid polymer has been developed for 3D tissue engineering scaffold, based on the unique cube structure of POSS unit and the characters of polymers.

These hybrid polymers can be well fabricated with mechanical and biological properties, and they should have the good cell adherence, subsequent proliferation, and differentiation. Guo and coworkers fabricated POSS-containing poly(ester urethanes) (PEU) into 3D tissue engineering scaffold with the desired porosity, mechanical property, and biodegradability, which could provide excellent support for embryonic stem cell (ESC), so POSS-containing PEU can be potentially applied in ESC-based tissue engineering and regenerative medicine [79, 80]. Wang and coworkers synthesized a series of novel injectable, photocrosslinkable, and biodegradable POSS-containing poly(propylene fumarate) (PPF) hybrid copolymers (PPF-*co*-POSS) via polycondensation [81]. The cell responses of these POSS-containing PPF were evaluated *in vitro* MC3T3-E1 cell experiments including cell attachment, spreading, proliferation, differentiation, and gene expression maximized at the content of POSS of 10%, indicating nonmonotonic or parabolic dependence on the content of POSS. The results revealed that POSS-containing PPF could provide a series of injectable and biodegradable tissue engineering scaffolds with better mechanical properties and osteoconductivity for bone repair and regeneration.

13.3.3 Dental Materials

Dental composites have been developed for over several decades. Dental composites based on acrylate have been extensively used with glass, metallic, or ceramic fillers as the fillers. The resin matrix can be cured (hardened) by photoinitiated free-radical polymerization. For example, the monomer 2,2-bis-[4-(methacryloxypropoxy)-phenyl]-propane (Bis-GMA) was first applied in construction of dental materials by Bowen [82]. Nowadays, it is still commonly used in this field. Despite much effort has been done to improve the properties of dental materials, there is still some problems in practical applications such as polymerization shrinkage, wear resistance, biocompatibility, toxicity of monomers, and modulus of elasticity [83]. Recently, POSS-containing hybrid polymers have been introduced into dental materials, which efficiently enhanced the properties of dental materials such as decreasing polymer shrinkage, enhanced mechanical properties, and oxygen permeability [84, 85]. Wu and coworkers synthesized POSS-containing dental composite resins using multifunctional methacryl POSS, Bis-GMA, and tri(ethylene glycol) dimethacrylate (TEGDMA) [86]. The mechanical behavior and volumetric shrinkage of these dental resins were evaluated for their dependence on the weight fraction of POSS from 0 to 15 wt%. The results revealed that the dental resin with 2 wt% POSS led to the increase of 15% in flexural strength, 12% in compressive strength, 4% in compressive modulus, 15% in hardness, 56% in fracture energy, and the decrease of volumetric shrinkage

from 3.53 to 3.10. This indicated that the properties of POSS-containing dental materials could be significantly improved only at addition of 2 wt% POSS. Kim and coworkers also studied the biocompatibility of POSS-containing methacrylate dental composites, and found POSS-containing acrylic-based hybrid composites with improved biocompatibility. This further confirmed POSS-containing hybrid resin could be used as dental materials with reduced shrinkage, improved mechanical property, and biocompatibility.

13.3.4 Drug Delivery

To well improve the efficiency and relieve some clinical side effects of drugs, the drug delivery materials have been developed. POSS-containing hybrids can be well used as drug delivery materials, since these materials have potential advantages such as their biodegradation, thermodynamic stability, and biocompatibility, and their being easily transferred by transmembrane and vascular pores [4, 24, 87–90]. Rotello and coworkers designed cationic POSS compounds as carriers to evaluate the processes of drug delivery [91]. They labeled octa-ammonium-POSS with a fluorescent dye (boron-dipyrromethene, BODIPY) to produce POSS-containing BODIPY (POSS–BODIPY) by the neutralization of ammonium on the POSS subunits with triethylamine and substitution with a succinimidyl ester derivative. BODIPY is a commonly employed fluorescent cellular membrane marker, which can be readily conjugated to various systems to track cellular migration patterns *in vitro*. The results showed that POSS presented very low toxicity and efficient uptake in the cytoplasm of Cos-1 cells. The POSS–BODIPY conjugate entered into the cells by passive diffusion, not by endocytosis. POSS-containing PU hybrids also have used in drug release system. For instance, Guo and coworkers incorporated the anticancer drug of paclitaxel (PTx) into POSS-containing hybrid biodegradable thermoplastic polyurethanes (TPUs) to prepare PTx/POSS TPU blends using a solution casting method, and systematically studied the morphology, miscibility, and interactions of paclitaxel over all proportions of the drug concentrations in POSS-containing TPUs [87, 88]. These POSS TPUs with an alternating multiblock structure are composed of POSS hard segments and biodegradable soft segments of amorphous polylactide/caprolactone copolymer (P(DLLA-*co*-CL)) incorporating PEG covalently. The results demonstrated PTx was not only amorphous and well miscible in all proportions in the PTx/POSS TPU blends, but also served as an antiplasticizer by increasing T_g of the blends, since it has strong H-bonding interactions with the PEG segments. Although PEG can promote the favorable H-bonding interactions in the PTx/POSS TPU blends, the interaction between the PTx and the lactide/caprolactone (LA/CL) repeat units still existed at all the range of the drug concentration in the POSS TPUs, which can achieve molecular-level miscibility. In these blends, the stable POSS molecules as hard segments endowed the PTx/POSS TPU blends with high mechanical properties, which should have the great potential for the drug-loaded polymeric systems requiring high mechanical strength in processing and during drug release. The release of PTx from POSS TPU stent coatings incorporating 5 wt% PTx in PBS buffer solution was also well evaluated, and the release rate can be precisely controlled through variation in T_g of polymer, degradation rate, and thickness increment rate, with 90% of the drug releasing from within half a day to about 90 days. Dendrimers feature a symmetric monodisperse macromolecule with a well-defined structure, globular shape, and surface functionalities. These characteristics make them as ideal drug carriers. Recently dendrimer-based bioconjugates have been developed for the delivery of anticancer drugs, gene, imaging agent, etc. [92, 93]. POSS molecule has its own advantages in the construction of the dendrimers. The cage-shaped 3D POSS molecule surrounded by eight corner organic groups is better for the growth of more generations in three dimensions, and the eight corner organic groups can be modified into a wide variety of functional groups such as vinyl, amino, and hydroxyl, which are easy to expand for the preparation of different kinds of dendrimers with a POSS core. Gu and coworkers prepared a novel POSS-containing poly(L-glutamic acid) dendrimer. The succinic acid-terminated POSS (POSS-SA, or OAS-SA) was prepared using octa(3-aminopropyl) POSS hydrochloride (POSS·HCl) with succinic anhydride in the presence of triethylamine, and POSS·HCl was further used to prepare the *tert*-butyl ester-protected POSS-poly(L-glutamic acid) dendrimers with excess H-Glu(OBut)-OBut·HCl. Finally, novel POSS-containing poly(L-glutamic acid) (OAS-Glu or POSS-Glu) dendrimer was obtained by the deprotection reaction of tert-butyl ester-protected OAS-poly(L-glutamic acid) dendrimers. The second and third POSS-Glu dendrimers were prepared by the last steps of the first POSS-Glu dendrimers. The POSS-Glu dendrimers was further conjugated with doxorubicin (DOX) via pH-sensitive hydrazine bonds and targeting moiety (biotin) to fabricate the complex of POSS-G3-Glu(NHN-DOX)/biotin (Figure 13.4) [89]. DOX, as a mostly used antitumor drug, has been studied for its delivery process. The POSS-G3-Glu(NHN-DOX)/biotin complex can self-assemble to spherical aggregates in aqueous solution with the size of 30–50 nm, which were suitable for injectable anticancer drug carriers due to the effect of enhanced permeability and retention (EPR). DOX release was carried out at different pH at 37°C, and the result showed that the release rates of DOX at pH = 5.0 were much faster than those at pH = 7.0, due to the acid cleavage of the hydrazine bonds. After 20 hours, the totally released DOX reached about 90% at pH = 5.0. The POSS-G3-Glu(NHN-DOX)/biotin complex had a low cytotoxicity when the concentrations were below 15 μg/mL. Moreover, the cytotoxicity increased with the increase of dendrimer concentration. In addition, the results

of the confocal laser scanning microscopy (CLSM) and flow cytometry analysis indicated that the cellular uptake of DOX dendrimer complex can be enhanced with the introduction of biotin through receptor-mediated endocytosis. Nowadays, gene therapy has been a promising technique in clinical field, so the gene delivery also becomes a research topic with the aim of improved efficiency. POSS-containing polymers have been developed as gene delivery vectors by several groups [94–97]. For example, POSS-containing poly(L-lysine) dendrimers with POSS as the core have been constructed using octa(3-aminopropyl)silsesquioxane as the initiator via liquid-phase peptide synthesis technique, and further applied as the carrier of gene delivery [96]. In each preparation step of POSS-containing poly(L-lysine) dendrimers in generation 1–4, a large excess of N_α, N_ε-di-*t*-BOC-L-lysine dicyclohexylammonium and coupling agents were used to ensure complete reaction of the surface amino groups with protected L-lysine, and diisopropylethylamine (DIPEA) was used to obtain free amino groups on the dendrimer surface. These POSS-containing poly(L-lysine) dendrimers exhibited a size-dependent toxicity in the cytotoxicity evaluation, but it was much lower than that of linear poly-L-lysine. Gel retardation assays showed that these dendrimers effectively bound and retarded the migration of DNA at low *N/P* ratios. In *in vitro* gene delivery experiments, they demonstrated they can efficiently deliver plasmid DNA to MDA-MB-231 cells. Loh and coworkers reported the star-shaped POSS-containing hybrid poly(2-dimethylamino)ethyl methacrylate (PDMAEMA) was prepared by ATRP using POSS as initiator, and further applied as gene delivery (Figure 13.5) [97]. PDMAEMA is a very popular cationic polymer for gene delivery as it has relatively low toxicity and high buffer capacity. This polymer can self-assemble into micelles for the encapsulation of PTx in the aqueous solution and the cationic PDMAEMA arms can bind with the negatively charged DNA plasmids by charge interaction. The literature has reported cationic polymers such as PDMAEMA and PEI are harmful to cells, since they could permeabilize membranes and cause cell rupture. In this work, the biological experiments showed that these PTx-loaded micelles were less toxic than poly(ethylenimine) (PEI) and PDMAEMA homopolymers. On the other hand, gene delivery efficiency can be improved by using cell-sensitizing drugs to treat the cells before administering gene transfection. Anticancer drug, PTx, has been used to pretreat cells for enhancing the expression of reporter genes. When PTx mixed with the POSS-*g*-PDMAEMA polymer, the gene transfection efficiency was efficiently improved. The transfection efficiency of PTx-loaded POSS-*g*-PDMAEMA is about 1.6 times without serum and 7 times with serum, over 7 times better than PEI at the N/P ratio of 10. Thus, PTx-encapsulated POSS-containing PDMAEMA complex had a superior gene transfection efficiency in human breast cancer cells than the nondrug-loaded one.

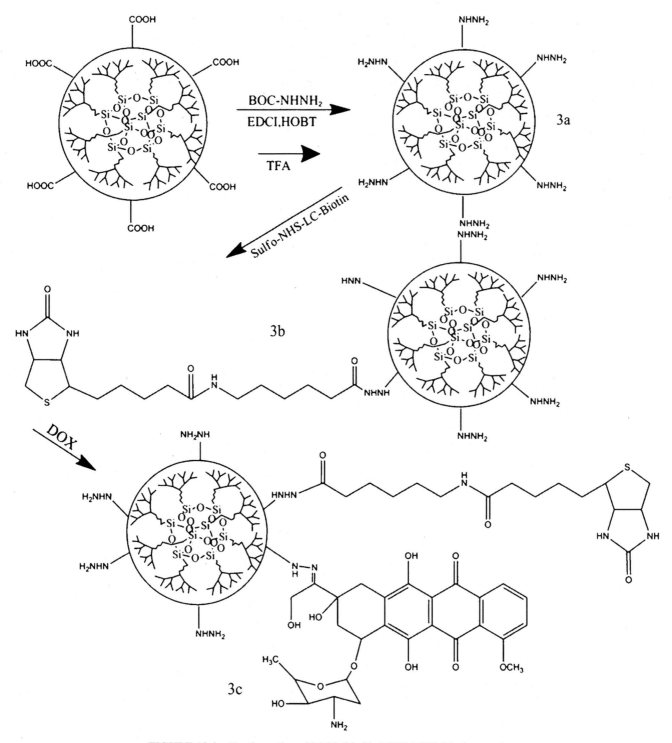

FIGURE 13.4   The formation of POSS-G3-Glu(NHN-DOX)/Biotin complexes.

13.3.5   Imaging Agents

Biological imaging also has attracted great attention, since it has been confirmed as an effective tool for the biology study. Among biological imaging techniques, the fluorescence-based materials have been widely used such as fluorescent proteins and small molecule-based fluorescent probes, due to their high resolution and easy operation. More recently, POSS molecule was developed in improvement of biological imaging technique [40, 98–101]. The arms of POSS molecules can be substituted by fluorescent molecules to produce the novel POSS-containing fluorescent materials with

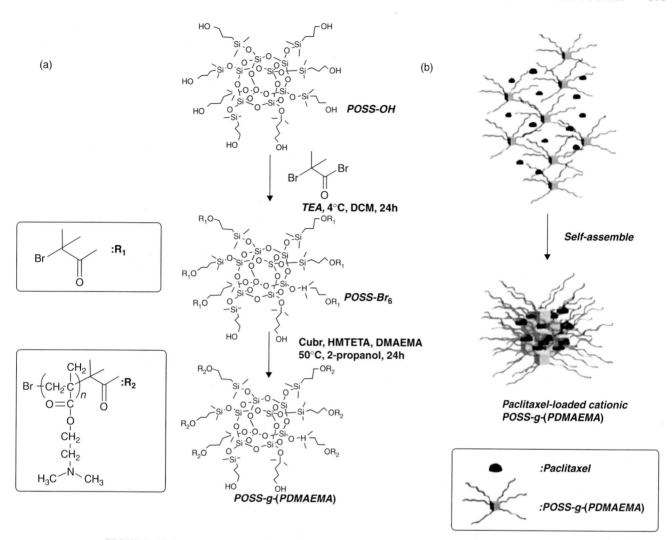

FIGURE 13.5   (a) Synthesis of POSS-g-PDMAEMA polymer by ATRP, (b) self-assembly of POSS-g-PDMAEMA polymer in aqueous solution for the encapsulation of paclitaxel.

three dimensions that possess higher PL quantum yields as compared to their linear counterparts. Liu and coworkers reported a bottom-up strategy to construct water-soluble fluorescent POSS-containing dendritic oligofluorene (OFP), which was used as the energy donor for fluorescence amplification in cellular imaging by Förster resonance energy transfer (FRET) approach with ethidium bromide (EB) as the energy acceptor (Figure 13.6) [98]. The OFP single molecular nanoparticle can be observed in water with an average diameter of 3.6 ± 0.3 nm, which easily can enter cellular nuclei. From the measurement of the optical properties, the absorption and emission peaks of OFP respectively appeared at 390 and 433 nm, which are red-shifted by 52 nm as compared to those of its arm molecule. A good overlap occurred between the emission spectrum of OFP and the absorption spectrum of EB, which favored FRET between them. The PL quantum yields of OFP and its arm are about 0.85 and 0.75 in water, and 0.80 and 0.28 in 150 mM phosphate-buffered

saline (PBS pH = 7.4), respectively. The decreased quantum yield of OFP in the buffer is ascribed to its 3D architecture that prevents the fluorescent arms from close packing to form excimers and ground-state aggregates. Thus, OFP with the strong fluorescence at high ionic strength could be a perfect energy donor for optical amplification in cells. The capability of OFP as an energy donor to amplify the signal of intercalating dye in buffer through FRET is evaluated, and the results showed that the fluorescence of intercalated EB was substantially amplified by 52-fold upon excitation of OFP in buffer, allowing naked-eye discrimination of double-stranded DNA from single-stranded DNA, and clearly visualize the entire cellular structure with weakly emissive dyes. Additionally, the cytotoxicity of OFP is also evaluated for mouse embryonic fibroblast cells (NIH 3T3) using a methylthiazolyldiphenyltetrazolium (MTT) cell-viability assay, and the results revealed that the *in vitro* NIH 3T3 cell viabilities are close to 100% in the tested period, which indicates that OFP

(a)

(b)

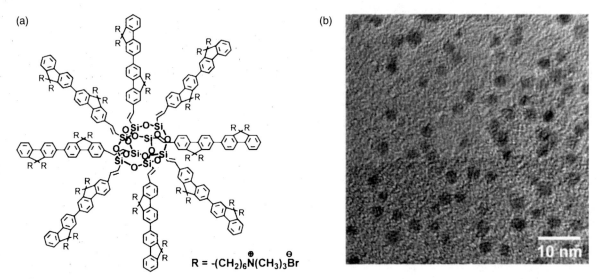

R = -(CH$_2$)$_6$N$^\oplus$(CH$_3$)$_3$Br$^\ominus$

FIGURE 13.6   (a) Chemical structure of cationic oligofluorene-substituted POSS (OFP). (b) HR-TEM image of OFP in aqueous solution.

exhibits low cytotoxicity and could be a potential candidate for long-term clinical applications. Fluorescence nucleus imaging is another important technique to gain insights into genomics and crucial diagnostic and prognostic information for pathologists, since the nucleus contains most of the cell's genetic materials in a eukaryotic cell, organized as multiple long linear DNA molecules in complex with a large variety of proteins, and can accommodate gene expression, replication, recombination, and repair. Liu and coworkers prepared a water-soluble POSS-containing conjugated oligoelectrolyte (POSS-COE) for two-photon excited fluorescence (TPEF) imaging of cellular nucleus [40]. POSS-COE with the 3D architecture was composed of a rigid silicon–oxygen inorganic core surrounded by cationic COE arms on its globular periphery, which takes advantage of a unique structure of POSS and the charge transfer-based optical properties of a cationic COE. The results showed that POSS-COE had a small size of about 3.3 nm in diameter, large Stokes shift (152 nm), high two-photon absorption (TPA) cross section (126 GM at 760 nm), low cytotoxicity, and efficient nucleus permeability. In conjunction with these desirable properties, the strong light-up response of COE-POSS toward nucleic acids rendered it an effective stain for nucleus imaging. POSS-COE as a novel TPEF has demonstrated it was more effective in illuminating and distinguishing the nucleoli from other internal compartments of the nucleus than OPEF and some mostly used commercial dye. Thus, POSS-COE could have great potential in clinical diagnosis and modern biological research. Subsequently, they further prepared the OFP-loaded chitosan/PEG nanoparticles with folic acid functionalization for targeted imaging of cancer cell nucleus, since chitosan is a biocompatible and pH-sensitive biopolymer, which can be degraded by cellular lysosomes [102]. These POSS-COE/CS/PEG-FA nanoparticles could be internalized in acidic lysosomes of MCF-7 breast cancer cells, trigger rapid release of COE-POSS, and exhibited low cytotoxicity with specific targeting capability for the nuclei of MCF-7 cancer cells using NIH/3T3 fibroblast normal cells as the control. So these pH-responsive NPs could be promising for targeted imaging of cancer cell nucleus.

13.4   CONCLUSIONS AND OUTLOOK

Although POSS-containing hybrid polymers have been well studied almost for two decades, and much effort has been contributed to their precious architecture, improved properties, and promising applications, only more recently, they were extended their application in biological fields. POSS molecules could be introduced into polymer matrices to produce a variety of POSS-containing bioconjugates, which is determined by their unique structure and properties. The POSS-containing hybrid polymers with enhanced mechanical properties can be well potentially applied as tissue engineering scaffolds and dental materials. For instance, POSS-containing hybrid resin could be used as dental materials with reduced shrinkage and improved mechanical property and biocompatibility. POSS molecules contain inorganic cages composed of Si-O-Si and Si-C bonds, which are inert and nontoxic, and they can be readily used to produce POSS-containing bioconjugates via the functionalization of corner groups, so POSS molecules have been developed to POSS-containing bioconjugated dendrimers, which exhibited great potential applications as drug delivery and biological imagines. Although some strategies have been concentrated on the construction of the POSS-containing bioconjugates, these works were only done in recent several years; moreover, most of works are first involved in field.

Besides tissue engineering and dental materials, the application of POSS-containing bioconjugates was mostly limited to POSS-containing dendrimeric bioconjugates. The preparation and application of well-defined POSS-containing bioconjugates with other architectures should be also explored in the future. Additionally, POSS-containing bioconjugates could be used as biological materials, which were still evaluated by cellular experiment and animal models. However, in the final clinical application, it needs to be addressed whether the properties of these materials such as stability, biocompatibility, nontoxicity can be still preserved. Thus much effort is still needed to focus on the construction and application of POSS-containing bioconjugates, and we expected that more and more excellent POSS-containing bioconjugates will be well developed and applied in the future.

REFERENCES

1. Fina A, Monticelli O, Camino G. *J Mater Chem* 2010;20: 9297–9305.
2. Gnanasekaran D, Madhavan K, Reddy BSR. *J Sci Ind Res* 2009;68:437–464.
3. Janowski B, Pielichowski K. *Polymer* 2008;53:87–98.
4. Kuo S-W, Chang F-C. *Prog Polym Sci* 2011;36:1649–1696.
5. Li GZ, Wang LC, Ni HL, Pittman CU. *J Inorg Organomet P* 2001;11:123–54.
6. Madbouly SA, Otaigbe JU. *Prog Polym Sci* 2009;34:1283–1332.
7. Phillips SH, Haddad TS, Tomczak SJ. *Curr Opin Solid State Mater Sci* 2004;8:21–29.
8. Pielichowski K, Njuguna J, Janowski B, Pielichowski J. Polyhedral oligomeric silsesquioxanes (POSS)-containing nanohybrid polymers. Abe, Akihiro; Dušek, Karel; Kobayashi, Shiro (Eds.), *Supramolecular Polymers Polymeric Betains Oligomers*. Berlin: Springer-Verlag; 2006. pp 225–296.
9. Tanaka K, Jeon JH, Inafuku K, Chujo Y. *Bioorgan Med Chem* 2012;20:915–919.
10. Wu J, Mather PT. *Polym Rev* 2009;49:25–63.
11. Brown JF. *J Am Chem Soc* 1965;87:4317–4324.
12. Sprung MM, Guenther FO. *J Am Chem Soc* 1955;77:3990–3996.
13. Scott DW. *J Am Chem Soc* 1946;68:356–358.
14. Baney RH, Itoh M, Sakakibara A, Suzuki T. *Chem Rev* 1995;95:1409–1430.
15. Voronkov MG, Lavrentyev VI. *Top Curr Chem* 1982;102:199–236.
16. Feher FJ. *J Am Chem Soc* 1986;108:3850–3852.
17. Zhang CX, Laine RM. *J Am Chem Soc* 2000;122:6979–6988.
18. Mather PT, Jeon HG, Romo-Uribe A, Haddad TS, Lichtenhan JD. *Macromolecules* 1999;32:1194–1203.
19. Xu HY, Kuo SW, Lee JS, Chang FC. *Macromolecules* 2002;35:8788–8793.
20. Leu CM, Chang YT, Wei KH. *Macromolecules* 2003;36:9122–9127.
21. Hsu Sh, Tseng Hj, Wu Ms. *Artif Organs* 2000;24:119–128.
22. Yu B, Jiang X, Qin N, Yin J. *Chem Commun* 2011;47:12110–12112.
23. Neumann D, Fisher M, Tran L, Matisons JG. *J Am Chem Soc* 2002;124:13998–13999.
24. Cordes DB, Lickiss PD, Rataboul F. *Chem Rev* 2010;110:2081–2173.
25. Shockey EG, Bolf AG, Jones PF, Schwab JJ, Chaffee KP, Haddad TS, Lichtenhan JD. *Appl Organomet Chem* 1999;13:311–327.
26. Li GZ, Pittman CU. "Polyhedral Oligomeric Silsesquioxane (POSS) Polymers, Copolymers, and Resin Nanocomposites," Chapter 5, in Abd-El-Aziz AS, Carraher CE, Pittman CU, Zeldin M, (eds), Macromolecules Containing Metal and Metal-Like Elements: Vol. 4, Group IVA Polymers, John Wiley and Sons, Hoboken, NY; 2005. pp. 79–132.
27. Laine RM. *J Mater Chem* 2005;15:3725–3744.
28. Pittman CU, Li GZ, Ni HL. *Macromol Symp* 2003;196:301–325.
29. Pu KY, Fan Q, Wang LH, Huang W. *Prog Chem* 2006;18:609–615.
30. Zhao JQ, Fu Y, Liu SM. *Polym Polym Compos* 2008;16:483–500.
31. Pan H, Qiu Z. *Macromolecules* 2010;43:1499–1506.
32. Song XY, Geng HP, Li QF. *Polymer* 2006;47:3049–3056.
33. Ishida Y, Tada Y, Hirai T, Goseki R, Kakimoto M-a, Yoshida H, Hayakawa T. *J Photopolym Sci Tech* 2010;23:155–159.
34. Hirai T, Leolukman M, Liu CC, Han E, Kim YJ, Ishida Y, Hayakawa T, Kakimoto M-a, Nealey PF, Gopalan P. *Adv Mater* 2009;21:4334–4338.
35. Song J, Filion TM, Xu JW, Prasad ML. *Biomaterials* 2011;32:985–991.
36. Wheeler PA, Fu BX, Lichtenhan JD, Jia WT, Mathias LJ. *J Appl Polym Sci* 2006;102:2856–2862.
37. Chen MH. *J Dent Res* 2010;89:549–560.
38. Yuan H, Luo K, Lai YS, Pu YJ, He B, Wang G, Wu Y, Gu Z. *Mol Pharm* 2010;7:953–962.
39. Gao Y, Eguchi A, Kakehi K, Lee YC. *Org Lett* 2004;6:3457–3460.
40. Pu K-Y, Li K, Zhang X, Liu B. *Adv Mater* 2010;22:4186–4189.
41. Sarkar S, Burriesci G, Wojcik A, Aresti N, Hamilton G, Seifalian AM. *J Biomech* 2009;42:722–730.
42. Bae Y, Fukushima S, Harada A, Kataoka K. *Angew Chem Int Ed* 2003;42:4640–4643.
43. Klok HA, *Adv Mater* 2001;13:1217–1229.
44. Braunecker WA, Matyjaszewski K. *Prog Polym Sci* 2007;32:93–146.
45. Higashihara T, Hayashi M, Hirao A. *Prog Polym Sci* 2011;36:323–375.
46. Moad G, Rizzardo E, Thang SH. *Aust J Chem* 2005;58:379–410.

47. Lin YC, Kuo SW. *J Polym Sci Pol Chem* 2011;49:2127–2137.

48. Zhang W, Fang B, Walther A, Mueller AHE. *Macromolecules* 2009;42:2563–2569.

49. Costa ROR, Vasconcelos WL, Tamaki R, Laine RM. *Macromolecules* 2001;34:5398–5407.

50. Zhang WA, Wang SH, Li XH, Yuan JY, Wang SL. *Eur Polym J* 2012;48:720–729.

51. Naik SS, Savin DA. *Macromolecules* 2009;42:7114–7121.

52. Kolb HC, Finn MG, Sharpless KB. *Angew Chem Int Ed* 2001;40:2004–2021.

53. Lin Y-C, Kuo S-W. *Polym Chem* 2012;3:882–891.

54. Lin Y-C, Kuo S-W. *Polym Chem* 2012;3:162–171.

55. Kuo S-W, Tsai H-T. *Polymer* 2010;51:5695–5704.

56. Lo Conte M, Staderini S, Chambery A, Berthet N, Dumy P, Renaudet O, Marra A, Dondoni A. *Org Biomol Chem* 2012;10:3269–3277.

57. Ray JG, Ly JT, Savin DA. *Polym Chem* 2011;2:1536–1541.

58. Fabritz S, Hörner S, Könning D, Empting M, Reinwarth M, Dietz C, Glotzbach B, Frauendorf H, Kolmar H, Avrutina O. *Org Biomol Chem* 2012;10:6287–6293.

59. Feher FJ, Wyndham KD, Knauer DJ. *Chem Commun* 1998:2393–2394.

60. Fabritz S, Heyl D, Bagutski V, Empting M, Rikowski E, Frauendorf H, Balog I, Fessner WD, Schneider JJ, Avrutina O, Kolmar H. *Org Biomol Chem* 2010;8:2212–2218.

61. Heyl D, Rikowski E, Hoffmann RC, Schneider JJ, Fessner W-D. *Chemistry* 2010;16:5544–5548.

62. Trastoy B, Eugenia Perez-Ojeda M, Sastre R, Luis Chiara J. *Chemistry* 2010;16:3833–3841.

63. Cui L, Zhu L. *Langmuir* 2006;22:5982–5985.

64. Cui L, Chen D, Zhu L. *ACS Nano* 2008;2:921–927.

65. Motwani MS, Rafiei Y, Tzifa A, Seifalian AM. *Biotechnol Appl Biochem* 2011;58:2–13.

66. Lickiss PD, Cordes DB, Rataboul F. *Chem Rev* 2010;110:2081–2173.

67. Soh MS, Sellinger A, Yap AUJ. *Curr Nanosci* 2006;2:373–381.

68. Kannan RY, Salacinski HJ, Butler PE, Seifalian AM. *Accounts Chem Res* 2005;38:879–884.

69. Seifalian AM, Ghanbari H, Cousins BG. *Macromol Rapid Commun* 2011;32:1032–1046.

70. Salacinski HJ, Goldner S, Giudiceandrea A, Hamilton G, Seifalian AM, Edwards A, Carson RJ. *J Biomater Appl* 2001;15:241–278.

71. Hu S, Ren X, Bachman M, Sims CE, Li GP, Allbritton N. *Anal Chem* 2002;74:4117–4123.

72. Werner C, Jacobasch HJ. *Int J Artif Organs* 1999;22:160–176.

73. Kannan RY, Salacinski HJ, Odlyhab M, Butler PE, Seifalian AM. *Biomaterials* 2006;27:1971–1979.

74. Kannan RY, Salacinski HJ, Sales K, Butler P, Seifalian AM. *Biomaterials* 2005;26:1857–1875.

75. Ghanbari H, Burriesci G, Ramesh B, Darbyshire A, Seifalian AM. *Acta Biomater* 2010;6:4249–4260.

76. Shanklin DR, Smalley DL. *Exp Mol Pathol* 1999;67:26–39.

77. McCarthy DJ, Chapman HL. *J Foot Surg* 1988 27:418–427.

78. Kannan RY, Salacinski HJ, Ghanavi JE, Narula A, Odlyhab M, Peirovi H, Butler PE, Seifalian AM. *Plast Reconstr Surg* 2007;119:1653–1662.

79. Guo YL, Wang W, Otaigbe JU. *J Tissue Eng Regen Med* 2010 4:553–564.

80. Wang WS, Guo YL, Otaigbe JU. *Polymer* 2009 50:5749–5757.

81. Cai L, Chen J, Rondinone AJ, Wang S. *Adv Funct Mater* 2012;22:3181–3190.

82. Bowen RL. *J Am Dent Assoc* 1963;66:57–64.

83. Dickens SH, Stansbury JW, Choi KM, Floyd CJE. *Macromolecules* 2003;36:6043–6053.

84. Soh MS, Yap AUJ, Sellinger A. *Eur Polym J* 2007;43:315–327.

85. Fong H, Dickensb SH, Flaim GM. *Dent Mater* 2005;21:520–529.

86. Wu X, Sun Y, Xie W, Liu Y, Song X. *Dent Mater* 2010;26:456–462.

87. Guo Q, Knight PT, Mather PT. *J Control Release* 2009;137:224–233.

88. Guo Q, Knight PT, Wu J, Mather PT. *Macromolecules* 2010;43:4991–4999.

89. Yuan H, Luo K, Lai Y, Pu Y, He B, Wang G, Wu Y, Gu Z. *Mol Pharm* 2010;7:953–962.

90. Ghanbari H, Cousins BG, Seifalian AM. *Macromol Rapid Commun* 2011;32:1032–1046.

91. McCusker C, Carroll JB, Rotello VM. *Chem Commun* 2005:996–998.

92. Ambade AV, Savariar EN, Thayumanavan S. *Mol Pharm* 2005;2:264–272.

93. Gupta U, Agashe HB, Asthana A, Jain NK. *Biomacromolecules* 2006;7:649–658.

94. Chen DY, Cui L, Zhu L. *ACS Nano* 2008;2:921–927.

95. He CB, Wang FK, Lu XH. *J Mater Chem* 2011;21:2775–2782.

96. Lu ZR, Kaneshiro TL, Wang X. *Mol Pharm* 2007;4:759–768.

97. Loh XJ, Zhang ZX, Mya KY, Wu YL, He CB, Li J. *J Mater Chem* 2010;20:10634–10642.

98. Pu KY, Li K, Liu B. *Adv Mater* 2010;22:643–646.

99. Olivero F, Renò F, Carniato F, Rizzi M, Cannasb M, Marchese L. *Dalton Trans* 2012;41:7467–7473.

100. Tan MQ, Ye Z, Jeong EK, Wu XM, Parker DL, Lu ZR. *Bioconjug Chem* 2011;22:931–937.

101. Mi Y, Li K, Liu YT, Pu KY, Liu B, Feng SS. *Biomaterials* 2011;32 8226–8233.

102. Ding D, Pu K-Y, Li K, Liu B. *Chem Commun* 2011;47:9837–9839.

SECTION V

CELL-BASED, HYDROGELS/MICROGELS AND GLYCO-BIOCONJUGATES

14

CELL-BASED BIOCONJUGATES

MITSUHIRO EBARA

Biomaterials Unit, International Center for Materials Nanoarchitectonics (WPI-MANA), National Institute for Materials Science (NIMS), 1-1 Namiki, Tsukuba, Ibaraki, Japan

14.1 INTRODUCTION

Modifying cell surfaces by conjugating them with bioactive molecules or synthetic polymers has been a versatile way to add new value, advanced features, and unique properties to inert cells. Creating a nanoscale layer on a cell surface, for example, significantly improve or even completely change its biological properties as well as introduce new unique properties, such as chemical functionality, surface roughness, surface tension, morphology, surface charge, surface reflectivity, surface conductivity, and optical properties. Recently, surface modification of living cells has been the subject of study for a variety of biological applications such as imaging, transfection, and control of cell surface interactions. The ability to visualize cell surfaces both *in vitro* and *in vivo* environments is essential to gaining further insight into the function of specific molecules or the entire entity. For example, cell labeling has been used for flow cytometric analysis and fluorescent microscopy. The cell surface modification has also been employed in masking surface antigens to prevent immune reaction. The surface modification with polyethylene glycol (PEG) called "PEGylation" can protect them from the adsorption of foreign proteins, making cells antigenically silent (stealth cells). This technology can be used as a means to silence the antigenic response of red blood cells (RBCs) toward the development of universal blood transfusion. The surface modification with adhesion ligands, on the other hand, can be used to systemically deliver a large quantity of cells to a specific tissue. Cell membrane modification can also be used to anchor cells to specified locations on surfaces in a sequence-dependent fashion. The ability to pattern cells on surfaces provides

a new platform for the study of cell–cell interactions, cell fusion, and three-dimensional (3D) tissue constructions. In addition to mammalian cells, surface modifications of other types of vesicles such as viruses, bacteria, and yeast cells have seen a dramatic growth of interest in recent years. The viral surface can potentially be modified to provide better means of preparation, purification, concentration, detection, tracking, imaging, and targeting for diagnostics, vaccine development, and drug/gene therapy. The surface modification of virus-based gene delivery vectors, for example, facilitates more efficient delivery of the gene of interest to desired sites of expression. The surface modification with biomimetic mineral layer has also been developed to enhance the viability of yeast cells for a longer storage time by protecting from harsh environments. These artificial shell structures can also provide protection of bacteria against deleterious conditions.

14.2 CLASSIFICATION OF CELL MEMBRANE CONJUGATIONS ON THE BASIS OF MODIFICATION METHODS

Nowadays, a variety of different bioconjugation techniques are known for the purpose of both *in vitro* and *in vivo* studies. In general, surface modification of cells or viruses can be broadly separated into four categories: (a) covalent immobilization, (b) electrostatic interactions, (c) hydrophobic anchors, and (d) ligand–receptor interactions. The covalent immobilization has been achieved through chemical or enzymatic treatment or by metabolic introduction (Figure 14.1a). Common types of covalent immobilization use amine coupling of lysine amino acid residues through

Chemistry of Bioconjugates: Synthesis, Characterization, and Biomedical Applications, First Edition. Edited by Ravin Narain.
© 2014 John Wiley & Sons, Inc. Published 2014 by John Wiley & Sons, Inc.

(a) Covalent immobilization   (b) Electrostatic interactions

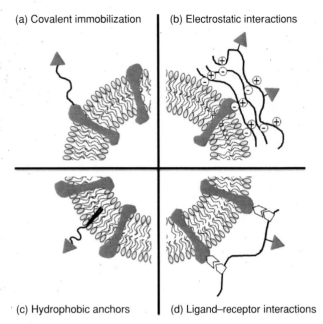

(c) Hydrophobic anchors   (d) Ligand–receptor interactions

FIGURE 14.1   Schematic illustration of cell surface modification methods by covalent immobilization (a), electrostatic interactions (b), hydrophobic anchors (c), and ligand–receptor interactions (d).

amine-reactive succinimidyl esters, sulfhydryl coupling of cysteine residues either other sulfhydryl groups or in a Michael addition via a sulfhydryl-reactive maleimide, and photochemically initiated free radical reactions. The use of *N*-hydroxysuccinimidyl ester (NHS) is one of the most common methods for condensation reaction between a carboxylic acid and an amino group of membrane proteins. Most problems, however, lie with the reagents and the byproducts which have the potential to be toxic to cells. The pair of functional groups chosen for cell membrane modification must be mutually reactive under physiological condition. "Staudinger ligation" has been developed to meet these criteria. The Staudinger reaction occurs between a phosphine and an azide to produce an aza-ylide rapidly in water at room temperature in high yield [1]. Both functional groups are abiotic and essentially unreactive toward biomolecules inside or on the surfaces of cells. Nowadays, "click chemistry" represented by the condensation reaction between an azido group and a triple bond is also used extensively because the reactions occur at room temperatures, give high yields, and are tolerant to the media since they can be carried out in the presence of oxygen and even in water [2]. In addition, both azide and alkyne groups can be appended to biomolecules without altering their function or metabolic processing. Another strategy to modify cell membrane surface is to coat the surface through nonspecific interactions represented by a layer-by-layer (LBL) technique of anionic and cationic polymers (Figure 14.1b) [3]. The versatility of this technique allows for the construction of thin polymer layer on the surface,

which interact electrostatically, through hydrogen bonding or hydrophobic interaction. In addition, the surface property can be controlled by the outermost layer of polymer. A drawback of this technique is cytotoxicity when polycations are used as the first layer which directly interacts with the cell surface. Another option for cell surface modification is the insertion of hydrophobic anchors into the cellular membranes, such as a fatty acid, steroid, lipophilic peptide, or other synthetic moieties (Figure 14.1c). This approach allows the controlled addition of structurally defined components to live cells. Another option is to use ligand–receptor interactions (Figure 14.1d). The pairs of CD44-hyaluronic acid (HA) and integrin-RGD peptides are commonly used for the modification. In these cases, signal transduction can also be controlled due to the activation of cell membrane receptors.

Although many methods have been developed for cell surface modification, it is also important to understand the dynamic features of conjugated polymers on the cell membranes including their uptake and exclusion. It has been reported that lipid molecules rapidly cycle between the membrane and the cytoplasm compartment and macrophages interiorize the equivalent of their cell surface area every 33 minutes [4]. The surface dynamics of the conjugated polymers on cell membranes can be examined using fluorescently labeled polymers. Iwata et al. compared the surface dynamics of three kinds of polymer conjugations as shown in Figure 14.2 [5]. When high concentration of cationic polymers such as poly(ethyleneimine) (PEI) was coated on the cell membrane, cell membrane was immediately destroyed, while anionic polymer did not interact with cells. When hydrophobic anchors were used such as PEG–conjugated phospholipid (PEG-lipid) and poly(vinyl alcohol) (PVA) with alkyl side chains (PVA-alkyl), they interacted with cells uniformly without

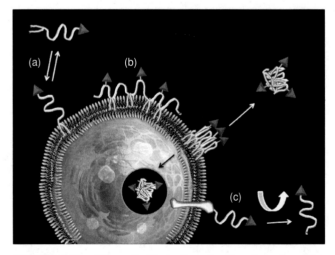

FIGURE 14.2   Schematic illustration of surface dynamics of the conjugated polymers on the cell membrane. For a color version, see the color plate section.

cytotoxicity. PEG-lipid was, however, rapidly released from the cell surface without uptake into the cytoplasm if the hydrophobic interaction between the lipid bilayer and alkyl chains was not strong enough (Figure 14.2a). In case of PVA-alkyl which has polyvalent hydrophobic anchors, they gradually assembled over time in an area of the membrane (Figure 14.2b). This phenomenon is similar to the capping phenomena observed on cells when treated with polyvalent antibodies [6]. The PVA-alkyl was then excluded from the cell surface or was uptaken into the cytoplasm. Of particular interest is that PEG with activated ester end-group (PEG-NHS) also disappeared from the living cell surface even though they were covalently bound to the membrane protein (Figure 14.2c). This may be caused by hydrolysis of the amide bond between PEG and protein or membrane proteins themselves by proteases because PEG-NHS did not disappear from the cells when cells were fixed and killed by paraformaldehyde (PFA) treatment.

14.3   IMMUNOCAMOUFLAGE OF RED BLOOD CELLS

Biomedical applications of cell surface conjugation have been found in the area of cell or tissue transplantation [7]. Transplant rejection is a process in which a transplant recipient's immune system attacks the transplanted cells or tissues. As soon as donor cells enter the body, the immune system recognizes as foreign and attacks them [8, 9]. There are three types of rejection: (1) Hyperacute rejection occurs a few minutes after the transplant, if the antigens are completely unmatched. The tissue must be removed right away so the recipient does not die. This type of rejection is seen when a recipient is given the wrong type of blood. (2) Acute rejection may occur any time from the first week after the transplant to 3 months afterward. (3) Chronic rejection takes place over many years. The body's constant immune response against the new organ slowly damages the transplanted tissues or organ. Therefore, masking of cell surface antigens and tissue offers a rational approach for attenuating deleterious host responses toward transplanted cells.

The immunologically protective strategy was first applied to the surface modification of RBCs using nonantigenic groups, mostly PEG to camouflage the antigenic determinants of RBC. Blood transfusion is the most widely used cell transplantation procedure in clinical practice and ABO blood matching tests are carried out routinely prior to blood transfusion in order to avoid incompatibilities and immunological reactions on the part of the patients (Figure 14.3a) [10]. In some cases, patients develop alloantibodies against antigens of several blood group systems after the transfusion, resulting in an increased risk of blood incompatibility in the event of future transfusions. This problem has prompted research toward the production of a "universal blood" in which the antigens are silenced. Methoxy-PEG, for example, has been employed to covalently modify the surface proteins of RBCs (almost exclusively lysine residues) via cyanuric chloride coupling [11]. Scott et al. have reported that the

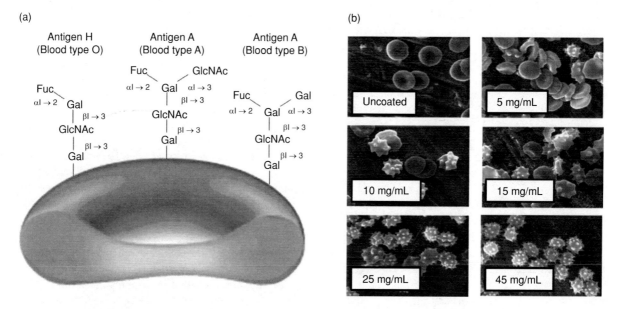

FIGURE 14.3   Structures of the ABO blood group antigens (a). Fuc, fucose; Gal, galactose; GalNAc, *N*-acetylgalactosamine. Schematic illustration of surface dynamics of the conjugated polymers on the cell membrane. The morphology of PEG-modified RBCs with different PEG concentrations (b).

PEG-modified RBCs effectively obscured antigenic determinants while leaving the RGB structurally and functionally normal and nonimmunogenic. Furthermore, they are resistant to phagocytosis and exhibited significantly prolonged survival when transfused into mice [12]. A highest recommended concentration of 15 mg/mL was also reported for a 5 kDa of cyanuric chloride coupled PEG due to the observation of increasing echinocytosis with increasing concentration (Figure 14.3b) [13]. The dependence of immunocamouflage on the target size for modification was also demonstrated. The larger polymers (20 kDa) were more effective at preventing protein adsorption on bigger latex particles (8 mm), while shorter polymers (2 kDa) were more effective on smaller particles (1.2 mm) [14]. Significantly, these findings were found to be relevant to the immunocamouflage of cells (8–10 mm in size) [15].

The other approach involves the formation of a polymeric shield around RBCs by electrostatic binding of polymers. LBL self-assembly method has been used to grow polyelectrolyte layers to modify RBCs. When the LBL method is applied on the erythrocyte surface, a polycation-based layer is formed first, then the excess of free positive charges is used to adsorb a subsequent anionic-based layer. This process can be repeated to increase the thickness of the coating. The multilayered film consisting of a protecting shell (five bilayers of chitosan-*graft*-phosphorylcholine (CH-PC) and sodium hyaluronate (HA)) and a camouflage shell (five bilayers of poly(L-lysine)-*graft*-poly(ethylene glycol) (PLL-*g*-PEG) and alginate (AL)) was successfully created on RBCs through LBL technique by Tabrizian et al. [16]. By modulation of the composition and buildup conditions, adjusting the film thickness, and the permeability, the antibody recognition was successfully prevented. Moreover, this method preserved the RBCs' viability and functionality such as the ability to take up oxygen [17]. Because LBL assembly may alter the membrane flexibility, a single polymer layer has also been developed by a simple one-step method using PLL-*g*-PEG [18] or poly(2-dimethylamino-ethylmethacrylate) (PDMAEMA)-*co*-PEG copolymer [19]. The molecular weight and architecture of the copolymers influenced the immunomasking efficiency.

14.4 ISLET TRANSPLANTATION

Except blood transfusion, one of the most reported examples of cell transplantation is islet transplantation for diabetes. Type 1 diabetes mellitus (T1DM) is a disease characterized by the lack of pancreatic islet β-cell function. β-cells are responsible for insulin production, and the lack of insulin results in unregulated glucose concentrations in the blood. Clinical islet transplantation is a promising treatment to restore β-cell function in type 1 diabetic patients [20]. However, widespread clinical application of islet transplantation

remains limited by the deleterious side effects of immunosuppressive therapy necessary to prevent host rejection of transplanted islets. Therefore, the desire to transplant islet without the need for immunosuppression has led to the development of semipermeable microcapsules capable of protecting donor cells from the host immune system while allowing transport of glucose, insulin, oxygen, and other essential nutrients. From these perspectives, several approaches have been proposed for thin coverage of the surface antigenic sites on islets. One of the most common coverage techniques involves cell encapsulation such as a coacervation between AL polyanions and poly-L-lysine (PLL) polycations [21]. Encapsulated islets remained morphologically and functionally intact over 15 weeks. Another encapsulation technique involves interfacial photopolymerization around an islet using prepolymer PEG [22]. Although the LBL method has also been widely used to encapsulate islets due to its ability to generate films of nanometer thickness on diverse substrates, direct contact between cationic polymers and islets significantly decreased viability. Incubation of islets with 1 mg/mL PLL for 15 minutes, for example, resulted in approximately 60% decrease in islet viability relative to untreated controls [23]. Similar results have also been reported when other polycations were used, such as poly(allylamine hydrochloride) (PAH), poly(ethyleneimine) (PEI), etc. [24]. Because polycation cytotoxicity is dependent on charge density, they can be attenuated by grafting PEG chains to a critical number of amine residues. Cytotoxicity of PLL-*g*-PEG, for example, decreases as the extent of PEG grafting increases. Chaikof et al. modified PLL-*g*-PEG copolymers to display biotin, hydrazide, and azide moieties by functionalizing the terminal group of the grafted PEG, which selectively captures streptavidin (SA), aldehyde, and cyclooctyne, respectively (Figure 14.4) [25]. PEG-rich layer was successfully assembled via LBL of biotinylated PLL-*g*-PEG and coated islets performed comparable viability and function to untreated control *in vivo* in a murine model of allogenic intraportal islet transplantation [23].

Amphiphilic polymers, such as PEG-conjugated phospholipid (PEG-lipid), have also been used to mask the surface antigens of islets. PEG-lipid is synthesized from NHS-PEG and 1,2-dipalmitoyl-*sn*-glycero-3-phosphatidylethanolamine (DPPE) [26]. The PEG-lipid spontaneously formed a thin layer on cells when they were added to islet suspension. Iwata et al. further modified the PEG-lipid layer on the islets with PVA using LBL method with thiol/disulfide exchange reaction [27]. This method effectively encapsulated islets in a multilayered ultra-thin PVA membrane without influencing cell viability or insulin release. This technology can be extended to immobilize bioactive substances such as heparin and enzymes on the islet surface. A plasminogen activator, urokinase, for example, was immobilized on the PVA-anchored islets [28]. PVA was first modified with sodium monochloroacetate to synthesize

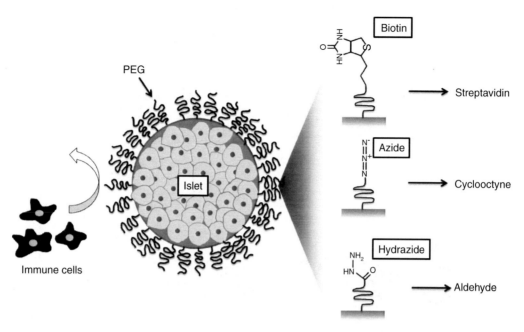

FIGURE 14.4 The concept of immunological protection of transplanted islets using PEGylation. The functional end-groups of PEG are used to display biotin, hydrazide, and azide moieties, which selectively captured SA, aldehyde, and cyclooctyne, respectively.

PVA–COOH and hexadecanal was then introduced to synthesize PVA-alkyl. Urokinase was then immobilized onto the islet surface through the thiol/maleimide reaction. Urokinase-immobilized islets exhibited fibrinolytic properties, which could help to improve graft survival by preventing thrombosis on the islet surface. In recent years, the surface coating of islets with living cells have also been attempted using biotin/SA reaction (Figure 14.5) [29]. Biotin-PEG-lipid was first incorporated into the islet surface by anchoring into the lipid bilayer. SA-immobilized human endoderm kidney cell line (HEK293 cells) was then immobilized on the biotin-PEG-lipid-modified islets. The surface of the islets has been covered with a cell layer for 5 days without central necrosis of the islet cells. This technique will enable to improve compatibility of islets with the recipient significantly and evade the immune-rejection reaction.

Several groups have also reported the surface coverage of antigenic sites on islets with polymers via covalent immobilization. PEG with activated ester, for example, has been employed to cover the surface antigen of islets. The activated ester group reacted with the membrane proteins or collagen layer on the islet surface. Russell et al. reported that PEG-modified islets released insulin in the same manner as unmodified islets [30]. Byun et al. reported that the immuno-camouflage of islets with PEG-protected transplanted islets in diabetic rats and normoglycemia could be maintained for at least 1 year by combination with cyclosporine A, whereas unmodified islets were completely eliminated within 2 weeks [31].

14.5 CELL SURFACE DECORATION FOR LABELING, MANIPULATION, AND PROGRAMMABLE ADHESION

Selective chemical modifications of cell membranes have also been powerful experimental tools for elucidating biological processes or engineering novel interactions. Cell surface oligosaccharides have been engineered to display unusual functional groups for the selective chemical remodeling of cell surfaces. In order to achieve this, the two participating functional groups must have finely tuned reactivity in a biological environment. Bertozzi et al. incorporated a ketone group into cell surface oligosaccharides using *N*-acetylmannosamine which was converted to the corresponding sialic acid metabolically [32]. The ketone group on the cell surface was then covalently ligated with molecules carrying a complementary reactive functional group such as hydrazide. This reaction provided a unique chemical target for covalent reaction with an aminooxy-functionalized magnetic resonance imaging (MRI) contrast reagent [33]. Tsien et al. reported another chemoselective ligation reaction, condensation reaction of a cysteine-rich peptide with a bis-dithioarsolane [34]. This enabled the labeling of a fluorescent dye to a single protein within the environs of living cells. In contrast to other bioconjugation moieties, azides are versatile and bio-orthogonal chemical moieties for labeling many classes of biomolecules in any biological context because they do not react with amines or other nucleophiles abundantly represented in biological systems. Fitting with

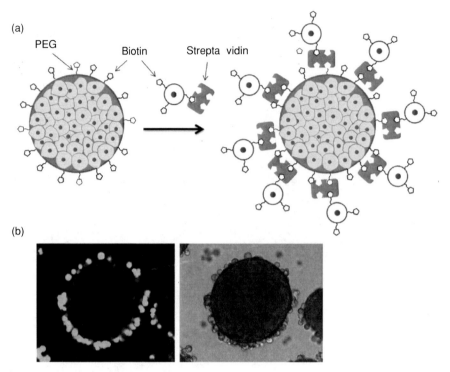

FIGURE 14.5 Schematic illustration of the immobilization of SA-immobilized HEK293 cells on the surface of biotin-PEG-modified islets (a). Confocal laser scanning (left) and differential interference (right) microscope images of surface-modified islets with HEK293 cells (b). For a color version, see the color plate section.

this definition, azido groups are highly suitable for either Staudinger ligation or click reaction. The Staudinger reaction occurs between a phosphine and an azide to produce an aza-ylide [1]. The phosphine and the azide react with each other rapidly in water at room temperature in high yield. Bertozzi et al. pioneered the field of Staudinger ligation in biological system. They demonstrated the modification of cell surface azides with phosphine reagents, yielding stable amide linkage [35]. The azide groups were installed within cell surface glycoconjugates metabolically (Figure 14.6a). First, cells are incubated with azide-modified monosaccharide building blocks which are uptaken in deacylation by cellular esterases in the endosomes. They enter the glycostructure salvage pathway, which includes activation of the monosaccharide by UDP-transfer. The UDP-activated monosaccharide is further incorporated into the core position of O-linked glycans. The complex O-linked glycans are finally presented at the cell surface with the azido groups intact. The cell surface azido groups can be labeled with a variety of phosphine probes such as fluorophores by Staudinger ligation (Figure 14.6b). This technique was further executed to tag cell surface glycans *in vivo* [36]. The ability to tag cell surface glycans *in vivo* may enable therapeutic targeting and noninvasive imaging of changes in glycosylation during disease progression.

The Staudinger ligation paved the way for a conceptually new way of thinking, involving the translation of knowledge of chemical reactions to reactions in living systems. Although Staudinger ligation has the potential for applications in noninvasive imaging and therapeutic targeting, the requisite phosphines are susceptible to air oxidation, and their optimization for improved water solubility and increased reaction rate has proven to be synthetically challenging. In this respect, particular inspiration is found in chemistry explored by Huisgen et al. involving cycloaddition reactions of unsaturated systems with 1,3-dipoles [37]. For example, Sharpless et al. identified Huisgen reactions as one of the key tools for the concept of "click chemistry" because participating functional groups can be appended to biomolecules without altering their function or metabolic processing [2]. Finn et al. demonstrated a robust method for the surface labeling of living Jurkat cells in culture using the Cu(I)-catalyzed alkyne–azide click reaction [38]. Alkynyl probe reagents were covalently attached to the azide derivative of N-acetylmannosamine on cell membrane. Although they optimized reaction conditions to preserve cell viability, the Cu(I) catalyst is generally regarded as toxic and incompatible with living cells. In order to circumvent the use of copper ions, Bertozzi et al. have devised a strain-promoted [3 + 2] cycloaddition reaction that involves azides and a strained cyclooctyne derivative [39]. Strain-alkyne, for example, provides copper-free alkyne–azide cycloaddition with cyclooctynes. Reactivity of cyclooctynes can be greatly enhanced by structural modification. The reaction rate increased according

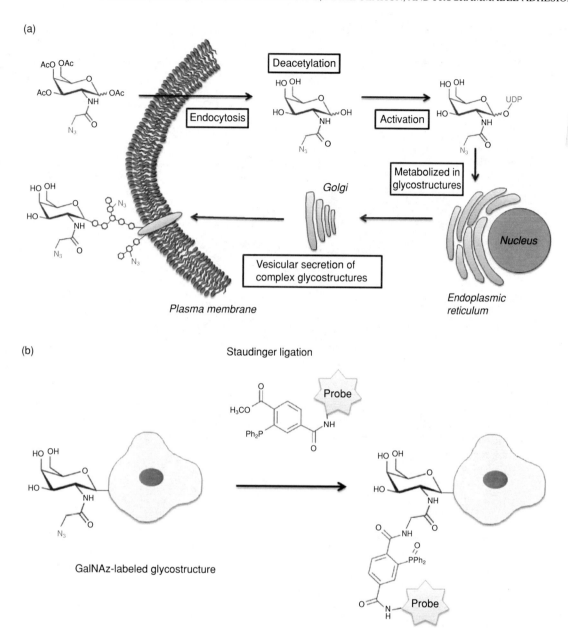

FIGURE 14.6 Schematic view of azide-modified monosaccharide metabolism (a). Staudinger ligation with GalNAz-modified glycoprotein (b).

to the trend: dibenzocyclooctyne (DIBO) < bicyclononyne (BCN) < dibenzoazacyclooctyne (DIBAC) < biarylazacyclooctynone (BARAC), which correlates the number of sp²-hybridized atoms as listed in Table 14.1 [40]. Van Delft et al. applied BCN for ligation to azido-containing cell surface glycans on MV3-melanoma cells [41]. Secondary labeling with SA-Alexa Fluor 488 revealed fine, subcellular details, which can discriminate surface glycan distribution states on individual living cells.

These cell membrane modification techniques can also be expanded to the development of new technologies for selective cell patterning for cell-based devices including

biosensors, drug-screening platforms, artificial tissues, and designed networks of neurons. To date, the patterning of cells on surfaces has been achieved with integrin-binding ligands such as fibronectin, laminin, or RGD (arginine–glycine–aspartic acid) containing peptides [42]. A particular advantage of this approach is its generality as most adherent cell types use integrins as their primary adhesion receptors [43]. However, the commonality of this mechanism makes it difficult to form patterns in which multiple types of cells are arrayed with precision on a single surface. As a result, considerable attention has been directed toward the development of new technologies for the programmable attachment of cells

TABLE 14.1   Reactivity of Functionalized Cyclooctynes

Name		$-\log(k \times 10^3)$
Biarylazacyclooctynone (BARAC)	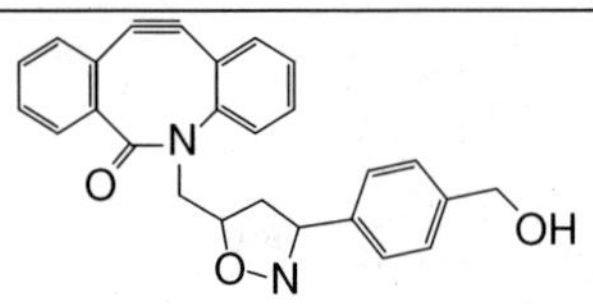	3.0
Dibenzoazacyclooctyne (DIBAC)	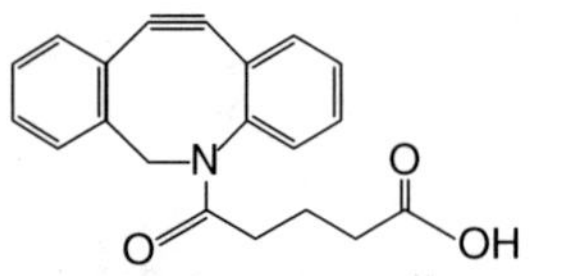	2.5
Bicyclononyne (BCN)	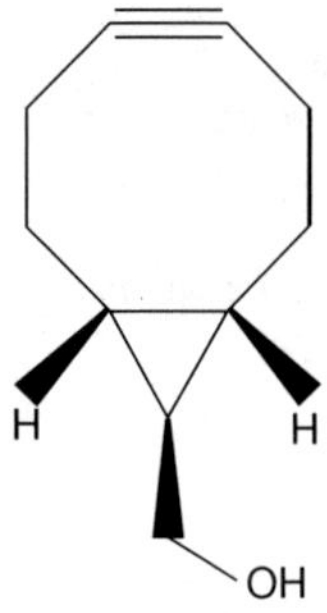	2.1
Dibenzocyclooctyne (DIBO)	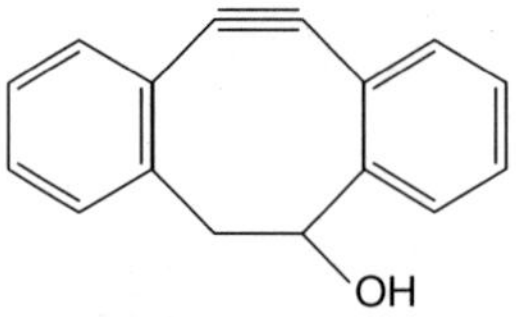	2.1
Difluorocyclooctyne (DIFO)	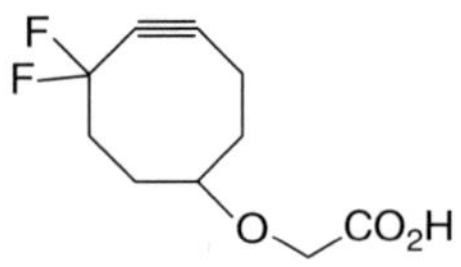	1.9
Monofluorinated cycloocyne (MOFO)	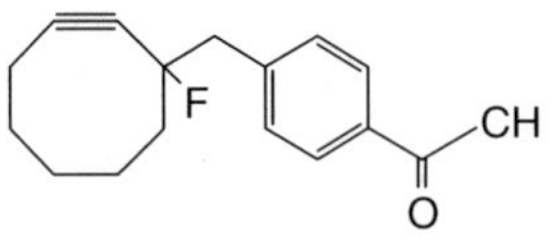	0.6
Cyclooctyne (OCT)	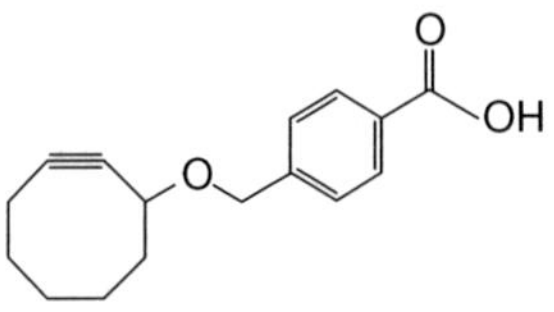	0.4
Dimethoxy azacylooctyne (DIMAC)	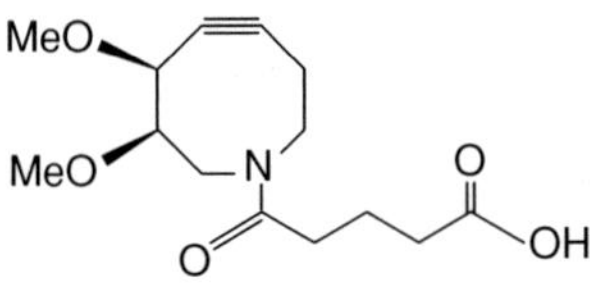	0.5

(*continued*)

TABLE 14.1 *(Continued)*

Name		$-\log(k \times 10^3)$
Nonfluorocyclooctyne (NOFO)		0.1
Aryl-less cyclooctyne (ALO)		0.1

to material surfaces. From these perspectives, Bertozzi and Francis et al. have expanded Staudinger ligation to include oligonucleotide attachment [44].

They prepared modified single-stranded DNA (ssDNA) strands with a phosphine group through the reaction of 5′-amine-modified ssDNA with a phosphine pentafluorophenyl (PFP) ester (Figure 14.7a). These strands were then used to anchor cells to specified locations on surfaces in a sequence-dependent fashion. Only cells bearing ssDNA strands that were complementary to the immobilized DNA were observed to bind to the surface, whereas otherwise identical cells bearing mismatched sequences were washed away (Figure 14.7b). This DNA-based strategy is sequence specific and applicable to a variety of devices and cell types such as T cells and cardiac myoblasts [45]. DNA hybridization techniques have also been utilized to control the attachment between heterogeneous and homogeneous cells. Iwata et al. introduced polyA and polyT onto the surface of the islets

(a)   (b)

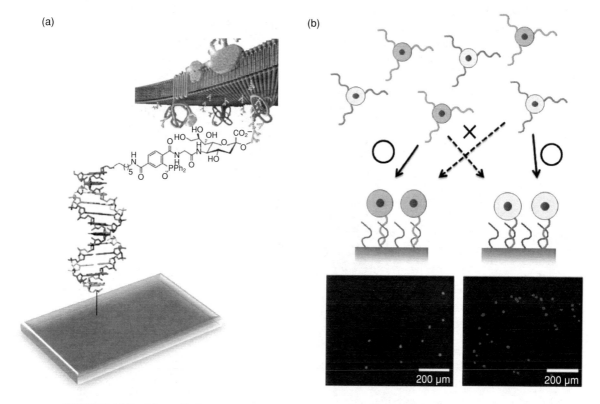

FIGURE 14.7 Live cells functionalized with single-stranded DNA oligonucleotides bind to substrates that bear complementary DNA strands in a sequence-specific manner (a). Only cells bearing ssDNA strands that were complementary to the immobilized DNA were observed to bind to the surface, whereas otherwise identical cells bearing mismatched sequences were washed away (b). For a color version, see the color plate section.

and HEK293 cells, respectively [46]. Other methods also exist for pattern cells on surfaces, such as polymer multilayers which offer a variety of favorable properties for cell surface modification. Rubner et al. created multilayer on the cell surface by photolithographic patterning technique [47]. First, a patterned array of heterostructured patches is created using a photolithographic lift-off process. Second, live cells are seeded onto the patterned assay and selectively bind to the surface of each patch. Finally, the patches are released from the surface while remaining attached to the cell membrane. When the patches carry magnetic nanoparticles, the cells can be spatially manipulated using a magnetic field.

14.6   VIRUSES, BACTERIA, YEAST CELL CONJUGATION

In addition to mammalian cells, selective functionalization of surface proteins via bio-orthogonal chemistry has been evolved into a powerful tool to engineer the surface of viruses, bacteria, yeast cells, etc. A myriad of viruses and virus-like particles have been genetically and chemically reprogrammed to function as drug/gene delivery vehicles, vaccines, and nanomaterials (Figure 14.8). For example, gene delivery vectors based on Adenoviral (Ad) vectors have enormous potential for the treatment of both hereditary and acquired diseases [48]. However, many of the therapeutically relevant target cells for gene therapy are refractory to Ad transduction due to low expression of primary receptors. Multiple strategies have been pursued to improve the efficacy of Ad vectors by targeting their transduction to specific

cell and tissue types. The chemical modification of the Ad capsid is one of the most direct approaches to modify vector tropism.

Seymour et al. modified the surface of Ads with a multivalent reactive 16.5 kDa poly(N-(2-hydroxypropyl) methacrylamide) (PHPMA)-based copolymer to shield them from recognition by antibodies [49]. Direct attachment of ligands like fibroblast growth factor-2 (FGF-2) through bifunctional PEG has been shown to augment coxsackie and adenovirus receptor (CAR)-independent gene transfer [50]. Reaction to PEG has also been shown to improve the *in vivo* pharmacokinetics of the vector by increasing vector persistence in the blood, preventing antibody neutralization [51]. Curiel et al. reported that coating of Ad with PEG reduced the clearance rate but also reduced infectivity [52]. Wilson et al. reported that PEG coating prolonged transgene expression and allowed partial readministration with native virus, although the activation of cytotoxic T lymphocytes and helper T cells of the type 1 subset (Th1 cells) against native viral antigens were diminished [53].

Of the many biomaterials used for building nanomaterials, viruses are of particular interest to the nanotechnology field because of their highly uniform structures, small size, and ability to self-assemble. Cowpea mosaic virus (CPMV) and tobacco mosaic virus (TMV), for example, have recently shown great potential with applications in nanoelectronics and energy harvesting devices. TMV is a classic example of rod-like plant viruses consisting of 2130 identical protein subunits arranged helically around genomic single RNA strand. Recent reports have demonstrated the addition of new functionality to the exterior and interior surfaces of viruses,

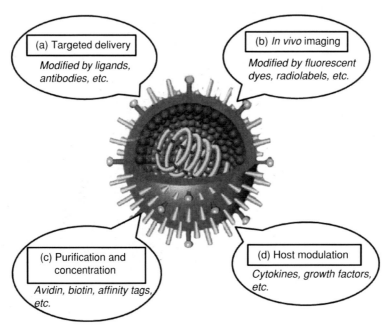

FIGURE 14.8   Overview of applications for surface modifications of viral vectors in gene therapy. For a color version, see the color plate section.

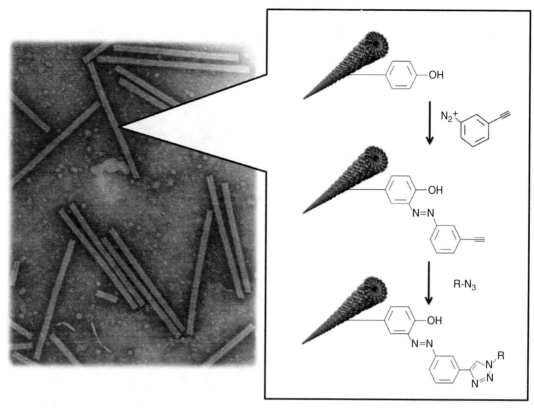

FIGURE 14.9   Surface modification of tobacco mosaic virus (TMV) by diazonium coupling and a sequential click reaction with azides.

yielding efficient routes to spherical core/shell materials [54]. Conductive polymers have also been coated on one-dimensional (1D) assembled TMV to produce conductive nanowires [55]. Lin has engineered cysteine residue on the CPMV because there are no native sulfhydryl groups on the exterior surface of the CPMV [56]. Compared to the native virus, the new inserted cysteines demonstrated higher reactivity with nearly all of the inserted thiol groups being chemically modified at very low concentration of a maleimide electrophile at neutral pH. In another system, the heat shock protein from *Methanococcus jannaschii* (MjHsp) has been engineered with a cysteine residue housed within the interior and by coupling the reactivity of the cysteine with a pH-sensitive maleimide derivative, antitumor drug was linked to the interior surface and selectively released upon decrease in pH [57]. Culver et al. functionalized cysteine-substituted TMV with fluorescent dyes, and the modified TMV was then partially disassembled to expose the single-stranded viral RNA. The exposed ssRNA strand was then utilized to hybridize to complementary DNA sequences patterned on surfaces [58]. It has been reported that tyrosine residues (Y139) of TMV are viable for chemical ligation using the electrophilic substitution reaction at the *ortho* position of the phenol ring with diazonium salts. Wang et al. demonstrated surface modification of TMV

by diazonium coupling and a sequential click reaction with azides (Figure 14.9) [59]. These studies highlight an important feature of viruses, namely that chemically reactive groups can be chemically engineered to selectively position drug molecules, imaging agent, and biologically relevant molecules on the 3D templates, which is extremely difficult realized using synthetic nanoparticles.

Surface-coating techniques have been also devised to protect bacteria from adverse environmental and processing conditions including *in vivo* conditions. Probiotics have recently received increased attention because of the multitude of health benefits they incur on the host system when taken in adequate amounts. However, the required stability of the microencapsulated organism is not always achieved and the number of viable bacteria becomes deficient to attain the intended effect. Encapsulation in microspheres using various polymers has been shown to provide improved survival in gastric/bile solutions (Figure 14.10). AL is a widely used encapsulating material that forms a gel in the presence of divalent cations. The limitation of using AL is its inability to withstand low pH encountered in the stomach. Raichur et al. demonstrated the encapsulation of probiotic *Lactobacillus acidophilus* through LBL self-assembly of chitosan (CH) and carboxymethyl cellulose (CMC). The survival rate of LBL-coated cells was enhanced during its gastrointestinal

Non coated   Coated

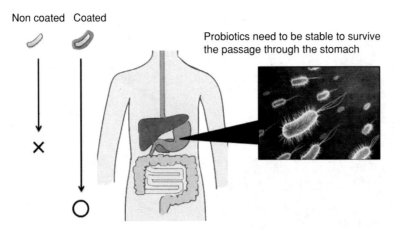

Probiotics need to be stable to survive
the passage through the stomach

FIGURE 14.10   Encapsulation of probiotics in microspheres can protect them from adverse environmental conditions and improve survival in gastric/bile solutions.

transit. About 33 log% of encapsulated cells survived when exposed to simulated gastric fluid for 2 hours followed by sequential exposure to simulated intestinal fluid for 2 hours, whereas uncoated cells were completely destroyed in identical conditions [60]. Lin et al. reported the selective functionalization of *o*-allyl-tyrosine genetically encoded in a Z-domain protein using a photoactivated, nitrile imine-mediated 1,3-dipolar cycloaddition reaction in *Escherichia coli*. The reaction procedure is very simple, straightforward, and nontoxic to cells. Fluorescent cycloadducts were formed which enabled a facile monitoring of the reaction *in vivo* [61].

14.7   CONCLUSIONS

Cell-based therapy is becoming the most powerful tool available to improve the human condition. Viral vectors are also proving to be useful delivery vehicles for gene therapy application. Modifying the surface of cells or viruses by conjugating them with bioactive molecules or synthetic polymers, therefore, has been a versatile way to add new value, advanced features, and unique properties to inert cells or viruses. Creating a nanoscale layer on a cell surface, for example, significantly improve or even completely change its biological properties as well as introduce new unique properties, such as chemical functionality, surface roughness, surface tension, morphology, surface charge, surface reflectivity, surface conductivity, and optical properties. Nowadays, a variety of different bioconjugation techniques are known for the purpose of both *in vitro* and *in vivo* studies. The range of opportunities provided by the bio-orthogonal reactions that can be carried out in whole organisms underscore the living cell surface modification. An ongoing challenge is to identify new transformations with the requisite qualities of selectivity and biocompatibility. Therefore, novel strategies for surface design of cells including mammalian cells, bacteria, yeast

cells, and viruses will facilitate the creation of new class of biomaterials in the future.

REFERENCES

1. Staudinger H, Meyer J. *Helv Chim Acta* 1919;2:635–646.
2. Rostovtsev VV, Green LG, Fokin VV, Sharpless KB. *Angew Chem Int Ed* 2002;41:2596–2599.
3. Decher G. *Science* 1997;277:1232–1237.
4. Steinman RM, Brodie SE, Cohn ZA. *J Cell Biol* 1976;68:665–687.
5. Teramura Y, Kaned Y, Totani T, Iwata H. *Biomaterials* 2008;29:1345–1355.
6. Harris AK. *Nature* 1976;263:781–783.
7. Teramura Y, Iwata H. *Soft Matter* 2010;6:1081–1091.
8. Frohn C, Fricke L, Puchta, JC, Kirchner H. *Nephrol Dial Transplant* 2001;16:355–360.
9. Bennet W, Sundberg B, Lundgren T, Tibell A, Groth CG, Richards A, White DJ, Elgue G, Larsson R, Nilsson B, Korsgren O. *Transplantation* 2000;69:711–719.
10. van Kim CL, Colin Y, Cartron JP. *Blood Rev* 2006;20:93–110.
11. Sehon AH. *Prog Immunol* 1983;5:483–491.
12. Scott MD, Murad KL, Koumpouras F, Talbot M, Eaton JW. *Proc Natl Acad Sci USA* 1997;94:7566–7571.
13. Hashemi-Najafabadi S, Vasheghani-Farahani E, Shojaosadati SA, Rasaee MJ, Armstrong JK, Moin M, Pourpak Z. *Bioconjugate Chem* 2006;17:1288–1293.
14. Le Y, Scott MD. *Acta Biomaterialia* 2010;6:2631–2641.
15. Chen AM, Scott MD. *Artif Cells Blood Substit Immobil Biotechnol* 2006;34:305–22.
16. Mansouri S, Fatisson J, Miao Z, Merhi Y, Winnik FM, Tabrizian M. *Langmuir* 2009;25:14071–14078.
17. Mansouri S, Merhi Y, Winnik FM, Tabrizian M. *Biomacromolecules* 2011;12:585–592.
18. Elbert DL, Hubbell JA. *Chem Biol* 1998;5:177–183.

19. Cerda-Cristerna BI, Cottin S, Flebus L, Pozos-Guillén A, Flores H, Heinen E, Jolois O, Gérard C, Maggipinto G, Sevrin C, Grandfils C. *Biomacromolecules* 2012;13, 1172–1180.

20. Shapiro AM, Lakey JR, Ryan EA, Korbutt GS, Toth E, Warnock GL, Kneteman NM, Rajotte RV. *N Engl J Med* 2000;343, 230–238.

21. Lim F, Sun AM. *Science* 1980;210:908–910.

22. Sawhney AS, Pathak CP, Hubbell JA. *Biotechnol Bioeng* 1994;44:383–836.

23. Wilson JT, Cui W, Chaikof, EL. *Nano Lett* 2008;8:1940–1948.

24. Lee DY, Park SJ, Lee S, Nam JH, Byun Y. *Tissue Eng* 2007;13:2133–2141.

25. Wilson JT, Krishnamurthy VR, Cui W, Qu Z, Chaikof EL. *J Am Chem Soc* 2009;131:18228–18229.

26. Miura S, Teramur Y, Iwata H. *Biomaterials* 2006;27:5828–5835.

27. Teramura Y, Kaneda Y, Iwata H. *Biomaterials* 2007;28:4818–4825.

28. Totani T, Teramura Y, Iwata H. *Biomaterials* 2008;29:2878–2883.

29. Teramura Y, Iwata H. *Biomaterials* 2009;30:2270–2275.

30. Panza JL, Wagner WR, Rilo HLR., Rao RH, Beckman EJ, Russell AJ. *Biomaterials* 2000;21:1155–1164.

31. Lee DY, Nam JH, Byun Y. *Biomaterials* 2007;28:1957–1966.

32. Mahal LK, Yarema KJ, Bertozzi CR. *Science* 1997;276:1125–1128.

33. Lemieux GA, Yarema KJ, Jacobs CL, Bertozzi CR. *J Am Chem Soc* 1999;121:4278–4279.

34. Griffin BA, Adams SR, Tsien RY. *Science* 1998;281:269–272.

35. Saxon E, Bertozzi CR. *Science* 2000;287:2007–2010.

36. Prescher JA, Dube DH, Bertozzi CR. *Nature* 2004;430:873–877.

37. Huisgen R, Padwa A (eds). 1,3-Dipolar Cycloaddition Chemistry. New York: John Wiley & Sons, Inc.; 1984.

38. Hong V, Steinmetz NF, Manchester M, Finn MG. *Bioconjugate Chem* 2010;21:1912–1916.

39. Agard NJ, Prescher JA, Bertozzi CR. *J Am Chem Soc* 2004;126:15046–15047.

40. Debets MF, van Berkel SS, Dommerholt J, Dirks AJ, Rutjes FPJT, van Delft FL. *Acc Chem Res* 2011;44:805–815.

41. Dommerholt J, Schmidt S, Temming R, Hendriks LJA, Rutjes FPJT, van Hest JCM, Lefeber DJ, Friedl P, van Delft FL. *Angew Chem Int Ed* 2010;49:9422–9425.

42. Singhvi R, Kumar A, Lopez GP, Stephanopoulos GN, Wang DIC, Whitesides GM, Ingber DE. *Science* 1994;264: 696–698.

43. Hynes RO. Integrins: versatility, modulation, and signaling in cell adhesion. *Cell* 1992;69:11–25.

44. Chandra RA, Douglas ES, Mathies RA, Bertozzi CR, Francis MB. *Angew Chem Int Ed* 2006;45:896–901.

45. Hsiao SC, Shum BJ, Onoe H, Douglas ES, Gartner ZJ, Mathies RA, Bertozzi CR, Francis MB. *Langmuir* 2009;25:6985–6991.

46. Teramura Y, Minh LN, Kawamoto T, Iwata H. *Bioconjugate Chem* 2010;21:792–796.

47. Swiston AJ, Cheng C, Um SH, Irvine DJ, Cohen RE, Rubner MF. *Nano lett* 2008;8:4446–4453.

48. Campos SK, Barry MA. *Curr Gene Ther* 2007;7:189–204.

49. Fisher KD, Stallwood Y, Green NK, Ulbrich K, Mautner V, Seymour LW. *Gene Ther* 2001;8:341–348.

50. Lanciotti J, Song A, Doukas J, Sosnowski B, Pierce G, Gregory R, Wadsworth SC, O'Riordan C. *Mol Ther* 2003;8:99–107.

51. Mok H, Palmer DJ, Ng P, Barry MA. *Mol Ther* 2005;11:66–79.

52. Alemany R, Suzuki K, Curiel DT. *J Gen Virol* 2000;81:2605–2609.

53. Croyle MA, Chirmule N, Zhang Y, Wilson JM. *Hum Gene Ther* 2002;13:1887–1900.

54. Schlick TL, Ding Z, Kovacs EW, Francis MB. *J Am Chem Soc* 2005;127:3718–3723.

55. Niu Z, Bruckman M, Kotakadi VS, He J, Emrick T, Russell TP, Yang L, Wang Q. *Chem Commun* 2006;28:3019–3021.

56. Lin T. *J Mater Chem* 2006;16:3673–3681.

57. Flenniken ML, Liepold LO, Crowley BE, Willits DA, Young MJ, Douglas T. *Chem Commun.* 2005;4:447–449.

58. Yi H, Rubloff GW, Culver JN. *Langmuir* 2007;23:2663–2667.

59. Bruckman MA, Kaur G, Lee LA, Xie F, Sepulveda J, Breitenkamp R, Zhang X, Joralemon M, Russell TP, Emrick T, Wang Q. *Chem Bio Chem* 2008;9:519–523.

60. Priya AJ, Vijayalakshmi SP, Raichur AM. *J Agric Food Chem* 2011;59:11838–11845.

61. Song W, Wang Y, Qu J, Lin Q. *J Am Chem Soc* 2008;130:9654–9655.

15

BIORESPONSIVE HYDROGELS AND MICROGELS

XUE LI AND MICHAEL J. SERPE
Department of Chemistry, University of Alberta, Edmonton, AB, Canada

15.1 INTRODUCTION

Polymers are among the most diverse class of materials employed and can be readily processed, chemically modified, and made degradable, which can be advantageous for biosensing [1–4], regenerative medicine [5–7], drug delivery [8–12], and tissue engineering [13–17]. Hydrogels are three-dimensional (3D), water-swollen, cross-linked polymer networks [18, 19]. The cross-links can be made via covalent bonds [20, 21], metal–ligand coordination [22], hydrogen bonds [23], van der Waals interactions [24], and other physical interactions such as entanglements and crystallites [25, 26]. Combinations of these cross-links are also frequently used [27–29]. Both synthetically prepared polymers and natural polymers have been used for polymeric hydrogels. Among the synthetic polymers are polyethylene oxide (PEO), poly(ethylene glycol) (PEG), poly(ethylene glycol)-diacrylate (PEGDA), polyvinyl alcohol (PVA), poly(acrylic acid) (PAA), poly(2-hydroxyethyl methacrylate) (pHEMA), and poly(N-isopropylacrylamide) (pNIPAm) [30]. Some of the examples of natural polymers include collagen, gelatin, glycosaminoglycans, and their derivatives [31]. Due to the mechanical and chemical properties of the hydrogel's polymer network, they can be used for sensing temperature [29, 32, 33], pH [34–36], biologically relevant analytes [37, 38], and ions [39, 40].

While the hydrogel's dimensions can be macroscopic, colloidally stable hydrogels—typically referred to as microgels (MGs) or nanogels (depending on their diameter)—can be synthesized. Additionally, MGs or nanogels can be synthesized to have tunable diameters from tens of nanometers to several micrometers and a high surface area due to their porosity. This is advantageous for coupling a high density

of biomolecules to the MG or nanogel structure, that is, bioconjugation [41–44]. In addition to covalent bioconjugation of biomolecules to the nanogel/MG network, they can also be entrapped via physical interactions. This has been used to entrap biomolecules such as drugs, proteins, carbohydrate, and DNA in these polymeric networks for *in vitro* small molecule release and various biomedical applications [45].

In this review, we highlight recent developments related to hydrogels and MGs. We highlight a number of examples of the use of these materials for biosensing, drug delivery, and tissue engineering. For many of these applications, the materials can change properties in response to the presence of selective biological targets. This often requires the material to be modified with biological recognition elements via covalent bonds. We will highlight a number of the coupling chemistries used to couple the biomolecules to the hydrogels.

15.2 BIOSENSING

15.2.1 Oligonucleotide-responsive Gels

Highly sensitive and selective DNA detection plays a central role in many fields of research such as disease diagnostics and forensic science [46]. DNA is a natural copolymer with four possible "monomers" (A, T, G, C) linked by a phosphodiester bond. Each phosphate carries one negative charge and therefore DNA is a polyanion. Being incorporated in the MG or hydrogel network, DNA can serve as a cross-linker controlling the mechanical properties of gel materials. Interestingly, it can serve as a probe to selectively bind to a variety of molecules. DNA can be chemically modified with many functional groups at specific sites, for example, thiol,

Chemistry of Bioconjugates: Synthesis, Characterization, and Biomedical Applications, First Edition. Edited by Ravin Narain.
© 2014 John Wiley & Sons, Inc. Published 2014 by John Wiley & Sons, Inc.

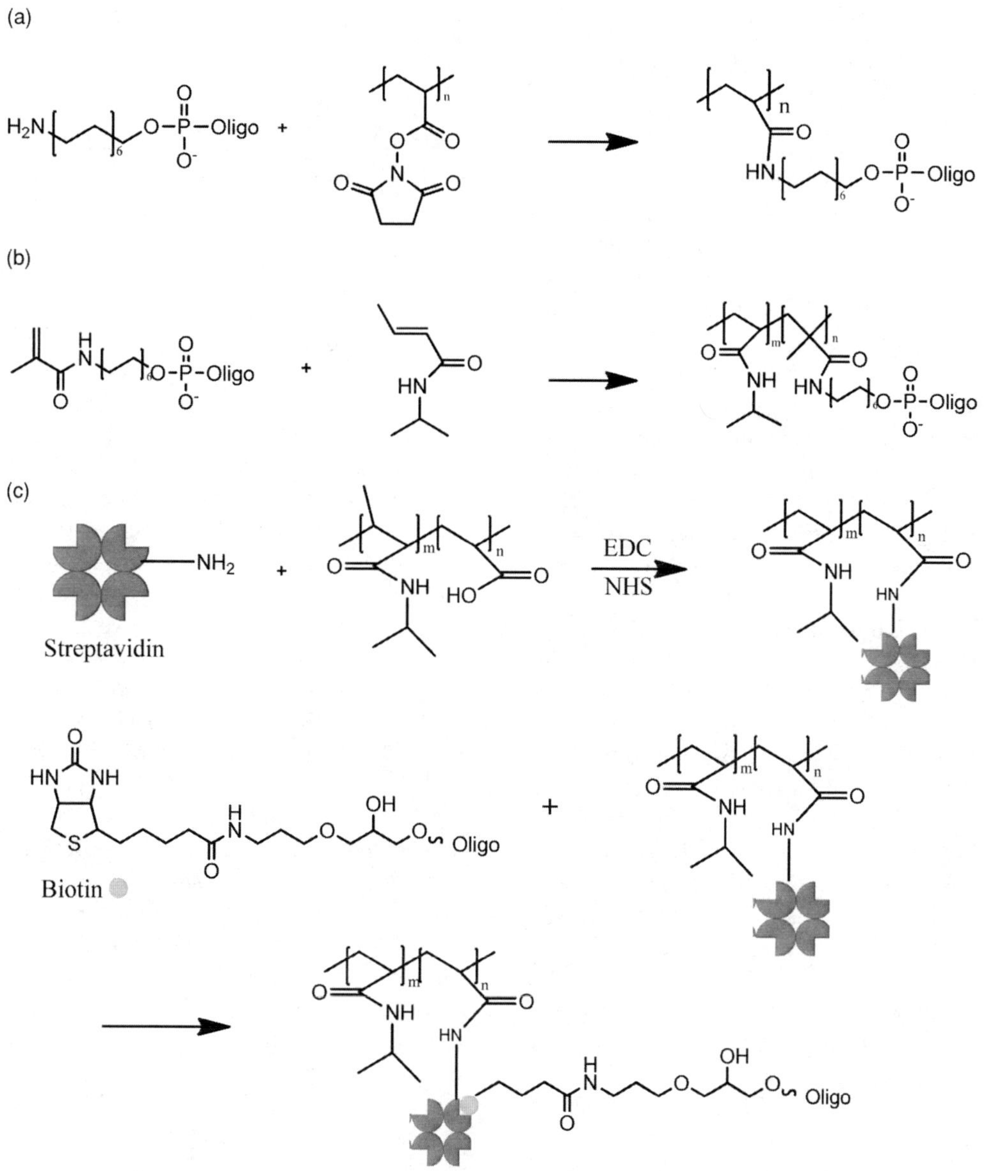

(a)

(b)

(c)

FIGURE 15.1   Three bioconjugate methods to attach DNA onto the gel system.

amino, biotin, and acrydite to accommodate various attachment chemistries. In gels, most DNA is functionalized with acrylic or acrylamide groups first in order for them to be incorporated in gel systems via copolymerization [47–50].

To date, there have been three prevalent bioconjugation chemistries used to covalently attach DNA to gels. First, as depicted in Figure 15.1a, amino-modified DNA reacts with a monomer or a polymer containing the reactive succinimidyl ester group to form an amide bond. In this way, the DNA is incorporated into the gel backbone. The DNA can be uniformly dispersed inside the gel networks by addition before the gel formation, otherwise the DNA will be mainly attached on surface of the gel systems [51]. Acrydite-modified DNA can be directly incorporated into the gel system as shown in

Figure 15.1b. In this case an acrydite-modified DNA serves as a comonomer and high incorporation efficiency can be achieved [52]. Another route is to use biotin–streptavidin (SP) interaction. Streptavidin is a type of tetrameric protein, which has four binding sites for biotin with high affinity ($K_d = 10^{-14}$ mol/L). Biotin-modified DNA can be directly attached to the SP-modified gel system with high affinity, as shown in Figure 15.1c. The streptavidin is attached onto the gel by coupling an amine group of the streptavidin to the carboxylate group of the gels by N-hydroxysuccinimide (NHS) and 1-ethyl-3-(3-dimethylaminopropyl) carbodiimide (EDC) coupling. Then the biotin-modified DNA can be immobilized onto the SP-gel by the interactions between the streptavidin and the biotin.

In addition to forming covalent bonds between DNA and the gel system, DNA can also be incorporated inside gels via physical interactions. For example, a new light-harvesting NIPAm-based MG, conjugated with polyelectrolyte (PE), was formed and fluorescein-modified ssDNA (DNA$_F$) was absorbed on the surface of the gel electrostatically [53]. The electrostatic interactions between positively charged MGs and negatively charged DNA$_F$ kept them in close proximity facilitating the energy transfer from the positively charged MG to DNA$_F$ causing the MG to fluoresce green under UV irradiation. Upon addition of another ssDNA (DNAc) complementary to DNA$_F$, double-stranded DNA was formed, which caused the MGs to fluoresce a salmon color instead of green. In another work, three different fluorescence signaling DNA enzymes were successfully entrapped within a series of sol–gel-derived matrices and used for sensing of various metal ions[54]. It was determined that the maximum sensitivity toward metal ions was obtained when the DNA enzymes were entrapped in composite materials containing approximately 40% methyltrimethoxysilane and approximately 60% tetramethoxysilane. These sol–gel technologies can provide new opportunities for the development of DNAzyme-based biosensors.

A new system referred to as a bioresponsive hydrogel suspension, which is based on pNIPAm hydrogel beads was studied to detect DNA [55]. Photonic structures were made from the hydrogel beads, which could convert physicochemical changes from DNA binding into spectral signals. In addition to using N,N'-methylenebisacrylamide (BIS) as cross-linkers in the hydrogels, they also employ single-stranded DNA to cross-link the polymer chains. They prepared a pregel solution composed of acrylamide, BIS, 5'- and 3'-acryloyl-modified ssDNA and photoinitiator. These monomers can be copolymerized to form hydrogel after exposing the pre-gel solution to the UV light. Specific hybridization of target DNA results in hydrogel shrinking, which can be detected from the blue shift of the Bragg diffraction peak position, as shown in Figure 15.2. Stokke and his coworkers [56, 57] studied the effect of the number of base pairs between two ssDNA on the physicochemical properties of hydrogels. The DNA linkages can be disrupted when a competitive DNA strand

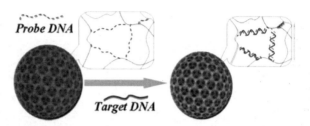

FIGURE 15.2 Schematic diagram of the label-free DNA detection based on DNA-responsive hydrogel photonic beads. Reprinted with permission from Reference 55, Copyright (2010), Copyright (2006), John Wiley & Sons, Inc.

is present. For this study, hydrogels were synthesized using acrylamide, BIS as the covalent cross-linker and hybridized oligonucleotides with attached methacrylic group at the 5' end acting as reversible physical cross-links. In the presence of target ssDNA, which can bind the cross-linked DNA more completely, the hydrogel structure swells, as shown in Figure 15.3.

Recently, Pelton and coworkers studied DNA oligonucleotide-conjugated MGs and their applications. They were trying to develop DNA–MG-based bioassays to investigate whether DNA–MG conjugates were compatible with enzymatic reactions [58]. They examined two enzymes: T4 DNA ligase and Phi29DNA polymerase. The mechanism is shown in Figure 15.4. They covalently coupled DNA and poly(isopropylacrylamide-co-vinylacetic acid) MG via EDC and (N-hydroxysulfosuccinimide) (NHSS) and they found that DNA molecules on the MG can still be manipulated by DNA-processing enzymes, which lay a solid foundation for the development of DNA–MG-based bioassays. They also studied the ability of these MGs for biosensing applications [59, 60]. Cao et al. [61] recently developed a type of biopolymer MG based on tetrameric protein avidin possessing four discrete and high affinity binding sites of biotins, DNA strands, and peptide nucleic acid (PNA), which is a synthetic analog of DNA in which the natural phosphodiester backbone is replaced with a polyamide [62, 63]. Firstly, three partially complementary DNA strands assemble into a three-way junction (TWJ)

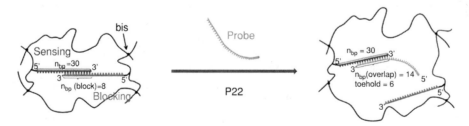

FIGURE 15.3 Schematic illustration of polymer–dsDNA hybrid hydrogels and processes of dsDNA exchange reactions. Reprinted with permission from Reference 56, Copyright (2011), Royal Society of Chemistry.

through Watson–Crick hydrogen bonding and at the end of each arm of the junction is a single-stranded overhang, as shown in Figure 15.5. These overhangs serve as the attaching point of the second components, which is a biotinylated PNA oligomer. PNA can bind with high affinity to the complementary DNA sequences according to the Watson–Crick rules [64]. Ultimately, the third component is avidin, which binds with biotins on PNAs. Thus, the biotinylated PNA links each avidin protein to as many as four TWJs, ultimately leading to formation of cross-linked network structure.

15.2.2 Protein-responsive Gels

pNIPAm displays temperature-dependent and reversible solubility behavior. This responsivity is a result of the lower critical solution temperature (LCST) of pNIPAm. This has been used by Hoffman and coworkers to design biohybrids [65]. As shown in Figure 15.6, the Michael-type addition reactions were applied to site-specifically conjugate maleimide-terminated oligo-NIPAm (oligomer) to cytochrome b5 which is genetically engineered to have a unique cysteine residue. The protein–oligomer conjugate was shown to also exhibit the LCST behavior, similar to the free oligomer. In another piece of Hoffman's work, streptavidin was engineered to contain a single cysteine near the biotin-binding site for site-specific conjugation of a vinyl-sulfone end-modified pNIPAm [66]. Normal binding of biotins and the pNIPAm-modified streptavidin occurs below 32°C, whereas above this temperature the polymer collapses and blocks the binding. In

a similar vein, Pennadam and coworkers conjugated pNIPAm to a large DNA "motor" to control methyltransferase activity of an enzyme complex [67]. Furthermore, bioconjugates based on pNIPAm have been studied for affinity precipitation of polysaccharides, proteins, and peptides [68, 69]. PEG is another highly investigated polymer for the covalent modification of proteins and peptides. The process of covalent attachment of PEG onto another molecule is called PEGylation. Reasons for PEGylation of peptides and proteins include shielding antigenic epitopes and receptor-mediated uptake and preventing recognition and degradation by proteolytic enzymes.

Most of conjugation chemistry was reviewed by Roberts and coworkers [70], as shown in Figure 15.7.

A reversible antigen-responsive hydrogel was obtained by immobilizing antigen and corresponding antibody on the semi-IPN hydrogel networks [71, 72]. Antigen (rabbit Immunoglobulin G (IgG)) and antibody (goat anti-rabbit IgG) were chemically modified by coupling the antigen and the antibody with N-succinimidylacrylamide (NSA) in pH 7.4 phosphate buffer solution to introduce vinyl group into IgG, as shown in Figure 15.8b. The resulting vinyl rabbit IgG was copolymerized with vinyl goat anti-rabbit IgG, acrylamide in the presence of cross-linker N,N'-methylenebisacrylamide (BIS). In this network system antigen–antibody bonding serves as additional cross-linkers. The polymerized antibody shows higher affinity with free antigen than polymerized antigen which already exists in the hydrogel system, then the free antigen replaced the binding of antibody from the polymerized antigen to the free antigen,

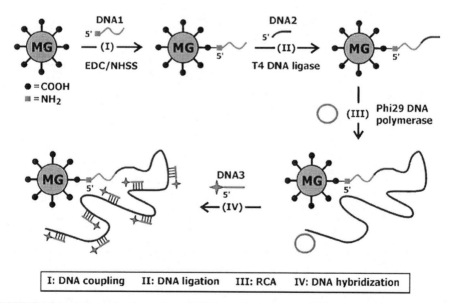

FIGURE 15.4 Schematic illustration of DNA manipulations on MG examined in this study. (I) covalent coupling of DNA with MG by NHSS/EDC. (II) DNA ligation. (III) Rolling circle amplification. (IV) Signal generation by hybridization with a fluorescent DNA probe. Reprinted with permission from Reference 58, Copyright (2007), Royal Society of Chemistry.

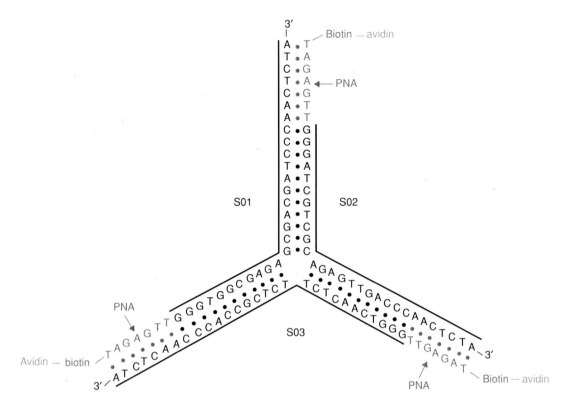

FIGURE 15.5 Molecular design of a cross-linked network structure based on a DNA three-way junction (TWJ). Note that each avidin protein can bind four biotinylated TWJ assemblies, leading to the network structure. Reprinted with permission from Reference 61, Copyright (2005), Nature Publishing Group.

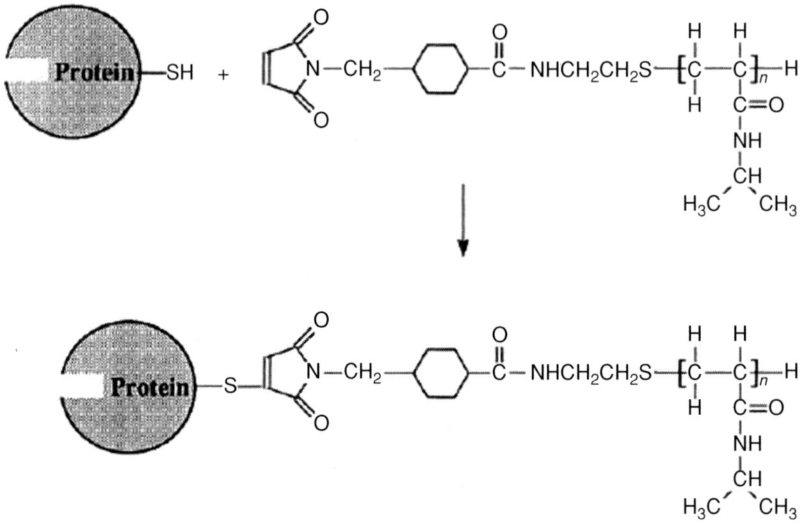

FIGURE 15.6 Reaction scheme of stoichiometrically precise conjugation of a protein to activated pNIPAm by reaction of the protein sulfhydryl group with a maleimide group, located at one end of the polymer chain. Reprinted with permission from Reference 65, Copyright (1994), American Chemical Society.

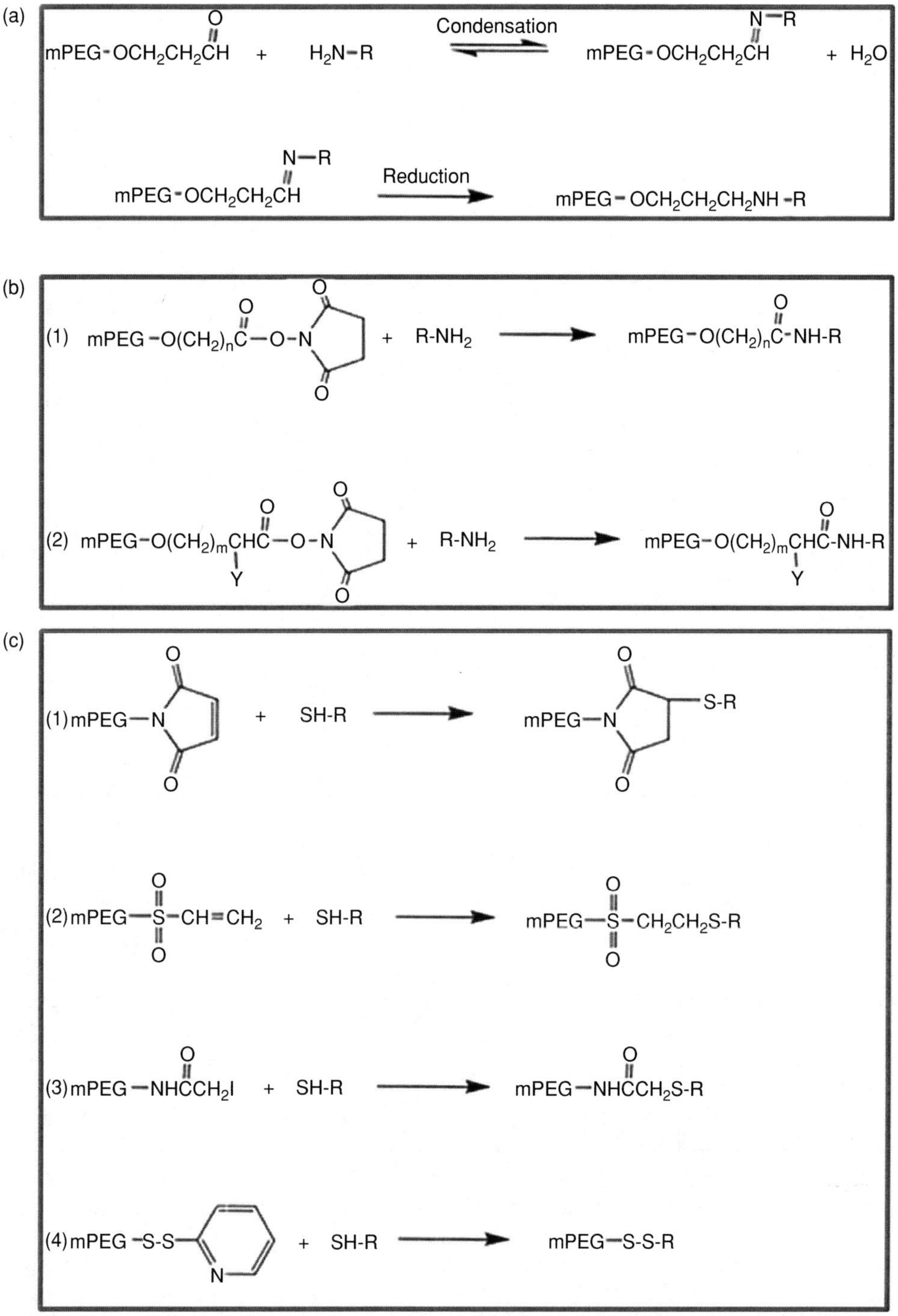

FIGURE 15.7 (a) Reductive amination using PEG-propionaldehyde, (b) PEG NHS esters: (1) PEG NHS esters based on propionic and butanoic acids and (2) α-branched PEG NHS esters based on propionic and butanoic acids, (c) thiol-reactive PEGs: (1) PEG maleimide, (2) PEG vinyl sulfone, (3) PEG iodoacetamide, and (4) PEG orthopyridyl disulfide. Reprinted with permission from Reference 70, Copyright (2002), Elsevier B.V.

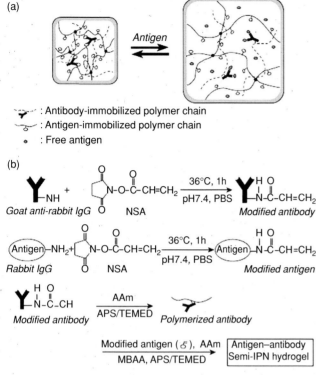

FIGURE 15.8 Strategy for the preparation of an antigen-responsive hydrogel. (a) Diagram of a suggested mechanism for the swelling of an antigen–antibody semi-IPN hydrogel in response to a free antigen. (b) Synthesis of the antigen–antibody semi-IPN hydrogel. Reprinted with permission from Reference 71, Copyright (1999), Nature Publishing Group.

which results in the swelling of the hydrogel, the suggested mechanism for the swelling of this semi-IPN hydrogel, as shown in Figure 15.8a.

Kim et al. [73] recently reported a whole-cell sensing system using interactions between antibody and specific antigen. The PEG hydrogel surface was functionalized with CD44 (a biotin) and the anti-CD44 antibody was conjugated to hydrogels via biotin–streptavidin interaction and this antibody allowed the specific immobilization of T-cells in a regular pattern at the hydrogel surface. And antigen-capturing B-cells overlaid on top acting as receptors for target molecules, as shown in Figure 15.9. Genetically engineered protein was attached on hydrogels by Ehrick [62] to study the stimuli-responsive characteristics of the hydrogels. Calmodulin (CaM), a calcium-binding protein was selected as it undergoes two conformational changes, one in the presence of Ca^{2+}, and the other in the presence of phenothiazines. These two changes induce reversible swelling of hydrogels. An allylamine moiety was attached to the free sulfhydryl residue on CaM after genetic engineering. Then the free amine group on the allylamine moiety was reacted with NSA, which results in the attachment of CaM on an acrylamide

moiety. Then the stimuli-responsive hydrogels were synthesized by free-radical polymerization of the modified CaM, N-{3-[2-(trifluoromethyl)-10H-phenothiazin-10-yl] propyl} acrylamide, acrylamide and N,N'-methylenebisacrylamide. The phenothiazine derivative and CaM were preincubated before polymerization, which resulted in phenothiazine non-covalently binding to CaM.

From the above examples, we found that NSA has been widely used to modify protein in order for them to be attached onto the gel systems. Recently Debord [74] studied the stability of succinimidyl esters on pNIPAm-co-(acrylic acid) (AAc) MGs. They used the well-known aqueous carbodiimide coupling method [75]. The coupling reagents EDC and sulfo-NHS were added to the MG solution and they found that microgel can only be recovered by aggressive hydrolysis protocols, which indicated that unusually stable succinimidyl ester has been formed in the MG during coupling. In the next few years they widely studied bioresponsive hydrogel microlenses prepared from stimuli-responsive poly(N-isopropylacrylamide-co-acrylic acid) MG functionalized with biotin via carbodiimide chemistry [76–78].

15.2.3 Glucose-sensitive Lectin-loaded Gels

Lectins are capable of binding carbohydrates with high affinity and specificity. These systems have been used for a number of applications, for example, for glucose sensing. Concanavalin A (ConA), which was first studied by Brownlee and Cerami [79] can complex with saccharide residues combined to polymer chains. Glycopolymers, which contain concentrated saccharide moieties on the polymer chains exhibit strong affinity interactions with ConA. Nakamae et al. [80] studied the complex formation between ConA and a polymer containing pendant glucose group poly(glucosyloxyethyl methacrylate) (pGEMA) in tris-HCl buffer solution. The solution became turbid due to the multiple interactions between ConA and pGEMA. When free glucose was added to the solution, the solution became transparent again due to the dissociation of the pGEMA–ConA complex. Novel glucose-sensitive hydrogels were prepared using complex between ConA and the pendant glucose groups of pGEMA to produce cross-linked points in the hydrogels [81, 82]. The hydrogel can serve as a glucose sensor, which relies on the dissociation of the complex formation between pGEMA and ConA in the presence of free glucose, as shown in Figure 15.10. The first step is to introduce vinyl group into the ConA to make the ConA monomer. The carboxyl group of AAc was bound to the amino group of ConA by condensation reaction with EDC. Since ConA modified by AAc can act as a cross-linking agent, the hydrogel was prepared without other cross-linking agents. The swelling of the hydrogel increased in response to free glucose and the swelling ratio was dependent on the glucose concentration. Kokufuta [83] also employed the carbohydrate-binding properties of ConA

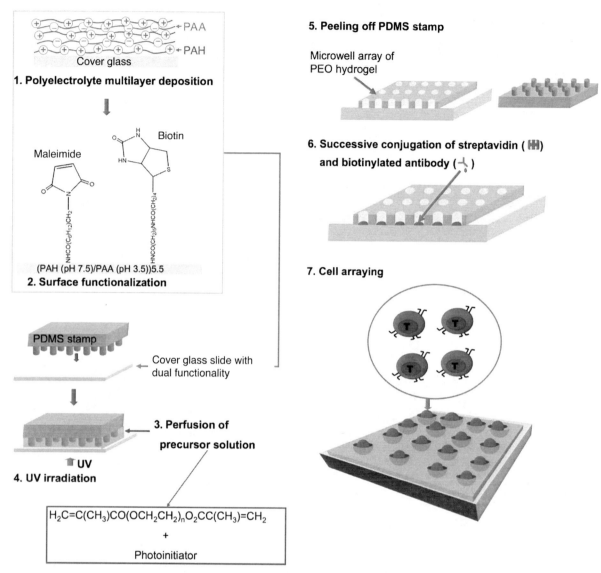

FIGURE 15.9   Schematic illustration of poly(ethylene glycol) dimethacrylate (PEGDMA) hydrogel microwell array fabrication and T-cell arraying. PAH, poly(allylamine hydrochloride); PAA, poly(acrylic acid); PEO, poly(ethylene oxide). Reprinted with permission from Reference 73, Copyright (2006), John Wiley & Sons, Inc.

and combined it with the temperature-sensitive properties of pNIPAm to prepare saccharide-sensitive hydrogels. This hydrogel swelled abruptly in the presence of the dextran sulfate at temperatures close to the volume phase transition point of the hydrogel and this signal is due to the ionic osmotic pressure exerted by the ionic saccharide.

15.2.4   Drug Delivery

The highly porous structure of hydrogels enables the introduction of relatively large amounts of small molecule drugs into them. The porosity of hydrogels can be tuned by controlling the density of cross-links [84] and by changing their

environmental conditions, which can facilitate the release of drugs through the matrix of hydrogels and into the environment.

A major advantage of polymer-based systems for drug delivery is that they may protect the active drugs from premature degradation. Ehrbar and coworkers [85] designed antibiotic-sensing hydrogels for the triggered release of human vascular endothelial growth factor. As shown in Figure 15.11, polyacrylamide was functionalized with nitrilotriacetic acid (NTA) for chelating Ni^{2+} ions to bind hexahistidine-tagged bacterial gyrase subunit (GyrB), which has been dimerized by the addition of the aminocoumarin antibiotic coumermycin. Synthesis of coumermycin

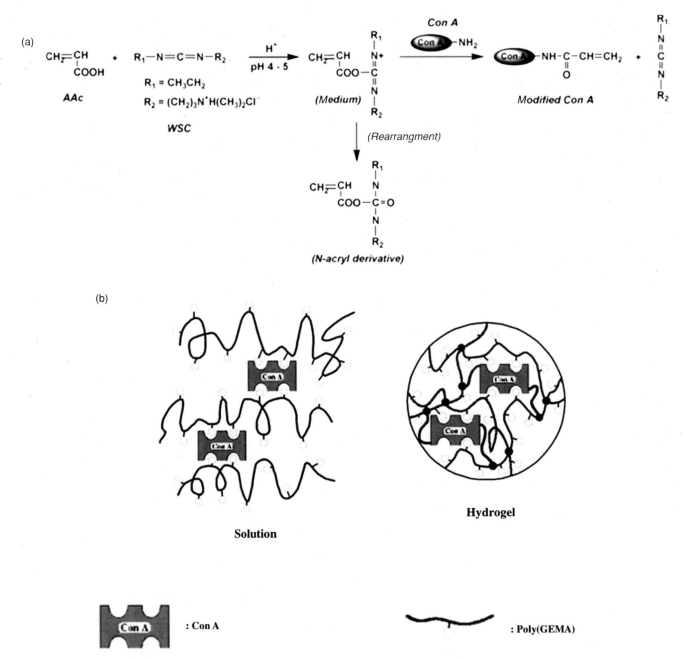

FIGURE 15.10 (a) Synthesis of modified ConA having vinyl groups. (b) Tentative models of complex formation between ConA and pendant glucose of GEMA in an aqueous solution and the ConA-copolymerized GEMA hydrogel. Reproduced with permission from Reference 80, Copyright (2004), Informa plc.

cross-linked hydrogels was carried out by incubating coumermycin and polyacrylamide, which was functionalized with dimerized GyrB. In the presence of novobiocin, which has higher affinity with GyrB than coumermycin, the GyrB subunits will disassociate with the hydrogels. This pharmacologically controlled hydrogels have the potential to be the smart devices for controlled delivery of drugs within the patient.

Enzyme-triggered drug release has been studied using drug molecules tethered to a hydrogel motif via enzyme cleavable linkers [86] and chemically cross-linked hydrogels whose cross-links contain peptide that is sensitive to an enzyme [87]. Thornton et al. [88, 89] designed an enzyme-responsive chemically cross-linked hydrogel that undergoes a macroscopic transition when triggered by a target protease, resulting in the release of entrapped molecules. As

Gel + Coumermycin   Sol + Novobiocin

(a)

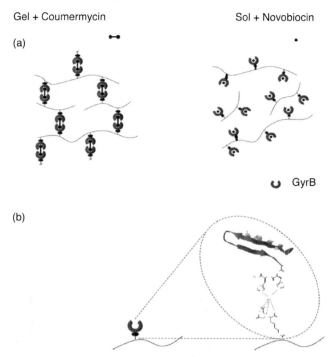

(b)

ᴗ GyrB

FIGURE 15.11   Design of pharmacologically controlled hydrogels. (a) GyrB coupled to an acrylamide polymer is dimerized by coumermycin, resulting in gelation of the hydrogel. In the presence of novobiocin, GyrB is dissociated, resulting in dissolution of the hydrogel. (b) Coupling of proteins to the acrylamide polymer. Polyacrylamide is functionalized with nitrilotriacetic acid chelating a Ni^{2+} ion to which GyrB can bind through a hexahistidine sequence. Reproduced with permission from Reference 85, Copyright (2008), Nature Publishing Group.

shown in Figure 15.12, PEG-based cross-linked polymers were modified with peptides providing the functions of sensing and actuation. The sensing part contains an enzyme cleavable linker (ECL) and the actuation part consists of two oppositely charged amino acids on either side, thus creating a zwitterionic peptide that has no overall charge when coupled to the hydrogel. When the ECL was selectively cleaved by the enzyme, a doubly negatively charged carboxylic acid fragment was removed leaving a doubly cationic amine fragment tethered to the hydrogel, which results in the swelling of the hydrogel. Polyacrylamide-based microparticles have also been used to be the carriers for the delivery of plasmid DNA for vaccine development [90]. Plasmid DNA was directly encapsulated into biocompatible polymer microparticles via radical polymerization in an inverse emulsion system, as shown in Figure 15.13. The DNA will be completely released under acidic conditions at lysosomal pH. Much work has been done on bioresponsive hydrogels related to the release of insulin in response to raised sugar levels. In one case, glucose oxidase molecules are immobilized onto a polymeric carrier [91, 92]. Glucose oxidase was employed in an oxidation-state responsive system based on

polysulfide nanoparticles modified with biocompatible polymer chains. Kim [93] and Miyata [82] prepared glucose-sensitive hydrogels by mixing glucose-containing polymers and PEGylation ConA. As shown in Figure 15.14, insulin release through glucose-sensitive hydrogel membrane and from the glucose-sensitive hydrogel matrix was dependent on the glucose concentration in the receptor chamber.

15.2.5   Tissue Engineering

Tissue engineering has been described as the process of creating artificial 3D tissue scaffolds or devices facilitating cell adhesion, growth, organization, and differentiation [94]. Scaffolds made from natural or synthetic polymers have been extensively used to drive the formation and maintenance of 3D tissue structures, which can be tailored to specific applications for the repair of replacement of tissues and organs [95]. In particular, hydrogels have been intensively studied and used as tissue engineering scaffolds as they can provide a highly swollen 3D environment similar to soft tissues and allow diffusion of nutrients and cellular waste through elastic networks [96, 97]. Compared with natural hydrogels, synthetic hydrogels have many advantages such as the ability of photopolymerization, adjustable mechanical properties, and convenient control of architecture and chemical compositions. Synthetic hydrogels have emerged as a preferable choice for tissue engineering scaffolds.

The environment surrounding cells in tissue is a complex network composed of proteins, carbohydrates, and growth factors called the extracellular matrix (ECM). ECM components play a crucial role in mediating cell functions, and possess critical biological functions like cell adhesion, proteolytic degradation, and growth factor binding [98]. Thus the natural ECM is an attractive model for design and fabrication of bioactive hydrogel scaffolds for tissue engineering.

The requirements of hydrogel scaffolds for tissue engineering include biocompatibility, biodegradability, high porosity, and no immunogenic reactions. Owing to the design flexibility, PEG is the most widely studied synthetic polymer used for scaffold fabrication. To meet the requirements of hydrogel scaffolds, PEG hydrogels have been modified with bioactive molecules, such as cell-adhesive peptide (CAP), enzyme-sensitive peptide (ESP), and growth factors (GF) to mimic ECM biofunctions such as cell adhesion, enzyme-sensitive degradation, and growth factors binding. These specific modifications are summarized as follows: the most commonly used CAP for cell-adhesive modification is Arg-Gly-Asp (RGD) [99] the cell-binding domain derived from Fibronectin (FN) and collagen [100]. To provide reactive sites for RGD on hydrogel surface, acrylic acid has been copolymerized with PEGDA to make hydrogels with carboxyl group, followed by conjugation with amine groups of peptide RGD [101]. Hynd and coworkers also applied streptavidin–avidin chemistry to modify hydrogel surface

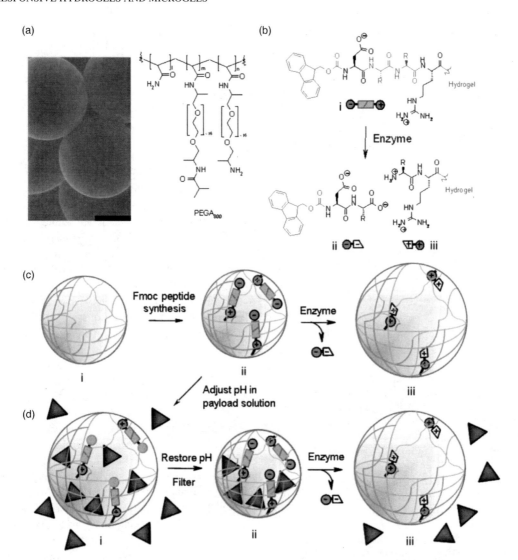

FIGURE 15.12   (a) The molecular structure of PEGA (copolymer of polyethylene glycol and acrylamide) and an environmental scanning electron microscopy (SEM) image of a PEGA hydrogel particle (scale bar is 100 μm). (b) Chemical structures of an ECL and consequent cleavage products. (c) Pendant amine groups may be readily functionalized with zwitterionic peptide linkers that confer no overall charge on the hydrogel. Enzyme-catalyzed hydrolysis reveals doubly charged peptide fragments causing the hydrogel particle to swell. (d) Hydrogel particles may be loaded with a macromolecular payload (represented by triangles) by lowering the pH of a dextran solution to less than 5. This pH drop leads to protonation of the side chains of aspartic acid residues, leaving a net positive charge that causes the hydrogel particle to swell and admit dextran molecules (i). A subsequent increase in solution pH regenerates the zwitterion, causing the hydrogel to collapse, reducing the mesh size, and entrapping the payload (ii). Upon enzymatic hydrolysis of the ECL, the dextran is released as the particle swells (iii). Reproduced with permission from Reference 89, Copyright (2007), John Wiley & Sons, Inc.

with CAPs [102]. This post-grafting approach is limited to attaching peptides on the hydrogel surface. Recently copolymerization of PEGDA with CAPs has become the major approach to make bulk cell-adhesive hydrogels. Hern and Hubbell [103] reported the incorporation of RGD into the PEG-based hydrogels. They synthesized the monoacrylate RGD by functionalizing the N-terminal amines of RGD

peptides with NHS ester of acrylic acid to produce monoacrylamidoyl RGD (RGD-MA), and subsequently RGD-MA was copolymerized with PEGDA upon photopolymerization to create cell-adhesive hydrogels, as shown in Figure 15.15. This method has been extensively studied to simulate cell adhesion, spreading, and growth on the nonadhesive surface of PEG hydrogels [104, 105]. For this method, PEGDA

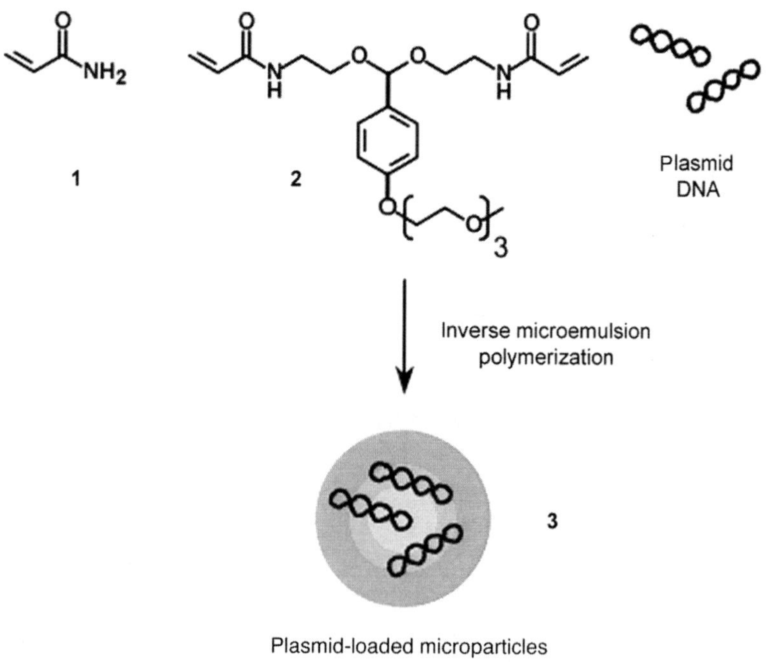

FIGURE 15.13   Synthesis of plasmid-loaded microparticles. Reprinted with permission from Reference 90, Copyright (2004), American Chemical Society.

polymerizes better than monoacrylated peptide because PEGDA is reactive on two sites. To solve this problem, Zhu and his coworkers described another method [106]. They used a hexapeptide, GRGDSP peptide, with specific binding to the integrin receptors to be the peptide sequence for the bioactive modification of PEGDA. The scheme is shown in Figure 15.16. To attach the RGD peptide in the middle of PEG chain, the GRGDSP was capped with diaminopropionic acid (Dap) to generate Dap-GRGDSP (1) with two free amine groups at the N-terminus, followed by reacting with acryloyl-PEG-NHS (Acr-PEG-NHS) (2) to generate the final product (3). Yang et al. reported on synthesis of cell-adhesive PEG hydrogels by click chemistry between 4-arm PEG acetylene (4-PEG-Ace) and RGD diazide (RGD-2N$_3$), as shown in Figure 15.17 [107]. RGD-2N3 was prepared by solid-phase peptide synthesis and 4-PEG-Ace wad synthesized by acetylation of tetrahydroxyl-terminated 4-arm PEG. PEG hydrogel network was formed by the copper sulfate as the catalyst between RGD-N3 and 4-PEG-Ace.

On the other hand, using the principle of the thermoresponsive phase transition, sol-to-gel phase transformation has been realized for potential tissue engineering application. pNIPAm, the most popular thermoresponsive polymer has been used in reversible cell attachment and detachment matrix [108], which is a pNIPAm copolymer partially derivatized with a cell-adhesion peptidyl moiety containing the RGD (Arg-Gly-Asp) sequence. Based on that, Ohya et al. [109] elaborated a novel strategy to synthesize aminated hyaluronic acid (HA)-g-pNIPAm (AHA-G-pNIPAm) to study its cell-adhesion behavior and the potential application on the tissue engineering. Hyaluronic acid (HA) is the nonsulfated glycosaminoglycan, which is widely distributed in the ECM of skin, cartilage, and the vitreous humor. It is composed of alternate disaccharide units of D-glucuronic acid and D-N-acetylglucosamine, which are linked together

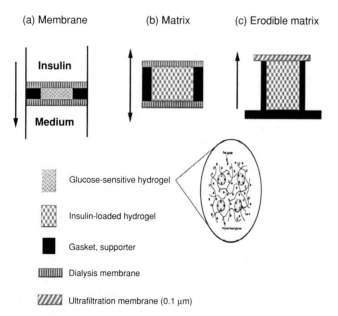

FIGURE 15.14   Three systems for modulated insulin delivery systems using phase-reversible glucose-sensitive hydrogels. Reproduced with permission from Reference 93, Copyright (2001), Elsevier B.V.

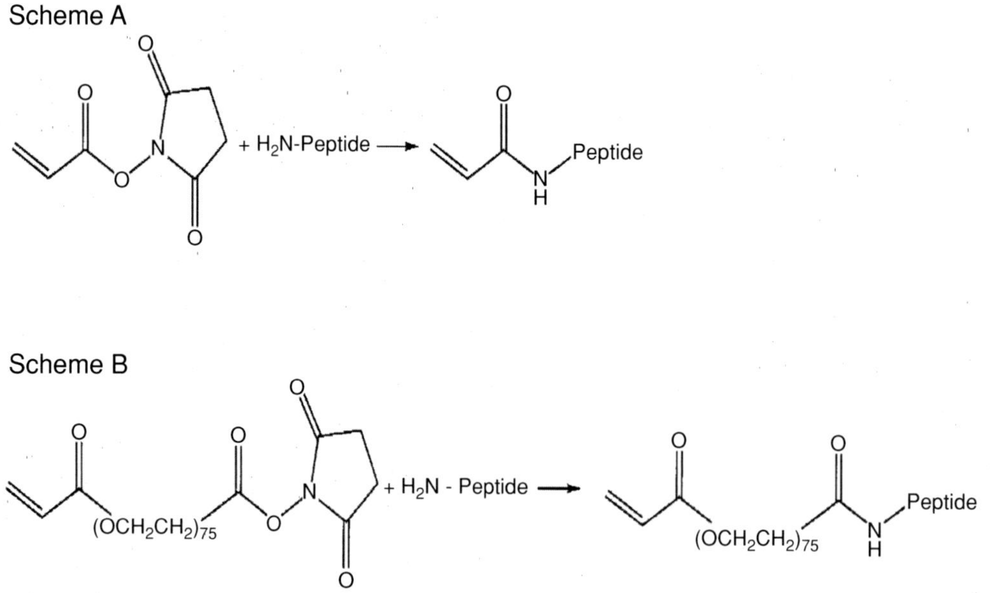

Scheme A

Scheme B

FIGURE 15.15 *N*-hydroxysuccinimidyl-activated esters were used to couple the N-terminal α-amine of the peptide to an acrylate moiety, either directly to acrylic acid or indirectly to an MW 3400 PEG spacer. Reproduced with permission from Reference 103, Copyright (1998), John Wiley & Sons, Inc.

via alternating β-1, 4 and β-1, 3 glycosidic bonds. AHA was prepared by grafting adipic dihydrazide to the HA backbone, and it was coupled to carboxylic end-capped pNIPAm (pNIPAm-COOH), which was produced via radical polymerization using 4,4'-azobis (4-cyanovaleric acid) as an initiator. The synthetic route of pNIPAm-grafted hyaluronan is shown in Figure 15.18. Chitosan is a linear natural polysaccharide composed of randomly distributed β-(1, 4)-linked D-glucosamine and N-acetyl-D-glucosamine units and has excellent biocompatibility and immunostimulatory activities [110]. pNIPAm has also been used to modify chitosan and other natural polymers to adjust their gelation temperature to the physiological temperature and improve their biocompatibility and mechanical strength [111]. Chitosan-g-pNIPAm copolymer was formed by conjugating the carboxylic acid group of pNIPAm-COOH to the amine group of chitosan in the presence of EDC and NHS. Then the CPN copolymers were evaluated for their potential use as injectable scaffolds for the culture of articular chondrocytes and meniscus cells.

Poly(α-hydroxyacid), another widely used synthetic polymeric material in tissue engineering, has also been fabricated into 3D scaffolds via a number of techniques. However, there are no functional groups available on the poly(α-hydroxyacid) chains. One method is to copolymerize the α-hydroxyacids with other monomers with pendant groups such as amino and carboxyl groups. He et al. [112] copolymerized L-lactide and (RS)-β-benzyl by ring-opening polymerization and pendant carboxyl group was obtained by removing the benzyl groups. Langer's group [113, 114]

synthesized poly[(L-lactic acid)-co-(L-lysine)] with functional lysine residue and was further coupled with RGD peptide, which can potentially control mammalian cell behavior.

Cells make up tissues and organs and they exist and communicate within a complex, 3D environment. Proper cell–cell communication is therefore crucial for a range of

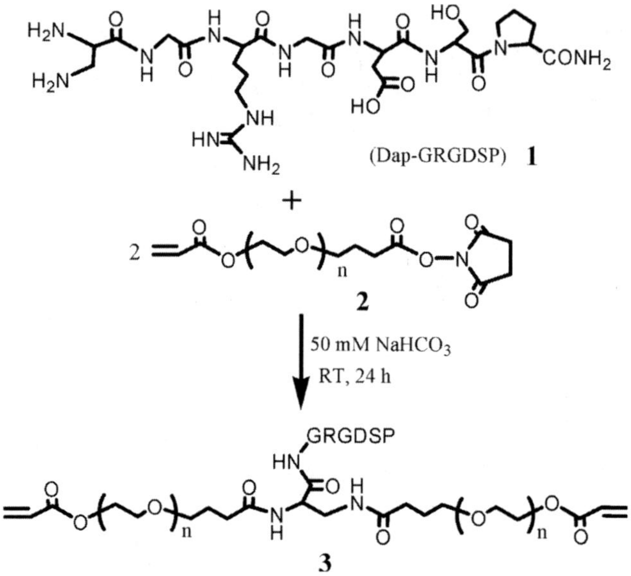

FIGURE 15.16 Synthesis of RGD-PEGDA. Reproduced with permission from Reference 106, Copyright (2006), American Chemical Society.

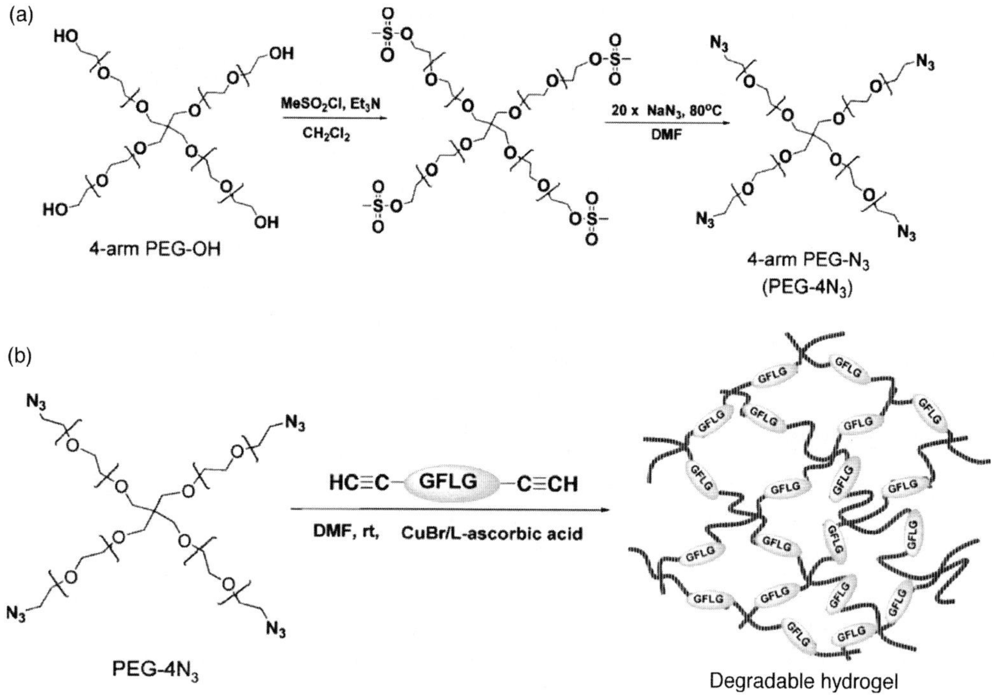

FIGURE 15.17 (a) Synthesis of 4-arm azido-terminated PEG. (b) Synthesis of degradable hydrogels by click reaction of 4-arm PEG-N3 with dialkyne-modified peptide. Reproduced with permission from Reference 107, Copyright (2010), John Wiley & Sons, Inc.

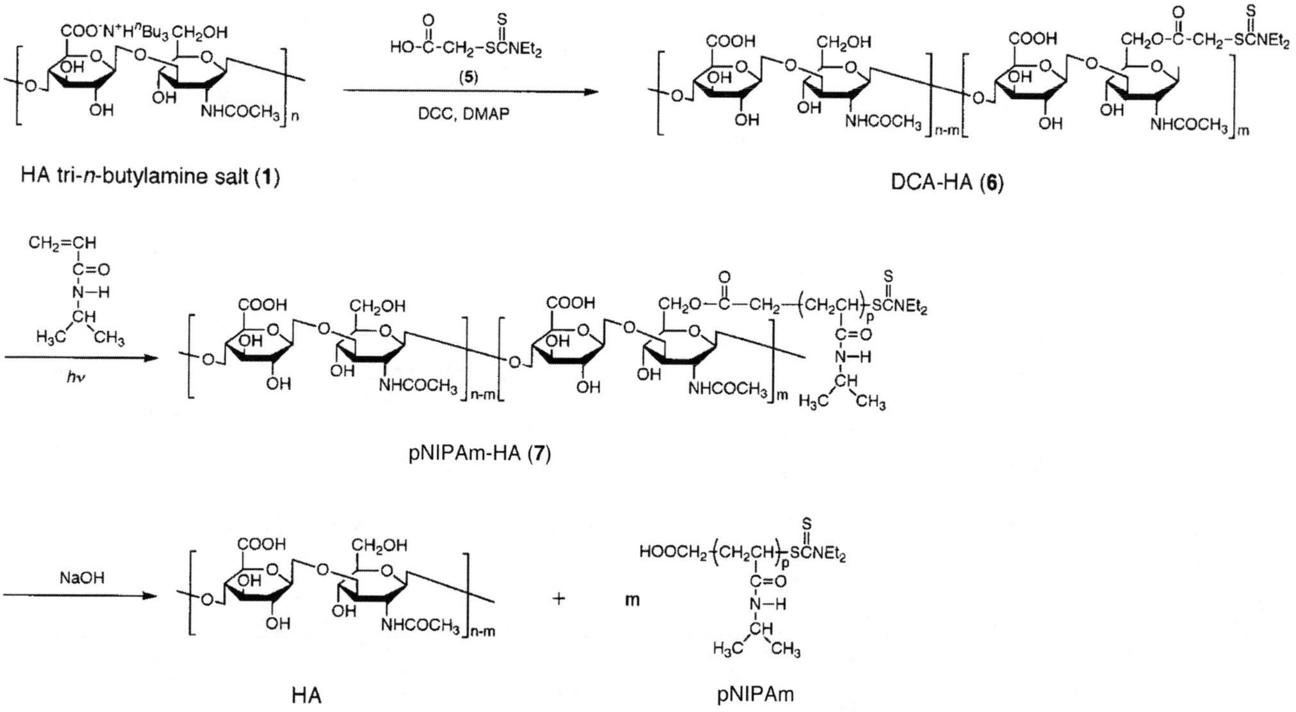

FIGURE 15.18 Synthetic route of pNIPAm-HA in which HA and pNIPAm molecules are connected via an ester bond. Reproduced with permission from Reference 109, Copyright (2001), American Chemical Society.

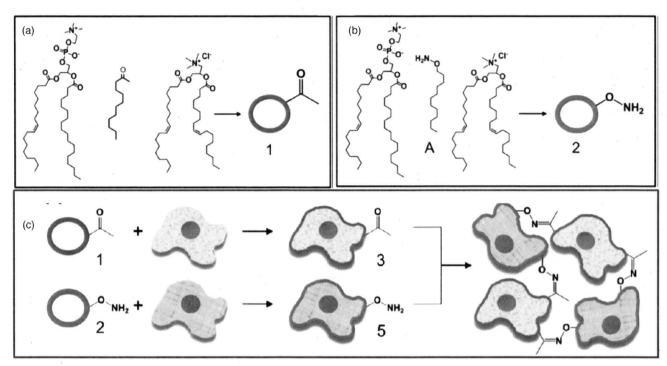

FIGURE 15.19 (a) Dodecanone molecules were incorporated into neutral, eggpalmitoyl-oleoyl phosphatidylcholine (POPC) and cationic, 1,2-dioleoyl-3-trimethylammonium-propane (DOTAP) at a ratio of 5:93:2 to form ketone-presenting liposomes (1). (b) O-Dodecyloxyamine molecules were incorporated into POPC and DOTAP at a ratio of 5:93:2 to form oxyamine-presenting liposomes (2). (c) Two fibroblast populations were cultured separately with ketone-containing (1) or oxyamine-containing (2) liposomes. Because of the presence of a positively charged liposome, fusion occurred, producing ketone-tethered (3) and oxyamine-tethered (5) cells. Upon mixing these cell populations, clustering and tissue-like formation, based on chemoselective oxime conjugation, occurred. Reproduced with permission from Reference 115, Copyright (2011), American Chemical Society. For a color version, see the color plate section.

fundamental biological processes. The development of 3D, *in vitro* model systems to investigate this complexity would provide a new platform for tissue engineering technologies. Yousef developed a strategy to induce specific and stable cell–cell contacts in 3D through chemoselective cell-surface engineering based on liposome delivery and fusion to display bio-orthogonal functional groups from cell membranes [115]. They use liposome fusion for the delivery of ketone or oxyamine groups to different populations of cells for subsequent cell assembly via oxime ligation, as shown in Figure 15.19. These methods can be used in many applications like bio-orthogonal ligand conjugation, rewiring cell adhesion, and the generation of multilayered tissue-like structures.

15.3 CONCLUSION

As mentioned above, the introduction of physical or synthetic elements into the natural or synthetic hydrogels or MGs can create materials with characteristics for biosensing, drug delivery, and tissue engineering. Probe-conjugated gels exhibit corresponding signal readout when exposed to the targeted molecules. Drug release from the polymeric gels is based on the drug diffusion and gel erosion. Therefore, the biodegradability of the polymeric gels is important for them to be used *in vivo*, and should be considered in the process of designing gels and choosing the synthetic method of the nondegradable elements. However, there is also a need for a continued improvement in the drug delivery of more sensitive molecules such as proteins, antibodies, or nucleic acids, which can readily be deactivated or unfolded by interactions with the gel-delivery vehicle. The development of tissue engineering has entered a new phase and rational design has been used to produce functionalized biomaterials and scaffolds tailored to specific applications. The polymer synthesis, scaffold fabrication, and growth factor delivery can be combined to create more advanced scaffolds for tissue engineering application.

REFERENCES

1. Hu Z, Chen Y, Wang C, Zheng Y, Li Y. *Nature* 1998;393:149.

2. Alarcon CdlH, Pennadam S, Alexander C. *Chem Soc Rev* 2005;34:276–285.

3. Galaev IY, Mattiasson B. *Trends Biotechnol* 1999;17:335–340.

4. Hoffman AS, Stayton PS. *Prog Polym Sci* 2007;32:922–932.

5. Cummins JM, Rago C, Kohli M, Kinzler KW, Lengauer C, Vogelstein B. *Nature* 2004;428.

6. Ratner BD, Bryant SJ. *Annu Rev Biomed Eng* 2004;6:41–75.

7. Bettinger CJ. *Macromol Biosci* 2011;11:467–482.

8. Das M, Mardyani S, Chan WCW, Kumacheva E. *Adv Mater* 2006;18:80–83.

9. Soppimath KS, Tan DCW, Yang YY. *Adv Mater* 2005;17:318–323.

10. Thornton PD, Mart RJ, Ulijn RV. *Adv Mater* 2007;19:1252–1256.

11. Soppimath KS, Kulkarni AR, Aminabhavi TM. *J. Controlled Release* 2001;75:331–345.

12. Gu J, Xia F, Wu Y, Qu X, Yang Z, Jiang L. *J. Controlled Release* 2007;117:396–402.

13. Yamaguchi N, Zhang L, Chae B-S, Palla CS, Furst EM, Kiick KL. *JACS* 2007;129:3040–3041.

14. Tae G, Kim Y-J, Choi W-I, Kim M, Stayton PS, Hoffman AS. *Biomacromolecules* 2007;8:1979–1986.

15. Moschou EA, Peteu SF, Bachas LG, Madou MJ, Daunert S. *Chem Mater* 2004;16:2499–2502.

16. Burdick JA, Prestwich GD. *Adv Mater* 2011;23:H41–H56.

17. Khetan S, Burdick JA. *Soft Matter* 2011;7:830–838.

18. Zhang S. *Nat Mater* 2004;3:7–8.

19. Kiyonaka S, Sada K, Yoshimura I, Shinkai S, Kato N, Hamachi I. *Nat Mater* 2004;3:58–64.

20. Cretu A, Kipping M, Adler H-J, Kuckling D. *Polym Int* 2008;57:905–911.

21. Bader RA. *Acta Biomaterialia* 2008;4:967–975.

22. Serpe MJ, Craig SL. *Langmuir* 2006;23:1626–1634.

23. Mitsumata T, Hasegawa C, Kawada H, Kaneko T, Takimoto J-i. *React Funct Polym* 2008;68:133–140.

24. Xiao C, Gao Y. *J Appl Polym Sci* 2008;107:1568–1572.

25. Peppas NA, Bures P, Leobandung W, Ichikawa H. *Euro J Pharm Biopharm* 2000;50:27–46.

26. Peppas NA, Mongia NK. *Euro J Pharm Biopharm* 1997;43:51–58.

27. Yang X, Zhu Z, Liu Q, Chen X. *J Appl Polym Sci* 2008;109:3825–3830.

28. Robb SA, Lee BH, McLemore R, Vernon BL. *Biomacromolecules* 2007;8:2294–2300.

29. Cheng V, Lee BH, Pauken C, Vernon BL. *J Appl Polym Sci* 2007;106:1201–1207.

30. Jagur-Grodzinski J. *Polym Adv Technol* 2010;21:27–47.

31. van Vlierberghe S, Dubruel P, Schacht E. *Biomacromolecules* 2011;12:1387–1408.

32. Geever LM, Nugent MJD, Higginbotham CL. *J Mater Sci* 2007;42:9845–9854.

33. Geever LM, Devine DM, Nugent MJD, Kennedy JE, Lyons JG, Hanley A, Higginbotham CL. *Eur Polym J* 2006;42:2540–2548.

34. Zhao Y, Yang Y, Yang X, Xu H. *J Appl Polym Sci* 2006;102:3857–3861.

35. Patel A, Mequanint K. *Macromol Biosci* 2007;7:727–737.

36. Hayashi H, Iijima M, Kataoka K, Nagasaki Y. *Macromolecules* 2004;37:5389–5396.

37. Lee M-C, Kabilan S, Hussain A, Yang X, Blyth J, Lowe CR. *Anal Chem* 2004;76:5748–5755.

38. Kang SI, Bae YH. *J Control Release* 2003;86:115–121.

39. Ju X-J, Chu L-Y, Liu L, Mi P, Lee YM. *J Phys Chem B* 2008;112:1112–1118.

40. Ahn S-k, Kasi RM, Kim S-C, Sharma N, Zhou Y. *Soft Matter* 2008;4:1151–1157.

41. Harris JM, Chess RB. *Nat Rev Drug Discov* 2003;2:214–221.

42. Vicent MJ, Greco F, Nicholson RI, Paul A, Griffiths PC, Duncan R. *Angew Chem Int Ed* 2005;44:4061–4066.

43. Khandare J, Minko T. *Prog Polym Sci* 2006;31:359–397.

44. Jung T, Kamm W, Breitenbach A, Kaiserling E, Xiao JX, Kissel T. *Euro J Pharm Biopharm* 2000;50:147–160.

45. Oh JK, Drumright R, Siegwart DJ, Matyjaszewski K. *Prog Polym Sci* 2008;33:448–477.

46. Rosi NL, Mirkin CA. *Chem Rev* 2005;105:1547–1562.

47. Wei B, Cheng I, Luo KQ, Mi Y. *Angew Chem Int Ed* 2008;47:331–333.

48. He X, Wei B, Mi Y. *Chem Commun* 2010;46:6308–6310.

49. Murakami Y, Maeda M. *Biomacromolecules* 2005;6:2927–2929.

50. Murakami Y, Maeda M. *Macromolecules* 2005;38:1535–1537.

51. Liu J. *Soft Matter* 2011;7:6757–6767.

52. Rehman FN, Audeh M, Abrams ES, Hammond PW, Kenney M, Boles TC. *Nucleic Acids Res* 1999;27:649–655.

53. Feng X, Xu Q, Liu L, Wang S. *Langmuir* 2009;25:13737–13741.

54. Shen Y, Mackey G, Rupcich N, Gloster D, Chiuman W, Li Y, Brennan JD. *Anal Chem* 2007;79:3494–3503.

55. Zhao Y, Zhao X, Tang B, Xu W, Li J, Hu J, Gu Z. *Adv Funct Mater* 2010;20:976–982.

56. Gao M, Gawel K, Stokke BT. *Soft Matter* 2011;7:1741–1746.

57. Tierney S, Stokke BT. *Biomacromolecules* 2009;10:1619–1626.

58. Ali MM, Su S, Filipe CDM, Pelton R, Li Y. *Chem Commun* 2007:4459–4461.

59. Su S, Ali MM, Filipe CDM, Li Y, Pelton R. *Biomacromolecules* 2008;9:935–941.

60. Ali MM, Aguirre SD, Xu Y, Filipe CDM, Pelton R, Li Y. *Chem Commun* 2009:6640–6642.

61. Cao R, Gu Z, Hsu L, Patterson GD, Armitage BA. *JACS* 2003;125:10250–10256.

62. Ehrick JD, Deo SK, Browning TW, Bachas LG, Madou MJ, Daunert S. *Nat Mater* 2005;4:298–302.

63. Nielsen PE, Haaima G. *Chem Soc Rev* 1997;26:73–78.

64. GREEN NM. *Biochem J* 1963;89:585–580.

65. Chilkoti A, Chen GH, Stayton PS, Hoffman AS. *Bioconjugate Chem* 1994;5:504–507.

66. Stayton PS, Shimoboji T, Long C, Chilkoti A, Chen GH, Harris JM, Hoffman AS. *Nature* 1995;378:472–474.

67. Pennadam SS, Lavigne MD, Dutta CF, Firman K, Mernagh D, Górecki DC, Alexander C. *JACS* 2004;126:13208–13209.

68. Takei YG, Aoki T, Sanui K, Ogata N, Sakurai Y, Okano T, Matsukata M, Kikuchi A. *Bioconjugate Chem* 1994;5:577–582.

69. Zheng Shu X, Liu Y, Palumbo FS, Luo Y, Prestwich GD. *Biomaterials* 2004;25:1339–1348.

70. Roberts MJ, Bentley MD, Harris JM. *Adv Drug Deliv Rev* 2002;54:459–476.

71. Miyata T, Asami N, Uragami T. *Nature* 1999;399:766–769.

72. Miyata T, Asami N, Uragami T. *J Polym Sci Part B: Polym Phys* 2009;47:2144–2157.

73. Kim H, Cohen RE, Hammond PT, Irvine DJ. *Adv Funct Mater* 2006;16:1313–1323.

74. Debord JD, Lyon LA. *Bioconjug Chem* 2007;18:601–604.

75. Grabarek Z, Gergely J. *Anal Biochem* 1990;185:131–135.

76. Kim J, Nayak S, Lyon LA. *JACS* 2005;127:9588–9592.

77. Kim J, Singh N, Lyon LA. *Biomacromolecules* 2007;8:1157–1161.

78. Kim J, Singh N, Lyon LA. *Angew Chem Int Ed* 2006;45:1446–1449.

79. Brownlee M, Cerami A. *Science* 1979;206:1190–1191.

80. Nakamae K, Miyata T, Jikihara A, Hoffman AS. *J Biomater Sci Polym Ed* 1994;6:79–90.

81. Miyata T, Jikihara A, Nakamae K, Hoffman AS. *Macromol Chem Phys* 1996;197:1135–1146.

82. Miyata T, Jikihara A, Nakamae K, Hoffman AS. *J Biomater Sci Polym Ed* 2004;15:1085–1098.

83. Kokufata E, Zhang YQ, Tanaka T. *Nature* 1991;351:302–304.

84. Hoare TR, Kohane DS. *Polymer* 2008;49:1993–2007.

85. Ehrbar M, Schoenmakers R, Christen EH, Fussenegger M, Weber W. *Nat Mater* 2008;7:800–804.

86. Law B, Weissleder R, Tung C-H. *Biomacromolecules* 2006;7:1261–1265.

87. Lutolf MP, Hubbell JA. *Nat Biotech* 2005;23:47–55.

88. Thornton PD, Mart RJ, Webb SJ, Ulijn RV. *Soft Matter* 2008;4:821–827.

89. Thornton PD, Mart RJ, Ulijn RV. *Adv Mater* 2007;19:1252–+.

90. Goh SL, Murthy N, Xu MC, Frechet JMJ. *Bioconjug Chem* 2004;15:467–474.

91. Fischel-Ghodsian F, Brown L, Mathiowitz E, Brandenburg D, Langer R. *Proc Natl Acad Sci* 1988;85:2403–2406.

92. Rehor A, Botterhuis NE, Hubbell JA, Sommerdijk NAJM, Tirelli N. *J Mater Chem* 2005;15:4006–4009.

93. Kim JJ, Park K. *J. Control Release* 2001;77:39–47.

94. Langer R, Vacanti JP. *Science* 1993;260:920–926.

95. Shoichet MS. *Macromolecules* 2009;43:581–591.

96. Lee KY, Mooney DJ. *Chem Rev* 2001;101:1869–1879.

97. Hoffman AS. In: Hunkeler D, Cherrington A, Prokop A, Rajotte R (eds), *Bioartificial Organs III: Tissue Sourcing, Immunoisolation, and Clinical Trials*, Part III, Immunoisolation and Encapsulation New York: New York Academy of Sciences; 2001. pp 62–73.

98. Badylak SF. *Biomaterials* 2007;28:3587–3593.

99. Hersel U, Dahmen C, Kessler H. *Biomaterials* 2003;24:4385–4415.

100. Leahy DJ, Aukhil I, Erickson HP. *Cell* 1996;84:155–164.

101. Drumheller PD, Hubbell JA. *Anal Biochem* 1994;222:380–388.

102. Hynd MR, Frampton JP, Dowell-Mesfin N, Turner JN, Shain W. *J Neurosci Methods* 2007;162:255–263.

103. Hern DL, Hubbell JA. *J Biomed Mater Res* 1998;39:266–276.

104. Gunn JW, Turner SD, Mann BK. *J Biomed Mater Res A* 2005;72:91–97.

105. Yang F, Williams CG, Wang D-a, Lee H, Manson PN, Elisseeff J. *Biomaterials* 2005;26:5991–5998.

106. Zhu JM, Beamish JA, Tang C, Kottke-Marchant K, Marchant RE. *Macromolecules* 2006;39:1305–1307.

107. Yang J, Jacobsen MT, Pan H, Kopeček J. *Macromol Biosci* 2010;10:445–454.

108. Yamada N, Okano T, Sakai H, Karikusa F, Sawasaki Y, Sakurai Y. *Die Makromolekulare Chemie, Rapid Communications* 1990;11:571–576.

109. Ohya S, Nakayama Y, Matsuda T. *Biomacromolecules* 2001;2:856–863.

110. De Souza R, Zahedi P, Allen CJ, Piquette-Miller M. *Biomaterials* 2009;30:3818–3824.

111. Chen J-P, Cheng T-H. *Macromol Biosci* 2006;6:1026–1039.

112. He B, Bei J, Wang S. *Polymer* 2003;44:989–994.

113. Barrera DA, Zylstra E, Lansbury PT, Langer R. *JACS* 1993;115:11010–11011.

114. Cook AD, Hrkach JS, Gao NN, Johnson IM, Pajvani UB, Cannizzaro SM, Langer R. *J Biomed Mater Res* 1997;35:513–523.

115. Dutta D, Pulsipher A, Luo W, Yousaf MN. *JACS* 2011;133:8704–8713.

16

CONJUGATION STRATEGIES USED FOR THE PREPARATION OF CARBOHYDRATE-CONJUGATE VACCINES

RACHEL HEVEY AND CHANG-CHUN LING
Alberta Glycomics Center and Department of Chemistry, University of Calgary, Calgary, AB, Canada

16.1 INTRODUCTION

Since the success of Edward Jenner's cowpox vaccine was first published in 1798, immunization has been used over the years as a preventative measure against the disease [1]. Immunization effectively prepares the immune system to respond against foreign pathogens, thereby preventing infection before it occurs. Resulting from the discovery of penicillin in 1929, the use of antibiotics against infection became increasingly popular throughout the years, causing the magnitude of research directed toward the development of vaccines to subside [2, 3]. Even with this great discovery, it has become increasingly evident over time that not all diseases respond to chemotherapeutic treatment. As well, with the growing concern regarding antibiotic-resistant strains of pathogens, such as the versatile methicillin-resistant *Staphylococcus aureus*, the development of vaccines has reemerged as a prominent subject in preventative disease medicine and research [4–8]. As the quantity of antibiotic-resistant bacterial strains rises, the want increases for alternative treatment options against infectious disease.

Vaccines can be classified into two broad categories: prophylactic vaccines and therapeutic vaccines. Prophylactic vaccines are administered prior to pathogen contact and effectively prepare the immune system to respond when exposure to a foreign invader occurs (active immunity). Adversely, therapeutic vaccines are used in patients postinfection. They can aid the immune system in attacking pathogenic organisms by introducing functional antibodies (passive immunity), which have been synthesized in a host organism prior to administration, into the serum of the infected individual. Therefore, the latter is a form of therapy used to treat infectious disease, whereas the former is meant to prevent it.

The immune system can be divided into two different types of responses: acquired (adaptive) responses and innate (natural or nonspecific) responses. Acquired immune responses can be either humoral (antibody-mediated) or cellular (cell-mediated). Alternatively, the innate immune response includes the process of phagocytosis, involving the engulfment and destruction of foreign pathogens and toxins, which can only occur in particular immune cells of the body such as monocytes, macrophages, or neutrophils [9]. Phagocytosis is also important for activating the acquired immune response. A second type of innate immune response includes complement activation; complement proteins exist as inactive zymogens circulating in the plasma until pathogen exposure occurs. Other types of nonspecific immune responses include cytokine release, which results in the aggregation of immune cells at the site of infection, and also inflammation.

Antigens are small biomolecules which are unique to individual pathogens, and can be recognized by antibodies produced by the host immune system. They are important because their presence signals to the host immune system when infectious organisms are present. Immunogens, on the other hand, are antigens which can stimulate an immune response. They are typically superior in size and molecular complexity. Immunogens are also more effective when in an appropriate physical form, most commonly as insoluble aggregates, since this allows them to be more readily taken

Chemistry of Bioconjugates: Synthesis, Characterization, and Biomedical Applications, First Edition. Edited by Ravin Narain.
© 2014 John Wiley & Sons, Inc. Published 2014 by John Wiley & Sons, Inc.

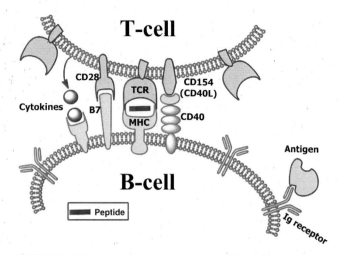

FIGURE 16.1 Immune responses involved in antigen presentation (peptide) by class II MHC proteins. For a color version, see the color plate section.

up by phagocytic cells [10]. Antigens can be immunogenic, although this is not necessarily the case. A subset of antigens, referred to as haptens, is unable to effectively elicit an immune response.

When a phagocytic cell encounters a foreign pathogen or toxin, it engulfs this nonself material and processes it into smaller fragments. These antigenic fragments are then bound to a family of receptor proteins within the cell, referred to as the major histocompatibility complex (MHC). The antigen–MHC complex is then transported across the cellular membrane and presented on the extracellular surface of the cell (Figure 16.1).

A family of lymphocytes, referred to as T-cells, have receptors expressed on their extracellular surface called T-cell receptors (TCRs). Each TCR binds specifically to a particular antigen–MHC combination, and each T-cell expresses thousands of the same TCR on its surface [10]. Upon binding, TCRs confirm the antigen as either self or nonself material. Upon detecting a nonself antigen, T-cells secrete cytokines that can target a cell for apoptosis, amplify the activity of phagocytes, or stimulate B-cells to produce and secrete soluble antibodies. Costimulation is also achieved through intercellular interactions involving pairs of CD28:B7 and CD154:CD40 receptors.

Different subclasses of T-cells exist, upon whose activation differentiates the process into either an antibody-mediated or a cellular-mediated immune response. The type of T-cell that becomes activated is directly dependent upon the class of MHC receptor that is presenting the nonself antigen. All nucleated cells have the ability to express class I MHC molecules [10]. Cytotoxic T-cells (T_C-cells) recognize antigen–MHC class I complexes, and upon binding the T_C-cell releases molecules (perforin and granzymes) which penetrate the cell membrane of the antigen-presenting cell, and eventually cause cell death. Class II MHC receptors

can be expressed on the surfaces of B lymphocytes and macrophages only, and selectively bind to helper T-cells (T_H-cells) [10]. T_H-cells have the ability to either activate macrophages or stimulate B-cells.

Upon stimulation, B-cells multiply and further differentiate into either plasma cells or memory cells. Plasma cells secrete high levels of soluble immunoglobulin M (IgM) antibodies into the serum, although have a lifetime limited to approximately 1 week [10]. Memory cells can persist for much longer, from several years to the entire host lifetime, and upon a second exposure to foreign antigen, they rapidly transform into plasma cells and secrete IgG antibodies [9]. IgG antibodies have a higher antigen-binding affinity and specificity than does IgM [9]. The transformation of memory cells into plasma cells does not require T-cell activation, and therefore, subsequent antigen exposures typically result in a much more rapid and exaggerated immune response.

Vaccines function by presenting particular target antigens to the immune cells. The immune system responds by producing B-cells which are specific for these target antigens. The antigens in the vaccine can be presented in a number of ways, either as a subset of the cellular surface or as peptides and other small biomolecules unique to that pathogen. Therefore, vaccines can be prepared from a variety of sources, such as killed microorganisms, attenuated microorganisms, inactivated toxins, or subunit vaccines [11].

The majority of subunit vaccines being used clinically have incorporated antigenic proteins or peptide fragments. Although the use of vaccines has significantly benefited preventative medicine, a large number of diseases still have no viable vaccine candidates. As alternative antigens, carbohydrates are extremely promising due to their high degree of molecular complexity afforded by multiple stereocenters, types of glycosidic linkages, ring size, and functional group substitution, although the extent of carbohydrate variety can be overwhelming [9]. Their expression on the extracellular surface also allows for easy recognition by host immune cells. The extracellular surface composition is often unique due to either the overexpression or selective expression of particular surface carbohydrates, which supplies this outer layer with the potential to be a novel immunotherapy target for disease prevention [12].

16.1.1 Early Carbohydrate-based Vaccines

The first carbohydrate vaccines were established against *Haemophilus influenzae* type b (Hib), comprising the purified capsular polysaccharide layer of the pathogen. The vaccines monopolized on a repeating polyribosylribitol phosphate (PRP) unit as the intended target antigen in the vaccine preparation, since earlier studies had demonstrated that an antibody response could be elicited against this polymer and was, more importantly, protective against Hib infection [13]. Unfortunately, the immune response against capsular polysaccharide vaccines was not as successful as was initially hoped for. When antigens are presented to the immune system with an adequate mass and hapten valency, carbohydrate vaccines are able to stimulate B-cell mitosis and antibody production without the involvement of T_H-cells [14, 15]. They are not presented via an MHC class II complex, resulting in poor long-term immunogenicity due to the carbohydrate antigen's inability to activate T_H-cells, referred to as a T-cell independent response.

This type of response results in poor long-term protection against diseases. When a sufficient number of antigens aggregate and bind to the cell surface receptors, B-cells can be activated to produce and release soluble antibodies [14]. Due to the absence of T_H-cell involvement, the required cytokines are not released to allow for the differentiation of B-cells into memory cells, or to facilitate the switch from IgM to IgG antibody production [15, 16]. As a result, booster injections of capsular polysaccharide vaccines do not induce an increase in antibody serum levels, but instead only return the IgM levels to the same value that was achieved after primary vaccination [17, 18].

Although effective at reducing the prevalence of Hib-related illness, there are drawbacks to using carbohydrate-based vaccines which have been formulated from crude capsular polysaccharides. These types of vaccines are not chemically homogeneous and can contain endotoxins, which are often the cause of adverse, and sometimes very serious, reactions in immunized individuals [19]. Local reactions at the site of immunization can occur if the vaccinated individual already contains circulating antibodies specific for one of the many antigens present in the capsular polysaccharide mixture [20]. In addition, although these vaccines have proven to be effective in healthy adults, they are not effective at conferring immunity in those who are at a higher risk for contracting the disease, such as neonates, children under 2 years of age, the elderly, chronically ill, splenectomized patients, and immunocompromised individuals [17, 21–23]. This ineffectiveness at immunizing the "high risk" population appears to be a common trend in vaccines comprising only carbohydrate antigens.

Smaller carbohydrate fragments are ineffective at eliciting any immune response. The molecules in the vaccine preparation must have an adequate molecular mass and antigen presentation in order for a T-cell independent immune response to be observed, further limiting the use of carbohydrate vaccines [14]. In order to overcome this obstacle, small carbohydrate fragments can be chemically conjugated to various scaffolds and immunogens in an attempt to trigger the desired immune response.

16.1.2 Carbohydrate-conjugate Vaccines

Landsteiner, van der Scheer, Avery, and Goebel were among the first to report that conjugation of small carbohydrates to

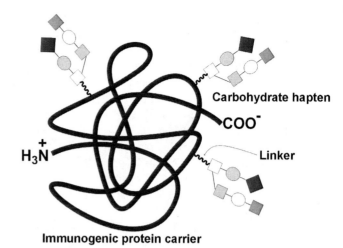

FIGURE 16.2 Schematic representation of carbohydrate-based conjugate vaccines.

an immunogenic protein confers an immune response to the carbohydrate epitope (Figure 16.2) [24–27]. The carbohydrates were conjugated to serum proteins, and the antibodies elicited against the carbohydrates were specific for particular stereoisomers and not cross-reactive with other isomeric antigens. Immunization studies involving these conjugates have demonstrated that a T-cell dependent immune response can occur against both the carbohydrate and protein regions of the molecule, to which Ig class switching, booster responses, and higher antibody titers have all been associated [23, 28–30]. It has also been demonstrated that the simultaneous administration of nonconjugated carbohydrate and protein does not elicit a T-cell dependent response with respect to the carbohydrate antigen, thereby supporting the requirement for a covalent bond [31]. This chapter will provide an overview of the different types of covalent bond formation that have been used recently in the preparation of carbohydrate-conjugate vaccines, in order to supplement earlier reviews published on the conjugate vaccines themselves [3, 9, 23, 32–34].

The first carbohydrate-conjugate vaccine was licensed for use in 1987 [23]. This Hib vaccine was composed of the same PRP polymer used in the earlier capsular polysaccharide vaccines, but instead the polymer was covalently bound to an immunogenic protein. The development of alternative Hib-conjugate vaccines ensued shortly after. Three different chemical methods were established to prepare the vaccines; all three were effective at preventing infection and were eventually commercialized by different pharmaceutical companies [3]. Altogether, these vaccines were responsible for a substantial reduction in prevalence of the disease in Finland, the United Kingdom, and the United States, thus providing the first piece of evidence to illustrate the promising nature of carbohydrate-based conjugate vaccines [35–37].

Unfortunately, the carbohydrate portion of these molecules was still being isolated and purified from natural sources, where quantities are often limited [1]. These vaccines also contained unregulated endotoxin content that occasionally resulted in adverse reactions to immunization [19, 38]. Studies with synthetic peptide haptens had been shown to induce a specific immune response, and that the elicited antibodies were cross-reactive with native antigens, suggesting that synthetic carbohydrate fragments could also be used to elicit an effective immune response [39–43].

Research into the synthesis of carbohydrate antigens ensued, and in 2004, the first clinical trial results were published on a conjugate vaccine containing a synthetic carbohydrate capsular polysaccharide antigen [44]. Since vaccinations are often administered parenterally in order to be most effective, safety regulations require that the chemical composition of the vaccine be known in precise detail. This makes synthetic antigens a more attractive target, since a chemical uniformity can be achieved which is not possible in natural product sources. Unfortunately, some studies have illustrated that a decrease in the magnitude of the immune response may be associated with more chemically pure vaccines, ultimately suggesting that a compromise may be necessary [45, 46].

16.1.3 Selecting the Immunogenic Carrier

When the carbohydrate–protein conjugate enters the body, it can be engulfed by B-cells, and then the individual peptide fragments from the immunogen are displayed on the MHC class II molecules at the extracellular surface. Upon presentation, the MHC-peptide fragments are able to bind to the receptors on T-cells, which eventually stimulate the T-cells to release the cytokines responsible for B-cell activation. As the immune response progresses and matures due to the presence of the T-cell dependent peptide fragments, the identity of the cytokines released gradually changes. If the carbohydrate antigens are bound to B-cells in the same microenvironment where these cytokines are being released, this may explain how the carbohydrate antigens also shift to a T-cell dependent immune response [47]. As a result, the B-cells which bind the carbohydrate antigens could differentiate into memory cells and further undergo an antibody class switch.

The efficacy of these conjugate vaccines is strongly dependent on the choice of immunogenic carrier [31, 48, 49]. It has been observed that both T-cell independent carriers and weakly immunogenic T-cell dependent carriers can induce immunogenic tolerance toward the conjugated antigen [15, 50]. Keyhole limpet hemocyanin (KLH) has been widely used in preclinical and clinical studies due to the strong T-cell dependent immune response that it is capable of eliciting, and has even been approved for use in areas of Europe and Asia [49, 51–54]. Unfortunately, it also suffers from some serious drawbacks such as a poorly defined structure

(making standardization difficult), poor solubility, and the ability to incorporate trace metals which as of 2011 has hindered its approval by the U.S. Food and Drug Administration [51, 53, 54].

Although several different proteins have been used in research for conjugation and subsequent immunological studies, only a small number of proteins have obtained widespread approval for use in humans [7]. Since the parenteral administration of foreign compounds into the body poses a high risk, this list is generally limited to proteins which have already been implicated for vaccine use in the past, such as tetanus toxoid (TT) and diphtheria toxoid (DT). Recombinant forms of these toxoids are also available, which display reduced toxicity yet still retain their immunogenicity. A frequently used example is CRM_{197}, a DT mutant which contains only a single amino acid substitution: a G52E point mutation in the gene results in the substitution of glutamic acid for a glycine residue [55]. A recombinant TT has also been formulated which contains only the receptor-binding domain of the toxoid ($\sim$50 kDa) instead of the entire protein ($\sim$159 kDa); this TT fragment has been shown to effectively elicit a protective antibody response in animals, yet its small size makes it easier to characterize which ultimately leads to improved homogeneity and reproducibility in the elicited immune responses [56].

There has been some evidence that vaccinating the host with the carrier molecule prior to immunization with the conjugate vaccine enhances the level of serum antibodies produced against the target, carbohydrate hapten [48, 57]. Unfortunately, contradictory results have also been published which suggest that this process, referred to as carrier priming, can actually suppress the immune response instead [57–61]. It appears that when the immunogenic protein is processed within the phagocytic cell, the peptides that are produced can either activate T_H-cells or instead activate suppressor T-cells, the latter resulting in a reduced immune response [62]. Therefore, particular carrier proteins can be more effective at suppressing antibody production than at enhancing it. Since the carrier proteins approved for human use are used in other vaccines, the chances are high that an individual will have already been primed to the carrier as a result of a previous vaccine regimen. In an attempt to overcome this obstacle, it has recently been observed that carbohydrate antigens conjugated to smaller peptides can also elicit an immune response. This allows for the selective use of those T-cell dependent peptides which enhance the immune response as opposed to suppress it [63, 64].

In addition, the conjugation of carbohydrates to proteins has proven difficult due to the large number of potential reactive sites on some proteins. Since the regiospecificity of these chemical conjugation reactions is very difficult to control, ambiguities can exist in the vaccine structure due to variations in the carbohydrate–protein ratio (i.e., number of protein functional groups reacted) and the particular sites that the carbohydrate is coupled to [11]. These variations in the structure may have implications on the host immune response; therefore, the reproducibility of experimental host responses to immunization can be complicated. Although progress has been made in terms of limiting this variability observed in conjugation to large proteins, the use of smaller peptides with a much lower number of reactive sites allows for the vaccine chemical structure to be more precisely defined, potentially allowing for increasingly reproducible immunological responses [65]. Efforts have also been directed at improving the methods used for characterization: one promising approach involves the use of matrix-assisted laser desorption/ionization time-of-flight/time-of-flight mass spectrometry (MALDI-TOF/TOF-MS) which has successfully been used to determine hapten/carrier ratios in bovine serum albumin (BSA) conjugates, and has also provided preliminary details on the specific locations of the glycosylation sites by comparing the fragmentation patterns of unconjugated and conjugated BSA after tryptic digestion [66].

One example of a peptide fragment used to stimulate T_H-cells is the 13 amino acid pan DR epitope (PADRE), comprising a sequence containing several nonnatural residues, which has been used in conjunction with both peptide and carbohydrate haptens in order to induce an Ig class-switching event (Figure 16.3) [67]. The IgG response of two different PADRE glycoconjugates was tested in mice, and was observed to be comparable to the antibody response observed in the related human serum albumin conjugates [68].

16.1.4   Current Obstacles in Conjugation Strategies

The variety of possible conjugation reactions has been restricted due to the limited number of appropriate functional groups available. The functional groups present often suffer from low reactivities, and achieving chemoselective control can be challenging [33, 69]. The majority of conjugation reactions target cysteine, lysine, aspartic acid, and glutamic acid residues due to the desirable reactivity of these free amine, sulfhydryl, and carboxylic acid functionalities. The modification of these or other particular functional groups on the substrates can improve the selectivity of the reaction, but can also negatively impact the binding specificity of elicited antibodies [70, 71].

Together with some of the previously mentioned obstacles (location and extent of conjugation, immunogenicity of carrier molecule, etc.), many other variables exist in the design and development of a conjugate vaccine. Other factors include (but are not exclusive to) the size of the epitope, or the distance between the epitope and the carrier, both of which can affect immunogenicity of the potential vaccine [72]. It was believed early on that carbohydrate epitopes had to contain at least two repeating oligosaccharide units in order to elicit an effective immune response, although an early study by Goebel as well as more recent studies have

contradicted this [72–74]. Even now, the ideal number of monomeric units is unknown and is likely dependent upon the particular identity and size of the carbohydrate antigen in question.

O-glycosidic linkages present in the carbohydrate epitopes could be labile against strong acids and sometimes strong bases, and thereby limit the conditions available for preparation of the conjugate [32, 75]. In addition, going to extreme pHs or less polar solvents could also denature proteins, which can eliminate the immunogenic properties of the protein carrier. Hence, the conjugation must be performed in aqueous solution at near neutral pH and at ambient temperature [76, 77].

It was also initially thought that the valency of the conjugate product and the elicited immune response were directly correlated. Although this is often the case at lower valencies, it has been observed that the relationship often holds only up until a maximum limit, after which point the immune response begins to decline [78, 79]. Even when the ideal valency is known, it is often difficult to achieve due to the steric effects imposed by bulky substituents, and in fact sterics often limit the number of carbohydrate antigens that can be conjugated to a carrier [32, 75].

For vaccines with a low number of monomeric epitope units, it is essential that the host immune cell receptors be able to bind the epitopic regions. Therefore, it is believed that the steric repulsion afforded by large protein carriers can have a detrimental effect on the immunogenicity of the conjugate [72]. Interference from the carrier could potentially inhibit the antigen–antibody complex binding, inactivating the immunizing properties of the vaccine. In order to overcome this obstacle, different linker molecules have been used as spacers during the chemical coupling of the carbohydrate and carrier components.

The linker molecule decreases the amount of steric interference around the carbohydrate antigen, thereby allowing easier access of the host antibodies to the carbohydrate [72]. Unfortunately, it is possible that the host can develop an immunogenic response to the linker itself [80]. This commonly results in decreased immunogenicity of the vaccine against the carbohydrate antigen, possibly a result of epitopic overload [81–83]. As a result, immunogenic linkers are often detrimental when being used in vaccines comprising self-antigens, such as those used for the prevention or treatment of cancer, since these antigens already possess an inherently low immunogenicity [84]. Immunogenicity of the linker is less of a concern in bacterial and viral vaccines comprising exogenic antigens. Therefore, an ideal linker molecule is likely nonimmunogenic and should minimize any steric changes that could result from conjugation, in either the immunogen or antigen. In order to act as a successful linker, it has been suggested that the molecule should contain a long, linear aliphatic chain in order to reduce its immunogenicity, yet be soluble enough in aqueous solution to

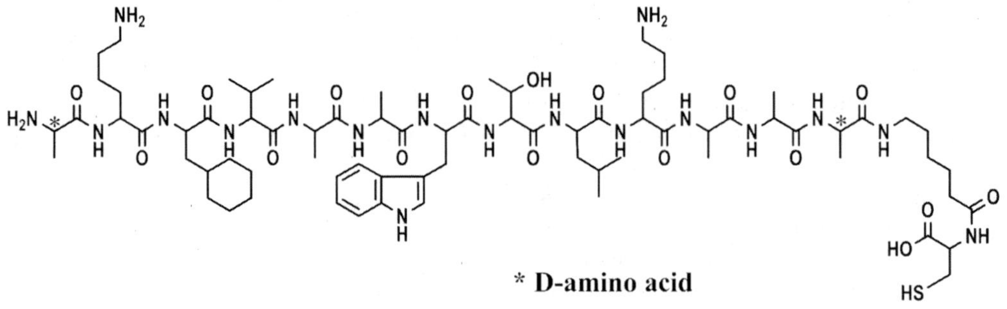

FIGURE 16.3 Pan DR epitope (PADRE) sequence used as a T-cell epitope in conjugate vaccines.

facilitate the conjugation reaction with water-soluble protein carriers [11, 85].

Alternative methods have been developed to overcome the low immunogenicity of cancer-specific antigens. For example, first generation cancer vaccines contained a single antigen type, although the different antigenic molecules were often administered simultaneously in order to widen the target cell population [86–91]. Later generation vaccines incorporated multiple antigens covalently bound to a scaffold, of which the entire molecular unit was subsequently conjugated to an immunogenic carrier [92]. This latter approach has advantages over the former, including a decreased amount of required immunogen which minimizes the risk of epitopic overload, and only a single, low yielding conjugation step [93].

Instead of using linker molecules, the assembly of lipopolysaccharide conjugates into liposomes has been suggested as an alternative [84, 94–97]. Liposomes are typically formed by conjugating the carbohydrate antigen to a large hydrophobic moiety, and then providing the conditions to allow these conjugate products to form micelles. Liposomes typically have very low immunogenicity and the liposome itself has been shown to have adjuvant-like properties, eliminating the need for the addition of auxiliary adjuvants which have often been linked to adverse reactions postimmunization [7, 84, 98, 99].

16.1.5 Direct Conjugation Techniques

Direct conjugation methods are useful because they avoid introducing potentially immunogenic linker molecules, and also reduce the number of synthetic steps required (resulting in improved efficiency, yield, etc.). Unfortunately, the steric hindrance caused by larger oligosaccharides often prohibits the direct conjugation of carbohydrate epitopes to an immunogenic protein carrier in significant yields [32]. To overcome this effect, synthetic chemists have developed a technique referred to as "glycoprotein remodeling," which involves conjugation of smaller mono- or disaccharides to the carrier, followed by subsequent glycosylation of the carbohydrate functionality in order to produce a more complex

carbohydrate [100]. A slightly different approach involves the carbohydrate modification of a single amino acid. Following this carbohydrate modification, solid-phase peptide synthesis (SPPS) can be used, and then further modification to the carbohydrate portion can be made after peptide assembly. SPPS is capable of synthesizing peptide chains of lengths up to approximately 50 amino acid residues [32].

16.1.5.1 Koenigs–Knorr Condensation
The conjugation between a glycosyl halide and a serine or threonine amino acid can produce a glycosylated amino acid suitable for incorporation into a peptide by SPPS methodology [101, 102]. This technique was used to produce both α- and β-linked N-acetyl lactosamine threonine derivatives [103]. This sequence is known to be added during the posttranslational modification stages of glycoprotein synthesis in the pathogen *Trypanosoma cruzi* [104, 105]. The α-analog was formed by coupling a 2-azido-2-deoxy-lactose derivative to a partially protected threonine residue (Figure 16.4a). Unfortunately, conjugation of the disaccharide was ineffective for the synthesis of the β-derivative. Instead, an alternative approach was developed where an N-acetyl glucosamine unit was conjugated to the amino acid, and then a bovine glycosyltransferase was subsequently used to enzymatically add a galactose unit to the carbohydrate-functionalized amino acid (Figure 16.4b). A review on other enzymatic applications in glycoprotein synthesis was recently published in 2007 [32].

16.1.5.2 Lansbury Aspartylation
Conjugation via reductive amination between a glycosyl amine epitope and an aspartic acid residue, referred to as Lansbury aspartylation, forms a conjugate product with linkage through an asparagine residue (Figure 16.5) [106]. This reaction is frequently used for the formation of N-linked glycosides, although a major drawback is the requirement for an exceedingly protected peptide [32]. The coupling requires a glycosyl amine, which can be prepared by either (i) reduction of the glycosyl azide, or (ii) Kochetkov amination, which involves treating the oligosaccharide with a saturated solution of aqueous ammonium bicarbonate [107]. In order to perform the coupling reaction, *tert*-butyloxycarbonyl-aspartic

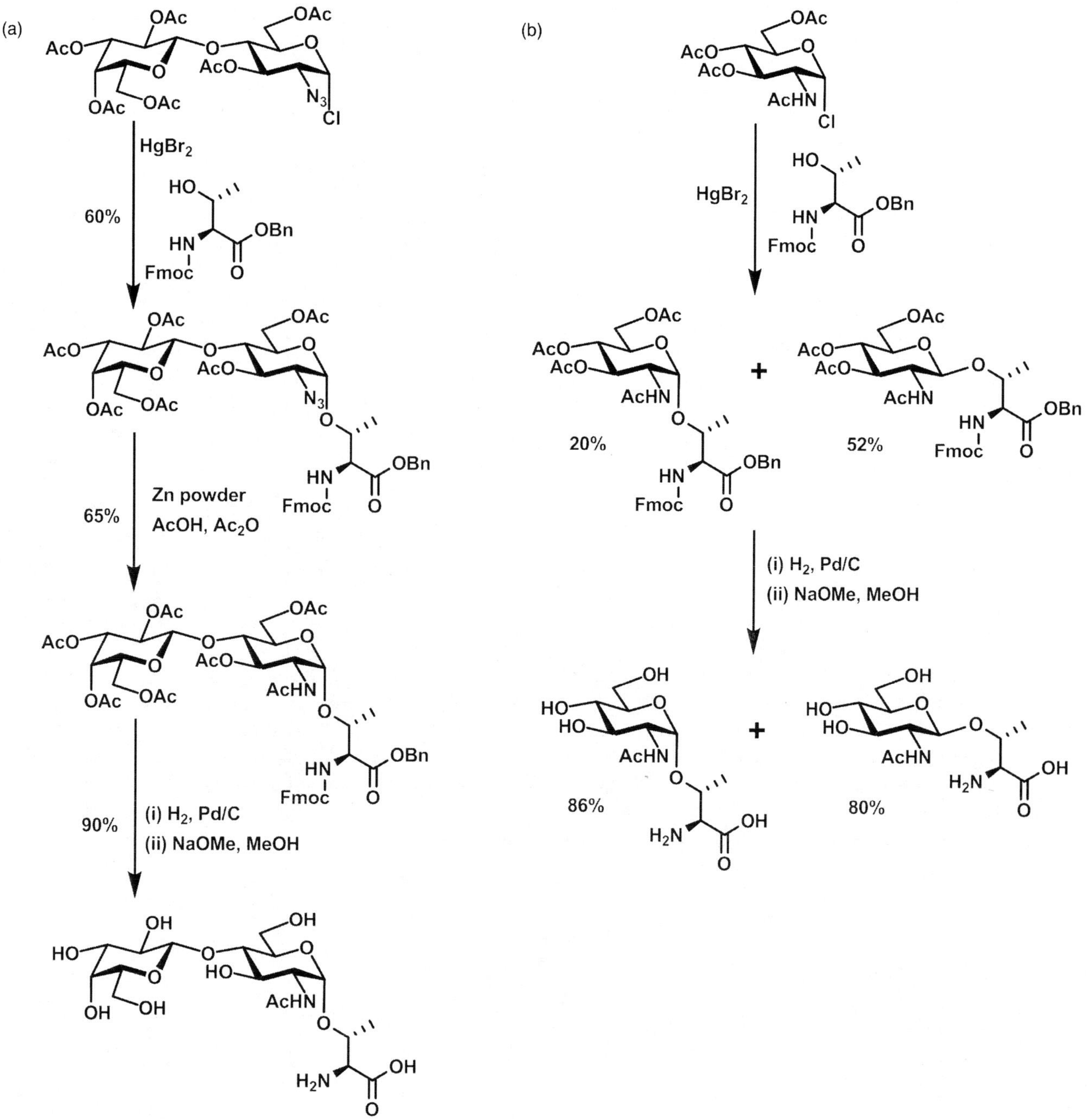

FIGURE 16.4 Koenigs–Knorr condensation in the synthesis of threonine sugar derivatives. (a) α-linked N-acetyl lactosamine derivatives, (b) α- and β-linked N-acetyl glucosamine derivatives.

acid benzyl ester (Boc-Asp-OBn) and hydroxybenzotriazole (5 eq.) were dissolved into dimethyl sulfoxide (DMSO), followed by the addition of glycosyl amine (1 eq.) and O-(benzotriazol-1-yl)-N,N,N',N'-tetramethyluronium hexafluorophosphate (HBTU) (5 eq.) [106]. For compounds prone to aspartimide formation, an additional equivalent of glycosyl amine was found to be optimal. The reaction mixture was stirred at room temperature for several days, and once the

coupling was complete the solution could be filtered and purified using high performance liquid chromatography (HPLC).

The reaction has been used for the attachment of a variety of prostate-specific antigens to peptide carriers [108]. The carbohydrate epitopes were synthesized in full, with an amino group at the anomeric position of the reducing end of the oligosaccharide chain, and then combined with the peptide fragments to form the conjugate product. This method has

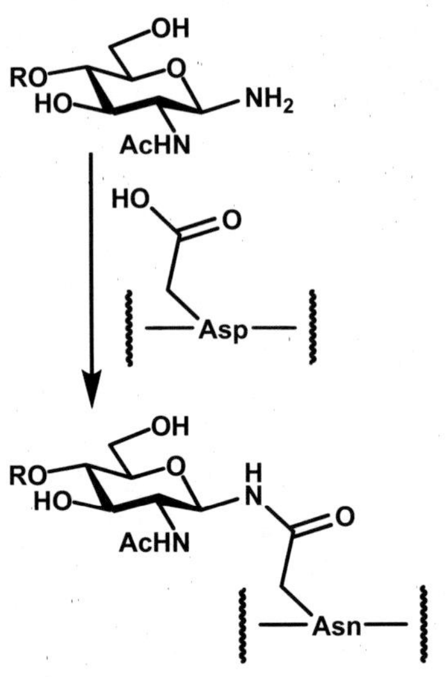

FIGURE 16.5 Lansbury aspartylation for direct conjugation to asparagine residues.

also been used in the preparation of multivalent constructs [109]. Lansbury aspartylation was performed on a natural short peptide, and then individual peptide fragments were fused together to form a more complex conjugate product.

Lansbury aspartylation was also used for the preparation of a potential carbohydrate-based human immunodeficiency virus (HIV) vaccine [79]. The target oligosaccharide epitope was synthesized and subsequently functionalized with an amino group (via Kochetkov amination) at the anomeric carbon of the reducing end sugar [107]. The glycosyl amine was then conjugated to aspartic acid residues in a small cyclic peptide scaffold. A cysteine residue in the scaffold construct

was used in the final synthetic steps to connect the antigen-containing cyclic peptide to an immunogenic carrier.

16.1.5.3 Nonnatural Amino Acids It has been suggested that the use of a nonnatural amino acid could potentially increase the immunologic response induced by a conjugate product [100, 110]. In addition to incorporating these unnatural residues directly into the immunogenic peptide, as in the earlier PADRE example (Figure 16.3), these unnatural amino acids can be installed into the linker portion itself. One such example involved the modification of individual amino acids with five different tumor-associated carbohydrate antigens [100]. The antigens employed in the vaccine construct were derivatives of Thomsen-nouvelle (Tn), Thomsen-Friedenreich (TF), sialyl-Tn, Lewis^y, and Globo H, which are known to be expressed in a variety of cancer types, such as those of the mammary glands, stomach, and colon [84, 111–114].

The synthesis began with a monosaccharide modified at the anomeric carbon with a trichloroacetimidate group, which was then conjugated to a free alcohol site on an unnatural amino acid containing a butanol side chain, fluorenylmethyloxycarbonyl (Fmoc)-L-hydroxynorleucine benzyl ester (Figure 16.6) [100]. Typically, the carbohydrate donor and amino acid acceptor (1.5 eq.) were dissolved into tetrahydrofuran (THF) containing molecular sieves; the glycosylation was promoted with trimethylsilyl triflate (0.1 eq.) at –78°C, and after 2 hours the reaction was quenched via addition of solid sodium bicarbonate. The reaction mixture was worked up and then purified using flash column chromatography on silica gel. These conditions were observed to be selective for the α-anomer. Glycoprotein remodeling was

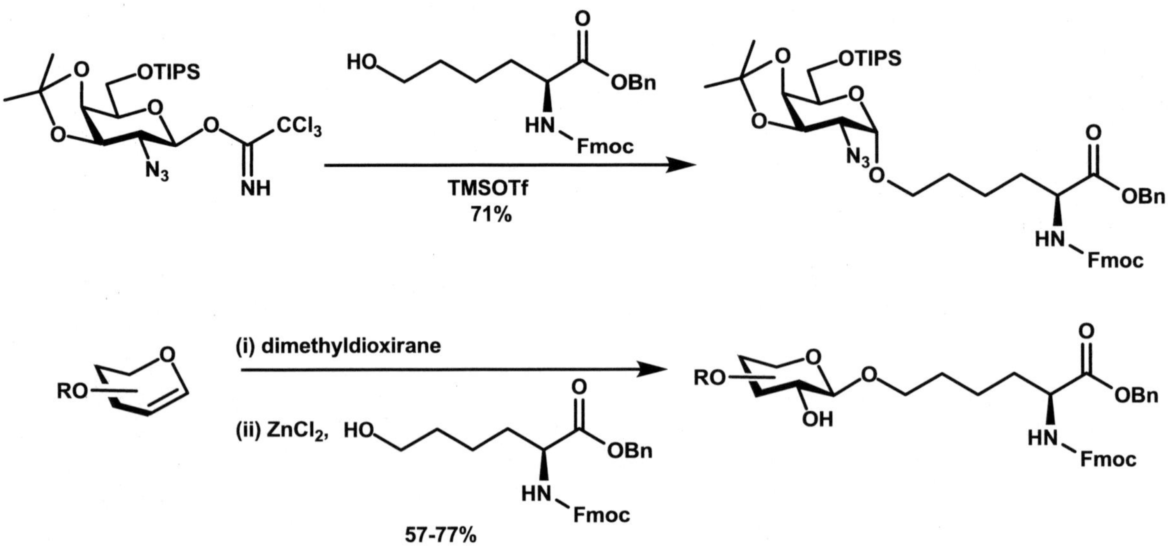

FIGURE 16.6 Conjugation to nonnatural amino acids could potentially improve the immunologic response. Coupling of an imidate to a butanol side chain affords the α-anomer, while addition of the same butanol side chain to an epoxide intermediate formed from a glycal affords the β-anomer.

performed on the monosaccharide derivative, resulting in the different target carbohydrate epitopes. In order to obtain the majority β-anomer an alternative approach was used, utilizing the same butanol-functionalized amino acid, but involving a different preparation of the glycosyl donor: a glycal was oxidized and then reacted with the amino acid moiety to form the conjugate product (Figure 16.6). A general procedure involves reacting the glycal (0.15M solution) with dimethyldioxirane (2 eq.) in dichloromethane (DCM) at 0°C for 30 minutes. After evaporation, the crude epoxide was reacted with the alcohol acceptor (5 eq.) in THF at –78°C in the presence of zinc(II) chloride (4 eq.). The formed β-glycoside was isolated by flash column chromatography on silica gel, and subsequently converted into the final target epitope through several synthetic steps.

16.1.6   Homobifunctional Linkers

16.1.6.1 Squarate Compounds Instead of performing direct conjugation, a linker molecule can be used to create a set distance between a carbohydrate hapten and an immunogen. Homobifunctional linkers contain the same functional groups at either end of the linking molecule. A commonly, well-studied example includes the class of squarate, or 3,4-dialkoxy-3-cyclobutene-1,2-dione compounds, which are often chosen due to the ability of the linkage chemistry to be performed under mild conditions, with high yields, and relatively small amounts of starting materials [115]. These molecules can conjugate to free amine sites, which can be present on the carbohydrate or protein moieties, or indirectly present via the use of another small linker molecule.

Bioconjugation through squarate coupling was first described by Tietze et al. in 1991, with the discovery that controlling the pH of the reaction solution enabled each side of the squaric acid diester to be conjugated to a different primary or secondary amine [116]. Shortly after this approach was used to generate a carbohydrate conjugate, where *p*-aminophenyl glycosides had been coupled via squarate to ε-amino groups at surface accessible lysine residues on BSA [117]. In order to perform this regioselective synthesis, a squarate diester and the amine (1.1 eq.) were coupled in an alcoholic solution at pH 7 [116]. The formed monoamide was then combined with a second amine (typically the protein) at elevated pH levels (pH ∼ 9). Upon completion, the reaction mixture was dialyzed to remove buffers and unreacted carbohydrates, and the desired protein glycoconjugate obtained upon lyophilization.

This technique has also been applied in the synthesis of *Vibrio cholerae* vaccines, which is the bacterial species responsible for the disease cholera [118–122]. The carbohydrate epitope was initially functionalized with an ester, which then selectively underwent amidation with ethylenediamine to produce a terminal free amine (Figure 16.7). The free amine was then available to conjugate to the squarate linker

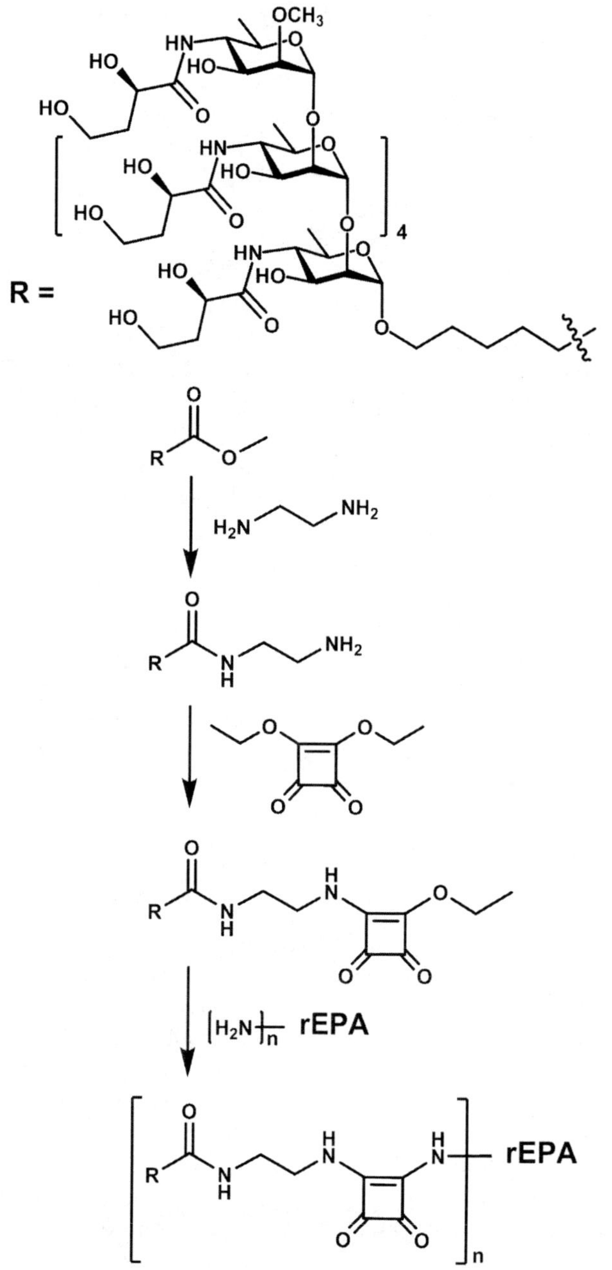

FIGURE 16.7   Utilization of the squarate linker for synthesizing a series of potential vaccines against the pathogen *Vibrio cholerae*.

to form the half-ester intermediate. The intermediate was subsequently conjugated to an immunogen, such as BSA, or recombinant exotoxin A (rEPA) from the bacterial species *Pseudomonas aeruginosa*. A combination of different carbohydrate epitopes was used to prepare a library of vaccines, one of which upon intraperitoneal injection demonstrated a protective response in mice [121].

A homobifunctional squarate linker was also used in the preparation of a vaccine against *Shigella flexneri* Y [123]. The oligosaccharide epitope was functionalized at the anomeric carbon via reductive amination with 1,3-diaminopropane in the presence of sodium cyanoborohydride. The product containing a free amine site was then attached to the diethyl squarate linker molecule. The half-ester product was reacted with available free amine functionalities found on the lysine residues of the immunogenic protein TT; approximately 44 carbohydrate epitopes were incorporated into each TT protein during the conjugation process.

Derivatives of the squarate linker were studied and compared over time, in order to determine the most effective method of conjugation involving squarate chemistry. One of the first studies involved examination of the importance of which secondary linker was used to introduce the amine functionality onto the anomeric carbon of the epitope [124]. A carbohydrate ester was treated with hydrazine hydrate, ethylenediamine, or hexamethylenediamine, and was afterward reacted with squarate at pH 7 to form the half-ester product. It was observed that the rate of monoester formation was most rapid for the hydrazide compound, while the other reaction rates for formation of the aminoethaneamide and the aminohexaneamide were not significantly different from one another. The half-esters were then conjugated to BSA at pH 9, and it was observed that the ethylenediamine underwent this final conjugation reaction the fastest, and also in the highest yield. Therefore, the authors concluded that ethylenediamine was the most suitable choice of modifying agent for the carbohydrate between the three studied compounds. Follow-up immunological studies on these conjugate products indicated no significant difference in the immune responses for the hydrazide and the aminoethaneamide linkers, although the longer aminohexaneamide linker was unable to elicit protective antibodies [125].

The effects of the length of the ester alkyl chain on the dialkyl squarate moiety were recently examined by Hou et al., along with a variation in reaction conditions, in order to observe how the maximum yield of desired conjugate product could be obtained [126]. It was found that both the product yield and the conjugation efficiency could be increased by increasing the concentration of both hapten and buffer used for the reaction. The conjugation reaction was most efficient at pH 9, and a decrease in pH had the ability to even terminate the reaction in some cases. The hydrazide-derivatized haptens were found to undergo a much slower and less efficient conjugation to the immunogen as compared to the amino amide haptens, which agrees with their conclusions drawn from earlier studies [124]. Overall, no significant advantage was evident based on the identity of the half-ester alkyl group used and therefore, the authors concluded that the dimethyl squarate linker was the best, since it was the easiest reagent to handle and is commercially available.

16.1.6.2 Adipic Acid Diester Derivatives

An oligosaccharide from *Streptococcus pneumoniae* was conjugated to an immunogenic CRM$_{197}$ protein (a derivative of the DT immunogen) for the preparation of a vaccine [127]. The two entities were attached by an N-hydroxysuccinimide-activated adipic acid diester. The linker molecule was first reacted with a free amine site on the carbohydrate hapten, and then subsequently conjugated to available free amine sites on the CRM$_{197}$ peptide (Figure 16.8). It was found that the antibody response against the oligosaccharide was greater with this adipic acid derivative than the corresponding antibody response against the same carbohydrate and protein combination that had been conjugated using a squarate linker.

p-Nitrophenol-based adipic acid diesters have also been used for the preparation of bioconjugate vaccines; they are easily prepared in a single step from the reaction of p-nitrophenol and adipoyl chloride in pyridine [128, 129]. The succinimide-activated half-ester intermediate used previously was found to be exceedingly reactive, and often not stable enough to withstand purification techniques. Conversely, the electron-withdrawing properties of the p-nitrophenol group were able to deactivate the intermediate enough so that purification of the half-ester product could be performed [128]. It was observed that the half-ester intermediates formed from p-nitrophenol adipic acid diesters were stable to the acidic conditions applied in both silica gel and reversed-phase chromatography [115].

In order to perform this conjugation method, a glycosyl amine was first combined with the adipic acid p-nitrophenyl diester linker (5 eq.) in dry N,N-dimethylformamide (DMF) [128]. After 5 hours the mixture was evaporated and the excess reagent removed using either regular or reversed-phase silica to obtain the pure amide half-ester, which was then coupled to protein in phosphate buffer (pH 7.5).

One series of examples utilizing the p-nitrophenol-based compounds includes the synthesis of a number of potential *Candida albicans* glycoconjugate cluster vaccines [78, 115, 130]. One of the vaccine constructs was synthesized from an O-alkenyl glycoside [115]. Photoaddition of 2-amino-ethanethiol to the alkene moiety was used to introduce an amine-containing functional group onto the carbohydrate epitope (Figure 16.9). The transformation retained the stereochemical configuration at the anomeric carbon center. The linear p-nitrophenol ester bifunctional linker then underwent conjugation with the free amine to form the half-ester intermediate product, which was subsequently purified using chromatographic techniques. The half-ester was then conjugated to either BSA or TT with an efficiency ranging from 32% to 45%, corresponding to roughly 12 carbohydrate ligands per protein.

A different amine-functionalized carbohydrate was used in the preparation of another potential *C. albicans* vaccine [78, 130]. The epitope was reacted with 5 equivalents of the p-nitrophenol-based linker molecule to form the half-ester intermediate product. The half-ester was then reacted with an immunogenic protein (either BSA or TT) and conjugation

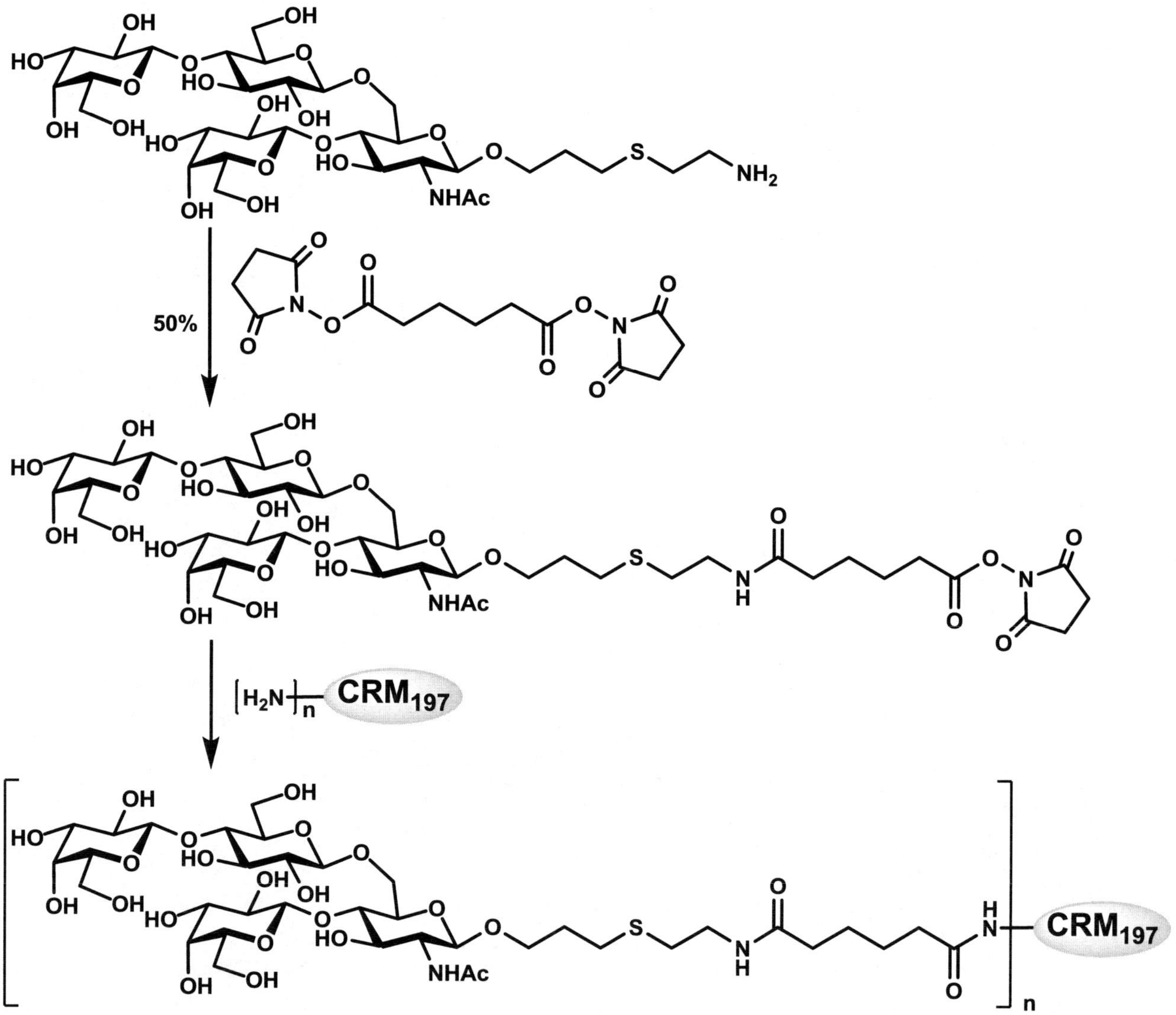

FIGURE 16.8 The use of *N*-hydroxysuccinimide adipic acid to conjugate a *Streptococcus pneumoniae* oligosaccharide to the immunogenic peptide CRM$_{197}$.

occurred with 18%–23% efficiency between the carbohydrate hapten and the available free amine sites on the protein.

The same technique was also applied to the synthesis of anticancer vaccines [131]. GM$_2$ and GM$_3$ ganglioside haptens were derivatized with a truncated ceramide functionality containing a free amine site on the ceramide moiety. The free amine was again reacted with *p*-nitrophenol adipic acid diester to form the half-ester compound, which was subsequently reacted with TT in buffer solution to form the conjugate product.

16.1.7 Heterobifunctional Linkers

16.1.7.1 Lemieux's Linker An early method for conjugation developed by Raymond Lemieux and colleagues utilized a heterobifunctional methyl 9-hydroxynonanoate linker

molecule [132]. To date, this acyl azide linking methodology has shown a wide variety of applications, including the investigation of novel cytokines, blood group immunochemistry studies, diagnostic assays, and the binding of fluorophores to assist with structural analysis [132–138]. The linker is first attached to the carbohydrate epitope by reaction with a per-acetylated glycosyl halide in the presence of mercuric cyanide in dry nitromethane (Figure 16.10) [134]. The formed product containing the Lemieux linker was purified and could be subjected to a number of subsequent transformations; the linker is compatible with acid-catalyzed acetal formation, acid-catalyzed glycosylation, acetylation, tritylation, catalytic hydrogenolysis, and reductive regioselective acetal opening. Upon deprotection of the sugar moiety, the ester functionality of the Lemieux linker was usually reacted with hydrazine in methanol to form the intermediate hydrazide;

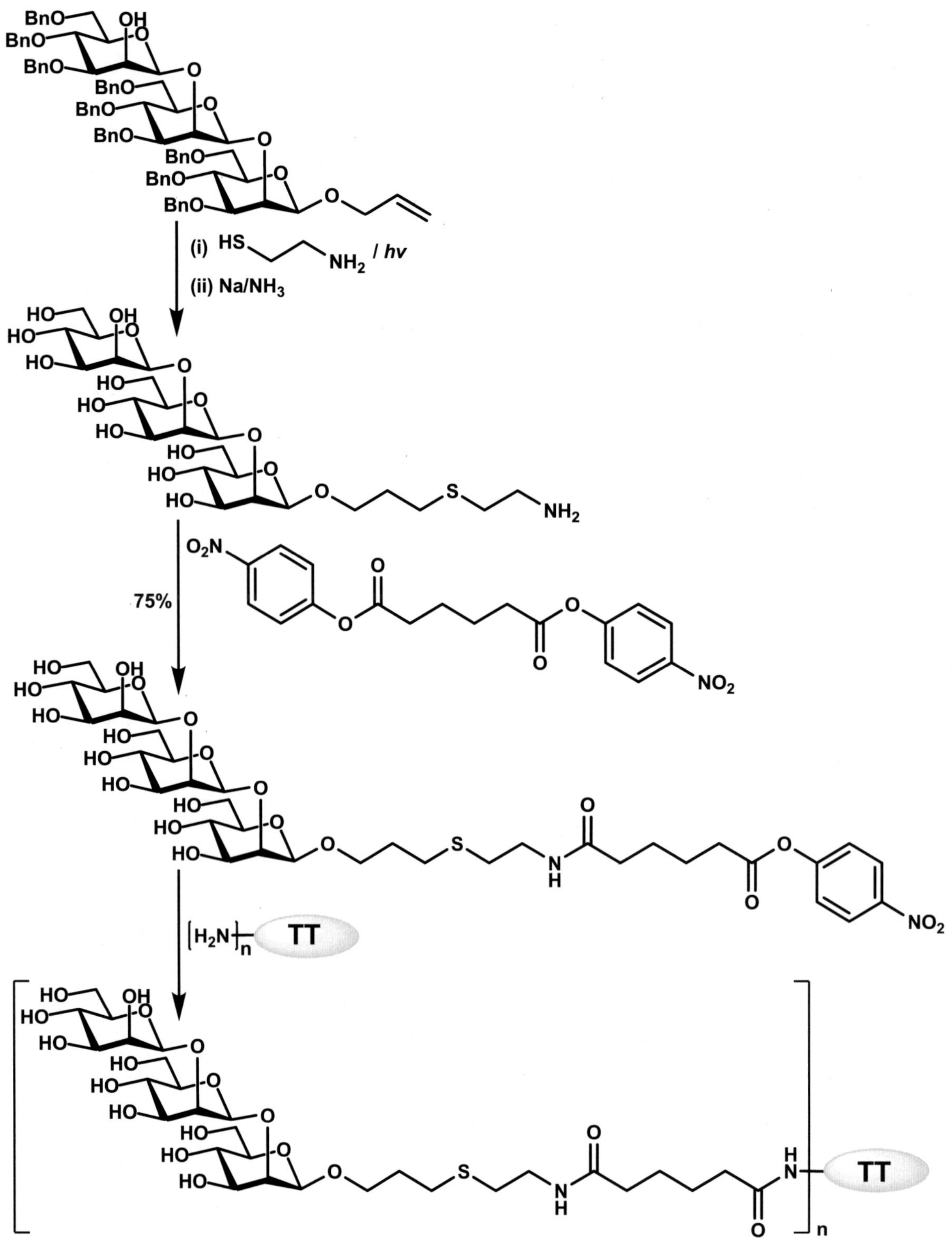

FIGURE 16.9 Attachment of the alkenyl derivative of an oligosaccharide from *Candida albicans* to tetanus toxoid (TT) with 2-amino-ethanethiol and a *p*-nitrophenol-based adipic acid linker.

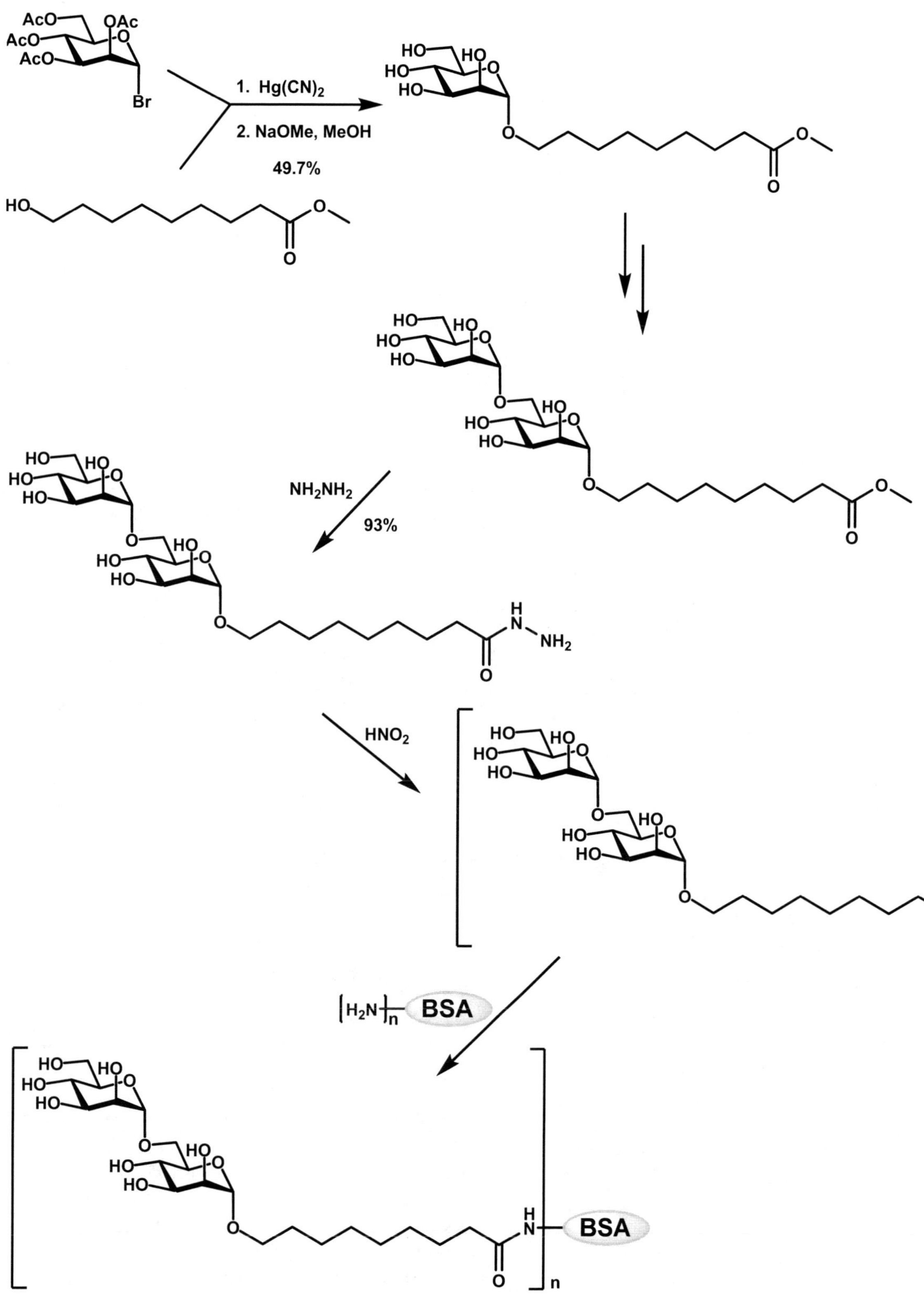

FIGURE 16.10   The use of Lemieux's linker for attaching target carbohydrates onto a protein.

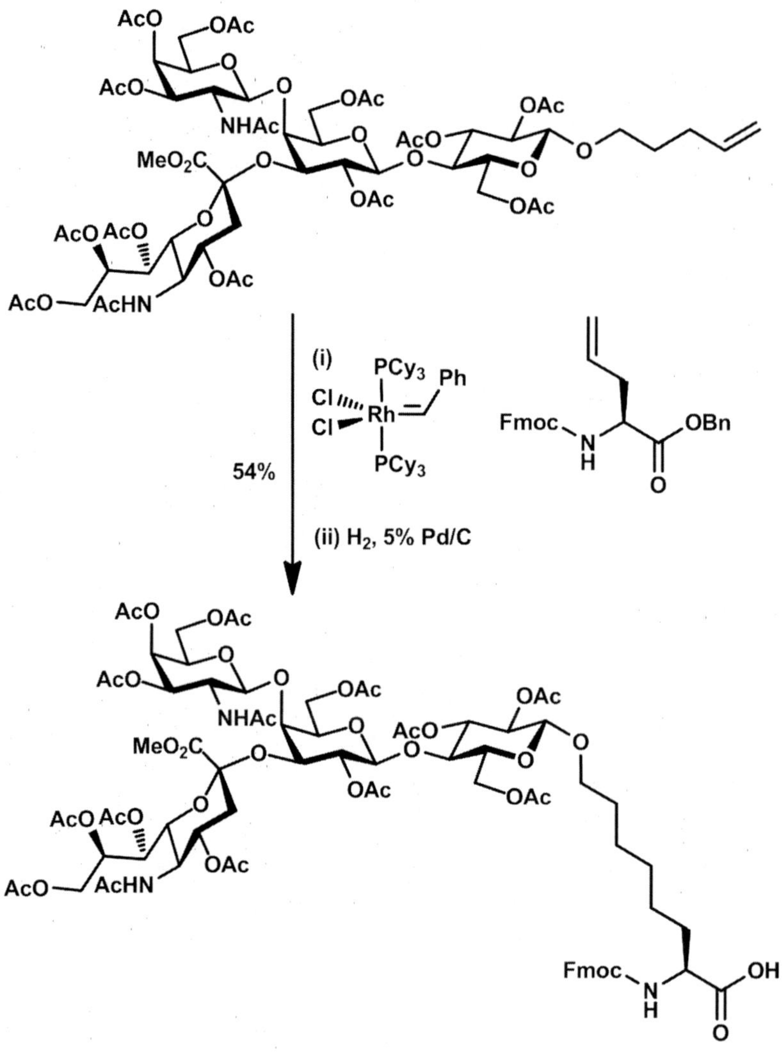

FIGURE 16.11   Utilization of a Grubb's catalyst and subsequent hydrogenation in the synthesis of a potential carbohydrate-based vaccine against prostate cancer.

by subsequent exposure to nitrous acid, the hydrazide was then converted to a reactive acyl azide intermediate, which could react with amino groups on the surface of a carrier protein to form the stable amide linkages.

16.1.7.2 Olefin Cross-metathesis   A potential prostate cancer vaccine was prepared by Cho et al. using olefin cross-metathesis [110]. Since multiple antigens can be selectively presented or overexpressed on an affected cell, the vaccine design involved incorporating several of these antigens into a single, multivalent construct. A peptide segment was synthesized that contained a distribution of unnatural amino acids, of which the alkyl side chain was terminated by a double bond. The anomeric carbon at the reducing end of the carbohydrate antigen was functionalized with a long, aliphatic chain that was also terminated by a double bond. The conjugation was performed using a Grubb's catalyst, and was

subsequently followed by the reduction of the double bond using a palladium–carbon catalyst (Figure 16.11). A hexavalent vaccine was produced, designed against prostate cancer, which contained the Globo H, Lewis^y, Tn, sialyl-Tn, GM$_2$, and TF antigens.

The same approach was used by the same group to construct an alternative multivalent vaccine against small cell lung cancer [139]. The trimeric fucosyl GM$_1$ vaccine was constructed using pentenyl-functionalized carbohydrate epitopes and the unnatural allyl glycine amino acid residues that had been previously utilized, which were incorporated into short peptides. The conjugation was again performed using a Grubb's catalyst, followed by hydrogenation to reduce the residual double bond in the linker region. The glycosylated polypeptide chain was eventually conjugated to the protein KLH to enhance the immunological response upon administration of the vaccine.

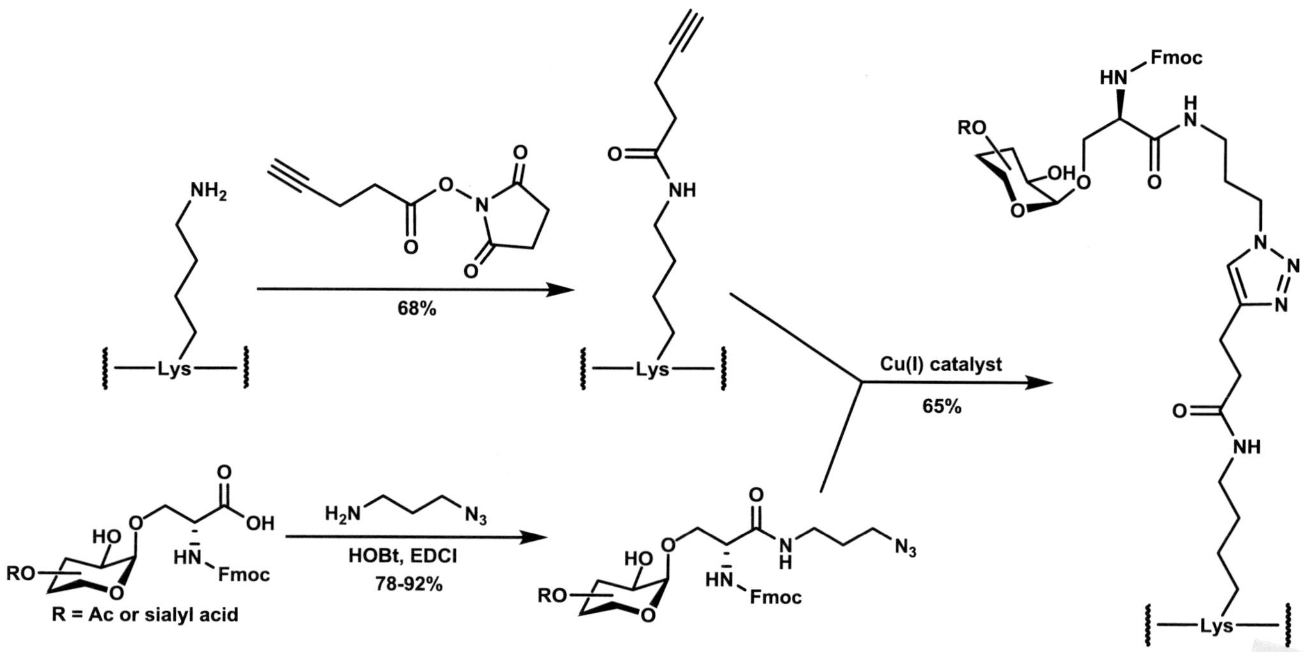

FIGURE 16.12   The modification of lysine residues with an alkyne moiety to subsequently enable Huisgen 1,3-dipolar cycloaddition conjugation to carbohydrate cancer antigens.

16.1.7.3 Cycloaddition Wan et al. used cycloaddition chemistry to perform conjugation during the synthesis of an anticancer vaccine [51]. An azide-containing oligosaccharide was reacted with an alkyne-functionalized polypeptide chain to form a five-membered ring (Figure 16.12). The copper(I)-catalyzed cycloaddition reaction was based on the Huisgen 1,3-dipolar cycloaddition reaction involving azides and terminal alkynes [140, 141]. The lysine residues of a peptide fragment were modified with an N-succinimidyl-4-pentynoate linker (in DMF), which introduced terminal alkyne functionalities onto the peptide fragment of the construct [51]. In addition, the anomeric position on the carbohydrate epitope was modified with Fmoc-protected serine, of which the carboxylic acid group was then further extended with a three-carbon linker containing a terminal azido functionality. A 1,3-dipolar cycloaddition was performed allowing for efficient conjugation of the carbohydrate and peptide portions of the vaccine construct. To afford the coupled product the glycosyl azide (4.5 eq.) and the alkyne-terminated peptide were reacted under the catalysis of Cu(0) nanosize powder (50% wt/wt) in phosphate-buffered saline [51]. This cycloaddition strategy is favorable since the introduction of functional groups is relatively easy, the method involves a high degree of chemoselectivity and functional group tolerance, and the reaction proceeds under aqueous conditions, although there are imperative toxicological considerations with using a copper catalyst at such a late stage in the synthesis of a potential therapeutic agent [51].

16.1.7.4 Traut's Reagent A vaccine construct was synthesized which contained a glycophosphatidylinositol (GPI) glycan from *Plasmodium falciparum*, conjugated to a maleimide-activated protein carrier [142, 143]. *P. falciparum* has unusually low levels of posttranslational N- and O-linked glycosylation; instead, the GPI glycan comprises greater than 95% of the total posttranslational modification in this species [144]. The nontoxicity of deacetylated GPIs made this epitope an attractive target for the construction of an antimalarial vaccine [145]. In order to couple the GPI to an immunogenic protein, the ethanolamine functionality on the GPI was treated with 2-iminothiolane, commonly referred to as Traut's reagent, which effectively introduced a sulfhydryl handle onto the GPI compound (Figure 16.13) [143]. The reduced thiol was then conjugated to maleimide-activated ovalbumin, KLH, or BSA to form the target product.

Similarly, an identical approach was used in the development of a related antimalarial vaccine, involving conjugation to an alternative site on the same GPI epitope [143]. A reduced sulfhydryl group connected to a phosphate oxygen at the reducing end of the GPI analog was again reacted with maleimide-activated BSA to form the target vaccine conjugate product. This approach has also been used for the preparation of a potential vaccine against *Mycobacterium tuberculosis* [146].

16.1.7.5 S-Acetyl Thioglycolic Acid Pentafluorophenyl Ester An alternative strategy was used by Vasan and colleagues to perform carbohydrate–protein conjugation [147].

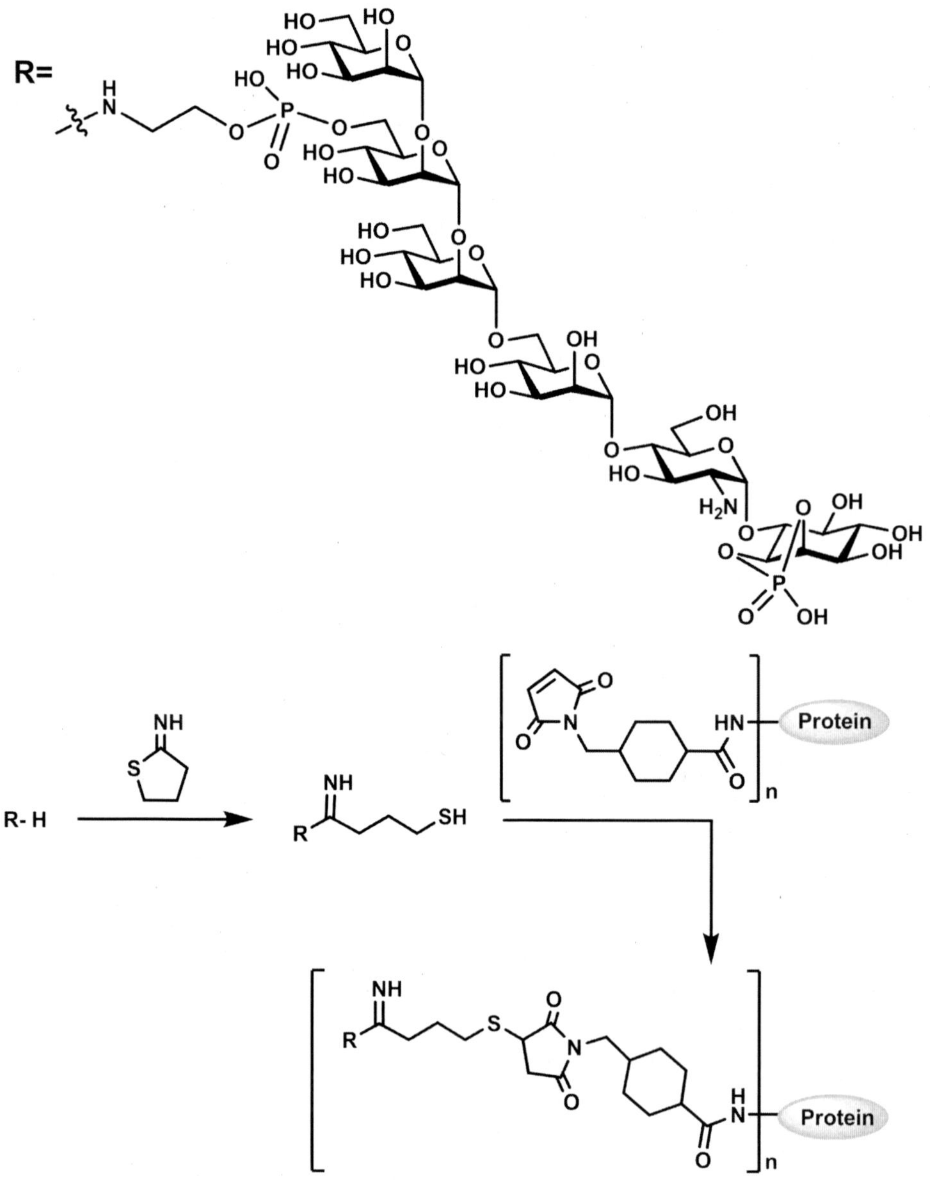

FIGURE 16.13 Utilization of Traut's reagent to couple an immunogen (BSA) to a carbohydrate epitope in the synthesis of a potential antimalarial vaccine.

The goal was to synthesize a potential vaccine against the pathogen *Bacillus anthracis*, otherwise referred to as anthrax. Two target antigens were proposed, both derived from carbohydrates found on the pathogen's vegetative cell wall, and both containing a free amine site at the anomeric carbon position. The free amine was treated with *S*-acetyl thioglycolic acid pentafluorophenyl ester (SAMA-OPfp) (1.5 eq.) and Hünig's base (3 eq.) in dry DMF to afford the corresponding thioacetate. De-*S*-acetylation was then achieved with neat 7% ammonia in DMF under argon, and the crude thiol was used directly for subsequent conjugation with maleimide-activated BSA in aqueous buffer (pH 7.2) (Figure 16.14). The produced conjugates displayed an average of 11–19 antigens

on each BSA, depending on amounts of the particular epitope used for conjugation.

The same technique was utilized to form another type of potential anthrax vaccine, but comprised a different epitope [148]. Several analogs of the carbohydrate portion of a glycoprotein (BclA) from *B. anthracis* were synthesized, and then the anomeric carbon on the reducing end was modified with an aminopropyl spacer. The free amine was again reacted with SAMA-OPfp, followed by deacetylation with ammonia. The reduced thiol was then reacted with either maleimide-activated BSA or succinimidyl-activated KLH. The latter conjugate had been prepared by reacting succinimidyl 3-(bromoacetamido) propionate with KLH initially,

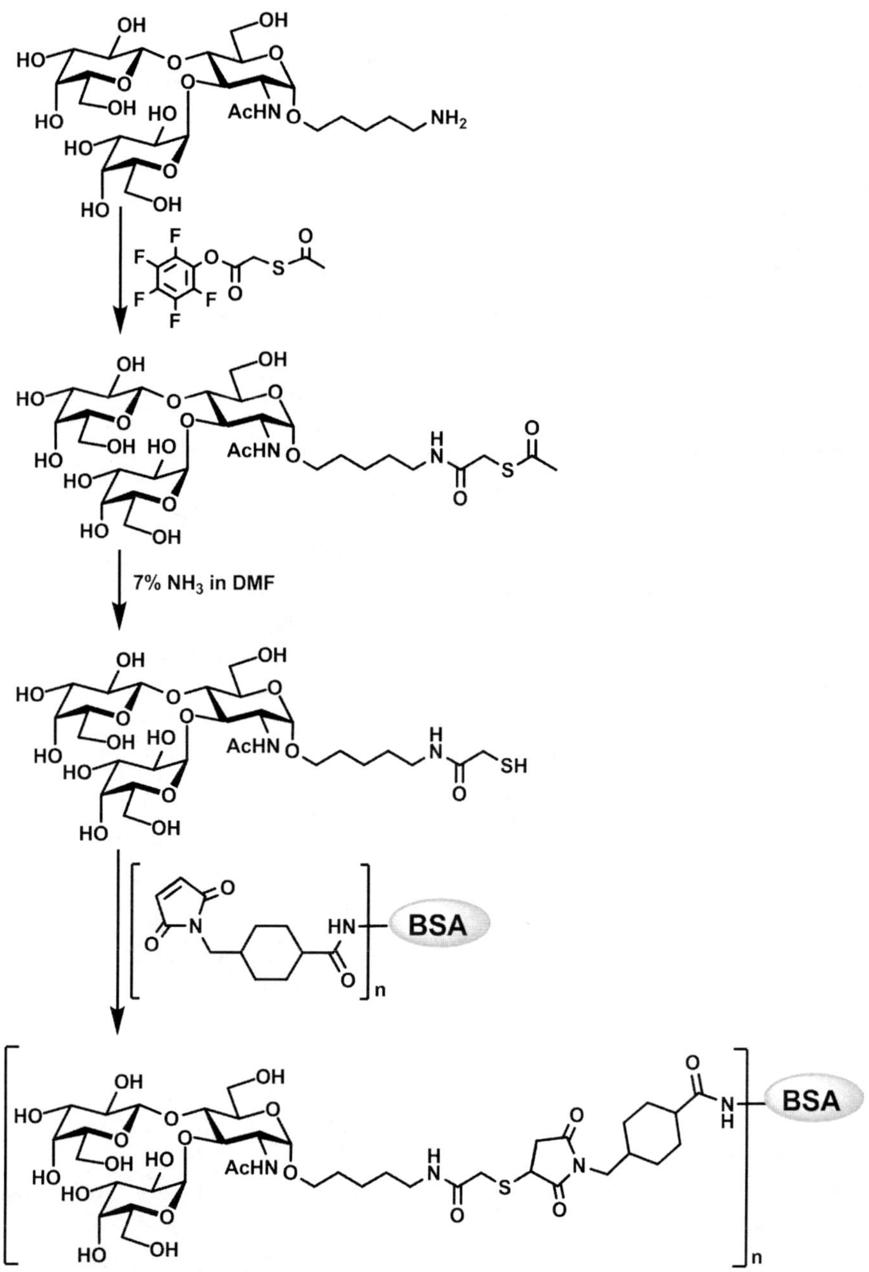

FIGURE 16.14 Treatment of a free amine with *S*-acetyl thioglycolic acid pentafluorophenyl ester and subsequent reaction with maleimide-functionalized protein (BSA) during the synthesis of a potential anthrax vaccine.

followed by incubation of the KLH derivative with the trisaccharide thiol.

The preparation of a cancer vaccine also utilized the identical technique [149]. A Lewis[y]-Lewis[x] heptasaccharide epitope was prepared and then modified with an aminopropyl linker. The amino group was modified by a thioacetyl group, using SAMA-OPfp, followed by deacetylation with ammonia, and then conjugation of this functional group to the protein KLH, which had been modified with

3-(bromoacetamido) propionyl groups. 1190 copies of heptasaccharide per KLH molecule were bound in the final conjugate product.

The above group had reported a year earlier that this same bifunctional cross-linker that was being used for conjugation was thought to be suppressing the immune responses toward the carbohydrate epitope that were being generated by these vaccines [150]. In the same study they also tried an alternative linker, generated by the addition of 3-(bromoacetamido)

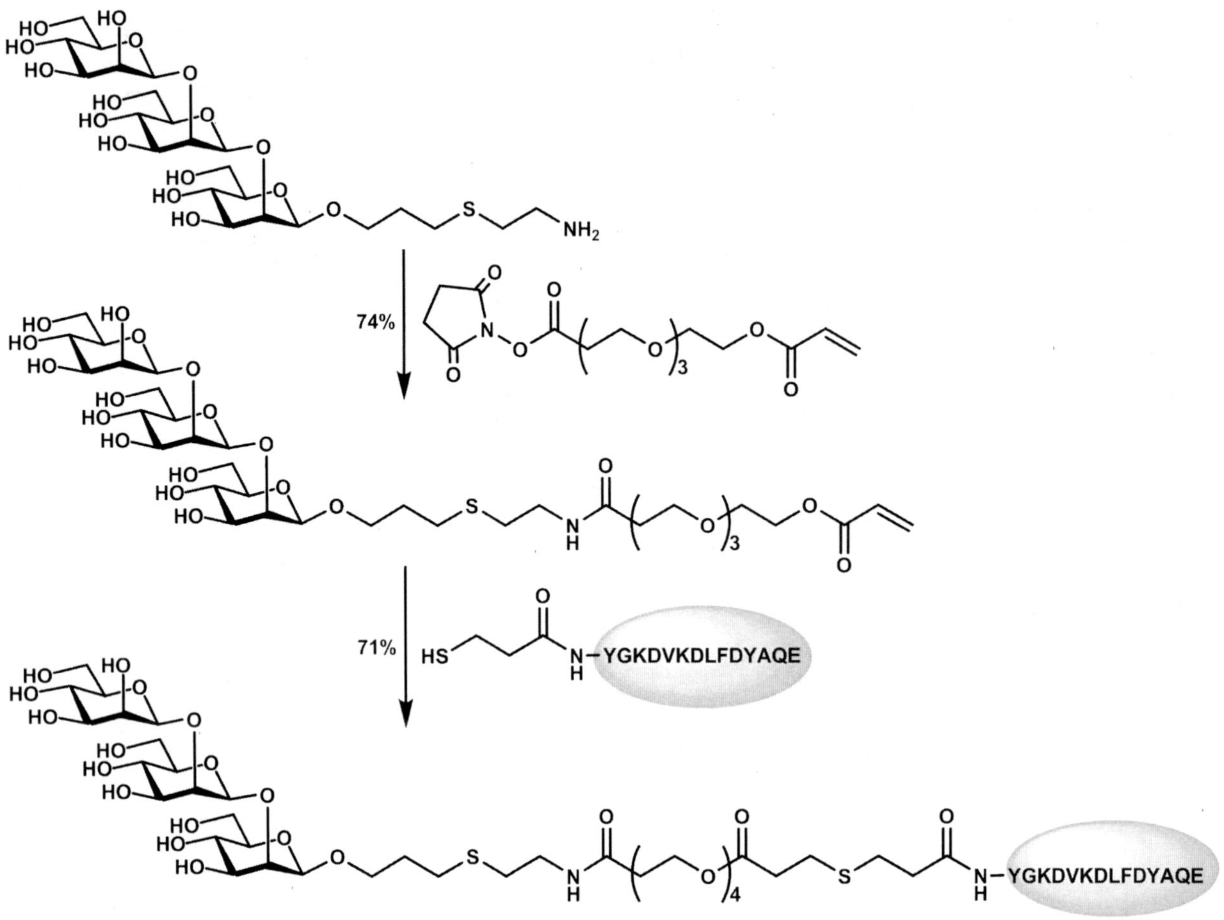

FIGURE 16.15 Conjugation strategy involving a linker with an *N*-hydroxysuccinimide ester at one end and a terminal acrylate moiety at the other, incorporated into the design of a *Candida albicans* conjugate vaccine.

propionate, which is smaller and more flexible. The use of this latter linker resulted in a decreased immune response against the linker itself, as compared to the response against the SAMA-linker, which correlated to an increased immunogenic response to the epitopic carbohydrate moieties.

16.1.7.6 Triethylene Glycol Chain Containing Terminal Acrylate and N-hydroxysuccinimide Ester

Bioconjugate vaccines against the fungal disease candidiasis were prepared, based on the conjugation of a β-1,2-mannan trisaccharide to a T-cell immunogenic peptide, both of which are found on the cell wall of the pathogen *C. albicans* [11, 151]. The β-1,2-mannopyranosyl trisaccharide was synthesized with a terminal amine, which was further functionalized through the conjugation of a heterobifunctional coupling reagent, containing at one end an amine-reactive *N*-hydroxysuccinimide ester and at the other a sulfhydryl-reactive acrylate group, separated by a core spacer region based on triethylene glycol (Figure 16.15). After conjugation to the oligosaccharide,

the acrylate was reacted with a cysteine functionality on the immunogenic peptide to form the glycopeptide-conjugate product.

The identical strategy was used to synthesize an anticancer vaccine containing the tumor-associated ganglioside GM$_2$ [11]. GM$_2$ is significantly overexpressed on certain types of cancer cells, and in this vaccine construct was conjugated to the T-helper immunogen. A heptadecapeptide fragment from a murine heat shock protein was chosen for this immunogenic fragment, since it had previously been shown to be effective in earlier conjugate vaccines against *S. pneumoniae* and *Neisseria meningitidis* [63, 152, 153]. The *N*-terminal end of the fragment was modified using *S*-trityl protected 3-mercaptopropionic acid in order to convert the free amine into the sulfhydryl group required for the conjugation reaction. After modification of the T-helper immunogen, conjugation between the carbohydrate linker portion of the molecule and the peptide was performed.

A modified approach, based on the above linker molecule, was used by Liu et al. to create a potential vaccine against

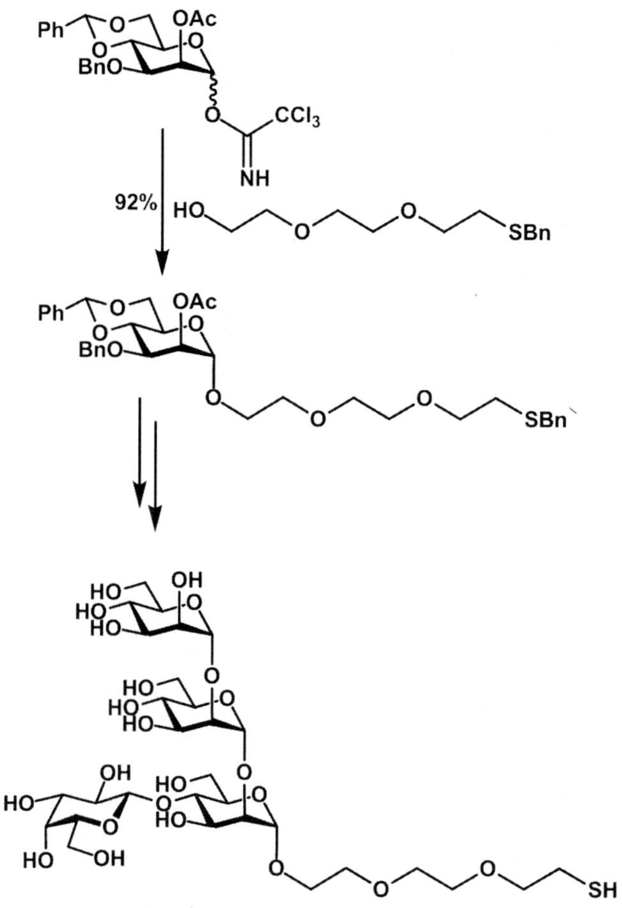

FIGURE 16.16 Covalent attachment of a secondary linker molecule to introduce a terminal sulfhydryl moiety necessary for subsequent conjugation to the primary linker immunogen fragment.

pathogens of the genus *Leishmania* [154]. The vaccine candidate contained a tetrasaccharide epitope essential for the entry of these parasites into host macrophages [155]. During synthesis of the target carbohydrate, a sulfhydryl-terminated triethylene glycol chain was incorporated onto the anomeric carbon at the reducing end of the sugar (Figure 16.16). The thiol on the carbohydrate epitope was conjugated to either maleimide-activated KLH or hemagglutinin protein in the presence of tris(2-carboxyethyl)phosphine, a reducing agent which prevents the formation of disulfide bonds.

This same terminal thiol was also conjugated to a phospholipid molecule that had been modified at the hydrophilic end with a maleimide functionality [154]. The glycolipid conjugates were synthesized and then eventually incorporated into virosomes by exploiting the disparate hydrophilic and hydrophobic regions of the conjugate. To prepare the virosomes, the conjugate product was suspended in octaethyleneglycol, and then a mixture of egg phosphatidylcholine, and various surface glycoproteins and phospholipids from a strain of inactivated influenza were added. Upon the

removal of detergent from the mixture virosomes formed which could then be used for subsequent immunological testing. The use of these immunostimulating influenza virosomes replaces the need for the addition of other adjuvants to the vaccine preparation, which reduces the potential negative side effects that are often induced by such adjuvants [154].

16.1.7.7 Succinimidyl 3-(bromoacetamido) Propionate and O-(3-thiopropyl) Hydroxylamine

An approach by Pozsgay et al. utilizes an oxime linkage, which was achieved by the use of two small linker molecules [45]. The first succinimidyl 3-(bromoacetamido) propionate linker was reacted with free amine sites on BSA, which was the immunogen that had been chosen in designing this particular vaccine against *Shigella dysenteriae* type 1 (Figure 16.17). A terminal bromide leaving group was replaced when the derivatized protein was mixed with the second linker, O-(3-thiopropyl) hydroxylamine, which functionalized the terminal end of the linker region with a free and reactive amine group. This amine was then treated with a synthetic keto-derivatized carbohydrate epitope from the pathogen to form the oligosaccharide–protein conjugates via the oxime linkage.

The same set of linker molecules was also used in the preparation of a potential *Bordetella* vaccine [156]. The vaccine was based on a lipopolysaccharide antigen that is presented on the extracellular surface of the pathogen. A ketone function was introduced onto the antigen by degrading the lipopolysaccharide using either mild acid hydrolysis or deamination, resulting in two distinct derivatives of the target antigen. BSA was chosen as the immunogen, and after functionalization with the two linker molecules, the protein was subsequently conjugated to the antigenic derivatives through a condensation reaction, forming an oxime linkage. The coupling reaction of both antigens to BSA proceeded near room temperature at neutral pH. Another study revealed that the same conjugation strategy could be used for aldehyde-functionalized antigens, although the ideal conditions at which conjugation occurs for aldehydes are slightly more acidic (~pH 5.5) [76].

The aldehyde or ketone functionality can be introduced into the carbohydrate through the use of an additional linker molecule. One example involves a Koenigs–Knorr type reaction of the target bromoacetylated carbohydrate with 5-methoxycarbonylpentanol, similar to the earlier processes used with Lemieux's linker, followed by deacetylation (Figure 16.18) [76]. The intermediate product was then reacted with either 1,2-diaminoethane or hydrazine to afford an amine-functionalized derivative. Finally, N-acylation was performed to afford the carbonyl-containing derivative, which could then undergo conjugation under mild conditions using the above strategy.

Teichoic acids are D-ribitol phosphate polymers that are present on the extracellular surface of many gram-positive bacterial species. Octa- and dodecamers of the teichoic

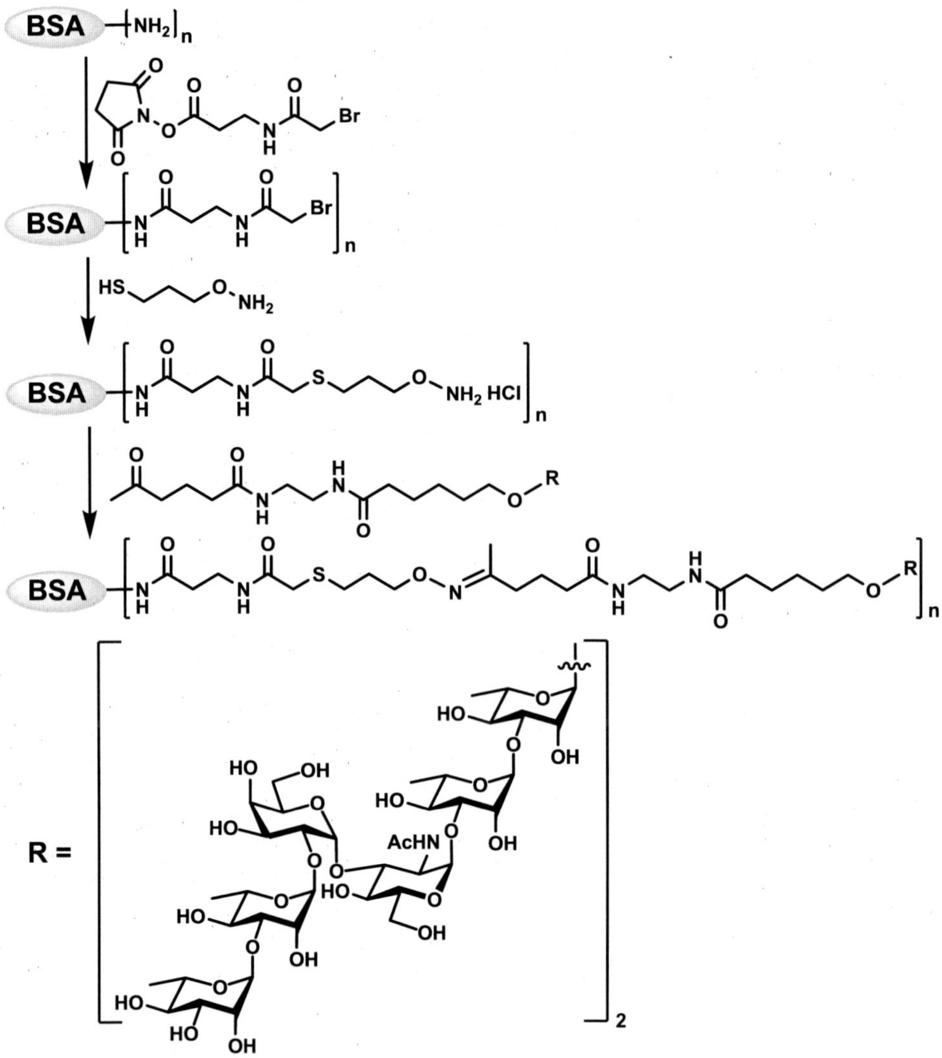

FIGURE 16.17 The use of multiple linkers to conjugate hapten to an immunogen during the preparation of a *Shigella dysenteriae* vaccine.

acid polymers were synthesized and then keto-functionalized using 5-ketohexanoic anhydride [157]. The carbohydrate derivatives were subsequently conjugated to aminooxylated BSA in a buffer solution to generate potential vaccine constructs. An average of 10–18 carbohydrate haptens were bound to each BSA, depending on the particular polymer employed.

16.1.7.8 Reductive Amination

Reductive amination was well studied and used early on by Jennings in his preparation of prospective vaccines against the pathogen *N. meningitidis* [158–160]. Polysaccharides purified from meningococcal strains were selectively oxidized by using sodium periodate in order to introduce aldehyde functionalities. The oxidized polysaccharides were then directly added to either TT or BSA, so that reductive amination conjugation could

occur between the aldehydes and the free amine groups on the proteins.

Ozonolysis of a linker molecule and reductive amination were used together to create an anticancer-conjugate vaccine [161]. The conjugation procedure began with an azide-functionalized carbohydrate, of which the azide group (used to protect the anomeric carbon during synthesis of the oligosaccharide epitope) was selectively reduced to an amine using 10% palladium/carbon and hydrogen gas in methanol (Figure 16.19). The obtained amine was acylated with 4-pentenoic anhydride, and the double bond of the 4-pentenoyl group was then cleaved by bubbling ozone gas through methanol at low temperature to afford the intermediate dimethyl acetal, which required a hydrolytic step with trifluoroacetic acid to convert it into the corresponding aldehyde; the oligosaccharide was finally conjugated to an

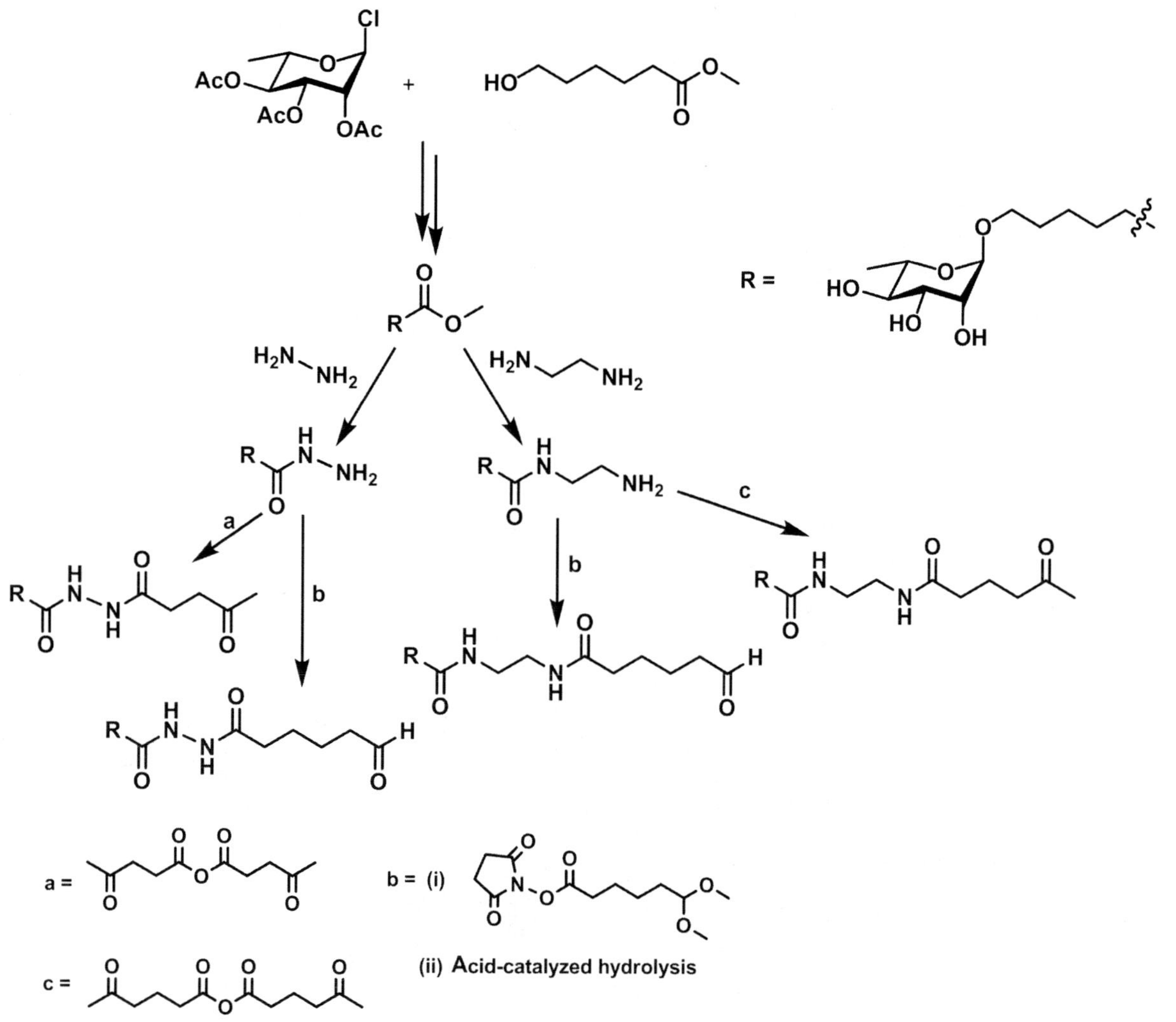

FIGURE 16.18   A Koenigs–Knorr glycosylation strategy for introducing a linker molecule, followed by multiple amide bond formations to obtain the final conjugated target.

immunogenic protein (human serum albumin or KLH) in aqueous carbonate buffer with sodium cyanoborohydride as a reducing reagent.

Another family of potential anticancer vaccines was similarly prepared, but the 4-pentenoyl linker was alternatively bound indirectly to the carbohydrate antigen [162–164]. The synthesis began with a galactosamine monomer, which was initially modified via a dehydration reaction with the reagent 2-chloroethanol (Figure 16.20) [162]. Sodium azide was subsequently added, resulting in substitution of the terminal halide. This azide-terminated alkyl chain then acted as a protecting group while the remainder of the carbohydrate epitope unit was synthesized from the monomeric starting material. The conjugation procedure was eventually completed in a manner comparable to the previous example.

A pentenyl linker was also used in the preparation of a target *B. anthracis* antigen [165]. The monosaccharide at the reducing end of the sugar was initially functionalized

at the anomeric site with a trichloroacetimidate group. The anomeric site was converted with 4-pentenol and the remainder of the carbohydrate antigen was subsequently synthesized from this product.

A vaccine against small cell lung cancer was similarly prepared [86]. The carbohydrate epitope was a ganglioside (fucosyl GM_1) which is both specific to and profusely expressed on the target cancer cells. The epitope was synthesized as the protected glycal, which was then functionalized with the pentenyl linker and subsequently deprotected. Ozonolysis was performed, followed by reductive amination with sodium cyanoborohydride and a second amine-functionalized linker molecule, 4-(4-*N*-maleimidomethyl)cyclohexane-1-carboxyl hydrazide (Figure 16.21). The maleimide-functionalized carbohydrate derivative was then conjugated to accessible sulfhydryl groups on the immunogenic protein KLH.

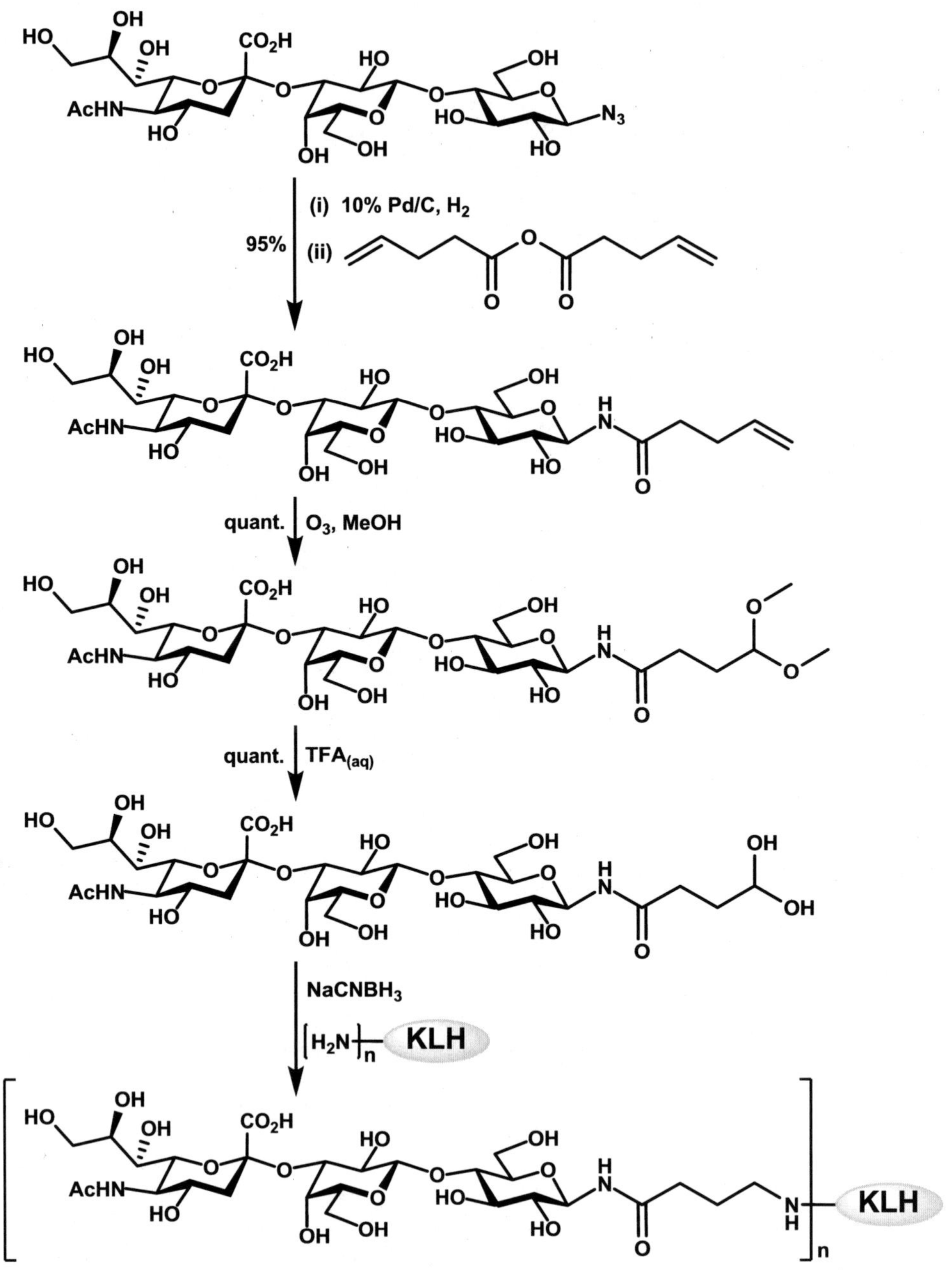

FIGURE 16.19   The use of an anhydride linker moiety and subsequent ozonolysis step for covalent conjugation in an anticancer vaccine.

Recently, it was suggested that conjugation via reductive amination may not be a suitable choice for coupling synthetic haptens [166]. Although it has shown great utility for the bacterial polysaccharides extracted from natural sources, conjugation reactions typically require a large excess of hapten and are often very low yielding. For most conjugation methods, the excess hapten remains unmodified and can be recovered from the reaction mixture in order to recycle the material. Conversely, reductive amination involves an intermediate imine (Schiff's base) which requires a reducing

agent in the reaction mixture in order to obtain the desired product. Unfortunately, the reducing reagents used often display poor chemoselectivity and concomitantly reduce the aldehyde functionality present on excess hapten. Due to the polyhydroxylated nature of the carbohydrates, this position cannot be selectively oxidized back to the aldehyde for recycling in subsequent conjugation reactions. Therefore, since this material cannot be recovered, this is not the wisest conjugation method to use for coupling complex synthetic haptens which are often very labor intensive and expensive to prepare.

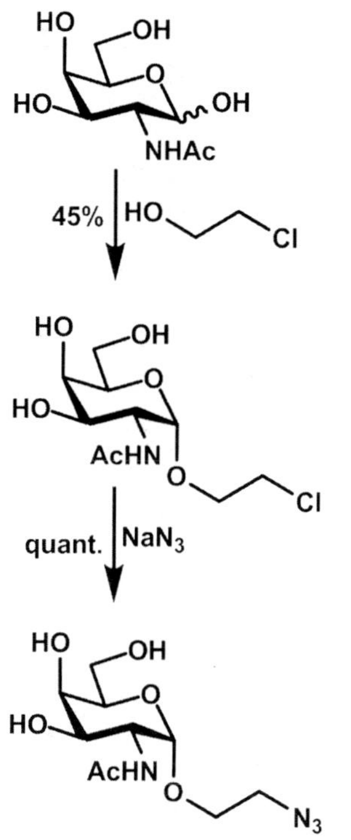

FIGURE 16.20 The use of a secondary 2-chloroethanol linker for conjugation.

16.2 FUTURE DIRECTIONS

Over the past decade, the conjugation chemistry available to connect individual vaccine subunits has expanded tremendously and will likely continue to do so. A lack of steric bulk surrounding the carbohydrate hapten has been recognized as important in order to improve binding interactions between the hapten and the immune cells, which effectively maximizes the antibody response generated postinoculation. As a result of this observation a number of linkers have been developed, and improvements to these moieties will likely continue in order to facilitate higher efficiency chemical reactions and decrease the immune response generated against some linkers *in situ*.

The types of linkages available will grow to facilitate conjugation to amino acids other than the commonly utilized residues containing terminal thiol, amine, or carboxylate functionalities. A recent example of this has already been illustrated where an electrophilic aromatic substitution reaction was performed on tyrosine residues [167]. Also, the increasing repertoire of synthetic peptides incorporating nonnatural amino acids will play a significant role in expanding the types of conjugation available for vaccine preparation. A radical reaction utilizing an anomeric sulfhydryl group and a terminal alkene has recently been explored, and although not yet examined as a conjugation method in vaccine preparation, the chemical stability and poor immunogenicity of the

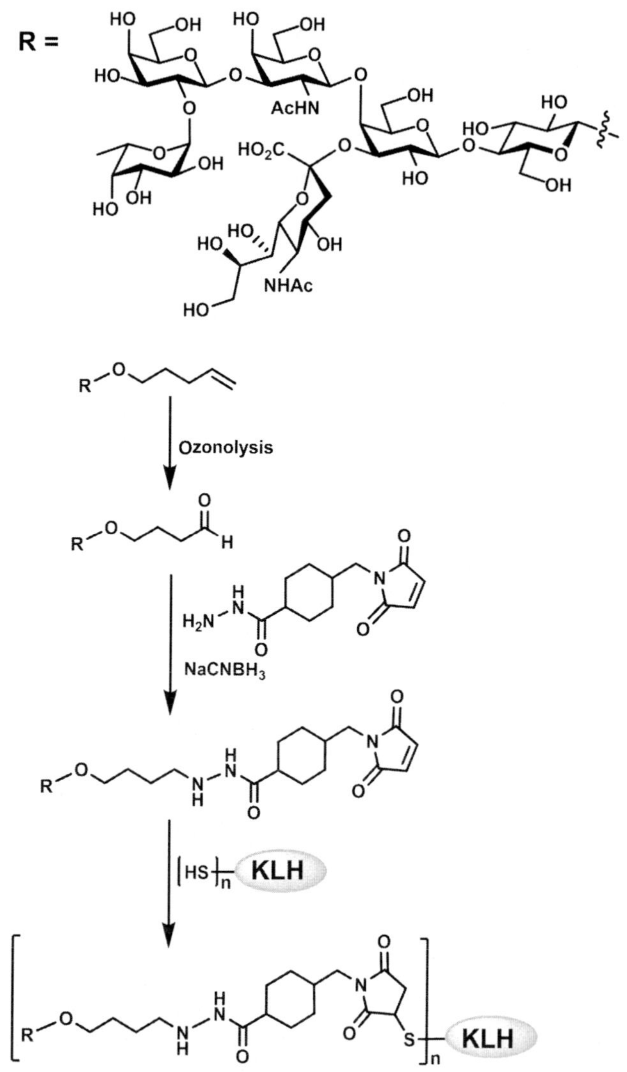

FIGURE 16.21 Ozonolysis, reductive amination, and subsequent conjugate addition to afford a conjugate anticancer vaccine against small cell lung cancer.

resultant thioether linkage holds much promise in this area for the future (Figure 16.22) [168].

In addition, a movement from protein immunogens to smaller peptide immunogens has already begun and will likely continue over the next decade. This strategy avoids the immunosuppressive epitopes that are often found in the larger proteins and also allows for better controlled conjugation reactions, resulting in more reproducible host immune responses [11, 63].

A final strategy employs a combination of several different carbohydrate haptens attached to the same scaffold [100]. An immunogen is also covalently bound to the same scaffold, and inoculation with this multivalent structure allows for the generation of antibodies against the different haptens simultaneously. This should improve the overall immune response and prohibit cellular resistance from evolving with time.

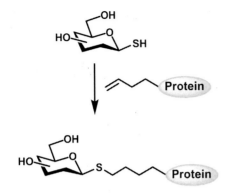

FIGURE 16.22   Conjugation of a reducing thioglycoside to an alkene-terminated nonnatural amino acid.

16.3   LIST OF ABBREVIATIONS

BSA	Bovine serum albumin
DCM	Dichloromethane
DMF	*N,N*-dimethylformamide
DMSO	Dimethyl sulfoxide
DT	Diphtheria toxoid
eq.	equivalent(s)
Fmoc	Fluorenylmethyloxycarbonyl
GPI	Glycophosphatidylinositol
HBTU	*O*-(benzotriazol-1-yl)-*N,N,N′,N′*-tetramethyluronium hexafluorophosphate
Hib	*Haemophilus influenzae* type b
HIV	Human immunodeficiency virus
HPLC	High-performance liquid chromatography
Ig	Immunoglobulin
KLH	Keyhole limpet hemocyanin
PRP	Polyribosylribitol phosphate
SAMA-OPfp	*S*-acetylthioglycolic acid pentafluorophenyl ester
SPPS	Solid-phase peptide synthesis
MALDI-TOF/TOF-MS	Matrix-assisted laser desorption/ionization time-of-flight/time-of-flight mass spectrometry
MHC	Major histocompatibility complex
PADRE	Pan DR epitope
T_C-cell	Cytotoxic T-cell
T_H-cell	Helper T-cell
TCR	T-cell receptor
TF	Thomsen-Friedenreich
THF	Tetrahydrofuran
Tn	Thomsen-nouvelle
TT	Tetanus toxoid

ACKNOWLEDGMENTS

We thank the Natural Sciences and Engineering Research Council of Canada (NSERC) and the Alberta Glycomics Center for financial support. Rachel Hevey is the recipient of a Queen Elizabeth II scholarship.

REFERENCES

1. Danishefsky SJ, Allen JR. *Angew Chem Int Ed* 2000;39:836–863.
2. Fleming A. *Brit J Exp Path* 1929;10:226–236.
3. Lindberg AA. *Vaccine* 1999;17: S28–S36.
4. Enright MC. *Curr Opin Pharmacol* 2003;3:474–479.
5. Ash C. *Trends Microbiol* 1996;4:371–372.
6. Spellberg B, Guidos R, Gilbert D, Bradley J, Boucher HW, Scheld WM, Bartlett JG, Edwards J. *Clin Infect Dis* 2008;46:155–164.
7. Dintzis RZ. *Pediatr Res* 1992;32:376–385.
8. Robbins JB. *Immunochemistry* 1978;15:839–854.
9. Kuberan B, Linhardt RJ. *Curr Org Chem* 2000;4:653–677.
10. Madigan MT, Martinko JM, Parker J. *Brock biology of microorganisms*. New Jersey: Prentice Hall, Upper Saddle River; 2003.
11. Dziadek S, Jacques S, Bundle DR. *Chem Eur J* 2008;14:5908–5917.
12. Hakomori S. *Cancer Res* 1985;45:2405–2414.
13. Alexander HE, Heidelberger M, Leidy G. *Yale J Biol Med* 1944;16:425–434.
14. Dintzis RZ, Okajima M, Middleton MH, Greene G, Dintzis HM. *J Immunol* 1989;143:1239–1244.
15. Lerman SP, Romano TJ, Mond JJ, Heidelberger M, Thorbecke GJ. *Cell Immunol* 1975;15:321–335.
16. Howard JG, Courtena BM. *Eur J Immunol* 1974;4:603–608.
17. Kayhty H, Karanko V, Peltola H, Makela PH. *Pediatrics* 1984;74:857–865.
18. Anderson P, Johnston RB, Smith DH, Wetterlo LH, Peter G. *J Clin Invest* 1972 ;51:39–44.
19. Peltola H, Kayhty H, Kuronen T, Haque N, Sarna S, Makela PH. *J Pediatr* 1978;92:818–822.
20. Borgono JM, McLean AA, Vella PP, Woodhour AF, Canepa I, Davidson WL, Hilleman MR. *Proc Soc Exp Biol Med* 1978;157:148–154.
21. Vliegenthart JFG. *Febs Lett* 2006;580:2945–2950.
22. Peltola H, Kayhty H, Sivonen A, Makela PH. *Pediatrics* 1977;60:730–737.
23. Sood RK, Fattom A, Pavliak V, Naso RB. *Drug Discovery Today* 1996;1:381–387.
24. Landsteiner K, van der Scheer J. *J Exp Med* 1928;48:315–320.
25. Landsteiner K, van der Scheer J. *J Exp Med* 1929;50:407–417.

26. Goebel WF, Avery OT. *J Exp Med* 1929;50:521–531.

27. Avery OT, Goebel WF. *J Exp Med* 1929;50:533–550.

28. Granoff DM, Weinberg GA, Shackelford PG. *Pediatr Res* 1988;24:180–185.

29. Lepow M, Randolph M, Cimma R, Larsen D, Rogan M, Schumacher J, Lent B, Gaintner S, Samuelson J, Gordon L. *J Pediatr* 1986;108:882–886.

30. Weinberg GA, Einhorn MS, Lenoir AA, Granoff PD, Granoff DM. *J Pediatr* 1987;111:22–27.

31. Kagan E, Ragupathi G, Yi SS, Reis CA, Gildersleeve J, Kahne D, Clausen H, Danishefsky SJ, Livingston PO. *Cancer Immunol Immunother* 2005;54:424–430.

32. Bennett CS, Wong CH. *Chem Soc Rev* 2007;36:1227–1238.

33. Hecht ML, Stallforth P, Silva DV, Adibekian A, Seeberger PH. *Curr Opin Chem Biol* 2009;13:354–359.

34. Pozsgay V. *Curr Top Med Chem* 2008;8:126–140.

35. Adams WG, Deaver KA, Cochi SL, Plikaytis BD, Zell ER, Broome CV, Wenger JD. *Jama-J Am Med Assoc* 1993;269:221–226.

36. Peltola H, Kilpi T, Anttila M. *Lancet* 1992;340:592–594.

37. Booy R, Heath PT, Slack MPE, Begg N, Moxon ER. *Lancet* 1997;349:1197–1202.

38. Ochiai M, Kataoka M, Toyoizumi H, Yamamoto A, Kamachi K, Arakawa Y, Kurata T, Horiuchi Y. *Jap J Infect Dis* 2004;57:58–59.

39. Audibert F, Jolivet M, Chedid L, Arnon R, Sela M. *Proc Natl Acad Sci USA* 1982;79:5042–5046.

40. Brown SE, Howard CR, Zuckerman AJ, Steward MW. *Lancet* 1984;2:184–187.

41. Houghten RA, Engert RF, Ostresh JM, Hoffman SR, Klipstein FA. *Infect Immun* 1985;48:735–740.

42. Shapira M, Jolivet M, Arnon R. *Int J Immunopharmacol* 1985;7:719–723.

43. Lopez JA, Weilenman C, Audran R, Roggero MA, Bonelo A, Tiercy JM, Spertini F, Corradin G. *Eur J Immunol* 2001;31:1989–1998.

44. Verez-Bencomo V, Fernandez-Santana V, Hardy E, Toledo ME, Rodriguez MC, Heynngnezz L, Rodriguez A, Baly A, Herrera L, Izquierdo M, Villar A, Valdes Y, Cosme K, Deler ML, Montane M, Garcia E, Ramos A, Aguilar A, Medina E, Torano G, Sosa I, Hernandez I, Martinez R, Muzachio A, Carmenates A, Costa L, Cardoso F, Campa C, Diaz M, Roy R. *Science* 2004;305:522–525.

45. Pozsgay V, Kubler-Kielb J, Schneerson R, Robbins JB. *Proc Natl Acad Sci USA* 2007; 104: 14478–14482.

46. Kumar A, Kaul S, Manivel V, Rao KVS. *Vaccine* 1992; 10:814–816.

47. Livingston PO. *Immunol Rev* 1995;145:147–166.

48. Schneerson R, Barrera O, Sutton A, Robbins JB. *J Exp Med* 1980;152:361–376.

49. Slovin SF, Ragupathi G, Musselli C, Olkiewicz K, Verbel D, Kuduk SD, Schwarz JB, Sames D, Danishefsky S, Livingston PO, Scher HI. *J Clin Oncol* 2003;21:4292–4298.

50. Mitchell GF, Williams AR, Humphrey JH. *Eur J Immunol* 1972, 2, 460–467.

51. Wan Q, Chen JH, Chen G, Danishefsky SJ. *J Org Chem* 2006;71:8244–8249.

52. Galonic DP, Gin DY. *Nature* 2007;446:1000–1007.

53. Helling F, Shang A, Calves M, Zhang SL, Ren SL, Yu RK, Oettgen HF, Livingston PO. *Cancer Res* 1994;54:197–203.

54. Goldman B, DeFrancesco L. *Nat Biotechnol* 2009;27:129–139.

55. Giannini G, Rappuoli R, Ratti G. *Nucleic Acids Res* 1984;12:4063–4069.

56. Bongat AFG, Saksena R, Adamo R, Fujimoto Y, Shiokawa Z, Peterson DC, Fukase K, Vann WF, Kovac P. *Glycoconj J* 2010;27:69–77.

57. Peeters C, Tenbergenmeekes AM, Poolman JT, Beurret M, Zegers BJM, Rijkers GT. *Infect Immun* 1991;59: 3504–3510.

58. Jacob CO, Arnon R, Sela M. *Mol Immunol* 1985;22:1333–1339.

59. Herzenberg LA, Tokuhisa T. *J Exp Med* 1982;155:1730–1740.

60. Herzenberg LA, Tokuhisa T, Herzenberg LA. *Nature* 1980;285:664–667.

61. Schutze MP, Leclerc C, Jolivet M, Audibert F, Chedid L. *J Immunol* 1985;135:2319–2322.

62. Adorini L, Harvey MA, Miller A, Sercarz EE. *J Exp Med* 1979;150: 293–306.

63. Amir-Kroll H, Nussbaum G, Cohen IR. *J Immunol* 2003;170:6165–6171.

64. Ingale S, AWolfert M, Gaekwad J, Buskas T, Boons GJ. *Nat Chem Biol* 2007;3:663–667.

65. Van Kasteren SI, Kramer HB, Gamblin DP, Davis BG. *Nat Protoc* 2007;2:3185–3194.

66. Jahouh F, Saksena R, Aiello D, Napoli A, Sindona G, Kovac P, Banoub JH. *J Mass Spectrom* 2010;45:1148–1159.

67. delGuercio MF, Alexander J, Kubo RT, Arrhenius T, Maewal A, Appella E, Hoffman SL, Jones T, Valmori D, Sakaguchi K, Grey HM, Sette A. *Vaccine* 1997;15:441–448.

68. Alexander J, del Guercio MF, Maewal A, Qiao L, Fikes J, Chesnut RW, Paulson J, Bundle DR, DeFrees S, Sette A. *J Immunol* 2000;164:1625–1633.

69. Davis BG. *Chem Rev* 2002;102:579–601.

70. Wessels MR, Paoletti LC, Kasper DL, Difabio JL, Michon F, Holme K, Jennings HJ. *J Clin Invest* 1990;86:1428–1433.

71. Ritter G, Boosfeld E, Adluri R, Calves M, Oettgen HF, Old LJ, Livingston P. *Int J Cancer* 1991;48:379–385.

72. Pozsgay V. *J Org Chem* 1998;63:5983–5999.

73. Goebel WF. *J Exp Med* 1940;72: 33–48.

74. Peeters C, Evenberg D, Hoogerhout P, Kayhty H, Saarinen L, Vanboeckel CAA, Vandermarel GA, Vanboom JH, Poolman JT. *Infect Immun* 1992;60:1826–1833.

75. Davis BG. *J Chem Soc Perkin Trans* 1999;1:3215–3237.

76. Kubler-Kielb J, Pozsgay V. *J Org Chem* 2005;70:6987–6990.

77. Langenhan JM, Thorson JS. *Curr Org Syn* 2005;2:59–81.

78. Wu XY, Lipinski T, Carrel FR, Bailey JJ, Bundle DR. *Org Biomol Chem* 2007;5:3477–3485.

79. Krauss IJ, Joyce JG, Finnefrock AC, Song HC, Dudkin VY, Geng X, Warren JD, Chastain M, Shiver JW, Danishefsky SJ. *J Am Chem Soc* 2007;129:11042–11044.

80. Ni JH, Song HJ, Wang YD, Stamatos NM, Wang LX. *Bioconjugate Chem* 2006;17:493–500.

81. Fattom A, Cho YH, Chu CY, Fuller S, Fries L, Naso R. *Vaccine* 1999;17:126–133.

82. Sarnaik S, Kaplan J, Schiffman G, Bryla D, Robbins JB, Schneerson R. *Pediatr Infect Dis J* 1990;9:181–186.

83. Barington T, Skettrup M, Juul L, Heilmann C. *Infect Immun* 1993;61:432–438.

84. Buskas T, Ingale S, Boons GJ. *Angew Chem Int Ed* 2005;44:5985–5988.

85. Wittrock S, Becker T, Kunz H. *Angew Chem Int Ed* 2007;46:5226–5230.

86. Krug LM, Ragupathi G, Hood C, Kris MG, Miller VA, Allen JR, Keding SJ, Danishefsky SJ, Gomez J, Tyson L, Pizzo B, Baez V, Livingston PO. *Clin Cancer Res* 2004;10:6094–6100.

87. Gilewski T, Ragupathi G, Bhuta S, Williams LJ, Musselli C, Zhang XF, Bencsath KP, Panageas KS, Chin J, Hudis CA, Norton L, Houghton AN, Livingston PO, Danishefsky SJ. *Proc Natl Acad Sci USA* 2001;98:3270–3275.

88. Krug LM, Ragupathi G, Ng KK, Hood C, Jennings HJ, Guo ZW, Kris MG, Miller V, Pizzo B, Tyson L, Baez V, Livingston PO. *Clin Cancer Res* 2004;10:916–923.

89. Ragupathi G, Park TK, Zhang SL, Kim IJ, Graber L, Adluri S, Lloyd KO, Danishefsky SJ, Livingston PO. *Angew Chem Int Ed in English* 1997;36:125–128.

90. Sabbatini PJ, Kudryashov V, Ragupathi G, Danishefsky SJ, Livingston PO, Bornmann W, Spassova M, Zatorski A, Spriggs D, Aghajanian C, Soignet S, Peyton M, O'Flaherty C, Curtin J, Lloyd KO. *Int J Cancer* 2000;87:79–85.

91. Slovin SF, Ragupathi G, Adluri S, Ungers G, Terry K, Kim S, Spassova M, Bornmann WG, Fazzari M, Dantis L, Olkiewicz K, Lloyd KO, Livingston PO, Danishefsky SJ, Scher HI. *Proc Natl Acad Sci USA* 1999;96:5710–5715.

92. Grigalevicius S, Chierici S, Renaudet O, Lo-Man R, Deriaud E, Leclerc C, Dumy P. *Bioconjugate Chem* 2005;16:1149–1159.

93. Ragupathi G, Koide F, Livingston PO, Cho YS, Endo A, Wan Q, Spassova MK, Keding SJ, Allen J, Ouerfelli O, Wilson RM, Danishefsky SJ. *J Am Chem Soc* 2006;128:2715–2725.

94. Chaicumpa W, Chongsa-nguan M, Kalambaheti T, Wilairatana PA, Srimanote P, Makakunkijcharoen Y, Looareesuwan S, Sakolvaree Y. *Vaccine* 1998;16:678–684.

95. Kalambaheti T, Chaisri U, Srimanote P, Pongponratn E, Chaicumpa W. *Vaccine* 1998;16:201–207.

96. Toth I, Simerska P, Fujita Y. *Int J Pept Res Thr* 2008;14:333–340.

97. Simerska P, Abdel-Aal ABM, Fujita Y, Batzloff MR, Good MF, Toth I. *Biopolymers* 2008;90:611–616.

98. Slovin SF, Ragupathi G, Fernandez C, Jefferson MP, Diani M, Wilton AS, Powell S, Spassova M, Reis C, Clausen H, Danishefsky S, Livingston P, Scher HI. *Vaccine* 2005;23:3114–3122.

99. Bettahi I, Dasgupta G, Renaudet O, Chentoufi AA, Zhang XL, Carpenter D, Yoon S, Dumy P, BenMohamed L. *Cancer Immunol Immunother* 2008;58:187–200.

100. Keding SJ, Endo A, Danishefsky SJ. *Tetrahedron* 2003;59:7023–7031.

101. Liebe B, Kunz H. *Helv Chim Acta* 1997;80:1473–1482.

102. Dziadek S, Kunz H. *Chem Rec* 2004;3:308–321.

103. Campo VL, Carvalho I, Allman S, Davis BG, Field RA. *Org Biomol Chem* 2007;5:2645–2657.

104. Acosta-Serrano A, Almeida IC, Freitas LH, Yoshida N, Schenkman S. *Mol Biochem Parasitol* 2001;114:143–150.

105. Buscaglia CA, Campo VA, Frasch ACC, Di Noia JM. *Nat Rev Microbiol* 2006;4:229–236.

106. Cohen-Anisfeld ST, Lansbury PT. *J Am Chem Soc* 1993;115:10531–10537.

107. Likhosherstov LM, Novikova OS, Derevitskaja VA, Kochetkov NK. *Carbohydr Res* 1986;146, C1–C5.

108. Dudkin VY, Miller JS, Danishefsky SJ. *J Am Chem Soc* 2004;126:736–738.

109. Warren JD, Miller JS, Keding SJ, Danishefsky SJ. *J Am Chem Soc* 2004;126:6576–6578.

110. Cho YS, Wan Q, Danishefsky SJ. *Bioorg Med Chem* 2005;13:5259–5266.

111. Westerlind U, Hobel A, Gaidzik N, Schmitt E, Kunz H. *Angew Chem Int Ed* 2008;47:7551–7556.

112. Cremer GA, Bureaud N, Piller V, Kunz H, Piller F, Delmas AF. *ChemMedChem* 2006;1:965–968.

113. Itzkowitz SH, Yuan M, Montgomery CK, Kjeldsen T, Takahashi HK, Bigbee WL, Kim YS. *Cancer Res* 1989;49:197–204.

114. Takahashi I, Maehara Y, Kusumoto T, Yoshida M, Kakeji Y, Kusumoto H, Furusawa M, Sugimachi K. *Cancer* 1993;72:1836–1840.

115. Wu XY, Bundle DR. *J Org Chem* 2005;70:7381–7388.

116. Tietze LF, Arlt M, Beller M, Glusenkamp KH, Jahde E, Rajewsky MF. *Chem Ber* 1991;124:1215–1221.

117. Tietze LF, Schroter C, Gabius S, Brinck U, Goerlachgraw A, Gabius HJ. *Bioconjugate Chem* 1991;2:148–153.

118. Wade TK, Saksena R, Shiloach J, Kovac P, Wade WF. *Med Microbiol* 2006;48:237–251.

119. Ma XQ, Saksena R, Chernyak A, Kovac P. *Org Biomol Chem* 2003;1:775–784.

120. Xu P, Alam MM, Kalsy A, Charles RC, Calderwood SB, Qadri F, Ryan ET, Kovac P. *Bioconjugate Chem* 2011;22:2179–2185.

121. Chernyak A, Kondo S, Wade TK, Meeks MD, Alzari PM, Fournier JM, Taylor RK, Kovac P, Wade WF. *J Infect Dis* 2002;185:950–962.

122. Rollenhagen JE, Kalsy A, Saksena R, Sheikh A, Alam MM, Qadri F, Calderwood SB, Kovac P, Ryan ET. *Vaccine* 2009;27:4917–4922.

123. Hossany RB, Johnson MA, Eniade AA, Pinto BM. *Bioorg Med Chem* 2004;12:3743–3754.

124. Saksena R, Zhang H, Kovac, P. *Tetrahedron-Asymmetry* 2005;16:187–197.

125. Saksena R, Ma XQ, Wade TK, Kovac P, Wade WF. *Fems Immunol Med Microbiol* 2006;47:116–128.

126. Hou SJ, Saksena R, Kovac P. *Carbohydr Res* 2008;343:196–210.

127. Mawas F, Niggemann J, Jones C, Corbel MJ, Kamerling JP, Vliegenthart JFG. *Infect Immun* 2002;70:5107–5114.

128. Wu XY, Ling CC, Bundle DR. *Org Lett* 2004;6:4407–4410.

129. Bundle DR, Rich JR, Jacques S, Yu HN, Nitz M, Ling CC. *Angew Chem Int Ed* 2005;44:7725–7729.

130. Wu XY, Lipinski T, Paszkiewicz E, Bundle DR. *Chem Eur J* 2008;14:6474–6482.

131. Jacques S, Rich JR, Ling CC, Bundle DR. *Org Biomol Chem* 2006;4:142–154.

132. Lemieux RU, Bundle DR, Baker DA. *J Am Chem Soc* 1975;97:4076–4083.

133. Becker B, Furneaux RH, Reck F, Zubkov OA. *Carbohydr Res* 1999;315:148–158.

134. Wada K, Chiba T, Takei Y, Ishihara H, Hayashi H, Onozaki K. *J Carbohydr Chem* 1994;13:941–965.

135. Zhao, JY Dovichi NJ, Hindsgaul O, Gosselin S, Palcic MM. *Glycobiology* 1994;4:239–242.

136. Khare DP, Hindsgaul O, Lemieux RU. *Carbohydr Res* 1985;136:285–308.

137. Chatterjee D, Douglas JT, Cho SN, Rea TH, Gelber RH, Aspinall GO, Brennan PJ. *Glycoconjugate J* 1985;2:187–208.

138. Chatterjee D, Cho SN, Stewart C, Douglas JT, Fujiwara T, Brennan PJ. *Carbohydr Res* 1988;183:241–260.

139. Wan Q, Cho YS, Lambert TH, Danishefsky SJ. *J Carbohydr Chem* 2005;24:425–440.

140. Rostovtsev VV, Green LG, Fokin VV, Sharpless KB. *Angew Chem Int Ed* 2002;41:2596–2599.

141. Tornoe CW, Christensen C, Meldal M. *J Org Chem* 2002;67:3057–3064.

142. Schofield L, Hewitt MC, Evans K, Siomos MA, *Nature* 2002;418:785–789.

143. Seeberger PH, Soucy RL, Kwon YU, Snyder DA, Kanemitsu T. *Chem Commun* 2004;15:1706–1707.

144. Gowda DC, Gupta P, Davidson EA. *J Biol Chem* 1997;272:6428–6439.

145. Schofield L, Hackett F. *J Exp Med* 1993;177, 145–153.

146. Kallenius G, Pawlowski A, Hamasur B, Svenson SB. *Trends Microbiol* 2008;16:456–462.

147. Vasan M, Rauvolfova J, Wolfert MA, Leoff C, Kannenberg EL, Quinn CP, Carlson RW, Boons GJ. *ChemBioChem* 2008;9:1716–1720.

148. Mehta AS, Saile E, Zhong W, Buskas T, Carlson R, Kannenberg E, Reed Y, Quinn CP, Boons GJ. *Chem Eur J* 2006;12:9136–9149.

149. Buskas T, Li YH, Boons GJ. *Chem Eur J* 2005;11:5457–5467.

150. Buskas T, Li YH, Boons GJ. *Chem Eur J* 2004;10:3517–3524.

151. Xin H, Dziadek S, Bundle DR, Cutler JE. *Proc Natl Acad Sci USA* 2008;105:13526–13531.

152. Konen-Waisman S, Cohen A, Fridkin M, Cohen IR. *J Infect Dis* 1999;179:403–413.

153. Amir-Kroll H, Riveron L, Sarmiento ME, Sierra G, Acosta A, Cohen IR. *Vaccine* 2006;24:6555–6563.

154. Liu XY, Siegrist S, Amacker M, Zurbriggen R, Pluschke G, Seeberger PH. *Acs Chem Biol* 2006;1:161–164.

155. Descoteaux A, Turco SJ. *Microbes Infect* 2002;4:975–981.

156. Kubler-Kielb J, Vinogradov E, Ben-Menachem G, Pozsgay V, Robbins JB, Schneerson R. *Vaccine* 2008;26:3587–3593.

157. Fekete A, Hoogerhout P, Zomer G, Kubler-Kielb J, Schneerson R, Robbins JB, Pozsgay V. *Carbohydr Res* 2006;341:2037–2048.

158. Jennings HJ, Lugowski C. *J Immunol* 1981;127:1011–1018.

159. Garcia-Ojeda PA, Monser ME, Rubinstein LJ, Jennings HJ, Stein KE. *Infect Immun* 2000;68:239–246.

160. Rubinstein LJ, Garcia-Ojeda PA, Michon F, Jennings HJ, Stein KE. *Infect Immun* 1998;66:5450–5456.

161. Xue J, Pan YB, Guo ZW. *Tetrahedron Lett* 2002;43:1599–1602.

162. Wang QL, Ekanayaka SA, Wu J, Zhang JP, Guo ZW. *Bioconjugate Chem* 2008;19:2060–2067.

163. Wu J, Guo ZW. *Bioconjugate Chem* 2006;17:1537–1544.

164. Chefalo P, Pan YB, Nagy N, Guo ZW, Harding CV. *Biochemistry* 2006;45:3733–3739.

165. Werz DB, Seeberger PH. *Angew Chem Int Ed* 2005;44:6315–6318.

166. Saksena R, Chernyak A, Poirot E, Kovac P. *Methods Enzymol* 2003;362:140–159.

167. Ban H, Gavrilyuk J, Barbas CF. *J Am Chem Soc* 2010;132:1523–1525.

168. Floyd N, Vijayakrishnan B, Koeppe AR, Davis BG. *Angew Chem Int Ed* 2009;48:7798–7802.

SECTION VI

CHARACTERIZATION, PHYSICO-(BIO)CHEMICAL PROPERTIES, AND APPLICATIONS OF BIOCONJUGATES

17

PROPERTIES AND CHARACTERIZATION OF BIOCONJUGATES

ALI FAGHIHNEJAD, JUN HUANG, AND HONGBO ZENG

Department of Chemical and Materials Engineering, Alberta Glycomics Centre, University of Alberta, Edmonton, AB, Canada

17.1 INTRODUCTION

Bioconjugation is the process of joining two biomaterials, or a synthetic material and a biomaterial, via covalent bonds or noncovalent intermolecular interactions in order to design a new material with properties and applications different from those of the original materials. Bioconjugates can be made with biomaterials such as amino acids, nucleotides, sugars, or complex moieties such as proteins, enzymes, and cells, and/or other organic materials such as polymers, nanoparticles, fullerenes, dendrimers, liposomes, and microgels. In this chapter, a review of different bioconjugate systems is provided followed by a brief description of intermolecular forces in biological systems.

17.2 POLYMER BIOCONJUGATES

The conjugation of synthetic polymers with biological molecules has been a focus in pharmacy research and in the development of biomaterials for many years [1–3]. Polymer bioconjugates are being used in new applications such as biosensors, electronic nanodevices, biometrics, and artificial enzymes [1]. Polymer bioconjugates are developed to obtain (1) highly ordered synthetic nanomaterials and nanostructures and (2) materials with biological properties and functionalities. A highly ordered synthetic structure can be developed by coupling a synthetic polymer with low molecular weight biological building blocks such as amino acids, nucleotides, or oligopeptides that are capable of forming self-organized structures. A synthetic polymer conjugated with complex biomolecules such as proteins or enzymes can achieve advanced functionalities such as specific recognition and selective catalytic activity [1].

Polymer–peptide bioconjugation has been used to improve solubility. It was reported that the incorporation of a single amino acid into the repeating unit of polybutadiene-*block*-poly (ethylene oxide) made the polymer soluble in mixtures of water and alcohol and changed the hydrophobicity of the synthetic polymer [4]. Single amino acids and short peptides were conjugated with synthetic polymers to enhance the solubility of hydrophobic drugs and the produced bioconjugates were less antigenic [4].

On the next level of complexity, conjugation of oligopeptides to polymers enables the formation of hierarchical structures [5–8]. Meredith et al. studied the conformation of a poly(ethylene glycol) (PEG)-peptide block copolymer in aqueous solutions, where the peptide was based on β-amyloid, a 40–43 amino acid peptide associated with Alzheimer's disease. The authors used small-angle neutron scattering and transmission electron microscopy (TEM) to confirm the formation of fibrils and found that a PEG coating which formed around the fibrils tended to stabilize them through steric forces [7, 8]. Conjugation of short polypeptides such as (Thr-Val)5, that is, five repeats of a threonine-valine diad, to poly(ethylene oxide) (PEO) led to well-defined microtape structures that were potentially suitable for material science applications [5, 6]. The secondary structure of peptides can also be utilized for assembly of synthetic polymers via specific interactions. Oligoalanine–PEG

Chemistry of Bioconjugates: Synthesis, Characterization, and Biomedical Applications, First Edition. Edited by Ravin Narain.
© 2014 John Wiley & Sons, Inc. Published 2014 by John Wiley & Sons, Inc.

were used to mimic silk and, due to the high affinity of alanine oligomers to adopt β-sheet secondary structure, β-sheet crossed-linked domains were formed, as characterized by IR spectroscopy and X-ray diffraction [9, 10]. Bioconjugates can also be designed to be stimuli responsive and capable of reacting to changes in environmental conditions such as temperature, pH, and chemicals. Such bioconjugates are of great interest in biomedical applications [11–14]. Hamley et al. studied the self-assembly of PEG conjugated with an amphiphilic β-strand peptide [12]. The researchers found that conjugation of the peptide with PEG induced structural transitions with changes in pH and thus made the peptide responsive to external stimuli (i.e., pH) [12]. The bioconjugates were characterized by circular dichroism (CD) which confirmed that conjugation of PEG with the peptide enhanced the stability of the β-sheet secondary structure of the peptide [5]. Several other studies showed enhanced thermal stability of the secondary structure of PEG-conjugated peptides as relative to unconjugated peptides and their enhanced ability to form different nanostructures [15, 16].

Protein–polymer conjugates were first introduced when van Hest et al. used genetically engineered proteins to control self-assembly and the material structure of protein-based polymer bioconjugates [17]. The bioconjugate comprised PEG chains attached to a genetically engineered protein that allowed self-assembly into well-defined fiber-like structures with a protein core and a PEG shell. The PEG chains prevented lateral aggregation of proteins through steric repulsion and led the assembly process toward the formation of needle-like lamellar crystals that were characterized by atomic force microscopy (AFM) and TEM [5, 17]. Another category includes protein-based block copolymers that are capable of forming ordered conformations with unique chemical and biological properties and with potential applications such as therapeutic delivery, tissue engineering, and medical imaging [18]. Some of the examples of engineered protein-based block copolymers include silk-like, resilin-like, and elastin-like polymers. Elastin is an extracellular matrix protein which has high elasticity and high fatigue lifetime. Elastin-based biopolymers have been found to be soft, flexible, and compatible with other biomaterials and are capable of forming spherical nanoparticles [18].

Nucleobases are nitrogen-based molecules that form the building blocks of nucleotides which in turn are the structural units of nucleic acids (i.e., DNA and RNA). Nucleic acids are vital in the transmission and expression of genetic information and protein synthesis. Therefore, conjugation of nucleobases and nucleic acids with synthetic polymers can potentially lead to the design of highly specialized biosensors and diagnostic kits [1]. Synthetic polymer–nucleobase conjugates in which the nucleobases sit on the polymer backbones or at the polymer chain ends, have been developed and found to be capable of forming ordered structures [1]. Recently, the self-assembly of DNA was used to form nanoscale shapes and patterns [19, 20] which suggested the use of nucleic acids as block copolymers for designing nanostructures and nanoarchitectures [21]. DNA block copolymers are virtually supramolecular polymers because the reversible bonds that hold the polymeric network together allow unique mechanical behavior. Oligonucleotide–polymer conjugates with unique recognition properties have been synthesized that can be used for diagnostic applications [22–24]. Polynorbornene conjugated with an oligonucleotide was found to be capable of the electrochemical detection of DNA [24]. Nucleic acid units have also been conjugated to light-emitting polymers for DNA detection and the range of detection was found to be from 10^{-3} to 10^{-8} M [21]. Nucleic acid–polymer conjugates are also used in the development of new drugs. Oligonucleotides are very selective in target recognition but have a short lifetime *in vivo* as they are not very stable against nucleases and are excreted quickly due to their low molecular weight. Conjugation of oligonucleotides with PEG stabilized oligonucleotides against nucleases and enhanced cell penetration by masking their negative charges [21]. Oligonucleotides were conjugated with polyelectrolytes and PEG to form micelles with the nucleotides and polyelectrolytes at the core and PEG chains constituting the shell. These polyelectrolyte/nucleotide/PEG complexes were used successfully to deliver drugs and genes to cancer cells [21].

Carbohydrates and polysaccharides—biomolecules that consist essentially of carbon, hydrogen, and oxygen—function biologically in storage and the transport of energy (e.g., starch, glycogen), formation of structural components (e.g., cellulose, chitin), lubrication in joints and extracellular matrix (e.g., hyaluronan), anti-inflammation (e.g., hyaluronan, chondroitin sulfate), and blood coagulation (e.g., heparin) [1, 25]. Carbohydrates also participate in cell-surface binding interactions and recognition processes. The wide range of carbohydrate functionalities make synthetic polymer–carbohydrate bioconjugates promising candidates for biomedical and biomedicine applications.

The binding strength of a saccharide with its corresponding protein receptor is not very strong. Nevertheless, the main mechanism behind recognition processes of cell-surface carbohydrates is through multivalency [25]. Multivalency is the simultaneous occurrence of hundreds of such carbohydrate–protein bonding interactions that create complex biological events such as cell–cell recognition and cell adhesion processes. Thus the development of enhanced carbohydrate-based biosensors for diagnostics of pathogens relies not only on knowledge of specific interactions between carbohydrate and pathogen but also on the multivalency of such interactions. Polymers consist of hundreds to thousands of repeating units that, when conjugated with carbohydrates, can potentially generate the required multivalency for cell recognition. For example, detection of *Escherichia coli* (*E. coli*) bacteria by poly(para-phenylene ethynylene) (PPE) conjugated with galactosides or mannosides was characterized by laser scanning confocal microscopy [26]. However, the ability of a carbohydrate to interact with different types of pathogens and proteins at the same time can result in poor selectivity in a biosensor. Different methods have been proposed to overcome this problem; conjugation of different types of carbohydrates with polymers allowed simultaneous analysis and detection of multiple pathogens, and addition of PEG chains to the carbohydrate–polymer conjugate minimized nonspecific adsorption of proteins [26, 27]. It was also shown that immobilization of carbohydrate–polymer conjugates on a surface enhances the binding affinity of cells and proteins and allows surface probing methods to be used in biosensor applications. For instance, surface plasmon resonance (SPR) spectroscopy characterizes a surface, enhances the detection sensitivity over conventional fluorescence microscopy, and eliminates the need to use fluorescently labeled proteins [26].

The biodegradability and biocompatibility of carbohydrate–polymer conjugates make them potential candidates for environment-friendly detergents, cosmetics, and therapeutic applications. The reader is referred to recent reviews on synthesis, properties, and applications of carbohydrate–polymer and polysaccharide–polymer conjugates for more details [28, 29].

17.3 NANOPARTICLE BIOCONJUGATES

Nanoparticles generally have a diameter of less than 100 nm. They can exist as spheres, tubes, or rods but the most common shape for biological applications is sphere. Nanoparticles can be made of different materials such as polymers (e.g., polystyrene, poly(methyl methacrylate) (PMMA), poly(hydroxyethyl methacrylate) (pHEMA)), metals (e.g., Au, Ag, and Pt), semiconductors (e.g., PbS, Ag_2S, CdS, and TiO_2), inorganic compounds (e.g., silica and $BaTiO_3$), and superparamagnetic composites [30]. Gas-phase, liquid-phase, and solid-phase processes have been developed for the synthesis of nanoparticles. Techniques for the production of nanoparticles include sol–gel, reduction, hot-soap, and ultrasonic methods [31]. Average particle diameter, particle size distribution, morphology, shape, and surface properties affect the behavior of nanoparticles in solution; thus, these factors must be considered in the preparation of nanoparticles. The Brownian motion that prevents particles from gravitational segregation and intermolecular and surface forces between particles in solution becomes important for particles smaller than 100 nm. The main interfacial forces between nanoparticles in a liquid medium include van der Waals and electrostatic forces, as described by the DLVO theory proposed by Derjaguin, Landau, Verwey, and Overbeek, hydrophobic interactions, structural and hydration forces, and steric repulsive forces [32]. Hydrophobic nanoparticles tend to segregate from aqueous solutions due to attractive hydrophobic interactions, while hydrophilic

particles form relatively stable solutions due to the formation of a hydration layer around each particle. Stabilization of nanoparticles in solution is essential for successful conjugation and is usually achieved by adjusting the solution pH/ionic strength or the surface charge of the nanoparticles.

Functionalization of nanoparticles is achieved by the association of molecular groups on the conjugate partner to the surface of the nanoparticles. Conjugation of biomolecules to nanoparticles can be accomplished by electrostatic adsorption, covalent binding, and specific biointeractions.

The functionalization of nanoparticles through electrostatic adsorption was investigated for both small biomolecules and large proteins/enzymes. Gold or silver nanoparticles that were stabilized by anionic ligands were functionalized with immunoglobulin G (IgG) through the electrostatic interactions between the positively charged amino acid chains of the protein and the negatively charged ligands of the nanoparticles. Electrostatic adsorption was also employed to generate multilayers of protein polymer or polyelectrolytes on nanoparticles in a shell–core structure. Multilayer adsorption of fluorescent functionalized polyelectrolytes on polystyrene nanoparticles allowed higher fluorescence output signals which can be used for detection of immunoassays with higher sensitivity [33].

The main drawback of physically adsorbed biomolecules on nanoparticles is that biomolecules, especially proteins, can denature or undergo conformational changes when adsorbed on a surface and thus lose their biological activities. Therefore, the covalent binding of biomolecules to nanoparticles is most often performed using thiol groups and bifunctional linkers. Proteins that have cysteine (an amino acid containing a thiol group) can be chemically attached to gold nanoparticles; alternatively, proteins can be genetically engineered to incorporate thiol groups in their structure, making them capable of chemisorption.

Biorecognition—a technique employed in antibody–antigen interactions, microbial diagnosis, and bioresponsive devices—is an important application of nanoparticles conjugated with proteins, nucleotides, or carbohydrates. Nanoparticle bioconjugates immobilized on a surface can act as biosensors. Metal and semiconductor nanoparticles are widely used in biosensor applications because of their unique optical and electronic characteristics. Depending on their size, gold and silver nanoparticles absorb plasmon (quantum oscillations) within the visible spectrum. This principle was used for detection of different DNA molecules since functionalized nanoparticles tend to aggregate as they bind to the DNA molecule, which increases the size and shifts the plasmon absorbance wavelength. Semiconductor nanoparticles have also been used as biosensors because of their size-dependent fluorescence properties. Other characterization methods used in nanoparticle bioconjugate biosensors include electrochemical reactions and surface-enhanced Raman scattering (SERS) [33, 34]. Gold nanoparticles were functionalized with antibodies and encoded with specific Raman dyes for the detection of different proteins using the SER scattering technique [33].

Magnetic nanoparticles (often Fe_3O_4 or γ-Fe_2O_3) conjugated with a biomolecular "shell" are used for therapeutic imaging and RNA, DNA, and cell purification purposes. Nanoparticle bioconjugates have also been used to generate nanostructures and nanoparticle networks, assembled structures on surfaces, and nanowires for nanoelectronic devices [33].

17.4 FULLERENE AND CARBON NANOTUBE BIOCONJUGATES

Carbon is a fundamental element in all forms of life and is a vital source of energy. Carbon exists in nature as different forms or allotropes such as diamond, graphite, and amorphous carbon with a wide range of different properties. Carbon atoms are oriented and bonded together in different ways in each allotrope. In diamond, carbon atoms form a tetrahedral lattice while in graphite they form flat hexagonal sheets. In fullerene (C_{60}), carbon atoms form a spherical cage comprised of 12 pentagons and 20 hexagons which is similar in shape to a modern soccer ball [30, 35]. Fullerene structures of C_{30}, C_{50}, C_{84}, and C_{540} have also been discovered, all having unique physical, electrochemical, and optical properties. Because fullerenes are extremely hydrophobic and insoluble in water, they are not very suitable for biological applications in their pristine form. Methods have been developed to make fullerenes water soluble through the covalent binding of hydrophilic groups. Fullerenes have been conjugated with molecules appropriate for biological applications such as drug and gene delivery and diagnostics. C_{60} conjugated with a carboxylic acid was observed to cleave DNA strands in the presence of visible light [36]. Several studies have shown the antiviral activity of fullerenes conjugated with ammonium groups (fulleropyrrolidines) which is attributed to the molecular structure and antioxidant properties of these conjugates. It was found that the structural position of ammonium groups on fullerenes has a vital role in determining its antiviral activity such that *trans* configuration is more active than *cis* isomer [37]. C_{60}/amino acid conjugates characterized by nuclear magnetic resonance (^{1}H-NMR) spectroscopy and electrospray mass spectroscopy (ES-MS) were found to be anti-HIV. It was suggested that the antiviral activity was due to hydrophobic interactions between C_{60} and a hydrophobic site on the HIV protease [37, 38]. Gene and drug delivery are other possible purposes for functionalized fullerenes. DNA-functionalized fullerenes were found to be capable of entering cells and had a relatively long lifetime. It was suggested that the stabilization of the DNA molecule by the fullerene increased the DNA lifetime in the cell [37]. The spherical cage

structure of fullerenes has been used to encapsulate magnetic and metallic agents for imaging and diagnostic applications. For instance, two titanium atoms encapsulated in a C_{84} fullerene (symbolically written as $Ti_2@C_{84}$) were characterized by laser-desorption time-of-flight mass spectroscopy (LD-TOF MS), UV–vis–NIR absorption, and electron energy loss spectroscopy (EELS) [37, 39].

Carbon nanotubes are allotropes of carbon with cylindrical structure and belong to the family of fullerenes. In carbon nanotubes, carbon atoms are connected to each other in hexagonal patterns and the configuration of these patterns determines the properties of the nanotube. Carbon nanotubes can be single-walled (SWNTs) or multi-walled (MWNTs). SWNTs are essentially a 1-atom-thick layer of carbon which is wrapped into a cylinder with a diameter of approximately 0.4–3 nm. A MWNT is a bundle of concentric SWNTs connected to each other; it has a diameter of approximately 1.4 to more than 100 nm. The unique properties of carbon nanotubes include a high length-to-diameter ratio (often more than 10,000), high tensile strength, and electrical conductivity [30]. Similar to fullerenes, carbon nanotubes are very hydrophobic and insoluble in water and thus easily form aggregates. Bare carbon nanotubes are susceptible to nonspecific protein adsorption because of their hydrophobic nature [40]. There are two main strategies to make carbon nanotubes biocompatible. The first method is based on the covalent binding of polar groups on the nanotube surface that makes them hydrophilic and capable of binding to biomolecules. For instance treatment of carbon nanotubes with strong acids (i.e., hydrochloric or nitric acid) introduces carboxyl and hydroxyl groups on the surface of carbon nanotubes. In the second strategy, surfactant and amphiphilic molecules are added to the carbon nanotube solution. Surfactant molecules are attached to the carbon nanotube surface through their hydrophobic tails leaving their hydrophilic parts exposed to the aqueous solution which increases the solubility of the carbon nanotube, enhancing its availability to interact with other molecules [30, 40, 41]. Carbon nanotubes with polar groups attached can be conjugated with proteins, carbohydrates, and nucleic acids for different applications. Carbon nanotubes functionalized with peptides can enter cells and can potentially be used for drug delivery or for diagnostic purposes. Characterization of peptide–carbon nanotube conjugates was performed by TEM and ^{1}H-NMR spectroscopy and the structures of synthesized bioconjugates were studied [42]. Functionalized carbon nanotubes were conjugated with DNA molecules for gene delivery and found to enhance the level of gene expression [43]. Functionalized carbon nanotubes were also used as biosensors through specific ligand–protein interactions. Carbon nanotubes conjugated to biotin, staphylococcal protein A (SpA), and U1A antigen could detect streptavidin, IgG, and mouse antibody 10E6, respectively, in a nanomolar concentration range as measured by a quartz crystal microbalance (QCM) [40]. In two separate studies, PEG-functionalized carbon nanotubes wer̶e̶ conjugated with an arg-gly-asp (RGD) peptide and pacl̶i̶t̶a̶x̶e̶l̶ (a known anticancer drug) and were found to be more successful at targeting tumor cells than available drugs. Carbon nanotubes loaded with paclitaxel were characterized using UV–vis–NIR spectroscopy and dynamic light scattering (DLS) and no significant aggregation of nanotubes was shown upon conjugation with the drug [40, 44].

17.5 DENDRIMER BIOCONJUGATES

Dendrimers are branched synthetic polymers with a multi-layer architecture where each layer is called a generation (G). The first dendrimers (from the Greek word "dendra" which means tree) were synthesized in early 1980s. A schematic drawing of a dendrimer with three generations is shown in Figure 17.1.

Poly(amidoamine) (PAMAM), the first synthesized and the most well-known dendrimer, has a diamine core. Trifunctional aromatic units, polyethers, and polyhydroxyls have been used to synthesize different types of dendrimers. A divergent method in which each generation is built upon the previous one, a convergent method in which dendrons are synthesized separately and then linked to each other, and self-assembly are methods used to create different types of dendrimers [30]. Size is an important factor in determining dendrimer properties. In general, small dendrimers (less than G-3) are flat and all the internal areas are accessible. As dendrimers grow in size (G-4–G-6) they tend to become spherical in shape with some accessible internal void areas

that can be used for encapsulation of guest molecules (i.e., drugs). Larger dendrimers (G-7–G-10) behave like spherical nanoparticles with no accessible interiors. As the generation of a dendrimer increases, the number of end-groups in its outer layer increases exponentially resulting in multivalency. Dendrimer multivalency has been exploited to enhance cell–ligand interactions for cancer therapy and HIV treatment [45]. The "core–shell"-like structure of dendrimers allows the design of dendrimers with a hydrophobic core and hydrophilic end-groups and vice versa that can potentially be used for controlled drug delivery [30, 45].

Dendrimers can be conjugated with a variety of biomolecules (e.g., amino acid-based dendrimers, glycodendrimers) which can be used in biological applications such as drug and gene delivery, imaging, and cancer therapy [45–47]. Stevelmans et al. modified the end-groups of hydrophilic poly(propylene imine) dendrimers with hydrophobic alkyl chains to produce dendritic micelles capable of encapsulating host molecules. Structural characterization of dendritic micelles was performed using ^{1}H-NMR, ^{13}C-NMR, and IR spectroscopies [48]. Polyester-based dendrimers were conjugated with the anticancer drug doxorubicin for drug delivery and the conjugates were characterized using ^{1}H-NMR, ^{13}C-NMR, size exclusion chromatography, and ES-MS [49].

17.6 LIPOSOME-BASED BIOCONJUGATES

17.6.1 Composition and Properties of Liposomes

Liposomes are artificial vesicle structures composed primarily of phospholipid bilayers that exhibit amphiphilic properties, as shown in Figure 17.2. Other molecules such as cholesterol or fatty acids can be included in the bilayer construction. The hydrophobic interaction between the fatty acid tails is the primary driving force for creating liposomal bilayers and holding the vesicles together in aqueous solution [30]. The phospholipid composition and bilayer structure of liposomes are similar to the composition and structure of the cell membrane. Lipid bilayers fuse well with other bilayers such as cell membranes and the physicochemical properties of liposomes were investigated extensively as models of membrane morphology. Liposomes are often used as delivery devices to encapsulate cosmetics, drugs, or fluorescent detection reagents, and to transport nucleic acids, peptides, and proteins to cellular sites *in vivo* [30]. Liposome-based anticancer drug delivery systems have been developed and widely used in clinical tumor therapy.

The most important component of a liposome is a phospholipid derivative that consists of a glycerol backbone that links two fatty acid molecules with a polar head-group. Phospholipids can contain mixed lipid chains with surfactant properties [50]. In addition to the phospholipid derivative, the presence of cholesterol in liposome membranes has a great effect on the properties of the liposome. The cholesterol

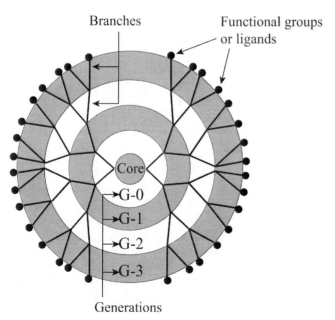

FIGURE 17.1 Schematic drawing of a dendrimer with three generations (layers).

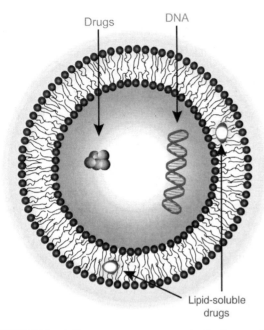

FIGURE 17.2   A liposome system as a carrier for drugs and DNA in bioengineering.

concentration can modulate the permeability and fluidity of an associated membrane, and decrease or even abolish (at high cholesterol concentrations) the phase transition from the gel state to the fluid or liquid-crystal state that occurs with increasing temperature [30]. The release of drugs can be controlled by adjusting liposome components.

Liposomes are roughly categorized by the diameter (small, large) of the vesicle and the number of bilayers surrounding each aqueous compartment (unilayer, multilamellar). Mechanical dispersion methods (sonication, shaking, freeze–thaw, etc.) are commonly used to prepare liposomes for research and scale production.

Liposomes are good nanocarriers for drugs because (1) the liposome bilayer structure can fuse well with cell membranes, (2) liposomes are nontoxic and (generally) nonimmunogenic to living organisms, (3) surface-modified, long-circulating, and drug-containing liposomes can slowly accumulate at pathological sites with affected and leaky vasculature, that is, they have an enhanced permeability and retention (EPR) effect [50], (4) liposomes can encapsulate water-soluble and lipophilic drugs, and (5) a liposome carrier can extend the half-life ($t_{1/2}$) of the drug, enhancing the bioavailability of the drug and allowing the drug to be released slowly increasing its absorption efficiency. Also, the toxicity of a drug has been shown to be reduced by liposome encapsulation. The properties and performance of liposomes need to be improved in the following aspects. (1) Unmodified liposomes are normally not stable enough for storage: small liposome vesicles often aggregate to form larger, more complex structures. Therefore, long-term storage in aqueous

solution is usually not possible without major transformations in liposome morphology [30, 51]. (2) Liposome encapsulating efficiency needs improvement for drug delivery. (3) The circulation time in blood for unmodified liposomes is short; liposomes are easily phagocytosed by the reticuloendothelial system (RES) and conjugated with plasma proteins. (4) Liposome targeting activity needs to be improved as it is hard to penetrate through biological barriers (i.e., skins, blood–brain barrier, etc.) for drug delivery. Surface modification by ligands, polymers, or other biomacromolecules could address these challenges [52].

17.6.2   PEG-conjugated Liposomes

PEG is often used as a surface modifier for liposome in drug delivery. PEG has excellent solubility in aqueous solutions; it is highly flexible and has low toxicity, immunogenicity, and antigenicity; it does not accumulate in RES cells and has a minimum influence on the specific biological properties of modified pharmaceuticals. The presence of PEG on the exterior surface of liposomes, typically achieved by formulation with PEG-modified lipids, significantly enhances the circulation lifetime of the liposomes. It is hypothesized that PEG prolongs the circulation time of liposomes through "steric stabilization" [53, 54] that reduces liposome aggregation [53] and plasma–protein adsorption [55]. Dos et al. studied the influence of PEG-grafting density and polymer length on the properties of liposomes. Characterization results indicated that the addition of 1,2-distearoyl-sn-glycero-3-phosphatidylethanolamine (DSPE)-PEG2000 prevented the aggregation of 1,2-distearoyl-sn-glycero-3-phosphatidylcholine (DSPC) liposomes and was correlated with prolonged circulation longevity, but the protein-binding profiles for these formulations were very similar. The primary effect of PEG-conjugated lipids in these liposomes was mediated via a PEG-dependent steric barrier that did not influence protein binding [56]. The mechanism of the PEG influence on liposomes is still under debate.

17.6.3   Nucleic Acid-conjugated Liposomes

Lipoplexes—biopolymer complexes made of cationic liposomes and negatively charged nucleic acids—are potential nonviral vectors for gene therapy. Introducing these biopolymers into cells *in vitro* and *in vivo* is a breakthrough for liposome application [52]. Development of nonviral vectors with high transfection efficiency and good pharmacokinetic properties *in vivo* is a major challenge for gene therapy. Chan et al. studied the influence of PEG-lipids (PEGylation) on the properties of lipoplexes. An acylhydrazone-based acid-labile PEG-lipid (HPEG2K-lipid, PEG MW 2000), stable at physiological pH but whose PEG chain would be cleaved from the tail at lower pH, was designed and synthesized [57]. Chan et al. also compared the properties of PEGylated complexes containing PEG2K-lipid versus HPEG2K-lipid. DLS was employed to identify the time-dependent and pH-dependent colloidal stability of these PEGylated complexes. Live-cell imaging was used to assess intracellular particle distributions directly [57]. Shirazi et al. designed and synthesized a series of cleavable multivalent lipids (CMVLs) with a disulfide bond in the linker between the cationic headgroup and the hydrophobic tails. Small-angle X-ray scattering (SAXS) and wide-angle X-ray scattering (WAXS) were employed to investigate the phase structure and phase transformation of the complexes. Complexes of the CMVLs and DNA were found to be environmentally responsive materials that underwent extensive structural rearrangements when exposed to reducing agents [58]. Kullberg et al. investigated conjugation of epidermal growth factor (EGF)—a stable protein with well-defined reaction sites for conjugation—to liposomes using a micelle-transfer method. EGF was conjugated to the distal end of PEG-DSPE lipid molecules in a micellar solution and the EGF-PEG-DSPE lipids were then transferred to preformed liposomes, with or without water-soluble boronated acridine-1 (WSA), a candidate drug for boron neutron capture therapy. The conjugate was shown to have EGF-receptor-specific cellular binding in cultured human glioma cells [59].

The high cost of lipid raw materials used for liposome preparation slows the development of liposomes for industrial applications [52], but liposome-based bioconjugates are expected to play an important role in cancer therapy.

17.7 MICROGEL AND HYDROGEL BIOCONJUGATES

17.7.1 Basic Properties of Microgels and Hydrogels

Gels are ubiquitous. From ordinary milk to interstitial spaces between cells and organs, gels play important roles in everyday life. Microgels are polymer gels with properties that are useful in scientific research and industry applications. Baker first defined a microgel in 1949 as a latex particle with a large gel network that was still only of supermolecular size and weight [60]. The definition of microgel has expanded with further research in this area, and dispersions with average diameters between 50 nm and 5 μm are generally considered to be microgels [61]. Hydrogels, that is, aqueous microgels, are an important part of water-borne polymer technologies, having properties in common with water-soluble polymers, water-swollen macrogels, and water-insoluble latex particles [61]. Polymer gels composed of polymer chains that swell in water due to cross-linking but do not dissolve are generally known as hydrogels [62].

In 1978, Tanaka observed the collapse of the polymer network in poly(acrylamide) gels upon changing temperature or fluid composition. This phenomenon led to a large increase in

gel swelling ability [63]. Further research demonstrated that the volume collapse of polymer gels can be induced by varying the pH of the gel fluid, and a thermodynamics theory of phase transition in ionic gels was proposed [64]. Philip et al. prepared the first reported temperature-sensitive hydrogel. It was composed of poly(N-isopropylacrylamide) (PNIPAM), which is composed of the monomer N-isopropylacrylamide [65]. Based on that work, a new concept named "smart material" or "switchable material" was proposed because these gels were sensitive to changes in environmental factors (e.g., temperature, pH, metal ions, antigens, glucose, saccharides). Pelton observed that if most of the polymer in the gel network displayed temperature-sensitive phase behavior in the swelling solvent, the microgel would be temperature sensitive [61]. Although other polymers show thermoresponsive behavior, PNIPAM is unique in that the phase transition occurs in an experimentally useful temperature range. PNIPAM exhibits a coil–globule transition in aqueous solution at its lower critical solution temperature (LCST), also known as its volume-phase transition temperature (VPTT), of 32°C in water, and this temperature is very close to human body temperature (37°C). Colloidal gels have high surface-to-volume ratios and the LCST of these thermoresponsive microgels can be modulated via copolymerization of various hydrophobic and hydrophilic monomers, proteins, or even DNA. In general, the incorporation of hydrophobic comonomers leads to a lower LCST and hydrophilic comonomers to a higher LCST [66]. These features make PNIPAM an excellent choice for bioconjugates research. In the last decade there has been a steady increase in the number of publications describing the preparation, characterization, and application of hydrogels, mostly based on PNIPAM or related polymers.

The N-isopropylacrylamide (NIPAM) structure is similar to that of acrylamide, thus many NIPAM properties resemble those of acrylamide. Like acrylamide, NIPAM is a suspected carcinogen and neurotoxin, but it has an intense odor so monomer contamination is easy to detect [61]. As shown in Figure 17.3, at room temperature PNIPAM hydrogels have high water content, a refractive index close to that of water, and a few electrically charged groups on the chain ends. In contrast, at elevated temperatures the particle volume of PNIPAM gels is 10 times lower than the volume at room temperature, the density of electrically charged groups is higher, and the difference in refractive index from that of water is greater. As there is a refractive index contrast between the medium and the microgels, deswelling is typically accompanied by an increase in the turbidity of the solution. This phenomenon is a sign of phase transition in the microgels, that is, the gels reach their LCST.

17.7.2 Characterization of Microgels

Microgels can be characterized by the standard techniques of electrophoresis, DLS, rheology, and electron microscopy.

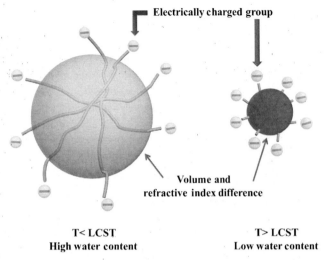

Electrically charged group

Volume and refractive index difference

| T< LCST | T> LCST |
| High water content | Low water content |

FIGURE 17.3 Schematic diagram of the volume difference in a microgel below (left) and above (right) its LCST.

Polymer/polymer interactions compete with polymer/water to determine the properties of the hydrogels. Like macroscopic aqueous gels, colloidal microgels are characterized by their degree of swelling, their average cross-link density, and their characteristic time constants for swelling and shrinking [61]. DLS, also known as photon correlation spectroscopy (PCS) or quasielastic light scattering (QELS), is sensitive to the diffusion of colloids in dispersion. This technique is capable of determining the size distribution of particles in solution and also gives perfectly acceptable VPTTs for hydrogels the DLS technique allows calculation of the change in microgel hydrodynamic radius with respect to temperature via correlation with the change in light scattering intensity [62]. Differential scanning calorimetry (DSC) is commonly used to characterize the water content of gels, and can be used to determine the LCSTs of bioconjugate copolymers. Imaging techniques such as scanning electron microscopy (SEM), TEM, and AFM are often employed to determine the morphology of microgels.

17.7.3 Bioconjugated Hydrogels

The first report of interactions between biological molecules and PNIPAM microgels was by Kawaguchi et al., who described the temperature-dependent sorption and desorption of human γ-globulin [67]. Shiroya et al. showed that the LCST of PNIPAM–trypsin conjugates changed with the molecular weight of coupled PNIPAM. Also, the enzymatic activity of the conjugates showed a significant temperature dependence which was affected by the molecular weight and the amount of coupled PNIPAM.[68]. Conjugates of polymers and DNA have a wide variety of applications in diagnostics and therapeutics such as gene delivery and DNA detection. The PNIPAM–DNA conjugate system also has been

used in the separation/purification of biomacromolecules, including DNA, by thermal stimulus [69]. Maeda et al. employed DLS to study the particle size for graft copolymers consisting of PNIPAM and single-stranded DNA [66, 70]. The results showed that the LCST of graft copolymers did not change significantly upon single-stranded DNA hybridization. However, the nanoparticles spontaneously aggregated above the LCST upon hybridization with complementary DNA. In addition to DLS, Ooi et al. used synchrotron radiation SAXS to follow the dynamical changes as well as to obtain detailed information about the particle structure for PNIPAM-g-(single-stranded DNA) hydrogel [71]. It was found that as the DNA fraction increased, the core size of the core–shell gel particle decreased significantly, whereas the shell thickness changed little. The thickness of the coronal layer was found to be close to the theoretical length of single-stranded DNA and the layer thickness reduced slightly on hybridization with its complementary DNA. Wu et al. studied the water content and volume-phase transition in swollen PNIPAM microgels with a combination of static and dynamic laser light scattering. Analysis of the results indicated that the volume change of the microgels was practically continuous, in contrast to the discontinuous volume-phase transition observed in bulk PNIPAM gels [72, 73].

Other PNIPAM derived materials such as Au-PNIPAM core–shell microgels can be used for SERS and biosensors [74, 75]. Zhou et al. demonstrated that PNIPAM combined with semiconductor quantum dots can be used as photoluminescent hybrids for sensing. Cadmium sulfide nanoparticles were loaded into glucose-sensitive poly (N-isopropylacrylamide-acrylamide-2-acrylamidomethyl-5-fluorophenylboronic acid), that is, PNIPAM-AAm-FPBA copolymer microgels, and the photoluminescence of the hybrid particles changed with changes in glucose concentration. This change was related to structural changes of the microgel network that swelled with increasing glucose concentration [76].

Microgel particles are attractive support materials that are readily available and nontoxic. These particles can provide enzyme stabilization; they have large surface areas for enzyme reactions and macropores for substrate and product transport with low diffusion restrictions. Such intelligent colloidal hydrogel systems could possibly be used as carriers for drug delivery systems, because the change in the colloidal assembly might result in the release of encapsulated drugs. The idea of using polymeric particles as biomolecular carriers has given rise to extensive research on the adsorption/grafting of protein molecules onto such dispersed materials [50, 62]. The recent work with composite microcolloids and hydrocolloids indicates a trend toward creating complex materials for diverse functions. Complex copolymer microgels show stimuli response under physiological conditions. The combination of different inorganic nanoparticles into one material can help obtain better performance by combining different organic and inorganic nanoparticles in a single material [75].

17.8   CELL-BASED BIOCOJUGATES

Cells are the basic structural and functional unit of living organisms. Cell-based therapies have recently found utility in the treatment of heart and vascular diseases, stroke, spinal injury, musculoskeletal disorders, cancer, and diabetes. Cell surface engineering—the introduction of exogenous molecules to the native cell surface—affords opportunities to control biochemical and cellular responses. These methods have important implications for drug delivery, cell-based therapy, and tissue engineering [77] and have the potential to improve clinical outcomes.

17.8.1   Cell Surface Engineering

The membrane of the mammalian cell is a complex composite of lipids, proteins, and carbohydrates. The well-known lipid bilayer forms a barrier between the cell cytoplasm and the environment. Cell surfaces are particularly available for manipulations that might make them useful in medical applications. Conjugation through covalent and noncovalent bonding, or through biomolecular recognition such as antibody/antigen and biotin/streptavidin interactions, can introduce targeting molecules, molecular and nanoparticle probes, polymer patches, and nanostructures onto the cell surface [78–80]. Chemical approaches or covalent conjugation methods for cell surface engineering have emerged as powerful tools for resurfacing the molecular landscape of cells and tissues [77, 81, 82]

Modification of the cell surface with PEG and PEG-related derivatives can decrease immunological recognition. For example, during cellular transplantation (such as blood transfusion), covalent modification of the surface of red blood cells with methoxy-PEG significantly diminished the immunological recognition of the surface antigens on these cells [83, 84]. Wilson et al. reported a versatile and facile noncovalent approach to cell surface engineering through electrostatic adsorption of designed cationic graft copolymers, poly(L-lysine)-*graft*-poly(ethylene glycol) (PLL-*g*-PEG), to cellular interfaces. It was found that grafting short PEG chains to PLL abrogated cytotoxicity and ultimately yielded cytocompatible polycations that presented functional groups upon adsorption to cell surfaces [77]. Based on this cytocompatible polycation polymer, PEG-dependent conformational changes in polycation structure were further exploited to unveil a narrow window in PLL-*g*-PEG copolymer structure space. Through these windows, cytocompatible polycations can facilitate the assembly of a uniss of polyelectrolyte multilayer (PEM) films wbiological and physicochemical properties [85].

Kozlovskaya et al. reported that cytocompatible synthetic shells from highly permeable, hydrogen-bonded multilayers could be engineered to cell surfaces while preserving long-term cell functioning. Nontoxic, nonionic, and biocompatible components such as poly(*N*-vinylpyrrolidone) (PVPON) and tannic acid (TA) were assembled on cell surfaces. These layer-by-layer shells facilitated outstanding (up to 6 days) cell survivability reaching 79% in contrast to only about 20% viability level for poly(allylamine hydrochloride)/poly(styrene sulfonate) coating [86].

17.8.2   Cell Encapsulation

The technology of cell microencapsulation is based on the immobilization of therapeutically active cells within a polymer matrix surrounded by a semipermeable membrane. Alginate is the most frequently used biomaterial for cell microencapsulation as it is biocompatible, easily manipulated, and has good gel-forming capacity and *in vivo* performance [87, 88]. The injection of insulin-secreting islet cells to enable continuous insulin treatment has been extensively investigated. Islet transplantation has been shown to control glucose levels successfully but improvement is needed in several areas: the availability of islets is limited, it is difficult to maintain islet functions such as cell growth and survival, and hosts reject implanted islets. Islet encapsulation is a possible solution to these challenges [89]. Cell surface engineering can improve cell interactions with drugs and drug delivery systems. Lee et al. decorated cell surfaces with a synthetic adenovirus receptor using a metabolic engineering approach [90]. Martin and Peterson used the concept of surface engineering to insert synthetic receptors into cell surfaces for the uptake of exogenous proteins, controlling the selective permeability to large drug molecules of the cell membrane [91].

The goal of cell surface engineering is to use externally applied chemical moieties and materials to control cell biology. It appears possible to modify cell membranes with specific ligands, micron-sized patches, and nanoparticles. Cell-based bioconjugation (including surface engineering, cell encapsulation, cell–matrix interactions) will likely find broad utility in tissue engineering, cell therapy, cell-based assays and devices, drug delivery and discovery, biosensing/bioimaging, cell separation, and manipulation of cell biological fates [78, 79].

17.9   BIOCONJUGATES ON SURFACES

Surface-modified bioconjugates can be built with inorganic materials (nanonoble metals such as Au, Ag, Pt), semiconductor quantum dots, graphene, carbon nanotubes, organic materials (polymers such as PEG, branched molecules such as dendrimers, lipid bilayers such as liposomes), biomaterials

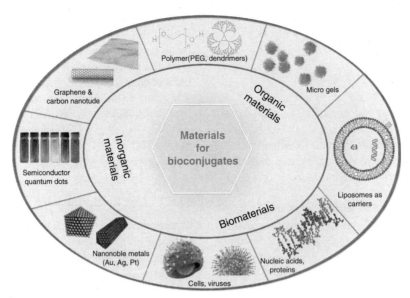

FIGURE 17.4   Bioconjugates achieved by modifying the surfaces of different materials. For a color version, see the color plate section.

(nucleic acids, proteins, cells, viruses), and many other substances, as shown in Figure 17.4.

The surface composition of a cell determines the cell's interactions with the environment, its ability to communicate with other cells, and its trafficking to tissues. Swiston et al. demonstrated that surface modification of cells can help the cells realize new functions. A photolithographic-patterning technique was used to engineer multilayer polymer patches containing a payload component (i.e., superparamagnetic nanoparticles) onto *T*-cell surfaces. It was also found that functional PEM patches can be attached to a fraction of the surface area of living, individual lymphocytes. To achieve this goal, a thermally responsive hydrogen-bonded multilayer system based on poly(methacrylic acid)/poly(N-isopropylacrylamide) (PMAA/PNIPAM) was developed. The complex patches carrying magnetic nanoparticles allowed the cells to be spatially manipulated using a magnetic field [92]. Dutta et al. rewired cell surfaces with ketone and oxyamine molecules based on liposome fusion for applications in cell surface engineering. The liposome fusion process was characterized by matrix-assisted laser-desorption/ionization mass spectrometry (MALDI-MS), DLS, fluorescence resonance energy transfer (FRET), and TEM. When cultured with cells, ketone- and oxyamine-containing liposomes underwent spontaneous membrane fusion to present the respective molecules from cell surfaces. The synthetic ketone and oxyamine molecules fused on the cell membrane serve as cell surface receptors, which provide another method to attach other functional materials to cell surface without using any cell-adhesive ligands [93]. Schmidt et al. demonstrated the switchability of cell adhesion on thermoresponsive PNIPAM microgel films. The films

allowed the researchers to control the detachment of adsorbed cells via temperature stimuli. Microgels attached noncovalently to surfaces allowing the gels to be used on a broad variety of (charged) surfaces. The noncovalent binding of the gels made it easier to attach them to surfaces compared to approaches that relied on covalent attachment of active films. In addition, the water content, mechanical properties, and adhesion forces of the microgel films were studied as a function of temperature by AFM [94].

Surface-modified nanoparticles and liposomes are now being tested as agents for drug delivery into regional lymph nodes and for diagnostic imaging purposes [95]. In cancer research, liposomes can be made in a particular size range that makes them viable targets for natural macrophage phagocytosis. Surface modifications can improve the targetability of liposomes. Components such as antibodies can be attached to liposomal surfaces to create large antigen-specific complexes that are good targets for macrophages [51, 52, 54, 55]. Surface modifications of pharmaceutical nanocarriers such as liposomes, micelles, nanocapsules, polymeric nanoparticles, solid lipid particles, and niosomes can control their biological properties and make these carriers perform various therapeutically or diagnostically important functions simultaneously [50]. The modification of hydrophobic polymeric nanoparticles can be performed by physical adsorption of a protecting polymer on a particle surface, or by chemical grafting of polymer chains onto a particle [50, 75]. The most significant biological consequence of nanoparticle modification with protecting polymers is a sharp increase in circulation time and a decrease in nanoparticle accumulation in the RES. For example, surface modification of polymer latexes with PEG (see also Section 17.6.2) allowed particles to circulate longer compared to unmodified particles, making it easier to study the particle penetration effects in tumors.

17.10   INTERMOLECULAR FORCES IN BIOLOGY

The successful operation and function of all bioconjugate systems relies on an understanding of their interactions at the molecular level within the biological environment. Intermolecular and surface interactions between two macromolecules or biological entities can be either attractive or repulsive as schematically shown in Figure 17.5. The type of interaction that exists between two molecules or surfaces depends on the nature of the molecules and the medium between them. In general, there are two categories of intermolecular and surface interactions in colloid and biological systems: (1) nonspecific interactions and (2) specific interactions.

Nonspecific interactions include van der Waals forces, electrostatic forces, solvation (or structural) forces, hydrophobic forces, and steric (polymer-mediated) forces. Van der Waals forces exist between all bodies and are usually attractive but can also become repulsive between dissimilar surfaces. Van der Waals forces originate from interactions of electric dipole moments between molecules and comprise three major forces: (i) the force between two permanent dipoles (Keesom interaction), (ii) the force between a permanent dipole and a corresponding induced dipole (Debye interaction), and (iii) the force between two instantly induced dipoles (London dispersion forces).

Electrostatic forces exist between charged molecules and surfaces and can be either attractive or repulsive. The electrostatic interaction that exists between charged surfaces in liquid solutions is known as the electric double-layer force.

Double layer and van der Waals forces are combined in DLVO theory (see Section 17.3) [32]. The double-layer force depends exponentially on separation distance while the van der Waals force is a power law function of the distance between two surfaces. The double-layer force strongly depends on solution conditions (i.e., electrolyte type and concentration) while van der Waals forces are almost insensitive to solution conditions. Therefore, DLVO forces can be either attractive or repulsive depending on the separation distance, nature of the surfaces, and solution conditions.

Steric forces exist between surfaces covered by large molecules such as polymers or proteins. Steric forces are repulsive and originate from the unfavorable entropy associated with confining large molecular chains between the two surfaces. The magnitude of steric forces depends on factors such as configuration and surface coverage of the polymer chains.

The long-range attractive forces between hydrophobic molecules, particles, or surfaces in aqueous solutions are the so-called hydrophobic forces. The magnitude of the hydrophobic force is much larger than the theoretical van der Waals force. The origin of the hydrophobic force is not well established; some of the mechanisms proposed include rearrangement of water molecules near hydrophobic entities and bridging of nanobubbles [96–98].

Solvation or structural forces are short-range oscillatory forces that exist between two surfaces or in strictly confined spaces with spherical liquid molecules between them. Oscillatory forces arise as the liquid molecules are forced to form discrete layers during the approach of the two surfaces. Therefore, oscillatory forces are mainly geometric in nature and depend on factors such as geometric shape of the liquid molecules and physical and chemical properties of the surfaces.

Specific interactions are essentially a combination of nonspecific forces that give rise to complex biological interactions such as receptor–ligand, lock-and-key, cell–cell, and cell–protein interactions. Biological interactions are never at thermodynamic equilibrium and thus are "dynamic" rather than "static" [32, 99] and are time and rate dependent. To pull apart a biological bond, energy is needed in proportion to the strength of the bond as well as the rate of pulling. Two highly specific receptor–ligand bonds are streptavidin–biotin and avidin–biotin interactions. Using AFM it was shown that for both of these ligand–receptor interactions, the rupture force increased dramatically with the rate when the ligand–receptor was pulled apart [32, 100].

Intermolecular forces in colloidal and biological systems can be measured with techniques such as AFM, surface forces apparatus (SFA), optical tweezers (OTs), micropipette aspiration (MPA), shear flow detachment (SFD), and bioforce probe (BFP). AFM has been extensively used to measure interfacial forces, nanoindentation, and for imaging surfaces in the nanoscale range. An SFA can measure forces between

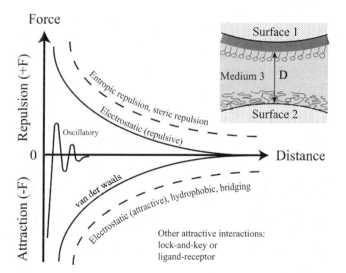

FIGURE 17.5   Generic interactions between two macromolecular or biological surfaces.

two macroscopic surfaces with a force resolution of 10^{-9} N and a distance resolution of 1 Å [101–106]. SFA and AFM techniques are used to measure long-range forces and separation distances; MPA, BFP, AFM, and SFD techniques are used to measure interactions between microscopic particles and cells; OT, AFM, SFD, and BFP are used to measure single-bond interactions [99].

The field of bioconjugates is an active research field and related products are already becoming an important part of daily life. By combining diverse materials, bioconjugate research not only discovers new materials, but also invents new biochemical communications. The fundamental understanding of the molecular and surface interactions of bioconjugate systems enhances the development of more advanced and functionalized bioconjugates. These inspiring results suggest that research of bioconjugates will lead to a new era of human life.

ACKNOWLEDGMENT

This work was supported by the Natural Sciences and Engineering Research Council of Canada (NSERC).

REFERENCES

1. Lutz J, Borner HG. *Prog Polym Sci* 2008;33:1–39.

2. Langer R, Tirrell D. *Nature* 2004;428:487–492.

3. Duncan R. *Nat Rev Drug Discov* 2003;2:347–360.

4. Geng Y, Discher DE, Justynska J, Schlaad H. *Angew Chem Int Ed Engl* 2006;45:7578–7581.

5. Borner HG.Schlaad H. *Soft Matter* 2007;3:394–408.

6. Hentschel J, Krause E, Borner HG. *J Am Chem Soc* 2006;128:7722–7723.

7. Burkoth TS, Benzinger TLS, Jones DNM, Hallenga K, Meredith SC, Lynn DG. *J Am Chem Soc* 1998;120:7655–7656.

8. Burkoth TS, Benzinger TLS, Urban V, Lynn DG, Meredith SC, Thiyagarajan P. *J Am Chem Soc* 1999;121:7429–7430.

9. Winningham MJ, Sogah DY. *Macromolecules* 1997;30:862–876.

10. Rathore O, Sogah DY. *Macromolecules* 2001;34:1477–1486.

11. Krishna OD, Wiss KT, Luo T, Pochan DJ, Theato P, Kiick KL. *Soft Matter* 2012;8:3832–3840.

12. Hamley IW, Ansari IA, Castelletto V. *Biomacromolecules* 2005;6:1310–1315.

13. Hamley IW, Cheng G, Castelletto V. *Macromol Biosci* 2011;11:1068–1078.

14. Top A, Roberts CJ, Kiick KL. *Biomacromolecules* 2011;12:2184–2192.

15. Vandermeulen GWM, Tziatzios C, Klok H. *Macromolecules* 2003;36:4107–4114.

16. Klok H, Vandermeulen GWM, Nuhn H, Rosler A, Hamley IW, Castelletto V, Xu H, Sheiko SS. *Faraday Discuss* 2005;128:29–41.

17. Smeenk JM, Otten MBJ, Thies J, Tirrell DA, Stunnenberg HG, van Hest JCM. *Angew Chem Int Ed Engl* 2005;44:1968–1971.

18. Rabotyagova OS, Cebe P, Kaplan DL. *Biomacromolecules* 2011;12:269–289.

19. Seeman NC. *Nature* 2003;421:427–431.

20. Rothemund PWK. *Nature* 2006;440:297–302.

21. Kwak M, Herrmann A. *Angew Chem Int Ed Engl* 2010;49:8574–8587.

22. Watson KJ, Park S, Im J, Nguyen ST, Mirkin CA. *J Am Chem Soc* 2001;123:5592–5593.

23. McLaughlin CK, Hamblin GD, Haenni KD, Conway JW, Nayak MK, Carneiro KMM, Bazzi HS, Sleiman HF. *J Am Chem Soc* 2012;134:4280–4286.

24. Gibbs JM, Park SJ, Anderson DR, Watson KJ, Mirkin CA, Nguyen ST. *J Am Chem Soc* 2005;127:1170–1178.

25. Ladmiral V, Melia E, Haddleton DM. *Eur Polym J* 2004;40:431–449.

26. Coullerez G, Seeberger PH, Textor M. *Macromol Biosci* 2006;6:634–647.

27. Disney MD, Zheng J, Swager TM, Seeberger PH. *J Am Chem Soc* 2004;126:13343–13346.

28. Spain SG, Cameron NR. *Polym Chem* 2011;2:60–68.

29. Schatz C, Lecommandoux S. *Macromol Rapid Commun* 2010;31:1664–1684.

30. Hermanson G. *Bioconjugate Techniques*. Amsterdam: Academic Press; 2008.

31. Nogi K, Naito M, Yokoyama T (eds) *Nanoparticle Technology Handbok*. Elsevier B. V; 2012.

32. Israelachvili JN. In: *Intermolecular and Surface Forces*: Academic Press: 2011.

33. Katz E, Willner I. *Angew Chem Int Ed Engl* 2004;43:6042–6108.

34. De M, Ghosh PS, Rotello VM. *Adv Mater* 2008;20:4225–4241.

35. Kroto HW, Heath JR, O'Brian SC, Curl RF, Smalley RE. *Nature* 1985;318:162–163.

36. Nakamura E, Isobe H. *Acc Chem Re* 2003;36:807–815.

37. Bakry R, Vallant RM, Najam-ul-Haq M, Rainer M, Szabo Z, Huck CW, Bonn GK. *Int J Nanomedicine* 2007;2:639–649.

38. Marcorin GL, Da Ros T, Castellano S, Stefancich G, Bonin I, Miertus S, Prato M. *Org Lett* 2000;2:3955–3958.

39. Cao B, Suenaga K, Okazaki T, Shinohara H. *J Phys Chem B* 2002;106:9295–9298.

40. Liu Z, Tabakman S, Welsher K, Dai H. *Nano Res* 2009;2:85–120.

41. Yang W, Thordarson P, Gooding JJ, Ringer SP, Braet F. *Nanotechnology* 2007;18:412001.

42. Pantarotto D, Partidos CD, Graff R, Hoebeke J, Briand J, Prato M, Bianco A. *J Am Chem Soc* 2003;125:6160–6164.

43. Bianco A, Kostarelos K, Partidos CD, Prato M. *Chem Commun* 2005:571–577.

44. Liu Z, Chen K, Davis C, Sherlock S, Cao Q, Chen X, Dai H. *Cancer Res* 2008;68:6652–6660.

45. Lee CC, MacKay JA, Fréchet J, Szoka FC. *Nat Biotechnol* 2005;23:1517–1526.

46. Liu J, Gray WD, Davis ME, Luo Y. *Interface Focus* 2012;2:307–324.

47. Medina SH, El-Sayed ME. *Chem Rev* 2009;109:3141–3157.

48. Stevelmans S, van Hest JCM, Jansen JF, van Boxtel DA, de Brabander-van den Berg EM, Meijer EW. *J Am Chem Soc* 1996;118:7398–7399.

49. Ihre HR, Padilla De Jesus OL, Szoka FC, Frechet JM. *Bioconjug Chem* 2002;13:443–452.

50. Torchilin VP. *Adv Drug Deliv Rev* 2006;58:1532–1555.

51. Woodle MC, Strom G (eds). *Long circulating liposomes: Old drugs, new therapeutics.* Springer; 1997.

52. Barenholz Y. *Curr Opin Colloid Interface Sci* 2001;6:66–77.

53. Hanahan D, Weinberg RA. *Cell* 2000;100:57–70.

54. Woodle MC, Lasic DD. *Biochim Biophys Acta* 1992;1113:171–199.

55. Du H, Chandaroy P, Hui SW. *Biochim Biophys Acta* 1997;1326:236–248.

56. Dos Santos N, Allen C, Doppen A, Anantha M, Cox K, Gallagher RC, Karlsson G, Edwards K, Kenner G, Samuels L, Webb MS, Bally MB. *Biochim Biophys Acta* 2007;1768:1367–1377.

57. Chan C, Majzoub RN, Shirazi RS, Ewert KK, Chen Y, Liang KS, Safinya CR. *Biomaterials* 2012;33:4928–4935.

58. Shirazi RS, Ewert KK, Silva BFB, Leal C, Li Y, Safinya CR. *Langmuir* 2012;28:10495–10503.

59. Kullberg EB, Bergstrand N, Carlsson J, Edwards K, Johnsson M, Sjoberg S, Gedda L. *Bioconjug Chem* 2002;13:737–743.

60. Baker WO. *Ind Eng Chem* 1949;41:511–520.

61. Pelton R. *Adv Colloid Interface Sci* 2000;85:1–33.

62. Debord J. Synthesis, characterization and properties of bioconjugated hydrogel nanoparticles, Dissertation, Georgia Institute of Technology, 2004.

63. Tanaka T. *Phys Rev Lett* 1978;40:820–823.

64. Tanaka T, Fillmore D, Sun ST, Nishio I, Swislow G, Shah A. *Phys Rev Lett* 1980;45:1636–1639.

65. Pelton RH, Chibante P. *Colloids Surf* 1986;20:247–256.

66. Maeda M. *Polym J* 2006;38:1099–1104.

67. Kawaguchi H, Fujimoto K, Mizuhara Y. *Colloid Polym Sci* 1992;270:53–57.

68. Shiroya T, Yasui M, Fujimoto K, Kawaguchi H. *Colloids Surf B* 1995;4:275–285.

69. Umeno D, Kawasaki M, Maeda M. *Bioconjug Chem* 1998;9:719–724.

70. Mori T, Maeda M. *Langmuir* 2004;20:313–319.

71. Ooi WY, Fujita M, Pan PJ, Tang HY, Sudesh K, Ito K, Kanayama N, Takarada T, Maeda M. *J Colloid Interface Sci* 2012;374:315–320.

72. Wu C, Zhou S, Au-yeung SCF, Jiang S. *Die Angewandte Makromolekulare Chemie* 1996;240:123–136.

73. Wu C. *Polymer* 1998;39:4609–4619.

74. Contreras-Cáceres R, Pacifico J, Pastoriza-Santos I, Pérez-Juste J, Fernández-Barbero A, Liz-Marzán LM. *Adv Funct Mater* 2009;19:3070–3076.

75. Karg M. *Colloid Polym Sci* 2012;290:673–688.

76. Wu WT, Zhou T, Aiello M, Zhou SQ. *Biosens Bioelectron* 2010;25:2603–2610.

77. Wilson JT, Krishnamurthy VR, Cui WX, Qu Z, Chaikof EL. *J Am Chem Soc* 2009;131:18228–18229.

78. Zhao WA, Teo GS, Kumar N, Karp JM. *Mater Today (Kidlington)* 2010;13:14–21.

79. Stephan MT, Irvine DJ. *Nano Today* 2011;6:309–325.

80. Stevens MM, George JH. *Science* 2005;310:1135–1138.

81. Prescher JA, Bertozzi CR. *Nat Chem Biol* 2005;1:13–21.

82. Chen I, Howarth M, Lin WY, Ting AY. *Nat Methods* 2005;2:99–104.

83. Murad KT, Mahany KL, Brugnara C, Kuypers FA, Eaton JW, Scott MD. *Blood* 1999;93:2121–2127.

84. Kellam B, De Bank PA, Shakesheff KM. *Chem Soc Rev* 2003;32:327–337.

85. Wilson JT, Cui WX, Kozovskaya V, Kharlampieva E, Pan D, Qu Z, Krishnamurthy VR, Mets J, Kumar V, Wen J, Song YH, Tsukruk VV, Chaikof EL. *J Am Chem Soc* 2011;133:7054–7064.

86. Kozlovskaya V, Harbaugh S, Drachuk I, Shchepelina O, Kelley-Loughnane N, Stone M, Tsukruk VV. *Soft Matter* 2011;7:2364–2372.

87. Orive G, De Castro M, Kong HJ, Hernandez R, Ponce S, Mooney DJ, Pedraz JL. *J Control Release* 2009;135:203–210.

88. Santos E, Zarate J, Orive G, Hernandez RM, Pedraz JL. *Adv Exp Med Biol* 2010;670:5–21.

89. Beck J, Angus R, Madsen B, Britt D, Vernon B, Nguyen KT. *Tissue Eng* 2007;13:589–599.

90. Lee JH, Baker TJ, Mahal LK, Zabner J, Bertozzi CR, Wiemer DF, Welsh MJ. *J Biol Chem* 1999;274:21878–21884.

91. Martin SE, Peterson BR. *Bioconjug Chem* 2003;14:67–74.

92. Swiston AJ, Cheng C, Um SH, Irvine DJ, Cohen RE, Rubner MF. *Nano Lett* 2008;8:4446–4453.

93. Dutta D, Pulsipher A, Luo W, Mak H, Yousaf MN. *Bioconjug Chem* 2011;22:2423–2433.

94. Schmidt S, Zeiser M, Hellweg T, Duschl C, Fery A, Mohwald H. *Adv Funct Mater* 2010;20:3235–3243.

95. Gupta AK, Gupta M. *Biomaterials* 2005;26:3995–4021.

96. Meyer EE, Rosenberg KJ, Israelachvili J. *Proc Natl Acad Sci U S A* 2006;103:15739–15746.

97. Faghihnejad A, Zeng H. *Soft Matter* 2012;8:2746–2759.

98. Christenson HK, Claesson PM. *Adv Colloid Interface Sci* 2001;91:391–436.

99. Leckband D, Israelachvili J. *Q Rev Biophys* 2001;34:105–267.

100. Evans E, Ritchie K. *Biophys J* 1997;72:1541–1555.

101. Israelachvili J, Min Y, Akbulut M, Alig A, Carver G, Greene W, Kristiansen K, Meyer E, Pesika N, Rosenberg K, Zeng H. *Rep Prog Phys* 2010;73:1–16.

102. Israelachvili JN, Adams GE. *J Chem Soc Faraday Trans 1* 1978;74:975–1001.

103. Israelachvili JN, McGuiggan PM. *J Mater Res* 1990;5:2223–2231.

104. Lu Q, Hwang DS, Liu Y, Zeng H. *Biomaterials* 2012;33:1903–1911.

105. Lu Q, Wang J, Faghihnejad A, Zeng H, Liu Y. *Soft Matter* 2011;7:9366–9379.

106. Zeng H, Hwang DS, Israelachvili JN, Waite JH. *Proc Natl Acad Sci U S A* 2010;107:12850–12853.

18

PHYSICO-CHEMICAL AND BIOCHEMICAL PROPERTIES OF BIOCONJUGATES

MARYA AHMED AND RAVIN NARAIN

Department of Chemical and Materials Engineering, Alberta Glycomics Centre, University of Alberta, Edmonton, AB, Canada

18.1 INTRODUCTION

The importance of bioconjugates in biomedicines, pharmaceuticals, and engineering is well established. The previous chapters provide a brief overview of novel bioconjugates; the synthetic approaches employed in the synthesis of bioconjugates and their applications are discussed in detail. This chapter focuses on physiochemical and biochemical properties of bioconjugates. The physical properties of conjugates include their response to temperature, external field (magnetic field, electric field, ultrasound), and light. The chemical properties of conjugates such as their response to change in pH of solution, and ionic strength are discussed. The properties of conjugates in response to biomolecules such as glutathione (GSH), hydrogen peroxide (H_2O_2), and glucose are also discussed.

18.2 PHYSICAL PROPERTIES

18.2.1 Temperature Responsive

Temperature is one of the critical factors that affect the biological activity of proteins [1–3]. Thermal inactivation is a major issue with enzyme stability. The bioconjugation of proteins with polymers is one way to improve the thermal stability of the proteins. The covalent attachment of polymers with enzymes reduces the mobility of the enzyme hence protecting the enzyme from environment-induced conformational changes [1–3]. poly-N-isopropylacrylamide (PNIPAM) is a

well-studied thermoresponsive polymer that shows a lower critical solution temperature (LCST) of approximately 32°C [1–4]. PNIPAM–trypsin conjugates are tested for their thermal stability, in precipitated form as well as in solution form [5]. These conjugates showed significantly high stability as compared to native trypsin. The polymer–protein conjugates were highly stable than native trypsin at high temperatures (60°C). However, low stability of polymer–protein conjugates was observed when temperature was cycled through LCST of PNIPAM. The conjugates were more susceptible to temperature change than native protein when the temperature was repeatedly cycled between 0° and 37°C (above and below LCST). This high vulnerability of conjugates around LCST is attributed toward conformational changes of polymer between dense globule and coil formation. This creates *"micro-environmental stress"* around protein conjugated to the polymer, hence causing its denaturation, as shown in Figure 18.1 [5]. In contrast, at constant high temperature (60°C) high stability of conjugates are due to the collapsed state of polymer around the protein [5].

The conjugation of enzyme with polymers can be done either by single-point or multiple-point covalent attachments. It is reported that single-point attachment of enzyme with scaffold or polymers offers good flexibility and operational stability to the conjugate. However, it does not provide conformational rigidity to the protein [6]. In contrast, multipoint attachment of scaffold provides enhanced rigidity to protein and minimizes the denaturation of proteins caused by conformational changes of the environment [7]. Chen et al. has

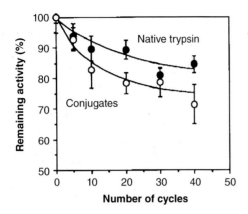

FIGURE 18.1   Retained activity of trypsin, as a function of number of cycles through LCST of PNIPAM [5].

synthesized thermoresponsive conjugates of α-amylase with PNIPAM [8]. The conjugation of PNIPAM with α-amylase was done either by single-point attachment or by multipoint attachment. It was found that multipoint attachment linkages provided high rigidity and protection to the enzyme against drastic environmental changes, as compared to the single-point attachment [8].

18.2.2   Field Responsive

The stimulation of conjugates by externally implied electric, magnetic, or electromagnetic field is another approach to attain controlled release of cargo materials. In addition, field-responsive nanomaterials have also been implied in biology for their unique responses to the stimuli [2, 3]. Hyperthermia treatment of tumor is an effective strategy to cure metastatic tissues, as neoplastic cells are more sensitive to hyperthermia than normal cells. Hyperthermia is obtained by applying alternating magnetic field of appropriate frequency, which causes heat dissipation due to oscillation of internal magnetic moment of small superparamagnetic iron oxide nanoparticles (SPIONs). Dextran-coated, folate-conjugated PEGylated SPION were prepared for the intracellular delivery of heat to the targeted cells, causing damage to their cytoplasm [9]. The nanoparticles were of 60–80 nm in diameter. The dextran-coated nanoparticles showed slight negative charge, which was not affected by pH of the media. The strongly negative charge of nanoparticles after PEGylation and folate conjugation at high pH values may be due to the deprotonation of carboxylic groups of folate. The changes in superparamagnetic properties of SPIONs before and after the surface modification were tested. The surface modification of SPIONs did not alter their paramagnetic properties and the particles were able to increase the temperature of aqueous solution at a constant rate under appropriate oscillating magnetic field [9].

Magnetic resonance imaging (MRI) is a noninvasive diagnostic technique for detailed imaging of internal organs [2, 3, 10–12]. Gadolinium (Gd)-based conjugates are most often used as contrast agents for MRI technique [2, 3, 10]. MRI contrast agents are usually evaluated by their relaxivity (r_{1p}) properties. The relaxivity of contrast agents can be increased by a number of ways, such as by optimizing the water residence time (τM), increasing the number of bound water molecules (q), and by conjugation with macromolecules or nanoparticles, hence increasing the rotational correlation time (τR) [13]. Gd(hydroxypropionate), Gd(HOPO), are well known and efficient contrast agents for MRI technique. The conjugates of Gd complexes with MS2 viral capsids were prepared, by selective conjugation of Gd^{3+} chelating ligand (Gd-TREN-*bis*-HOPO-TAM) to the exterior or interior of capsid, as shown in Figure 18.2. The number of ligands immobilized on a single capsid were 90 (50% functionalization), due to limited solubility of conjugates at high ligand concentration. The resultant complexes were modified with Gd^{3+} to obtain desired contrast agents. The relaxivity measurements of Gd–capsid conjugates showed significantly high r_{1p} values (41.6 mM^{-1} s^{-1} at 25°C and 30 MHz), as compared to Gd–HOPO complexes which show r_{1p} values between 8 and 14 mM^{-1} s^{-1} at 25°C and 20 MHz [14–16]. The magnetic coupling of the solvent with the system can be determined by measuring the change in relaxivity as a function of magnetic field. This gives $1/T1$ (proton nuclear magnetic relaxation dispersion) NMRD profile, which can be fitted to an appropriate model to obtain q, τM, and τR. NMRD profiles of Gd–capsid conjugates were compared with small molecule chelate of Gd-TREN-*bis*-HOPO-TAM-CO$_2$H at 298 K and 310 K. [10].

In comparison to Gd-TREN-*bis*-HOPO-TAM-CO$_2$H, four- to fivefold increase in relaxivity of Gd–capsid conjugates was observed (Figure 18.3). A slight decrease in relaxivity values at high temperature (37°C) suggests that relaxivity values of conjugates are not dependent on slow rate of water exchange from the inner coordination sphere of Gd^{3+} but they depend on the tumbling rate of Gd^{3+}. The relaxation rate of MS2 capsid alone was 0.4–0.5 s^{-1} and showed negligible dependence on magnetic field. The effect of temperature on relaxivity of externally and internally modified capsid conjugates was studied. The increase in temperature caused an exponential decrease in relaxivity of both conjugates, due to slow tumbling and fast exchanging inner sphere water molecules [10].

18.2.3   Light Responsive

Near-infrared (NIR) light-responsive materials are important for gene and drug delivery applications, due to the high penetration ability of NIR in tissue along with low absorbance in plasma [17, 18]. NIR absorbing properties of gold nanorods are utilized as a tool to construct light-responsive

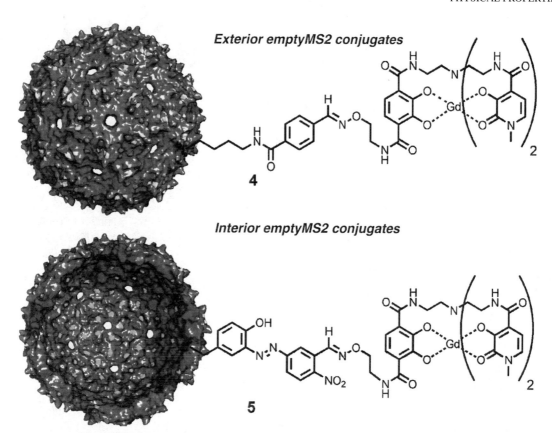

FIGURE 18.2 Gd complex modified externally or internally with MS2 virus capsid [10]. For a color version, see the color plate section.

nanomaterials, as drug delivery carriers [19–22]. Polymer-encapsulated gold nanorods are used for site-specific drug delivery and photothermal therapy by exploiting glass transition temperature of polymeric nanocarriers [19]. Glass transition temperature (T_g) of the cross-linked polymers is a characteristic property, which indicates the transition of

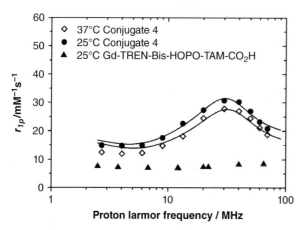

FIGURE 18.3 NMRD profiles of capsid conjugate at 25° and 37°C, in comparison with Gd-TREN-*bis*-HOPO-TAM-CO$_2$H NMRD profile at 25°C [10].

polymer from glassy to rubbery state, hence indicating the mobility of polymer chains. When T < T_g (glassy state), the drug remains encapsulated in nanocarrier, in contrast at $T >$ T_g (rubbery state) triggered release of drug from conjugate occurs. Gold nanorods embedded polymer film (1 mm thick) of degradable poly(β-amino esters), 2-hydroxyethylacrylate, and tertiary butyl acrylate was prepared. T_g of the film was controlled by controlling the cross-linking density and by the hydrophobicity of functional end-groups. T_g of the film was determined by dynamic mechanical analysis (DMA) and was found to be 40.5°C. The addition of gold nanorods in film caused a further increase (∼3°C) in T_g of the film. T_g higher than body temperature ensured that passive diffusion of drug from polymer film at body temperature will not occur. DOX-loaded polymeric discs were prepared from the film. The amount of DOX loaded per disc was 13.5 μg. The effect of heat on DOX-loaded nanorods embedded polymer discs was evaluated by obtaining the laser on and laser off profile. The discs were incubated at 37°C in PBS in laser off mode for 24 hours, followed by the incubation for 30 minutes in the presence of NIR laser (Figure 18.4). The cycle was repeated five times to determine the degradability of polymer disc, hence causing the release of DOX. The temperature of the disc 30 minutes post-exposure to NIR laser was 69.9°C. The samples were compared with negative control (in the absence

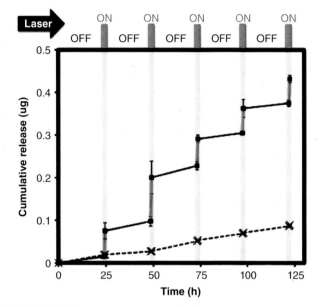

FIGURE 18.4   Average amount of DOX released from polymer discs in the presence (-■-) and absence (-X-) of NIR laser light. Five cycles of on and off laser were evaluated for the release of DOX in triplicates [19].

of laser). The samples treated with laser showed significant increase in DOX levels (~5 times), as compared to negative control. The amount of DOX released in every cycle was consistent and was repeatable [19].

Lanthanide-doped upconverting nanoparticles (UCNPs) have emerged as alternative to NIR-responsive materials [23, 24]. NIR lights are inefficient and slow to activate a chemical reaction due to low two-photon absorbing cross sections of chromophores and requirement of high power laser for simultaneous excitation of two photons [24]. UCNPs absorb NIR light and convert it to photons of higher energy in UV–visible region. Block copolymer (BCP) micelles encapsulated UCNPs were prepared [23]. The core–shell NaYF4:TmYb nanoparticles of 30 nm diameter were encapsulated into block copolymeric micelles made from poly(ethylene oxide) (PEO) and hydrophobic polymethacrylate bearing photolabile *O*-nitrobenzyl groups (PNBMA). The exposure of UCNP-loaded micelles to NIR light (980 nm) emits photon of approximately 350 nm. BCPs contain two methoxy groups on nitrobezyl moiety, which shift the absorption maximum from 300 to 350 nm hence cleaving the photoresponsive groups upon absorption of UV light emitted by UCNPs. This photoreaction converts PNBMA into hydrophilic polymethacrylic acid, and destabilizes the micelles. The photo-dissociation of micelles in response to NIR light was confirmed by co-encapsulation of Nile Red (NR) into micelles along with UCNPs. The control experiments were performed by loading NR in micelles in the absence of UCNPs. The absorption spectra were recorded by conjugates in the presence and absence of NIR light. In

the absence of NIR light, no change in spectra of water was recorded even after 24 hours of incubation. In contrast, in the presence of NIR light, the absorption bands of 350–400 nm were detected, indicating the cleavage of nitrosobenzaldehyde. The change in fluorescence emission spectra of NR-encapsulated micellar solution was detected at 550 nm, as another method to measure NR-encapsulated micellar disruption. The disruption in micellar conformation releases NR in aqueous solution and quenches the fluorescence of NR. In the absence of UCNPs no disruption in micellar confirmation was detected even after 4 hours of exposure with NIR. However, release of NR was detected for UCNP co-loaded micelles within 20 minutes of exposure. The continuous decrease in fluorescence emission upon exposure to NIR light indicates constant release of NR in aqueous solution. The effect of laser intensity on the disintegration of UCNPs and NR-loaded micelles was then studied. The normalized fluorescence intensity of NR as a function of exposure time of NIR light was plotted. Under all studied concentrations, the fluorescence intensity decreases in first 40–50 minutes of exposure time and then plateaus. The high concentration of UCNPs or high laser power showed faster release of NR from BCP micelles (Figure 18.5) [23].

18.3   CHEMICAL PROPERTIES

18.3.1   pH Responsive

Stimuli-responsive systems have been used for temporal and spatial delivery of biomolecules at the site of interest. pH-responsive systems are of significant interest due to acidic pH of tumor tissue, as well as for gene and drug delivery applications via endocytosis [3, 25, 26]. Multilayered multifunctional drug delivery system (DDS) was studied for their pH-responsive de-shielding ability and specificity [27]. The low pH values are characteristics of inflamed and neoplastic tissues. Myosin specific monoclonal antibody (mAb 2G4) modified, pH-responsive PEGylated micelles were prepared. PEGylated conjugates of phosphatidylethanolamine spontaneously form micelles in aqueous solution. The critical micelle concentration of conjugates was determined by pyrene method and was found to be 0.1 μM. The brief incubation of micelles at low pH leads to the removal of outer PEG layer, hence exposing the cell-specific antibody and cell-penetrating molecules on the surface of DDS as shown in Figure 18.6 [27].

The stability of DDS as a function of time at varying pH values was studied by size exclusion high permeation liquid chromatography (HPLC) analysis. The degradation kinetics of PEGylated micelles is summarized in Table 18.1. As shown, the micelles are stable at high pH values (pH 8 and above) and rapidly dissociate into polymer chains at pH 5 [27].

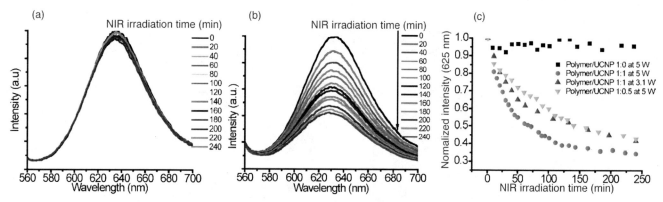

FIGURE 18.5   Fluorescence emission spectra of BCP micelles measured at $\lambda_{exc} = 550$ nm loaded with UCNPs and NR, and are exposed to NIR light, (a) emission spectra of BCPs loaded with NR, (b) emission spectra of BCPs loaded with both UCNPs and NR, (c) normalized fluorescence intensity of NR versus NIR exposure time studied at varying ratios of UCNPs loading and light intensity [23]. For a color version, see the color plate section.

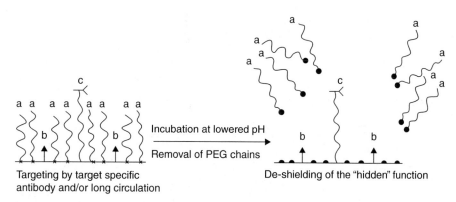

FIGURE 18.6   A multifunctional DDS, segment "a" represents pH cleavable PEG-Hz-PE, segment "b" represents temporarily shielded biotin or Tat peptide, segment "c" represents monoclonal antibody [27].

In another study, water-soluble drug conjugates of PHPMA with monoclonal and polyclonal antibodies were prepared to obtain an optimum design for the targeted drug delivery in living systems [28]. PHPMA backbone was modified by degradable tetrapeptide chain, Gly-DL-Phe-$_L$-Leu-Gly, bearing terminal drug moieties and antibodies, distributed in random fashion. The structure of polymer–drug–antibody conjugates is shown in Figure 18.7 [28]. The samples of PHPMA drug conjugates prepared differ from each other in configuration of phenylalanine (Phe) ($_{D, L, DL}$) residues in tetrapeptide chain. The anticancer activity of poly-

mer conjugates depends on the release of free drug (DOX) from bioconjugates. The rate of DOX released from bioconjugates containing Phe in tetrapeptide spacer decreased in the following order; $_L$-Phe $>$ $_{DL}$-Phe $>$ $_D$-Phe. However, the toxicity results obtained by *in vitro* and *in vivo* analyses did not show a direct relationship between the amount of DOX released from conjugates and tumor suppression. Although, $_D$-Phe-based tetrapeptide spacer-containing drug conjugates showed efficient release of DOX, the inhibition of T-cell proliferation did not occur efficiently in comparison to $_{D,L}$-Phe-containing analog. Similarly, the treatment of murine EL4 lymphoma in mice showed effective and prolonged inhibition of tumors after the treatment with conjugates bearing $_{D,L}$-Phe tetrapeptide spacer, and the least effect was observed for $_L$-Phe tetrapeptide spacer analogs [28].

18.3.2   Ionic Strength

PEGylation of proteins, drugs, or biomolecules was first approved by FDA in 1990 [29]. PEGylation of molecules is known to increase their blood circulation, and to decrease

TABLE 18.1   PEG-Hz-PE micelles' stability at different pH values

Incubation time (min)	pH = 5	pH = 7	pH = 10
20	3	56	99
40	2.5	28	99
60	2	10	99

From Reference 27.

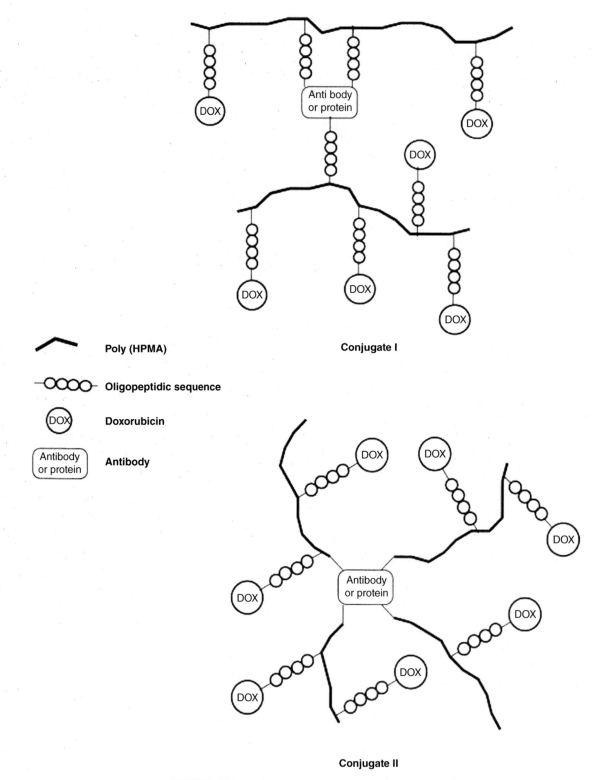

FIGURE 18.7   Design of polymer drug conjugates [28].

their renal filtration by masking their immunogenicity [30]. PEGylation of proteins increases the steric hindrance of the products and decreases the adsorption of proteins per unit surface area. The effect of linear or branched PEGylation (PEG-L and PEG-B) on protein adsorption was studied [31]. A 348 Dalton peptide "glucagon" was used as a model protein. The effect of architecture of PEG on glucagon adsorption on hydrophobic surfaces was evaluated and was compared with native glucagon. The adsorption of glucagon, glucagon-PEG-L, and glucagon-PEG-B on polystyrene (PS) beads was compared by reflectometry and isothermal calorimetry (ITC). It was found that adsorption of glucagon-PEG-L and glucagon-PEG-B was significantly lower on PS surface, as compared to glucagon alone. However, no difference in adsorption of glucagon-PEG-L or glucagon-PEG-B on hydrophobic surface was observed, which indicates that the protein modified by PEG of two different architectures occupy same space on the surface. This is further supported by the studies that modification of glucagon with PEG-L or PEG-B of similar weight results in same increase in the protein size. In contrast, results obtained from ITC indicate that there is a difference in heat flow, upon adsorption of glucagon-PEG-L or glucagon-PEG-B on hydrophobic PS beads. The difference in conformational changes of PEG-L and PEG-B and variations in flexibility of polymer chains of two different architectures are thought to contribute to the difference in heat flow [31].

Mitomycin C-dextran conjugates (MMC-D) of varying molecular weights (10–500 kDa) and of cationic (MMC-D_{cat}) and anionic (MMC-D_{an}) charges were prepared [32]. The physiochemical properties of bioconjugates as a function of their charge and molecular weight were studied. The interactions of MMC-D_{an} and MMC-D_{cat} with Ehrlich ascites carcinoma (EAC) cells were compared. MMC-D_{cat} showed significantly high cellular interaction and cytotoxicity, possibly due to electrostatic interactions between cationic MMC-D and anionic cell surface, as compared to MMC-D_{an} *in vitro*. In contrast, exceptional tumor inhibition activity of MMC-D_{an} of 500 kDa was observed in tumor bearing mice post IP injection [32]. In another study, MMC-D were studied for their regeneration rates. MMC-D samples containing different spacer lengths ranging from 4 to 8 carbon atoms, namely MMC(C4)D, MMC(C6)D, and MMC(C8)D were prepared. All the MMC-D samples contained net positive charge. The rate of MMC release from conjugates was studied *in vitro* and was directly related to the spacer length. MMC(C8)D showed longer half-life, that is, slow release of MMC than MMC(C6)D and MMC(C4)D. The interactions of drug conjugates with cells were studied *in vitro*. MMC(C8)D and MMC(C6)D showed significant interactions with cell surface as compared to MMC(C4)D. In addition, superior antitumor activities of MMC(C8)D and MMC(C6)D were obtained during *in vitro* and *in vivo* analyses [33].

Polysaccharide–protein conjugates show high thermal stability due to co-operative contribution of a number of factors, which maintain native conformation of enzymes at high temperatures [34–38]. Polyanionic nature of polysaccharides is one of the factors that helps in the formation of multipoint salt bridges on the surface of conjugated protein. This helps in reducing the conformational flexibility of conjugated protein, hence maintaining its native structure at high temperatures. The cationic and anionic polyelectrolytes have also been used for the modification of proteins, to maintain their structural integrity at high temperatures [34–38].

The glycosylation of proteins impart thermal stability to bioconjugates due to several factors:

1. Hydration and dehydration of protein functional groups
2. Formation of hydrogen bonds
3. Intermolecular and intramolecular cross-linking between carbohydrates and proteins

The native conformation of protein in polysaccharide–protein conjugates is maintained due to hydrophilic nature of carbohydrates, which inhibits their aggregation and denaturation. In addition, covalent linkage of carbohydrates with proteins provides conformational stability to the proteins, due to the formation of hydrogen bonding between proteins and carbohydrates [1].

DNA-functionalized nanoparticle conjugates are used for gene delivery applications as well as the formation of biosensors. DNA-nanoparticle assemblies are affected by several parameters, such as ionic strength, DNA loading, and spacer length [39–44]. The disassembly or melting properties of DNA-silver nanocrystals (DNA-AgNCs) were studied thermodynamically at varying salt concentrations by Park et al. [22]. A series of hybridized DNA-AgNCs were prepared by mixing two types of DNA-AgNCs at varying NaCl concentrations (0.1–0.6M), and their melting properties were studied by monitoring their extinction at $\lambda = 420\,nm$, as a function of temperature. Melting transitions of hybridized NCs are shown in Figure 18.8, as a function of NaCl concentration. In all studied cases, a sharp melting transition of approximately $2.3°C$ was observed. However, a gradual increase in melting temperatures (T_m) of conjugates (46–60°C) as a function of salt concentration was observed. The rapid increase in T_m was observed when salt concentration changed from 0.1 to 0.2M, in contrast relatively slow and steady increase in T_m occurred with further increase in salt concentration in solution [22].

18.4 BIOCHEMICAL PROPERTIES

18.4.1 Glucose Responsive

Glucose-responsive materials are "intelligent" nanomaterials for the effective diabetes therapies. The two main applications of glucose-responsive nanomaterials are controlled release of insulin and diagnosis of elevated sugar levels in

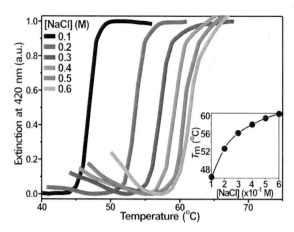

FIGURE 18.8   Melting transitions of DNA-AgNCs studied at varying salt concentrations. The inset shows the melting temperature (T_m) of the DNA-AgNCs plotted against salt concentrations [22].

the body [45]. Phenyl boric acid and its analogs are synthetic glucose sensors, which make them promising candidates for glucose sensing and controlled drug delivery [46]. Sugar-responsive homopolymer of boroxole-containing styrenic monomers (PBOx) and their block copolymers with PEG were synthesized via RAFT polymerization approach by Kim et al. [47]. PBOx and their PEGylated block copolymers were insoluble in water at physiological pH, and self assembled to form polymersomes in water. Organoboronic acids can bind to hydroxyl groups (1,2 and 1,3 diols) of monosaccharides and nucleic acids, reversibly in solution, switching the solubility of boronic acid-containing polymers from insoluble to soluble in water, hence triggering the disassembly of polymersomes [48, 49]. The addition of 0.2M fructose or 0.5M glucose facilitates the dissolution of boroxole-containing polymers in aqueous solution at physiological conditions. The binding of boroxole with monosaccharides was characterized by Wang's competitive binding assay. Alizarin red S (ARS)–PBOx complexes were formulated in PBS/dioxane mixture (9:1 v/v ratio), which showed the absorption maxima at 425 nm. The addition of monosaccharides shifted the absorption maxima to 520 nm indicating the replacement of boroxole-bound ARS with sugars. The association constant K_a of PBOx was measured by the decrease in fluorescence emission of PBOx–ARS complexes, due to the replacement of ARS with monosaccharides. The sugar-responsive behavior of PBOx at neutral pH makes them ideal candidates for the formulation of stimuli-responsive carriers for the delivery of insulin [47].

Quantum dot (QD)–protein/receptor bioconjugates were characterized for their fluorescence resonance energy transfer (FRET) ability in recognition-based sensing [50]. The conjugates were prepared by electrostatic interactions of maltose-binding protein (a variant containing pentahistidine segment

at its C-terminus (MBP-5HIS)) on anionic CdSe-ZnS core–shell QDs. The number of proteins per QD were calculated, and there were 10 proteins/560 nm emitting QDs. These protein-modified QDs showed enhanced photoluminescence (PL) as compared to unconjugated analogs (Figure 18.9) [50]. The QD–MBP sugar-sensing nanoassemblies were modified with displaceable β-CD-QSY9 at sugar binding pocket of MBP, as shown in Figure 18.9a. The overlap between QD-emission spectrum and β-CD-QSY9 excitation spectrum results in an efficient quenching (∼50%) of QD conjugates, hence making a FRET system; see Figure 18.9b. The high luminescent properties of QD560 in the presence of 20 MBP (∼300%) and their interaction with 1 μM β-CD-QSY9 are shown in Figure 18.9c. MBP has similar binding affinities for β-CD-QSY9 and maltose, hence titration of maltose with bioconjugate displaces β-CD-QSY9 quencher, and increases the PL of QDs in concentration-dependent manner (Figure 18.9d). The amount of maltose required to displace β-CD-QSY9 was determined quantitatively and apparent dissociation constant K_{app} for maltose was calculated (6.9 ± 0.2 μM) (Figure 18.9e). This QD nanoassembly was further tested for a variety of sugars for their FRET response. It was found that only sugars with α-1,4-glycosidic linkages elicit FRET response [50].

18.4.2   Antigen Responsive

Antigen–antibody interactions are probably the most specific responses associated with immune response of the body. The binding of antigens with antibodies occur via a number of interactions such as noncovalent and hydrophobic interactions. These specific interactions of antigens with antibodies have led to the production of antigen-responsive materials, containing physically entrapped or chemically conjugated network of antibodies. These polymers or nanomaterials are then investigated for the delivery of biomolecules [1, 51, 52]. This antigen-responsive characteristic of nanoparticles has also been employed to obtain targeting properties. Paclitaxel (PTX) was linked to epidermal growth factor receptor (EGFR) specific monoclonal antibody (mAb-CC25) with either hydrophilic (PEG or PAMAM) linkers or hydrophobic short chain linkers such as succinic acid and glutaric acid [53]. EGF is a transmembrane protein of tyrosine-kinase family, and EGFRs are overexpressed in a number of human malignancies such as ovarian, head, and neck cancers [54–58]. The linear or branched antibody–drug conjugates of varying molecular weights and antibody:PTX ratios were prepared and were studied for their physiochemical properties. It was found that for linear conjugates, regardless of the type of linker used, conjugates with PTX:CC25 > 3–5 lost their solubility and were precipitated out of solution. The incorporation of linker in the middle of PTX and mAb leads to loss of solubility beyond 3-PTX substitution number. However, use of hydrophilic branched linkers as a pendant

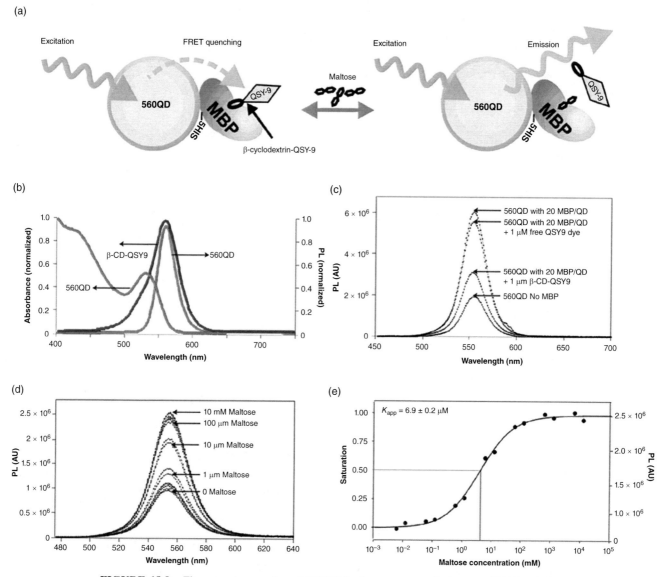

FIGURE 18.9 Fluorescent properties of 560QD in the presence and absence of MBPs, β-CD-QSY9, and maltose. (a) A depiction of FRET system, (b) spectral properties of 560QDs, and β-CD-QSY9, samples were excited at 400 nm and emission spectra were recorded from 420 to 750 nm, (c) FRET quenching of 560QD/MBP in the presence of β-CD-QSY9, (d) maltose sensing of Qd/MBP system, (e) determination of apparent dissociation constant of maltose, by assuming 10% and 90% saturation into a sensing range of 500 nm to approximately 100 μM maltose [50].

moiety was found to effectively improve the solubility of the conjugates. However, no effect on immunoreactivity of conjugates, as a function of number of mAb-CC25 or cross-linker type was observed [53].

Wheat germ agglutinin (WGA) is a plant lectin which possesses high muco-adhesive properties [59]. Insulin-loaded methacrylic acid and PEG-based hydrogels (PMAA-g-PEG) were surface functionalized with WGA to increase the local concentration of drug and bioavailability [60]. The release profile of insulin from WGA-modified hydrogel (WGA-(PMAA-g-PEG)) and control hydrogel (PMAA-g-PEG) was

compared at different pH values. It was found that both hydrogels showed limited release (~10%) of insulin at acidic pH, indicating that bioconjugates can withstand the acidic conditions of the stomach. However, a significant release of insulin from bioconjugates occurred at pH = 7. The specific binding of WGA with pig gastric mucin (PGM) treated microplates was tested and compared with control hydrogels, under physiological conditions. WGA-(PMAA-g-PEG) showed improved adhesion (~33%) at 1 mg/mL concentration in PBS at pH = 7.4, as compared to (PMAA-g-PEG) hydrogels (~15% adhesion). Similarly the titration of competitive carbohydrates to PGM-functionalized plate in the presence of WGA-(PMAA-g-PEG) or (PMAA-g-PEG) hydrogels significantly reduced the interactions of former with the plate. Furthermore, interactions of both hydrogels with Caco-2 cells were studied. Caco-2 contains glycocalyx layer; WGA-modified hydrogels specifically interact with *N*-acetyl glucosamine and sialic acid residues of glycocalyx. WGA-(PMAA-g-PEG) hydrogels showed significant binding ability with Caco-2 cells (39%) as compared to control hydrogels which showed limited binding ability (22%) to Caco-2 cells [60].

18.4.3   Glutathione Responsive

GSH/GSSG is a major redox-responsive system in mammalian cells, which determine their antioxidative capacity. This redox-responsive system has been implied by researchers to effectively deliver and unpack cargo material in the cytosol of the targeted cell. The significant differences in intracellular and extracellular levels of GSH make GSH-responsive vectors appealing for gene and drug delivery applications [61, 62]. Poly(amidoamine)-*N*-acetyl-L-cysteine (NAC) (PAMAM-s-s-NAC) conjugates containing cleavable disulfide linkages were designed for intracellular release of NAC in response to GSH levels [63]. GSH levels are significantly higher in intracellular compartments (cytosol; 1–10 mM), as compared to extracellular compartments (plasma; 0.2 μM) [64]. The release of NAC from conjugates occurs in the presence of high levels of GSH via a disulfide exchange reaction. Navath et al. studied the release of NAC at high (intracellular) and low levels (extracellular) of GSH by reverse-phase high permeation liquid chromatography (RP-HPLC) analysis and UV absorbance at 220 nm [63]. The dendrimer–NAC conjugates were found to be highly stable at 0.2 μM concentration, a limited amount of NAC (1%) was released within the first hour of incubation period. The remaining conjugates stayed intact even after 17 hours of incubation period, due to the depletion of GSH from solution. In contrast, a rapid release of NAC from conjugates occurred in the presence of 10 mM GSH. PAMAM-s-s-NAC conjugates released 47% of free NAC in solution within 1 hour. The analysis of products and by-products formed during the reaction showed that NAC-GS-S-G was formed in solution (19%) after 1 hour. The extent of NAC release did not change significantly after 1 hour of incubation time. However, no NAC–NAC conjugates were formed during the reaction, possibly due to the excess amount of GSH in solution. The reaction products were compared with known reaction species for their elution times. Free NAC and GSH has elution time of 4.7 minutes and 3.8 minutes respectively, as shown in chromatogram (Figure 18.10) [63]. The oxidized form of GSH appears at 3.9 minutes, whereas oxidized form of NAC (NAC–NAC) appears at 8.3 minutes. NAC–NAC molecules are more hydrophobic than GSSG and NAC alone, hence

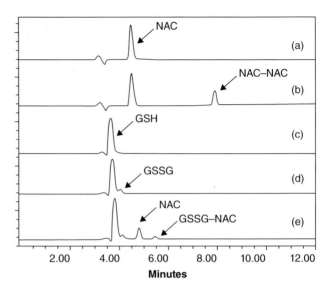

FIGURE 18.10   Reverse-phase (RP)-HPLC, UV absorbance chromatograms obtained at 220 nm (a) NAC, (b) NAC and NAC-NAC, (c) GSH, (d) GSH and GSSG, (e) GSH, GSSG, NAC, and GSSG–NAC [63].

show longer elution times on chromatogram. The elution of GSSG–NAC takes longer time than GSH and NAC (5.3 min); this indicates formation of disulfide bond between NAC and GSH, and reduces the hydrophilicity of the reactants [63].

GSHs are also used as stabilizers during the synthesis of photoluminescent nanomaterials, called quantum dots (QDs) [65, 66]. Li et al. have prepared cystamine-protected cationic QDs and complexed them with DNA through electrostatic interactions [67]. These DNA–QD complexes showed inhibitory gene expression in the absence of GSH. However, the addition of GSH at intracellular levels led to release of DNA, and superior ability to express the gene in HEK 293T cells. The release of DNA from complexes in the presence of GSH was studied by dye exclusion assay. Dye displacement assay detected the DNA-binding affinity of polymers/nanomaterials by a simple nondestructive way. GeneFinder dye was used for this purpose, as shown in Figure 18.11 [67]. Point "a" represents background fluorescence of tris-buffer. Point "b" shows the fluorescence of GeneFinder in the presence of tris-buffer alone. The addition of 200 ng of DNA causes significant increase in fluorescence as shown by point "c." The step-by-step addition of cationic QDs extrude the inserted dye from DNA coils, hence decreasing the fluorescence of the system, as shown in point "d." The addition of GSH in cationic-DNA QD system releases DNA from the complexes, hence restoring the fluorescence intensity as shown in point "e." The addition of low concentration of GSH (10 μM) releases only 10% of pDNA, the increase in concentration of GSH (1 mM, similar to intracellular environment) causes 79% release of plasmid in solution [67].

This dye displacement method demonstrated that GSH has the ability to release DNA from QD–DNA complexes in

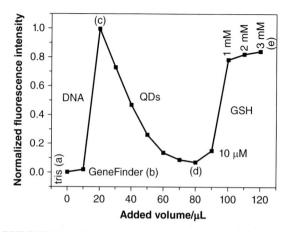

FIGURE 18.11 Dye exclusion assay showing the relative fluorescence of DNA, complexation of DNA at different concentrations on QDs and interactions of QD–DNA complexes with varying levels of GSH [67].

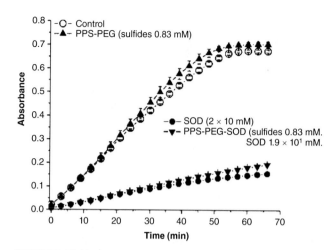

FIGURE 18.12 Formation of formazan crystals due to oxidation of MTS by SOD alone and SOD-PPS-PEG conjugates. The absorbance was read at 490 nm [72].

a concentration-dependent manner. The negatively charged GSH has strong affinity for CdTe core; hence it displaces DNA by counteracting positive charge of QDs, and releasing DNA in solution. This speculation was further affirmed by ζ-potential measurements. The QD–DNA complexes show net charge of $+26 \pm 4$ mV, the addition of 1 mM GSH reverse the cationic charge of complexes, giving net charge of -12 ± 2 mV. The interactions of GSH with QD–DNA complexes were also characterized by EMSA, TEM, and XPS, all of these evidences confirmed the release of DNA in the presence of GSH from QD–DNA complexes [67].

18.4.4 Hydrogen Peroxide Responsive

Reactive oxygen species (ROS) such as hydrogen peroxide (H_2O_2), superoxide anions, and hydroxyl radicals are by-products of cell metabolism, and are hazardous agents as they can perform significant cellular damage upon intracellular accumulation [68–71]. The measurement of H_2O_2 in patient's body is used as a diagnostic marker for a variety of pathological states. Oxidation-responsive nanoparticles are promising tools for the detection of H_2O_2 in patient's body. Polysulfide-containing micelles such as (poly-propylsulfide-*b*-PEG) (PPS-*b*-PEG) are studied for their morphological changes, which occur in the presence of H_2O_2 [72]. The reaction of PPS with oxidants produce hydrophilic molecules, hence dissolution of PPS-*b*-PEG micelles in water occurs [73]. These morphological changes in micellar conformation in the presence of H_2O_2 are monitored by using Nile red (NR) as a probe. Superoxide dismutase (SOD) is a ubiquitous enzyme involved in the defense of prokaryotic and eukaryotic organisms [74–77]. SOD-PPS-*b*-PEG conjugates were prepared and were characterized for their superoxide scavenging ability. In contrast to PPS-*b*-PEG, PPS-*b*-PEG-SOD

micelles show significant superoxide scavenging ability, as shown by their capacity to inhibit the oxidation of MTS in Figure 18.12 [72].

The high molar activity of free SOD as compared to SOD micelles may be due to the steric hindrance or partial denaturation of the enzyme during conjugation. The activity of conjugated enzyme was quantified by measuring the rate of WST-1 formazan production, as a function of molar concentration of free or conjugated SOD. SOD micelles showed reduced activity (by magnitude of order 1) as compared to free SOD. The Nile red fluorescence assay showed that SOD conjugates caused quicker oxidation of PPS, as compared to unconjugated analogs, as activity of SOD increases the local concentration of H_2O_2 hence causing rapid dissolution of micelles in water [72].

Electrochemical sensors are another type of convenient tools constructed for H_2O_2 detection [77]. The performance of biosensors depends on the process used for the immobilization of enzyme, as well as on the matrix used for immobilization. Sol–gel technology is an excellent tool to produce biosensors of high performance. Ferrocene monocarboxylic acid/bovine serum albumin (FMC-BSA) conjugate-doped multiwalled carbon nanotubes (MWNTs)/ormosil conjugates were studied for the entrapment of horseradish peroxidase (HRP). BSA is an inert biomolecule that is used to entrap ferrocene mediators, hence improving the signal intensity. The surface of electrode was analyzed by SEM to ensure the homogeneity and smoothness of the surface. The cyclic voltammogram (CV) of electrode showed some redox peaks due to the oxidation and reduction of immobilized ferrocene. Moreover, shape of the CV of biosensor was better than MWNTS alone, due to the close interaction of MWNT with mediators. The continuous cyclic scans of voltammeter confirmed the stability of biosensor. The addition of H_2O_2

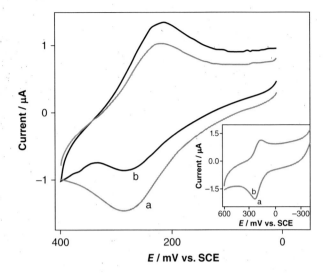

FIGURE 18.13   Cyclic voltammogram of biosensors in the presence and absence of HRP in PBS at pH = 6.8, and 100 mV/s [77].

(6 mM) in HRP-modified biosensor caused significant differences in CV response as shown in Figure 18.13. The reduction peak current of biosensor increased, while oxidation peak current decreased. In contrast, no change was observed for biosensors (BSA-FMC-MWNTs)/ormosil constructed in the absence of HRP (inset in Figure 18.13), indicating the reduction of H_2O_2 in the presence of HRP. The amperometric response of biosensor was evaluated by the application of potential from $+600$ mV to $+100$ mV. The reduction of H_2O_2 occurred at $+400$ mV. The decrease in applied potential from $+400$ mV to $+220$ mV increases the current due to fast reduction of mediators at low potential values, the response plateau at $+220$ mV. The concentration of HRP was an important factor in amperometric response. The increase in concentration of HRP up to 0.55 IU on biosensor surface increases the catalytic current. Further increase in HRP concentration on biosensor decreases the current by reducing the conductivity of the composite or by restricting the diffusion of substrate to immobilized enzyme. Under optimum conditions, addition of H_2O_2 increases the amperometric response to a maximum value within 5 seconds. The decrease in current at low H_2O_2 concentrations is due to fast catalysis ability of HRP, which leads to uneven concentrations on electrode surface. The increase in H_2O_2 concentration leads to a linear increase in amperometric response, in the range of 0.02–4 mM. The detection limit of H_2O_2 was found to be 5 μM [77].

REFERENCES

1. Shakya AK, Sami H, Srivastva A, Kumar A. *Prog Polym Sci* 2010;35:495–486.

2. Roy D, Cambre JN, Sumerlin BS. *Prog Polym Sci* 2010;35:278–301.

3. Ganta S, Devalapally H, Shahiwala A, Amiji N. *J Control Release* 2008;126:187–204.

4. Shimoboji T, Larenas E, Fowler T, Hoffman AS, Stayton PS. *Bioconjugate Chem* 2003;14:517–525.

5. Ding Z, Chen G, Hoffman AS. *J Biomed Mater Res* 1998;39:498–505.

6. Gupta MN. *Biotechnol Appl Biochem* 1991;14:1–11.

7. Guisan JM. *Enzyme Microb Technol* 1988;10:375–382.

8. Chen JP, Chu DH, Sun YM. *J Chem Technol Biotechnol.* 1997;69:421–428.

9. Sonvico F, Mornet S, Vasseur S, Dubernet C, Jaillard D, Degrouard J, Hoebeke J, Duguet E, Colombo P, Couvreur P. *Bioconjugate Chem* 2005;16:1181–1188.

10. Datta A, Hooker JM, Botta M, Francis MB, Aime S, Raymond KN. *J Am Chem Soc* 2008;130:2546–2552.

11. Zarabi B, Nan A, Zhou J, Gullapalli R, Ghandehari H. *Mol Pharmaceutics* 2006;3:550–557.

12. Rudovsky J, Botta M, Hermann P, Hardcastle KI, Lukes I, Aime S. *Bioconjugate Chem* 2006;17:975–987.

13. Aime S, Botta M, Terreno E. *Gd(III)-based contrast agents for MRI.* San Diego, CA: Elsevier; 2005. Vol. 57, pp 173–237.

14. Doble DMJ, Melchior M, O'Sullivan B, Siering C, Xu JD, Pierre VC, Raymond KN. *Inorg Chem* 2003;42:4930–4937.

15. Pierre VC, Botta M, Aime S, Raymond KN. *Inorg Chem* 2006;45:8355–8364.

16. Pierre VC, Melchior M, Doble DMJ, Raymond KN. *Inorg Chem* 2004;43:8520–8525.

17. Zhao Y. *Macromolecules* 2012;45:3647–3657.

18. Rapoport N. *Prog Polym Sci* 2007;32:962–990.

19. Hribar KC, Lee MH, Lee D, Burdick JA. *ACS Nano* 2011;5:2948–2956.

20. Chen C-C, Lin Y-P, Wang C-W, Tzeng H-C, Wu C-H, Chen Y-C, Chen C-P, Chen L-C, Wu Y-C. *J Am Chem Soc* 2006;128:3709–3715.

21. Saha K, Agasti SS, Kim C, Li X, Rotello VM. *Chem Rev* 2012;112:2739–2779.

22. Park H-G, Joo JH, Kim H-G, Lee J-S. *J Phys Chem C* 2012;16:2278–2284.

23. Yan B, Boyer J-C, Branda NR, Zhao Y. *J Am Chem Soc.* 2011;133:19714–19717.

24. Wang F, Liu X. *Chem Soc Rev* 2009;38:976–989.

25. Crownover EF, Convertine AJ, Stayton PS. *Polym Chem* 2011;2:1499–1504.

26. Wan X, Zhang G, Ge Z, Narain R, Liu S. *Chem Asian J* 2011;6:2835–2845.

27. Sawant RM, Hurley JP, Salmaso S, Kale A, Tolcheva E, Levchenko TS, Torchilin VP. *Bioconjugate Chem* 2006;17:943–949.

28. Ulbrich K, Subr V, Strohalm J, Polcova D, Jelinkova M, Rihova B. *J Control Release* 2000;64:63–79.

29. Kang JS, Deluca PP, Lee KC. *Expert Opin Emerg Drugs* 2009;14:363–380.

30. Veronese FM, Mero A. *Biodrugs* 2008;22:315–329.

31. Pinholt C, Bukrinsky JT, Hostrup S, Frokjaer S, Norde W, Jorgensen L. *Eur J Pharm Biopharm* 2011;77:139–147.

32. Takakura Y, Kitajima M, Matsumoto S, Hashida M, Sezaki H. *Intl J Pharm* 1987;37:135–143.

33. Takakura Y, Matsumoto S, Hashida M, Sezaki H. *J Control Release* 1989;10:97–105.

34. Villalonga R, Villalonga ML, Gomez L. *J Mol Catal B Enzym* 2000;10:483–490.

35. Gomez L, Villalonga R. *Biotechnol Lett* 2000;22:1191–1195.

36. Gomez L, Ramirez HL, Villalonga R. *Biotechnol Lett* 2000;22:347–350.

37. Fagain CO. *Enzyme Microb Technol* 2003;13:137–149.

38. Arica MY, Yavuz H, Patir S, Denizili A. *J Mol Catal B Enzym* 2000;11:127–138.

39. Jin R, Wu G, Li Z, Mirkin CA, Schatz GC. *Am Chem Soc* 2003;125:1643–1654.

40. Storhoff JJ, Lazarides AA, Mucic RC, Mirkin CA, Letsinger RL, Schatz GC. *J Am Chem Soc* 2000;122:4640–4650.

41. Storhoff JJ, Mirkin CA. *Chem Rev* 1999;99:1849–1862.

42. Park SY, Lytton-Jean AKR, Lee B, Weigand S, Schatz GC, Mirkin CA. *Nature* 2008;451:553–556.

43. Oh J-H, Lee J-S. *Chem Commun* 2010;46:6382–6384.

44. Gibbs-Davis JM, Schatz GC, Nguyen ST. *J Am Chem Soc* 2007;129:15535–15540.

45. Wu Q, Wang L, Yu H, Wang J, Chen Z. *Chem Rev* 2011;111:7855–7875.

46. Springsteen G, Wang B. *Tetrahedron* 2002;58:5291–5300.

47. Kim H, Kang YJ, Kang S, Kim KT. *J Am Chem Soc* 2012;134:4030–4033.

48. Nishiyabu R, Kubo Y, James TD, Fossey JS. *Chem Commun* 2011;47:1106–1123.

49. Cambre JN; Sumerlin BS. *Polymer* 2011;52:4631–4643.

50. Medintz IL, Clapp AR, Mattoussi H, Goldman ER, Fisher B, Mauro JM. *Nature Mat* 2003;2:630–638.

51. Lele BS, Murata H, Matyjaszewski K, Russell AJ. *Biomacromolecules* 2005;6:3380–3387.

52. Yamamoto Y, Tsutsumi Y, Yoshioka Y, Nishibata T, Kobayashi K, Okamoto T, Mukai Y, Shimizu T, Nakagawa S, Nagata S, Mayumi T. *Nat Biotechnol* 2003;21:546–552.

53. Quiles S, Raisch KP, Sanford LL, Bonner JA, Safavy A. *J Med Chem* 2010;53:586–594.

54. Carpenter G, Cohen S. *Annu Rev Biochem* 1979;48:193–216.

55. Ma WW, Adjei AA. *CA Cancer J Clin* 2009;59:111–137.

56. Singletary SE, Baker FL, Spitzer G, Tucker SL, Tomasovic B, Brock WA, Ajani JA, Kelly AM. *Cancer Res* 1987;47:403–406.

57. Real PJ, Benito A, Cuevas J, Berciano MT, de Juan A, Coffer P, Gomez-Roman J, Lafarga M, Lopez-Vega JM, Fernandez-Luna JL. *Cancer Res* 2005;65:8151–8157.

58. Reilly RM, Chen P, Wang J, Scollard D, Cameron R, Vallis KA. *J Nucl Med* 2006;47:1023–1031.

59. Meng FH, Hennink WE, Zhong ZY. *Biomaterials* 2009;30:2180–2198.

60. Wood KM, Stone GM, Peppas NA. *Biomacromolecules* 2008;9:1293–1298.

61. Cheng R, Feng F, Meng F, Deng C, Feijn J, Zhong Z. *J Control Release* 2011;152:2–12.

62. Schafer FQ, Buettner GR. *Free Radic Biol Med* 2001;30:1191–1212.

63. Navath RS, Kurtoglu YE, Wang B, Kannan S, Romero R, Kannan RM. *Bioconjugate Chem* 2008;19:2446–2455.

64. Hong R, Han G, Fernandez MJ, Kim JB, Forbes SN, Rotello MV. *J Am Chem Soc* 2006;128:1078–1079.

65. Zheng YG, Gao SJ, Ying JY. *Adv Mater* 2007;19:376–380.

66. Qian HF, Dong CQ, Weng JF, Ren JC. *Small* 2006;2:747–751.

67. Li D, Li G, Guo W, Li P, Wang E, Wang J. *Biomaterials* 2008;29:2276–2282.

68. Griendling KK, Sorescu D, Ushio-Fukai M. *Circ Res* 2000;86:494–501.

69. Halliwell B. *J Neurochem* 2006;97:1634–1658.

70. Madamanchi NR, Vendrov A, Runge MS. *Arterioscler Thromb Vasc Biol* 2005;25:29–38.

71. Pavlick KP, Laroux FS, Fuseler J, Wolf RE, Gray L, Hoffman J, Grisham MB. *Free Radic Biol Med.* 2002;33:311–322.

72. Hu P, Tirelli N. *Bioconjugate Chem* 2012;23:438–449.

73. Vo CD, Kilcher G, Tirelli N. *Macromol Rapid Commun* 2009;30:299–315.

74. Jolly SR, Kane WJ, Bailie MB, Abrams GD, Lucchesi BR. *Circ Res* 1984;54:277–285.

75. Mates JM, Perez-Gomez C, De Castro IN. *Clin Biochem* 1999;32:595–603.

76. Orr WC, Sohal RS. *Science* 1994;263:1128–1130.

77. Tripathi VS, Kandimalla VB, Ju H. *Biosens Bioelectron* 2006;21:1529–1535.

19

APPLICATIONS OF BIOCONJUGATES

MARYA AHMED AND RAVIN NARAIN

Department of Chemical and Materials Engineering, Alberta Glycomics Centre, University of Alberta, Edmonton, AB, Canada

19.1 INTRODUCTION

Bioconjugates of synthetic polymers, natural polymers, hydrogels, and nanoparticles (NPs) have received a lot of attention in the decade [1–10]. These bioconjugates have been implied for a variety of biological applications, including drug and gene delivery applications, biological assays, imaging, and biosensors [1–10]. The success of these bioconjugates in research laboratories, as compared to their precursor biomolecules, have further encouraged for their use in industrial applications [11]. Some of the gene and drug delivery agents are now in clinical trials. The chapter describes the applications of these bioconjugates in detail.

19.2 LABELING TAGS AND PROBES

DNA microarrays are set of known oligodinucleotide (ODN) sequences, immobilized on a solid support, on a precisely defined location. These DNA microarrays can selectively hybridize with a complementary ODN sequence. The hybridization of ODN with its complementary set produces a signal which can be detected either by using fluorescence probes or by an enzymatic reaction (which can convert a substrate into a fluorescent product) [12]. The sensitivity of this DNA-microarray technology can be enhanced by immobilizing the polymer–ODN conjugates on the surface of a solid support. De Lamber et al. studied the role of polymers for improving the sensitivity of ODN microarrays. The polymer–ODN conjugates of poly(tert-butylacrylamide-*b*-(*N*-acryloylmorpholine-*N*-acryloxysuccinimide)) (p(TBAm-*b*-(NAM/NAS)) with dT25-ODN are prepared and are studied for DNA hybridization assay in a 96 well microtiter plate,

as shown in Figure 19.1. The copolymer conjugates are functionalized in each well by 16 different spots, followed by the deposition of biological species solution by using droplet spotting technology. The presence of specific DNA sequence (Cy-3′-labeled dA-25) is detected by the bright spot, due to the release of Cy-3′ dye. The experiments reveal that sensitivity of microhybridization assay can be amplified by five times by using polymer-conjugated ODNs, as compared to ODN alone [12].

Fluorophore-labeled proteins play an important role in many vital cellular processes including protein dynamics, localization, and interactions with cell matrix. The specific labeling of proteins in their native environment has been done using various treatments. One possible approach is labelling of enzymes with smart polymers, which can undergo self-immolation in the presence of external stimulus [13]. The self-immolative polymers (SIPs) have also been used for signal amplification and drug release. The disassembly of SIPs upon reaction in aqueous solution releases a fluorescent probe, which is azaquinone-methide intermediate. If the external stimulus is an enzyme, SIP disassembly can produce fluorescent enzymes upon the entrapment of azaquinone-methide intermediates by nucleophilic residues of protein. The mechanism of enzyme labeling using SIPs is depicted in Figure 19.2. The potential of SIPs, as labeling probes is studied by using water-soluble self-immolative trimers, equipped with phenylacetamide trigger, the trigger can be cleaved by penicillin G amidase (PGA). The labeling process of the enzyme is monitored by spectral changes as a function of time. The labeled enzyme is dialyzed and is characterized by fluorescence spectroscopy. To further explore the scope of SIPs, catalytic antibody (Ab38C2) is fluorescently labeled using SIPs specific for Ab38C2. The labeled antibody and

Chemistry of Bioconjugates: Synthesis, Characterization, and Biomedical Applications, First Edition. Edited by Ravin Narain.

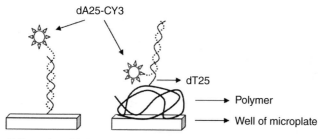

CY3 : carbonyl-indocyanine

FIGURE 19.1 Hybridization of target Cy-3′ labeled ODN on dT25 alone or dT-25-polymer conjugates, on a microarray system [12].

PGA are then tested for their physiological activity using chromogenic substrate, *p*-nitroaniline [13].

19.3 BIOLOGICAL ASSAYS

The conjugation of stimuli-responsive polymers with proteins and vitamins can be detected by using various covalent and noncovalent approaches and assays [14]. Biotin–avidin interaction are one of the strongest noncovalent interactions ($K_d = 10^{-15}$) in the universe [15]. These biotin–avidin interactions have been exploited in the form of biological assays to determine the extent of biotinylation of a polymer. For example, thermoresponsive biotinylated polymers are subjected to HABA/biotin assay to show the availability of biotin on the surface of polymers. Poly-*N*-isopropylacrylamide

(PNIPAM) and biotin conjugates are injected in aqueous solution in Dot-lab on the surface of avidin-coated sensorchip. The response obtained due to the binding of biotinylated polymer on the surface of avidin chip is recorded as a function of time. It is concluded from the results that diblock copolymer of PNIPAM (PNIPAM–biotin–PNIPAM) shows lower response than PNIPAM–biotin conjugate, possibly due to the steric hindrance of biotinylated polymer chains, in the case of diblock architecture [14].

The immobilization of enzymes on the surface of nanomaterials is an important tool to produce functional materials and devices. However, the retention of activity of enzymes on the surface of nanomaterials is a major challenge. A thin layer of enzyme–NP conjugates with high surface to volume ratio can provide ideal geometry to construct biocatalysts for various applications [16]. The assembly of enzyme–NP conjugates on oil–water interface emulsions has been used as a technique to produce catalytic microcapsules. The cationic gold NPs immobilized with β-galactosidase enzyme are decorated on oil–water interface to make microcapsules. These microcapsules containing β-galactosidase enzyme are then studied for their enzymatic activity using chlorophenol red β-D-galactopyranoside (CPRG) substrate (Figure 19.3a). The hydrolysis of CPRG produces red color in solution, which is quantified to measure the enzymatic activity of microcapsules as compared to β-galactosidase alone, as shown in Figure 19.3b. It is found that immobilization of enzymes on NP surface and their orientation in the form of microcapsules maintained their enzymatic activity. The enzymatic activity of microcapsules was determined and was 76% as compared to β-galactosidase [16].

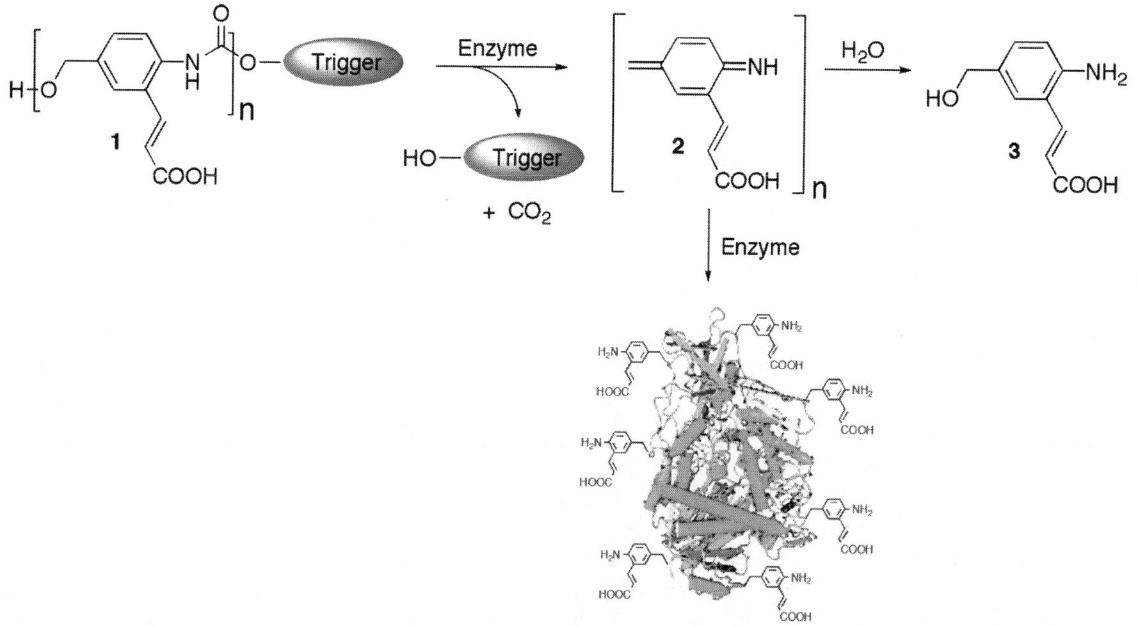

FIGURE 19.2 Labeling of enzymes using self-immolative polymers (SIPs) [13].

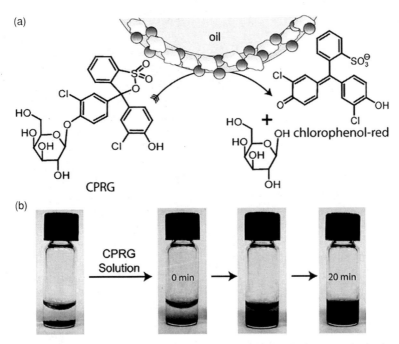

FIGURE 19.3 (a) Enzymatic reaction of β-galactosidase decorated microcapsules in the presence of chlorophenol red β-ᴅ-galactopyranoside (CPRG), as a substrate, (b) development of red color solution as a function of time, due to the hydrolysis of CPRG [16].

19.4 IMAGING *in vivo*

In vivo imaging plays a critical role in detection of diseases and surgery. The clinical trials of therapeutics are greatly dependent on imaging data to provide a noninvasive measure of the effect of therapy. *In vivo* imaging of bioconjugated systems can be done by using a variety of methods such as optical imaging, single photon emission computed tomography (SPECT), positron emission tomography (PET), fluorescent microscopy, magnetic resonance imaging (MRI), and ultrasound [3]. The imaging technologies are extensively applied to the field of oncology in an effort to develop new analysis techniques and quantitative imaging software tools to understand the cancer-related signal pathways [4]. Near-infrared (NIR) fluorescent carbocyanine probes are ideal tool for optical imaging, due to their high selectivity, sensitivity, and stability. The excitation of NIR probes causes low autofluorescence and deep tissue imaging, as the light in 500–700 nm range can propagate several centimeters in depth in tissues. The multivalent NIR conjugates of glucosamine dendrimers are prepared and are studied for their *in vivo* distribution in proliferating pancreatic tumor bearing mice. The *in vivo* and *ex vivo* distribution of the fluorescent molecules is determined by NIR fluorescence imaging, using an excitation wavelength of 780 nm. The localization of carbohydrate-based fluorescent dendrimers in tumor site is obtained 24 hours postinjection, possibly due to the noninvasive passive targeting of carbohydrates-based vectors at tumor site [17].

Radiolabeled amphiphilic polymer-based nanocapsules "polymersomes" are studied for their *in vivo* biodistribution, as a function of vesicle size in Balb/C mice using SPECT/CT imaging. ^{111}In-radiolabeled PEG-butadiene polymersomes of 90 nm showed longer circulation times and longer blood half-life, as compared to polymersomes of 120 nm (Figure 19.4). The polymersomes of 120 nm or larger sizes show rapid renal clearance with blood half-life less than 4 hours [18].

Semiconductor NPs of nanometer size possess unique optical and electronic properties with tunable fluorescent spectra ranging from visible to NIR wavelengths [19]. The broad excitation profiles, narrow emission spectra, and photostability of quantum dots (QDs) make them ideal candidates for *in vivo* imaging. Bioconjugated QDs of 10–15 nm are studied for human prostate cancer imaging in mice. Polymer-encapsulated bioconjugated QDs are engineered for *in vivo* cancer targeting and imaging. Prostate-specific membrane antigen (PSMA)-conjugated PEGylated QDs are compared with COOH-coated QDs and PEGlyated QDs for their *in vivo* stability and active targeting to prostate tumor in mice. It is found that high nonspecific uptake of COOH-coated QDs and PEGylated QDs occur in liver and spleen of treated mice. Although, the passive targeting of PEGylated QDs at tumor site is achieved over a time period of 24 hours, the active targeting of PSMA-PEGylated QDs to tumor site is observed within 2 hours of injection time. The toxicity of QDs is found to be a function of stability of polymer layer on

Biodistribution and SPECT imaging of polymersomes

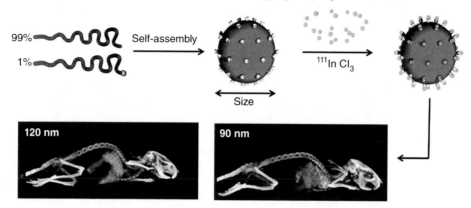

FIGURE 19.4 ^{111}In-radiolabeled PEGylated polymersome self-assembly and their biodistribution in mice [18].

the surface of QDs, the stable polymer layer on the surface of QDs eliminates the toxicity of QDs by increasing their stability under physiological condition [20].

19.5 BIOSENSORS

The synthesis of novel nanomaterials allow the construction of biosensors with low cost and improved sensing techniques. The light weight, flexibility, corrosion resistance, chemical inertness, and facile processing and handling of polymers make them ideal for biosensor applications. The immobilization of biological moieties (probes) on polymer surface is the key step for the construction of biosensors [21]. DNA–polymer hybrids are used for the detection of DNA probes, using three-component sandwich-type electrochemical detection strategy, as shown in Figure 19.5 [22]. In a typical experiment, DNA (a) is immobilized on the surface of gold electrode. The synthetic target-DNA (a'b') and DNA-polymer hybrid are then incubated with glass electrode for 4 hours at room temperature. In a control experiment, modified glass electrode is incubated with DNA-polymer hybrid in the

absence of target-DNA. Alternating current (AC) voltammograms are obtained for each sample to detect the presence of target-DNA at very low concentration (100 pM) [22].

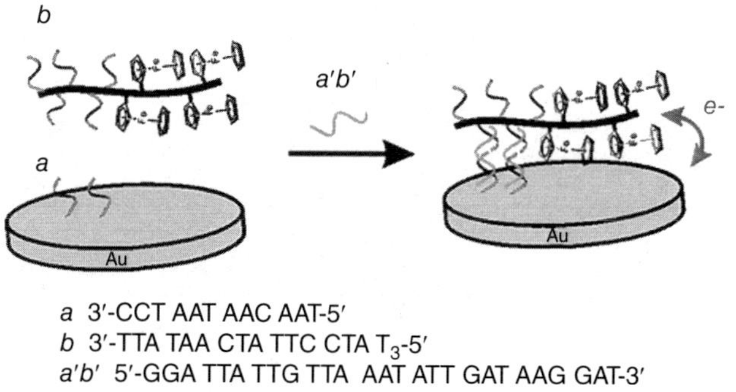

Aflatoxin B (AfB1) is a highly toxic, immunogenic hepatotoxin which occurs in nuts, rice, and oil seed [23]. The high levels of AfB1 in body can produce liver cirrhosis and liver carcinoma. AfB1 has been classified as a class I toxin by the international agency for research on cancer (IARC). The conventional methods for the detection of AfB1, such as thin layer chromatography, are expensive and time consuming. NPs have attracted a lot of attention for biosensing applications, due to their large surface-to-volume ratio, facile biofunctionalization, selectivity, shelf life, and fast electron transfer ability. Ring-like nickel NPs (RnNi) of 10–20 nm in diameter are prepared and are self-assembled on DMSO-functionalized indium tin oxide (ITO) substrate for the covalent immobilization of anti-aflatoxin B monoclonal antibodies (a-AfB1). The bioelectrode thus produced (a-AfB1/DMSO/RnNi/ITO) is studied for the electrochemical response as a function of AfB1 concentration. The bioelectrode showed broad detection range of toxin (5–100 ng/dL) along with high sensitivity and shelf life of 60 days [23].

Aptamers are novel nucleic acid structures which are largely used for the development of biosensors for proteins, DNAs, and small molecules [24]. Thrombin-specific aptamers contain binding sites for the recognition of fibrinogen. The sensitive biosensors for the detection of thrombin are prepared by the immobilization of thrombin-specific aptamers on gold NPs. The aptamers are conjugated on gold NPs by three different approaches: (i) aptamers are covalently immobilized (Ap-Im-GNPs); (ii) aptamers are hybridized (Ap-Hy-GNPs); and (iii) aptamers are physically adsorbed (Ap-Ad-GNPs), on the surface of gold NPs, as shown in Figure 19.6 [25].

The aptamer (P1) is either chemically immobilized or physically hybridized on the surface of 10 nm GNPs, as

a 3'-CCT AAT AAC AAT-5'
b 3'-TTA TAA CTA TTC CTA T$_3$-5'
a'b' 5'-GGA TTA TTG TTA AAT ATT GAT AAG GAT-3'

FIGURE 19.5 Polymer–DNA conjugates for the DNA detection technology [22].

(a) Ap-Im-GNPs

(b) Ap-Hy-GNPs

(c) Ap-Ad-GNPs

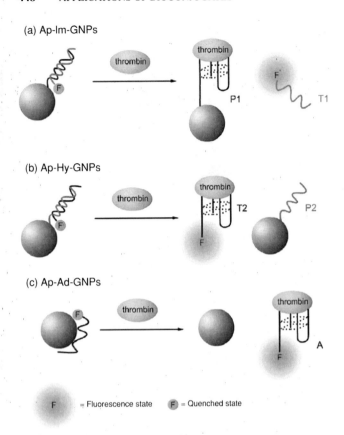

F = Fluorescence state   F = Quenched state

FIGURE 19.6 Thrombin-specific aptamer-conjugated gold nanoparticles prepared by (a) covalent immobilization, (b) hybridization, and (c) physical adsorption, and detection of thrombin [25].

shown in the case of A and B of Figure 19.6. The dye-labeled complementary DNA (T1) is then hybridized with aptamer, so that the fluorescence of the complex is quenched. The addition of thrombin interacts with aptamer causing the dissociation of duplex and the release of fluorescent T1 from GNP surface. The fluorescence intensity from two experiments is compared and it is found that although the two methods show similar fluorescence intensities, the equilibrium constant (K) is much lower for Ap-Im-GNPs than Ap-Hy-GNPs. K is used to measure the affinity of biosensors for different methods. The nonspecific adsorption of nonthiolated DNA on the surface of gold NPs is done and its interactions with thrombin are studied and compared with Ap-Im-GNPs and Ap-Hy-GNPs. It is concluded that covalent immobilization of DNA on GNPs is the optimum approach to produce biosensors for the detection of proteins with high accuracy and sensitivity [25].

19.6 DRUG DELIVERY

Synthetic drug delivery systems represent an excellent example of contribution of scientific development toward human

health and well-being. As compared to the conventional dosage of drugs, drug conjugates offer various advantages, including longer blood circulation time, reduced toxicity, and improved efficacy. The first controlled drug delivery system was introduced in 1997 [26]. Since then, a large number of drug conjugates have been prepared and are studied to treat various disorders. The conjugation of drugs with synthetic polymers or natural polysaccharides, nanomaterials, or their encapsulation in hydrogels is the commonly used approach to obtain controlled release of drugs *in vitro* and *in vivo* [26]. Doxorubicin (DOX) and cisplatin are effective chemotherapeutic agents for the treatment of various cancers [27–29]. The conjugates of these anticancer drugs with a variety of water-soluble polymers are prepared in order to improve the solubility and efficacy of the drug along with decreasing its toxicity [27–29]. DOX-conjugated, polyester dendritic scaffolds are prepared and are studied for its biodistribution *in vivo*. The conjugation of DOX on a polymeric scaffold significantly increases its half-life (72 minutes vs. 8 minutes) in the blood of treated mice. In addition, no significant accumulation of DOX conjugates in vital organs is observed, as compared to DOX alone, which shows high nonspecific accumulation in liver and heart. Hence, conjugation of anticancer drugs with polymeric scaffolds can significantly alter their pharmacokinetics [27]. The encapsulation of DOX in hydrogels is used as another approach to obtain controlled release of the drug *in vitro* and *in vivo*. Hyaluronic acid-based click-gels are synthesized in the presence of DOX and drug release is studied over a period of 10 days. The percentage release of DOX is studied at 37°C by UV spectrophotometer ($\lambda = 486$ nm), as a function of time [30].

DOX-loaded, transferrin-conjugated PEGylated-polycarpolactone Tf-(PEG-PCL) polymersomes are studied as drug delivery carrier to glioma bearing rats (Figure 19.7) [31]. The biodistribution of DOX-loaded polymersome is compared to DOX alone, and it is found that DOX-loaded polymersomes show longer circulation time in blood than DOX alone (24 hours vs. 15 minutes). The ability of polymersomes to surpass blood–brain barrier is also studied

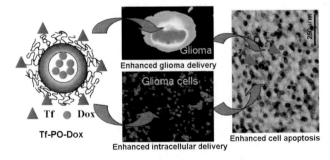

FIGURE 19.7 Uptake of transferring-conjugated, DOX-loaded polymersomes in rat glioma cells and their potential as anticancer agent [31]. For a color version, see the color plate section.

and it is found that transferrin-conjugated polymersomes show high uptake in brain cells as compared to unconjugated polymersomes or free DOX. The antitumor effect of polymersomes in tumor bearing rats show that DOX-loaded, transferrin-conjugated polymersomes significantly reduce the tumor sizes in rats, 14 days posttreatment, as compared to the control groups. The survival time of rats is also significantly increased (from 70 to 100%), upon treatment with transferrin-conjugated, drug-loaded polymersomes, showing the potent antitumor activity of the compound [31].

An important characteristic of tumor tissues is their relatively low pH (pH = 6.8) as compared to the blood stream (pH = 7.4). Anticancer drugs such as DOX usually show reduced activity at acidic pH, hence causing the low efficacy of the drug. pH-responsive micelles are designed to selectively deliver the drug in cancer tissues. The antitumor activity of DOX at acidic pH (pH = 6) is improved by designing pH-sensitive, DOX-loaded p(HEMA)-b-p(His) micelles. These DOX-loaded micelles show high growth inhibition of human colon carcinoma cells (HCT-116) at all studied concentrations of drug, as compared to DOX alone [32].

19.7   TISSUE ENGINEERING

Biomaterials play a critical role in tissue engineering by providing scaffolds and matrices for the growth and differentiation of living cells. The first biologically active scaffold was produced in 1974, since then variety of polymers have been produced and are studied for tissue engineering applications. The ideal properties of a biological scaffold are (i) to provide cell interactions and adhesion with the biomaterial surface, (ii) to allow the perfusion of nutrients and gases to allow cell survival and proliferation, (iii) degradation of biomaterials occurs at the rate constant with tissue growth, (iv) show limited toxicity and inflammation. The natural polymers, synthetic polymers, ceramics, and nanofibers have been studied for tissue engineering applications [33]. Liver transplantation is an established therapy for liver failure or for inherited or acquired liver diseases. However, the clinical trials show limited efficacy of the approach due to a number of factors. The direct injection of hepatocyte suspension into host liver, into spleen, or into pleural cavity show limited efficacies either due to a small number of active liver cells successfully transplanted into liver, or due to low cell survival and cell functioning after transplantation. Tissue engineering provides a feasible solution to introduce a large number of cells into the host body on biodegradable scaffold. Tissue engineering involves the development of biological structures to maintain or to improve the function of certain organs in living system. Poly(L-lactic acid) (PLLA) provides a biodegradable, three-dimensional porous scaffold for the optimum growth and handling of hepatocytes. Hepatocytes are cultured on PLLA matrix in a flow bioreactor. The growing cells in tissue scaffolds are tested for their cell viability and albumin production, as a function of time, to determine the differentiation ability of hepatocytes. The high cell viability and sustained production of albumin indicate that tissue engineering scaffold is ready for implantation after 2 days of culturing. The hepatocytes continued to grow and produce albumin for a period of 6 days, producing spheroids of 80–100 μm in diameters [34].

In another study, incorporation of plasmid DNA into tissue engineering matrices is used as an approach to obtain a large number of transfected cells *in vivo*, at a localized site. Platelet-derived growth factor (PDGF) is a potent factor, which enhances vascularization and tissue repair. PDGF-loaded highly porous, three-dimensional poly(lactide-co-glycolide) (PLG) scaffolds are prepared and are tested for the sustained release of plasmid DNA and protein expression. The efficacies of plasmid-loaded tissue scaffold for sustained gene delivery efficacies and quality of protein production are studied *in vivo*. PGDF-loaded matrices show increased granulation and vascularization of the tissue over a time period of 2–4 weeks. This suggests the prolonged release of plasmid from polymeric scaffold over a long period of time [35].

19.8   GENE DELIVERY

The adverse side effects of viral gene delivery vectors led to the design and evaluation of a variety of nonviral vectors. Since the advent of gene delivery in 1965, a number of synthetic and natural polymers are studied for gene delivery application. The cationic polymers such as polyethyleneimine, poly(L-lysine), and cationic lipoplexes such as lipofectamine are commercially available gene delivery vectors, which are used as standards due to their high gene expression efficacies. In order to reduce the toxicity of cationic vectors, their analogs with nonionic moieties such as poly(ethylene glycol) (PEG) and poly(hydroxypropyl methacrylamide) (PHPMA) have been prepared. However, the modifications of cationic polymers with nonionic moieties are reported to reduce their gene expression. With the advances in the field of polymerization, well-defined cationic polymers and their analogs are prepared without modifying the cationic content of the gene therapy vectors. The libraries of synthetic cationic glycopolymers, peptide-functionalized copolymers, and lipids modified copolymers have been produced and their gene transfection efficacies are studied [36]. Moreover, the modification of cationic vectors with nuclear localization signals and peptides are reported to enhance their nuclear localization ability and hence gene expression [37].

The cationic glycocopolymer of linear (block vs. statistical and *"block–statistical"*) and hyperbranched architecture bearing glucose-derived or galactose-based pendant moieties are studied for their gene expression in different cell lines [38–40]. The gene expressions of copolymers are found to

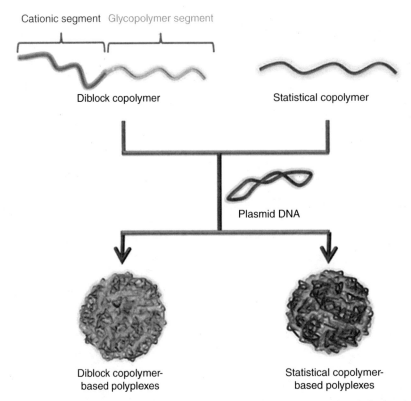

FIGURE 19.8 Formulation of polyplexes using cationic glycopolymers of statistical and diblock architectures and β-galactosidase plasmid DNA [38]. For a color version, see the color plate section.

be dependent on the amine content of the copolymers, their molecular weights, and their architecture. The copolymers of statistical and *"block–statistical"* architectures show high gene expression and low toxicities *in vitro* in the presence and absence of serum proteins. In contrast, their diblock analogs show low gene expression in the presence of serum protein along with high toxicities [38]. The formulation of polyplexes using copolymers of diblock and statistical configurations is depicted in Figure 19.8 [38]. The interaction of hyperbranched cationic glycopolymer-based polyplexes with cell surface receptors is further studied in hepatocytes, as a function of sugar moieties on the surface of polyplexes. It is found that galactose-based hyperbranched cationic glycopolymer-based polyplexes show high cellular uptake and gene expression in hepatocytes, in comparison to glucose-based analogs possibly due to their specific interactions with asialoglycoprotein receptors (ASGPRs) on the surface of HepG2 cells [39].

HPMA copolymers of statistical architecture bearing pendant lysine and histidine residues are studied for their gene delivery efficacies. The lysine-containing copolymers of HPMA (Lys residues 20 mol%) show high transfection efficacies and low toxicities; the gene expression of copolymers is similar to positive control (PEI) [41].

Protein–polymer conjugates bearing synthetic DNA-binding domains are constructed. Different generations of

dendrons are modified with bovine serum albumin (BSA) and class II hydrophobin protein (HFBI) to produce self-assembling protein–dendron conjugates. HFBI is a small surface-active protein from *Trichoderma reesei*, which is known to form structures by self-assembly. HFBI–dendrons are shown to self-assemble into hexagonally ordered arrays onto hydrophobic surfaces. HFBI-conjugated DNA domains are thought to serve as cationic surfactant which can pass the cell membrane and deliver DNA. The gene delivery of protein–dendron conjugates are studied in kidney fibroblast cells (CV1-P) using β-galactosidase expression [42].

19.9 BIOCOMPATIBILITY

Biocompatibility of a material is defined as the ability of the materials to produce suitable host response for a specific application. For example, biocompatibility of blood-contacting materials is related to their thrombotic response. The blood biocompatibility is a critical requirement for the cardiovascular devices, delivery vehicles, and for other foreign materials required to implant inside the body. A variety of synthetic polymers have been prepared and their blood compatibility is studied [43]. Hyperbranched glycerols are studied for their blood compatibility and cytotoxicity. These hyperbranched polymers showed no red blood

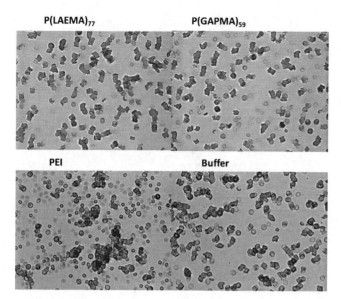

FIGURE 19.9 Interactions of hyperbranched glycopolymers with red blood cells, as compared to positive control (PEI) and negative control buffer [45].

cell aggregation, red blood cell lysis, or complement activation [44]. Hyperbranched glycopolymers of different molecular weights and pendant sugar moieties are also prepared and are studied for their blood biocompatibility [45]. It was found that both glucose- and galactose-based hyperbranched polymers of varying molecular weights (40–80 kDa) show high compatibility toward serum proteins, and do not activate immune response at all studied concentrations. The interaction of glucose-derived and galactose-based hyperbranched glycopolymers with red blood cells lead to *rouleaux* formation, in contrast aggregation and lysis of red blood cells occurs in the presence of cationic polymer (PEI) as shown in Figure 19.9 [45].

Hydrophobically derived hyperbranched polyglycerols are then used as human serum albumin (HSA) substitutes, due to their high biocompatibility during *in vitro* and *in vivo* studies [46].

Biodegradable and biocompatible PEO-*b*-PCL- and PEO-*b*-PLA-based polymersomes are prepared and are used as hemoglobin-based oxygen carriers (HbOCs). Hb is encapsulated in polymersomes at Hb/polymer ratio of 1.2–1.5. The oxygen-binding capacity of Hb-loaded polymersomes is studied and is directly related to the membrane thickness of vesicles, as oxygen will take longer time to cross the thick polymersome membrane. The oxygen affinity (P_{50}) and cooperativity coefficient (n) are studied for Hb-loaded polymersomes and it is found that loading of Hb in polymersomes do not alter their oxygen-binding ability. The cooperativity coefficient is the ability of Hb to change from low affinity to high affinity state, as a function of surrounding oxygen concentration. Hb-loaded polymersomes showed low P_{50} values,

indicating that these materials are suitable candidates to carry oxygen to ischemic tissues [47].

The conjugation of enzymes with biocompatible polymers, such as poly(methacryloxy ethylphosphorylcholine) (PMPC), is shown to maintain the stability and hence the activity of the enzymes over a period of time. Papain is modified with PMPC of varying chain lengths (5–40 kDa). The activity of papain enzyme is studied as a function of polymer chain length and degree of modification. The polymers with smaller degree of modification and longer chain length and polymers with high degree of modification and shorter chain length show high activity of enzyme over a period of 37 days [48]. Similarly, PEGylation of trypsin is shown to maintain the activity of trypsin enzyme after conjugation. The PEGylated enzyme is incubated with its substrate *N*-α-benzoyl-$_L$-arginine 4-nitroanilide hydrochloride (BAPNA). The release of *p*-nitroaniline is studied spectrophotometrically (absorption at 410 nm, extinction coefficient = 8800) [49]. α-Linked disaccharide, trehalose is known to impart the stability to organisms living under extreme conditions by protecting their cells and proteins. The conjugation of trehalose polymers directly to the proteins, such as lysozyme, is found to maintain the biological activity of the protein–polymer conjugates as compared to the wild-type protein, when exposed to a variety of stresses such as heat and lyophilization [50].

19.10 VACCINES

Vaccination is a well-known approach to induce adaptive immune response against a variety of pathogens and diseases. One of the critical issues with vaccination is the need for new immunostimulants or adjuvants and an effective delivery system. Polymeric bioconjugates and their NPs offer a feasible solution to coencapsulate multiple antigens and adjuvants as well as to reduce the systemic toxicity of vaccines. In another strategy, polymeric NPs have been used to deliver antigen to dendritic cells (DCs) for a prolonged time [7]. Antigen-loaded polymeric NPs are studied for the enhancement of antigen-specific humoral response by their selective presentation to antigen presenting cells (APCs). DCs are the major component of immune response, and are capable of processing antigens by major histocompatibility complexes (MHCs) I and II. The delivery of antigens to DC is obtained by using amphiphilic NPs. Ovalbumin (OVA)-encapsulated γPGA-Phe NPs of 25 nm in diameter are studied for their uptake in bone marrow-derived DCs (Figure 19.10). It is found that as compared to OVA alone, OVA–NPs are efficiently uptaken by DCs [51].

The activation of DCs by NPs is dependent on their sizes, surface charge, and geometries. PLA NPs of 200–600 nm in diameter show higher uptake by DCs as compared to 2–6 μm particles [52]. The NPs with net cationic charge show high

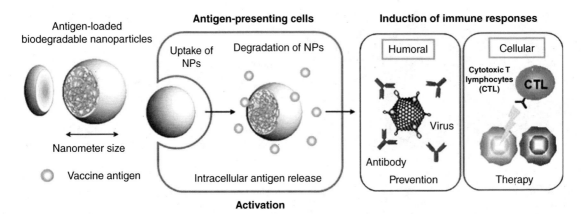

FIGURE 19.10   Formulation of nanoparticle-based vaccines and induction of immune response [51]. For a color version, see the color plate section.

cellular uptake due to negatively charged cell surface of DCs [53]. In addition, spherical particles show high uptake by DCs as compared to the worm-like particles of high aspect ratios [54, 55].

The peptide–polymer conjugates that can competitively block the entry of human immunodeficiency type-1 virus (HIV-1) into human cells are designed and studied for their antiviral properties [56]. These polypeptide-based anti-inhibitors block the functions of viral glycoproteins (gp 120), which facilitates the fusion of virion into the cells. The peptide–polymer conjugates presenting multiple copies of HIV-neutralizing antibody IgG b12 are prepared. IgG b12 interacts with CD4 of gp120 and possesses weak antiviral activity. The synthesis of peptide–polymer bioconjugates are shown to improve the antiviral activities of the antibody as a function of polymer chain length [56].

Well-defined polymethacrylates bearing sulfated malto-heptose units are studied for their anti-HIV activities and as blood anticoagulant. Anti-HIV activities of polymers containing varying degrees of sulfate content are studied by the inhibitory concentration (EC_{50}) of the polymers on the infection of HIV in MT-4 cells. It is found that the degree of sulfation is an important parameter in improving the anti-HIV activities of polymers. The increase in the degree of sulfation of polymers increases their anti-HIV properties. In addition, the spatial distance between sulfated chains of the polymers is also a detrimental factor to their anti-HIV properties [57].

The introduction of xenobiotics in living organisms facilitates their detoxification by their conjugation with UDP-D-glucuronic acid in the liver. The resulting β-D-glucuronide is excreted through the kidney and the liver. β-D-glucuronidase enzyme in small intestine however cleaves these conjugates, causing the absorption of xenobiotics in intestine. The inhibition activity of β-D-glucuronidase has been observed in the presence of aldaric acid, D-glucaric acid, and their derivatives. The glycopolymers bearing pendant D-glucaric acid and D-gluconic acid are synthesized and are studied for the inhibition activity of β-D-glucuronidase. The hydrolysis of substrate (*p*-nitrophenyl β-D-glucuronide) by the enzyme in the presence of glycopolymers is kinetically determined by studying the absorbance of *p*-nitrophenol from glucuronide at $\lambda = 400$ nm. Indeed, the copolymers bearing high concentration of D-glucaric acid showed high inhibition activity of enzyme, as compared to the polymers with low concentration of D-glucaric acid [58].

Hybrid hydrogels of PNIPAM containing silver NPs (AgNPs) are studied for their antibacterial activities as a function of size of AgNPs embedded in the hydrogels. It is shown that small AgNPs (2.67 nm) loaded hydrogels show improved antibacterial activities as compared to the larger-sized particles. This is possibly due to the facile diffusion of smaller AgNPs out of the hydrogel meshwork, which allows rapid interaction of NPs with *Escherichia coli* bacteria and hence higher antibacterial activities [59].

19.11   *In vitro* AND *in vivo* TARGETING

Site-specific delivery of macromolecules is an interesting approach to overcome the toxicity and side effects of cargo materials. The three important steps for targeting a macromolecule are (i) successful targeting to cell markers, (ii) intracellular trafficking; delivery of macromolecules inside the cytoplasm from intracellular compartments, and (iii) delivery of macromolecules to the nucleus [9]. The intracellular trafficking of biomolecules and their delivery in cytosol has been significantly improved by the incorporation of endocytosis-mediating agents on pH-responsive cargo carriers. The delivery of biomolecules to the desired cell type or tissue is still a major challenge for clinical applications of therapeutics. The delivery of cancer therapeutics to the tumor site is one of the most studied applications *in vitro* and *in vivo*. The active targeting to tumor vasculature has been used as an approach for selective imaging and delivery of therapeutics to tumor site.

HPMA-based copolymers containing $\alpha_v\beta_3$ integrin-targeting peptide and radioactively labeled ^{99m}Tc or ^{90}Y are synthesized and are studied for their tumor-targeting efficacies *in vitro* and *in vivo*. The injection of copolymers in SCID mice shows time-dependent accumulation of these compounds in tumor tissue, over a period of 72 hours. The treatment of SCID mice carrying DU145, a human prostate carcinoma xenografts with radiolabeled polymer–peptide conjugates, shows significant reduction in tumor volume (63% decrease in tumor volume upon treatment with 250 μCi dose) as compared to the untreated controls [60].

The overexpression of folate receptors on the surface of a variety of cancer cells makes them an interesting candidate for targeting drug and gene delivery in carcinomas. DOX-loaded folate-conjugated poly(N-isopropylacrylamide-*co*-N,N-dimethylacrylamide-*co*-2-aminoethyl methacrylate)-*b*-poly(undecenoic acid) (P(NIPAAm-*co*-DMAAm-*co*-AMA)-*b*-PUA) micelles are studied for the targeted drug delivery in folate expressing 4T1 mouse breast cancer cells and human epidermal carcinoma KB cells [61]. These folate-functionalized drug carriers show higher accumulation in 4T1 cells *in vitro* as compared to controls. The *in vivo* distribution of folate conjugates is then studied in a mouse breast cancer model induced by 4T1 cells. The results show high accumulation of DOX-loaded folate-functionalized micelles in tumor tissue, as compared to DOX alone [61].

Curcuma Longa commonly referred to as "turmeric" is used as spice, as cosmetic, and as a medicine in ancient ayurvedic system of medicine. The primary active ingredient in turmeric [(1E,6E)-1,7-bis(4-hydroxy-3-methoxyphenyl)hepta-1,6-diene-3,5-dione] "curcumin" is interesting because of its antioxidant, anti-inflammatory, and anticancer properties [62]. Curcumin is also known for its potent anti-Alzheimer's disease activity, due to its ability to bind and dissolve amyloid fibrils. The poor solubility of curcumin is a major drawback in its use for medicinal purposes. Dendrimer–curcumin conjugates prepared by click chemistry significantly improve the water solubility of curcumin. The conjugates are studied for their interactions with amyloid fibrils. Congo red-stained amyloid fibrils are formed by incubating Aβ 1–40 peptides, in the presence or absence of curcumin and its conjugates. It is found that amyloid fibril formation occurs in the absence of curcumin, while curcumin and its derivatives inhibit the fibril formation by dissolving the peptides in solution, as shown in Figure 19.11 [62]. Hence, conjugates of curcumin maintain their physiological activity. Curcumin conjugates are then studied for their anticancer activity by measuring caspase-3 activation in metastatic neurotumor cells. The conjugates of curcumin show significant apoptosis in metastatic cells, and the apoptosis was similar to curcumin drug itself [62].

The growth of new blood vessels from existing vasculature is termed as angiogenesis. Angiogenesis is a pathological procedure involved in tumor progression, hence the synthesis

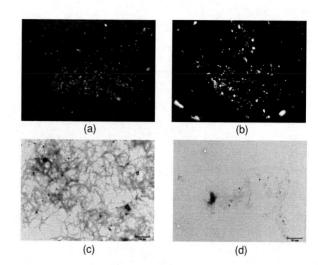

(a)　　　　(b)

(c)　　　　(d)

FIGURE 19.11 Congo red-stained heart tissue containing intracellular amyloid in the absence (a, c) and presence (b, d) of curcumin, studied using polarized light microscope (a, b) and transmission electron microscope (c, d) [62].

of angiogenesis inhibitors has become an important part of cancer therapy. TNP-470 is a potent angiogenesis inhibitor, and has shown a broad anticancer spectrum in animal studies. TNP-470 is a synthetic analog of a compound called, *fumagillin*, secreted by fungus *Aspergillus fumigatus fresenius* [63]. However, severe nephrotoxicity associated with TNP-470 greatly limits its applications as anticancer agent. In order to improve the biodistribution and accumulation of TNP-470 in cancer tissue, enzymatically active, HPMA polymer-conjugated TNP-470 analogs are tested for their anticancer activity and nephrotoxicity in mice. The conjugation of TNP-470 with HPMA polymers is done using a degradable tetrapeptide linker (Gly-Phe-Leu-Gly). Gly-Phe-Leu-Gly linker is cleavable by lysosomal cysteine proteases, cathepsin B. Cathepsin B is overexpressed in cancer cells hence releasing the drug in lysosome of malignant cells. The conjugation of TNP-470 to HPMA polymer significantly decreases the penetration of drug through the blood–brain barrier, while increasing its solubility and circulation time in blood. These conjugates also show improved antitumor activity *in vitro* and *in vivo* [63].

19.12 DIAGNOSTIC AND AFFINITY SEPARATIONS

A range of chemical and biochemical assays are used in health care system for the early detection and diagnosis of diseases. These diagnostic tests rely on biological receptors, such as protein, DNA, and antibodies as recognition elements. The instability, poor reproducibility, and low yield of biological receptors are some of the problems associated with diagnostic tests. One way to solve these problems is the conjugation of recognition elements with polymeric or nanomaterial scaffolds [10]. Screening of protein biomarkers for early cancer detection is still a challenge in the field of biomedicine. The commercial immunoassays require 10–100 pg/mL of proteins, which is a significant amount for the human tissues. A novel amplification strategy is used to detect cancer biomarkers using SWNT-based immunosensors. Prostate-specific antigen (PSA)-functionalized SWNTS are prepared. PSA is a biomarker for prostate cancer and is used as a clinical tool to diagnose and monitor the disease. For the detection, calf serum-containing PSA is incubated on modified SWNT surface, followed by the addition of horse radish peroxidase (HRP)-labeled secondary anti-PSA. The washed immunosensor is put into electrochemical cells containing H_2O_2 for signal detection. This method can successfully detect PSA in serum in a range of 0.4–40 ng/mL. The current detection range for human PSA in serum is 4–10 ng/mL. To further improve the sensitivity of the system, HRP-modified secondary anti-PSA is conjugated to the surface of MWCNT. The conjugates are then used for the detection of PSA in serum, instead of using HRP–anti-PSA itself. The MWCNT conjugates of HRP–anti-PSA significantly improve (∼800 times higher than HRP–anti-PSA) the sensitivity of the system. Moreover, the system shows superior detection ability (4 pg/mL) than all the present analysis techniques available for the detection of PSA. The schematics of SWNT-based immunosensors is depicted in Figure 19.12 [64].

"Smart" polymers are interesting due to their stimuli-responsive (temperature- and pH-responsive) properties in solution. The conjugation of smart polymers with biomolecules produces a biohybrid, which combine the individual properties of two components to produce "doubly smart conjugates" with distinct properties. These bioconjugates are extensively used in the field of biotechnology and medicines. Temperature-sensitive polymer conjugates are well studied for their affinity phase separation applications [6]. PNIPAM is a well-known temperature-sensitive polymer which undergoes phase change at its LCST (32°C).

PNIPAM-based microgels are used for affinity separation of immunoglobulin G (IgG). IgG is a 150 kDa protein, which is extensively used in research and diagnostics. IgG exists as tetramer, having two identical halves that form Y-shaped fork. The affinity purification of IgG is usually done by using proteins A/G. Proteins A/G are recombinant proteins, which require tedious purification methods. The alternative techniques for the screening of IgG have been used by employing Fc domain of IgG as a target for its capture. The affinity between protein–protein interactions arises by the cumulative effect of individually weak interactions including van der waals interactions, electrostatic interactions, and hydrogen bonding. The library of synthetic polymeric NPs containing a combination of charged, hydrophobic, and hydrophilic monomers is produced. These multifunctional NPs are screened for their interactions with IgG. The NPs showing some interactions with Fc component of IgG are combined to produce second generation of NPs which show enhanced affinity toward IgG. This iterative process helped in the identification of critical components for the synthesis of Fc-specific NPs. It is found that PNIPAM-based polymeric NPs of 50–65 nm diameter, containing 40% hydrophobic component and 20% anionic component, show high affinity toward Fc component of IgG. The binding of these polymeric NPs is a pH-dependent process. The amount of NPs require to bind Fc component of IgG at acidic pH is twice than the amount required to bind at neutral pH. This difference is attributed to the charged state of histidine residues at Fc component, at different pH of solution [65].

19.13 INDUSTRIAL APPLICATIONS

The development of food preservatives with superior antimicrobial activities and low toxicities have been the focus of food industries. Hen egg lysozymes are the main ingredients for antimicrobial products, due to their effective antibacterial properties. However, the activities of lysozymes are limited on gram-negative bacteria due to their inability to penetrate the lipopolysaccharide (LPS) membrane. The detergents and high temperatures have been used to inhibit the activity of gram-negative bacteria [11]. The surfactant activity of polysaccharides and antimicrobial activities of lysozyme are combined by preparing chitosan–lysozyme conjugates. High molecular weight chitosan–lysozyme conjugates show potent antimicrobial activities, specifically against gram-negative bacteria at room temperature. The excellent emulsifying properties and bactericidal action make chitosan–lysozyme conjugates promising candidates for industrial applications [66].

The allergens of proteins can be masked by their conjugation with biocompatible moieties. Soy protein contains a 34 kDa protein which is a well-identified allergen in soy. The patients sensitive to soy are due to the presence of 34 kDa

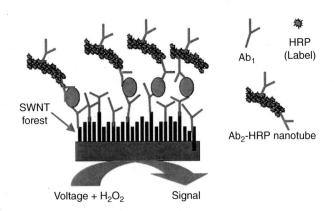

FIGURE 19.12 Depiction of SWNT-based immunosensors for the detection of PSA [64].

protein in soy. This allergen protein can be conjugated to polysaccharides by Maillard reaction. These conjugates are found to be an excellent source to reduce the allergenicity of the proteins [67]. For example, during *in vivo* studies in mice lysozyme induction produced significant amount of IgE, in contrast production of IgE was significantly decreased upon the injection of polysaccharide–lysozyme conjugates [68].

19.14 CONCLUSION

The advancements in the field of biology, biomedicines, and biotechnology today largely depend on the contribution of bioconjugates. A variety of bioconjugates have been prepared and are studied for drug and gene delivery, diagnostics, affinity separation, imaging, targeting, and as biosensors. Significant efforts have been made to synthesize novel material to overcome the barriers in targeted drug delivery and tumor imaging. The versatile biosensors with improved sensitivity and reproducibility are synthesized to efficiently diagnose several health disorders. The *in vivo* imaging techniques are greatly improved by the availability of nanoparticles and polymeric conjugates, hence providing a real-time imaging and evaluation of therapeutics. The blood compatible polymers are also prepared and are shown to be an efficient candidate for systemic applications. The well-defined cationic polymers of varying architectures with pendant biocompatible moieties are prepared and are shown to serve as potent gene delivery agents. Further *in vivo* studies are required to evaluate the role of these polymers for systemic applications. The polysaccharide bioconjugates are shown to contain antibacterial properties and are used in industrial applications as food preservatives.

REFERENCES

1. Wagner V, Dullaart A, Bock AK, Zweck A. *Nat Biotechnol* 2006;24:1211–1217.
2. Williams DF. *Biomaterials* 2009;30:5897–5909.
3. Koo OM, Rubinstein I, Onyuksel H. *Nanomedicine: Nanotech Bio Med* 2005;1:193–212.
4. Kherlopian AR, Song T, Duan Q, Niemark MA, Po MJ, Gohagan JK, Laine AF. *BMC Syst Biol* 2008;2:74. doi:10.1186/1752-0509-2-74
5. West JL, Halas NJ. *Annu Rev Biomed Eng* 2003;5:285–292.
6. Hoffman AS. *Clinl Chem* 2000;46:1478–1486.
7. Akagi T, Baba M, Akashi M. *Adv Polym Sci* 2012;247:31–64.
8. Rice-Ficht AC, Arenas-Gamboa AM, Kahl-Macdonagh MM, Ficht TA. *Curr Opin Microbio* 2010;13:106–112.
9. Stayton PS, Hoffman AS, Murth N, Lackey C, Cheung C, Tan P, Klumb LA, Chilkoti A, Wilbur FS, Press OW. *J Control Release* 2000;65:203–220.
10. Piletsky SA, Turner NW, Laitenberger P. *Med Eng Phys* 2006;28:971–977.
11. Kato A. *Food Sci Technol Res* 2002;8:193–99.
12. De Lambert B, Chaix C, Charreyre M-T, Laurent A, Aigoui A, Perrin-Rubins A, Pichot C. *Bioconjug Chem* 2005;16:265–274.
13. Weinstain R, Baran PS, Shabat D. *Bioconjug Chem* 2009;20:1783–1791.
14. Wan X, Zhang G, Ge Z, Narain R, Liu S. *Chem Asian J* 2011;6:2835–2845.
15. Wong J, Chilkoti A, Moy VT. *Biomol Eng* 1999;16:45–55.
16. Samanta B, Yang X-C, Ofir Y, Park M-H, Patra D, Agasti SS, Miranda OR, Mo Z-H, Rotello VM. *Angew Chem Intl Ed* 2009;48:5341–5344.
17. Ye Y, Bloch S, Kao J, Achilefu S. *Bioconjug Chem* 2005;16:51–61.
18. Brinkhuis RP, Stojanov K, Laverman P, Eilander J, Zuhorn IS, Rutjes FPJT, Van Hest JCM. *Bioconjug Chem* 2012;23:958–965.
19. Jamieson T, Bakshi R, Petrova D, Pocock R, Imani M, Seifalian AM. *Biomaterials* 2007;28:4717–4732.
20. Gao X, Cui Y, Levenson RM, Chung LWK, Nie S. *Nat Biotechnol* 2004;22:969–976.
21. Teles FRR, Fonseca LP. *Mater Sci Eng C* 2008;28:1530–1543.
22. Gibbs JM, Park SJ, Anderson DR, Watson KJ, Mirkin CA, Nguyen ST. *J Am Chem Soc* 2005;127:1170–1178.
23. Kalita P, Singh J, Singh MK, Solanki PR, Sumana G, Malhotra BD. *Appl Phys Lett* 2012;100:0937021–0937024.
24. Bouchard PR, Hutabarat RM, Thompson KM. *Annu Rev Pharmacol Toxicol* 2010;50:237–257.
25. Wang W, Chen C, Qian M, Zhao XS. *Analyt Biochem* 2008;373:213–219.
26. Uhrich KE, Cannizzaro SM, Langer RS, Shakesheff KM. *Chem Rev* 1999;99:3181–3198.
27. De Jesus OLP, Ihre HR, Gagne L, Frechet JMJ, Szoka FC Jr. *Bioconjug Chem* 2002;13:453–461.
28. Huynh VT, Binauld S, De Souza PL, Stenzel MH. *Chem Mater* 2012;24:3197–3211.
29. Huynh VT, De Souza P, Stenzel MH. *Macromolecules* 2011;44:7888–7900.
30. Crescenzi V, Cornelio L, Meo CD, Nardecchia S, Lamanna R. *Biomacromolecules* 2007;8:1844–1850.
31. Pang Z, Gao H, Yu Y, Guo L, Chen J, Pan S, Ren J, Wen Z, Jiang X. *Bioconjug Chem* 2011;22:1171–1180.
32. Johnson RP, Jeong Y-I, Choi E, Chung CW, Kang DH, Oh SO, Kim I. *Adv Funct Mater* 2012;22:1058–1068.
33. Dhandayuthapani B, Yoshida Y, Maekawa T, Kumar DS. *Intl J Polym Sci* 2011;2011. doi: 10.1155/2011/290602
34. Torok E, Pollok J-M, Ma PX, Vogel C, Dandri M, Petersen J, Burda MR, Kaufmann PM, Kluth D, Rogiers X. *Dig Surg* 2001;18:196–203.
35. Shea LD, Smiley E, Bonadio J, Mooney DJ. *Nat Biotechnol* 1999;17:551–554.
36. Ahmed M, Narain R. *Prog Poly Sci* 2012;38:767–790. doi:10.1016/j.progpolymsci.2012.09.008

37. Noor F, Wustholz A, Kinscherf R, Metzler-Nolte N. *Angew Chem Int Ed* 2005;44:2429–2432.

38. Ahmed M, Narain R. *Biomaterials* 2011;32:5279–5290.

39. Ahmed M, Narain R. *Biomaterials* 2012;33:3990–4001.

40. Ahmed M, Jawanda M, Ishihara K, Narain R. *Biomaterial* 2012;33:7858–7870.

41. Johnson RN, Burke RS, Convertine AJ, Hoffman AS, Stayton PS, Pun SH. *Biomacromolecules* 2010;11:3007–3013.

42. Kostiainen MA, Szilavy GR, Lehtinen J, Smith DK, Linder MB, Urtti A, Ikkala O. *ACS Nano* 2007;1:103–113.

43. Gorbet MB, Sefton MV. *Biomaterials* 2004;25:5681–5703.

44. Kainthan RK, Hester SR, Levin E, Devine DV, Brooks DE. *Biomaterials* 2007;28:4581–4590.

45. Ahmed M, Lai FLB, Kizhakkedathu JN, Narain R. *Bioconjug Chem* 2012;23:1050–1058.

46. Kainthan RK, Janzen J, Kizhakkedathu JN, Devine DV, Brooks DE. *Biomaterials* 2008;29:1693–1704.

47. Rameez S, Alosta H, Palmer AF. *Bioconjug Chem* 2008;19:1025–1032.

48. Miyamoto D, Watanabe J, Ishihara K. *Biomaterials* 2004;25:71–76.

49. Zarafashani Z, Obata T, Lutz JF. *Biomacromolecules* 2010;11:2130–2135.

50. Mancini RJ, Lee J, Maynard HD. *J Am Chem Soc* 2012;134:8474–8479.

51. Akagi T, Wang X, Uto T, Baba M, Akashi M. *Biomaterials* 2007;28:3427–3436.

52. Kanchan V, Panda AK. *Biomaterials* 2007;28:5344–5357.

53. Foged C, Brodin B, Frokjaer S, Sundblad A. *Int J Pharm* 2005;298:315–322.

54. Champion JA, Mitragotri S. *Proc Natl Acad Sci USA* 2006;103:4930–4934.

55. Champion JA, Mitragotri S. *Pharm Res* 2009;26:244–249.

56. Danial M, Root MJ, Klok HA. *Biomacromolecules* 2012;13:1438–1447.

57. Yoshida T, Akasaka T, Choi Y, Hattori K, Yu B, Mimura T, Kaneko Y, Nakashima H, Aragaki E, Premanathan M, Yamamoto N, Uryu T. *J Polym Sci Part A: Polym Chem* 1999;37:789–800.

58. Hashimoto K, Siato H, Ohsawa R. *J Polym Sci Part A Polym Chem* 2006;44:4895–4903.

59. Mohan YM, Lee K, Premkumar T, Geckeler KE. *Polymer* 2007;48:158–164.

60. Mitra A, Nan A, Papadimitriou JC, Ghandehari H, Line BR. *Nucl Med Biol* 2006;33:43–52.

61. Liu SQ, Wiradharma N, Gao SJ, Tong YW, Yang YY. *Biomaterials* 2007;28:1423–1433.

62. Shi W, Dolai S, Rizk S, Hussain A, Tariq H, Averick S, Amoreaux WL, Idrissi AE, Banerjee P, Raja K. *Org let* 2007;9:5461–5464.

63. Satchi-Finaro R, Puder M, Davies JW, Tran HT, Sampson DA, Greene AK, Corfas G, Folkman J. *Nat Med* 2004;10:255–261.

64. Yu X, Munge B, Patel V, Jensen G, Bhirde A, Gong JD, Kim SN, Gillespie J, Gutkind JS, Papadimitrakopoulos F, Rusling JF. *J Am Chem Soc* 2006;128:11199–11205.

65. Lee SH, Hoshino Y, Randall A, Zeng Z, Baldi P, Doong RA, Shea KJ. *J Am Chem Soc* 2012;134:15765–15772.

66. Tsai GJ, Su WH. *J Food Protect* 1999;62:239–243.

67. Babiker EE, Azakami H, Matsudomi N, Iwata H, Ogawa T, Bando N, Kato A. *J Agric Food Chem* 1998;46:866–871.

68. Arita K, Babiker EE, Azakami H, Kato A. *J Agric Food Chem* 2001;49:2030–2036.

INDEX

Chemistry of Bioconjugates: Synthesis, Characterization, and Biomedical Applications, First Edition. Edited by Ravin Narain.
© 2014 John Wiley & Sons, Inc. Published 2014 by John Wiley & Sons, Inc.